T0176765

Hydrogeology

Fresh Water to Drink

Laura Ingalls Wilder
1867–1957

In a little while the well was almost full of water. A circle of blue sky lay not far down in the ground, and when Laura looked at it, a little girl's head looked up at her. When she waved her hand, a hand on the water's surface waved, too.

The water was clear and cold and good. Laura thought she had never tasted anything so good as those long, cold drinks of water. Pa hauled no more stale, warm water from the creek. He built a solid platform over the well, and a heavy cover for the hole that let the water-bucket through. Laura must never touch that cover. But whenever she or Mary was thirsty, Ma lifted the cover and drew a dripping bucket of cold, fresh water from that well.

Little House on the Prairie, 1935
© HarperCollins Publishers Inc., New York.

Hydrogeology

Principles and Practice

Third Edition

Kevin M. Hiscock and Victor F. Bense

WILEY Blackwell

This third edition first published 2021
© 2021 John Wiley & Sons Ltd

Edition History
Blackwell Science Ltd (1e, 2005); John Wiley & Sons Ltd (2e, 2014)

All rights reserved. No part of this publication may be reproduced, stored in a retrieval system, or transmitted, in any form or by any means, electronic, mechanical, photocopying, recording or otherwise, except as permitted by law. Advice on how to obtain permission to reuse material from this title is available at http://www.wiley.com/go/permissions.

The right of Kevin M. Hiscock and Victor F. Bense to be identified as the authors of this work has been asserted in accordance with law.

Registered Offices
John Wiley & Sons, Inc., 111 River Street, Hoboken, NJ 07030, USA
John Wiley & Sons Ltd, The Atrium, Southern Gate, Chichester, West Sussex, PO19 8SQ, UK

Editorial Office
9600 Garsington Road, Oxford, OX4 2DQ, UK

For details of our global editorial offices, customer services, and more information about Wiley products visit us at www.wiley.com.

Wiley also publishes its books in a variety of electronic formats and by print-on-demand. Some content that appears in standard print versions of this book may not be available in other formats.

Limit of Liability/Disclaimer of Warranty
The contents of this work are intended to further general scientific research, understanding, and discussion only and are not intended and should not be relied upon as recommending or promoting scientific method, diagnosis, or treatment by physicians for any particular patient. In view of ongoing research, equipment modifications, changes in governmental regulations, and the constant flow of information relating to the use of medicines, equipment, and devices, the reader is urged to review and evaluate the information provided in the package insert or instructions for each medicine, equipment, or device for, among other things, any changes in the instructions or indication of usage and for added warnings and precautions. While the publisher and authors have used their best efforts in preparing this work, they make no representations or warranties with respect to the accuracy or completeness of the contents of this work and specifically disclaim all warranties, including without limitation any implied warranties of merchantability or fitness for a particular purpose. No warranty may be created or extended by sales representatives, written sales materials or promotional statements for this work. The fact that an organization, website, or product is referred to in this work as a citation and/or potential source of further information does not mean that the publisher and authors endorse the information or services the organization, website, or product may provide or recommendations it may make. This work is sold with the understanding that the publisher is not engaged in rendering professional services. The advice and strategies contained herein may not be suitable for your situation. You should consult with a specialist where appropriate. Further, readers should be aware that websites listed in this work may have changed or disappeared between when this work was written and when it is read. Neither the publisher nor authors shall be liable for any loss of profit or any other commercial damages, including but not limited to special, incidental, consequential, or other damages.

Library of Congress Cataloging-in-Publication Data

Names: Hiscock, K. M. (Kevin M.), author. | Bense, V. F. (Victor
  Franciscus), author. | John Wiley & Sons, publisher.
Title: Hydrogeology : principles and practice / Kevin M. Hiscock and Victor
  F. Bense.
Description: Third edition. | Hoboken, NJ : Wiley, 2021. | Includes
  bibliographical references and index.
Identifiers: LCCN 2021003266 (print) | LCCN 2021003267 (ebook) | ISBN
  9781119569534 (paperback) | ISBN 9781119569527 (adobe pdf) | ISBN
  9781119569510 (epub)
Subjects: LCSH: Hydrogeology.
Classification: LCC GB1003.2 .H57 2021 (print) | LCC GB1003.2 (ebook) |
  DDC 551.49–dc23
LC record available at https://lccn.loc.gov/2021003266
LC ebook record available at https://lccn.loc.gov/2021003267

Cover Design: Wiley
Cover Image: © Kevin Hiscock

Cover caption: Sluice overflow at Cheddar Rising, Somerset, England. With a catchment area of $40\,km^2$, the Cheddar resurgence is the largest of the Mendip Hills springs, with flow varying from $0.6–10\,m^3s^{-1}$. The main conduit is blocked by scree through which water emerges from many outlets in the bed of the artificial lake.

Set in 9.5/12.5pt STIXTwoText by Straive, Pondicherry, India

Printed in Singapore
M095793_160821

Contents

List of colour plates

Plate 1.1 Examples of now redundant village pumps once widespread in their use in Britain: (a) the wooden pump on Queen's Square, Attleborough, Norfolk, enclosed 1897; and (b) the large Shalders pump used in the days before tarmac for dust-laying on the old turnpike (Newmarket Road) at Cringleford, near Norwich, Norfolk.

Plate 1.2 Arcade of the aqueduct Aqua Claudia situated in the Parco degli Acquedotti, 8 km east of Rome. The aqueduct, which is built of cut stone masonry, also carries the brick-faced concrete Anio Novus, added later on top of the Aqua Claudia.

Plate 1.3 The baroque *mostra* of the Trevi Fountain in Rome. Designed by Nicola Salvi in 1732, and fed by the Vergine aqueduct, it depicts Neptune's chariots being led by Tritons with sea horses, one wild and one docile, representing the various moods of the sea.

Plate 1.4 Global map of epithermal neutron currents measured on the planet Mars obtained by the NASA Odyssey Neutron Spectrometer orbiter. Epithermal neutrons provide the most sensitive measure of hydrogen in surface soils. Inspection of the global epithermal map shows high hydrogen content (blue colour) in surface soils south of about 60° latitude and in a ring that almost surrounds the north polar cap. The maximum intensity in the northern ring coincides with a region of high albedo and low thermal inertia, which are both required for near-surface water ice to be stable. Also seen are large regions near the equator that contain enhanced near-surface hydrogen, which is most likely in the form of chemically and/or physically bound water and/or hydroxyl radicals since water ice is not stable near the equator. (Source: Reproduced from Los Alamos National Laboratory. © Copyright 2011 Los Alamos National Security, LLC. All rights reserved.)

Plate 1.5 Hose reel and rain gun irrigation system applied to a potato field in Norfolk, eastern England on 30 May 2020 and supplied by groundwater from the underlying Cretaceous Chalk aquifer.

Plate 1.6 NASA Landsat-7 satellite image of the Ouargla Oasis, Algeria on 20 December, 2000. In this false-colour image, red indicates vegetation (the brighter the red, the more dominant the vegetation). Pale pink and orange tones show the desert landscape of sand and rock outcrops. The satellite image shows date palms surrounding the urban area of Ouargla and Chott Aïn el Beïda in the south-west, a saline depression that has traditionally collected irrigation runoff, as well as the proliferation of irrigated land to the north and east of Ouargla in the vicinity of Chott Oum el Raneb. The width of the image shown is approximately 40 km. (Source: Reproduced from http://earthobservatory.nasa.gov/IOTD. NASA images created by Jesse Allen and Rob Simmon, using Landsat data provided by the United States Geological Survey. Caption by Michon Scott.)

Plate 1.7 Estimated depth in metres below ground level (m bgl) to groundwater in Africa (Bonsor and MacDonald 2011). (Source: Bonsor, H.C. and MacDonald, A.M. (2011). *An initial estimate of depth to groundwater across Africa, 2011*, British Geological Survey. © 2011, British Geological Survey.)

Plate 1.8 Aquifer productivity in litres per second ($L\,s^{-1}$) for Africa showing the likely interquartile range for boreholes drilled and sited using appropriate techniques and expertise (Bonsor and MacDonald 2011). (Source: Bonsor, H.C. and MacDonald, A.M. (2011). *An initial estimate of depth to groundwater across Africa, 2011*, British Geological Survey. © 2011, British Geological Survey.)

Plate 2.1 Aerial view of artesian springs and spring mounds west of Lake Eyre South (137° E, 29°S) in the Great Artesian Basin in northern South Australia (see Box 2.11) showing the flowing artesian Beresford Spring (A in foreground), the large, 45 m high Beresford Hill with an extinct spring vent (C), the flowing artesian Warburton Spring (E), and the flat topped hill (F) capped by spring carbonate deposits (tufa) overlying Bulldog Shale. The diameter of the upper part of the circular Beresford Hill, above the rim, is about 400 m. Luminescence ages of 13.9 ± 1 ka were determined for samples from the carbonate mound of the actively flowing Beresford Spring (B) and of 128 ± 33 ka from the north west side of the dry extinct Beresford Hill spring carbonate mound deposits (D) (Prescott and Habermehl 2008). (Source: Prescott, J.R. and Habermehl, M.A. (2008) Luminescence dating of spring mound deposits in the southwestern Great Artesian Basin, northern South Australia. *Australian Journal of Earth Sciences* **55**, 167–181. Reproduced with permission from Taylor & Francis.)

Plate 2.2 Big Bubbler Spring, with its spring outlet on top of an elevated mound, located west of Lake Eyre South in the Great Artesian Basin in northern South Australia

(see Box 2.11). The spring outflow runs into a small channel and forms small wetlands (to the right). Hamilton Hill and its cap of spring carbonate deposits is visible in the background and is similar to Beresford Hill (**Plate 2.1**) located approximately 30 km to the north-west (Prescott and Habermehl 2008). (Source: Prescott, J.R. and Habermehl, M.A. (2008) Luminescence dating of spring mound deposits in the southwestern Great Artesian Basin, northern South Australia. *Australian Journal of Earth Sciences* **55**, 167–181. Reproduced with permission from Taylor & Francis.)

Plate 2.3 1 : 50 000 000 Transboundary Aquifers of the World (Special Edition for the 7th World Water Forum 2015) map. There are 592 identified transboundary aquifers, including transboundary 'groundwater bodies' as defined by the European Union Water Framework Directive, underlying almost every nation. Areas of transboundary aquifer extent are shown with brown shading and areas of transboundary groundwater body extent are shown with green shading, with overlapping aquifers and groundwater bodies shown in gold shading. Individual blue squares and green circles, respectively, indicate small aquifers and groundwater bodies ($<6000\,km^2$). The thematic inset maps combine, from left to right, respectively, the delineations of transboundary aquifers of the world with maps of climate zones, groundwater resources and recharge, and population at 1 : 135 000 000. For more information on individual transboundary aquifers and groundwater bodies and an extended view of the small aquifers and groundwater bodies, visit IGRAC's online Global Groundwater Information System: (https://ggis.un-igrac.org/ggis-viewer/viewer/tbamap/public/default). (Source: Transboundary Aquifers of the World, Special Edition for the 7 World Water Forum 2015, IGRAC. © 2015, IGRAC.)

Plate 2.4 Groundwater discharge in the intertidal zone of Kinvara Bay on 15 September

2010 at Dunguaire Castle, County Galway, Ireland.

Plate 2.5 Lake Caherglassaun (for location see Box 2.14, Fig. 2.66) responding to high tide as observed at 14.53 h on 13 September 2006 in the karst aquifer of the Gort Lowlands, County Galway, Ireland.

Plate 2.6 The Carran Depression and turlough (a fluctuating, groundwater level-controlled ephemeral lake) on 13 September 2006, The Burren, County Clare, Ireland.

Plate 2.7 1 : 25 000 000 Groundwater Resources of the World (2008 edition) map showing the distribution of large aquifer systems (excluding Antarctica). Blue shading represents major groundwater basins, green shading areas with complex hydrogeological structure and brown shading areas with local and shallow aquifers. Darker and lighter colours represent areas with high and low groundwater recharge rates, respectively, generally above and below 100 mm a^{-1}. For further discussion see Section 2.17. (Source: Wall map "Groundwater Resources of the World", Global groundwater wall map, 2008. © 2008, WHYMAP.)

Plate 2.8 1 : 120 000 000 Groundwater Recharge (1961–1990) per Capita (2000) map showing groundwater recharge in m^3 capita^{-1} a^{-1} aggregated for countries or sub-national units (excluding Antarctica). (Source: Wall map "Groundwater Resources of the World", Global groundwater wall map, 2008. © 2008, WHYMAP.)

Plate 3.1 Variable-density groundwater flow simulations to evaluate the efficiency of different styles of salinization processes in layered aquifer systems on the continental shelf during and after transgression of the sea (see Section 3.6.1 and Fig. 3.12). The upper panel shows the model set-up representing a slightly seaward dipping layered aquifer system in which the left-hand boundary represents fresh, meteoric water originating as recharge in the hinterland. The right-hand boundary represents coastal seawater. In the initial steady-state situation ($t = 0$ years), the sideways sag of the saline water underneath the sea floor results in a tongue of saline water in the deeper inland aquifers. When sea-level rises, seawater starts to sink into the upper aquifer with a characteristic finger pattern indicative of free-convection replacing fresh water. This process of salinization is rapid compared to the salinization process in the deeper aquifer which only proceeds slowly by transverse movement of the saline-fresh interface. The simulation shows that it takes millennia for these processes to result in complete salinization of sub-seafloor aquifers which explains the current occurrence of fresh water in many parts of the continental shelf (e.g Fig. 3.13).

Plate 3.2 Example of a numerical simulation illustrating aspects of the hydrodynamics within sedimentary basins during glaciation. (a) A bowl-shaped sedimentary basin is conceptualized consisting of several thick aquifers and aquitards. This basin is overridden by an ice-sheet, which results in a complex hydrodynamic response. A deformation of the finite-element mesh accommodates the flexure of the sedimentary basin caused by the weight of the ice-sheet. (b) The high hydraulic head at the ice-sheet base is propagated into the aquifer units in the basin and results in a strong groundwater flow component away from the base of the ice-sheet. At the same time, the increasing weight exerted as the ice-sheet advances results in a build-up of hydraulic head in the aquitard units in the basin which is considerably more compressible than the aquifers. Consequently, groundwater is moving away from these aquitard units. In this model simulation, the lower aquitard is more compressible than the upper aquitard (Bense and Person 2008). (Source: Adapted from Bense, V.F. and Person, M.A. (2008) Transient hydrodynamics in inter-cratonic sedimentary basins during glacial cycles. *Journal of Geophysical Research* **113**, F04005.)

Plate 6.1 Global map of the groundwater footprint of aquifers. Six aquifers that are important to agriculture are shown at the bottom of the map (at the same scale as the global map) with the surrounding grey areas indicating the groundwater footprint proportionally at the same scale. The ratio GF/A$_A$ indicates widespread stress of groundwater resources and/or groundwater-dependent ecosystems. The inset histogram shows that GF is less than A$_A$ for most aquifers (Gleeson *et al.* 2012). (Source: Gleeson, T., Wada, Y., Bierkens, M.F.P. and van Beek, L.P. H. (2012) Water balance of global aquifers revealed by groundwater footprint. *Nature* **488**, 197–200.)

Plate 7.1 Temperature and fluid electrical conductivity (EC) logs in the Outokumpu Deep Drill Hole, eastern Finland. The 2516 m deep research borehole was drilled in 2004–2005 into a Palaeoproterozoic metasedimentary, igneous and ophiolite-related sequence of rocks in a classical ore province with massive Cu-Co-Zn sulphide deposits. The 'Sample EC' column shows the results of drill borehole water sampling in 2008. Arrows pointing to the left indicate interpreted depths of saline formation fluid flowing into the borehole and arrows to the right indicate fluid flowing out of the borehole. Arrows pointing up and down indicate the flow direction in the borehole. The 'Fractures' column indicates the interpreted fractures from sonic, electrical potential and calliper logs. The 'Hydraulic tests' column shows the test intervals and hydraulic permeabilities from packer experiments during drilling breaks. The 'Lithology' column shows the rock types (blue: metasediments; green and orange: ophiolite-derived serpentinite and skarn rocks; pink: pegmatitic granite). (Source: Adapted from Ahonen *et al.* 2004.)

Plate 7.2 Automated time-lapse electrical resistivity tomography (ALERT) monitoring results during an interruption in groundwater pumping in an operational Lower Cretaceous

sand and gravel quarry in West Sussex, England. Two times are shown: (a) t_a and (b) t_b imaged 15 days apart, as well as (c) the log resistivity ratio (t_b/t_a) plot showing sub-surface change. Water levels shown are for piezometers P1 and P6. Dashed lines show the minimum and maximum water levels estimated from the log resistivity ratio section (Chambers *et al.* 2015). (Source: Adapted from Chambers, J.E., Meldrum, P.I., Wilkinson, P.B. *et al.* (2015) Spatial monitoring of groundwater drawdown and rebound associated with quarry dewatering using automated time-lapse electrical resistivity tomography and distribution guided clustering. *Engineering Geology* **193**, 412–420.)

Plate 7.3 Satellite-derived images of (a) shallow groundwater storage and (b) root zone soil moisture content in Europe on 22 June 2020 as measured by the Gravity Recovery and Climate Experiment Follow On (GRACE-FO). GRACE-FO employs a pair of satellites that detect the movement of water based on variations in the Earth's gravity field by measuring subtle shifts in gravity from month to month. Variations in land topography, ocean tides and the addition or subtraction of water change the distribution of the Earth's mass and gravity field. Measurements are integrated with data from the original GRACE mission (2002–2017), together with current and historical ground-based observations using a sophisticated numerical model of water and energy processes at the land surface. The colours depict the wetness percentile to illustrate the status of groundwater storage and soil moisture content compared to long-term records for the month. Blue areas have more abundant water than usual, while orange and red areas have less. The darkest red areas represent dry conditions that should occur only 2% of the time (a return period of about once every 50 years). Much of Europe experienced drought in the summers of 2018 and 2019, followed by little snow in the winter of 2019–2020, the warmest on record. As a consequence, much of the continent began 2020 with a significant water deficit, with the threat of a groundwater drought and implications for maize and wheat yields compared to the five-year average in a number of countries. (Source: Signs of Drought in European Groundwater, NASA Earth Observatory, https://earthobservatory.nasa.gov/images/146888/signs-of-drought-in-european-groundwater?src=eoa-iotd.)

Plate 8.1 Replica pump with missing handle (see Plate 1b for comparison) in present-day Broadwick Street, Soho, London. The handle from the original Broad Street pump was famously removed on 8 September 1854 on the recommendation of Dr John Snow (1813–1858) who had concluded that the outbreaks of deaths from cholera among residents of the parishes of St James and St Anne were due to drinking contaminated water from the Broad Street well. From his investigation into the epidemiology of the cholera outbreak around the well, Snow gained valuable evidence that cholera is spread by contamination of drinking water. Subsequent research by others showed that the well was contaminated by sewage from an adjacent cess pool at 40 Broad Street entering the 1.83 m diameter, 8.8 m deep, brick-lined well sunk in sand above London Clay. This case represents one of the first, if not the first, study of an incident of groundwater contamination in Great Britain (Price 2004).

Plate 9.1 (a) and (b) Location of the Kaibab Plateau in the Colorado Plateau physiographic province (maximum elevation of 2807 m) north of the Grand Canyon, Arizona, United States, including the outline of the Grand Canyon National Park. (c) Shaded relief image of the Kaibab Plateau and surrounding region with approximate locations of major faults in the area. (d) and (e) Two karst aquifer vulnerability maps of the deep (approximately between 650 and 1000 m below ground surface), semi-confined Kaibab Plateau R (Redwall-Muav) aquifer system created, respectively, with the original concentration-overburden-precipitation (COP) method described by Vías *et al.* (2006) and the modified COP method of Jones *et al.* (2019) that uses sinkhole density as well as the location of faulted and fractured rock to model intrinsic vulnerability. Note that the modified model has a reduced overall intrinsic vulnerability to contamination and greater spatial variation of vulnerability (Jones *et al.* 2019). (Source: Adapted from Jones, N.A., Hansen, J., Springer, A.E. *et al.* (2019) Modeling intrinsic vulnerability of complex karst aquifers: modifying the COP method to account for sinkhole density and fault location. *Hydrogeology Journal* **27**, 2857–2868.)

Plate 10.1 An example of a dune slack at Winterton Dunes National Nature Reserve on the east coast of Norfolk, eastern England, observed in September 2020. Dune slack (or pond) habitats are a type of wetland that appear as damp or wet hollows left between sand dunes where, as here, the groundwater reaches or approaches the surface of the sand. The unusual acidic dunes and heaths at Winterton are internationally important for the rare groups of plants and animals which they support. The temporary pools in the dune slacks provide breeding sites for nationally important colonies of natterjack toads. The natterjack toad *Epidalea calamita* is often associated with dune slacks. To breed successfully, natterjacks require warmer water such as found in shallow dune slacks.

Plate 10.2 Multi-model mean changes in: (a) precipitation (mm/day), (b) soil moisture content (%), (c) runoff (mm/day) and (d) evaporation (mm/day). To indicate consistency in the sign of change, regions are stippled where at least 80% of models agree on the sign of the mean change. Changes are annual means for the medium, A1B scenario 'greenhouse gas' emissions scenario for the period 2080–2099 relative to 1980–1999. Soil moisture and runoff changes are shown at land points with valid data from at least 10 models

(Collins *et al.* 2007). (Source: Collins, W.D., Friedlingstein, P., Gaye, A.T. *et al.* (2007) Global climate projections, Chapter 10. In: *Climate Change 2007: The Physical Science Basis. Contribution of Working Group I to the Fourth Assessment Report of the Intergovernmental Panel on Climate Change* (eds S. Solomon, D. Qin, M. Manning *et al.*). Cambridge University Press, Cambridge, pp. 747–846. © 2007, Cambridge University Press.)

Plate 10.3 (a) Map showing global land-ocean temperature anomalies in 2019. Regional temperature anomalies are compared with the average base period (1951–1980)

(Source: NASA (2020). *2019 was the second warmest year on record.* https://earthobservatory.nasa.gov/images/146154/2019-was-the-second-warmest-year-on-record (accessed 13 September 2020).) (b) NASA Goddard Institute for Space Studies (GISS) graph showing global surface temperature anomalies from 1880 through to 2013 compared to the base period from 1951 to 1980. The thin red line shows the annual temperature anomaly, while the thicker red line shows the five-year running average. (Source: Global Temperature Anomaly, 1880–2013, NASA Earth Observatory, NASA.)

List of boxes

Preface to the third edition

The scope of this book remains the same in presenting the study of hydrogeology and the significance of groundwater in the terrestrial aquatic environment. This new edition reflects developments in hydrogeology since publication of the second edition in 2014 and aims to capture contemporary topics linking groundwater resources to global challenges such as climate change, energy resources and groundwater depletion embedded in global crop production and trade. Advances in the availability of digital data and techniques for modelling global-scale datasets have revealed the current state of groundwater resources worldwide, only further emphasizing the societal importance of groundwater as a natural resource. This third edition continues to be aimed at both students and practitioners in hydrogeology and related subjects. As in the previous two editions, this edition includes a set of exercises in Appendix 10, together with some new numerical problems, to assist students in gaining practice in problem solving in hydrogeology.

The overall structure and order of the book chapters remain largely the same, with Chapter 1 having undergone revision to present new insights into the distribution of groundwater in the Earth's upper continental crust, groundwater-related tipping points, the role of groundwater as an agent of global material and elemental fluxes, and the influence of humans on the water cycle. As an illustration of global groundwater challenges, a new box

is included on groundwater quality and depletion in the Indo-Gangetic Basin.

Chapters 2 and 3 cover basic principles of physical hydrogeology and the response of groundwater systems to geological processes. There is further presentation of the problems of land subsidence from groundwater over-abstraction as observed using remote sensing methods in the Tehran Plain, Iran, and a new section on the types of sinkholes and triggers for their collapse.

Chapters 4 and 5 have been extended with new material on the relationships between geology, hydrogeological environment and groundwater chemistry and how these factors have influenced the location of British spa towns and the history of malting and brewing in Europe. The application of the stable isotopes of nitrogen and sulphur in groundwater investigations is presented in the context of regional hydrochemical processes. A new box illustrates the application of multiple environmental tracers in revealing groundwater processes with the example of fluid flow along faults in the Lower Rhine Embayment, Germany.

Chapter 6 has a new section on global-scale surface water-groundwater modelling and Chapter 7 includes new material on the phenomenon of terrestrial water loading by monsoon rains on groundwater level fluctuation in the Bengal Basin. Chapter 7 also extends the section on downhole and surface geophysical techniques as applied in hydrogeology.

Chapter 8 on contaminant hydrogeology has a new section on emerging contamination from microplastic pollution and the section on agricultural contaminants is extended to include a global assessment of nitrate stored in the vadose zone. A section on the problem of saline water intrusion on small oceanic islands is included, illustrated with the example of groundwater supply on the island of Malta in the Mediterranean Sea. Given the importance of karst groundwater resources in many areas of the world, Chapter 9 now considers methods for groundwater vulnerability assessment and mapping of carbonate aquifers with an example from the Grand Canyon, Arizona.

Chapter 10 has been expanded with new sections on the role of managed aquifer recharge in sustainable water management and on the linkages between groundwater and climate change. The response time of groundwater to climate change, the connection between groundwater pumping and global greenhouse gas emissions and the influence of global warming on cold-region hydrogeology are all explored. The challenge of achieving net-zero carbon emissions by 2050 is introduced with new material discussing the link between groundwater and energy resources with an introduction to geothermal energy resources, ground source heat pumps and shale gas exploration. Finally, and with a view to looking forward, this third edition concludes with an updated discussion of future challenges for groundwater governance and management.

Kevin Hiscock, Norwichand
Victor Bense, Wageningen
July 2021

Preface to the second edition

Reflecting on the first edition of this book, written a decade ago, it has become increasingly evident that groundwater will play an essential part in meeting the water resources demands of the twenty-first century, with groundwater already supplying an estimated 2 billion people worldwide with access to freshwater. Furthermore, the challenge of feeding a projected population of nine billion people by 2030 is likely to require an ever greater demand for water in growing crops, with a large fraction of irrigation water supplied by groundwater. Combined with other global environmental pressures resulting from altered patterns of temperature and precipitation as driven by climate change, adaptation responses to water use and management will become critical if demands for water are to be met. Hence, in order to set the scene for students and practitioners in hydrogeology, the second edition of this book includes new sections on the distribution and exploitation of global groundwater resources and possible approaches to adapting to climate change. A longer term view is also presented in which processes that act over geological timescales, such as the formation of sedimentary basins and crustal deformation during ice ages, are shown to have a profound influence on our understanding of groundwater flow patterns and the distribution of fresh groundwater resources today.

As with the first edition, the main emphasis of this second edition is to present the principles and practice of hydrogeology, without which the appropriate investigation, development and protection of groundwater resources is not feasible. An important addition to the current edition is Chapter 3 in which regional characteristics such as topography, compaction and variable fluid density are introduced and explained in terms of geological processes affecting the past, present and future groundwater flow regimes. In support of the new material presented in this chapter and throughout this second edition, and given the positive reception to the case studies published in the first edition, a further 13 boxes are included, as well as a set of colour plates, that are drawn from our teaching and research experience. The case studies illustrate international examples ranging from transboundary aquifers and submarine groundwater discharge to the over-pressuring of groundwater in sedimentary basins and, as a special topic, the question of whether there is groundwater on the planet Mars. To help with a more rational presentation, some reorganization of material has occurred to separate investigation of catchment processes required to understand the role of groundwater as part of a catchment water balance (Chapter 6) from groundwater investigation techniques used to determine aquifer properties (Chapter 7). Also, Appendix 10 now includes a set of answers to the review questions in order to assist the reader consolidate his or her hydrogeological knowledge and understanding.

Kevin Hiscock & Victor Bense, Norwich
July 2013

Preface to the first edition

In embarking on writing this book on the principles and practice of hydrogeology, I have pur-
posely aimed to reflect the development of
hydrogeology as a science and its relevance to
the environment. As a science, hydrogeology
requires an interdisciplinary approach with
applications to water resources investigations,
pollution studies and environmental management. The skills of hydrogeologists are required
as much by scientists and engineers as by planners and decision-makers. Within the current
era of integrated river basin management, the
chance to combine hydrogeology with wider
catchment or watershed issues, including the
challenge of adapting to climate change, has
never been greater. Hence, to equip students
to meet these and future challenges, the
purpose of this book is to demonstrate the principles of hydrogeology and illustrate the importance of groundwater as a finite and vulnerable
resource. By including fundamental material

in physical, chemical, environmental isotope
and contaminant hydrogeology together with
practical techniques of groundwater investigation, development and protection, the content
of this book should appeal to students and
practising professionals in hydrogeology and
environmental management. Much of the
material contained here is informed by my
own research interests in hydrogeology and
also from teaching undergraduate and postgraduate courses in hydrology and hydrogeology within the context of environmental
sciences. This experience is reflected in the
choice of case studies, both European and
international, used to illustrate the many
aspects of hydrogeology and its connection
with the natural and human environments.

Kevin Hiscock, Norwich
May 2004

Acknowledgements

No book is produced without the assistance of others, and we are no exception in recognizing the input of colleagues, family and friends. Several people have provided help with proof reading sections and in supplying references and additional material. These people are Julian Andrews, Alison Bateman, Ros Boar, Lewis Clark, Sarah Cornell, Kate Dennis, Alan Dutton, Jerry Fairley, Aidan Foley, Tom Gleeson, Thomas Grischek, Rien Habermehl, Norm Henderson, Mike Leeder, Beth Moon, Lorraine Rajasooriyar, Peter Ravenscroft, Mike Rivett, Raphael Schneeberger, Wilhelm Struckmeier and John Tellam. An enormous thank you is owed to Phillip Judge, Sheila Davies and Laura Hiscock for their patient and expert preparation of the majority of the figures and Rosie Cullington for typing the many tables contained throughout. The staff, facilities and electronic resources of the Library at the University of East Anglia are appreciated for providing the necessary literature with which to compile this book. If this were not enough, we are indebted to Tim Atkinson, Richard Hey, Alan Kendall and Helen He for helping form the content of this book through the years spent together teaching and examining undergraduate and postgraduate students in hydrology and hydrogeology in the School of Environmental Sciences at the University of East Anglia. We are also grateful to the editorial team at John Wiley & Sons Limited for their guidance and support during the publication process. Last but not least, we again especially thank Cathy, Laura, Rebecca, Sylvia, Ronja, Kailash and Nico in supporting our endeavours in hydrogeology and for their patience during the time spent preparing this new edition.

Symbols and abbreviations

Multiples and submultiples

Symbol	Name	Equivalent
E	exa	10^{18}
P	peta	10^{15}
T	tera	10^{12}
G	giga	10^{9}
M	mega	10^{6}
k	kilo	10^{3}
h	hecto	10^{2}
da	deca	10^{1}
d	deci	10^{-1}
c	centi	10^{-2}
m	milli	10^{-3}
μ	micro	10^{-6}
n	nano	10^{-9}
p	pico	10^{-12}

Symbols and abbreviations

Symbol	Description	Units
[–]	activity	$mol\,kg^{-1}$
(–)	concentration (see Box 4.1)	$mol\,L^{-1}$ or $mg\,L^{-1}$
A	area	m^2
A	radionuclide activity	
AE	actual evapotranspiration	mm
ASR	artificial storage and recovery	
(aq)	aqueous species	
atm	atmosphere (pressure)	
B	barometric efficiency	
BOD	biological oxygen demand	$mg\,L^{-1}$
Bq	becquerel (unit of radioactivity; 1 Bq = 1 disintegration per second)	

Symbol	Description	Units
b	aquifer thickness	m
$2b$	fracture aperture	m
C	Sediment loading efficiency	
C	shape factor for determining k_i	
C	specific moisture capacity of a soil	(m of water)$^{-1}$
C, c	concentration	
$°C$	degrees Celsius (temperature)	
CEC	cation exchange capacity	meq (100 g)$^{-1}$
CFC	chlorofluorocarbon	
Ci	curie (older unit of radioactivity; 1 Ci = 3.7 × 10^{10} disintegrations per second)	
COD	chemical oxygen demand	$mg\,L^{-1}$
D	hydraulic diffusivity	$m^2\,s^{-1}$
D	hydrodynamic dispersion coefficient	$m^2\,s^{-1}$
$D*$	molecular diffusion coefficient	$m^2\,s^{-1}$
DIC	dissolved inorganic carbon	$mg\,L^{-1}$
DNAPL	dense, non-aqueous phase liquid	
DOC	dissolved organic carbon	$mg\,L^{-1}$
d	mean pore diameter	m
$\left(\dfrac{dh}{dl}\right)$	hydraulic gradient	
E^o	standard electrode potential	V
EC	electrical conductivity	$S\,cm^{-1}$
Eh	redox potential	V
e	void ratio	

Symbol	Description	Units	Symbol	Description	Units
e-	electron		k_i	intrinsic permeability	m^2
eq	chemical equivalent (see Box 4.1)	$eq\,L^{-1}$	L	litre (volume)	
F	Faraday constant $(9.65 \times 10^4\,C\,mol^{-1})$		LNAPL	light, non-aqueous phase liquid	
F	Darcy–Weisbach friction factor		l	length	m
f_c	infiltration capacity	$cm\,h^{-1}$	MNA	monitored natural attenuation	
f_t	infiltration rate	$cm\,h^{-1}$	m	mass	kg
f_{oc}	weight fraction organic carbon content		mol	amount of substance (see Box 4.1)	
G	Gibbs free energy	$kJ\,mol^{-1}$	n	an integer	
g	gravitational acceleration	$m\,s^{-2}$	n	roughness coefficient (Manning's n)	
g	gram (mass)		n	porosity	
(g)	gas		n_e	effective porosity	
H	depth (head) of water measured at a flow gauging structure	m	P	Peclet number	
			P	precipitation amount	mm
H	enthalpy	$kJ\,mol^{-1}$	P	pressure	Pa (or $N\,m^{-2}$)
h	hydraulic head	m	P	partial pressure	atm or Pa
I	ionic strength	$mol\,L^{-1}$	P_A, P_o	atmospheric pressure	atm or Pa
i	hydraulic gradient $\left(\dfrac{dh}{dl}\right)$		Pa	pascal (pressure)	
IAEA	International Atomic Energy Agency		P_w	porewater pressure	Pa or m of water
IAP	ion activity product	$mol^n\,L^{-n}$	PAH	polycyclic aromatic hydrocarbon	
J	joule (energy, quantity of heat)		PDB	Pee Dee Belemnite	
K	equilibrium constant	$mol^n\,L^{-n}$	PE	potential evapotranspiration	mm
K (hydraulics)	hydraulic conductivity	$m\,s^{-1}$	p	$-\log_{10}$	
K (temperature)	kelvin		ppm	parts per million	
K_d	partition or distribution coefficient	$mL\,g^{-1}$	Q	discharge	$m^3\,s^{-1}$
			Q_f	fracture flow discharge	$m^3\,s^{-1}$
K_f	fracture hydraulic conductivity	$m\,s^{-1}$	q	specific discharge or darcy velocity	$m\,s^{-1}$
K_H	Henry's law constant	Pa $m^3\,mol^{-1}$	R	hydraulic radius	m
			R (R_d)	recharge (direct recharge)	$mm\,a^{-1}$
K_{oc}	organic carbon-water partition coefficient		R	universal gas constant (8.314 $J\,mol^{-1}\,K^{-1}$)	
K_{ow}	octanol-water partition coefficient		RC	root constant	mm
K_s	selectivity coefficient		R_d	retardation factor	
			R_e	Reynolds number	
K_{sp}	solubility product	$mol^n\,L^{-n}$	rem	roentgen equivalent man (older unit of dose equivalent; 1 rem = 0.01 Sv)	

(Continued)

Symbol	Description	Units
S	entropy	J mol^{-1} K^{-1}
S	sorptivity of soil	cm (min)$^{-1/2}$
S	storativity	
S	slope	
S_p	specific retention	
S_s	specific storage	s^{-1}
S_{sp}	specific surface area	m^{-1}
S_y	specific yield	
SMD	soil moisture deficit	mm
STP	standard temperature and pressure of gases (0 °C, 1 atmosphere pressure)	
Sv	sievert (unit of dose equivalent that accounts for the relative biological effects of different types of radiation; 1 Sv = 100 rem)	
s	groundwater level drawdown	m
s	solubility	g (100 g)$^{-1}$ solvent
T	transmissivity	m^2 s^{-1}
T	absolute temperature	K
TDS	total dissolved solids	mg L^{-1}
TOC	total organic carbon	mg L^{-1}
t	time	s
$t_{1/2}$	radionuclide half-life	s
V	volume	m^3
V	volt (electrical potential)	
V_s	volume of solid material	m^3
VOC	volatile organic compound	
V_{SMOW}	Vienna Standard Mean Ocean Water	
v	groundwater or river water velocity	m s^{-1}
$\bar{v}$	average linear velocity (groundwater)	m s^{-1}
$\bar{v}_c$	average linear contaminant velocity	m s^{-1}
$\bar{v}_w$	average linear water velocity	m s^{-1}
WHO	World Health Organization	

Symbol	Description	Units
WMO	World Meteorological Organization	
WMWL	World Meteoric Water Line	
$W(u)$	well function	
w	width	m
ZFP	zero flux plane	
z (hydraulics)	elevation	m
z (chemistry)	electrical charge	

Greek symbols

Symbol	Description	Units
α	aquifer compressibility	m^2 N^{-1} (or Pa^{-1})
α	aquifer dispersivity	m
α	isotope fractionation factor	
β	water compressibility	m^2 N^{-1} (or Pa^{-1})
γ (chemistry)	activity coefficient	kg mol^{-1}
γ (hydraulics)	specific weight	N
δ	stable isotope notation	
$\delta-$	partial negative charge	
$\delta+$	partial positive charge	
θ	volumetric moisture content	
λ	radionuclide decay constant	s^{-1}
μ	viscosity	N s m^{-2}
μ/ρ	kinematic viscosity	m^2 s^{-1}
ρ	fluid density	kg m^{-3}
ρ_b	bulk mass density	kg m^{-3}
ρ_s	particle mass density	kg m^{-3}
σ_e	effective stress	Pa (or N m^{-2})
σ_T	total stress	Pa (or N m^{-2})
T	total competing cation concentration	meq (100 g)$^{-1}$
Σ	sum of	
Φ	fluid potential (=h)	m
ψ	fluid pressure	m of water
ψ_a	air entry pressure	m of water
Ω	saturation index	

About the companion website

This book is accompanied by a companion website:

www.wiley.com/go/hiscock/hydrogeology3e

The website includes:

- Powerpoints of all figures from the book for downloading
- PDFs of tables from the book

1

Introduction

1.1   Scope of this book

This book is about the study of hydrogeology and the significance of groundwater in the terrestrial aquatic environment. Water is a precious natural resource, without which there would be no life on Earth. We, ourselves, are comprised of two-thirds water by body weight. Our everyday lives depend on the availability of inexpensive, clean water and safe ways to dispose of it after use. Water supplies are also essential in supporting food production and industrial activity. As a source of water, groundwater obtained from beneath the Earth's surface is often cheaper, more convenient and less vulnerable to pollution than surface water.

Groundwater, because it is unnoticed underground, is often unacknowledged and undervalued resulting in adverse environmental, economic and social consequences. The over-exploitation of groundwater by uncontrolled pumping can cause detrimental effects on neighbouring boreholes and wells, land subsidence, saline water intrusion and the drying out of surface waters and wetlands. Without proper consideration for groundwater resources, groundwater pollution from uncontrolled uses of chemicals and the careless disposal of wastes on land cause serious impacts requiring difficult and expensive remediation over long periods of time. Major sources of contamination include agrochemicals, industrial and municipal wastes, tailings and process wastewater

from mines, oil field brine pits, leaking underground storage tanks and pipelines, and sewage sludge and septic systems.

Achieving sustainable development of groundwater resources by the future avoidance of over-exploitation and contamination is an underlying theme of this book. By studying topics such as the properties of porous material, groundwater flow theory and geological processes, well hydraulics, groundwater chemistry, environmental isotopes, contaminant hydrogeology and techniques of groundwater remediation and aquifer management, it is our responsibility to manage groundwater resources to balance environmental, economic and social requirements and achieve sustainable groundwater development (Fig. 1.1).

The 10 chapters of this book aim to provide an introduction to the principles and practice of hydrogeology and to explain the role of groundwater in the aquatic environment. Chapter 1 provides a definition of hydrogeology and charts the history of the development of hydrogeology as a science. The water cycle is described and the importance of groundwater as a natural resource is explained. The legislative framework for the protection of groundwater resources is introduced with reference to developed and developing countries. Chapters 2–4 discuss the principles of physical and chemical hydrogeology that are fundamental to an understanding of the occurrence, movement and chemistry of groundwater in the Earth's crust. The relationships between

Hydrogeology: Principles and Practice, Third Edition. Kevin M. Hiscock and Victor F. Bense.
© 2021 John Wiley & Sons Ltd. Published 2021 by John Wiley & Sons Ltd.
Companion website: www.wiley.com/go/hiscock/hydrogeology3e

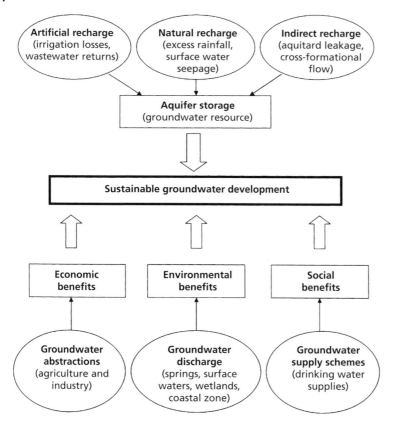

Fig. 1.1 The achievement of sustainable groundwater development through the balance of recharge inputs to aquifer storage (the groundwater resource) against discharge outputs for economic, environmental and human (social) benefits (Hiscock *et al.* 2002). (*Source:* Hiscock, K.M., Rivett, M.O. and Davison, R.M. (2002) Sustainable groundwater development. In: *Sustainable Groundwater Development* (eds K.M. Hiscock, M.O. Rivett and R.M. Davison). Geological Society, London, Special Publications 193, pp. 1–14. © 2002, Geological Society of London.)

geology and aquifer conditions are demonstrated both in terms of flow through porous material and rock-water interactions. Chapter 5 provides an introduction to the application of environmental isotopes in hydrogeological investigations for assessing the age of groundwater recharge and includes a section on noble gases to illustrate the identification of palaeowaters and aquifer evolution.

In the second half of this book, Chapters 6 and 7 provide an introduction to the range of field investigation techniques used in the assessment of catchment water resources and includes stream gauging methods, well hydraulics and tracer techniques. The protection of

groundwater from surface contamination requires knowledge of solute transport processes, and Chapter 8 introduces the principles of contaminant hydrogeology. Chapter 8 also covers water quality criteria and discusses the nature of contamination arising from a variety of urban, industrial and agricultural sources and also the causes and effects of saline intrusion in coastal regions and oceanic islands. The following Chapter 9 discusses methods of groundwater pollution remediation and protection, and includes sections that introduce risk assessment methods and spatial planning techniques. The final chapter, Chapter 10, returns to the topic of catchment water resources and demonstrates integrated methods for the management of groundwater together with consideration of groundwater interactions with rivers and wetlands, as well as the potential impacts of climate change on groundwater. Given the drive to net-zero carbon emissions by 2050, the role and interaction of groundwater in the exploitation of energy resources, including renewable resources and shale gas, is reviewed. Finally, Chapter 10 concludes with a description of approaches to groundwater governance and management for the long-term sustainability of groundwater resources.

Each chapter in this book concludes with recommended further reading to help extend the reader's knowledge of hydrogeology. In addition, for students of hydrogeology, a set of discursive and numerical exercises are provided in Appendix 10 to provide practice in solving groundwater problems. The remaining appendices include data and information in support of the main chapters of this book and will be of wider application in Earth and environmental sciences.

1.2 What is hydrogeology?

Typical definitions of hydrogeology emphasize the occurrence, distribution, movement and geological interaction of water in the Earth's crust. Hydrogeology is an interdisciplinary subject and also encompasses aspects of hydrology. Hydrology has been defined as the study of the occurrence and movement of water on and over the Earth's surface independent of the seepage of groundwater and springs which sustain river flows during seasonal dry periods. However, too strict a division between the two subjects is unhelpful, particularly when trying to decipher the impact of human activities on the aquatic environment. How well we respond to the challenges of pollution of surface water and groundwater, the impacts of over-exploitation of water resources, and

the potential impact of climate change will depend largely on our ability to take a holistic view of the aquatic environment.

1.3 Early examples of groundwater exploitation

The vast store of water beneath the ground surface has long been realized as an invaluable source of water for human consumption and use. Throughout the world, wells and springs fed by groundwater are revered for their life-giving or curative properties (see Fig. 1.2), and utilization of groundwater long preceded understanding of its origin, occurrence and movement (Bord and Bord 1985).

Fig. 1.2 Lady's Well in Coquetdale, northern England (National Grid Reference NT 953 028). Groundwater seeping from glacial deposits at the foot of a gently sloping hillside is contained within an ornamental pool floored with loose gravel. The site has been used since Roman times as a roadside watering place and was walled round and given its present shape in either Roman or medieval times. Anglo Saxon Saint Ninian, the fifth-century apostle, is associated with the site, and with other 'wells' beside Roman roads in Northumberland, and marks the spot where Saint Paulinus supposedly baptized 3000 Celtic heathens in its holy water during Easter week, 627 AD. The name of the well, Lady's Well, was adopted in the second half of the twelfth century when the nearby village of Holystone became the home of a priory of Augustinian canonesses. The well was repaired and adorned with a cross, and the statue brought from Alnwick, in the eighteenth and nineteenth centuries. Today, groundwater overflowing from the pool supplies the village of Holystone.

A holy or sacred well is commonly a well or spring at which religious devotions are, or have been, practised. In Ireland, for example, there are more than 3000 holy wells, many of which are sites of devotion, especially on the saint's day. Many of these wells have reputations for healing, with commonly cited cures being eye problems, toothache and warts (Misstear *et al.* 2018).

Springs are significant cultural places, embodying traditional folklore and mythology (Idris 1996; Park and Ha 2012; Powell *et al.* 2015) and supporting settlements along ancient trade routes (Aldumairy 2005). Indeed, the very survival and dispersal of early hominins, and later *Homo*, in the East African Rift System may have been influenced by springs. Hundreds of springs and groundwater-fed perennial streams currently distributed across East Africa are likely to have functioned as persistent hydro-refugia during dry periods of orbital-scale climate cycles in the Plio-Pleistocene and may have facilitated unexpected variations in isolation and dispersal of hominin populations (Cuthbert *et al.* 2017).

Evidence for some of the first wells to be used by modern humans is found in the far west of the Levant on the island of Cyprus. It is likely that Cyprus was first colonized by farming communities in the Neolithic, probably sailing from the Syrian coast about 9000 BC (Mithen 2012). Several Neolithic wells have been excavated from known settlements in the region of Mylouthkia on the west coast of Cyprus (Peltenberg *et al.* 2000). The wells are 2 m in diameter and had been sunk at least 8 m through sediment to reach groundwater in the bedrock. The wells lacked any internal structures or linings other than small niches within the walls, interpreted as hand- and foot-holds to allow access during construction and for cleaning. When abandoned, the wells were filled with domestic rubbish which dates from 8300 BC, indicating that the wells had been built at or just before this date (Mithen 2012).

Wells from the Neolithic period are also recorded in China, a notable example being the wooden Hemudu well in Yuyao County, Zhejiang Province, in the lower Yangtze River coastal plain. Based on carbon-14 dating of the well wood, it is inferred that the well was built in 3710 ± 125 BC (Zhou *et al.* 2011). The depth of the well was only 1.35 m with over 200 wooden components used in its construction comprising an outer part of 28 piles surrounding a pond, and an inner part, the wooden well itself, in the centre of the pond. The walls of the well were lined with close-set timber piles reinforced by a square wooden frame. The 28 piles in the outer part of the site may have been part of a shelter for the well, suggesting awareness by the people of the Hemudu culture that their water source required protection (Zhou *et al.* 2011).

According to archaeological research, the Chinese are credited with developing the percussion method of well construction, a technique that has been in continuous use now for 4000 years. The 'rope and drop' method involved a steel rod or piston that was raised and dropped vertically via a rope supported by a bamboo framework. Using this percussion system with a heavy chiselling or crushing tool, wells were drilled to depths of 130 m around 3000 years ago, although construction took years to complete (Zhou *et al.* 2011). The cable tool drilling rig used today (see Section 7.2.2) is directly descended from the bamboo framework percussion drilling techniques developed in China.

Evidence for the appearance of dams, wells and terraced walls, three methods of water management, is widespread by the Early Bronze Age from 3600 BC, as part of what has been termed a 'Water Revolution' (Mithen 2012). The recognisable development of groundwater as part of a water management system also dates from ancient times, as manifest by the wells and horizontal tunnels known as qanats (ghanats) or aflaj (singular, falaj), both Arabic terms describing a small, artificial channel excavated as part of a water distribution system, which appear to have originated in Persia about 3000 years ago. Examples of such systems are found in a band across the arid regions extending from Afghanistan to Morocco. In Oman, the rural villages and aflaj-supplied oases lie at the heart of Omani culture and tradition. The system of participatory management of communal aflaj is an ancient tradition in Oman by which common-property flows are channelled and distributed to irrigation plots on a time-based system, under the management of a local community (Young 2002).

Figure 1.3 shows a cross-section along a qanat with its typical horizontal or gently sloping gallery laboriously dug through alluvial material, occasionally up to 30 km in length, and with vertical shafts dug at closely spaced intervals to provide access to the tunnel. Groundwater recharging the alluvium in the mountain foothills is fed by gravity flow from beneath the water table at the upper end of the qanat to a ground surface outlet and

irrigation canal on the arid plain at its lower end (Fig. 1.4). The depth of the mother well (Fig. 1.3) is normally less than 50 m. Discharges, which vary seasonally with water table fluctuations, seldom exceeding 3 m^3 s^{-1}.

Such early exploitation of groundwater as part of a sophisticated engineered system is also evident in the supply of water that fed the fountains of Rome (see Box 1.1). Less sophisticated but none the less significant, hand-operated pumps installed in wells and boreholes have been used for centuries to obtain water supplies from groundwater found in surface geological deposits. The fundamental design of hand pumps of a plunger (or piston) in a barrel (or cylinder) is recorded in evidence from Greece in about 250 BC (Williams 2009). It is assumed that wooden pumps were in continuous use after the end of the Roman period, although examples are difficult to find given that wooden components perish in time. In Britain, the majority of existing hand-operated pumps are cast iron, dating

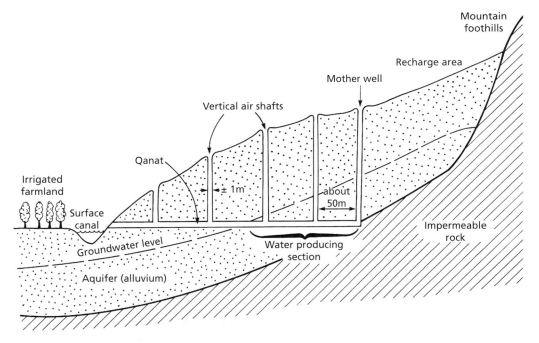

Fig. 1.3   Longitudinal section of a qanat (Beaumont 1968 and Biswas 1972). (*Sources:* Based on Beaumont, P. (1968) Qanats on the Varamin plain, Iran. *Transactions of the Institute of British Geographers* **45**, 169–179; Biswas, A.K. (1972) *History of Hydrology*. North-Holland, Amsterdam.)

Fig. 1.4 Irrigation canal supplied with water by a qanat or falaj in Oman. (Photograph provided courtesy of M.R. Leeder.)

from the latter part of the nineteenth century (see Plate 1.1). Although entirely redundant now due to issues of unreliability in dry weather and the risk of surface-derived pollution, private and domestic pumps were once widely used for supplying houses, farms, inns, almshouses, hospitals, schools and other institutions in cities, towns and villages. Ultimately, as mains water was introduced across Britain from the nineteenth century onwards following the Public Health (Water) Act of 1878, the village pump was superseded by the communal outdoor tap or water pillar, itself made redundant when piped water was provided to individual houses.

1.4 History of hydrogeology

It is evident from the examples mentioned previously that exploitation of groundwater resources long preceded the founding of geology, let alone hydrogeology. Western science was very slow in achieving an understanding of the Earth's hydrological cycle. Even as late as the seventeenth century it was generally

---

Box 1.1 The aqueducts of Rome

The aqueducts of ancient Rome are often associated with Roman expertise in civil engineering, and the fact that most of the aqueducts are supplied by springs is a tribute to the importance of groundwater in sustaining human civilization (Deming 2020). The remarkable organization and engineering skills of the Roman civilization are demonstrated in the book written by Sextus Julius Frontinus and translated into English by Bennett (1969). In the year 97 AD, Frontinus was appointed to the post of water commissioner, during the tenure of which he wrote the *De Aquis*. The work is of a technical nature, written partly for his own instruction, and partly for the benefit of others. In it, Frontinus painstakingly details every aspect of the

construction and maintenance of the aqueducts existing in his day.

For more than 400 years, the city of Rome was supplied with water drawn from the River Tiber, and from wells and springs. Springs were held in high esteem, and treated with veneration. Many were believed to have healing properties, such as the springs of Juturna, part of a fountain known from the south side of the Roman Forum. As shown in Fig. 1.5 and illustrated in Plate 1.2, by the time of Frontinus, these supplies were augmented by several aqueducts, presumably giving a reliable supply of good quality water, in many cases dependent on groundwater. For example, the Vergine aqueduct brought water from the estate of Lucullus

Box 1.1 (Continued)

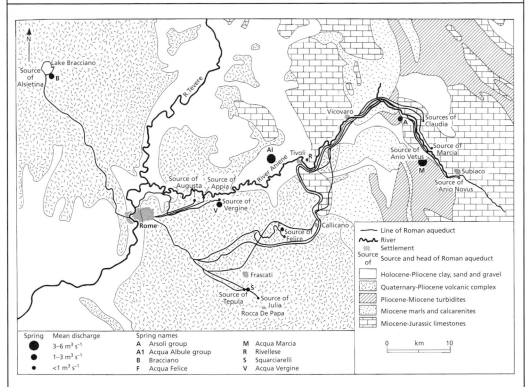

Spring	Mean discharge	Spring names
●	3–6 m³ s⁻¹	A Arsoli group	M Acqua Marcia
●	1–3 m³ s⁻¹	A1 Acqua Albule group	R Rivellese
●	<1 m³ s⁻¹	B Bracciano	S Squarciarelli
	F Acqua Felice	V Acqua Vergine

Fig. 1.5 Map of the general geology in the vicinity of Rome showing the location of the spring sources and routes of Roman aqueducts (Bennett 1969 and Boni *et al.* 1986). (*Sources:* Based on Bennett, C.E. (1969) *Frontinus: The Stratagems and the Aqueducts of Rome.* Harvard University Press, Cambridge, Massachusetts; Boni, C., Bono, P. and Capelli, G. (1986) *Hydrogeological Scheme of Central Italy. Sheet 1 (A. Hydrogeological Map).* Memoir of the Geological Society of Italy **XXXV**, Geological Society of Italy, Rome.)

where soldiers, out hunting for water, were shown springs which, when dug out, yielded a copious supply. Frontinus records that the intake of Vergine is located in a marshy spot, surrounded by a concrete enclosure for the purpose of confining the gushing waters. The length of the water course was 14 105 paces (20.9 km). For 19.1 km of this distance the water was carried in an underground channel, and for 1.8 km above ground, of which 0.8 km was on substructures at various points, and 1.0 km on arches. The source of the Vergine spring, located approximately 13 km east of Rome in the small town of Sal-one, is shown on a modern hydrogeological map (Boni *et al.* 1986) as issuing from

permeable volcanic rocks with a mean discharge of 1.0 m³ s⁻¹ (Fig. 1.5). Frontinus also describes the Marcia aqueduct with its intake issuing from a tranquil pool of deep green hue. The length of the water-carrying conduit is 61 710½ paces (91.5 km), with 10.3 km on arches. Today, the source of the Marcia spring is known to issue from extensively fractured limestone rocks with a copious mean discharge of 5.4 m³ s⁻¹.

After enumerating the lengths and courses of the several aqueducts, Frontinus enthuses: 'with such an array of indispensable structures carrying so many waters, compare, if you will, the idle Pyramids or the useless, though famous, works of the Greeks!' To

(Continued)

Box 1.1 **(Continued)**

protect the aqueducts from wilful pollution, a law was introduced such that: 'No one shall with malice pollute the waters where they issue publicly. Should any one pollute them, his fine shall be 10 000 sestertii' which, at the time, was a very large fine. Clearly, the 'polluter pays' principle was readily adopted by the Romans! Further historical, architectural and engineering details of the ancient aqueducts of Rome are given by Bono and Boni (2001) and Hodge (2008).

The Vergine aqueduct is one of only two of the original aqueducts still in use. The name derives from its predecessor, the Aqua Virgo, constructed by Marcus Agrippa in 19 BC. The main channels were renovated and numerous

secondary channels and end-most points (*mostre*) added during the Renaissance and Baroque periods, culminating in several fountains, including the famous Trevi fountain completed in 1762 (Plate 1.3). The total discharge of the ancient aqueducts was in excess of 10 m^3 s^{-1} supplying a population at the end of the first century AD of about 0.5 million. Today, Rome is supplied with 23 m^3 s^{-1} of groundwater, mainly from karst limestone aquifers, and serving a population of 3.5 million (Bono and Boni 2001). Many of the groundwater sources are springs from the karst system of the Simbruini Mountains east of Rome.

---

assumed that water emerging from springs could not be derived from rainfall, in that it was believed that the quantity was inadequate and the Earth too impervious to permit infiltration of rain water far below the surface. For example, Athanasius Kircher (1602–1680) erroneously considered in his publication *Mundus Subterraneus* of 1664 that the tides were caused by water moving to and from a subterranean ocean (Fig. 1.6). In contrast, Eastern philosophical writings had long considered that the Earth's water flowed as part of a great cycle involving the atmosphere. For example, ancient China had explicit concepts about water circulation as early as the mid-fourth to early-third centuries BC and also documented relationships between topography, soil type and groundwater depth in the book *Guan Zi* authored in the early Warring States Period (475 BC–221 BC) (Zhou *et al.* 2011). Even earlier, about 3000 years ago, the sacred Hindu Vedas texts of India explained the Earth's water movements in terms of cyclical processes of evaporation, condensation, cloud formation, rainfall, river flow and water storage (Chandra 1990).

A clear understanding of the hydrological cycle was achieved by the end of the seventeenth century. The English experimentalist Robert Hooke (1635–1703) made a number of observations on whether precipitation was sufficient to fully account for terrestrial stream flow (Deming 2019) and the French experimentalists Pierre Perrault (1611–1680) and Edme Mariotte (ca. 1620–1684) recorded measurements of rainfall and runoff in the River Seine drainage basin. In addition, the English astronomer Edmond Halley (1656–1742) demonstrated that evaporation of seawater was sufficient to account for all springs and stream flow (Halley 1691). Over 100 years later, the famous chemist John Dalton (1766–1844) made further observations of the water cycle, including a consideration of the origin of springs (Dalton 1799).

One of the earliest applications of the principles of geology to the solution of hydrological problems was made by the Englishman William Smith (1769–1839), the 'father of English geology' and originator of the epoch-making Map of England (1815). During his work as a canal and colliery workings

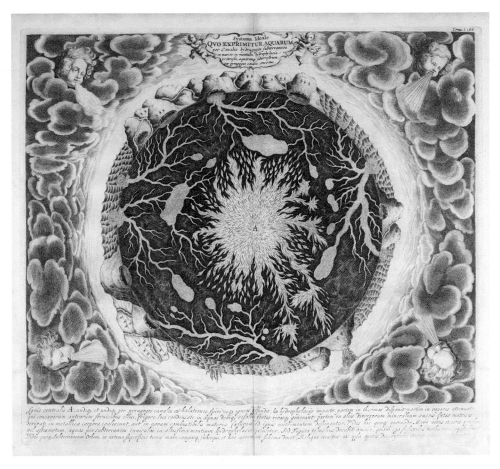

Fig. 1.6 Baroque-style depiction of the interlaced systems of air, fire and water within the Earth as conceived by the German Jesuit scholar Athanasius Kircher (1602–1680) in his book *Mundus Subterraneus* (1664). *Source:* AF Fotografie/Alamy Stock Photo.

surveyor in the west of England, Smith noted the various soils and the character of the rocks from which they were derived and used his knowledge of rock succession to locate groundwater resources to feed the summit levels of canals and supply individual houses and towns (Mather 1998).

In Britain, the industrial revolution led to a huge demand for water resources to supply new towns and cities, with Nottingham, Liverpool, Sunderland and parts of London all relying on groundwater. This explosion in demand for water gave impetus to the study of the economic aspects of geology. It was at this time that Lucas (1874) introduced the term 'hydrogeology' and produced the first real

hydrogeological map (Lucas 1877). Towards the end of the nineteenth century, William Whitaker, sometimes described as the 'father of English hydrogeology,' and an avid collector of well records, produced the first water supply memoir of the Geological Survey (Whitaker and Reid 1899) in which the water supply of Sussex is systematically recorded.

The drilling of many artesian wells stimulated parallel activity in France during the first half of the nineteenth century. The French municipal hydraulic engineer Henry Darcy (1803–1858) studied the movement of water through sand and from empirical observations defined the basic equation, universally known as Darcy's Law that governs groundwater flow in most alluvial and sedimentary formations (Freeze 1994). The equation was published in one of eight appendices in a volume that is partly a consulting report on the water supply for the City of Dijon, France, and partly an encyclopaedia of mid-nineteenth century water knowledge (Bobeck 2006) and can be found in the entire translation of Darcy's report by Bobeck (2004). Darcy's Law is the foundation of the theoretical aspects of groundwater flow and his work was extended by another Frenchman, Arsène Dupuit (1804–1866), whose name is synonymous with the equation for axially-symmetric flow towards a well in a permeable, porous medium.

The pioneering work of Darcy and Dupuit was followed by the German civil engineer, Adolph Thiem (1836–1908), who made theoretical analyses of problems concerning groundwater flow towards wells and galleries, and by the Austrian Philip Forchheimer (1852–1933) who, for the first time, applied advanced mathematics to the study of hydraulics. One of his major contributions was a determination of the relationship between equipotential surfaces and flow lines. Inspired by earlier techniques used to understand heat flow problems, and starting with Darcy's Law and Dupuit's assumptions, Forchheimer derived a partial differential equation, the Laplace equation, for steady groundwater flow. Forchheimer was also the first to apply the method of mirror images to groundwater flow problems; for example, the case of a pumping well located adjacent to a river.

Much of Forchheimer's work was duplicated in the United States by Charles Slichter (1864–1946), apparently oblivious of Forchheimer's existence. However, Slichter's theoretical approach was vital to the advancement of groundwater hydrology in America at a time when the emphasis was on exploration and understanding the occurrence of groundwater. This era was consolidated by Meinzer (1923) in his book on the occurrence of groundwater in the United States. Meinzer (1928) was also the first to recognize the elastic storage

behaviour of artesian aquifers. From his study of the Dakota sandstone (Meinzer and Hard 1925), it appeared that more water was pumped from the region than could be explained by the quantity of recharge at outcrop, such that the water-bearing formation must possess some elastic behaviour in releasing water contained in storage. Seven years later, Theis (1935), again using the analogy between heat flow and water flow, presented the ground-breaking mathematical solution that describes the transient behaviour of water levels in the vicinity of a pumping well.

Two additional major contributions in the advancement of physical hydrogeology were made by Hubbert and Jacob in their 1940 publications. Hubbert (1940) detailed work on the theory of natural groundwater flow in large sedimentary basins, while Jacob (1940) derived a general partial differential equation describing transient groundwater flow. Significantly, the equation described the elastic behaviour of porous rocks introduced by Meinzer over a decade earlier. Today, much of the training in groundwater flow theory and well hydraulics, and the use of computer programmes to solve hydrogeological problems, is based on the work of these early hydrogeologists during the first half of the twentieth century.

The development of the chemical aspects of hydrogeology stemmed from the need to provide good quality water for drinking and agricultural purposes. The objective description of the hydrochemical properties of groundwater was assisted by Piper (1944) and Stiff (1951) who presented graphical procedures for the interpretation of water analyses. Later, notable contributions were made by Chebotarev (1955), who described the natural chemical evolution of groundwater in the direction of groundwater flow, and Hem (1959), who provided extensive guidance on the study and interpretation of the chemical characteristics of natural waters. Later texts by Garrels and Christ (1965) and Stumm and Morgan (1981) provided thorough, theoretical treatments of aquatic chemistry.

By the end of the twentieth century, the previous separation of hydrogeology into physical and chemical fields of study had merged with the need to understand the fate of contaminants in the sub-surface environment. Contaminants are advected and dispersed by groundwater movement and can simultaneously undergo chemical processes that act to reduce pollutant concentrations. More recently, the introduction of immiscible pollutants, such as petroleum products and organic solvents into aquifers, has led to intensive research and technical advances in the theoretical description, modelling and field investigation of multi-phase systems. At the same time, environmental legislation has proliferated, and has acted as a driver in contaminant hydrogeology and in the protection of groundwater-dependent ecosystems. Today, research efforts are directed towards understanding natural attenuation processes as part of a managed approach to restoring contaminated land and groundwater and also in developing approaches to manage groundwater resources in the face of global environmental change.

Hence, hydrogeology has now developed into a truly interdisciplinary subject, and students who aim to become hydrogeologists require a firm foundation in Earth sciences, physics, chemistry, biology, mathematics, statistics and computer science, together with an adequate understanding of environmental economics and law, and government policy. Indeed, the principles of hydrogeology can be extended to the exploration of water on other planetary systems. Finding water on other planets is of great interest to the scientific community, second only to, and as a prerequisite for, detecting evidence for extraterrestrial life. As an example, a discussion of the evidence for water on Mars is given in Box 1.2.

Box 1.2 Groundwater on Mars?

Significant amounts of global surface hydrogen as well as seasonally transient water and carbon dioxide ice at both the North and South Polar Regions of Mars have been detected and studied for several years. The presently observable cryosphere, with volumes of $1.2 - 1.7 \times 10^6$ km^3 and $2 - 3 \times 10^6$ km^3, respectively, at the north and south poles, contains an equivalent global layer of water (EGL), if melted, of a few tens of metres deep (Smith *et al.* 1999; Farrell *et al.* 2009). Surface conditions on Mars are currently cold and dry, with water ice unstable at the surface except near the poles. Geologically recent, glacier-like landforms have been identified in the tropics and the mid-latitudes of Mars and are thought to be the result of obliquity-driven climate change (Forget *et al.* 2006). The relatively low volume of the EGL, coupled with widespread indications of chemical and geological landforms shaped by areas of recent groundwater seepage (Malin and Edgett 2000) and extensive palaeohydrological activity (Andrews-Hanna *et al.* 2010; Michalski *et al.* 2013; Salese *et al.* 2019), has resulted in the search for other extant water resources, as well as evidence of how much water, hydrogen and oxygen was stripped from the Martian atmosphere about 4 Ga.

The most likely reservoir for extensive storage of water on Mars is groundwater and a global Martian aquifer was long been assumed to exist beneath the permafrost at a depth where crustal temperatures maintained by geothermal heating may support liquid water. Depending on latitude, this melting isotherm is tentatively estimated to be located between depths of 5–9 km and to be overlain by a layer of mixed soil and ice (Farrell *et al.* 2009; Harrison and Grimm 2009). It is thought that topographic and

(Continued)

Box 1.2 (Continued)

temperature gradients act to create a significant and prolonged difference in hydraulic head between the melt water-fed, polar groundwater 'mound' and the equatorial aquifer and this is assumed to facilitate significant subsurface flow over geological timescales to establish a global equilibrium depth to the melting isotherm (Baker *et al.* 1991; Clifford 1993).

Martian groundwater research advanced greatly in the 1980s and early 1990s when the currently accepted ideas regarding subterranean dynamics and subsurface structure were hypothesized. Contemporary investigations are examining these assumptions using the imagery and data now collected by the extensive array of Martian orbiters, landers and rovers, notably NASA's Mars Odyssey satellite, launched in 2001, and the ESA Mars Express, in orbit since 2003. As Mars has a very thin atmosphere and no planetary magnetic field, solar cosmic rays reach the planet's surface unimpeded where they interact with nuclei in subsurface layers up to 2 m in depth, producing gamma rays and neutrons of differing kinetic energies that leak from the surface. Instruments on board the Mars Odyssey orbiter can detect this nuclear radiation and use it to calculate the spatial and vertical distribution of soil water and ice in the upper permafrost layer (Plate 1.4) (Mitrofanov *et al.* 2004; Feldman *et al.* 2008). The results indicate water ice content ranging from 10 to 55% by mass, depending on latitude, with the highest concentrations in and around the southern subpolar region (Mitrofanov *et al.* 2004).

The Mars Advanced Radar for Subsurface and Ionospheric Sounding (MARSIS) instrument mounted on the Mars Express satellite analyses the reflection of active, low frequency radio waves to identify aquifers containing liquid water, since these have a significantly different radar signature to the surrounding rock. The initial findings of the MARSIS sensor effectively identified the basal interface of the ice-rich layered deposits in the South Polar Region with a maximum measured thickness of 3.7 km, with an estimated total volume of 1.6×10^6 km^3, equivalent to a global water layer of approximately 11 m thick (Plaut *et al.* 2007). However, more recent studies using the MARSIS instrument presented a lack of direct evidence for the existence of subsurface water resources on Mars, possibly as a result of the high conductivity of the overlying crustal material (a mix of water ice and rock) resulting in a radar echo below the detectable limit of the MARSIS sensor (Farrell *et al.* 2009).

Other studies based on groundwater modelling approaches to explain various topographic features on Mars, such as chaotic terrains thought to have formed owing to disruptions of a cryosphere under high aquifer pore pressure, have concluded that a global confined aquifer system, for example as proposed by Risner (1989), is unlikely to exist and, instead, regionally or locally compartmentalized groundwater flow is more probable (Harrison and Grimm 2009).

Interestingly, the discovery of recurrent slope lineae (RSL) may indicate the presence of seasonal brine water flow on the Martian surface (McEwen *et al.* 2014). From data acquired by the Mars Reconnaissance Orbiter, RSL appear as narrow, dark markings, typically extending downslope on steep slopes from bedrock areas, often associated with small gullies, and indicative of intermittent, flow-like features (McEwen *et al.* 2014). It is conjectured that RSL emanate from bedrock outcrops and progressively lengthen during warm seasons and fade during cooler seasons, preferably on equatorial- and west-facing slopes. From structural mapping using

Box 1.2 (Continued)

observations from the High Resolution Imaging Science Experiment (HiRISE) of RSL source regions along the walls of craters, together with heat flow modelling and comparison with terrestrial analogues, Abotalib and Heggy (2019) considered that the source of RSL could be natural discharge along geological structures from briny aquifers within the cryosphere at depths of 750 m. The conceptual model presented by Abotalib and Heggy (2019) suggested that deep groundwater occasionally surfaces on Mars under present-day conditions (Fig. 1.7). The presence of RSL suggests that there is potentially abundant liquid water in some near-surface equatorial regions of Mars (McEwen *et al.* 2014), a necessary requirement of habitability for any planet.

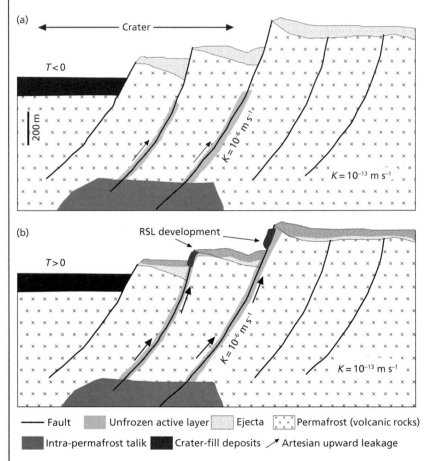

- —— Fault ▨ Unfrozen active layer ▦ Ejecta ⊞ Permafrost (volcanic rocks)
- ■ Intra-permafrost talik ■ Crater-fill deposits ↗ Artesian upward leakage

Fig. 1.7 Schematic diagram of the control of seasonal melting and freezing of shallow subsurface recurrent slope lineae (RSL) activity on Mars in which discharge of deep groundwater under high hydrostatic pressure occurs preferentially along fault-related ridges and scarps. (a) In winter, the system shuts down when ascending brines freeze within fault pathways in the near-surface. (b) In summer, the system resumes when the brine temperature rises above freezing point (Abotalib and Heggy 2019). (*Source:* Abotalib, A.Z. and Heggy, E. (2019) A deep groundwater origin for recurring slope lineae on Mars. *Nature Geoscience* **12**, 235–241. DOI: 10.1038/s41561-019-0327-5.)

1.5 The water cycle

A useful start in promoting a holistic approach to linking ground and surface waters is to adopt the hydrological cycle as a basic framework. The hydrological cycle, as depicted in Fig. 1.8, can be thought of as the continuous circulation of water near the surface of the Earth from the ocean to the atmosphere and then via precipitation, surface runoff and groundwater flow back to the ocean. Warming of the ocean by solar radiation causes water to be evaporated into the atmosphere and transported by winds to the land masses where the vapour condenses and falls as precipitation. The precipitation is either returned directly to the ocean, intercepted by vegetated surfaces and returned to the atmosphere by evapotranspiration, collected to form surface runoff, or infiltrated into the soil and underlying rocks to form groundwater. The surface runoff and groundwater flow contribute to surface streams and rivers that flow to the ocean, with pools and lakes providing temporary surface storage.

Of the total water in the global cycle, Table 1.1 shows that saline water in the oceans accounts for 97.25%. Land masses and the atmosphere therefore contain 2.75%. Ice caps and glaciers hold 2.05%, groundwater to a depth of 4 km accounts for 0.68%, freshwater lakes 0.01%, soil moisture 0.005% and rivers 0.0001%. About 75% of the water in land areas is locked in glacial ice or is saline (Fig. 1.9). The relative importance of groundwater can be realized when it is considered that, of the remaining quarter of water in land areas, around 98% is stored underground, and so making groundwater the second largest store of freshwater in the global cycle. In addition to the more accessible groundwater involved in the water cycle above a depth of 4 km, estimates of the volume of interstitial water in rock pores at even greater depths range from 53×10^6 km^3 (Ambroggi 1977) to 320×10^6 km^3 (Garrels *et al.* 1975).

Within the water cycle, and in order to conserve total water, evaporation must balance precipitation for the Earth as a whole. The average global precipitation rate, which is equal to the evaporation rate, is 496 000 km^3 a^{-1}. However, as Fig. 1.8 shows, for any one portion of the Earth, evaporation and precipitation generally do not balance. The differences comprise water transported from the oceans to the continents as atmospheric water vapour and water returned to the oceans as river runoff and a small amount (~6%) of direct

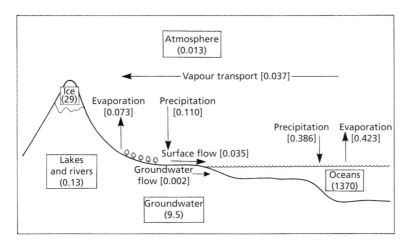

Fig. 1.8 The hydrological cycle. The global water cycle has three major pathways: precipitation, evaporation and water vapour transport. Vapour transport from sea to land is returned as runoff (surface water and groundwater flow). Numbers in ( ) represent inventories (in 10^6 km^3) for each reservoir. Fluxes in [ ] are in 10^6 km^3 a^{-1} (Berner and Berner 1987). (*Source: Berner, E.K. and Berner, R.A. (1987) The Global Water Cycle: Geochemistry and Environment.* Prentice-Hall, Inc., Englewood Cliffs, New Jersey. © 1987, Pearson Education.)

Table 1.1   Inventory of water at or near the Earth's surface (Berner and Berner 1987).

Reservoir	Volume (×10⁶ km³)	Percentage of total
Oceans	1370	97.25
Ice caps and glaciers	29	2.05
Deep groundwater (750–4000 m)	5.3	0.38
Shallow groundwater (<750 m)	4.2	0.30
Lakes	0.125	0.01
Soil moisture	0.065	0.005
Atmosphere[a]	0.013	0.001
Rivers	0.0017	0.0001
Biosphere	0.0006	0.00004
Total	1408.7	100

Note:
[a] As liquid equivalent of water vapour.
(*Source:* Berner, E.K. and Berner, R.A. (1987) *The Global Water Cycle: Geochemistry and Environment.* Prentice-Hall, Inc., Englewood Cliffs, New Jersey. © 1987, Pearson Education.)

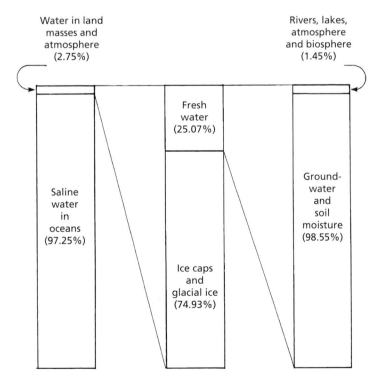

Fig. 1.9   The distribution of water at or near the Earth's surface. Only a very small amount of freshwater (<0.3% of total water) is readily available to humans and other biota (Maurits la Rivière 1989). (*Source:* Maurits la Rivière, J.W. (1989) Threats to the world's water. *Scientific American* **261**, 48–55.)

groundwater discharge to the oceans (Zektser and Loaiciga 1993).

The interfaces between hydrological compartments in the water cycle have important implications for water quantity and quality. Processes at the interfaces between the hydrological compartments (for example, soil-atmosphere or soil-groundwater) determine the age distribution of the water fluxes between these compartments and can, thus, greatly influence water travel and residence times (Sprenger *et al.* 2019). The age distribution of water spans over a wide range of temporal scales. In the 'critical zone', the Earth's boundary layer ranging from the top of the vegetation layer to the bottom of the groundwater storage, water ages range from hours to millennia (Fig. 1.10).

By taking the constant volume of water in a given reservoir and dividing by the rate of addition (or loss) of water to (from) it enables the calculation of a residence time for that reservoir. For the oceans, the volume of water present (1370×10^6 km^3; see Fig. 1.8) divided by the rate of river runoff to the oceans (0.037×10^6 km^3 a^{-1}) gives an average time that a water molecule spends in the ocean of about 37 000 years. Lakes, rivers, glaciers and shallow groundwater have residence times ranging between days and thousands of years. Because of extreme variability in volumes and precipitation and evaporation rates, no simple average residence time can be given for each of these reservoirs. As a rough calculation, and with reference to Fig. 1.8 and Table 1.1, if about 6% (2220 km^3 a^{-1}) of runoff from land is taken as active groundwater circulation, then the time taken to replenish the volume (4.2×10^6 km^3) of shallow groundwater stored below the Earth's surface is of the order of 2000 years. In reality, groundwater residence times vary from about 2 weeks to 10 000 years (Nace 1971), and longer (Edmunds 2001). A similar estimation for rivers provides a value of about 20 days. These estimates, although a gross simplification of the natural variability, do serve to emphasize the potential longevity

of groundwater pollution compared to more rapid flushing of contaminants from river systems.

1.5.1 Groundwater occurrence in the upper continental crust

Focusing on the upper 2 km of the continental crust in which most hydrogeological observations are made, Gleeson *et al.* (2015) combined multiple approaches using geospatial datasets, tritium age dating of groundwater and numerical modelling to show that less than 6% of the groundwater in the uppermost portion of the Earth's land mass is less than 50 years old, representing modern groundwater that is the most recently recharged. Gleeson *et al.* (2015) found that the total groundwater volume in the upper 2 km of continental crust is approximately 22.6×10^6 km^3, of which 0.1–5.0×10^6 km^3 is less than 50 years old. The distribution of this modern groundwater is spatially heterogeneous, with very little in arid regions. Although modern groundwater represents a small percentage of the total groundwater storage on Earth, the volume of this component is still very significant, equivalent to a water depth of about 3 m spread over the world's continents.

1.5.2 Groundwater-related tipping points

Several potential groundwater-related tipping points are associated with the storage function of groundwater (Gleeson *et al.* 2020). Most critical for aquatic ecosystems is the role of groundwater as a stable supply of baseflow, and therefore a key tipping point is when a stream transitions from perennial to intermittent due to groundwater depletion (see Section 6.8). Groundwater-related tipping points are also present for terrestrial groundwater-dependent ecosystems. Groundwater within or near the root zone provides a stable supply of water, particularly during drought, for many natural and agricultural crops via capillary rise and direct groundwater uptake (see Section 6.4.1). Since groundwater is

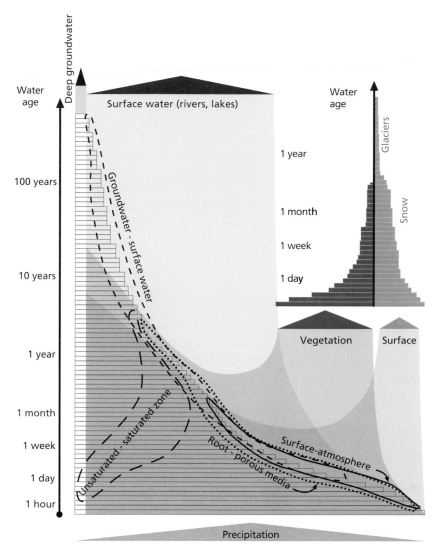

Fig. 1.10   Conceptual diagram showing hypothetical age distributions in the Earth's critical zone. The envelopes shown indicate the mixing of water with different ages at the interfaces between hydrological compartments (Sprenger *et al.* 2019). (*Source:* Adapted from Sprenger, M., Stumpp, C., Weiler, M *et al.* (2019) The demographics of water: a review of water ages in the critical zone. *Reviews of Geophysics* **57**, DOI: 10.1029/2018RG000633.)

estimated to influence terrestrial ecosystems over 7–17% of global land area (Fan *et al.* 2013) and can contribute substantially to evapotranspiration, it is likely that groundwater constitutes an important component of terrestrial evapotranspiration (Gleeson *et al.* 2020). For instance, groundwater is an essential contributor to evapotranspiration in the Amazon Basin (Fang *et al.* 2017).

1.5.3   Groundwater discharge to the oceans

The approximate breakdown of direct groundwater discharge from continents to adjacent oceans and seas was estimated by Zektser and Loaiciga (1993) as follows: Australia 24 km^3 a^{-1}; Europe 153 km^3 a^{-1}; Africa 236 km^3 a^{-1}; Asia 328 km^3 a^{-1}; the Americas

$729 \, \text{km}^3 \, \text{a}^{-1}$; and major islands $914 \, \text{km}^3 \, \text{a}^{-1}$. The low contribution from the Australian continent of direct groundwater discharge, despite its relatively large territory, is attributed to the widespread occurrence of low-permeability surface rocks that cover the continent. At the other extreme, the overall proximity of recharge areas to discharge areas is the reason why major islands of the world contribute over one-third of the world's direct groundwater discharge to the oceans. The largest direct groundwater flows to oceans are found in mountainous areas of tropical and humid zones and can reach $10–15 \times 10^{-3} \, \text{m}^3 \, \text{s}^{-1} \, \text{km}^{-2}$. The smallest direct groundwater discharge values of $0.2–0.5 \times 10^{-3} \, \text{m}^3 \, \text{s}^{-1} \, \text{km}^{-2}$ occur in arid and arctic regions that have unfavourable recharge and permeability conditions (Zektser and Loaiciga 1993).

In a later study presented by Luijendijk *et al.* (2020), the application of a spatially resolved, density-driven global model of coastal groundwater discharge showed that the contribution of fresh groundwater to the world's oceans is equal to 224 (range 1.4–500) $\text{km}^3 \, \text{a}^{-1}$, and accounts for approximately 0.6% (range 0.004–1.3%) of the total freshwater input and approximately 2% (range 0.003–7.7%) of the solute input of carbon, nitrogen, silica and strontium. The uncertainty ranges reported are mostly caused by the high uncertainty of the values of permeability that were used, which is on average two orders of magnitude. Additional sources of uncertainty are the representative topographic gradient of coastal watersheds, groundwater recharge, and the size of the area that contributes to coastal groundwater discharge.

The coastal discharge of freshwater showed a high spatial variability. For an estimated 26% (0.4–39%) of the world's estuaries, 17% (0.3–31%) of salt marshes and 14% (0.1–26%) of coral reefs, the flux of terrestrial groundwater exceeds 25% of the river flux and poses a risk for pollution and eutrophication. Catchments with hotspots of coastal groundwater discharge, where coastal groundwater discharge exceeds $100 \, \text{m}^2 \, \text{a}^{-1}$ and 25% of the river

discharge, were located predominantly in areas with a steep coastal topography due to glacio-isostatic rebound, active tectonics or volcanic activity, and in areas consisting of permeable unconsolidated sediments, carbonates or volcanic rocks. The distribution of these hotspots is consistent with reported sites of high fresh groundwater discharge found in North America, Europe and East Asia. However, at many hotspots, such as Iceland and parts of South America, Africa and South Asia, and many tropical islands, coastal groundwater discharge requires further exploration. In summary, Luijendijk *et al.* (2020) concluded that fresh groundwater discharge is insignificant for the world's oceans, but important for coastal ecosystems. For further discussion of groundwater discharge to the oceans, see Section 2.16.

1.5.4 Global groundwater material and elemental fluxes

As an agent of material transport to the oceans of products of weathering processes, groundwater probably represents only a small fraction of the total transport (see Table 1.2). Rivers (89% of total transport) represent an important pathway while groundwater accounts for a poorly constrained estimate of 2% of total transport in the form of dissolved materials (Garrels *et al.* 1975). Estimates by Zektser and Loaiciga (1993) indicated that globally the transport of salts via direct groundwater discharge is approximately $1.3 \times 10^9 \, \text{t a}^{-1}$, roughly equal to half of the quantity contributed by rivers to the oceans. Given a volumetric rate of direct groundwater discharge to the oceans of 2220 $\text{km}^3 \, \text{a}^{-1}$, the average dissolved solids concentration is about $585 \, \text{mg L}^{-1}$. This calculation illustrates the long residence time of groundwater in the Earth's crust, where its mineral content is concentrated by dissolution.

In an analysis of comprehensive datasets of the chemistry of groundwater and produced water (groundwater pumped during oil and gas extraction) compiled by the US Geological Survey (DeSimone *et al.* 2014; Blondes *et al.*

Table 1.2 Material transport and subsurface dissolved salts discharge from groundwater to the world's oceans (Garrels *et al.* 1975 and Zektser and Loaiciga 1993).

Agent or Ocean	% of total material transport (Remarks)	% of total dissolved salts transport	Subsurface dissolved salts discharge (10^6 t a^{-1})
Surface runoff	89 (Dissolved load 19%, suspended load 81%)	66	–
Glacier ice, coastal erosion, volcanic and wind-blown dust	~9 (Ice-ground rock debris, cliff erosion sediments, volcanic and desert-source dust)	–	–
Groundwater	2 (Dissolved salts similar to river water composition)	34	–
Pacific		–	520.5
Atlantic		–	427.8
Mediterranean Sea		–	42.5
Indian		–	295.5
Arctic		–	7.2
All oceans			1293.5

(*Sources:* Garrels, R.M., Mackenzie, F.T. and Hunt, C. (1975) *Chemical Cycles and the Global Environment: Assessing Human Influences.* Kaufman, Los Altos, California; Zektser, I.S. and Loaiciga, H.A. (1993) Groundwater fluxes in the global hydrologic cycle: past, present and future. *Journal of Hydrology* **144**, 405–427.)

2017), together with estimates of global groundwater usage, Stahl (2019) estimated elemental fluxes from global pumping and found that groundwater fluxes contribute appreciably to the overall cycles of a number of important elements and may provide a significant portion (more than 10%) of crop requirements of key nutrients (e.g. potassium and nitrogen) where groundwater is used for irrigation. Comparing the dissolved solute flux from groundwater pumping to the dissolved solute flux of approximately 4–5×10^9 t a^{-1} delivered annually to the ocean by rivers (Sen and Peucker-Ehrenbrink 2012), Stahl (2019) calculated that total pumping with and without produced waters gives fluxes of total dissolved solids (TDS) of 881×10^6 and 513×10^6 t a^{-1}, respectively, which represent 20 and 7% of the global dissolved solute flux carried by rivers. The fact that groundwater and produced water pumping are exclusively anthropogenic fluxes of water highlights the significance of groundwater pumping in global elemental cycles.

1.5.5 Human influence on the water cycle

Although rarely depicted in diagrams of the water cycle, human activity alters the water cycle in three distinct, but inter-related ways (Abbott *et al.* 2019). First, humans appropriate water through: (1) livestock, crop and forestry use of soil moisture that eventually flows back to the atmosphere as evapotranspiration (so-called 'green water' use with an estimated flux of 15–22×10^3 km^3 a^{-1}); (2) as surface water and groundwater abstraction ('blue water' use with an estimated flux of 3.8–6.0×10^3 km^3 a^{-1}); and (3) as water required to assimilate pollution ('grey water' use with an estimated flux of 1.0–2.0×10^3 km^3 a^{-1}) (Abbott *et al.* 2019; Schyns *et al.* 2019). Second, humans have disturbed approximately three-quarters of the Earth's ice-free land surface through activities that include agriculture, deforestation and wetland destruction (Ellis *et al.* 2010). These disturbances alter evapotranspiration, groundwater recharge, river discharge and precipitation at continental scales (Falken-mark *et al.* 2019). Third, climate change is disrupting patterns of water flow and storage at local to global scales (Haddeland *et al.* 2014). These human interferences with the water cycle have created problems for billions of individuals and many ecosystems worldwide. The estimates of human green, blue and grey water use (about $24 \times 10^3\,\mathrm{km}^3\,\mathrm{a}^{-1}$) indicates that human freshwater appropriation redistributes the equivalent of half of global river discharge or double global groundwater recharge each year (Abbott *et al.* 2019).

1.6 Global groundwater resources

Groundwater is an important natural resource. Worldwide, more than two billion people depend on groundwater for their daily supply (Kemper 2004). Total global fresh water use is estimated at about $4000\,\mathrm{km}^3\,\mathrm{a}^{-1}$ (Margat and Andréassian 2008) with 99% of the irrigation, domestic, industrial and energy use met by abstractions from renewable sources, either surface water or groundwater. Less than 1% (currently estimated at $30\,\mathrm{km}^3\,\mathrm{a}^{-1}$) is obtained from non-renewable (fossil groundwater) sources mainly in three countries: Algeria, Libya and Saudi Arabia.

The increase in global groundwater exploitation has been stimulated by the development of low-cost, power-driven pumps and by individual investment for irrigation (Plate 1.5) and urban uses. Currently, aquifers supply approximately 20% of total water used globally, with this share rising rapidly, particularly in dry areas (IWMI 2007). Globally, 65% of groundwater utilization is devoted to irrigation, 25% to the supply of drinking water and 10% to industry.

Groundwater resources are often the only source of supply in arid and semi-arid zones (for example, 100% in Saudi Arabia and Malta, 95% in Tunisia and 75% in Morocco). As demonstrated for the case of the North-west Sahara Aquifer System (Box 1.3), irrigated agriculture is the principal user of groundwater from the major sedimentary aquifers of the Middle East, North Africa, North America and the Asian alluvial plains of the Punjab and Terai (WWAP 2009). Groundwater has been most intensively developed in South Asia and North America, where it provides 57 and 54%, respectively, of all irrigation water. The rapid growth in the construction of irrigation water-wells now supplies $3.9 \times 10^6\,\mathrm{km}^2$ of irrigated land in India, $1.9 \times 10^6\,\mathrm{km}^2$ in China, $1.7 \times 10^6\,\mathrm{km}^2$ in the United States and also across large areas of Pakistan and Bangladesh (Konikow *et al.* 2015).

Whether groundwater or surface water is exploited for water supply is largely dependent on the location of aquifers relative to the point of demand. A large urban population with a high demand for water would only be able to exploit groundwater if the aquifer, typically a sedimentary rock, has favourable storage and transmission properties, whereas in a sparsely populated rural district more limited but essential water supplies might be found in poor aquifers, such as weathered basement rock.

Box 1.3 The North-west Sahara Aquifer System and the Ouargla Oasis, Algeria

The Sahara Basin covers an area of about 780 000 km^2 and includes two sub-basins separated by the M'zab High. The western sub-basin occupies about 280 000 km^2 and is covered by sand dunes of the Grand Erg Occidental and the eastern sub-basin extends over about 500 000 km^2 and is covered by the desert of the Grand Erg Oriental. The Sahara Basin is underlain by two major aquifers that comprise the Northwest Sahara Aquifer System (NWSAS) of Cretaceous age that extends below Algeria, Tunisia and Libya

Box 1.3 (Continued)

in North Africa (see Plate 2.7). The Lower Aquifer (the Continental Intercalaire) is composed of continental sandstone alternating with argillaceous layers and the Upper Aquifer (the Complex Terminal) is a multi-layered aquifer consisting of sandstones and limestones. The thickness of the thicker, more extensive Lower Aquifer ranges between 200 and 1000 m, decreasing north-eastwards to 125 m. The lower confining unit consists of argillaceous and marly formations of Devonian-Triassic age while the upper confining units consist of evaporites and clays of Upper Cretaceous age (Zektser and Everett 2004).

In western Algeria, the Lower and Upper Aquifers are almost independent but towards the Mediterranean coast the aquifers become interconnected or merge to form one aquifer system. Groundwater movement in the NWSAS is towards the south and south-west in the western sub-basin. In the eastern sub-basin, where the Complex Terminal aquifer is heavily exploited in Algeria and Tunisia, groundwater flows towards discharge areas, mainly desert depressions or oases known as 'chotts'. The chotts supply irrigation water through traditional qanat systems (foggaras), with some 570 foggaras discharging about 90×10^6 m^3 a^{-1} (Zektser and Everett 2004).

In the western sub-basin, the total dissolved solids content of groundwater in the Lower Aquifer ranges from 0.5 to 1 g L^{-1}. In the eastern sub-basin, salinity increases to 5 g L^{-1}. The concentration of total dissolved solids in the Upper Aquifer is about 2 g L^{-1} in southern areas of the Grand Erg Oriental, increasing in concentration north-eastwards from 2 to 5 g L^{-1} at Tozeur in Tunisia. At Ouargla in Algeria, concentrations reach 8 g L^{-1} in discharge areas (Zektser and Everett 2004).

Groundwater reserves in the NWSAS are estimated to be 60 000 km^3 although, given the low rainfall amount, the aquifer system is generally considered a non-renewable aquifer system. Use of the superficial water table of the NWSAS extends back to ancient times and, from the middle of the nineteenth century, boreholes were drilled to access deeper parts of the aquifer. In Algeria, exploitation of groundwater from the aquifer system was about 150×10^6 m^3 until 1940 since when pumping has increased to about 260×10^6 m^3 a^{-1} (Zektser and Everett 2004). By the 1970s, in the Ouargla Oasis of northern Algeria, there were approximately 2000 boreholes developed in the NWSAS in order to irrigate date palms (see Plate 1.6). In southern Tunisia, exploitation of the Complex Terminal aquifer increased from 9×10^6 m^3 a^{-1} in 1900 to about 190×10^6 m^3 a^{-1} in 1995. The impact of this intensive development on the aquifer system has been observed in discharge areas. In Algeria, the flow of springs decreased from 200 L s^{-1} in 1900 to 6 L s^{-1} in 1970, whereas in Tunisia, the flow from springs decreased from 2500 L s^{-1} in 1900 to virtually nil (less than 30 L s^{-1}) in 1990 (Zektser and Everett 2004). Traditional irrigation methods in the region used sustainable quantities of water, but the more intensive modern irrigation methods used at present have led to a degraded water quality, decreased water levels and loss of artesian pressure, as well as salinization of the superficial water table and soil zone due to the drainage conditions. This salinized water is typically at a depth of 0.5–1.5 m below the soil surface and is detrimental to date palms (UNEP 2008).

1.6.1 Global groundwater abstraction

With rapid population growth, groundwater abstractions have tripled over the last 40 years (Fig. 1.11), largely explained by the rapid increase in irrigation development stimulated by food demand in the 1970s and by the continued growth of agriculture-based economies (World Bank 2007). Emerging market economies such as China, India and Turkey, which

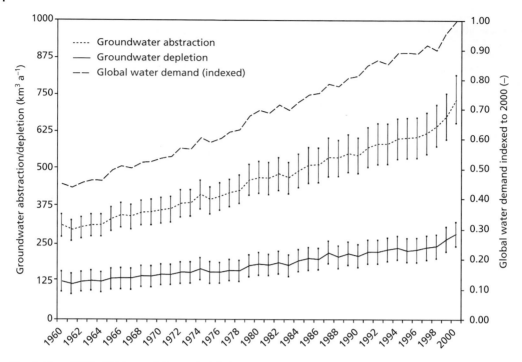

Fig. 1.11 1960–2000 trends in total global water demand (right axis, indexed for the year 2000), global groundwater abstraction (left axis, km^3 a^{-1}) and global groundwater depletion (left axis, defined as groundwater abstraction in excess of groundwater recharge in km^3 a^{-1}) for sub-humid and arid areas (Wada *et al.* 2010). (*Source:* Wada, Y., van Beek, L.P.H., van Kempen, C.M. *et al.* (2010). Global depletion of groundwater resources. *Geophysical Research Letters* **37**, L20402, 5 pp. © 2010, John Wiley & Sons.)

still have an important rural population dependent on water supply for food production, are also experiencing rapid growth in domestic and industrial demands linked to urbanization. Urbanized and industrial economies such as the European Union and the United States import increasing amounts of 'virtual water' (Allan 1998, 2003) in food (see Box 1.4) and manufactured products, while water use in industrial processes and urban environments has been declining, due to both technological changes in production processes and pollution mitigation efforts (WWAP 2009).

Table 1.3 shows that three countries, India, China and the United States, are estimated to abstract in excess of 100 km^3 a^{-1} of groundwater and that several countries, including Pakistan, Saudi Arabia, Syria, India, Iran and Bangladesh, use greater than 86% of groundwater abstraction for irrigation, while Indonesia, the Russian Federation and Thailand use greater than 60%

of groundwater abstraction for domestic use. Also, Saudi Arabia, Bangladesh, Syria and Iran abstract more than 57% of their total freshwater demand from groundwater.

In regions of the world where the rate of groundwater abstraction exceeds the rate of natural groundwater recharge over extensive areas and for long time periods, over-exploitation or persistent groundwater depletion is the consequence. As shown by the global overview of Wada *et al.* (2010), in the year 2000, the rate of total global groundwater depletion is estimated to have increased to 283 km^3 a^{-1} (Fig. 1.11). Groundwater depletion rates were found to be highest in some of the world's major agricultural regions including north-east Pakistan and north-west India, north-east China, the Ogallala aquifer in the central United States (see Section 10.2.1), the San Joaquin aquifer in the Central Valley of California (see Box 2.7), Iran, Yemen and south-east Spain.

Table 1.3 Countries with estimated groundwater abstraction greater than 10 km^3 a^{-1} (Margat and van der Gun 2013).

Country	Estimated groundwater abstraction (km^3 a^{-1})	Groundwater abstraction by sector			Groundwater share of total freshwater abstraction (%)
		Irrigation (%)	Domestic use (%)	Industry (%)	
India	251.00	89	9	2	33
China	111.95	54	20	26	18
United States of America	111.70	71	23	6	23
Pakistan	64.82	94	6	0	32
Iran	63.40	87	11	2	57
Bangladesh	30.21	86	13	1	79
Mexico	29.45	72	22	6	35
Saudi Arabia	24.24	92	5	3	95
Indonesia	14.93	2	93	5	11
Turkey	13.22	60	32	8	16
Russian Federation	11.62	3	79	18	18
Syria	11.29	90	5	5	65
Japan	10.94	23	29	48	
Thailand	10.74	14	60	26	17
Italy	10.40	67	23	10	25
Brazil	10.06	38	38	24	14

(*Source:* Adapted from Margat, J. and van der Gun, J. (2013) *Groundwater Around the World: A Geographic Synopsis.* CRC Press/Balkema, EH Leiden.)

Box 1.4   Groundwater depletion from global irrigated crop production and trade

The over-abstraction of groundwater in major food producing areas of the world such as north-west India, North China, the central United States and California is rapidly depleting groundwater storage in large aquifer systems. The global scale of groundwater depletion due to irrigation is difficult to assess given the lack of research integrating crop water use, groundwater depletion and the embedding of groundwater in international food trade. In the study by Dalin *et al.* (2017), groundwater depletion linked to irrigation (GWD) (in which GWD is defined as the volume of groundwater abstracted for irrigation use in excess of the natural recharge rate and irrigation return flow, and allowing for environmental flow requirements) is estimated based on 26 crop classes and bilateral trade flow for 360 commodities. The results shown in Table 1.4 for 2010, show the volume of GWD embedded in food production and trade for the top 10 countries with the most GWD.

(*Continued*)

Box 1.4 (Continued)

Table 1.4 Groundwater depletion (GWD) for irrigation embedded in national food production, imported and exported GWD, and corresponding fractions of GWD in global food production, national food consumption and national food production for the year 2010 (Dalin *et al.* 2017).

Country	GWD in production (km^3 a^{-1})	Fraction of global GWD (%)	GWD in imports (km^3 a^{-1})	Fraction of GWD in national consumption (%)	GWD in exports (km^3 a^{-1})	Fraction of GWD in national production (%)
India	73.5	33.9	0.2	0.3	3.0	4.0
Iran	33.3	15.4	1.4	4.2	1.2	3.5
Pakistan	27.5	12.7	0.2	1.2	7.3	26.4
China	24.0	11.1	2.2	8.5	0.3	1.1
USA	16.2	7.5	1.7	15.3	6.9	42.4
Saudi Arabia	12.5	5.7	0.8	6.0	0.4	3.5
Mexico	11.1	5.1	1.0	10.6	2.5	22.6
Libya	2.5	1.1	0.1	2.4	0	0.1
Turkey	2.0	0.9	0.5	22.6	0.4	18.0
Italy	2.0	0.9	0.5	27.9	0.8	39.2
Total top ten	204.6	84.8	8.6	4.5	22.8	11.1
Total world	241.4	100	25.6	NA	25.6	NA

Note: Also shown are totals for these ten countries and for the world. NA, not applicable.
(*Source:* Adapted from Dalin, C., Wada, Y., Kastner, T. and Puma, M.J. (2017) Groundwater depletion embedded in international food trade. *Nature* **543**: 700–704.)

Global GWD increased to 292 km^3 in 2010, mostly due to increases in India, China and the United States. The crops accounting for most depletion globally, both in terms of their large production and GWD intensity, are wheat (22% of global GWD, or 65 km^3), rice (17%), sugar crops (7%), cotton (7%) and maize (5%) (Dalin *et al.* 2017). The countries irrigating crops from over-exploited aquifers export these crops in various proportions: India retains most of its large GWD-based crop production for domestic use (only 4% of GWD exported), while the United States, Pakistan and Mexico export significant portions of their GWD-based crop production (Table 1.4). Globally, about 11% of GWD is embedded in international food trade,

of which exports from Pakistan, the United States and India alone account for more than two-thirds of all embedded GWD. Pakistan is the largest exporter, with 29% of the global GWD trade volume, followed by the United States (27%) and India (12%) (Dalin *et al.* 2017).

Five of the 10 countries shown in Table 1.4 with the most GWD (the United States, Mexico, Iran, Saudi Arabia and China) are also the top importers of GWD via food trade. Critically, these countries import or export crops irrigated from the world's most stressed aquifer systems. As demonstrated by Gleeson *et al.* (2012), food production relying on these aquifers is particularly unsustainable with extraction rates 20–50 times higher than required for sustainable groundwater use

Box 1.4 (Continued)

(see further Table 6.1 and Section 10.2). For example, the United States imports about 1.5 times as much GWD from Mexico (mainly via citrus and sugar crops) as it exports there (mainly via cotton and maize) (Dalin *et al.* 2017).

Therefore, it is of concern that exhaustion of aquifers in areas that are hotspots of water and food security related to GWD threaten the food supply both domestically and in their water-stressed trade partners. Clearly, solutions are required to improve the sustainability of water use and food production for those regions, crops and trade relationships

that are most reliant on over-exploited aquifers. In the food producing countries, solutions could include water-saving strategies such as improving irrigation efficiency and growing more drought-resistant crops, together with targeted measures such as metering and regulation of groundwater pumping, while accounting for local socio-economic, cultural and environmental requirements (Dalin *et al.* 2017). In addition, food importing countries can assist these solutions by promoting and supporting sustainable irrigation practices with their trade partners.

1.6.2 Global groundwater depletion and sea level rise

Using a global hydrological model, Wada *et al.* (2010) assessed the amount of groundwater depletion, defined as the excess of abstraction over recharge replenishment, and estimated that for sub-humid and arid areas the rate of total global groundwater depletion has increased from 126 ± 32 km^3 a^{-1} in 1960 to 283 ± 40 km^3 a^{-1} in 2000 (Fig. 1.11). Groundwater depletion in 2000 equalled about 40% of the global annual groundwater abstraction, about 2% of the global annual groundwater recharge and about 1% of the global annual continental runoff, contributing a considerable amount (about 25%) of 0.8 ± 0.1 mm a^{-1} to current sea level rise.

Using a similar approach in which groundwater depletion was directly calculated using calibrated groundwater models, analytical approaches or volumetric budget analyses for multiple aquifer systems, Konikow (2011) estimated an average global groundwater depletion rate of 145 km^3 a^{-1} during the period 2000–2008, equivalent to 0.4 mm a^{-1} of sea-level rise, or 13% of the reported rise of 3.1 mm a^{-1} during this period.

Using an integrated water resources assessment model to simulate global terrestrial water stocks and flows, Pokhrel *et al.* (2012) estimated that the sum of unsustainable groundwater use, artificial reservoir water impoundment, climate-driven changes in terrestrial water storage and the loss of water in closed basins, principally the Aral Sea, has contributed a sea-level rise of about 0.77 mm a^{-1} between 1961 and 2003, or about 42% of the observed sea-level rise. Considering a simulated mean annual unsustainable groundwater use during 1951–2000 of about 359 km^3 a^{-1}, Pokhrel *et al.* (2012) estimated, using the assumption of Wada *et al.* (2010) that 97% of unsustainable groundwater use ends up in the oceans, a cumulative sea-level rise due to groundwater over-abstraction during this period of 48 mm or about 1 mm a^{-1}.

1.7 Groundwater resources in developed countries

Approaches to the management and protection of groundwater resources have developed in parallel with our understanding of the economic and environmental implications of groundwater exploitation. In developed countries, with the availability of a dense network of field monitoring data and remotely sensed information for the quantification, modelling and reporting of water resources within existing regulatory frameworks, there is a good understanding of the distribution, size and utilization of groundwater resources. In this section, patterns of groundwater abstraction, as well as regulations for the management of groundwater resources, are described for several industrialized regions of the world (the United Kingdom, Europe, Canada and the United States), as well as the rapidly developing country of China.

1.7.1 Groundwater abstraction in the United Kingdom

The distribution of groundwater abstraction volumes in the United Kingdom is largely determined by the relationship between population and geology, with major producing aquifers in the densely populated south and east of the country and minor producing aquifers in the less-densely populated north and west. Table 1.5 provides a breakdown of water use by purpose and type (surface water and groundwater) for regions of England. Surface water abstraction for electricity generation is the largest category, but most of the freshwater abstracted for cooling purposes is returned to rivers and can be used again downstream. For England, groundwater accounts for 20% of total abstractions (surface water and groundwater) for all water use purposes and 31% of total abstractions for domestic use. In terms of abstractions for domestic water supply, groundwater is especially significant in the Southern (75% dependence on groundwater), Anglian (36%), Thames (34%) and Midlands (32%) regions. In these densely populated regions of south-east England and the English Midlands, good quality groundwater is obtained from the high-yielding Cretaceous Chalk and Triassic sandstone aquifers.

Groundwater is also important for spray irrigation. The Anglian region in the east of England, an area of intensive arable farming and horticulture, has the highest demand, with 45% of abstractions for spray irrigation obtained from groundwater (Plate 1.5).

Scotland, Wales and Northern Ireland have ample surface water resources that are important for public supplies. Aquifers are typically less productive and/or more localized in these countries, although groundwater is significant for private supplies where properties are not connected to the public supply network. In Scotland, groundwater contributes about 5% to public supplies, with approximately 100 boreholes and springs used to supply some major rural towns (Dochartaigh *et al.* 2015). The total volume of public supply from groundwater in Scotland is estimated as $235 \times 10^3 \, \mathrm{m}^3 \, \mathrm{day}^{-1}$ in 2004. In addition, groundwater is important for about 70% of private supplies in Scotland, serving at least 330 000 people. More than 4000 boreholes, as well as some large springs, are used for large private, industrial or agricultural supplies, and approximately 20 000 boreholes, small springs and wells provide private water supplies for at least 80 000 people (Dochartaigh *et al.* 2015).

Groundwater supplies about 3% of public supplies in Wales, which equates to approximately $40 \times 10^3 \, \mathrm{m}^3 \, \mathrm{day}^{-1}$, with most groundwater sources operated conjunctively with surface water sources. However, some groundwater sources are critical in supplying local areas that cannot be supplied by other means (Environment Agency 2015; Welsh Water 2019). In Northern Ireland, groundwater is a negligible component of the public water supply, contributing only 0.6% (Northern Ireland Water 2013).

1.7.1.1 Management and protection of groundwater resources in the United Kingdom

In the United Kingdom, it is interesting to follow the introduction of relevant legislation, and how this has increased hydrogeological

Table 1.5 Estimated abstractions from all surface water and groundwater in England by purpose and Environment Agency region for 2017. All data are given as 10^6 m^3.

Region	Domestic use[a]		Spray irrigation		Agriculture[b]		Industry[c]		Total	
	Surface water	Groundwater	Surface water	Groundwater	Surface water	Groundwater	Surface water	Groundwater	Surface water	Groundwater
England	3661	1669	46	41	770	140	3873	195	8350	2044
North West	624	65	0	1	13	3	863	23	1501	92
North East	640	104	1	3	65	24	1126	21	1834	153
Midlands	585	280	9	7	12	2	814	38	1429	327
Anglian	484	269	29	24	19	3	145	35	680	332
Thames	918	482	1	3	54	10	286	27	1262	523
Southern	112	341	4	2	257	47	23	32	397	421
South West	297	127	0	1	350	51	600	19	1248	197

[a] Category includes public and private water supplies.
[b] Category includes fish farming, cress growing and amenity ponds. Excludes spray irrigation.
[c] Category includes electricity supply and other industries.
(*Source:* Water abstraction data sets, Department for Environment, Food and Rural Affairs, 2017. Contains public sector information licensed under the Open Government Licence v3.0.)

knowledge, with overviews provided by Downing (1993) and Streetly and Heathcote (2018). Hydrogeological experience prior to 1945 rested on a general awareness of sites likely to provide favourable yields, changes in chemistry down-gradient from the point of recharge and hazards such as ground subsidence from groundwater over-exploitation. The Water Act 1945 provided legal control on water abstractions and this prompted an era of water resources assessment that included surveys of groundwater resources, the development of methods to assess recharge amounts (Section 6.5), and the initiation of groundwater studies. Increased abstraction from the Chalk aquifer during the 1950s and a drought in 1959 highlighted the effect of groundwater abstractions upon Chalk streams and stimulated the need for river baseflow studies (Section 6.7.1). Furthermore, the application of quantitative pumping test analysis techniques (Section 7.3.2) during this period revealed spatial variations in aquifer transmissivity and an association between transmissivity and topography.

The Water Resources Act 1963 led to the formation of 27 catchment-based authorities responsible for pollution prevention, fisheries, land drainage and water resources. The Act ushered in a decade of groundwater resources management that required the licensing of all abstractions in England and Wales. Under Section 14 of the Act, each authority was required to undertake a survey of resources and the Water Resources Board (abolished 1974) was established with the task of resource planning on a national scale. Regional groundwater schemes were developed in the context of river basin analysis for the purposes of river augmentation by groundwater, seasonal abstraction and artificial recharge. Scientific advancement in the application of numerical models to solve non-linear equations of groundwater flow permitted the prediction of future groundwater abstraction regimes.

The Water Act 1973 reflected the importance of water quality aspects and heralded the developing interest in groundwater quality. The Act led to the formation of 10 catchment-based regional water authorities with responsibility for all water and sewerage services and for all parts of the water cycle. The Control of Pollution Act 1974 extended the powers of the regional water authorities in controlling effluent discharge to underground strata and limited certain activities that could lead to polluting discharges. The first aquifer protection policies were developed at this time.

The Water Act 1989 separated the water supply and regulatory functions of the regional water authorities, and the new National Rivers Authority was set-up to manage water resources planning, abstraction control, pollution prevention and aquifer protection. A number of other Acts of Parliament followed including the Environmental Protection Act 1990 and the Water Resources Act 1991 that control the direct and indirect discharge of harmful substances into groundwater and are, in part, an enactment of the European Communities Directive on the Protection of Groundwater Against Certain Dangerous Substances (80/68/EEC). Further controls on discharges were implemented under the Groundwater Regulations 1998. In addition, the Water Resources Act 1991 consolidated all the provisions of the Water Resources Act 1963 in respect of the control of groundwater abstractions. In pursuing a strategy to protect both individual borehole sources and wider groundwater resources, the National Rivers Authority (1992) developed its practice and policy for the protection of groundwater with the aim of raising awareness of the vulnerability of groundwater to surface-derived pollution. Following the establishment of the Environment Agency under the Environment Act 1995 (when the National Rivers Authority, Her Majesty's Inspectorate of Pollution and the Waste Regulatory Authorities were brought together) the practice and policy document for the protection of groundwater was updated (Environment Agency 1998).

Currently, the Environment Agency for England and Wales promotes a national framework for water resources protection in the context of existing European initiatives, principally the Water Framework Directive (Section 1.7.2.1). The Water Act 2003 is one example of legislation to further the sustainable use of water resources and protect the environment. The Act links water abstraction licensing to local water resource availability and moves from a licensing system based on purpose of use to one based on volume consumed. The Act also introduces time-limited licences to give flexibility in making changes to abstraction rights in the face of climate change and increased demand. From 2012, licences without a time limit will be revoked, without a right to compensation, if an abstraction causes significant environmental damage.

1.7.2 Groundwater abstraction in Europe

In Europe, groundwater is again a significant economic resource. According to a report commissioned for the European Commission (RIVM and RIZA 1991), about 75% of the inhabitants of Europe depend on groundwater for their water supply. As Table 1.6 reveals, large quantities of groundwater are abstracted in Italy (10.40 km^3 a^{-1}), as well as Portugal, Germany, Spain and France (all in excess of 5.00 km^3 a^{-1}). Groundwater accounts for in excess of 97% of total freshwater abstraction in Malta, Montenegro, Denmark, Croatia and Iceland to less than 10% in Belgium, Bulgaria, Ukraine, The Netherlands, Romania, Hungary and Lithuania. Mediterranean countries rely heavily on groundwater for agricultural irrigation, and of the 29 countries with reported

Table 1.6 Estimated abstraction in European countries (Margat and van der Gun 2013).

| Country | Estimated groundwater abstraction (km^3 a^{-1}) | Groundwater abstraction by sector | | | Groundwater share of total freshwater abstraction (%) |
		Irrigation (%)	Domestic use (%)	Industry (%)	
Albania	0.90	61	33	6	53
Andorra	0.01	—	—	—	—
Austria	1.12	5	52	43	30
Belgium	0.65	4	55	41	10
Bosnia and Herzegovina	0.30	33	67	0	32
Bulgaria	0.58	—	—	—	10
Croatia	1.16	0	86	14	97
Cyprus	0.15	87	13	0	62
Czech Republic	0.38	—	—	—	19
Denmark	0.65	38	40	22	98
Estonia	0.33	0	20	80	18
Finland	0.28	24	65	11	17
France	5.71	14	63	23	18
Germany	5.83	4	48	48	18
Greece	3.65	86	14	0	39

(Continued)

Table 1.6 (Continued)

Country	Estimated groundwater abstraction (km³ a⁻¹)	Groundwater abstraction by sector			Groundwater share of total freshwater abstraction (%)
		Irrigation (%)	Domestic use (%)	Industry (%)	
Hungary	0.37	18	35	47	7
Iceland	0.16	—	—	—	97
Ireland	0.21	29	35	36	27
Italy	10.40	67	23	10	25
Latvia	0.11	—	—	—	26
Lithuania	0.17	—	—	—	7
Luxembourg	0.02	—	—	—	40
Macedonia	0.16	—	—	—	16
Malta	0.03	75	16	9	100
Moldova	0.60	—	—	—	31
Montenegro	0.05	0	100	0	100
Netherlands	0.97	23	32	45	9
Norway	0.41	0	27	73	14
Poland	2.59	0	70	30	22
Portugal	6.29	89	7	4	74
Romania	0.63	1	61	38	9
Serbia	0.53	—	—	—	—
Slovakia	0.36	3	84	13	52
Slovenia	0.19	1	83	16	15
Spain	5.70	72	23	5	18
Sweden	0.35	0	92	8	14
Switzerland	0.79	0	72	28	30
Ukraine	4.02	52	30	18	10
United Kingdom	2.16	9	77	14	17

(*Source:* Adapted from Margat, J. and van der Gun, J. (2013) *Groundwater Around the World: A Geographic Synopsis.* CRC Press/Balkema, EH Leiden.)

values in Table 1.6, 14 use over half of groundwater abstraction for domestic use.

1.7.2.1   European Union Water Framework Directive

The Water Framework Directive (WFD) establishing a framework for Community action in the field of water policy is a far-reaching piece of legislation governing water resources management and protection in the European Union (Council of the European Communities 2000). The Directive (2000/60/EC) was adopted in December 2000 and requires Member States to enforce appropriate measures to achieve good ecological and chemical status of all water bodies with a review of progress based on a six-year cycle. The purpose of the Directive is to establish a framework for the protection of inland surface waters, transitional waters (estuaries), coastal waters and groundwater to prevent further deterioration of aquatic eco-systems and, with regard to their water needs, terrestrial ecosystems and wetlands. In its implementation, the WFD requires an inte-grated approach to river basin management and promotes sustainable water use based on long-term protection of available water resources. A specific purpose of the WFD is to ensure the progressive reduction of pollution of groundwater and prevent its further pollution.

Article 17 of the WFD required a proposal (2003/0210(COD)) from the Commission for a Groundwater Daughter Directive leading to the adoption of specific measures to prevent and control groundwater pollution and achieve good groundwater chemical status (Commission of the European Communities 2003). In addition, the proposal introduced mea-sures for protecting groundwater from indirect pollution (discharges of pollutants into ground-water after percolation through the ground or subsoil). In the Groundwater Directive (Council of the European Union 2006), compli-ance with good chemical status is based on a comparison of monitoring data with quality standards existing in EU legislation on nitrates and plant protection and biocidal products which set threshold values (maximum permissi-ble concentrations) in groundwater for a num-ber of pollutants. With regard to pollutants that are not covered by EU legislation, the Direc-tive (2006/118/EC) requires Member States to establish threshold values defined at the national, river basin or groundwater body levels, thus taking into account the great diversity of groundwater characteristics across the EU.

The Groundwater Directive sets out specific criteria for the identification of significant and sustained upward trends in pollutant con-centrations, and for the definition of starting points for when action must be taken to reverse these trends. In this respect, significance is defined both on the basis of time series and environmental significance. Time series are periods of time during which a trend is detected through regular monitoring. Environmental

significance describes the point at which the concentration of a pollutant starts to threaten to worsen the quality of groundwater. This point is set at 75% of the quality standard or the threshold value defined by Member States. Under the WFD, a comprehensive programme of measures to prevent or limit pollution of water, including groundwater, became opera-tional. Monitoring results obtained through the application of the Groundwater Directive are used to design the measures to prevent or limit pollution of groundwater.

1.7.3 Groundwater abstraction in North America

A similar picture emerges of the importance of groundwater for the population of North America. In Canada, almost nine million peo-ple, or 30% of the population, rely on ground-water for domestic use (Government of Canada 2021). Approximately two-thirds of these users live in rural areas where groundwa-ter is a reliable and cheap water supply that can be conveniently abstracted close to the point of use. The remaining groundwater users are located primarily in smaller municipalities where groundwater provides the primary source for their water supply systems. For example, 100% of the population of Prince Edward Island and over 60% of the population of New Brunswick rely on groundwater for domestic supplies. In Ontario, a province where groundwater is also used predominantly for supplying municipalities, 29% of the popu-lation is reliant on groundwater. Furthermore, the predominant use of groundwater varies by province. In Ontario, Prince Edward Island, New Brunswick and the Yukon, the largest users of groundwater are municipalities; in Alberta, Saskatchewan and Manitoba, the main users are in the agricultural sector for livestock watering; in British Columbia, Que-bec and the Northwest Territories, the princi-pal users are in the industrial sector; and in Newfoundland and Nova Scotia, the predomi-nant use is for rural domestic supplies. Prince

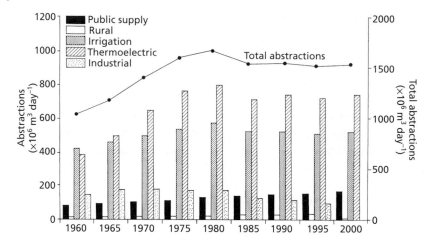

Fig. 1.12 Trends in water abstractions (fresh and saline) by water use category and total (fresh and saline) abstractions in the United States from 1960 to 2000 (Solley *et al.* 1998 and Hutson *et al.* 2004). (*Sources:* Adapted from Solley, W.B., Pierce, R.R. and Perlman, H.A. (1998) Estimated use of water in the United States in 1995. *United States Geological Survey Circular* **1200**, 71 pp; Hutson, S.S., Barber, N.L., Kenny, J.F. *et al.* (2004) Estimated use of water in the United States in 2000. *United States Geological Survey Circular* **1268**, 46 pp.)

Edward Island is almost totally dependent on groundwater for all its uses.

The abstraction of fresh and saline water in the United States from 1960 to 2000 as reported by Solley *et al.* (1998) and Hutson *et al.* (2004) is shown in Fig. 1.12. The estimated total abstraction for 1995 is 1522×10^6 m^3 day^{-1} for all off-stream uses (all uses except water used instream for hydroelectric power generation) and is 10% less than the 1980 peak estimate. This total has varied by less than 3% since 1985. In 2000, the estimated total water use in the United States is 1544×10^6 m^3 day^{-1}. Estimates of abstraction by source indicate that during 1995, total fresh surface water abstractions were 996×10^6 m^3 day^{-1} and total groundwater abstractions were 293×10^6 m^3 day^{-1} (or 23% of the combined freshwater abstractions). The respective figures for 2000 are 991×10^6 m^3 day^{-1} and 316×10^6 m^3 day^{-1}, with 24% of freshwater abstractions from groundwater.

Total water abstraction for public water supply in the United States in 2000 is estimated to have been 163×10^6 m^3 day^{-1}, an 8% increase since 1995. This increase compares with a 7% growth in the population for the same period.

Per capita public water supply use increased from about 678 L day^{-1} in 1995 to 683 L day^{-1} in 2000, but is still less than the per capita consumption of 696 L day^{-1} recorded for 1990.

The two largest water use categories in 2000 were cooling water for thermoelectric power generation (738×10^6 m^3 day^{-1} of fresh and saline water) and irrigation (518×10^6 m^3 day^{-1} of freshwater). Of these two categories, irrigation accounts for the greater abstraction of freshwater. The area of irrigated land increased nearly 7% between 1995 and 2000 with an increase in freshwater abstraction of 2% for this water use category. The area irrigated with sprinkler and micro-irrigation systems has continued to rise and now comprises more than half of the total. In 2000, surface water was the primary source of irrigation water in the arid West and the Mountain States and groundwater was the primary source in the Central States. California, Idaho, Colorado and Nebraska combined accounted for one-half of the total irrigation water abstractions. California and Idaho accounted for 40% of surface water abstractions and California and Nebraska accounted for one-third of groundwater abstractions. In general, groundwater abstractions for irrigation have increased significantly. In 1950, groundwater accounted for 23% of total irrigation water, while in 2000 it accounted for 42%.

1.7.3.1 Management and protection of groundwater resources in the United States

Groundwater management in the United States is highly fragmented, with responsibilities shared among a large number of federal, state and local programmes. At each level of government, unique legal authorities allow for the control of one or more threats to groundwater, such as groundwater contamination arising from municipal, industrial, mining and agricultural activities.

Beginning with the 1972 amendments to the federal Water Pollution Control Act, and followed by the Safe Drinking Water Act 1974, the federal government's role in groundwater management has increased. The introduction of the Resource Conservation and Recovery Act (RCRA) 1976 and the Comprehensive Environmental Response, Compensation and Liability Act (CERCLA) 1980, established the federal government's current focus on groundwater remediation. With these acts, the federal government has directed billions of dollars in public and private resources towards cleaning up contaminated groundwater at 'Superfund' sites, RCRA corrective action facilities and leaking underground storage tanks. In 1994, the National Academy of Sciences estimated that over a trillion dollars, or approximately $4000 per person in the United States, will be spent in the next 30 years on remediating contaminated soil and groundwater.

The approach to groundwater protection at the federal level has left the management of many contaminant threats, for example hazardous materials used by light industries (such as dry cleaners, printers or car maintenance workshops), to state and local government authorities. Other groundwater threats, such as over-abstraction, are not generally addressed under federal law, but left to states and local governments to manage.

In 1984, the US Environmental Protection Agency (USEPA) created the Office of Ground Water Protection to initiate a more comprehensive groundwater resource protection approach and to lead programmes aimed at resource protection. Such programmes include the Wellhead Protection and Sole Source Aquifer Programs, which were established by Amendments to the Safe Drinking Water Act 1986. The Wellhead Protection Program (WHPP) encourages communities to protect their groundwater resources used for drinking water. The Sole Source Aquifer Program limits federal activities that could contaminate important sources of groundwater.

State groundwater management programmes are seen as critical to the future achievement of effective and sustainable protection of groundwater resources. In 1991, the USEPA established a Ground Water Strategy to place greater emphasis on comprehensive state management of groundwater as a resource through the promotion of Comprehensive State Ground Water Protection Programs (CSGWPPs) together with better alignment of federal programmes with state groundwater resource protection priorities (USEPA 1992).

1.7.4 Groundwater abstraction in China

The rapidly growing economy of China, with about 20% of the world's population but only about 5–7% of global freshwater resources, has a high demand for groundwater. Groundwater is used to irrigate more than 40% of China's farmland and supplies about 70% of drinking water in the dry northern and north-western regions, with the past few decades having seen groundwater extraction increase by about $2.5 \times 10^9 \, \mathrm{m^3 \, a^{-1}}$ to meet these needs. Consequently, groundwater levels below the arid North China Plain have dropped by as much as $1 \, \mathrm{m \, a^{-1}}$ between 1974 and 2000 (Qiu 2010). Further discussion of the significance of groundwater leading to economic development in the rural and expanding urban areas underlain by the Quaternary Aquifer of the North China Plain is presented in Box 1.5.

Currently, the largest threat to sustainable water supplies in China is the growing geographical mismatch between agricultural development and water resources. The centre of grain production in China has moved from the humid south to the water-scarce north over the past 30 years, as southern cropland is urbanized and more land is irrigated further north. As the north has become drier, increased food production in this region has largely relied on unsustainable overuse of local water resources, especially groundwater. Wasteful irrigation infrastructure, poorly managed water use, as well as fast industrialization and urbanization, have led to a serious depletion of groundwater aquifers, loss of natural habitats and water pollution (Yu 2011).

To provide more sustainable management of groundwater resources, China needs to build an integrated network to monitor surface water and groundwater, and use it to assess and set water policies through an integrated water-resource management system, backed up by legislation that sets out clear policies on data sharing, and penalties for those who do not comply (Yu 2011). Arguably, the biggest improvement could come in the agriculture sector, which already uses 70% of the China's fresh water. For instance, to boost grain production and help maintain food security, China has a double-cropping system of growing wheat in winter and maize in summer, an unsustainable system that needs reconsidering. Meanwhile, the Chinese government hopes that a massive system of canals and pipes, to

---

Box 1.5 Groundwater Development of the Quaternary Aquifer of the North China Plain

The Quaternary Aquifer of the North China Plain represents one of the world's largest aquifer systems and underlies extensive tracts of the Hai River Basin and the catchments of the adjacent Huai and Huang (Yellow) River Systems (Fig. 1.13) and beyond. This densely populated area comprises a number of extensive plains, known collectively as the North China Plain, and includes three distinct hydrogeological settings within the Quaternary aquifer system (Fig. 1.14). The semi-arid climate of northeastern China is characterized by cold, dry winters (December–March) and hot, humid summers (July–September).

The Quaternary Aquifer supports an enormous exploitation of groundwater which has led to large socio-economic benefits in terms of irrigated grain production, farming employment and rural poverty alleviation, together with urban and industrial water supply provision. An estimated water supply of $27 \times 10^9 \, m^3 \, a^{-1}$ in the Hai River Basin alone was derived from wells and boreholes in 1988 (MWR 1992), but such large exploitation of groundwater has led to increasing difficulties in the last few years.

Given the heavy dependence on groundwater resources in the North China Plain, a number of concerns have been identified in recent years (Fig. 1.13) including a falling water table in the shallow freshwater aquifer, declining water levels in the deep freshwater aquifer, aquifer salinization as a result of inadequately controlled pumping and aquifer pollution from uncontrolled urban and industrial wastewater discharges. These issues are interlinked, but do not affect the three main hydrogeological settings equally (Table 1.7). A range of water resources management strategies are considered by Foster et al. (2004) that could contribute to reducing and

Box 1.5   (Continued)

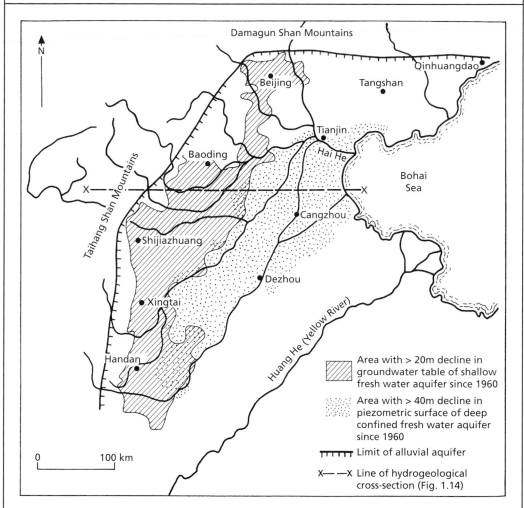

Fig. 1.13   Location map of the North China Plain showing the distribution of areas exhibiting marked groundwater depletion as a consequence of aquifer over-exploitation of the Quaternary aquifer system (Fig. 1.14) (Foster *et al.* 2004). (*Source:* Foster, S., Garduno, H., Evans, R. *et al.* (2004) Quaternary Aquifer of the North China Plain – assessing and achieving groundwater resource sustainability. *Hydrogeology Journal* **12**, 81–93. © 2004, Springer Nature.)

eventually eliminating the current aquifer depletion and include agricultural water-saving measures, changes in land use and crop regimes, artificial aquifer recharge of excess surface runoff, re-use of treated urban wastewater, and improved institutional arrangements that deliver these water savings and technologies while at the same time limiting further exploitation of groundwater for irrigated agriculture and industrial production (Foster *et al.* 2004).

(Continued)

Box 1.5 **(Continued)**

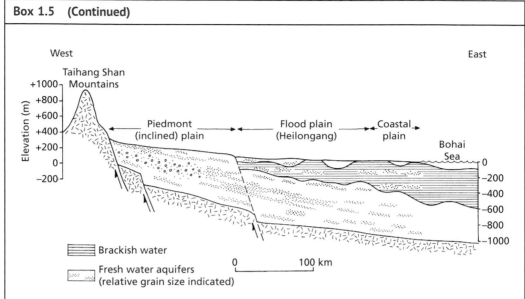

Fig. 1.14 Cross-section of the North China Plain showing the general hydrogeological setting of the Quaternary aquifer system which includes the gently sloping piedmont plain and associated major alluvial fans, the main alluvial plain (Heilongang) and the coastal plain around the margin of the Bohai Sea (Foster *et al.* 2004). (*Source:* Foster, S., Garduno, H., Evans, R. *et al.* (2004) Quaternary Aquifer of the North China Plain – assessing and achieving groundwater resource sustainability. *Hydrogeology Journal* **12**, 81–93. © 2004, Springer Nature.)

Table 1.7 Key groundwater issues in the North China Plain listed according to hydrogeological setting (Fig. 1.14) (Foster *et al.* 2004).

	Hydrogeological setting		
Groundwater issue	**Piedmont plain**	**Flood plain**	**Coastal plain**
Falling water table of shallow freshwater aquifer	+++[a]	+++	+
Depletion of deep freshwater aquifer	0[b]	+++	++
Risk of shallow aquifer and/or soil salinization	0	++	+++
Groundwater pollution from urban and industrial wastewater	+++	+	0

Notes:
[a] +++, very important; ++, important; +, minor importance; 0, not important.
[b] Effects of excessive abstraction may be reflected in the overlying shallow freshwater aquifer which is here in hydraulic continuity.
(*Source:* Foster, S., Garduno, H., Evans, R. *et al.* (2004) Quaternary Aquifer of the North China Plain – assessing and achieving groundwater resource sustainability. *Hydrogeology Journal* **12**, 81–93. © 2004, Springer Nature.)

transfer 45×10^9 m^3 a^{-1} from China's wetter south to its arid north, will alleviate groundwater depletion once completed in 2050 (Qiu 2010).

1.8 Groundwater resources in developing countries

It remains one of the greatest challenges for the future to provide the basic amenity of a safe and reliable supply of drinking water to the entire world's population. Despite the efforts of governments, charities and aid agencies, many villagers have to walk hundreds of metres to obtain drinking water from sources that may be unprotected from contamination (Fig. 1.15). Pollution sources include unsewered pit latrines to dispose of human wastes, inorganic fertilizers and pesticides used in an effort to secure self-sufficiency in food production and industrial wastes in urban areas.

In the developing world, groundwater is extensively used for drinking water supplies, especially in smaller towns and rural areas, where it is often the cheapest source. Groundwater schemes consist typically of large numbers of boreholes, often drilled on an uncontrolled basis, providing untreated, unmonitored and often unconnected supplies. Shallower dug wells continue to be constructed in some cases. Better yielding boreholes (100 L s^{-1}) are quite widely developed in larger towns to provide piped supplies. Even in these cases, raw water monitoring and treatment are often limited and intermittent.

The Third World Water Forum held in Osaka, Japan, in March 2003 emphasized

Fig. 1.15 Collection of water for domestic use from a hand-pumped tube well drilled in Precambrian metamorphic rock in the Uda Walawe Basin, Sri Lanka.

issues relating to the development and management of groundwater and recommended that many developing nations need to appreciate their social and economic dependency on groundwater and to invest in strengthening institutional provisions and building institutional capacity for its improved management. International development agencies and banks were urged to give higher priority to supporting realistic initiatives to strengthen governance of groundwater resources and local aquifer management. Future sustainable livelihoods, food security and key ecological systems in developing nations will be dependent on such initiatives.

As examples of the significance of groundwater in leading the economic development of rural and expanding urban areas in developing countries, Box 1.6 provides a description of the groundwater potential of the African continent and Box 1.7 describes groundwater resources in the Indo-Gangetic Basin in South Asia.

Box 1.6   Groundwater resources potential in Africa

Currently, there are more than 300 million people in Africa without access to safe drinking water, many of whom are amongst the poorest and most vulnerable in the world (JMP 2010; Hunter *et al.* 2010). Even for those with access to improved water sources, there is growing evidence that domestic water use will need to increase substantially to help lift people out of poverty (Grey and Sadoff 2007; Hunter *et al.* 2010). In Africa, groundwater is the major source of drinking water and its use for irrigation is forecast to increase substantially to counter growing food insecurity. At present, only 5% of arable land is irrigated (Siebert *et al.* 2010), and there is discussion of the need to increase irrigation to help meet rising demands for food production in the context of future, less reliable rainfall (UNEP 2010, Pfister *et al.* 2011).

Increasing reliable water supplies throughout Africa will depend on the development of groundwater (Giordano 2009; MacDonald and Calow 2009). However, quantitative, spatially explicit information on groundwater in Africa is required to characterize this resource in order to inform strategies to adapt to growing water demand associated not only with population growth but also climate variability and change. To address this significant knowledge gap, MacDonald *et al.* (2011, 2012) have developed the first quantitative continent-scale maps of groundwater storage and potential yields in Africa based on an extensive review of available maps, publications and data. From this analysis, MacDonald *et al.* (2012) estimated total groundwater storage in Africa to be $0.66 \times 10^6 \text{ km}^3$ (range $0.36-1.75 \times 10^6 \text{ km}^3$). Not all of this groundwater storage is available for abstraction, but the estimated volume is more than 100 times estimates of annual renewable freshwater resources in Africa.

Groundwater resources are unevenly distributed in Africa. The largest groundwater volumes are found in the large sedimentary aquifers in the North African countries of Libya, Algeria, Egypt and Sudan (see Box 1.3). Crystalline basement rocks have the lowest yields, generally less than 0.5 L s^{-1}, though a significant minority of areas has yields that are in excess of 1 L s^{-1}. Highest borehole yields ($>20 \text{ L s}^{-1}$) can be found in thick sedimentary aquifers, particularly in unconsolidated or poorly consolidated sediments. Depth to groundwater (Plate 1.7) is another important factor controlling accessibility and cost of developing groundwater resources. Water levels deeper than 50 m are not easily accessible by a hand pump. At depths >100 m, the cost of borehole drilling increases significantly due to the requirement for more advanced drilling equipment.

The aquifer productivity map (Plate 1.8) shows that for many African countries, appropriately sited and constructed boreholes will be able to sustain community hand pumps (yields of $0.1-0.3 \text{ L s}^{-1}$) and, for most of the populated areas of Africa, groundwater levels are likely to be sufficiently shallow to be accessed using a hand pump. The majority of large groundwater stores in the sedimentary basins which can accommodate high yielding boreholes are in northern Africa. These are often far from population centres and have deep water levels and are therefore costly to develop. Away from the large sedimentary aquifers, the potential for borehole yields exceeding 5 L s^{-1} is not widespread, though higher yielding boreholes may be successful in some areas if accompanied by detailed hydrogeological investigation. The potential for intermediate boreholes yields of $0.5-5 \text{ L s}^{-1}$, which could be suitable for small-scale household and community irrigation, or multiple-use water supply systems, is much higher, but will again require effective hydrogeological investigation and borehole

| **Box 1.6 (Continued)** |

siting. According to MacDonald *et al.* (2012), strategies for increasing the use of groundwater throughout Africa for irrigation and urban water supplies should not be based on the widespread expectation of high-yielding boreholes but recognize that high borehole yields may occasionally be realized where a detailed knowledge of the local groundwater conditions has been developed.

Of particular focus, the population of sub-Saharan Africa is currently about 1 billion, and is predicted to double by 2050, whereas the region's climate is predicted to become drier during the same period (Healy 2019). Groundwater in sub-Saharan Africa supports livelihoods and poverty alleviation and maintains vital ecosystems, owing to its widespread availability, generally high quality and intrinsic ability to buffer the impacts of episodic drought and pronounced climate variability that characterize this region (Cuthbert *et al.* 2019). Therefore, it is impotent to understand the renewability of groundwater under present and future climatic conditions. The sustainability of groundwater in sub-Saharan Africa in response to future climate variability depends critically on the relationship between precipitation and recharge. Cuthbert *et al.* (2019) presented an analysis of multi-decadal groundwater hydrographs across sub-Saharan Africa and showed that levels of aridity determine the recharge processes, whereas the local groundwater conditions influence the type and sensitivity of precipitation-recharge relationships. Cuthbert *et al.* (2019) found that intense precipitation events, even during years of lower overall precipitation, can produce years with the largest recharge in some dry subtropical locations, and so challenging the general consensus of decreasing water resources in such regions under climate change.

| **Box 1.7 Groundwater quality and depletion in the Indo-Gangetic Basin** |

The Indo-Gangetic Basin (IGB) alluvial aquifer system is one of the world's most important freshwater resources. The IGB encompasses more than 2.5×10^6 km^2 across Bangladesh, India, Pakistan and southern Nepal, with a population of over 750 million people and comprising over 100×10^6 km^2 of agricultural land (Fendorf and Benner 2016) Formed by sediments eroded from the Himalayas and redistributed by the Indus, Ganges and Brahmaputra River systems, this transboundary aquifer system forms a flat fertile plain across Pakistan, northern India, southern Nepal and Bangladesh (Fig. 1.16). Groundwater abstraction from the IGB aquifer system comprises 25% of global groundwater withdrawals in sustaining agricultural productivity in this region of South Asia (MacDonald *et al.* 2016). Fifteen to twenty million water wells abstract an estimated 205 km^3 a^{-1}, and this volume continues to increase at 2–5 km^3 a^{-1} as agricultural production intensifies (MacDonald *et al.* 2016). Although unevenly distributed, the IGB aquifer system also supplies drinking water for rural and urban populations across the whole area. The IGB aquifer system is complex and heterogeneous, with large spatial differences in permeability, storage, recharge and water chemistry that can also vary with depth (Bonsor *et al.* 2017). The IGB has the largest surface water irrigation system in the world, constructed during the nineteenth and early twentieth centuries to redistribute water from the Indus and Ganges through a canal network that is greater than 100 000 km long (MacDonald *et al.* 2016).

(Continued)

Box 1.7 **(Continued)**

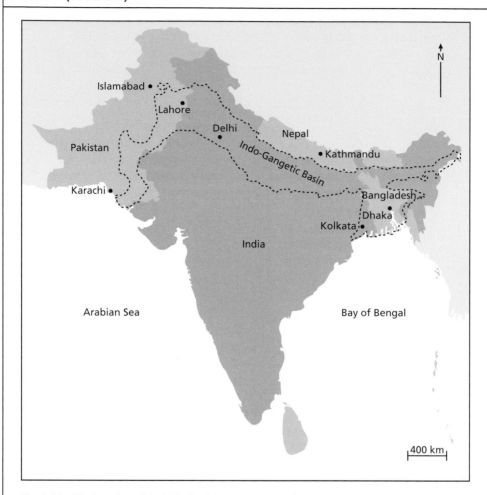

Fig. 1.16 The location of the IGB alluvial aquifer system (MacDonald *et al.* 2016). (*Source:* Adapted from MacDonald, A.M., Bonsor, H.C., Ahmed, K.M. *et al.* (2016) Groundwater quality and depletion in the Indo-Gangetic Basin mapped from *in situ* observations. *Nature Geoscience* **9**, 762–766.)

MacDonald *et al.* (2016) found that the water table within the IGB aquifer system is typically shallow (<5 m below ground level) and relatively stable since at least 2000 throughout much of the basin, with some important exceptions. In areas of high groundwater abstraction in north-west India and the Punjab in Pakistan (Regions 2 and 4, Fig. 1.17), the water table can be >20 m bgl (below ground level) and in some locations is falling at rates of >1 m a^{-1}. In areas of equivalent high irrigation abstraction within Bangladesh, the average water table remains shallow (<5 m bgl) due to greater direct recharge and a high capacity for induced recharge. Groundwater levels are deep and falling beneath many urban areas, and particularly in large groundwater-dependent cities such as Lahore, Dhaka and Delhi (Chatterjee *et al.* 2009). Shallow

Box 1.7   (Continued)

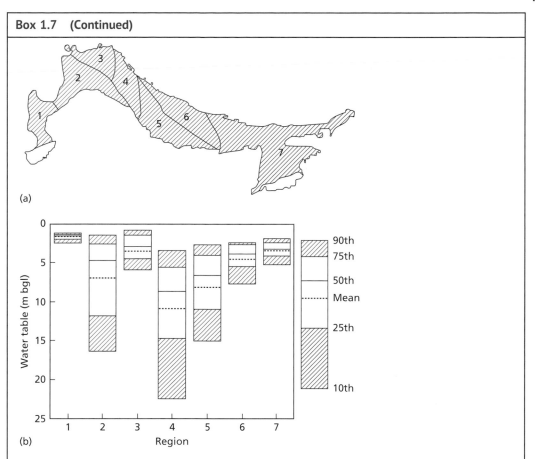

(a)

(b)

Fig. 1.17 Groundwater level variations in metres below ground level (m bgl) across the IGB alluvial aquifer system. (a) Location of analysis regions (divided by aquifer and climate): (1) Sindh; (2) middle Indus; (3, 4) upper Indus; (5) drier Uttar Pradesh; (6) wetter Uttar Pradesh; and (7) Lower Ganges and Bengal Basin. (b) Data from 3429 monitoring points showing mean water-table depths in individual wells for the period 2000–2012. Areas with high abstraction and lower rainfall show the deepest groundwater levels and a wide range in measured groundwater levels (MacDonald *et al.* 2016). (*Source:* Adapted from MacDonald, A.M., Bonsor, H.C., Ahmed, K.M. *et al.* (2016) Groundwater quality and depletion in the Indo-Gangetic Basin mapped from *in situ* observations. *Nature Geoscience* **9**, 762–766.)

and rising water tables are found in the Lower Indus, parts of the lower Bengal Basin and in places throughout the IGB aquifer system, as a consequence of leakage from canals, rivers and irrigation (MacDonald *et al.* 2016).

From mapping specific yield from lithological and hydrogeological data, MacDonald *et al.* (2016) estimated that the total volume in the top 200 m of the IGB aquifer system is 30 000 ± 14 000 km^3, equivalent to 20–30 times the combined mean annual flow in the rivers within the basin

(1000–1500 km^3 a^{-1}). Estimated trends in groundwater storage for the IGB aquifer system, derived from *in situ* measurements of water-table variations and estimates of specific yield derived across the basin, indicate a net average annual groundwater depletion within the period 2000–2012 of 8.0 km^3 a^{-1} (range 4.7–11.0 km^3 a^{-1}) with significant variation across the basin. The largest depletion occurred in areas of high abstraction and consumptive use in northern India and Pakistan, including Punjab, 2.6 ± 0.9 km^3 a^{-1};

(Continued)

Box 1.7 (Continued)

Haryana, 1.4 ± 0.5 km^3 a^{-1}; Uttar Pradesh, 1.2 ± 0.5 km^3 a^{-1}; and Punjab Region, Pakistan, 2.1 ± 0.8 km^3 a^{-1}. In the Lower Indus, within the Sindh, groundwater is accumulating at a rate of 0.3 ± 0.15 km^3 a^{-1}, which has led to increased waterlogging of land and significant reduction in the outflow of the River Indus (Basharat *et al.* 2014). Across the rest of the IGB aquifer system, changes in groundwater storage are generally modest (± 10 mm a^{-1}) (MacDonald *et al.* 2016).

From an analysis of satellite and well data, Asoka *et al.* (2017) found that groundwater storage in northern India has declined at a rate of 2 cm a^{-1} between 2002 and 2013 and that groundwater storage variability in north-western India is explained predominantly by variability in abstraction for irrigation, which itself is influenced by changes in precipitation. Asoka *et al.* (2017) suggested that declining precipitation in northern India is linked to Indian Ocean warming, in turn influencing groundwater storage either directly by changing recharge or indirectly by changing abstraction.

Based on national surveys on water quality, MacDonald *et al.* (2016) found that groundwater quality is highly variable and often stratified with depth. The two main water quality concerns are salinity and arsenic. Elevated arsenic is primarily a concern for drinking water, while salinity affects irrigation and also the acceptability of groundwater for drinking. Other pollutants are present and most areas are vulnerable to contamination from nitrate and faecal pathogens. Of the estimated 30 000 km^3 of groundwater storage in the basin, 7000 ± 3000 km^3 (23%) is estimated as having salinity greater than 1000 mg L^{-1}. A further $11\,000 \pm 5000$ km^3 (37%) of groundwater storage is affected by arsenic at toxic concentrations.

The origin of the saline groundwater is complex due to a variety of natural processes: saline intrusion, historic marine transgression, dissolution of evaporite layers and excessive evaporation of surface water or shallow groundwater. Natural salinity is exacerbated by the long-term impact of irrigation and shallow water tables. Only the lower Bengal Basin has been subject to Quaternary marine influence, together with the Pakistan coast. The widespread salinity in the Indus Basin and drier parts of the Upper Ganges is terrestrial in origin, formed by a combination of natural and anthropogenic activities (MacDonald *et al.* 2016).

Arsenic-rich groundwater occurs in chemically reducing, grey-coloured, Holocene sediments, mostly restricted to groundwater in the uppermost 100 m across the floodplains in the southern Bengal Basin, where arsenic is commonly present at >100 µg L^{-1} (Fendorf *et al.* 2010). Less extreme arsenic concentrations, though still >10 µg L^{-1} (the World Health Organization (1994) recommended limit), occur in other parts of the IGB aquifer system, including Assam; southern Nepal; the Sylhet trough in eastern Bangladesh; and within Holocene sediments along the course of the Ganges and Indus river systems (MacDonald *et al.* 2016). Intensive abstraction of shallow groundwater can flush aqueous arsenic from the aquifer (Shamsudduha *et al.* 2011), but there is concern that high-capacity deep pumping may draw arsenic down to levels in the Bengal aquifer system which are otherwise of good quality, although retardation is expected to delay vertical migration by centuries (Radloff *et al.* 2011). Age-depth profiles and hydrochemical data from monitoring wells in the coastal Bengal Basin aquifer system demonstrate the regional resilience of deep groundwater (>100 m) to the ingress of shallow, contaminated groundwater due to the high regional anisotropy of aquifer properties (Lapworth *et al.* 2018).

From their study, MacDonald *et al.* (2016) concluded that the complex and dynamic

Box 1.7 (Continued)

nature of the IGB aquifer system highlights the fundamental importance of regular and spatially distributed measurements of groundwater levels and water quality to acquire data of sufficient resolution to identify processes, monitor changes and adopt appropriate management strategies. Increasing groundwater use for irrigation poses legitimate questions about the future sustainability of abstraction from the basin.

Faced with this challenge, and in order to maintain water and food security, strong scientific and management capacities at local levels are required, together with the legal and policy frameworks necessary to design management strategies to match local groundwater and surface water conditions within the various groundwater typologies cross the IGB aquifer system (Bonsor *et al.* 2017).

Further reading

Anderson, M.P. ed. (2008) *Groundwater: Selection, Introduction and Commentary.* International Association of Hydrological Sciences, Benchmark Papers in Hydrology, **3**. IAHS Press, Wallingford, Oxfordshire, UK.

Appleton, J.D., Fuge, R. and McCall, G.J.H. (1996). *Environmental Geochemistry and Health with Special Reference to Developing Countries.* Geological Society, London, Special Publications, 113.

Deming, D. (2002) *Introduction to Hydrogeology.* McGraw-Hill Higher Education, New York.

Downing, R.A. and Wilkinson, W.B. (eds) (1991) *Applied Groundwater Hydrology: A British Perspective.* Clarendon Press, Oxford.

Hiscock, K.M., Rivett, M.O. and Davison, R.M. (eds) (2002). *Sustainable Groundwater Development.* Geological Society, London, Special Publications, 193.

IYPE (2005). *Groundwater – Reservoir for a Thirsty Planet? International Year of Planet Earth.* Earth Sciences for Society Foundation, Leiden.

Jones, J.A.A. (ed.) (2011) *Sustaining Groundwater Resources: A Critical Element in the Global Water Crisis.* Springer, Dordrecht.

Jones, J.A.A. (1997) *Global Hydrology: Processes, Resources and Environmental Management.* Addison Wesley Longman Ltd., Harlow.

Kemper, K.E. (ed.) (2004) Theme issue: groundwater – from development to management. *Hydrogeology Journal* **12**, 3–5.

Price, M. (1996) *Introducing Groundwater* (2nd edn). Chapman & Hall, London.

Younger, P.L. (2007) *Groundwater in the Environment.* Blackwell Publishing Ltd., Malden, Massachusetts.

References

Abbott, B.W., Bishop, K., Zarnetske, J.P. *et al.* (2019) Human domination of the global water cycle absent from depictions and perceptions. *Nature Geoscience* **12**, 533–540. DOI: 10.1038/s41561-019-0374-y.

Abotalib, A.Z. and Heggy, E. (2019) A deep groundwater origin for recurring slope lineae on Mars. *Nature Geoscience* **12**, 235–241. DOI: 10.1038/s41561-019-0327-5.

Aldumairy, A. (2005) *Siwa: Past and Present.* Yasso, Alexandria, Egypt.

Allan, J.A. (1998) Virtual water: a strategic resource – global solutions to regional deficits. *Ground Water* **36**, 545–546.

Allan, J.A. (2003) Integrated water resources management is more a political than a technical challenge. *Developments in Water Science* **50**, 9–23.

Ambroggi, R.P. (1977) Underground reservoirs to control the water cycle. *Scientific American* **236** (5), 21–27.

Andrews-Hanna, J.C., Zuber, M.T., Arvidson, R. E. and Wiseman, S.M. (2010) Early Mars hydrology: Meridiani playa deposits and the sedimentary record of Arabia Terra. *Journal of Geophysical Research, E: Planets* **115**, E06002.

Asoka, A., Gleeson, T., Wada, Y. and Mishra, V. (2017) Relative contribution of monsoon precipitation and pumping to changes in groundwater storage in India. *Nature Geoscience* **10**, 109–117. DOI: 10.1038/NGEO2869.

Baker, V.R., Strom, R.G., Gulick, V.C. *et al.* (1991) Ancient oceans, ice sheets and the hydrological cycle on Mars. *Nature* **362**, 589–594.

Basharat, M., Hassan, D., Bajkani, A.A. and Sultan, S.J. (2014) *Surface Water and Groundwater Nexus: Groundwater Management Options for Indus Basin Irrigation System*. International Waterlogging and Salinity Research Institute (IWASRI), Publication no. 299. Pakistan Water and Power Development Authority, Lahore, 136 pp.

Beaumont, P. (1968) Qanats on the Varamin plain, Iran. *Transactions of the Institute of British Geographers* **45**, 169–179.

Bennett, C.E. (1969) *Frontinus: The Stratagems and the Aqueducts of Rome*. Harvard University Press, Cambridge, Massachusetts.

Berner, E.K. and Berner, R.A. (1987) *The Global Water Cycle: Geochemistry and Environment*. Prentice-Hall, Inc., Englewood Cliffs, New Jersey.

Biswas, A.K. (1972) *History of Hydrology*. North-Holland, Amsterdam.

Blondes, M., Gans, K., Engle, M. *et al.* (2017). *National Produced Waters Geochemical Database v2.3*. United States Geological Survey, Energy Resources Program. https://www.usgs.gov/energy-and-minerals/energy-resources-program.

Bobeck, P. trans. (2004). *The Public Fountains of the City of Dijon by Henry Darcy (1856)*. Kendall/Hunt Publishing Company, Dubuque, Iowa, USA.

Bobeck, P. (2006) Henry Darcy in his own words. *Hydrogeology Journal* **14**, 998–1004.

Boni, C., Bono, P. and Capelli, G. (1986) *Hydrogeological Scheme of Central Italy. Sheet 1 (A. Hydrogeological Map)*. Memoir of the Geological Society of Italy **XXXV**, Geological Society of Italy, Rome.

Bono, P. and Boni, C. (2001) Water supply of Rome in antiquity and today. In: *Springs and Bottled Waters Of the World: Ancient History, Source, Occurrence, Quality and Use* (eds P.E. LaMoreaux and J.T. Tanner). Springer-Verlag, Berlin, pp. 200–210.

Bonsor, H.C. and MacDonald, A.M. (2011). *An initial estimate of depth to groundwater across Africa, 2011*, British Geological Survey.

Bonsor, H.C., MacDonald, A.M., Ahmed, K.M. et al. (2017) Hydrogeological typologies of the Indo-Gangetic basin alluvial aquifer, South Asia. *Hydrogeology Journal* **25**, 1377–1406.

Bord, J. and Bord, C. (1985) *Sacred Waters: Holy Wells and Water Lore in Britain and Ireland*. Granada Publishing Ltd, London.

Chandra, S. (1990) *Hydrology in Ancient India*. National Institute of Hydrology, Roorkee, India.

Chatterjee, R., Gupta, B.K., Mohiddin, S.K. *et al.* (2009) Dynamic groundwater resources of National Capital Territory, Delhi: assessment, development and management options. *Environmental Earth Sciences* **59**, 669–686.

Chebotarev, I.I. (1955) Metamorphism of natural water in the crust of weathering. *Geochimica et Cosmochimica Acta* **8**, 22–48, 137–170, 198–212.

Clifford, S.M. (1993) A model for the hydrologic and climatic behaviour of water on Mars. *Journal of Geophysical Research* **98**, 10973–11016.

Commission of the European Communities (2003). *Proposal for a Directive of the European Parliament and of the Council on the protection of groundwater against pollution (2003/0210 (COD)). COM(2003) 550 final,* Brussels, 19.9.2003, 20 pp.

Council of the European Communities (2000) Directive establishing a framework for Community action in the field of water policy (2000/60/EC). *Official Journal of the European Communities,* **L327**, 1–73. Brussels.

Council of the European Union (2006) Directive on the protection of groundwater against pollution and deterioration (2006/118/EC). *Official Journal of the European Union,* **L372**, 19–31. Brussels.

Cuthbert, M.O., Gleeson, T., Reynolds, S.C. *et al.* (2017) Modelling the role of groundwater hydro-refugia in East African hominin evolution and dispersal. *Nature Communications* **8**:15696. DOI: 10.1038/ncomms15696.

Cuthbert, M.O., Taylor, R.G., Favreau, G *et al.* (2019) Observed controls on resilience of groundwater to climate variability in sub-Saharan Africa. *Nature* **572**, 230–234.

Dalin, C., Wada, Y., Kastner, T. and Puma, M.J. (2017) Groundwater depletion embedded in international food trade. *Nature* **543**: 700–704.

Dalton, J. (1799) *Experiments and Observations to Determine Whether the Quantity of Rain and Dew is Equal to the Quantity of Water Carried Off by Rivers and Raised by Evaporation, with an Enquiry into the Origin of Springs.* Literary and Philosophical Society, Manchester.

Deming, D. (2019) Robert Hooke's contributions to hydrogeology. *Groundwater* **57**, 177–184.

Deming, D. (2020) The aqueducts and water supply of ancient Rome. *Groundwater* **58**, 152–161.

DeSimone, L.A., McMahon, P.B. and Rosen, M.R. (2014) The quality of our Nation's waters: Water quality in principal aquifers of the United States, 1991–2010. *United States Geological Survey Circular* **1360**. 10.3133/cir1360.

Dochartaigh, B.É.Ó, MacDonald, A.M., Fitzsimons, V. and Ward, R. (2015). *Scotland's aquifers and groundwater bodies.* British Geological Survey Open Report, OR/15/028. British Geological Survey, Keyworth, Nottingham, 76 pp.

Downing, R.A. (1993) Groundwater resources, their development and management in the UK: an historical perspective. *Quarterly Journal of Engineering Geology* **26**, 335–358.

Edmunds, W.M. (2001) Palaeowaters in European coastal aquifers – the goals and main conclusions of the PALAEAUX project. In: *Palaeowaters in Coastal Europe: Evolution of Groundwater since the Late Pleistocene* (eds W.M. Edmunds and C.J. Milne). Geological Society, London, Special Publications 189, pp. 1–16.

Ellis, E.C., Goldewijk, K.K., Siebert, S. et al. (2010) Anthropogenic transformation of the biomes, 1700 to 2000. *Global Ecology and Biogeography* **19**, 589–606.

Environment Agency (1998) *Policy and Practice for the Protection of Groundwater.* Environment Agency, Bristol.

Environment Agency (2015) *Underground, Under Threat.* Environment Agency, Bristol, 24 pp.

Falkenmark, M., Wang-Erlandsson, L. and Rockström, J. (2019) Understanding of water resilience in the Anthropocene. *Journal of Hydrology X* **2**, 100009. DOI: 10.1016/j.hydroa.2018.100009.

Fan, Y., Li, H. and Miguez-Macho, G. (2013) Global patterns of groundwater table depth. *Science* **339**, 940–943. DOI: 10.1126/science.1229881.

Fang, Y., Leung, L.R., Duan, Z. et al. (2017) Influence of landscape heterogeneity on water available to tropical forests in an Amazonian catchment and implications for modeling drought response. *Journal of Geophysical Research-Atmospheres* **122**, 8410–8426. DOI: 10.1002/2017JD027066.

Farrell, W.M., Plaut, J.J., Cummer, S.A. *et al.* (2009) Is the Martian water table hidden from radar view? *Geophysical Research Letters* **36**, L15206.

Feldman, W.C., Bandfield, J.L., Diez, B. *et al.* (2008) North to south asymmetries in the water-equivalent hydrogen distribution at high latitudes on Mars. *Journal of Geophysical Research* **113**, E08006.

Fendorf, S. and Benner, S.G. (2016) Indo-Gangetic groundwater threat. *Nature Geoscience* **9**, 732–733.

Fendorf, S., Michael, H.A. and van Geen, A. (2010) Spatial and temporal variations of groundwater arsenic in south and southeast Asia. *Science* **328**, 1123–1127.

Forget, F., Haberle, R.M., Montmessin, F. *et al.* (2006) Formation of glaciers on Mars by atmospheric precipitation at high obliquity. *Science* **311**, 368–371.

Foster, S., Garduno, H., Evans, R. *et al.* (2004) Quaternary Aquifer of the North China Plain – assessing and achieving groundwater resource sustainability. *Hydrogeology Journal* **12**, 81–93.

Freeze, R.A. (1994) Henry Darcy and the fountains of Dijon. *Ground Water* **32**, 23–30.

Garrels, R.M. and Christ, C.L. (1965) *Solutions, Minerals and Equilibria*. Harper and Row, New York.

Garrels, R.M., Mackenzie, F.T. and Hunt, C. (1975) *Chemical Cycles and the Global Environment: Assessing Human Influences*. Kaufman, Los Altos, California.

Giordano, M. (2009) Global groundwater? Issues and solutions. *Annual Review of Environment and Resources* **34**, 153–178.

Gleeson, T., Befus, K.M., Jasechko, S. *et al.* (2015) The global volume and distribution of modern groundwater. *Nature Geoscience* **9**, 161–167. DOI: 10.1038/NGEO2590.

Gleeson, T., Wada, Y., Bierkens, M.F.P. and van Beek, L.P.H. (2012) Water balance of global aquifers revealed by groundwater footprint. *Nature* **488**, 197–200.

Gleeson, T., Wang-Erlandsson, L., Porkka, M. *et al.* (2020) Illuminating water cycle modifications and Earth system resilience in the Anthropocene. *Water Resources Research* **56**, e2019WR024957. DOI: 10.1029/2019WR024957.

Government of Canada (2021). Water sources: groundwater. https://www.canada.ca/en/environment-climate-change/services/water-overview/sources/groundwater.html#protection (accessed 17 January 2021).

Grey, D. and Sadoff, C.W. (2007) Sink or swim? Water security for growth and development. *Water Policy* **9**, 545–571.

Haddeland, I., Heinke, J., Biemans, H. *et al.* (2014) Global water resources affected by human interventions and climate change. *Proceedings of the National Academy of Sciences* **111**, 3251–3256.

Halley, E. (1691) On the circulation of the vapors of the sea and the origin of springs. *Philosophical Transactions of the Royal Society* **17**(192), 468–473.

Harrison, K.P. and Grimm, R.E. (2009) Regionally compartmented groundwater flow on Mars. *Journal of Geophysical Research* **114**, E04004.

Healy, R.W. (2019) Groundwater resilience in sub-Saharan Africa. *Nature* **572**, 185–186.

Hem, J.D. (1959) Study and interpretation of the chemical characteristics of natural water. *United States Geological Survey Water Supply Paper* **1473**, 269 pp.

Hiscock, K.M., Rivett, M.O. and Davison, R.M. (2002) Sustainable groundwater development. In: *Sustainable Groundwater Development* (eds K.M. Hiscock, M.O. Rivett and R.M. Davison). Geological Society, London, Special Publications 193, pp. 1–14.

Hodge, T.A. (2008) *Roman Aqueducts and Water Supply* (2nd edn). Duckworth, London.

Hubbert, M.K. (1940) The theory of groundwater motion. *The Journal of Geology* **48**, 785–944.

Hunter, P.R., MacDonald, A.M. and Carter, R.C. (2010) Water supply and health. *PLoS Medicine* **7**, e1000361.

Hutson, S.S., Barber, N.L., Kenny, J.F. *et al.* (2004) Estimated use of water in the United States in 2000. *United States Geological Survey Circular* **1268**, 46 pp.

Idris, H. (1996) Springs in Egypt. *Environmental Geology* **27**, 99–104.

IWMI (2007) *Water for Food, Water for Life: A Comprehensive Assessment of Water Management in Agriculture.* Earthscan, London and International Water Management Institute, Colombo.

Jacob, C.E. (1940) On the flow of water in an elastic artesian aquifer. *Transactions of the American Geophysical Union* **22**, 574–586.

JMP (2010) *Progress on Sanitation and Drinking Water, 2010 Update.* WHO/UNICEF, Geneva and New York.

Kemper, K.E. (2004) Groundwater – from development to management. *Hydrogeology Journal* **12**, 3–5.

Konikow, L., Custodio, E., Villholth, K. *et al.* (2015). *Food Security and Groundwater.* International Association of Hydrogeologists, Strategic Overview Series, 6 pp. https://iah.org/wp-content/uploads/2015/11/IAH-Food-Security-Groundwater-Nov-2015.pdf (accessed 17 January 2021).

Konikow, L.F. (2011) Contribution of global groundwater depletion since 1900 to sea-level rise. *Geophysical Research Letters* **38**, L17401.

Lapworth, D.J., Zahid, A., Taylor, R.G. *et al.* (2018) Security of deep groundwater in the coastal Bengal basin revealed by tracers. *Geophysical Research Letters* **45**, 8241–8252.

Lucas, J. (1874) *Horizontal Wells. A New Application of Geological Principles to Effect the Solution of the Problem of Supplying London with Pure Water.* Edward Stanford, London.

Lucas, J. (1877) *Hydrogeological Survey. Sheet 1 (South London).* Edward Stanford, London.

Luijendijk, E., Gleeson, T. and Moosdorf, N. (2020) Fresh groundwater discharge insignificant for the world's oceans but important for coastal ecosystems. *Nature Communications.* DOI: 10.1038/s41467-020-15064-8.

MacDonald, A.M., Bonsor, H.C., Ahmed, K.M. *et al.* (2016) Groundwater quality and depletion in the Indo-Gangetic Basin mapped from *in situ* observations. *Nature Geoscience* **9**, 762–766.

MacDonald, A.M., Bonsor, H.C., Calow, R.C. *et al.* (2011). *Groundwater resilience to climate change in Africa.* British Geological Survey Open Report, OR/11/031. British Geological Survey, Keyworth, Nottingham, 25 pp.

MacDonald, A.M., Bonsor, H.C., Dochartaigh, B. É.Ó. and Taylor, R.G. (2012) Quantitative maps of groundwater resources in Africa. *Environmental Research Letters* **7**, 024009.

MacDonald, A.M. and Calow, R.C. (2009) Developing groundwater for secure water supplies in Africa. *Desalination* **248**, 546–556.

Malin, M.C. and Edgett, K.S. (2000) Evidence for recent groundwater seepage and surface runoff on Mars. *Science*, **288**, 2330–2335.

Margat, J. and Andréassian, V. (2008) *L'Eau, les Idées Reçues.* Editions le Cavalier Bleu, Paris.

Margat, J. and van der Gun, J. (2013) *Groundwater Around the World: A Geographic Synopsis.* CRC Press/Balkema, EH Leiden.

Mather, J. (1998) From William Smith to William Whitaker: the development of British hydrogeology in the nineteenth century. In: *Lyell: The Past is the Key to the Present* (eds D.J. Blundell and A.C. Scott). Geological Society, London, Special Publications 143, pp. 183–196.

Maurits la Rivière, J.W. (1989) Threats to the world's water. *Scientific American* **261**, 48–55.

McEwen, A.S., Dundas, C.M., Mattson, S.S. *et al.* (2014) Recurring slope lineae in equatorial regions of Mars. *Nature Geoscience* **7**, 53–58. DOI: 10.1038/ngeo2014.

Meinzer, O.E. (1923) The occurrence of groundwater in the United States with a discussion of principles. *United States Geological Survey Water Supply Paper* **489**, 329 pp.

Meinzer, O.E. (1928) Compressibility and elasticity of artesian aquifers. *Economic Geology* **23**, 263–291.

Meinzer, O.E. and Hard, H.H. (1925) The artesian water supply of the Dakota sandstone in North

Dakota, with special reference to the Edgeley Quadrangle. *United States Geological Survey Water Supply Paper* **520-E**, 73–95.

Michalski, J.R., Cuadros, J., Niles, P.B. *et al.* (2013) Groundwater activity on Mars and implications for a deep biosphere. *Nature Geoscience* **6**, 133–138.

Misstear, B., Gill, L., McKenna, C. and Foley, R. (2018). Hydrology and Communities: A Hydrogeological Study of Irish Holy Wells. In: *Proceedings of the Irish National Hydrology Conference*, Mullingar (20 November 2018), 39–47.

Mithen, S. (2012) *Thirst: Water and Power in the Ancient World*. Weidenfeld & Nicolson, London.

Mitrofanov, I.G., Litvak, M.L., Kozyrev, A.S. *et al.* (2004) Soil water content on Mars as estimated from neutron measurements by the HEND instrument onboard the 2001 Mars Odyssey spacecraft. *Solar System Research* **38**, 253–257.

MWR (1992) *Water Resources Assessment for China*. Ministry of Water Resources, China Water and Power Press, Beijing, China.

Nace, R.L. (ed.) (1971). *Scientific Framework of World Water Balance*. UNESCO Technical Papers in Hydrology **7**. UNESCO, Paris, 27 pp.

National Rivers Authority (1992) *Policy and Practice for the Protection of Groundwater*. National Rivers Authority, Bristol.

Northern Ireland Water (2013) *Drinking Water and Health: A Guide for Public and Environmental Health Professionals and for Those in the Water Industry in Northern Ireland*. Northern Ireland Water, Belfast, 134 pp.

Park, W.-B. and Ha, K. (2012) Spring water and water culture on Jeju Island. *Ground Water* **50**, 159–165.

Peltenberg, E., Colledge, S., Croft, P. *et al.* (2000) Agro-pastoralist colonization of Cyprus in the 10th millennium BP: initial assessments. *Antiquity* **74**, 844–853.

Pfister, S., Bayer, P., Koehler, A. and Hellweg, S. (2011) Projected water consumption in future global agriculture: scenarios and related impacts. *Science of the Total Environment* **409**, 4206–4216.

Piper, A.M. (1944) A graphic procedure in the geochemical interpretation of water analyses. *Transactions of the American Geophysical Union* **25**, 914–923.

Plaut, J.J., Picardi, G., Safaeinili, A. *et al.* (2007) Subsurface radar sounding of the South Polar layered deposits of Mars. *Science* **316**, 92–95.

Pokhrel, Y.N., Hanasaki, N., Yeh, P.J.-F., et al. (2012) Model estimates of sea-level change due to anthropogenic impacts on terrestrial water storage. *Nature Geoscience* **5**, 389–392.

Powell, O., Silcock, J. and Fensham, R. (2015) Oases to oblivion: The rapid demise of springs in the south-eastern Great Artesian Basin, Australia. *Groundwater* **53**, 171–178.

Qiu, J. (2010) China faces up to groundwater crisis. *Nature* **466**, 308.

Radloff, K.A., Zheng, Y., Michael, H.A. *et al.* (2011) Arsenic migration to deep groundwater in Bangladesh influenced by adsorption and water demand. *Nature Geoscience* **4**, 793–798.

Risner, J.K. (1989) The geohydrology of Mars. *Ground Water*, **27**, 184–192.

RIVM and RIZA (1991) *Sustainable Use of Groundwater: Problems and Threats in the European Communities*. National Institute of Public Health and Environmental Protection and Institute for Inland Water Management and Waste Water Treatment, Bilthoven, The Netherlands. Report no. 600025001.

Salese, F., Pondrelli, M., Neeseman, A. *et al.* (2019) Geological evidence of planet-wide groundwater system on Mars. *Journal of Geophysical Research, Planets* **124**. DOI: 10.1029/2018JE005802.

Schyns, J.F., Hoekstra, A.Y., Booij, M.J. *et al.* (2019) Limits to the world's green water resources for food, feed, fiber, timber, and bioenergy. *Proceedings of the National Academy of Sciences* **116**, 4893–4898.

Sen, I.S. and Peucker-Ehrenbrink, P. (2012) Anthropogenic disturbance of element cycles at the Earth's surface. *Environmental Science and Technology* **46**, 8601–8609. 10.1021/es301261x.

Shamsudduha, M., Taylor, R.G., Ahmed, K.M. and Zahid, A. (2011) The impact of intensive abstraction on recharge to a shallow regional aquifer system: evidence from Bangladesh. *Hydrogeology Journal* **19**, 901–916.

Siebert, S., Burke, J., Faures, J.M. *et al.* (2010) Groundwater use for irrigation – a global inventory. *Hydrology and Earth System Sciences* **14**, 1863–1880.

Smith, D.E., Zuber, M.T., Solomon, S.C. *et al.* (1999) The global topography of Mars and implications for surface evolution. *Science* **284**, 1495–1503.

Solley, W.B., Pierce, R.R. and Perlman, H.A. (1998) Estimated use of water in the United States in 1995. *United States Geological Survey Circular* **1200**, 71 pp.

Sprenger, M., Stumpp, C., Weiler, M *et al.* (2019) The demographics of water: a review of water ages in the critical zone. *Reviews of Geophysics* **57**, DOI: 10.1029/2018RG000633.

Stahl, M.O. (2019) Groundwater pumping is a significant unrecognized contributor to global anthropogenic element cycles. *Groundwater* **57**, 455–464.

Stiff, H.A. (1951) The interpretation of chemical water analysis by means of patterns. *Journal of Petroleum Technology* **3**, 15–17.

Streetly, M. and Heathcote, J.A. (2018) Advances in groundwater system measurement and monitoring documented in 50 years of *QJEG*H. *Quarterly Journal of Engineering Geology and Hydrogeology* **51**, 139–155.

Stumm, W. and Morgan, J.J. (1981) *Aquatic Chemistry: An Introduction Emphasizing Chemical Equilibria in Natural Waters* (2nd edn). Wiley, New York.

Theis, C.V. (1935) The relation between the lowering of the piezometric surface and rate and duration of discharge of a well using groundwater storage. *Transactions of the American Geophysical Union* **2**, 519–524.

UNEP (2008) *Africa: Atlas of Our Changing Environment*. Division of Early Warning and Assessment, United Nations Environment Programme, Nairobi, Kenya.

UNEP (2010) *Africa Water Atlas. Division of Early Warning and Assessment*. United Nations Environment Programme, Nairobi.

USEPA (1992). Final Comprehensive State Ground Water Protection Program Guidance. Report EPA 100-R-93-001. United States Environmental Protection Agency, 30 pp.

Wada, Y., van Beek, L.P.H., van Kempen, C.M. *et al.* (2010). Global depletion of groundwater resources. *Geophysical Research Letters* **37**, L20402, 5 pp.

WaterAid (2011) *Sustainability Framework*. WaterAid, London.

Welsh Water (2019). Final water resources management plan 2019, 257 pp. https://www.dwrcymru.com/en/our-services/water/water-resources (accessed 17 January 2021).

Whitaker, W. and Reid, C. (1899) *The Water Supply of Sussex from Underground Sources*. Memoir of the Geological Survey. HMSO, London.

Williams, R.K. (2009) *Village Pumps*. Shire Publications Ltd, Botley, Oxford.

World Bank (2007) *World Development Report 2008: Agriculture for Development*. World Bank, Washington, DC.

World Health Organisation (1994) *Guidelines for Drinking Water Quality. Volume 1: Recommendations* (2nd edn). World Health Organisation, Geneva.

WWAP (2009) *The United Nations World Water Development Report 3: Water in a Changing World*. World Water Assessment Programme, UNESCO, Paris and Earthscan, London, 318 pp.

Young, M.E. (2002) Institutional development for sustainable groundwater management – an Arabian perspective. In: *Sustainable Groundwater Development* (eds K.M. Hiscock, M. O. Rivett and R.M. Davison). Geological Society, London, Special Publications 193, pp. 63–74.

Yu, C. (2011) China's water crisis needs more than words. *Nature* **470**, 307.

Zektser, I.S. and Everett, L.G. (eds) (2004) *Groundwater Resources of the World and Their Use.* International Hydrological Programme-VI, Series on Groundwater No. 6. UNESCO, Paris.

Zektser, I.S. and Loaiciga, H.A. (1993) Groundwater fluxes in the global hydrologic cycle: past, present and future. *Journal of Hydrology* **144**, 405–427.

Zhou, Y., Zwahlen, F. and Wang, Y. (2011) The ancient Chinese notes on hydrogeology. *Hydrogeology Journal* **19**, 1103–1114.

2

Physical hydrogeology

2.1 Introduction

The occurrence of groundwater within the Earth's crust and the emergence of springs at the ground surface are determined by the lithology of geological materials, regional geological structure, geomorphology of landforms and the availability of recharge sources. The infiltration of rainfall to the water table and the flow of groundwater in an aquifer towards a discharge area are governed by physical laws that describe changes in energy of the groundwater. In this chapter, the physical properties of aquifer storage and permeability are discussed in relation to different rock types and hydrogeological conditions. Of particular concern, the physical mechanisms that cause aquifer compaction as a result of groundwater abstraction from confined aquifers leading to land subsidence is explained. Then, starting with Darcy's Law, the fundamental law of groundwater flow, the equations of steady-state and transient groundwater flow are derived for the hydraulic conditions encountered above and below the water table. Next, examples of analytical solutions to simple one-dimensional groundwater flow problems are presented, and this is followed by an explanation of the influence of topography in producing various scales of groundwater movement, including patterns of local, intermediate and regional flow. The last section of this chapter deals with the occurrence of groundwater resources of the world. The wide range of aquifer types and expected borehole yields associated with sedimentary, metamorphic and igneous rock types are described with reference to the hydrogeological units that occur in the United Kingdom.

2.2 Porosity

The porosity of a soil or rock is that fraction of a given volume of material that is occupied by void space, or interstices. Porosity, indicated by the symbol n, is usually expressed as the ratio of the volume of voids, V_v, to the total unit volume, V_t, of a soil or rock, such that $n = V_v/V_t$. Porosity can be determined in the laboratory from knowledge of the bulk mass density, ρ_b, and particle mass density, ρ_s, of the porous material (see Section 6.4.1) using the relationship:

$$n = 1 - \frac{\rho_b}{\rho_s} \qquad \text{(eq. 2.1)}$$

In fractured rocks, secondary or fracture porosity can be estimated by the field method of scan lines using the relation $n_f = Fa$, where F is the number of joints per unit distance intersecting a straight scan line across a rock outcrop, and a is the mean aperture of the fractures.

Porosity is closely associated with the void ratio, e, the ratio of the volume of voids to

Hydrogeology: Principles and Practice, Third Edition. Kevin M. Hiscock and Victor F. Bense.
© 2021 John Wiley & Sons Ltd. Published 2021 by John Wiley & Sons Ltd.
Companion website: www.wiley.com/go/hiscock/hydrogeology3e

the volume of the solid material, V_s, such that $e = V_v/V_s$. The relation between porosity and void ratio can be expressed as follows:

$$n = \frac{e}{(1 + e)} \qquad \text{(eq. 2.2)}$$

or

$$e = \frac{n}{(1 - n)} \qquad \text{(eq. 2.3)}$$

Void ratio displays a wide range of values. In soils and rocks with a total porosity ranging from 0.001 to 0.7, the corresponding void ratio range is 0.001−2.3.

In general, unconsolidated sediments, such as gravels, sands, silts and clays, which are composed of angular and rounded particles, have larger porosities than indurated, consolidated sediments such as sandstone and limestone. Crystalline igneous and metamorphic rocks have especially low porosities because the pores are merely within the inter-crystal surfaces. Conversely, formations rich in platy clay minerals with very fine grain size can achieve high porosity values.

As illustrated in Fig. 2.1, porosity is controlled by the shape and arrangement of constituent grains, the degree of sorting, compaction, cementation, fracturing and solutional weathering. Porosity values range from negligibly small (0%) for unfractured to 0.1 (10%) for weathered crystalline rocks to 0.4–0.7 (40–70%) for unconsolidated clay deposits (Table 2.1).

There is a distinction between primary porosity, which is the inherent character of a soil or rock matrix that developed during its formation, and secondary porosity. Secondary porosity may develop as a result of secondary physical and chemical weathering along the bedding planes and joints of indurated sediments such as limestones and sandstones, or as a result of structurally controlled regional fracturing and near-surface weathering in hard rocks such as igneous and metamorphic rocks. Where both primary and secondary porosities are present, a dual-porosity system is recognized, for example as a result of fracturing and fissuring in porous sandstone or limestone.

Not all the water contained in the pore space of a soil or rock can be viewed as being available to groundwater flow, particularly in fine grained or fractured aquifers. In an aquifer with a water table, the volume of water released from groundwater storage per unit surface area of aquifer per unit decline in the

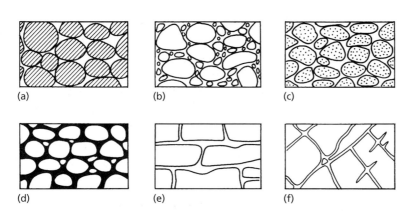

(a)   (b)   (c)

(d)   (e)   (f)

Fig. 2.1 Types of porosity with relation to rock texture: (a) well-sorted sedimentary deposit having high porosity; (b) poorly sorted sedimentary deposit having low porosity; (c) well-sorted sedimentary deposit consisting of pebbles that are themselves porous so that the whole deposit has a very high porosity; (d) well-sorted sedimentary deposit whose porosity has been reduced by the deposition of mineral matter (cementation) in the interstices; (e) soluble rock made porous by solution; (f) crystalline rock made porous by fracturing (Meinzer 1923). (*Source:* Adapted from Meinzer, O.E. (1923) The occurrence of groundwater in the United States with a discussion of principles. *United States Geological Survey Water Supply Paper* **489**.)

Table 2.1 Range of values of hydraulic conductivity and porosity for different geological materials (Freeze and Cherry 1979 and Back *et al.* 1988).

Geological material	Hydraulic conductivity, K (m s^{-1})	Porosity, n
Fluvial deposits (alluvium)	10^{-5}–10^{-2}	0.05–0.35
Glacial deposits		
– Basal till	10^{-11}–10^{-6}	0.30–0.35
– Lacustrine silt and clay	10^{-13}–10^{-9}	0.35–0.70
– Outwash sand and gravel	10^{-7}–10^{-3}	0.25–0.50
– Loess	10^{-11}–10^{-5}	0.35–0.50
Sandstone	10^{-10}–10^{-5}	0.05–0.35
Shales		
– Unfractured	10^{-13}–10^{-9}	0–0.10
– Fractured	10^{-9}–10^{-5}	0.05–0.50
Mudstone	10^{-12}–10^{-10}	0.35–0.45
Dolomite	10^{-9}–10^{-5}	0.001–0.20
Oolitic limestone	10^{-7}–10^{-6}	0.01–0.25
Chalk		
– Primary	10^{-8}–10^{-5}	0.15–0.45
– Secondary	10^{-5}–10^{-3}	0.005–0.02
Coral limestones	10^{-3}–10^{-1}	0.30–0.50
Karstified limestones	10^{-6}–10^{0}	0.05–0.50
Marble, fractured	10^{-8}–10^{-5}	0.001–0.02
Volcanic tuff	10^{-7}–10^{-5}	0.15–0.40
Basaltic lava	10^{-13}–10^{-2}	0–0.25
Igneous and metamorphic rocks – Unfractured and fractured	10^{-13}–10^{-5}	0–0.10

(*Sources:* Freeze, R.A. and Cherry, J.A. (1979) *Groundwater*. Prentice-Hall, Inc., Englewood Cliffs, New Jersey; Back, W., Rosenshein, J.S. and Seaber, P.R. (eds) (1988) *Hydrogeology. The Geology of North America*, Vol. **O-2**. The Geological Society of North America, Boulder, Colorado.)

water table is known as the specific yield, S_y (see Section 2.10). The fraction of water that is retained in the soil or rock against the force of gravity is termed the specific retention, S_r. As shown in Fig. 2.2, the sum of the specific yield and specific retention ($S_y + S_r$) is equal to the total porosity, n. It is useful to distinguish the total porosity from the effective porosity, n_e, of a porous material. The total porosity relates to the storage capability of the material, whereas the effective porosity relates to the transmissive capability of the material.

In coarse-grained rocks with large pores, the capillary films that surround the solid particles occupy only a small proportion of the pore space such that S_y and n_e will almost equal n. In fine-grained rocks and clay, capillary forces dominate such that S_r will almost equal n, but n_e will be much less than n. These variations can be described by the term specific surface area, S_{sp}, defined as the ratio of total surface area of the interstitial voids to total volume of the porous material. In sands, S_{sp} will be of the order of 1.5×10^4 m^{-1}, but in montmorillonite clay, it

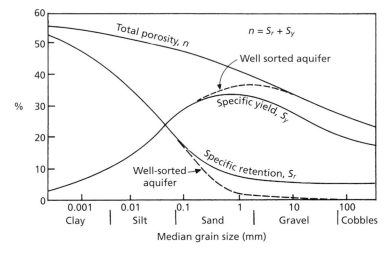

Fig. 2.2 Relation between median grain size and water storage properties of typical alluvial sediments (Davis and De Wiest 1966). (*Source:* Davis, S.N. and De Wiest, R.J.M. (1966) *Hydrogeology.* John Wiley & Sons, Inc., New York. © 1966, John Wiley & Sons.)

is about 1.5×10^9 m^{-1} (Marsily 1986). These properties are important in the adsorption of water molecules and dissolved ions on mineral surfaces, especially on clay.

In the case of a fissured or fractured aquifer, such as weathered limestone and crystalline rocks, water contained in the solid matrix is typically immobile and the only effective porosity is associated with the mobile water contained in the fissures and fractures. With increasing depth, the frequency of fissures and fractures decreases and the increasing overburden pressure closes any remaining openings such that the effective porosity of these formations substantially declines.

2.3 Hydraulic conductivity

Hydraulic conductivity or, as it is occasionally referred to in older publications, the coefficient of permeability, has dimensions of $[\mathrm{L\,T^{-1}}]$ and is a measure of the ease of movement of a water through a porous material. Values of hydraulic conductivity display a wide range in nature, spanning 13 orders of magnitude (Table 2.1). In general, coarse-grained and fractured

materials have high values of hydraulic conductivity, while fine-grained silts and clays have low values. An illustration of the relationship between hydraulic conductivity and grain size is shown for two alluvial aquifers in Fig. 2.3. In such aquifers, the sediment grain size commonly increases with depth such that the greatest hydraulic conductivity is generally deep in the aquifer (Sharp 1988). The properties of the geological material will significantly influence the isotropy and homogeneity of the hydraulic conductivity distribution (Section 2.4).

The hydraulic conductivity of geological materials is not only a function of the physical properties of the porous material but also the properties of the migrating fluid, including specific weight, γ ($= \rho g$, where ρ is the density of the fluid and g is the gravitational acceleration), and viscosity, μ, such that

$$K = k_i \frac{\gamma}{\mu} \qquad \text{(eq. 2.4)}$$

where the constant of proportionality, k_i, is termed the intrinsic permeability because it is a physical property intrinsic to the porous material alone.

The density and viscosity of water are functions of temperature and pressure, but these

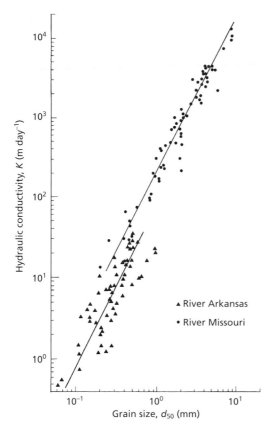

Fig. 2.3 Laboratory determined values of hydraulic conductivity as a function of grain size for alluvial aquifers in the Rivers Missouri and Arkansas. Note the log-log scales (Sharp 1988). (*Source:* Sharp, J.M. (1988) Alluvial aquifers along major rivers. In: *Hydrogeology. The Geology of North America*, Vol. **O-2** (eds W. Back, J.S. Rosenshein and P.R. Seaber). The Geological Society of North America, Boulder, Colorado, pp. 273–282. © 1988, Geological Society of America.)

effects are not great for the ranges of temperature and pressure encountered in most groundwater situations (see Appendix 2). A one-third increase in hydraulic conductivity is calculated using eq. 2.4 for a temperature increase from 5°C ($\rho = 999.965\,\text{kg m}^{-3}$, $\mu = 1.5188 \times 10^{-3}\,\text{N s m}^{-2}$) to 15°C ($\rho = 999.099\,\text{kg m}^{-3}$, $\mu = 1.1404 \times 10^{-3}\,\text{N s m}^{-2}$), although a groundwater flow system exhibiting such a temperature change would be considered unusual. An example is groundwater that penetrates

deep in the Earth's crust, becomes heated and returns rapidly to the surface as highly mineralized hot springs. Equally, in coastal areas, saline intrusion into fresh groundwater will cause variations in fluid density such that information about both k_i and K is required in any investigation.

The intrinsic permeability is representative of the properties of the porous material alone and is related to the size of the openings through which the fluid moves. For unconsolidated sand, Krumbein and Monk (1943) derived the following empirical relationship, where GM_d is the geometric mean of the grain diameter (mm) and σ is the standard deviation of the grain size in phi units ($-\log_2(\text{grain diameter in mm})$):

$$k_i = 760(GM_d)^2 e^{-1.3\sigma} = Cd^2 \quad (\text{eq. } 2.5)$$

As shown, eq. 2.5 is more generally expressed as $k_i = Cd^2$, where d is equal to the mean pore diameter and C represents a dimensionless 'shape factor' assessing the contribution made by the shape of the pore openings, as influenced by the relationship between the pore and grain sizes and their effect on the tortuosity of fluid flow. Intrinsic permeability has the dimensions of $[L^2]$ and, using nomenclature common in the petroleum industry, the unit of k_i is the darcy, where 1 darcy is equivalent to $9.87 \times 10^{-13}\,\text{m}^2$.

Statistical power least-squares regression analyses performed on 19 sets of published data on particle size and laboratory intrinsic permeability of unconsolidated sediments by Shepherd (1989) yielded the exponent of the grain diameter in eq. 2.5 to range from 1.11 to 2.05, with most values significantly less than 2.0. The results presented by Shepherd (1989) indicate that the permeability-grain size relation may be expressed alternatively, on an empirical basis, as $k_i = Cd^{1.65 \text{ to } 1.85}$. Values of C and the exponent of the grain diameter were shown to both generally decrease with decreased textural maturity (characterized by uniformly sized particles, better sorted samples and grains with higher roundness and sphericity) and increased induration (consolidation).

2.4 Isotropy and homogeneity

Aquifer properties, such as hydraulic conductivity, are unlikely to conform to the idealized, uniform porous material whether viewed at the microscopic or regional scales. The terms isotropy, anisotropy, homogeneity and heterogeneity are used to describe the spatial variation and directional trends in aquifer property values.

If the hydraulic conductivity, K, is independent of position within a geological formation, the formation is homogeneous. If the hydraulic conductivity varies from place to place, then the formation is heterogeneous. The type of heterogeneity will depend on the geological environment that gave rise to the deposit or rock type. As shown in Fig. 2.4, layered heterogeneity is common in sedimentary rocks where each bed comprising the formation has its own hydraulic conductivity value. Strong, layered heterogeneity will be present in interbedded deposits of clay and sand. Similarly large contrasts can arise in cases of discontinuous heterogeneity caused by the presence of faults or large-scale stratigraphic features. Trending heterogeneity exists in formations such as deltas, alluvial fans and glacial outwash plains, where there is sorting and grading of the material deposits. Vertical trends in hydraulic conductivity are also present in consolidated rocks where permeability is dependent on joint and fracture density.

It is widely accepted that the statistical distribution of hydraulic conductivity for a geological formation is described by a log-normal probability density function with the average hydraulic conductivity calculated as a geometric mean. Trending heterogeneity within a geological formation can be regarded as a trend in the mean hydraulic conductivity value.

An isotropic geological formation is one where the hydraulic conductivity is independent of the direction of measurement at a point in the formation. If the hydraulic conductivity varies with the direction of measurement at a point, the formation is anisotropic at that point. The principal directions of anisotropy correspond to the maximum and minimum values of hydraulic conductivity and are usually at right angles to each other. The primary cause of anisotropy on a small scale is the orientation of clay minerals in sedimentary rocks and unconsolidated sediments. In consolidated rocks, the direction of jointing or fracturing can impart strong anisotropy at various scales, from the local to regional.

Combining the previous definitions, and as shown in Fig. 2.5, it is possible to recognise four possible combinations of heterogeneity and anisotropy when describing the nature of the hydraulic conductivity of a formation.

2.5 Aquifers, aquitards and aquicludes

Natural variations in the permeability and ease of transmission of groundwater in different geological materials lead to the recognition of aquifers, aquitards and aquicludes. An aquifer is a layer or layered sequence of rock or sediment comprising one or more geological formations that contains water and is able to transmit significant quantities of water under an ordinary hydraulic gradient (see Section 2.6). Aquifers therefore have sufficient permeability to transmit groundwater that can be exploited economically from wells or springs. Good aquifers are usually developed in sands, gravels, solutionally weathered limestones and fractured sandstones.

The term 'aquitard' is used to describe a formation of lower permeability that may transmit quantities of water that are significant in terms of regional groundwater flow, but from which negligible supplies of groundwater can be obtained. Examples of aquitards include

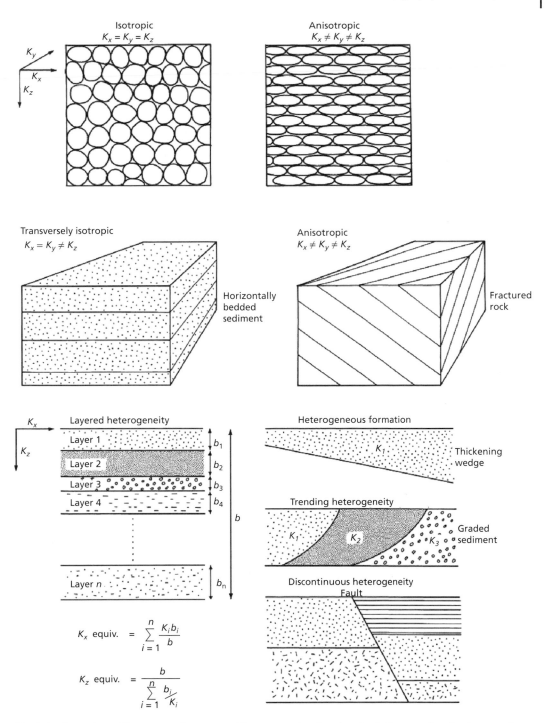

Fig. 2.4 Examples of isotropy, anisotropy and heterogeneity showing the influence of grain shape and orientation, sedimentary environment and geological structure on hydraulic conductivity (Fetter 2001). (*Source:* Adapted from Fetter, C.W. (2001) *Applied Hydrogeology* (4th edn). Pearson Higher Education, Upper Saddle River, New Jersey.)

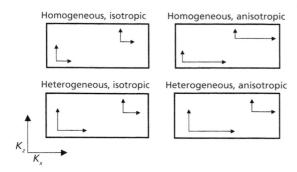

Homogeneous, isotropic   Homogeneous, anisotropic

Heterogeneous, isotropic   Heterogeneous, anisotropic

K_z

K_x

Fig. 2.5 Four possible combinations of heterogeneity and anisotropy describing the hydraulic conductivity of a porous material (Freeze and Cherry 1979). (*Source:* Freeze, R.A. and Cherry, J.A. (1979) *Groundwater*. Prentice-Hall, Inc., Englewood Cliffs, New Jersey. © 1979, Pearson Education.)

fluvial and glacio-fluvial silts and sandy clays, sedimentary rocks with few fractures and fractured crystalline rock.

Although a less commonly used term, an aquiclude is a saturated geological unit of such low permeability that is incapable of transmitting significant quantities of water under ordinary hydraulic gradients and can act as a barrier to regional groundwater flow. Aquiclude rocks include clays, shales and metamorphic rocks.

2.6 Darcy's Law

Water contained within the interconnected voids of soils and rocks is capable of moving, and the ability of a rock to store and transmit water constitutes its hydraulic properties. At the centre of the laws that govern the behaviour of groundwater flow in saturated material is that formulated empirically by the French municipal engineer for Dijon, Henry Darcy, in 1856. Using the type of experimental apparatus shown in Fig. 2.6, Darcy studied the flow of water through porous material contained in a column and found that the total flow, Q, is proportional to both the difference in water level, $h_1 - h_2$, measured in manometer tubes at either end of the column and the cross-sectional area of flow, A, and inversely proportional to the column length, L. When combined with the constant of proportionality, K, Darcy obtained:

$$Q = KA \frac{(h_1 - h_2)}{L} \qquad \text{(eq. 2.6)}$$

In general terms, Darcy's Law, as it is known, can be written as follows:

$$Q = -KA \frac{dh}{dl} \qquad \text{(eq. 2.7)}$$

where dh/dl represents the hydraulic gradient, with the negative sign indicating flow in the direction of decreasing hydraulic head. K is the hydraulic conductivity of the porous material. Adopting the shorthand of dh/dl equal to i, then eq. 2.7 can be written as follows:

$$Q = -AiK \qquad \text{(eq. 2.8)}$$

Now, combining eqs 2.4, 2.5 and 2.7 gives a full expression of the flow through a porous material as follows:

$$\frac{Q}{A} = q = -K \frac{dh}{dl} = -\frac{Cd^2 \rho g}{\mu} \frac{dh}{dl}$$

$$\text{(eq. 2.9)}$$

The quotient Q/A, or q, indicates the discharge per unit cross-sectional area of saturated porous material. The term q, referred to as the specific discharge, has the dimensions of velocity [L T^{-1}] and is also known as the Darcy velocity or Darcy flux. It is important to remember that the Darcy velocity is not the true, microscopic velocity of the water moving along winding flowpaths within the soil or rock. Instead, by dividing the specific discharge by the fraction of open space (in other words, effective porosity, n_e) through which groundwater flows across a given sectional area, this provides an average measure of groundwater velocity such that

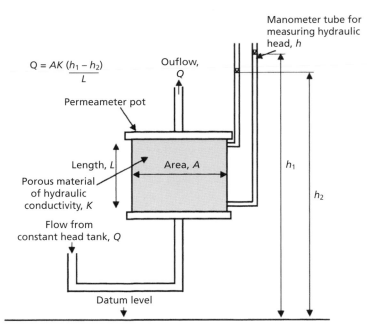

Fig. 2.6   Permeameter apparatus for determining the hydraulic conductivity of saturated porous material using Darcy's Law.

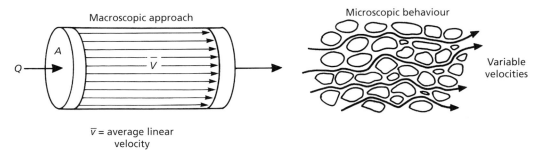

Fig. 2.7   Macroscopic (Darcian) approach to the analysis of groundwater flow contrasted with the true, microscopic behaviour of tortuous flowpaths.

$$\frac{Q}{An_e} = \frac{q}{n_e} = \bar{v} \qquad \text{(eq. 2.10)}$$

where $\bar{v}$ is the average linear velocity (Fig. 2.7).

As illustrated in Box 2.1, the application of eqs 2.7 and 2.10 to simple hydrogeological situations enables first estimates to be obtained for groundwater flow and velocity. More accurate calculations require the use of more advanced techniques such as flow net analysis (Box 2.3 later) and groundwater modelling (see Further Reading and Section 7.5).

As a result of introducing anisotropy, it is necessary to recognize that in a three-dimensional flow system the specific discharge or darcy velocity (eq. 2.9) as defined by Darcy's Law, is a vector quantity with components q_x, q_y and q_z given by:

$$q_x = -K_x \frac{\partial h}{\partial x}, \quad q_y = -K_y \frac{\partial h}{\partial y},$$

$$q_z = -K_z \frac{\partial h}{\partial z}$$

where K_x, K_y and K_z are the hydraulic conductivity values in the x, y and z directions.

Box 2.1   Application of Darcy's Law to simple hydrogeological situations

The following two worked examples illustrate the application of Darcy's Law to simple hydrogeological situations. In the first example, the alluvial aquifer shown in Fig. 2.8 is recharged by meltwater runoff from the adjacent impermeable mountains that run parallel to the axis of the valley. If the groundwater that collects in the aquifer discharges to the river, then it is possible to estimate the river flow at the exit from the valley. To solve this problem, and assuming that the river is entirely supported by groundwater discharge under steady, uniform flow conditions, the groundwater discharge (Q) can be calculated using eq. 2.7 and the information given in Fig. 2.8, as follows:

$$Q = -KA\frac{dh}{dl}$$
$$Q = 1 \times 10^{-3} \times 20 \times 5000 \times 4 \times 10^{-3}$$
$$Q = 0.4\,\text{m}^3\,\text{s}^{-1}$$

Accounting for both halves of the valley floodplain, the total discharge from the alluvial aquifer as river flow is 0.8 m³ s⁻¹.

In the second example (Fig. 2.9), a municipal waste disposal facility is situated in a former sand and gravel quarry. The waste is in contact with the water table and is directly contaminating the aquifer. The problem is to estimate the time taken for dissolved solutes to reach a spring discharge area located down-gradient of the waste tip. Assuming that the contaminant is unreactive and moves at the same rate as the steady, uniform groundwater flow, then from a consideration of eq. 2.10 and the information given in Fig. 2.9, the average linear velocity is calculated as follows:

$$\bar{v} = \frac{q}{n_e} = \frac{K\frac{dh}{dl}}{n_e}$$

$$\bar{v} = \frac{5 \times 10^{-4} \times 5 \times 10^{-3}}{0.25}$$

$$\bar{v} = 1 \times 10^{-5}\,\text{m s}^{-1}$$

Therefore, the time taken, t, to move a distance of 200 m in the direction of groundwater flow from the waste tip to the spring is

$$t = \frac{200}{1 \times 10^{-5}} = 2 \times 10^7\,\text{s} \approx 230\,\text{days}$$

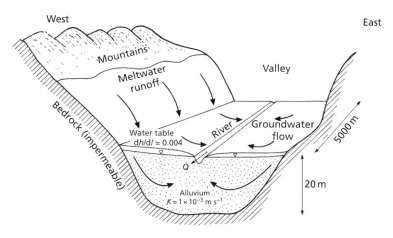

Fig. 2.8   Alluvial aquifer bounded by impermeable bedrock.

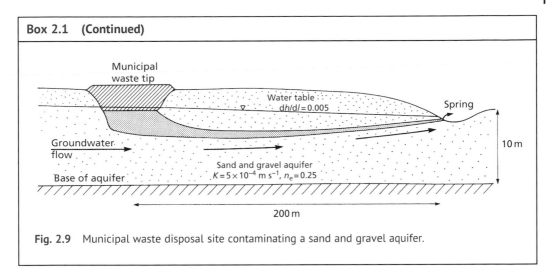

Box 2.1 (Continued)

Fig. 2.9 Municipal waste disposal site contaminating a sand and gravel aquifer.

2.6.1 Hydraulic properties of fractured rocks

By adopting the Darcian approach to the analysis of groundwater flow, it is implicit that the physical assemblage of grains that comprise the porous material are considered as a representative continuum, and that macroscopic laws, such as Darcy's Law, provide macroscopically averaged descriptions of the microscopic behaviour. In other words, Darcy's Law describes groundwater flow as a flux through a porous material that is imagined to have continuous, smoothly varying properties. In reality, intergranular and fractured porous materials are highly heterogeneous when examined at a scale similar to the spacing of the dominant pore size. The consequence of this is that Darcy's Law can be used successfully, but only at a scale large enough to contain a representative assemblage of pores. This is the continuum scale. At sub-continuum scales, the local pore network geometry strongly influences flow and the transport of contaminants. This is particularly relevant in fractured rocks where the dimension of the fracture spacing can impart a continuum scale that exceeds the size of many practical problems.

In fractured material such as carbonate and crystalline rocks and fissured clay sediments such as glacial tills, the conceptual model of groundwater flow can either be grossly simplified or a detailed description of the aquifer properties attempted as depicted in Fig. 2.10. With the exception of conduit flow in karst aquifers, fracture flow models generally assume that both fracture apertures and flow velocities are small such that Darcy's Law applies and flow is laminar (Box 2.2). In the example of the equivalent porous material shown in Fig. 2.10(b), the primary and secondary porosity and hydraulic conductivity distributions are represented as the equivalent or effective hydraulic properties of a continuous porous material. A drawback with this approach is that it is often difficult to determine the size of the representative elementary volume of material from which to define the effective hydraulic property values. Hence, the equivalent porous material approach may adequately represent the behaviour of a regional flow system but is likely to reproduce local conditions poorly.

More advanced approaches, such as the discrete fracture and dual-porosity models shown in Figs 2.10(c) and 2.10(d) represent groundwater movement through the fracture network. Flow through a single fracture may be idealized as occurring between two parallel plates with a

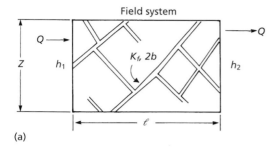

(a)

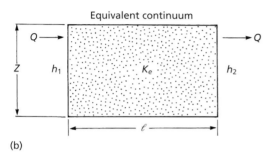

(b)

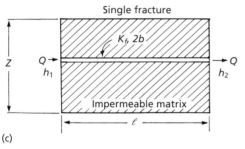

(c)

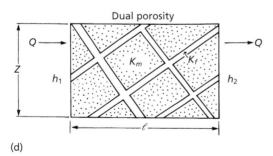

(d)

Fig. 2.10 Conceptual models to represent a fractured rock system. The fracture network of aperture 2*b* and with groundwater flow from left to right is shown in (a). The equivalent porous material, discrete fracture and dual porosity models representative of (a) are shown in (b), (c) and (d), respectively (Gale 1982). (*Source:* Gale, J.E. (1982) Assessing permeability characteristics of fractured rock. *Geological Society of America, Special Paper* **189**, 163–181, Boulder CO. © 1982, Geological Society of America.)

uniform separation or fracture aperture, 2*b*. The relation between flow and hydraulic gradient for individual fractures under laminar flow conditions is usually considered to be governed by the 'cubic law' presented by Snow (1969) and further validated by Witherspoon *et al.* (1980) and Gale (1982). In this treatment, the flow rate through a fracture, Q_f, may be expressed as follows:

$$Q_f = -2bwK_f \frac{dh}{dl} \qquad \text{(eq. 2.11)}$$

where w is the width of the fissure, K_f the hydraulic conductivity of the fracture and l the length over which the hydraulic gradient is measured. The hydraulic conductivity, K_f, is calculated from

$$K_f = \frac{\rho g(2b)^2}{12\mu} \qquad \text{(eq. 2.12)}$$

where ρ is fluid density, μ is fluid viscosity and g is the gravitational acceleration.

If the expression for K_f (eq. 2.12) is substituted in eq. 2.11, then

$$Q_f = -\frac{2}{3} \frac{w\rho g b^3}{\mu} \frac{dh}{dl} \qquad \text{(eq. 2.13)}$$

It can be seen from eq. 2.13 that the flow rate increases with the cube of the fracture aperture. Use of a model based on these equations requires a description of the fracture network, including the mapping of fracture apertures and geometry, that can only be determined by careful fieldwork.

In the case of the dual-porosity model, flow through the fractures is accompanied by exchange of water and solute to and from the surrounding porous rock matrix. Exchange between the fracture network and the porous blocks may be represented by a term that describes the rate of mass transfer. In this model, both the hydraulic properties of the fracture network and porous rock matrix need to be assessed, adding to the need for field mapping and hydraulic testing.

Box 2.2 Laminar and turbulent flows

Darcy's Law applies when flow is laminar but at high flow velocities, turbulent flow occurs and Darcy's Law breaks down. Under laminar conditions individual 'particles' of water move in paths parallel to the direction of flow, with no mixing or transverse component to the fluid motion. These conditions can be visualized by making an analogy between flow in a straight, cylindrical tube of constant diameter, and flow through porous granular or fissured material. At the edge of the tube, the flow velocity is zero rising to a maximum at the centre. As the flow velocity increases, so fluctuating eddies develop and transverse mixing occurs whereupon the flow becomes turbulent.

Flow rates that exceed the upper limit of Darcy's Law are common in karstic limestones (Section 2.6.2) and dolomites and highly permeable volcanic formations. Also, the high velocities experienced close to the well screen of a pumping borehole can also create turbulent conditions. The change from laminar flow at low velocities to turbulent flow at high velocities is usually related to the dimensionless Reynolds number, R_e, which expresses the ratio of inertial to viscous forces during flow. For flow through porous material, the Reynolds number is expressed as follows:

$$R_e = \frac{\rho q d}{\mu} \qquad \text{(eq. 1)}$$

where ρ is fluid density, μ is viscosity and q is the specific discharge (or characteristic velocity for fissured or fractured material). The characteristic length, d, can represent the mean pore diameter, mean grain diameter or, in the case of a fissure or fracture, either the hydraulic radius (cross-sectional area/wetted perimeter) or width of the fissure.

For laminar flow in granular material, Darcy's Law is valid as long as values of R_e do not exceed the range 1 to 10. Since fully turbulent flow does not occur until velocities are high and R_e is in the range 10^2 to 10^3, the transition between the linear laminar and turbulent regimes is characterized by non-linear laminar flow. In karst aquifers, conduit flow may remain in the laminar regime in pipes up to about 0.5 m in diameter provided the flow velocity does not exceed 1×10^{-3} m s^{-1} (Fig. 2.11).

The following example illustrates the application of the Reynolds number in determining whether groundwater flow is laminar or turbulent. A fissure in a limestone aquifer has a width, w, of 2 m and an aperture, $2b$, of 0.1 m. A tracer dye moves along the fissure at a velocity of 0.03 m s^{-1}. From this information, the characteristic length of the fissure (equal to the hydraulic radius) is $(2bw)/(2(2b + w)) = (0.1 \times 2)/(2(0.1 + 2)) = 0.05$ m. The characteristic velocity is equal to the tracer velocity. Hence, if the kinematic viscosity, μ/ρ, at 10°C is 1.31×10^{-6} m^2 s^{-1}, then using eq. 1:

$$R_e = \frac{0.03 \times 0.05}{1.31 \times 10^{-6}} = 1145 \qquad \text{(eq. 2)}$$

and the flow is transitional to turbulent.

(Continued)

Box 2.2   (Continued)

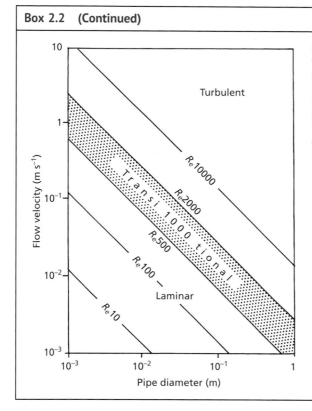

Fig. 2.11   Values of Reynolds number, R_e, at various velocities and conduit diameters and showing fields of different flow regimes (Smith *et al.* 1976). (*Source:* Adapted from Smith, D.I., Atkinson, T.C. and Drew, D.P. (1976) The hydrology of limestone terrains. In: *The Science of Speleology* (eds T.D. Ford and C.H.D. Cullingford). Academic Press, London.)

2.6.2   Karst aquifer properties

The term *karst* is used widely to describe the distinctive landforms that develop on rock types such as limestones, gypsum and halite that are readily dissolved by water. The name karst is derived from a word meaning stony ground used to describe the Kras region, now part of Slovenia and Croatia, where distinctive karst landforms are exceptionally well-developed. Karst areas are typically characterized by a lack of permanent surface streams and the presence of swallow holes (Fig. 2.12) and enclosed depressions. Rainfall runoff usually occurs underground in solutionally enlarged channels, some of which are large enough to form caves. Well-known karst areas include the pinnacle karst of the Guilin area, southern China (Guo *et al.* 2015); Mammoth Caves, Kentucky, USA (White and White 1989); the Yucatán Peninsula, Mexico (Bauer-Gottwein *et al.* 2011); Greek Islands (Panagopoulos and Lambrakis 2006); the Dordogne, Vercors and Tarn areas of France (Peyraube *et al.* 2012); Postojna Caves in Slovenia (Kogovsek and Petric 2014); and The Burren, County Clare, Ireland (Drew 1990). The karst of Ireland is further described by the Karst Working Group (2000) and Ford and Williams (2007) provide an extensive treatment of karst geomorphology and hydrology in general.

Karst aquifers are significant resources for human and agricultural uses across the world. Despite only covering 12% of the Earth's surface, karst aquifers contain some of the largest and most productive springs, and directly supply up to 25% of the world's population with water for drinking, agriculture and other water needs (Ford and Williams 2007). Carbonate rocks, many of which are karstic, underlie 35% of Europe and, over wide areas, karst waters form the only available natural resource for drinking water supply (Daly *et al.* 2002).

Fig. 2.12 The disappearance of the upper River Fergus at An Clab, south-east Burren, County Clare, Ireland, where surface runoff from Namurian shales disappears into a swallow hole at the contact with Carboniferous limestone.

In addition, karst aquifers provide extended seasonal storage for base flow support of rivers and subsequent water use downstream (Tobin *et al.* 2018).

Springs are among the most characteristic features of karst areas. Karst areas where springs with elevated temperatures occur are termed thermal karst. The largest naturally flowing thermal water system in Europe, the hot spring and wells that supply the baths of Budapest, Hungary, discharges from Triassic carbonate rocks. Many caves and related phenomena can be observed in the "Buda Thermal Karst" in the northeast extreme of the Transdanubian Range (Dublyansky 1995; Erhardt *et al.* 2017). The second-largest occurrence of mineral and thermal springs in Europe, in Stuttgart, Germany, is also associated with a karst aquifer, as are other thermal springs and spas in Germany and many other regions of the world (see Box 2.12; Goldscheider *et al.* 2010).

Figure 2.13 shows a typical model of groundwater flow and is used here to describe groundwater conditions in the Mendip Hills karst aquifer located in the west of England. The hills extend 50 km east-west and 10 km north-south. In the west, they rise above surrounding lowlands and form a broad karst plateau at about 260 m above sea level, developed in Carboniferous limestone. Structurally, the Mendips comprise four *en echelon* periclines with cores of Devonian sandstone. The dip of the limestones on the northern limbs of the folds is generally steep ($60°$ to $90°$), but to the south, it is more gentle ($20°$ to $40°$). The periclines emerge from beneath younger Triassic rocks in the west, but in the east are covered by Mesozoic strata which are in the process of being removed by erosion. As a result, the karstic features are better developed and probably older in the west than in the east. The principal aquifer is the Carboniferous limestone which has been extensively exploited for water supply, primarily by spring abstraction. Spring discharges are generally flashy, with a rapid response to storms, such that abstracted water is normally stored in surface reservoirs.

As shown in Fig. 2.13, groundwater discharge is via springs located at the lowest limestone outcrop, often where the limestones dip below Triassic mudstones. The larger springs are fed by conduits or flooded cave systems. The conduits act as drains within the saturated zone of the aquifer, and groundwater in fissures and fractures flows towards the conduits. Within the saturated zone, the conduit flow has a turbulent regime while the diffuse fissure flow obeys Darcy's Law (see Box 2.2).

Recharge to the aquifer can be characterized as allogenic and autogenic. Allogenic recharge comprises sinking streams which collect on sandstone and shale exposed in the core of the periclines. These streams pass directly into the conduit system through swallow holes. Autogenic recharge is either concentrated by closed depressions (dolines) or occurs as diffuse infiltration through the soil. Closed depressions are the first-order tributaries of the conduit system and focus concentrated recharge

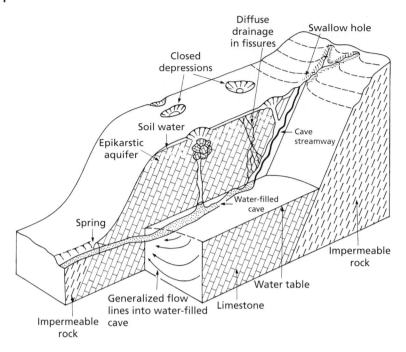

Fig. 2.13 Block diagram showing the occurrence of groundwater in karst aquifers. (*Source:* Adapted from Atkinson, T.C. (1977) Diffuse flow and conduit flow in limestone terrain in the Mendip Hills, Somerset (Great Britain). *Journal of Hydrology* **35**, 93–110.)

into shafts and caves. Weathering in the upper few metres of bedrock produces dense fissuring that provides storage for water in the unsaturated zone in what is sometimes referred to as the epikarstic aquifer. The epikarstic aquifer is recharged by infiltration and drains to the saturated zone via fractures and fissures, but with frequent concentration of drainage into shafts which form tributaries to cave systems.

Analysis of hydrographs (Section 6.7), baseflow recession curves, water balances (Section 6.2) and tracer tests (Section 7.3.3) indicates that the diffuse flow component of the saturated zone in the Mendip Hills has a storativity of about 1% and a hydraulic conductivity of 10^{-4} to 10^{-3} m s^{-1} (Atkinson 1977). About 70% of the flow in the saturated zone is via conduits, but these comprise less than one thirtieth of the active storage. From direct exploration, the depth of conduit circulation beneath the water table is known to exceed 60 m, implying a total storage in the saturated

zone of at least 600 mm of precipitation, roughly equivalent to one year's runoff. Significant storage also occurs in the epikarstic aquifer although the total amount is not known.

Karst aquifers can be classified according to the relative importance of diffuse flow and conduit flow, the degree of concentration of recharge and the amount of storage in the aquifer as shown in Fig. 2.14. The Mendip Hills karst aquifer has high storage, about 50% concentration of recharge into streams and closed depressions and 70% conduit flow in the saturated zone.

In karst aquifers where turbulent flow conditions can develop in solutionally developed conduits, representation of the hydraulic behaviour of the system is complicated by the difficulty in characterizing the hydraulic properties. A number of approaches are commonly used to model the behaviour of karst aquifers. The first is to assume that groundwater flow is governed by Darcy's Law and then to use one of

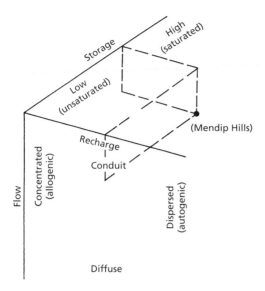

Fig. 2.14 Conceptual classification of karst aquifers from a consideration of recharge and groundwater flow mechanisms and the degree of saturated aquifer storage. (*Source:* Atkinson, T.C. (1977) Diffuse flow and conduit flow in limestone terrain in the Mendip Hills, Somerset (Great Britain). *Journal of Hydrology* **35**, 93–110.)

the models shown in Fig. 2.10. Further approaches to modelling flow in karst conduits is to adopt the Darcy–Weisbach pipe flow equation (eq. 6.21) or, for mature karst landscapes, to use a 'black box' model in which empirical functions are developed based on field observations of flow to reproduce input and output responses, in particular of recharge and spring flow. These functions may or may not include the usual aquifer parameters such as hydraulic conductivity, storativity and porosity. A third, hybrid approach is to use the aquifer response functions developed as for the 'black box' approach and then make use of these in an equivalent porous material model, although it must be recognized that large uncertainties remain, requiring careful field validation.

2.6.3 Sinkholes and land subsidence

Land subsidence includes both the gentle down-warping and sudden sinking of discrete segments of the ground surface. In areas underlain by soluble rocks such as limestone, dolomite, gypsum and halite, land subsidence is a common phenomenon and can result in the sudden collapse of the ground surface. Mining activities that remove materials (such as coal and salt) from below the surface can also result in a sudden collapse. Displacement of the ground as a result of a subsidence is principally vertical, although horizontal deformation often causes significant damage. The extraction of fluids such as groundwater, crude oil and natural gas from subsurface formations can play a direct role in land subsidence by causing compaction (see Section 2.9.1) that can disturb existing infrastructure, including buildings, roads, railways and pipelines (Galloway and Burbey 2011). Areas of the world with major land subsidence due to groundwater abstraction and fluid withdrawal from oil and gas fields are tabulated by Poland (1972).

Sinkholes are a form of instability and can be considered as a category of mass movement, even though they generally affect low-gradient or sub-horizontal slopes in soluble rocks. Sinkholes, also termed dolines, are a typical hazard in karst regions often related to the presence of underground caves. In addition to underground caves, sinkholes can also be linked to artificial cavities relating to human activity (Parise *et al.* 2015). Sinkholes not only threaten human lives and cause structural problems but also contaminate groundwater in that open sinkholes can create pathways for transmitting polluted surface water directly into groundwater (Xiao *et al.* 2018).

Sinkholes can resemble various shapes, including dishes, bowls, cones and cylinders and range in size from less than a metre deep and wide to over hundreds of metres deep and several hundred metres or even a kilometre wide. The large forms tend to be complex and grade into other classes of closed depressions. Several processes form sinkholes, although these processes frequently occur in combination and most sinkholes are polygenic (Huggett 2011). A six-fold classification of sinkhole formation is shown in Fig. 2.15. If the sinkhole collapse occurs into a water-filled cave, or if the water table rises after the collapse occurs, the collapse doline may contain a lake, often deep, covering its floor. Such lakes are called cenotes on the Yucatán Peninsula, Mexico and Obruk lakes on the Turkish Plateau (Huggett 2011).

The triggering mechanisms for sinkholes to collapse can be one, or a combination of mechanisms, including enlargement of caves due to subsurface dissolution, infiltration of water from the surface washing down fine materials from the covering deposits, fluctuations in groundwater levels rising to wet the covering materials and then falling leaving the cover saturated and without the hydraulic support offered by the water (Gutiérrez et al.

2014). For example, Florida has experienced numerous sinkhole incidents, a region with gently dipping to horizontal limestones. On 9 May 1981, a large collapse occurred in Winter Park. Over the period of a few hours, a house, several cars and half of a municipal swimming pool fell into a sinkhole. After several months, water rose to approximately the same level as lakes in the area. Geotechnical investigations indicated that the sinkhole was caused by ravelling of terra rossa, a karstic clayey silt, into a cave (Rahn 1996). The collapse, 100 m in diameter, occurred after a dry decade of low rainfall and a lowered water table in the area that most likely triggered the collapse. In a further case study, in urban areas in central Florida, the opposite effect of heavy rainfall and rapid increase of groundwater level within a

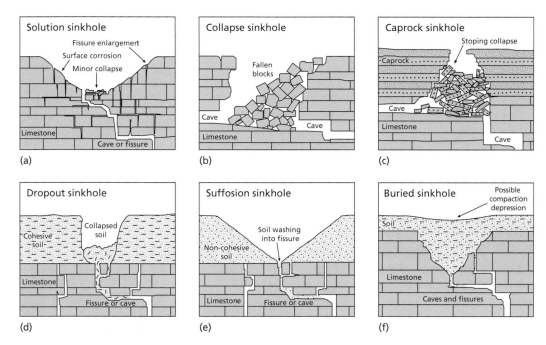

Fig. 2.15 Six types of sinkholes and their typical cross-sections. (a) Solution sinkhole: dissolutional lowering of the surface. (b) Collapse sinkhole: rock roof failure into an underlying cave. (c) Caprock sinkhole: failure of insoluble rock into a cave in soluble rock below. (d) Dropout sinkhole: cohesive soil collapse into a soil void formed over a bedrock fissure (see also Fig. 2.17). (e) Suffosion sinkhole: down-washing of non-cohesive soil into fissures in bedrock. (f) Buried sinkhole: sinkhole in rock, soil-filled after environmental change. Note that the dropout and suffosion sinkholes may be described as forms of subsidence sinkholes (Waltham *et al.* 2005). (*Source:* Adapted from Waltham, T., Bell, F. and Culshaw, M. (2005) *Sinkholes and Subsidence: Karst and Cavernous Rocks in Engineering and Construction.* Praxis Publishing Ltd, Chichester.)

relatively short period of time are considered major factors affecting the timing of sinkhole occurrences (Xiao *et al.* 2018).

Mining can also be a factor in causing sinkholes, either by dewatering and lowering of the water table, or by intercepting clay-filled voids which subsequently collapse (British Geological Survey 2020). A significant problem is that once created, underground cavities are later overlooked and become part of built-up areas that expand above the zones where the cavities are located, and so increasing the risk of damage in the event of a collapse (Fig. 2.16).

A famous sinkhole area associated with mining is found in the Far West Rand, about 65 km west of Johannesburg, Republic of South Africa, where gently dipping Precambrian dolomites overlie gold-bearing conglomerates. The dolomites have a thick terra rossa residual soil, or regolith, cover. As a result of groundwater pumping from the deep gold mines, soil choking a vertical pipe above a large cavern can start to desiccate and erode from the bottom upwards, creating a temporary arch above the pipe until a catastrophic collapse and sinkhole forms (Rahn 1996).

In the example of cohesive soil collapse into a soil void formed over carbonate rock (a dropout sinkhole, Fig. 2.15d), a solution feature begins with the opening out of natural joints in the carbonate rock by the process of dissolution where slightly acidic water percolating through overlying superficial deposits such as sand dissolves calcium carbonate (Fig. 2.17). The joints at the top of the carbonate rock become enlarged and form fissures that gradually fill with material washed down from above. This process can be either a sudden or geologically slow process, depending on drainage conditions. These sand-filled fissures tend to become preferred drainage pathways for percolating water. The process can continue over a long period, with progressive enlargement of the fissure and downward movement of sand, so that a zone of loosened sand develops above the

feature. The whole process is driven by water, so what might take tens of years during normal climatic conditions can occur over hours if there is a burst water main or major drain leak. Uneven slumping of the sand at depth can result in cavities forming. When these cavities eventually collapse, the overall effect is that the cavities migrate upwards (in other words, they swap places with the overlying sand) until they finally appear at the ground surface. A collapse of this type can cause unexpected and serious impacts on building foundations or on paved areas (Williams 2017).

Areas prone to sinkhole formation occur throughout the United Kingdom, although most are relatively small or are in upland rural locations. These include areas underlain by Carboniferous limestones, notably the Mendips, parts of Wales, the Peak District, and the northern Pennines including the Yorkshire Dales. The Cretaceous Chalk is also susceptible, especially where it is covered by Palaeogene clay and sand deposits, notably in parts of Dorset, Hampshire, and the Chilterns. However, the most susceptible area in the United Kingdom is the Permian (Zechstein Group) gypsum deposits of north-east England, particularly around Ripon where thick gypsum ($CaSO_4.2H_2O$) rests on porous and jointed dolomite. The groundwater flow and active gypsum karstifcation leads to the formation of sinkholes that can be up to 35 m across and up to 20 m deep, some of which have affected property and infrastructure (Cooper 1998; Cooper *et al.* 2013). Sinkholes also occur over salt deposits, commonly in areas such as Cheshire, where brine has been extracted making it difficult to separate naturally formed sinkholes from those created by mining. Natural subsidence as a result of groundwater flow and solution are believed to have been responsible for the formation of the Cheshire Meres (Bell *et al.* 1986). In Scotland, sinkholes are generally rare except in parts of Assynt underlain by the Cambrian Durness limestone (British Geological Survey 2020).

Fig. 2.16   Double-decker bus with its front wheels in the air that sank into a hole that opened suddenly on 3 March 1988 in Earlham Road, Norwich, eastern England. The ground collapse, or denehole, was caused by a set of tunnels related to chalk mining and thought to date from the sixteenth century. Deneholes are medieval chalk extraction pits, often comprising a narrow shaft with a number of chambers radiating from the base. The depth of such features reflects the depth to the underlying chalk bedrock with a shaft width commonly of 2–3 m, widening out into galleries at depth. (*Source:* Bryn Colton/Hulton Archive/Getty Images)

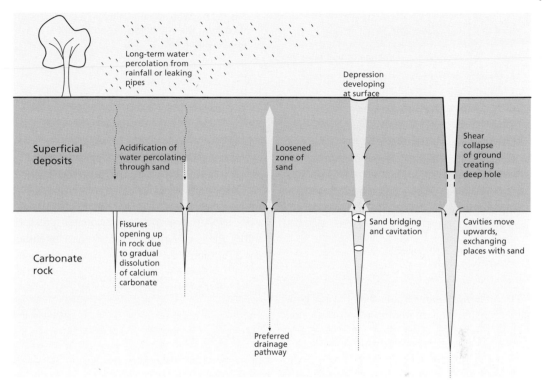

Fig. 2.17 Idealised diagram of the progressive development of a natural solution feature driven by water percolation resulting in a dropout sinkhole (Williams 2017). (*Source:* Adapted from Williams, M. (2017) *Subterranean Norwich: The Grain of the City.* Lasse Press, Norwich.)

2.7 Groundwater potential and hydraulic head

As described in the previous sections of this chapter, the porosity and hydraulic conductivity of porous material characterize the distribution and ease of movement of groundwater in geological formations. When analysing the physical process of groundwater flow, analogies are drawn with the flow of heat through solids from higher to lower temperatures and the flow of electrical current from higher to lower voltages. The rates of flow of heat and electricity are proportional to the potential gradients and, in a similar way, groundwater flow is also governed by a potential gradient.

Groundwater possesses energy in mechanical, thermal and chemical forms with flow controlled by the laws of physics and thermodynamics. With reference to Fig. 2.18, the work done in moving a unit mass of fluid from the standard state to a point, P, in the flow system is comprised of the following three components:

1) potential energy (mgz) required to lift the mass to elevation, z;
2) kinetic energy ($mv^2/2$) required to accelerate the fluid from zero velocity to velocity, v; and
3) elastic energy required to raise the fluid from pressure P_o to pressure P. The latter quantity can be thought of as the change in potential energy per unit volume of fluid and is found from

$$m\int_{P_o}^{P} \frac{V}{m}\,\mathrm{d}P = m\int_{P_o}^{P} \frac{\mathrm{d}P}{\rho} \qquad \text{(eq. 2.14)}$$

Given that groundwater velocities in porous material are very small, the kinetic energy term

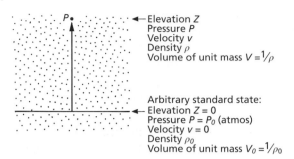

Fig. 2.18 Work done in moving a unit mass of fluid from the standard state to a point P in a groundwater flow system.

Elevation Z
Pressure P
Velocity v
Density ρ
Volume of unit mass $V = 1/\rho$

Arbitrary standard state:
Elevation $Z = 0$
Pressure $P = P_0$ (atmos)
Velocity $v = 0$
Density ρ_0
Volume of unit mass $V_0 = 1/\rho_0$

can be ignored such that at the new position, P, the fluid potential, Φ, or mechanical energy per unit mass ($m = 1$) is

$$\Phi = gz + \int_{P_o}^{P} \frac{dP}{\rho} \qquad \text{(eq. 2.15)}$$

For incompressible fluids that have a constant density, and therefore are not affected by a change in pressure, then

$$\Phi = gz + \frac{(P - P_o)}{\rho} \qquad \text{(eq. 2.16)}$$

To relate the fluid potential to the hydraulic head measured by Darcy in his experiment (Fig. 2.6), Fig. 2.19 demonstrates that the fluid pressure at position P in a column containing porous material is found as follows:

$$P = \rho g \psi + P_o \qquad \text{(eq. 2.17)}$$

where ψ is the height of the water column above P and P_o is atmospheric pressure (the pressure at the standard state).

It can be seen that $\psi = h - z$ and so, substituting in eq. 2.17:

$$P = \rho g(h - z) + P_o \qquad \text{(eq. 2.18)}$$

By substituting this expression for pressure into the equation for fluid potential, eq. 2.16, then

$$\Phi = gz + \frac{[\rho g(h - z) + P_o - P_o]}{\rho}$$
$$\text{(eq. 2.19)}$$

and, thus:

$$\Phi = [gz + gh - gz] = gh \qquad \text{(eq. 2.20)}$$

The result of eq. 2.20 provides a significant relationship in hydrogeology: the fluid potential, Φ, at any point in a porous material can simply be found from the product of hydraulic head and acceleration due to gravity. Since gravity is, for all practical purposes, almost constant near the Earth's surface, Φ is almost exactly correlated with h. The significance is that hydraulic head is a measurable, physical quantity and is therefore just as suitable a measure of fluid potential as Φ.

Returning to the analogy with heat and electricity, where rates of flow are governed by

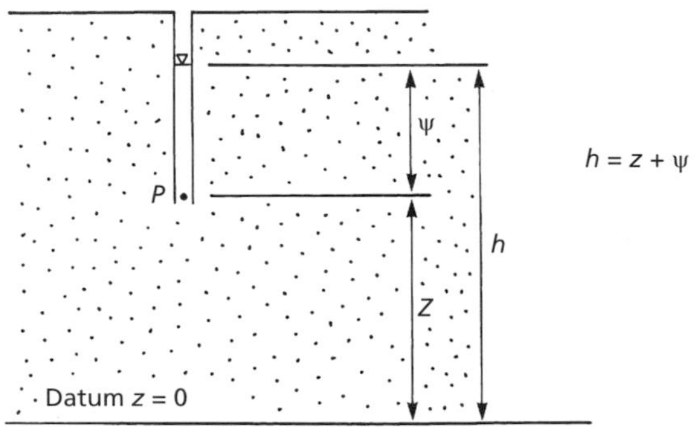

Fig. 2.19 Relation between hydraulic head, h, pressure head, ψ, and elevation head, z, at a point P in a column of porous material.

$h = z + \psi$

ψ

P

h

z

Datum $z = 0$

potential gradients, it is now shown that groundwater flow is driven by a fluid potential gradient equivalent to a hydraulic head gradient. In short, groundwater flows from regions of higher to lower hydraulic head.

With reference to eqs 2.16 and 2.20, and, by convention, setting the atmospheric pressure, P_o to zero, then

$$gh = gz + \frac{P}{\rho} \qquad \text{(eq. 2.21)}$$

The pressure at point P in Fig. 2.19 is equal to $\rho g \psi$, and so it can be shown by substitution in eq. 2.21 and by dividing through by g that

$$h = z + \psi \qquad \text{(eq. 2.22)}$$

Equation 2.22 confirms that the hydraulic head at a point within a saturated porous material is the sum of the elevation head, z, and pressure head, ψ, thus providing a relationship that is basic to an understanding of groundwater flow. This expression is equally valid for the unsaturated and saturated zones of porous material, but it is necessary to recognize, as shown in Fig. 2.20, that the pressure head term, ψ, is a negative quantity in the unsaturated zone as a result of adopting the convention of

setting atmospheric pressure to zero and working in gauge pressures. From this, it follows that at the level of the water table the water pressure is equal to zero (i.e. atmospheric pressure). In the capillary fringe above the water table, the aquifer material is completely saturated, but because of capillary suction drawing water up from the water table, the porewater pressure is negative, that is less than atmospheric pressure (for further discussion, see Section 6.4.1). The capillary fringe varies in thickness depending on the diameter of the pore space and ranges from a few centimetres for coarse-grained material to several metres for fine-grained deposits.

2.8 Interpretation of hydraulic head and groundwater conditions

2.8.1 Groundwater flow direction

Measurements of hydraulic head, normally achieved by the installation of a piezometer or well point, are useful for determining the

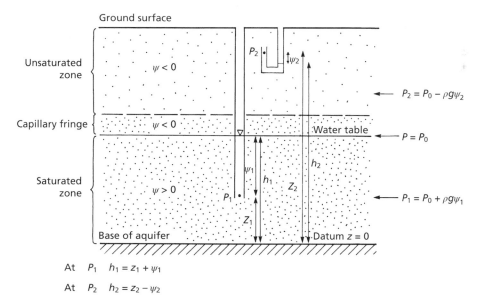

At P_1 $h_1 = z_1 + \psi_1$

At P_2 $h_2 = z_2 - \psi_2$

Fig. 2.20 Condition of pressure head, ψ, for the unsaturated and saturated zones of an aquifer. At the water table, fluid pressure is equal to atmospheric (P_o) and by convention is set equal to zero. Note also that the unsaturated or vadose zone is the region of a geological formation containing solid, water and air phases while in the saturated or phreatic zone, pore spaces of the solid material are all water-filled.

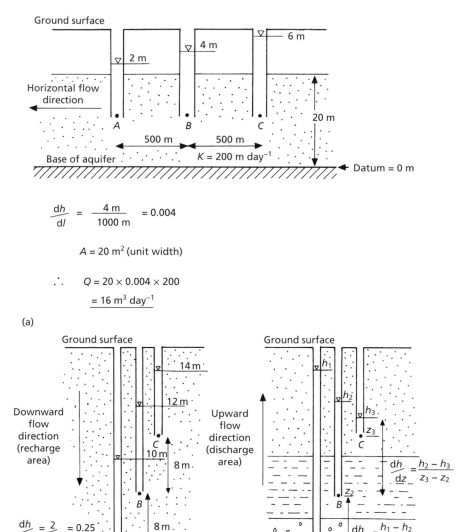

$$\frac{dh}{dl} = \frac{4\ m}{1000\ m} = 0.004$$

$$A = 20\ m^2\ \text{(unit width)}$$

$$\therefore \quad Q = 20 \times 0.004 \times 200$$

$$\underline{= 16\ m^3\ day^{-1}}$$

(a)

(b)

Fig. 2.21 Determination of groundwater flow direction and hydraulic head gradient from piezometer measurements for (a) horizontal flow and (b) vertical flow. The elevation of the water level indicating the hydraulic head at each of the points *A, B* and *C* is noted adjacent to each piezometer.

directions of groundwater flow in an aquifer system. In Fig. 2.21(a), three piezometers installed to the same depth enable the determination of the direction of groundwater flow and, with

the application of Darcy's Law (eq. 2.7), the calculation of the horizontal component of flow. In Fig. 2.21(b), two examples of piezometer nests are shown that allow the measurement of

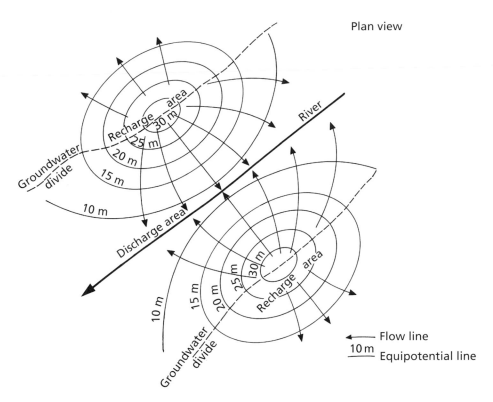

Plan view

Fig. 2.22 Sketch map of the surface of the water table in an unconfined aquifer showing recharge and discharge areas and the position of groundwater divides.

hydraulic head and the direction of groundwater flow in the vertical direction to be determined either at different levels in the same aquifer formation or in different formations.

2.8.2 Water table and potentiometric surface maps

Observation boreholes and piezometers located within a district provide a picture of the three-dimensional distribution of hydraulic head throughout an aquifer system. Lines drawn joining points of equal groundwater head, or groundwater potential, are termed equipotential lines. Lines perpendicular to the equipotential lines are flow lines and can be used in the construction of a flow net

(Box 2.3). In plan view, the construction of equipotential contours results in a map of the potentiometric surface. In an unconfined aquifer, the potentiometric surface is a map of the water table, where the groundwater is by definition at atmospheric pressure. In a confined aquifer, the potentiometric surface predicts the position that the water level would rise to in a borehole that penetrates the buried aquifer. As shown in Fig. 2.22, areas of high hydraulic head may be interpreted as groundwater recharge zones, while areas of low hydraulic head are typically in groundwater discharge zones. Box 2.4 provides an example of an actual potentiometric surface map for the Chalk aquifer underlying the London Basin.

Box 2.3 Flow nets and the tangent law

The construction of a flow net, for example a water table or potentiometric surface map, and the interpretation of groundwater flow lines, requires the implicit assumption that flow is perpendicular to the lines of equal hydraulic head (i.e. the porous material is isotropic), with flow in the direction of decreasing head. All flow nets, however, simple or advanced, can be drawn using a set of basic rules. When attempting to draw a two-dimensional flow net for isotropic porous material by trial and error, the following rules must be observed (Fig. 2.23):

1) Flow lines and equipotential lines should intersect at right angles throughout the groundwater flow system;
2) Equipotential lines should meet an impermeable boundary at right angles resulting in groundwater flow parallel to the boundary;
3) Equipotential lines should be parallel to a boundary that has a constant hydraulic head resulting in groundwater flow perpendicular to the boundary;
4) In a layered, heterogeneous groundwater flow system, the tangent law must be satisfied at geological boundaries;
5) If squares are created in one portion of one formation, then squares must exist throughout that formation and throughout all formations with the same hydraulic conductivity with the possible exception of partial stream tubes at the edge. Rectangles will be created in formations with different hydraulic conductivity.

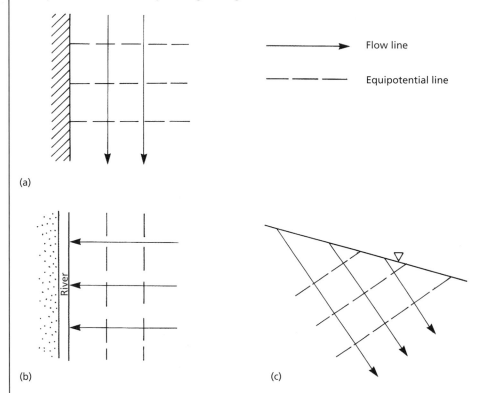

Fig. 2.23 Simple rules for flow net construction for the cases of (a) an impermeable boundary, (b) a constant head boundary (here shown as a river), and (c) a water table boundary.

Box 2.3 (Continued)

The last two rules are particularly difficult to observe when drawing a flow net by hand but even a qualitative flow net, in which orthogonality is preserved but with no attempt to create squares, can help provide a first understanding of a groundwater flow system. In simple flow nets, the squares are actually 'curvilinear squares' that have equal central dimensions able to enclose a circle that is tangent to all four bounding lines (Fig. 2.24).

In Fig. 2.24, a flow net is constructed for groundwater flow beneath a dam structure that is partially buried in an isotropic and homogeneous sand aquifer. To calculate the flow beneath the dam, consider the mass balance for box ABCD for an incompressible fluid. Under steady-state conditions, and assuming unit depth into the page, the flow into the box across face AB with width, Δw,

will equal the flow out of the box across face DC. From Darcy's Law (eq. 2.7) the best estimate of flow through box ABCD, ΔQ, is equal to

$$\Delta Q = \Delta w \times K \frac{\Delta h}{\Delta l} \qquad \text{(eq. 1)}$$

or, on rearrangement:

$$\Delta Q = K \times \Delta h \frac{\Delta w}{\Delta l} \qquad \text{(eq. 2)}$$

If the flow net is equi-dimensional (curvilinear squares), then $\Delta w/\Delta l$ is about equal to unity and eq. 2 becomes

$$\Delta Q = K \times \Delta h \qquad \text{(eq. 3)}$$

Δh is found from the total head drop $(h_1 - h_2)$ along the stream tube divided by the number of head divisions, n, in the flow net:

$$\Delta h = \frac{(h_1 - h_2)}{n} \qquad \text{(eq. 4)}$$

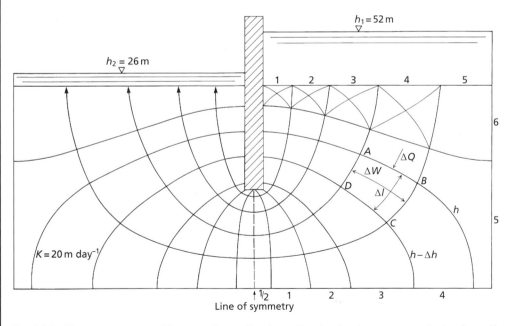

Fig. 2.24 Flow net constructed for groundwater flow beneath a dam in a homogeneous, isotropic aquifer.

(Continued)

Box 2.3 (Continued)

If the number of stream tubes in the region of flow is m, then the total flow below the dam is

$$Q = \frac{m}{n} \times K(h_1 - h_2) \qquad \text{(eq. 5)}$$

For the example of flow beneath a dam shown in Fig. 2.24, $m = 5$, $n = 13$ and $(h_1 - h_2) = 26$ m. If the hydraulic conductivity, K, is 20 m day^{-1}, then the total flow is found from

$$Q = \frac{5}{13} \times 20 \times 26$$
$$= 200 \text{ m}^3 \text{ day}^{-1}.$$

In homogeneous but anisotropic porous material, flow net construction is complicated by the fact that flow lines and equipotential lines are not orthogonal. To overcome this problem, a transformed section is prepared through the application of a hydraulic conductivity ellipse (Fig. 2.25). Considering

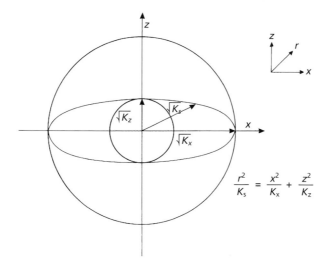

$$\frac{r^2}{K_s} = \frac{x^2}{K_x} + \frac{z^2}{K_z}$$

(a)

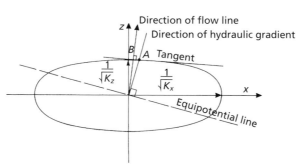

(b)

Fig. 2.25 In (a) the hydraulic conductivity ellipse for a homogeneous, anisotropic material is shown with principal hydraulic conductivities K_x and K_z. The hydraulic conductivity value K_s for any direction of flow in an anisotropic material can be found graphically if K_x and K_z are known. Also shown are two circles representing the possible isotropic transformations for flow net construction (see text for explanation). In (b) the method for determining the direction of flow in an anisotropic material at a specified point is shown represented as an inverse hydraulic conductivity ellipse. A line drawn in the direction of the hydraulic gradient intersects the ellipse at point A. If a tangent is drawn to the ellipse at A, the direction of flow is then found perpendicular to this tangent (point B).

Box 2.3   (Continued)

a two-dimensional region in a homogeneous, anisotropic aquifer with principal hydraulic conductivities K_x and K_z, the hydraulic conductivity ellipse will have semi-axes $\sqrt{K_x}$ and $\sqrt{K_z}$. The co-ordinates in the transformed region, X–Z, are related to the original x–z system by:

$$X = x$$
$$Z = \frac{z\sqrt{K_x}}{\sqrt{K_z}}$$ (eq. 6)

For $K_x > K_z$, this transformation will expand the vertical scale of the region of flow and also expand the hydraulic conductivity ellipse into a circle of radius $\sqrt{K_x}$. The fictitious, expanded region of flow will then act as if it were homogeneous with hydraulic conductivity K_x. The graphical construction of the flow net follows from the transformation of the co-ordinates and using the previous rules for homogeneous, isotropic material. The final step is to redraw the flow net by inverting the scaling ratio to the original dimensions. If discharge quantities or flow velocities are required, it is easiest to make these calculations in the transformed

section and applying the hydraulic conductivity value K', found from

$$K' = \sqrt{K_x \times K_z}$$ (eq. 7)

In the absence of a transformation of the co-ordinate system, the direction of groundwater flow at a point in an anisotropic material can be found using the construction shown in Fig. 2.25(b). A line drawn in the direction of the hydraulic gradient intersects the ellipse at point A. If a tangent is drawn to the ellipse at A, then the direction of flow is perpendicular to this tangent line. For a further treatment of the topic of flow net construction, refer to Cedergren (1967) and Freeze and Cherry (1979).

When groundwater flows across a geological boundary between two formations with different values of hydraulic conductivity, the flow lines refract in an analogous way to light passing between two materials. Unlike in the case of light that obeys a sine law, groundwater refraction obeys a tangent law, as explained next.

In Fig. 2.26, a stream tube is shown with flow from a region with hydraulic conductivity

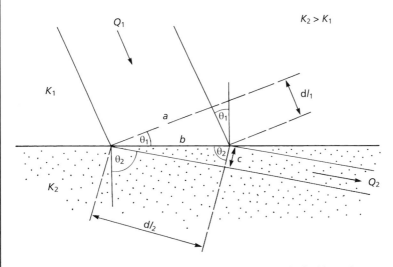

Fig. 2.26   Refraction of groundwater flow lines at a geological boundary.

(Continued)

Box 2.3 (Continued)

K_1 to a region with hydraulic conductivity K_2, where $K_2 > K_1$. Considering a stream tube of unit depth perpendicular to the page, for steady flow, the inflow Q_1 must equal the outflow Q_2; then, from Darcy's Law (eq. 2.7):

$$K_1 a \frac{dh_1}{dl_1} = K_2 c \frac{dh_2}{dl_2} \qquad \text{(eq. 8)}$$

where dh_1 is the decrease in head across distance dl_1 and dh_2 is the decrease in head across distance dl_2. In that dl_1 and dl_2 bound the same two equipotential lines, then dh_1 equals dh_2; and from a consideration of the geometry of Fig. 2.26, $a = b.\cos \theta_1$ and $c = b.\cos \theta_2$. Noting that $b/dl_1 = 1/\sin \theta_1$ and $b/dl_2 = 1/\sin \theta_2$, eq. 8 now becomes

$$K_1 \frac{\cos \theta_1}{\sin \theta_1} = K_2 \frac{\cos \theta_2}{\sin \theta_2} \qquad \text{(eq. 9)}$$

or

$$\frac{K_1}{K_2} = \frac{\tan \theta_1}{\tan \theta_2} \qquad \text{(eq. 10)}$$

Equation 10 is the tangent law for the refraction of groundwater flow lines at a geological boundary in heterogeneous material. In layered aquifer systems, as shown in Fig. 2.27, the outcome of the tangent law is that flow lines have longer, horizontal components of flow in aquifer layers and shorter, vertical components of flow across intervening aquitards. The aquifer layers act as conduits for groundwater flow. If the ratio of the aquifer to aquitard hydraulic conductivities is greater than 100, then flow lines are almost horizontal in aquifer layers and close to vertical across aquitards. This is commonly the case, as the values of hydraulic conductivity of natural geological materials range over many orders of magnitude (Table 2.1).

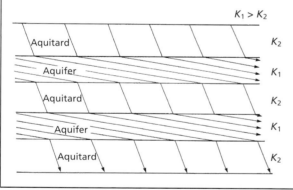

Fig. 2.27 Refraction of groundwater flow lines across a layered aquifer system.

Box 2.4 Potentiometric surface map of the London Basin

The recording of groundwater levels in wells and boreholes and their reference to a common datum such as sea level is a basic requirement in hydrogeology. Maps of the water table or, more correctly, the potentiometric surface assist in the management of groundwater resources by enabling the identification of recharge and discharge areas and groundwater conditions. The repeat mapping of an area enables the storage properties of the aquifer to be understood from an examination of observed fluctuations in groundwater levels (Section 7.2). Potentiometric surface maps can also be used as the basis for constructing a regional flow net (Box 2.3) and can provide useful information on the hydraulic conductivity of the aquifer from inspection of the gradients of the equipotential contour lines. A further important application is in groundwater modelling, where a high-quality

Box 2.4   (Continued)

potentiometric surface map and associated observation borehole hydrographs are necessary in developing a well-calibrated groundwater flow model.

A map of the potentiometric surface of the Chalk aquifer below the London Basin is shown in Fig. 2.28 and illustrates a number of the above points. Areas of high groundwater level in excess of 50 m above sea level are present in unconfined areas, where the Chalk is exposed on the northern and southern rims of the synclinal basin. Here, the Chiltern Hills and North Downs are the recharge areas for the London Basin, respectively. In the centre of the Basin, the residual drawdown in Chalk

groundwater levels due to earlier over-exploitation of the aquifer (see Box 2.5) is clearly visible in the wide area where the Chalk potentiometric surface is less than 10 m below sea level. Additional disturbance of the regional groundwater level is noticeable along the River Lea Valley to the north of London where large abstractions have disturbed the equipotential contours. To the east of Central London, the Chalk potentiometric surface is at about sea level along the estuary of the River Thames and here saline water can intrude the aquifer where the overlying Lower London Tertiaries and more recent deposits are thin or absent.

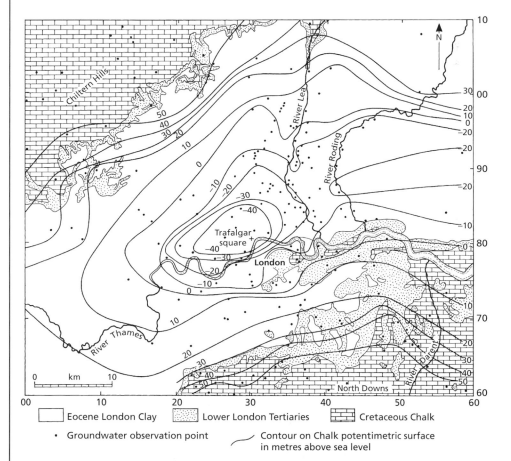

Fig. 2.28   Map of the potentiometric surface of the Chalk aquifer underlying the London Basin drawn from observations made in January 1994 (Lucas and Robinson 1995). (*Source:* Lucas, H.C. and Robinson, V.K. (1995) Modelling of rising groundwater levels in the Chalk aquifer of the London Basin. *Quarterly Journal of Engineering Geology* **28**, S51–S62.)

2.8.3 Types of groundwater conditions

Groundwater conditions are strongly influenced by the juxtaposition of lithological units and by geological structure. The nature of aquifer geometry can give rise to four basic types of groundwater conditions, as depicted in Fig. 2.29, and also determines the occurrence of springs. An unconfined aquifer exists when a water table is developed that separates the unsaturated zone above from the saturated zone below. It is possible for an unconfined aquifer to develop below the lower surface of an aquitard layer. In this case, a concealed unconfined aquifer is recognized.

In heterogeneous material, for example sedimentary units containing intercalated lenses or layers of clay, perched water table conditions can develop. As shown in Fig. 2.30, above the regional water table and within the unsaturated zone, a clay layer within a sand matrix causes water to be held above the lower permeability material creating a perched water table. Because water table conditions occur where groundwater is at atmospheric pressure, inverted water tables occur at the base of the perched lens of water in the clay layer and also below the ground surface following a rainfall event in which water infiltrates the soil zone.

A confined aquifer is contained between two aquitards or aquicludes. Water held in a

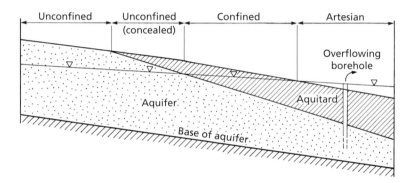

Fig. 2.29 Cross-section showing four types of groundwater conditions. A water table is developed where the aquifer is unconfined or concealed, and a potentiometric surface is present where the aquifer experiences confined or artesian conditions.

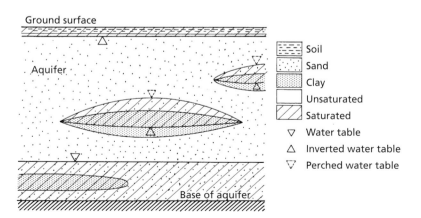

Fig. 2.30 Perched and inverted water table conditions developed within a sand aquifer containing clay lenses. An inverted water table is also shown below the wetted ground surface in the soil zone.

confined aquifer is under pressure, such that groundwater in a borehole penetrating a confined unit will rise to a level above the top of the aquifer. If the groundwater level rises to the top of the borehole above ground level and overflows, then an overflowing artesian groundwater condition is encountered. If over-abstraction of groundwater occurs from boreholes exploiting a confined aquifer, the groundwater level can be drawn down below the top of the aquifer such that it becomes unconfined. As shown in Box 2.5, this situation developed in the Chalk aquifer of the London Basin as a result of over-abstraction of groundwater to sustain the growth of London's industry and population from between the early 1800s and the early 1960s.

Box 2.5   History of groundwater exploitation in the Chalk aquifer of the London Basin

The industrial revolution and associated population growth of Greater London resulted in a large demand for water. With surface water resources becoming polluted, increasing use was made of the substantial storage in the Cretaceous Chalk aquifer that underlies the London Basin. The Chalk forms a gentle syncline with extensive outcrops to the south and north-west of London (Fig. 2.31) and is confined by Tertiary strata, mainly Eocene London Clay. The aquifer is recharged at outcrop by rainfall on the Chiltern Hills and North Downs. A component of this water flows through the aquifer towards and along the easterly dipping axis of the syncline. Prior to exploitation, this groundwater eventually discharged into the Thames Estuary in the Woolwich area via springs (Marsh and Davies 1983).

The first deep boreholes were sunk in the middle of the eighteenth century, and there was rapid development from about 1820. At this time, as shown in Fig. 2.32a, an artesian situation existed throughout much of the London Basin. The potentiometric surface of the Chalk aquifer was typically at shallow depth and in low lying areas, such as the Hackney Marshes, water seeped upwards to the surface. In Central London, the natural potentiometric surface was a few metres above sea level (+7.5 m; Fig. 2.32a) creating artesian groundwater conditions. At the start of the twentieth century, rest water levels in the borehole that at first supplied the fountains in Trafalgar Square under an artesian condition (Kirkaldy 1954) had declined to 60 m below sea level (Fig. 2.33), a drop of 40 m since the borehole was drilled in

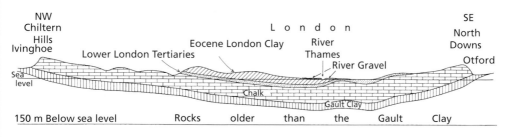

Fig. 2.31 Schematic geological section across the London Basin. The vertical scale is greatly exaggerated (Sherlock 1962). (*Source:* Sherlock, R.L. (1962) *British Regional Geology: London and Thames Valley* (3rd edn Reprint with Minor Additions). HMSO, London. © 1962, British Geological Survey.)

(Continued)

Box 2.5   (Continued)

1844, causing the groundwater condition in the Chalk aquifer to change from confined to unconfined. The decrease in hydraulic head caused a loss of yield from 2600 to 36 m^3 day^{-1} by 1900.

By 1950, and over an area extending from the centre of London south-west to Richmond (Fig. 2.32b), groundwater levels were more than 60 m below sea level over an area of 200 km^2. In the 100 years up to 1950,

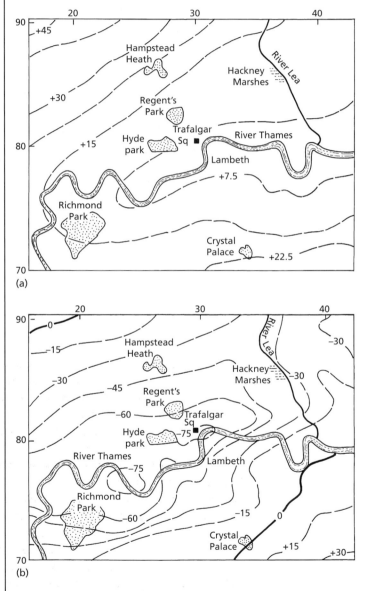

Fig. 2.32 Maps of the Chalk potentiometric surface in London showing groundwater levels in metres relative to mean sea level: (a) before groundwater exploitation commenced in the early 1800s; and (b) in 1950 (Marsh and Davies 1983). (*Source:* Marsh, T.J. and Davies, P.A. (1983) The decline and partial recovery of groundwater levels below London. *Proceedings of the Institution of Civil Engineers, Part 1* **74**, 263–276.)

Box 2.5   (Continued)

groundwater abstractions are estimated to have increased from 9.0×10^6 to 73.0×10^6 m^3 a^{-1} (Marsh and Davies 1983).

The general decline in groundwater levels had a number of impacts on groundwater resources in the London Basin, including a loss of yield and eventual failure of supply boreholes, and saline water intrusion into the Chalk aquifer. Saline water was recorded as far west as Lambeth in a zone 5–8 km wide centred on the River Thames. As discussed in Box 2.7, falling groundwater levels in a confined aquifer system can cause land subsidence, although the well-consolidated nature of the marine London Clay formation prevented substantial compaction in this

case. Nevertheless, the overall settlement approached 0.2 m in Central London over the period 1865–1931 (Wilson and Grace 1942) as a consequence of the combined effects of dewatering the aquifer system, heavy building development and the removal of fine sediment during pumping from sand and gravel horizons.

An increasing awareness of the problems of over-abstraction from the Chalk aquifer together with a reduction in industrial abstractions since 1945 has led to a virtual end to further aquifer exploitation in London. Water supplies switched to piped supplies drawn predominantly from reservoirs in the Thames and Lea reservoirs, or from major

Fig. 2.33   Variation in groundwater level at Trafalgar Square (borehole TQ28/119) showing the changes in groundwater conditions in the Chalk aquifer below London. In the period before groundwater exploitation, the natural potentiometric surface was near to ground surface at about 7.5 m above sea level. At this time, the Chalk aquifer was confined by the Eocene London Clay. By 1965, the groundwater level had declined to 83 m below sea level as a result of over-exploitation of the groundwater resource. The lower section of the Tertiary strata includes sands of the Lower London Tertiaries which are in hydraulic continuity with the Chalk and these extend the aquifer vertically. By 1900, the Chalk potentiometric surface was within these strata and the groundwater condition had become semi-confined. For most of the twentieth century until the mid-1960s when groundwater levels began to rise, the Chalk potentiometric surface was below the base of the Tertiary strata and the groundwater condition was unconfined (Marsh and Davies 1983). (*Source:* Marsh, T.J. and Davies, P.A. (1983) The decline and partial recovery of groundwater levels below London. *Proceedings of the Institution of Civil Engineers, Part 1* **74**, 263–276.)

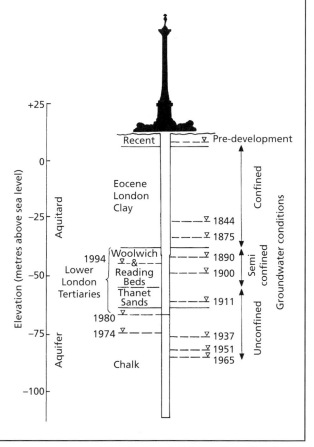

Box 2.5 (Continued)

abstraction boreholes located outside of Greater London. Since 1965, the reduction in groundwater abstraction has resulted in commencement of a recovery of groundwater levels at a rate of about $1\,\mathrm{m\,a}^{-1}$ in the centre of the cone of depression (Lucas and Robinson 1995; Fig. 7.4b). The re-saturation of the Tertiary strata has implications for building structures because of increased hydrostatic pressure on deep foundations and the flooding of tunnels constructed during the main period of depressed groundwater levels between 1870 and 1970. To combat this problem, it is recognized that pumping will be required to control groundwater levels. Regional groundwater modelling of the London Basin indicated that an additional $70 \times 10^3\,\mathrm{m}^3\,\mathrm{day}^{-1}$ of pumping would be necessary to control the rise in groundwater to an acceptable level assuming that existing abstractions continued at their current level (Lucas and Robinson 1995). To be effective, much of this abstraction would need to be concentrated in an area incorporating Central London together with an extension eastwards along the River Thames where structural engineering problems due to groundwater level rise are predicted (Simpson *et al.* 1989).

If the overlying geological unit behaves as an aquitard, then leakage of water to the underlying aquifer can occur if a vertical hydraulic gradient is developed across the aquitard-aquifer boundary. This situation is commonly encountered where fluvial or glacio-fluvial silts and sandy clays overlie an aquifer and results in a semi-confined aquifer condition.

A more complete regional hydrogeological interpretation requires the combination and analysis of mapped geological and geomorphological information, surveyed groundwater level data and hydrogeological field observations. As an example of such an integrated approach, groundwater conditions prevailing in the Qu'Appelle Valley of Saskatchewan are explained in Box 2.6. Interpretation of the relationship between geology, groundwater occurrence and potentiometric head distribution demonstrates the connection between groundwater and spring-fed lakes in this large glacial meltwater channel feature.

Box 2.6 Relationship between geology, geomorphology and groundwater in the Qu'Appelle Valley, Saskatchewan

The Qu'Appelle Valley in southern Saskatchewan in Canada is a major landscape feature that owes its origin to the continental glaciations of the Quaternary Period. During the advance of the first glacier, the Hatfield Valley was cut into bedrock by glacial meltwater to a width of 20 km from near the Manitoba border to Alberta. When the glacier advanced to the vicinity of the Hatfield Valley, sands of the Empress Group were deposited and these now form a major aquifer. A similar sequence of events occurred during the advance of the third glacier which deposited the Floral Formation till. Meltwaters from this glacier cut the Muscow Valley, which was then filled with silt, sand and gravel of the Echo Lake Gravel. The fourth and last ice advance finally retreated from the Qu'Appelle area about 14

Box 2.6 (Continued)

000 years ago, leaving its own distinctive till, the Battleford Formation, and other major landscape features.

The Qu'Appelle Valley was carved by meltwater issuing from the last retreating ice sheet, draining eastwards to glacial Lake Agassiz through the ice marginal Qu'Appelle Spillway. As the water continued eastwards through the spillway it cut a wide valley to a depth of 180 m into the underlying glacial deposits and bedrock. Where it crossed the buried Hatfield Valley it cut into the sand deposits. Since the retreat of the glacier, the present Qu'Appelle Valley has continued

to fill with alluvial material derived from down-valley transport of sediment and from the erosion of valley sides and adjacent uplands.

Presently, a number of freshwater lakes and the existence of tributary valleys owe their existence to groundwater discharge and demonstrate the relationship between geology, geomorphology and groundwater. To illustrate this relationship, the cross-section through the Qu'Appelle Valley at Katepwa Provincial Park (Fig. 2.34) shows recharge from rainfall and snowmelt on the adjacent prairie moving vertically

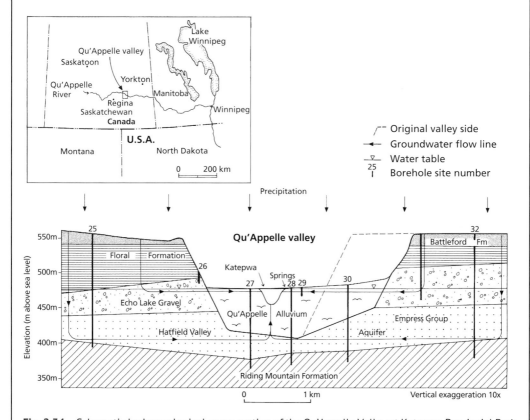

Fig. 2.34 Schematic hydrogeological cross-section of the Qu'Appelle Valley at Katepwa Provincial Park, Saskatchewan (Christiansen *et al.* 1981). (*Source:* Christiansen, E.A., Acton, D.F., Long, R.J. *et al.* (1981) *Fort Qu'Appelle Geolog. The Valleys – Past and Present.* Interpretive Report No. 2. The Saskatchewan Research Council, Canada.)

(*Continued*)

Box 2.6 (Continued)

downwards to the Echo Lake Gravel and Empress Group, then horizontally through these more permeable deposits, before finally moving vertically upwards to discharge into Katepwa Lake as underwater springs.

The evolution of the tributary valleys, as shown in Fig. 2.35, is linked to past and present groundwater flow regimes. At the time the Qu'Appelle Valley was cut by glacial meltwater, the water-bearing Echo Lake Gravel was penetrated, and large quantities of groundwater discharged from this aquifer into the Qu'Appelle Valley in the form of major springs. It is conjectured that discharging groundwater carried sand

and gravel from the Echo Lake Gravel and, to a lesser extent, from the Empress Group into the Qu'Appelle Valley, where part of it was swept away by meltwater flowing through the Qu'Appelle Spillway. This loss of sand and gravel by 'spring sapping' caused the overlying till to collapse, forming a tributary valley, which developed headwards along the path of maximum groundwater flow. Spring sapping forms short, wide tributaries and accounts for the short, well-developed gullies observed in the valley sides that deliver material to build the alluvial fans that today project into Katepwa Lake.

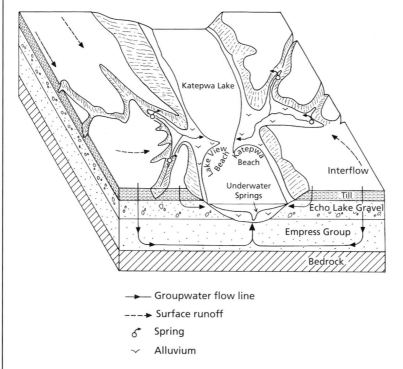

Groupwater flow line
Surface runoff
Spring
Alluvium

Fig. 2.35 Block diagram showing the final stage of evolution of tributary valleys and alluvial fans by spring sapping in the Qu'Appelle Valley, Saskatchewan (Christiansen *et al.* 1981). (*Source:* Christiansen, E.A., Acton, D.F., Long, R.J. *et al.* (1981) *Fort Qu'Appelle Geolog. The Valleys – Past and Present.* Interpretive Report No. 2. The Saskatchewan Research Council, Canada.)

2.9 Transmissivity and storativity of confined aquifers

For a confined aquifer of thickness, b, the transmissivity, T, is defined as follows:

$$T = Kb \qquad \text{(eq. 2.23)}$$

and represents the rate at which water of a given density and viscosity is transmitted through a unit width of aquifer or aquitard under a unit hydraulic gradient. Transmissivity has the units of $L^2\,T^{-1}$.

The storativity (or storage coefficient), S, of a confined aquifer is defined as follows:

$$S = S_s b \qquad \text{(eq. 2.24)}$$

where S_s is the specific storage term, and represents the volume of water that an aquifer releases from storage per unit surface area of aquifer per unit decline in the component of hydraulic head normal to that surface (Fig. 2.36a). Storativity values are dimensionless and range in value from 0.005 to 0.00005 such that large head changes over extensive areas are required to produce significant yields from confined aquifers.

2.9.1 Release of water from confined aquifers

At the beginning of the last century, Meinzer and Hard (1925) observed in a study of the

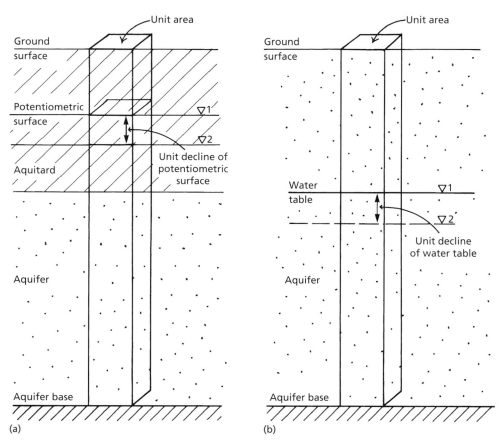

Fig. 2.36 Schematic representation of storativity in (a) a confined aquifer (the storage coefficient) and (b) an unconfined aquifer (the specific yield).

Dakota sandstone that more water was pumped from the region than could be accounted for (as water was pumped, a cone of depression developed and the rate of abstraction decreased, but with no apparent effect on groundwater levels in the recharge zone), such that the water-bearing formation was demonstrating elastic behaviour in releasing water from storage. Later, in deriving the general partial differential equation describing transient groundwater flow, Jacob (1940) formally described the elastic behaviour of porous rocks. There are two mechanisms that explain how water is produced by confined aquifers: the porosity of the aquifer is reduced by compaction and groundwater is released; and the water itself expands since water is slightly compressible.

As shown in Fig. 2.37, the total downward stress, σ_T, applied at the top of a confined aquifer is supported by an upward effective stress, σ_e, on the aquifer material and the water pressure contained in the pore space, P_w, such that

$$\sigma_T = \sigma_e + P_w \qquad \text{(eq. 2.25)}$$

If the porewater pressure is decreased by groundwater pumping or by natural groundwater outflow, the stress on the aquifer material will increase causing it to undergo compression.

Compressibility is a material property that describes the change in volume (strain) induced in a material under an applied stress and is the inverse of the modulus of elasticity (equal to a change in stress divided by a change in strain). The compressibility of water, β, is defined as follows:

$$\beta = \frac{\frac{-dV_w}{V_w}}{dP_w} \qquad \text{(eq. 2.26)}$$

where P_w is porewater pressure, V_w is volume of a given mass of water and dV_w/V_w is volumetric strain for an induced stress dP_w. For practical purposes, β can be taken as a constant equal to 4.4×10^{-10} m^2 N^{-1} (or Pa^{-1}).

The compressibility of aquifer material, α, is defined as follows:

$$\alpha = \frac{\frac{-dV_T}{V_T}}{d\sigma_e} \qquad \text{(eq. 2.27)}$$

where V_T is total volume of aquifer material and dV_T/V_T is volumetric strain for an induced change in effective stress $d\sigma_e$.

Now, with reference to eq. 2.27, for a reduction in the total volume of aquifer material, dV_T, the amount of water produced by compaction of the aquifer, dV_w, is

$$dV_w = -dV_T = \alpha V_T d\sigma_e \qquad \text{(eq. 2.28)}$$

If the total stress does not change ($d\sigma_T = 0$), then from a knowledge that $P_w = \rho g \psi$ and $\psi = h - z$ (eq. 2.22), with z remaining constant, then, using eq. 2.25:

$$d\sigma_e = 0 - \rho g d\psi = -\rho g dh \qquad \text{(eq. 2.29)}$$

For a unit decline in hydraulic head, $dh = -1$, and if unit volume is assumed ($V_T = 1$), then eq. 2.28 becomes

$$dV_w = \alpha(1)(-\rho g)(-1) = \alpha \rho g$$

$$\text{(eq. 2.30)}$$

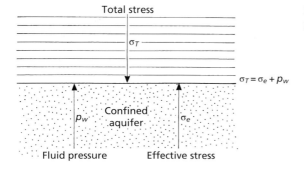

Fig. 2.37 Total stress, effective stress and fluid pressure at the top of a confined aquifer.

The water produced by the expansion of water is found from eq. 2.26 thus

$$dV_w = -\beta V_w dP_w \qquad \text{(eq. 2.31)}$$

Recognizing that the volume of water, V_w, in the total unit volume of aquifer material, V_T, is nV_T, where n is porosity, and that $dP = \rho g d\psi$ or $-\rho g$ for a unit decline in hydraulic head (where $\psi = h - z$ (eq. 2.22), with z remaining constant), then for unit volume, $V_T = 1$, eq. 2.31 gives

$$dV_w = -\beta n(1)(-\rho g) = \beta n \rho g$$
$$\text{(eq. 2.32)}$$

Finally, the volume of water that a unit volume of aquifer releases from storage under a unit decline in hydraulic head (the specific storage, S_s) is the sum of the volumes of water produced by the two mechanisms of compaction of the aquifer (eq. 2.30) and expansion of the water (eq. 2.32) thus

$$S_s = \alpha \rho g + \beta n \rho g = \rho g (\alpha + n\beta)$$
$$\text{(eq. 2.33)}$$

In other words, groundwater pumped from a confined aquifer does not represent a dewatering of the physical pore space in the aquifer but, instead, results from the secondary effects of aquifer compaction and water expansion. As a consequence, for an equivalent unit decline in hydraulic head, yields from confined aquifers are much less than from unconfined aquifers. Hence, storage coefficient values of

Table 2.2 Range of values of compressibility (Freeze and Cherry 1979).

Geological material	Compressibility, α (m^2 N^{-1} or Pa^{-1})
Clay	10^{-6}–10^{-8}
Sand	10^{-7}–10^{-9}
Gravel	10^{-8}–10^{-10}
Jointed rock	10^{-8}–10^{-10}
Sound rock	10^{-9}–10^{-11}
Water (β)	4.4×10^{-10}

(*Source:* Freeze, R.A. and Cherry, J.A. (1979) *Groundwater.* Prentice-Hall, Inc., Englewood Cliffs, New Jersey.)

confined aquifers are much smaller than for unconfined aquifers.

Values of material compressibility, α, range from 10^{-6} to 10^{-9} m^2 N^{-1} for clay and sand and from 10^{-8} to 10^{-10} m^2 N^{-1} for gravel and jointed rock (Table 2.2). These values indicate that a greater, largely irrecoverable compaction is expected in a previously unconsolidated clay aquitard, while smaller, elastic deformations are likely in gravel or indurated sedimentary aquifers. A possible consequence of groundwater abstraction from confined aquifers is land subsidence following aquifer compaction, especially in sand-clay aquifer-aquitard systems found in unconsolidated alluvial or basin-fill environments (Galloway and Burbey 2011). Notable examples are the Central Valley, California (Box 2.7) and the Tehran Plain, Iran (Box 2.8).

Box 2.7   Land subsidence in the Central Valley, California

Agricultural production in the Central Valley of California is dependent on the availability of water for irrigation. One-half of this irrigation water is supplied by groundwater and accounts for 74% of California's total abstractions and about 20% of the irrigation abstractions in the United States (Williamson *et al.* 1989). Groundwater abstraction is especially important in dry years when it supplements highly variable surface water supplies. In 1975, about 57% of the total land area (5.2×10^6 ha) in the Central Valley was irrigated. The intensive agricultural development during the past 100 years has had major impacts on the aquifer system.

(Continued)

Box 2.7 (Continued)

The Central Valley is a large structural trough filled with marine sediments overlain by continental deposits with an average total thickness of about 730 m (Fig. 2.38). More than half of the thickness of the continental deposits is composed of fine grained sediments. When development began in the 1880s, flowing wells and marshes were found throughout most of the Central Valley. The northern one-third of the valley, the Sacramento Valley, is considered to be an unconfined aquifer with a water table and the southern two-thirds, the San Joaquin Valley, as a two-layer aquifer system separated by a regional confining clay layer, the Pleistocene Corcoran Clay. This clay layer is highly susceptible to compaction. Figure 2.38 is a conceptual model of the hydrogeology of the Central Valley based on Williamson *et al.* (1989) who considered the entire thickness of continental deposits to be one aquifer system that has varying vertical leakance (ratio of vertical hydraulic conductivity to bed thickness) and confinement depending on the proportion of fine grained sediments encountered.

During 1961–1977, an average of 27×10^9 m^3 a^{-1} of water was used for irrigation with about one-half derived from groundwater. This amount of groundwater abstraction has caused water levels to decline by in excess of 120 m in places (Fig. 2.39) resulting in the largest volume of land subsidence in the world due to groundwater abstraction. Land subsidence has caused problems such as cracks in roads and canal linings, changing slopes of water channels and ruptured well casings.

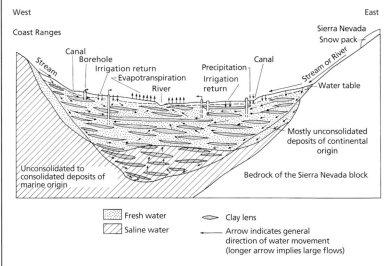

Fig. 2.38 Conceptual model of the hydrogeology of the Central Valley of California. Before groundwater development, water that recharged the aquifer at the valley margins moved downwards and laterally into the aquifer system and then moved upwards to discharge at rivers and marshes along the valley axis. The entire aquifer system is considered to be a single heterogeneous system in which vertically and horizontally scattered lenses of fine grained materials provide increasing confinement with depth (Williamson *et al.* 1989). (*Source:* Williamson, A.K., Prudic, D.E. and Swain, L.A. (1989) Ground-water flow in the Central Valley, California. *United States Geological Survey Professional Paper* **1401-D**.)

Box 2.7   (Continued)

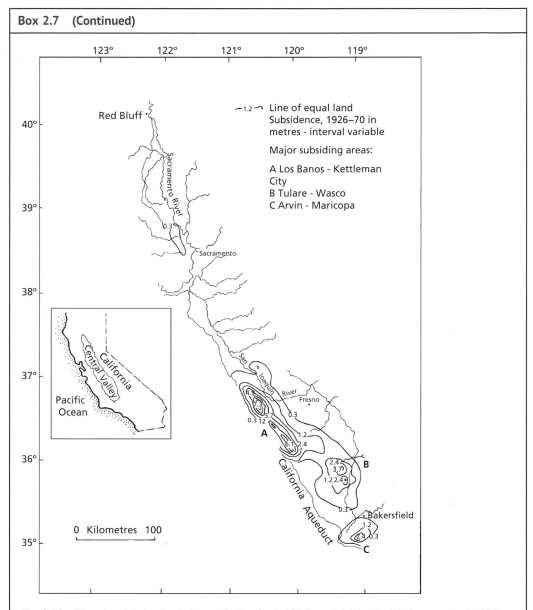

Fig. 2.39   Map showing land subsidence in the Central Valley of California (Williamson *et al.* 1989). (*Source:* Williamson, A.K., Prudic, D.E. and Swain, L.A. (1989) Ground-water flow in the Central Valley, California. *United States Geological Survey Professional Paper* **1401-D**.)

From pre-development until 1977, the volume of water in aquifer storage declined by about 74×10^9 m³, with 49×10^9 m³ from the water table zone, 21×10^9 m³ from inelastic compaction of fine grained sediments and 4×10^9 m³ from elastic storage. Elastic storage is a result of the expansion of water and the compression of sediments resulting from a change in fluid pressure (see Section 2.9.1). The estimated average elastic specific storage, S_s (eq. 2.33), is 1×10^{-5} m⁻¹. The inelastic compaction of fine grained

(Continued)

Box 2.7 (Continued)

sediments in the aquifer system caused by a decline in the hydraulic head results in a reorientation of the grains and a reduction in pore space within the compacted beds, thus releasing water. The volume of water released by compaction is approximately equal to the volume of land subsidence observed at the surface. The loss of pore space represents a permanent loss of storage capacity in the aquifer system. Even if water levels were to recover to their previous highest position, the amount of water stored in the aquifer system would be less than the amount stored prior to compaction. Inelastic compaction means permanent compaction. This type of land subsidence represents a once only abstraction of water from storage.

The cumulative volume of subsidence in the San Joaquin Valley is shown in Fig. 2.40. By 1970, the total volume of subsidence was 19×10^9 m³. Also included in Fig. 2.40 are cumulative volumes of subsidence for each of the three major subsiding areas. The volume of subsidence in the Los Banos-Kettleman City area (Area *A* in Fig. 2.39) accounted for nearly two-thirds of the total volume of subsidence as of 1970. From 1970 to 1975, there was little further subsidence in this area because surface water imports from the California Aqueduct greatly reduced the demand for groundwater. However, subsidence recurred during the drought of 1976–1977 owing to an increase in groundwater abstraction. The correlation

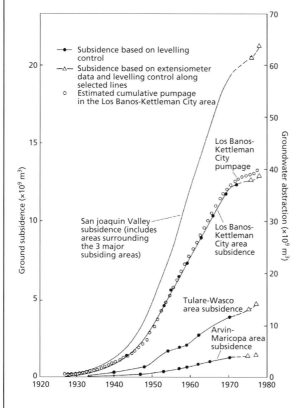

Fig. 2.40 Volumes of land subsidence in the major subsiding areas of the San Joaquin Valley and groundwater abstraction in the Los Banos-Kettleman City area, 1925–1977 (Williamson *et al.* 1989). (*Source:* Williamson, A.K., Prudic, D.E. and Swain, L.A. (1989) Ground-water flow in the Central Valley, California. *United States Geological Survey Professional Paper* **1401-D**.)

Box 2.7   (Continued)

between groundwater abstraction and the volume of subsidence in the Los Banos-Kettleman City area is high, indicating that about 43% of the water pumped from the lower pumped zone (at least 75–80% of the total) was derived from compaction of the fine grained sediments in the aquifer system.

Land subsidence continues to be a problem in some areas, although the areas of greater subsidence have been controlled by importing surface water. In the late 1960s, the surface-water delivery system began to route water from the wetter Sacramento Valley to the drier, more heavily pumped San Joaquin Valley. The surface-water delivery system was fully functional by the early 1970s, resulting in water-level recovery in the northern and western parts of the San Joaquin Valley. Overall, the Tulare Basin part of the San Joaquin Valley (Area *B* in Fig. 2.39) is still showing dramatic declines in groundwater levels and accompanying increased depletion of groundwater storage. Other than the large loss in storage in the Tulare Basin, on average there has been little overall change in storage throughout the rest of the Central Valley (Faunt 2009).

Climate variability has had profound effects on the Central Valley hydrologic system. For example, the droughts of 1976–1977 and 1987–1992 led to reduced surface-water deliveries and increased groundwater abstraction, thereby reversing the overall trend of groundwater-level recovery and re-initiating land subsidence in the San Joaquin Valley. In areas where groundwater pumping increases again, water levels can drop rapidly towards the previous lows because of the loss of aquifer storage capacity that resulted from the previous compaction of fine grained sediments. Since the mid-1990s, although annual surface-water deliveries generally have exceeded groundwater abstraction, water is still being removed from storage in most years in the Tulare Basin (Faunt 2009).

From an analysis of data from the NASA Gravity Recovery and Climate Experiment (GRACE) satellite mission (Section 7.4) to estimate water storage changes in California's Sacramento and San Joaquin River Basins for the 78-month period, October 2003–March 2010, Famiglietti *et al.* (2011) found that groundwater storage in the basins decreased by 31.0 ± 2.7 mm a^{-1}, which corresponds to a volume of 30.9 km^3 of water loss. As shown in Fig. 2.41, there is an apparent break in the behaviour of groundwater storage variations. Beginning with the drought in 2006, a steep decline in groundwater storage of 38.9 ± 9.5 mm a^{-1} (6.0 km^3 a^{-1}) occurred between April 2006 and March 2010, nearly as large as previous model-based estimates of groundwater losses (Faunt 2009) during the two major droughts of the last 50 years. Reported groundwater losses during those periods were approximately 12.3 km^3 a^{-1} from 1974 to 1976, and 8.2 km^3 a^{-1} from 1985 to 1989.

Fig. 2.41 Monthly groundwater storage anomalies (mm) for the Sacramento and San Joaquin River Basins for the period October 2003 to March 2010. Monthly errors are shown by grey shading. The overall trend in changes in groundwater storage for the 78-month period is shown as the single dashed line. The two dotted lines represent the trends from October 2003 to March 2006 and April 2006 to March 2010 (Famiglietti *et al.* 2011). (*Source:* Famiglietti, J.S., Lo, M., Ho, S.L. *et al.* (2011) Satellites measure recent rates of groundwater depletion in California's Central Valley. *Geophysical Research Letters* **38**, L03403, doi: 10.1029/2010GL046442. © 2011, John Wiley & Sons.)

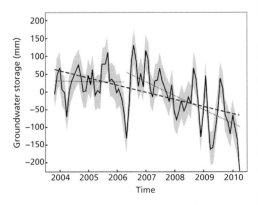

Box 2.8   Land subsidence in the Tehran Plain, Iran

Tehran, western Asia's most populous city, is situated on an alluvial plain immediately south of the Alborz Mountains that form an east-west chain stretching across northern Iran at an average elevation of 1300 m above mean sea level. The Tehran Province has experienced enormous population growth over the last 40 years and now hosts approximately 13 million people compared to less than 6 million in 1979. The urban area of Tehran has increased dramatically, particularly towards the west and southwest. As a result, Tehran has been experiencing an increasing demand for water to meet the needs of its growing population and economy. Water shortages have been experienced in recent years due to periods of sustained drought that have stressed existing groundwater resources. Most of the qanat galleries in the region (constructed underground water channels built centuries ago to transfer water from the mountains to the plains; see Section 1.3 and Knill and Jones 1968), have dried out. In 2012, only 167 branches of the 522 qanats in 1970 remained effective as a reliable means of water transport (Mahmoudpour *et al.*, 2016).

Several dams were constructed in the 1960s along the main watercourses in the area to channel surface water for farming, municipal and industrial uses. However, these supplies do not meet the water needs of the region, and groundwater is currently the primary source of water. The number of boreholes in the area has increased from 3906 in 1968 to 26 076 in 2003 and 32 518 in 2012. In 2012, about $1.9 \times 10^9\,m^3$ of groundwater were extracted from the Tehran aquifer for mainly agricultural (51%), domestic and drinking (34%), industrial (7%) and other (8%) purposes. Despite the increase in the number of boreholes, the total amount of abstraction has actually decreased as a result of the decline in groundwater levels and reduced water consumption. A comparison of the number of pumping wells and volume of groundwater abstracted from 1968 ($40–50 \times 10^6\,m^3\,a^{-1}$; Knill and Jones 1968) and 2003 indicates that the ability of the aquifer system to yield water has significantly decreased because of insufficient recharge (Mahmoudpour *et al.* 2016).

The Quaternary Tehran Alluvial Formation and underlying Kahrizak Formation dominate the central Tehran Plain and represent potential aquifers with good hydraulic conductivity. The folded beds of the underlying Pliocene Hezardarreh Formation dominate the northern part of the Plain but have poor aquifer characteristics resulting from high cementation with low conductivity (Mahmoudpour *et al.* 2016). Boreholes drilled to a maximum depth of 100 m south-west of Tehran identified three confined aquifers and three aquitards comprising a layered aquifer system (Fig. 2.42). From the examination of borehole drilling logs, a shallow aquifer extends down from 7 m to more than 30 m, a second aquifer extends down from 35 m to more than 65 m, and a third, deep aquifer extends downward from 70 m to at least 100 m. The deep aquifer is the major aquifer for groundwater extraction and is composed of alluvial fine sand and silty sand with a thickness of 2–20 m (Mahmoudpour *et al.* 2016).

The aquitards are formed of mostly silty clay and clayey soil, which play a significant role in land subsidence. The thickest aquitard (9–41 m) confining the deep aquifer (Fig. 2.42) plays a significant role in the compression of the aquifer system. Compressible sediments also occur in the aquifer system as discontinuous interbeds within aquifers.

The minimum and maximum thicknesses of the saturated zone in the region are 34 and 84 m, respectively. The groundwater

Box 2.8   (Continued)

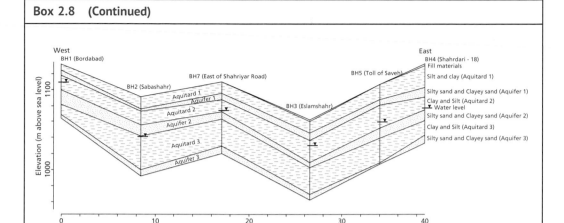

Fig. 2.42 Hydrogeological cross-section showing the multi-layered aquifer system of the Tehran Plain (Mahmoudpour *et al.* 2016). (*Source:* Adapted from Mahmoudpour, M., Khamehchiyan, M., Nikudel, M.R. and Ghassemi, M.R. (2016) Numerical simulation and prediction of regional land subsidence caused by groundwater exploitation in the southwest plain of Tehran, Iran. *Engineering Geology* **201**, 6–28.)

table generally follows the topographic gradient in the region and slopes from the north to south-east (Mahmoudpour *et al.* 2016). Partly due to increased exploitation and partly due to the indirect effects of dams, which have decreased the flow of surface water into the aquifers, Tehran's groundwater has been depleted over the last four decades. The average groundwater level in Tehran decreased by about 11.65 m from 1984 to 2012, representing an average decline of 42 cm a^{-1} (Fig. 2.43).

The decrease in the hydraulic head in the Tehran Plain is associated with local and regional land subsidence. Most of the developed groundwater basins in Iran are subject to land subsidence hazards resulting from the over-abstraction of groundwater. Earth fissures, damage to buildings, shifts in the ground and cracks in walls are evidence of groundwater-induced compaction that have been observed. Huge fissures, several kilometres long and up to 4 m wide and deep, have opened up in the land to the southeast of Tehran, and some are threatening

power-transmission lines and railways (Ravilious 2018).

In an example of the application of remote sensing techniques to hydrogeology (see Section 7.4), Haghshenas Haghighi and Motagh (2019) used InSAR (Interferometric Synthetic Aperture Radar) time series to investigate ground surface displacements in the Tehran Plain for the period of 2003–2017. InSAR is a powerful technique for measuring the topography of a surface and its changes over time (Bürgmann *et al.*, 2000). Given its broad spatial coverage and high accuracy, InSAR has become a preferred geodetic method for the study of land deformation in developed groundwater basins and provides insight into the geological and hydrological parameters that characterize underlying aquifer systems (Galloway and Burbey 2011). InSAR also provides valuable information about the consequences of rapid subsidence on infrastructure, facilities and urban areas.

A total area of approximately 1300 km^2 is affected by land subsidence in the Tehran

Box 2.8 (Continued)

Plain and Haghshenas Haghighi and Motagh (2019) identified three distinct subsidence features with rates exceeding 25 cm a^{-1} in the western area of the Plain and rates of approximately 5 cm a^{-1} in the immediate vicinity of Tehran international airport and 22 cm a^{-1} in the Varamin Plain to the south-east of Tehran city. The temporal pattern of land subsidence, which is dominated by a decreasing trend, generally follows the regional decline in groundwater level (Fig. 2.43). The subsidence has two main patterns: a long-term displacement as a result of declines in the groundwater level in recent decades; and short-term variations related to seasonal groundwater discharge and recharge. Analysis of the area south-west of Tehran shows that the non-recoverable portion of the deformation is likely to be dominant with an average elastic/inelastic ratio of approximately 0.4 (Haghshenas Haghighi and Motagh 2019).

Unless effective groundwater management is implemented, including the regulation of groundwater abstraction (Ravilious 2018), ongoing subsidence in Tehran is expected to cause further damage to infrastructure, particularly in the regions of high displacement gradients in the urban areas of Tehran and in the vicinity of the international airport.

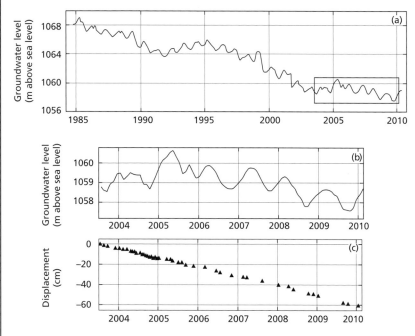

Fig. 2.43 (a) Average groundwater level for the Tehran Plain between 1984 and 2010. The groundwater level decreased by approximately 10 m during this period. Periods of rapid and gentle declines, small recoveries and seasonal fluctuations are recorded. (b) Close-up view of groundwater changes for the period since 2004 with InSAR measurements. (c) Average vertical ground displacement derived from InSAR time series of all points with subsidence rates greater than 2 cm a^{-1} south-west of Tehran (Haghshenas Haghighi and Motagh 2019). (*Source:* Adapted from Haghshenas Haghighi, M. and Motagh, M. (2019) Ground surface response to continuous compaction of aquifer system in Tehran, Iran: results from a long-term multi-sector InSAR analysis. *Remote Sensing of Environment* **221**, 534–550.)

2.10   Transmissivity and specific yield of unconfined aquifers

For an unconfined aquifer, the transmissivity is not as well defined as in a confined aquifer, but eq. 2.23 can be applied with b now representing the saturated thickness of the aquifer or the height of the water table above the top of a lower aquitard boundary. The transmissivity will, therefore, vary if there are large seasonal fluctuations in the elevation of the water table or if the saturated thickness of the aquifer shows lateral variation as a result of an irregular lower aquitard boundary or differences between recharge and discharge areas in the same aquifer.

The storage term for an unconfined aquifer is known as the specific yield, S_y, (or the unconfined storativity) and is that volume of water that an unconfined aquifer releases from storage per unit surface area of aquifer per unit decline in the water table (see Fig. 2.36b), and is approximately equivalent to the total porosity of a soil or rock (see Section 2.2). Specific yield is a dimensionless term and the normal range is from 0.01 to 0.30. Relative to confined aquifers, the higher values reflect the actual dewatering of pore space as the water table is lowered. Consequently, the same yield can be obtained from an unconfined aquifer with smaller head changes over less extensive areas than can be produced from a confined aquifer.

Although not commonly used, by combining the aquifer properties of transmissivity (T or K) and storativity (S or S_s) it is possible to define a single formation parameter, the hydraulic diffusivity, D, defined as either:

$$D = \frac{T}{S} \quad \text{or} \quad \frac{K}{S_s} \qquad \text{(eq. 2.34)}$$

Aquifer formations with a large hydraulic diffusivity respond quickly in transmitting changed hydraulic conditions at one location to other regions in an aquifer, for example in response to groundwater abstraction.

2.11   Equations of groundwater flow

In this section, the mathematical derivation of the steady-state and transient groundwater flow equations will be presented followed by a demonstration of simple analytical solutions to groundwater flow problems. Following from this, different scales of flow systems are shown to exist in regional aquifer systems.

Equations of groundwater flow are derived from a consideration of the basic flow law, Darcy's Law (eq. 2.7), and an equation of continuity that describes the conservation of fluid mass during flow through a porous material. In the following treatment, which derives from the classic paper by Jacob (1940), steady-state and transient saturated flow conditions are considered in turn. Under steady-state conditions, the magnitude and direction of the flow velocity at any point are constant with time. For transient conditions, either the magnitude or direction of the flow velocity at any point may change with time, or the potentiometric conditions may change as groundwater either enters into or is released from storage.

2.11.1   Steady-state saturated flow

First, consider the unit volume of a porous material (the elemental control volume) depicted in Fig. 2.44. The law of conservation of mass for steady-state flow requires that the rate of fluid mass flow into the control volume, ρq (fluid density multiplied by specific discharge across a unit cross-sectional area), will be equal to the rate of fluid mass flow out of the control volume, such that the incremental differences in fluid mass flow, in each of the directions x, y, z, sum to zero, thus:

$$\left(\rho q_x + \frac{\partial(\rho q_x)}{\partial x} - \rho q_x\right) + \left(\rho q_y + \frac{\partial\left(\rho q_y\right)}{\partial y} - \rho q_y\right)$$
$$+ \left(\rho q_z + \frac{\partial(\rho q_z)}{\partial z} - \rho q_z\right) = 0$$
$$\text{(eq. 2.35)}$$

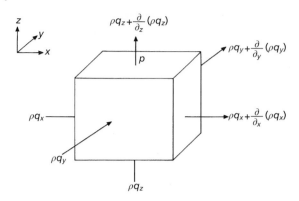

$$\rho q_z + \frac{\partial}{\partial z}(\rho q_z)$$

p

$$\rho q_y + \frac{\partial}{\partial y}(\rho q_y)$$

ρq_x

$$\rho q_x + \frac{\partial}{\partial x}(\rho q_x)$$

ρq_y

ρq_z

Fig. 2.44 Unit volume (elemental control volume) for flow through porous material.

From eq. 2.35, the resulting equation of continuity is

$$\frac{\partial(\rho q_x)}{\partial x} + \frac{\partial\left(\rho q_y\right)}{\partial y} + \frac{\partial(\rho q_z)}{\partial z} = 0$$

(eq. 2.36)

If the fluid is incompressible, then density, $\rho(x, y, z)$, is constant and eq. 2.36 becomes

$$\frac{\partial q_x}{\partial x} + \frac{\partial q_y}{\partial y} + \frac{\partial q_z}{\partial z} = 0 \qquad \text{(eq. 2.37)}$$

From Darcy's Law, each of the specific discharge terms can be expressed as follows:

$$q_x = -K_x \frac{\partial h}{\partial x}, \quad q_y = -K_y \frac{\partial h}{\partial y},$$

$$q_z = -K_z \frac{\partial h}{\partial z}$$

(eq. 2.38)

and upon substitution in eq. 2.37:

$$\frac{\partial}{\partial x}\left(K_x \frac{\partial h}{\partial x}\right) + \frac{\partial}{\partial y}\left(K_y \frac{\partial h}{\partial y}\right)$$
$$+ \frac{\partial}{\partial z}\left(K_z \frac{\partial h}{\partial z}\right) = 0$$

(eq. 2.39)

For an isotropic and homogeneous porous material, $K_x = K_y = K_z$ and $K(x,y,z) = $ constant, respectively. By substituting these two conditions in eq. 2.39 it can be shown that

$$\frac{\partial^2 h}{\partial x^2} + \frac{\partial^2 h}{\partial y^2} + \frac{\partial^2 h}{\partial z^2} = 0 \qquad \text{(eq. 2.40)}$$

Thus, the steady-state groundwater flow equation is the Laplace equation and the solution $h(x,y,z)$ describes the value of the hydraulic head at any point in a three-dimensional flow field. By solving eq. 2.40, either in one- two- or three-dimensions depending on the geometry of the groundwater flow problem under consideration, a contoured equipotential map can be produced and, with the addition of flow lines, a flow net drawn (Box 2.3).

2.11.2 Transient saturated flow

The law of conservation of mass for transient flow in a saturated porous material requires that the net rate of fluid mass flow into the control volume (Fig. 2.44) is equal to the time rate of change of fluid mass storage within the control volume. The equation of continuity is now

$$\frac{\partial(\rho q_x)}{\partial x} + \frac{\partial\left(\rho q_y\right)}{\partial y} + \frac{\partial(\rho q_z)}{\partial z}$$
$$= \frac{\partial(\rho n)}{\partial t} = n\frac{\partial \rho}{\partial t} + \rho\frac{\partial n}{\partial t}$$

(eq. 2.41)

The first term on the right-hand side of eq. 2.41 describes the mass rate of water produced by expansion of the water under a change in its density, ρ, and is controlled by the compressibility of the fluid, β. The second term is the mass rate of water produced by the compaction of the porous material as influenced by the change in its porosity, n, and is determined by the compressibility of the aquifer, α. Changes in fluid density and formation porosity are both produced by a change in hydraulic head and the volume of water produced by the two mechanisms for a unit decline in head is the specific storage, S_s. Hence, the time rate of change of fluid mass storage within the control volume is

$$\rho S_s \frac{\partial h}{\partial t}$$

and eq. 2.41 becomes

$$\frac{\partial (\rho q_x)}{\partial x} + \frac{\partial (\rho q_y)}{\partial y} + \frac{\partial (\rho q_z)}{\partial z} = \rho S_s \frac{\partial h}{\partial t}$$

(eq. 2.42)

By expanding the terms on the left-hand side of eq. 2.42 using the chain rule (eliminating the smaller density gradient terms compared with the larger specific discharge gradient terms) and, at the same time, inserting Darcy's Law to define the specific discharge terms, then

$$\frac{\partial}{\partial x}\left(K_x \frac{\partial h}{\partial x}\right) + \frac{\partial}{\partial y}\left(K_y \frac{\partial h}{\partial y}\right)$$
$$+ \frac{\partial}{\partial z}\left(K_z \frac{\partial h}{\partial z}\right) = S_s \frac{\partial h}{\partial t}$$

(eq. 2.43)

If the porous material is isotropic and homogeneous, eq. 2.43 reduces to

$$\frac{\partial^2 h}{\partial x^2} + \frac{\partial^2 h}{\partial y^2} + \frac{\partial^2 h}{\partial z^2} = \frac{S_s}{K} \frac{\partial h}{\partial t}$$

(eq. 2.44)

or, expanding the specific storage term, S_s (eq. 2.33):

$$\frac{\partial^2 h}{\partial x^2} + \frac{\partial^2 h}{\partial y^2} + \frac{\partial^2 h}{\partial z^2} = \frac{\rho g(\alpha + n\beta)}{K} \frac{\partial h}{\partial t}$$

(eq. 2.45)

Equations 2.43, 2.44 and 2.45 are all transient groundwater flow equations for saturated anisotropic (eq. 2.43) and homogeneous and isotropic (eqs 2.44 and 2.45) porous material. The solution $h(x,y,z,t)$ describes the value of hydraulic head at any point in a three-dimensional flow field at any time. A solution requires knowledge of the three

hydrogeological parameters, K, α and n, and the fluid parameters, ρ and β. A simplification is to take the special case of a horizontal confined aquifer of thickness, b, storativity, S ($= S_s b$), and transmissivity, T ($= Kb$), and substitute in eq. 2.44, thus

$$\frac{\partial^2 h}{\partial x^2} + \frac{\partial^2 h}{\partial v^2} = \frac{S}{T} \frac{\partial h}{\partial t}$$

(eq. 2.46)

The solution of this equation, $h(x,y,t)$, describes the hydraulic head at any point on a horizontal plane through a horizontal aquifer at any time. A solution requires knowledge of the aquifer parameters T and S, both of which are measurable from field pumping tests (see Section 7.3.2).

2.11.3 Transient unsaturated flow

A treatment of groundwater flow in unsaturated porous material must incorporate the presence of an air phase. The air phase will affect the degree of connectivity between water-filled pores, and will therefore influence the hydraulic conductivity. Unlike in saturated material where the pore space is completely water-filled, in unsaturated material the partial saturation of pore space, or moisture content (θ), means that the hydraulic conductivity is a function of the degree of saturation, $K(\theta)$. Alternatively, since the degree of moisture content will influence the pressure head (ψ), the hydraulic conductivity is also a function of the pressure head, $K(\psi)$. In soil physics, the degree of change in moisture content for a change in pressure head ($d\theta/d\psi$) is referred to as the specific moisture capacity, C (the unsaturated storage property of a soil), and can be empirically derived from the slope of a soil characteristic curve (see Section 6.4.1).

Returning to Fig. 2.44, for flow in an elemental control volume that is partially saturated, the equation of continuity must now express the time rate of change of moisture content as well as the time rate of change of storage due to water expansion and aquifer compaction. The fluid mass storage term (ρn) in eq. 2.41 now becomes $\rho\theta$ and:

$$\frac{\partial(\rho q_x)}{\partial x} + \frac{\partial(\rho q_y)}{\partial y} + \frac{\partial(\rho q_z)}{\partial z} = \frac{\partial(\rho\theta)}{\partial t}$$

$$= \theta\frac{\partial\rho}{\partial t} + \rho\frac{\partial\theta}{\partial t}$$

(eq. 2.47)

The first term on the right-hand side of eq. 2.47 is insignificantly small and by inserting the unsaturated form of Darcy's Law, in which the hydraulic conductivity is a function of the pressure head, $K(\psi)$, then eq. 2.47 becomes, upon cancelling the ρ terms:

$$\frac{\partial}{\partial x}\left(K(\psi)\frac{\partial h}{\partial x}\right) + \frac{\partial}{\partial y}\left(K(\psi)\frac{\partial h}{\partial y}\right)$$

$$+ \frac{\partial}{\partial z}\left(K(\psi)\frac{\partial h}{\partial z}\right) = \frac{\partial\theta}{\partial t}$$

(eq. 2.48)

It is usual to quote eq. 2.48 in a form where the independent variable is either θ or ψ. Hence, noting that $h = z + \psi$ (eq. 2.22) and defining the specific moisture capacity, C, as $d\theta/d\psi$, then

$$\frac{\partial}{\partial x}\left(K(\psi)\frac{\partial\psi}{\partial x}\right) + \frac{\partial}{\partial y}\left(K(\psi)\frac{\partial\psi}{\partial y}\right) +$$

$$\frac{\partial}{\partial z}\left(K(\psi)\left(\frac{\partial\psi}{\partial z} + 1\right)\right) = C(\psi)\frac{\partial\psi}{\partial t}$$

(eq. 2.49)

This equation (eq. 2.49) is the ψ-based transient unsaturated flow equation for porous material and is known as the Richards equation. The solution $\psi(x, y, z, t)$ describes the pressure head at any point in a flow field at any time. It can be easily converted into a hydraulic head solution $h(x, y, z, t)$ through the relation $h = z + \psi$ (eq. 2.22). To be able to provide a solution to the Richards equation, it is necessary to know the characteristic curves $K(\psi)$ and $C(\psi)$ or $\theta(\psi)$ (Section 6.4.1).

2.12 Analytical solution of one-dimensional groundwater flow problems

The three basic steps involved in the mathematical analysis of groundwater flow problems are the same whatever the level of mathematical difficulty and are (1) conceptualizing the problem; (2) finding a solution; and (3) evaluating the solution (Rushton 2003). Simple one-dimensional problems can be solved using ordinary differential and integral calculus. Two-dimensional problems or transient (time-variant) flow problems require the use of partial derivatives and more advanced calculus.

In conceptualizing the groundwater flow problem, the basic geometry should be sketched and the aquifers, aquitards and aquicludes defined. Simplifying assumptions, for example concerning isotropic and homogeneous hydraulic conductivity, should be stated and, if possible, the number of dimensions reduced (for example, consider only the horizontal component of flow or look for radial symmetry or approximately parallel flow). If the groundwater flow is confined, then a mathematically linear solution results which can be combined to represent more complex situations. Unconfined situations produce higher-order equations.

As a first step, an equation of continuity is written to express conservation of fluid mass. For incompressible fluids, this is equivalent to conservation of fluid volume. Water is only very slightly compressible, so conservation of volume is a reasonable approximation. By combining the equation of continuity with a flow law, normally Darcy's Law, and writing down equations that specify the known conditions at the boundaries of the aquifer, or at specified points (for example, a well), provides a general differential equation for the specified system. Solving the problem consists of finding an equation (or equations) which describes the system and satisfies both the differential equation and the boundary conditions. By

integrating the differential equation, the resulting equation is the general solution. If the constants of integration are found by applying the
boundary conditions, then a specific solution
to the problem is obtained (for examples,
see Box 2.9).

Box 2.9   Examples of analytical solutions to one-dimensional groundwater flow problems

To illustrate the basic steps involved in the
mathematical analysis of groundwater flow
problems, consider the one-dimensional
flow problem shown in Fig. 2.45 for a confined aquifer with thickness, b. The total flow
at any point in the horizontal (x) direction is
given by the equation of continuity of flow:

$$Q = q \times b \qquad \text{(eq. 1)}$$

where q, the flow per unit width (specific
discharge), is found from Darcy's Law:

$$q = -K\frac{dh}{dx} \qquad \text{(eq. 2)}$$

where x increases in the direction of flow.
Combining eqs 1 and 2 gives the general differential equation:

$$Q = -Kb\frac{dh}{dx} \qquad \text{(eq. 3)}$$

By integrating eq. 3, it is possible to
express the groundwater head, h, in terms
of x and Q:

$$\int dh = -\int \frac{Q}{Kb}dx$$
$$\therefore h = -\frac{Q}{Kb}x + c \qquad \text{(eq. 4)}$$

c is the constant of integration and can be
determined by use of a supplementary
equation expressing a known combination
of h and x. For example, for the boundary
condition $x = 0$, $h = h_0$ and by applying this
condition to eq. 4 gives:

$$h_0 = -\frac{Q}{Kb} \times 0 + c$$
$$\therefore c = h_0$$

which gives the solution:

$$h = h_0 - \frac{Q}{Kb}x \qquad \text{(eq. 5)}$$

The specific solution given in eq. 5 relates h
to location, x, in terms of two parameters, Q
and Kb (transmissivity), and one boundary
value, h_0. The solution is the equation of a
straight line and predicts the position of the
potentiometric surface as shown in Fig. 2.45.
Also note, by introducing a new pair of values
of x where h is known, for example $x = D$, $h = h_D$,
we can use the following equation to evaluate
the parameter combination Q/Kb since:

$$\frac{Q}{Kb} = \frac{h_0 - h_D}{D} \qquad \text{(eq. 6)}$$

If we know Kb, we can find Q, or vice versa.
As a further example, the following groundwater flow problem provides an analytical
solution to the situation of a confined aquifer
receiving constant recharge, or leakage.
A conceptualization of the problem is shown
in Fig. 2.46, and it should be noted that

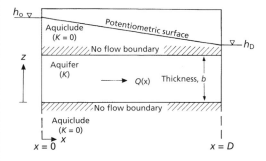

Section (x,z plane)

Fig. 2.45   Definition sketch of steady flow
through a uniform thickness, homogeneous
confined aquifer.

(Continued)

Box 2.9   (Continued)

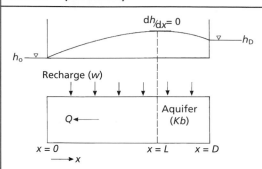

Fig. 2.46 Definition sketch of steady flow through a uniform thickness, homogeneous confined aquifer with constant recharge.

recharge, W, at the upper boundary of the aquifer is assumed to be constant everywhere.

Between $x = 0$ and $x = L$, continuity and flow equations can be written as follows:

$$Q = W(L-x) \qquad \text{(eq. 7)}$$

$$Q = Kb\frac{dh}{dx} \qquad \text{(eq. 8)}$$

and between $x = L$ and $x = D$:

$$Q = W(x-L) \qquad \text{(eq. 9)}$$

$$Q = -Kb\frac{dh}{dx} \qquad \text{(eq. 10)}$$

By combining the first pair of equations (eqs 7 and 8) between $x = 0$ and $x = L$ and integrating:

$$Kb\frac{dh}{dx} = W(L-x)$$

$$\int dh = \int \frac{W}{Kb}(L-x)dx$$

$$\therefore h = \frac{W}{Kb}\left(Lx - \frac{x^2}{2}\right) + c \qquad \text{(eq. 11)}$$

for the boundary condition $x = 0$, $h = h_0$, then the constant of integration $c = h_0$ and the following partial solution is found:

$$h - h_0 = \frac{W}{Kb}\left(Lx - \frac{x^2}{2}\right) \qquad \text{(eq. 12)}$$

Note that in eq. 12, the position of the divide, L, appears as a parameter. We can eliminate L by invoking a second boundary condition, $x = D$, $h = h_D$ and rearranging eq. 12 such that

$$L = \frac{Kb}{W}\frac{(h_D - h_0)}{D} + \frac{D}{2} \qquad \text{(eq. 13)}$$

By substituting L found from eq. 13 into eq. 12, we obtain the specific solution:

$$h - h_0 = x\frac{(h_D - h_0)}{D} + \frac{W}{2Kb}\left(Dx - x^2\right) \qquad \text{(eq. 14)}$$

Thus, the potentiometric surface in a confined aquifer with constant transmissivity and constant recharge is described by a parabola as shown in Fig. 2.46.

To obtain a solution to simple one-dimensional groundwater flow problems in unconfined aquifers, it is necessary to adopt the Dupuit assumptions that in any vertical section the flow is horizontal, the flow is uniform over the depth of flow and the flow velocities are proportional to the slope of the water table and the saturated depth. The last assumption is reasonable for small slopes of the water table. From a consideration of flow and continuity, and with reference to Fig. 2.47, the discharge per unit width, Q, for any given vertical section is found from

$$Q = -Kh\frac{dh}{dx} \qquad \text{(eq. 15)}$$

and integrating:

$$Qx = -\frac{K}{2}h^2 + c \qquad \text{(eq. 16)}$$

For the boundary condition $x = 0$, $h = h_0$, we obtain the following specific solution known as the Dupuit equation:

Box 2.9 (Continued)

$$Q = \frac{K}{2x}\left(h_0^2 - h^2\right) \qquad \text{(eq. 17)}$$

The Dupuit equation therefore predicts that the water table is a parabolic shape. In the direction of flow, the curvature of the water table, as predicted by eq. 17, increases. As a consequence, the two Dupuit assumptions become poor approximations to the actual groundwater flow and the actual water table increasingly deviates from the computed position as shown in Fig. 2.47. The reason for this difference is that the Dupuit flows are all assumed horizontal, whereas the actual velocities of the same magnitude have a vertical downward component so that a greater saturated thickness is required for the same discharge. As indicated in Fig. 2.47, the water table actually approaches the right-hand boundary tangentially above the water body surface and forms a seepage face. However, this discrepancy aside, the Dupuit equation accurately determines heads for given values of boundary heads, Q and K.

As a final example, Fig. 2.48 shows steady flow to two parallel stream channels from an unconfined aquifer with continuous recharge applied uniformly over the aquifer. The stream channels are idealized as two long parallel streams completely penetrating the aquifer. Adopting the Dupuit assumptions, the flow per unit width is given by eq. 15 and from continuity:

$$Q = Wx \qquad \text{(eq. 18)}$$

By combining eqs 15 and 18, then integrating and setting the boundary condition $x = a$, $h = h_a$, the following specific solution is obtained:

$$h^2 = h_a^2 + \frac{W}{K}\left(a^2 - x^2\right) \qquad \text{(eq. 19)}$$

From symmetry and continuity, the baseflow entering each stream per unit width of channel is equal to $2aW$. If h is known at any point, then the baseflow or recharge rate can be computed provided the hydraulic conductivity is known. This type of analysis has been applied to the design of parallel drains on agricultural soils to calculate the necessary spacing for specified soil, crop and irrigation conditions. A reappraisal of techniques used to analyse the horizontal flow of water to tile drains that also separate the external boundary of the water table from the internal boundary of the tile drain is given by Khan and Rushton (1996).

Fig. 2.47 Definition sketch of steady flow in a homogeneous unconfined aquifer between two water bodies with vertical boundaries.

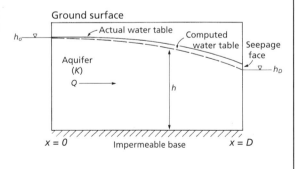

(Continued)

Box 2.9 (Continued)

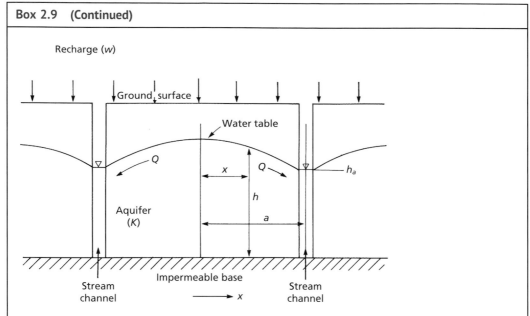

Fig. 2.48 Definition sketch of steady flow to two parallel streams from a homogeneous unconfined aquifer with constant recharge.

The final step in the mathematical analysis is the evaluation of the solution. The specific solution is normally an equation that relates groundwater head to position and to parameters contained in the problem such as hydraulic conductivity, or such factors as the discharge rate of wells. By inserting numerical values for the parameters, the solution can be used to evaluate groundwater head in terms of position in the co-ordinate system. The results might be expressed as a graph or contour diagram, or as a predicted value of head for a specified point.

2.13 Groundwater flow patterns

Preceding sections in this chapter have introduced the fundamental principles governing the existence and movement of groundwater, culminating in the derivation of the governing groundwater flow equations for steady-state, transient and unsaturated flow conditions. Within the water or hydrological cycle, groundwater flow patterns are influenced by geological factors such as differences in aquifer lithologies and structure of confining strata. A further influence of groundwater flow, other than aquifer heterogeneity, is the topography of the ground surface. Topography is a major influence on groundwater flow at local, intermediate and regional scales. The elevation of recharge areas in areas of aquifer outcrop, the degree to which river systems incise the landscape and the location and extent of lowland areas experiencing groundwater discharge determine the overall configuration of groundwater flow. As shown by Freeze and Witherspoon (1967), the relative positions and difference in elevation of recharge and discharge areas determine the hydraulic gradients and the length of groundwater flowpaths.

To understand the influence of topography, consider the groundwater flow net shown in Fig. 2.49 for a two-dimensional vertical

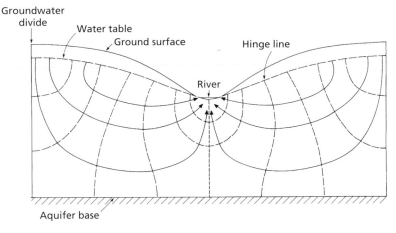

Fig. 2.49 Groundwater flow net for a two-dimensional vertical cross-section through a homogeneous, isotropic aquifer with a symmetrical valley topography.

cross-section through a homogeneous, isotropic aquifer. The section shows a single valley bounded by groundwater divides and an impermeable aquifer base. The water table is a subdued replica of the topography of the valley sides. The steady-state equipotential and groundwater flow lines are drawn using the rules for flow net analysis introduced in Box 2.3. It is obvious from the flow net that groundwater flow occurs from the recharge areas on the valley sides to the discharge area in the valley bottom. The hinge line separating the recharge from the discharge areas is also marked in Fig. 2.49. For most common topographic profiles, hinge lines are positioned closer to valley bottoms than to catchment divides with discharge areas commonly comprising only 5–30% of the catchment area (Freeze and Cherry 1979).

At the regional scale, the water table can be differentiated into two types: topography-controlled water tables where the water-table elevation is closely associated with topography; and recharge-controlled water tables that are largely disconnected from topography (Haitjema and Mitchell-Bruker 2005). Recharge-controlled water tables are expected in arid regions with mountainous topography and high hydraulic conductivity, whereas

topography-controlled water tables are expected in humid regions with subdued topography and low hydraulic conductivity (Haitjema and Mitchell-Bruker 2005). The water table depth is generally deeper and more variable in the case of recharge-controlled water tables. In mountainous regions, numerical model simulations suggest that regional groundwater flow (generally 10–1000 km in extent between surface water divides) is limited with topography-controlled water tables but can be significant in regions with recharge-controlled water tables (Gleeson and Manning 2008). In an analysis of the contiguous United States, Gleeson et al. (2011) identified regions with recharge-controlled water tables, for example the Southwest or Rocky Mountains, where water tables depths are generally greater and more variable, and where regional groundwater flow is more important as a percentage of the catchment water budget. Gleeson et al. (2011) also showed that water table depths are generally shallow and less variable, and regional groundwater flow limited in areas with topography-controlled water tables such as the north-east of the United States.

Returning to the boundary-value problem represented by the flow net shown in Fig. 2.49, Tóth (1963) determined an analytical

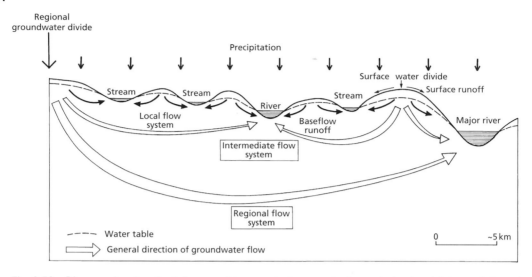

Fig. 2.50 Diagram showing the influence of hummocky topography in producing local, intermediate and regional groundwater flow systems (Tóth 1963). (*Source:* Modified from Tóth, J. (1963) A theoretical analysis of groundwater flow in small drainage basins. *Journal of Geophysical Research* **68**, 4795–4812.)

expression for the hydraulic head in the flow field for simple situations of an inclined water table of constant slope and cases in which a sine curve is superimposed on the incline to represent hummocky topography. Freeze and Witherspoon (1967) developed this mathematical approach further by employing numerical simulations to examine the effects of topography and geology on the nature of regional groundwater flow patterns. As illustrated in Fig. 2.50, Tóth (1963) proposed that it is possible to differentiate between local, intermediate and regional groundwater flow systems. Tóth (1963) further showed that as the ratio of depth to lateral extent of the entire aquifer system becomes smaller and as the amplitude of the hummocks becomes larger, the local flow systems are more likely to reach the aquifer base, thus creating a series of small groundwater flow cells. In general, where the local topography is subdued, only regional systems develop compared to areas of pronounced local relief where only local flow systems develop. Groundwater flow in the local and intermediate systems moves relatively quickly along short flowpaths and discharges as baseflow to

streams at the local scale and rivers at the intermediate scale. The regional component of groundwater flow has a relatively long residence time and follows long flowpaths before discharging to major rivers. Field examples of regional-scale and continental-scale groundwater flow systems are given in Boxes 2.10 and 2.11 and illustrate the combined influence of geology and topography on groundwater flow.

2.14 Classification of springs and intermittent streams

The classification of springs has been discussed from as early as Bryan (1919) who recognized the following types: volcanic, fissure, depression, contact, artesian and springs in impervious rock. Simply defined, springs represent the termination of underground flow systems and mark the point at which fluvial processes become dominant. The vertical position of the spring marks the elevation of the water table or a minimum elevation of the potentiometric surface at the point of discharge from

Box 2.10   Regional-scale groundwater flow in the Lincolnshire Limestone aquifer, England

The Lincolnshire Limestone aquifer in eastern England is part of a relatively uniform stratigraphical succession of Jurassic strata that dip 1–2° to the east. The pattern of recharge and groundwater flow in the aquifer is influenced by various geological and hydrological controls as shown in the schematic cross-section of Fig. 2.51. The Upper Lias Clay forms an effective impermeable base to the limestone aquifer. In the west of the section, the limestone is 6–8 km wide at outcrop. Formations overlying the limestone have a significant effect on surface water and groundwater flows since they form an alternating sequence of limestone aquifer units and confining beds of low permeability consisting of clays, shales and marls. The uppermost deposits of the confining strata include the Oxford Clay that extends to the east of the area where the continuation of the limestone below the Lincolnshire Fens is uncertain. Quaternary glacial deposits consisting mainly of boulder clay (glacial till) and sands and gravels occur in the west of the area and also in west-east drainage channels.

From a consideration of catchment water balances (Section 6.2) and the use of numerical modelling, Rushton and Tomlinson (1999) showed that recharge to the Lincolnshire Limestone aquifer occurs as direct recharge in the western outcrop area and as runoff recharge and downward leakage from the low permeability boulder clay and confining strata. The boulder clay produces large volumes of runoff which commonly recharges the limestone through the many swallow holes that occur at the edge of the boulder clay. In places, the West Glen River and East Glen River incise the confining strata and provide a mechanism for localized groundwater discharge from the limestone aquifer. Further east, the groundwater potentiometric surface is above the top of the aquifer, with overflowing artesian conditions developed in the Fens. In this region of the aquifer, slow upward leakage of the regionally extensive groundwater flow system is expected, although the large volumes of groundwater abstracted in recent years for public water supply has disrupted the natural flow patterns and drawn modern recharge water further into the confined aquifer.

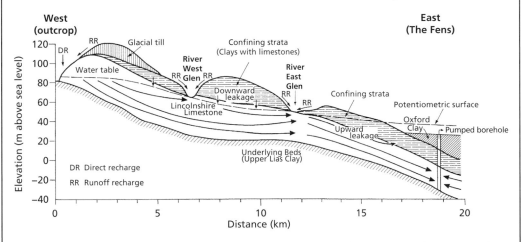

Fig. 2.51 Schematic cross-section of the Lincolnshire Limestone aquifer in eastern England showing mechanisms of groundwater recharge and directions of local and regional groundwater flow (Rushton and Tomlinson 1999). (*Source:* Rushton, K.R. and Tomlinson, L.M. (1999) Total catchment conditions in relation to the Lincolnshire Limestone in South Lincolnshire. *Quarterly Journal of Engineering Geology* **32**, 233–246.)

Box 2.11   Large-scale groundwater flow in the Great Artesian Basin, Australia

The Great Artesian Basin covers an area of 1.7×10^6 km^2 or about one-fifth of the Australian continent (Fig. 2.52) and is one of the world's largest artesian groundwater basins. The Basin underlies arid and semi-arid regions where surface water is sparse and unreliable, and extends across parts of the outback of Queensland, New South Wales, South Australia and the Northern Territory (Ransley *et al.* 2015). Discovery of the Basin's artesian groundwater resources in 1878 made settlement possible and led to the development of an important pastoral industry. Farming and public water supplies, and increasingly the mining and petroleum industries, are largely or totally dependent on the Basin's artesian groundwater (Habermehl 1980; Habermehl and Lau 1997; Habermehl 2020).

The Basin consists of a multi-layered confined aquifer system, with aquifers occurring in continental quartzose sandstones of Triassic, Jurassic and Cretaceous age and ranging in thickness from several metres to several hundreds of metres. The intervening confining beds consist of siltstone and mudstone, with the main confining unit formed by a sequence of argillaceous Cretaceous sediments of marine origin (Fig. 2.53). Basin sediments are up to 3000 m thick and form a large synclinal structure, uplifted and exposed

Fig. 2.52   Map of the location and extent of the Great Artesian Basin, Australia (Habermehl 1980). (*Source:* Habermehl, M.A. (1980) The Great Artesian Basin, Australia. *BMR Journal of Australian Geology and Geophysics* **5**, 9–38.)

Box 2.11 (Continued)

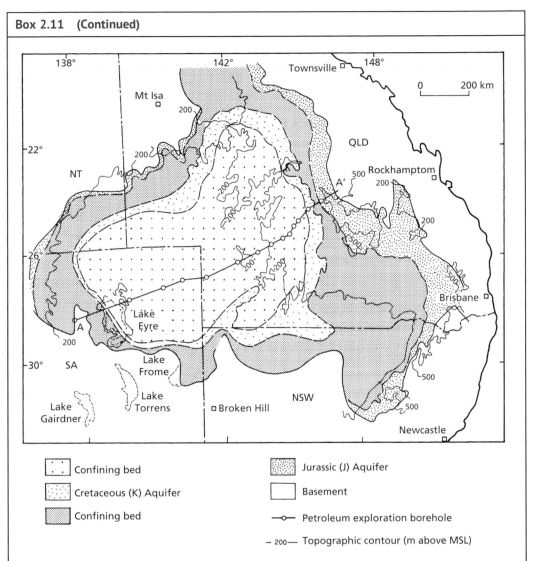

Fig. 2.53 Map showing lateral extent of hydrogeological units forming the Great Artesian Basin (Habermehl 1980). (*Source:* Habermehl, M.A. (1980) The Great Artesian Basin, Australia. *BMR Journal of Australian Geology and Geophysics* **5**, 9–38.)

along its eastern margin and tilted towards the southwest (Fig. 2.54).

Groundwater recharge occurs mainly in the eastern marginal zone on the western slope of the Great Dividing Range in an area of relatively high rainfall (median annual rainfall >500 mm) and large-scale regional groundwater movement is predominantly towards the south-western, western and southern discharge margins, as well as a component northwards (Fig. 2.55). Recharge also occurs along the western margin of the Basin and in the arid centre of the continent with groundwater flow towards the south-western discharge margin. Natural groundwater discharge occurs as groups of springs (Fig. 2.55; Plates 2.1 and

(Continued)

Box 2.11 (Continued)

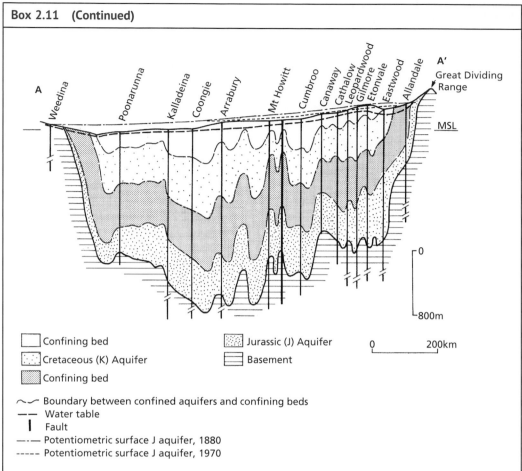

Confining bed
Cretaceous (K) Aquifer
Confining bed

Jurassic (J) Aquifer
Basement

0   200km

~⌣ Boundary between confined aquifers and confining beds
—— Water table
I Fault
—·— Potentiometric surface J aquifer, 1880
----- Potentiometric surface J aquifer, 1970

Fig. 2.54 Simplified cross-section of the Great Artesian Basin showing the position of major aquifer and confining units (see Fig. 2.53 for location of section A–A′) (Habermehl 1980). (*Source:* Habermehl, M.A. (1980) The Great Artesian Basin, Australia. *BMR Journal of Australian Geology and Geophysics* **5**, 9–38.)

2.2) that are commonly characterized by conical mounds of clayey and/or sandy sediments overlain by carbonates which range in height and diameter from a few metres to tens of metres and up to several metres high, some with water-filled craters (Habermehl 1980; Prescott and Habermehl 2008). Diffuse discharge also occurs from the artesian aquifers near the Basin margins where the overlying confining beds are thin. The location of many springs appears to be fault-controlled with others present where aquifers abut low permeability basement rocks or where only thin confining beds are present. The ages of spring mound deposits established by luminescence dating of quartz sand grains that have been incorporated in the mound deposits cover a time range from modern to more than 700 ka, with the luminescence ages corresponding with wetter hydrological conditions during Pleistocene interglacial periods (Prescott and Habermehl 2008).

The potentiometric surface of the Triassic, Jurassic and Lower Cretaceous confined aquifers is still above ground level in most areas (Fig. 2.54) despite considerable lowering of

Box 2.11 (Continued)

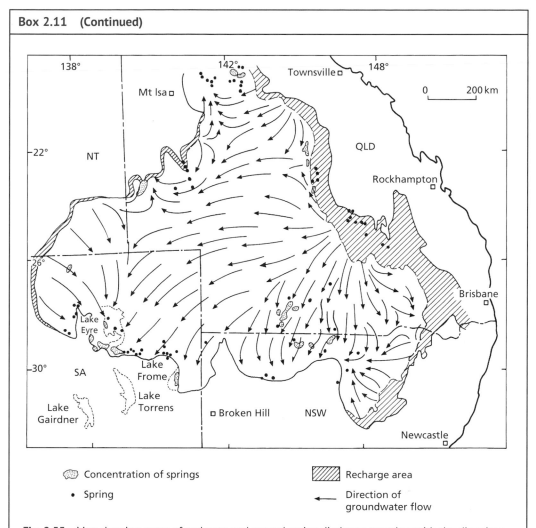

Fig. 2.55 Map showing areas of recharge and natural spring discharge together with the directions of regional groundwater flow (Habermehl 1980). (*Source:* Habermehl, M.A. (1980) The Great Artesian Basin, Australia. *BMR Journal of Australian Geology and Geophysics* **5**, 9–38.)

heads due to extensive groundwater development since the 1880s. Groundwater levels in the confined aquifer in the upper part of the Cretaceous sequence have always been below ground level throughout most of the Basin area and development of these non-flowing artesian aquifers requires the installation of pumping equipment. Transmissivity values of the main aquifers in the Lower Cretaceous-Jurassic sequence, from which most overflowing artesian groundwater is obtained, are typically several tens to several hundreds of m^2 day^{-1}. Average groundwater flow rates in the eastern and western parts of the Basin range from $1-5\,m\,a^{-1}$ based on hydraulic data and radiometric dating using carbon-14 and chlorine-36 radioisotopes, and yield residence times from several thousands

(*Continued*)

Box 2.11 (Continued)

of years near the marginal recharge areas to more than one million years near the centre of the Basin (for further discussion see Section 5.5.3 and Bentley *et al.* 1986; Love *et al.* 2000).

Groundwater in the most widely exploited artesian aquifers in the Lower Cretaceous-Jurassic sequence generally contains between 500 and 1000 mg L^{-1} of total dissolved solids, predominantly as sodium and bicarbonate ions. Water quality improves with increasing depth of aquifers in the sequence and on the whole the groundwater is suitable for domestic and stock use, although it is generally unsuitable for irrigation use due to the high sodium concentration and its chemical incompatibility with the montmorillonite clay soils. Groundwater from aquifers in the upper part of the Cretaceous sequence has a higher salinity.

The exploitation and management of the groundwater resources from the Great Artesian Basin must take account of the artesian springs and spring mound deposits given their ecological and biodiversity significance, as well as their Aboriginal and European cultural heritage values. For example, the springs in northern South Australia largely determined the location of early pastoral settlements and the location of the Overland Telegraph Line from Adelaide to Darwin. Also, the original railway line from Marree to Oodnadatta and Alice Springs

followed the springs, which were the only permanent sources of water in this desert region (Harris 1981).

Exploitation of groundwater resources of the Great Artesian Basin has impacted surface pressures, borehole yields and spring flows. For example, in the south-eastern portion of the Great Artesian Basin, around 55% of the Eulo, Bourke and Bogan River spring groups have been rendered inactive due to groundwater extraction, while many of the remaining springs have diminished in flow. Just 13 spring groups in New South Wales and 47 in Queensland remained active (Powell *et al.* 2015). To remedy the impacts of groundwater development, the Great Artesian Basin Sustainability Initiative (GABSI), completed in 2018, and other related bore-capping programmes have made considerable progress in restoring aquifer pressure. In 2004, the Great Artesian Basin Consultative Council (GABCC) was established with the primary role of providing advice to relevant Australian and state and territory ministers on the sustainable, whole-of-basin resource management, and to coordinate activity between stakeholders, with a new Strategic Management Plan developed in 2018 (Habermehl 2020). Further discussion of the history of the Great Artesian Basin, its exploration, scientific investigations and management is presented by Habermehl (2020).

---

the aquifer. The influence which springs exert on the aquifers they drain depends principally upon the topographic and structural context of the spring. Ford and Williams (1989) discussed hydrogeological controls on springs and recognized three principle types of springs (free draining, dammed and confined), principally in relation to karst aquifers in which some of the world's largest springs occur (Table 2.3).

With reference to Fig. 2.56, free draining springs experience groundwater discharge under the influence of gravity and are entirely or dominantly in the unsaturated (vadose) zone. Dammed springs are a common type and result from the location of a major barrier in the path of the underground drainage. The barrier may be caused by another lithology, either faulted or in conformable contact, or be caused by

Table 2.3 Discharges of 10 of the world's largest karst springs (Ford and Williams 1989).

Spring	Discharge (m³ s⁻¹)			Basin area (km²)
	Mean	Max	Min	
Matali, Papua New Guinea	90	>240	20	350
Bussento, Italy	–	117	76	–
Dumanli*, Turkey	50	–	25	2800
Trebišnijca, Bosnia-Herzegovina	50	250	3	–
Chingshui, China	33	390	4	1040
Vaucluse, France	29	200	4.5	2100
Frió, Mexico	28	515	6	>1000?
Silver, USA	23.25	36.5	15.3	1900
Waikoropupu, New Zealand	15	21	5.3	450
Maligne, Canada	13.5	45	1	730

* Dumanli spring is the largest of a group of springs that collectively yield a mean flow of 125–130 m³ s⁻¹ at the surface of the Manavgat River.
(*Source:* Ford, D.C. and Williams, P.W. (1989) *Karst Geomorphology and Hydrology.* Unwin Hyman Ltd., London.)

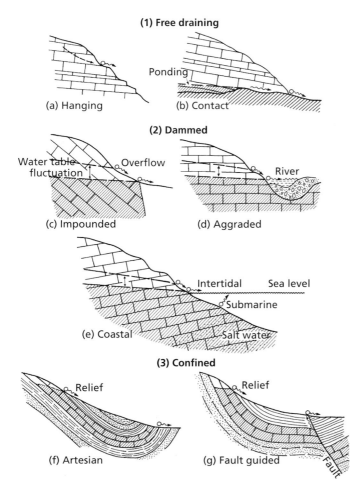

Fig. 2.56 Types of springs (Ford and Williams 1989). (*Source:* Modified from Ford, D.C. and Williams, P.W. (1989) *Karst Geomorphology and Hydrology.* Unwin Hyman Ltd., London.)

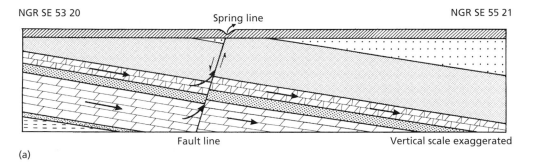

(a)

(b)

Fig. 2.57 (a) Geological section at Gale Common, South Yorkshire showing a spring line at the position of a normal fault that acts as a barrier to groundwater movement. The existence of the fault causes groundwater flow through the overlying Upper Permian Marl via a zone of enhanced permeability. (b) Springs from Permian Magnesian Limestones appearing through Quaternary deposits at Gale Common, South Yorkshire. For scale, a camera lens cap is shown to the right of the spring pool.

valley aggradation, such as by the deposition of glacial deposits. The denser salt water of the sea also forms a barrier to submarine groundwater discharge of freshwater. In each case, temporary overflow springs may form at times of higher water table elevation. Confined springs arise where artesian conditions are caused by an overlying impervious formation. Fault planes occasionally provide a discharge route for the confined groundwater as in the case of the Bath

thermal springs (Box 2.12) and the example of the Permian Magnesian Limestones in South Yorkshire (Fig. 2.57). Elsewhere, the groundwater escapes where the overlying strata are removed by erosion. Since the emerging water is usually rapidly equilibrating to atmospheric pressure, dissolved gases can create a 'boiling' appearance within the spring pool.

In rivers that flow over an aquifer outcrop, both influent and effluent conditions can

---

Box 2.12   The thermal springs of Bath, England

The thermal springs at Bath (the Roman town of Aquae Sulis) in the west of England are the principal occurrence of thermal springs in the British Isles and have been exploited for at least the past 2000 years. The Romans built the first baths around the shrine of the goddess Sulis Minerva about AD 76 (Bord and Bord 1985). A succession of buildings were constructed over the springs, beginning with the baths and a Roman temple (Fig. 2.58). The springs have temperatures of 44–47°C with an apparently constant flow of $15\,\mathrm{L\,s^{-1}}$. Three springs, the King's (Fig. 2.59), Cross and Hetling Springs, issue from what were probably once pools on a floodplain terrace on the River Avon, in the centre of Bath. Further details and an account of the hydrogeology of the thermal springs are given by Atkinson and Davison (2002).

The origin of the thermal waters has been subject to various investigations. Andrews *et al.* (1982) examined the geochemistry of the hot springs and other groundwaters in the region and demonstrated that they are

Fig. 2.58   The overflow of thermal spring water adjacent to the King's Bath shown in Fig. 2.59.

Fig. 2.59   The King's Spring (or Sacred Spring) emerging into the King's Bath at the Roman Baths, Bath, England.

(Continued)

Box 2.12 (Continued)

of meteoric origin. The silica content indicates that the thermal waters attain a maximum temperature between 64°C and 96°C, the uncertainty depending on whether chalcedony or quartz controls the silica solubility. Using an estimated geothermal gradient of 20°C km^{-1}, Andrews *et al.* (1982) calculated a circulation depth for the water of between 2.7 and 4.3 km from these temperatures. The natural groundwater head beneath central Bath is about 27–28 m above sea level, compared with normal spring pool levels at about 20 m. For this head to develop, Burgess *et al.* (1980) argued that the recharge area is most likely the Carboniferous limestone outcrop in the Mendip Hills (see Section 2.6.2), 15–20 km south and south-west of Bath, in order to drive recharge down along a permeable pathway and then up a possible thrust fault to the springs themselves. The structural basin containing the Carboniferous limestone lies at depths exceeding 2.7 km at the centre of the basin, sufficient for groundwater to acquire the necessary temperature indicated by its silica content. This 'Mendips Model' for the origin of the Bath hot springs is summarized by Andrews *et al.* (1982) and shown in Fig. 2.60.

In a later study of the hydrochemistry of two boreholes sampled in the centre of Bath, Edmunds *et al.* (2014) further developed the Mendips Model to give improved understanding of the residence times and flow regime

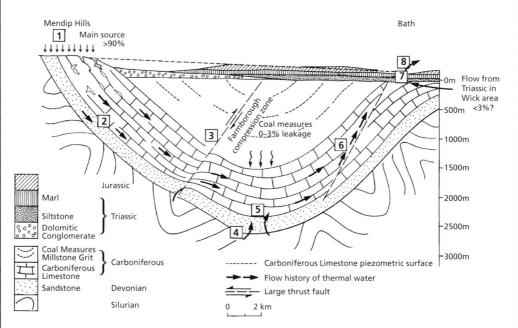

Fig. 2.60 Conceptual model for the origin of the Bath thermal springs. The numbers shown in squares indicate: (1) recharge (9–10°C) at the Carboniferous limestone/Devonian sandstone outcrop on the Mendip Hills; (2) flow down-dip and downgradient; (3) possible downward leakage from Upper Carboniferous Coal Measures; (4) possible leakage of very old ^{4}He-bearing groundwater from Devonian sandstone and Lower Palaeozoic strata; (5) storage and chemical equilibration within the Carboniferous limestone at 64–96°C; (6) rapid ascent, probably along Variscan thrust faults re-activated by Mesozoic tectonic extension; (7) lateral spread of thermal water into Permo-Triassic strata at Bath; and (8) discharge of the thermal springs at Bath (46.5°C) (Andrews *et al.* 1982). (*Source:* Andrews, J.N., Burgess, W.G., Edmunds, W.M. *et al.* (1982) The thermal springs of Bath. *Nature* **298**, 339–343.)

Box 2.12 (Continued)

of the thermal springs. Concentrations of the noble gas ^{39}Ar showed that the bulk of the thermal water has been in circulation within the Carboniferous limestone for at least 1000 years. Other stable isotope and noble gas measurements strongly suggested that recharge of the spring water occurred within the Holocene time period (i.e. the last 12 ka). Measurements of ^{85}Kr and chlorofluorocarbons (CFCs) corroborate previous understanding from tritium that a small proportion (<5%) of the thermal water originates as modern leakage into the spring pipe passing through the Mesozoic valley fill underlying Bath. Edmunds *et al.* (2014) concluded that the hydrochemical evidence confirms that the spring water evolved within the

Carboniferous limestone and that the groundwater, in a refinement of the Mendips Model, travels to the surface relatively slowly along a more complex fracture network during which it evolves geochemically and retains its thermal signature. However, the chemistry alone does not define the geometry of the recharge area or circulation route. For a likely residence time of 1–12 ka, volumetric calculations imply a large storage volume and circulation pathway, if typical porosities of the limestone at depth are assumed. These calculations indicate that much of the Bath-Bristol basin must be involved in the water storage, consistent with the geological model of Gallois (2006) and the 'base case' flow model of Atkinson and Davison (2002).

develop depending on the position of the water table in relation to the elevation of the river bed. With the seasonal fluctuation of the water table, the sections of river that receive groundwater discharge in addition to surface runoff will also vary. A good example of this type of river, are the intermittent streams that appear over areas of Chalk outcrop in southern England. In these areas, the low specific yield of the Chalk aquifer causes large fluctuations in the position of the water table between the summer and winter and, therefore, in the length of the intermittent streams. The intermittent streams, or Chalk bournes or winterbournes as they are known, flow for part of the year, usually during or after the season of most precipitation. An example is the River Bourne located in the north-east of Salisbury Plain as shown in Fig. 2.61(a). In this area of undulating Chalk downland in central southern England, the intermittent section of the River Bourne is 10 km in length until the point of the perennial stream head is met below which the Chalk water table permanently

intersects the river bed. In drought years, the intermittent section may remain dry while in wet years, such as shown in Fig. 2.61(b), the upper reaches sustain a bank full discharge.

2.15 Transboundary aquifer systems

A transboundary aquifer system is recognized when a natural groundwater flowpath is intersected by an international boundary, such that water transfers from one side of the boundary to the other (Fig. 2.62). In many cases, the aquifer may receive the majority of its recharge on one side of the boundary, and the majority of its discharge on the other. The subsurface flow system at the international boundary itself can be visualized to include regional, as well as the local movement of water. In hydrogeological terms, these shared resources can only be estimated through good observations and measurements of selected hydrologic

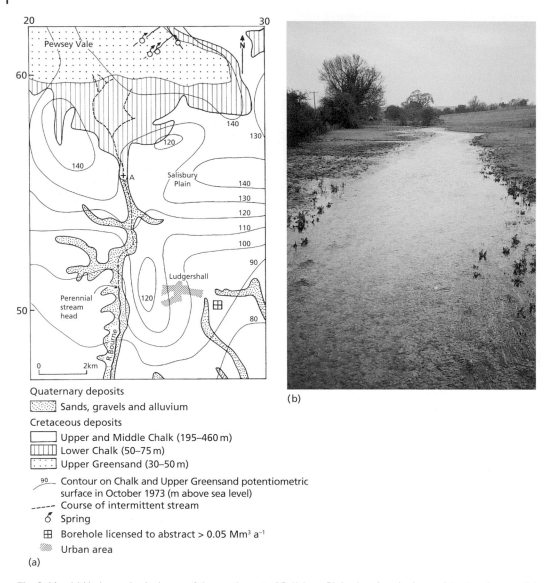

Quaternary deposits
[::::::] Sands, gravels and alluvium

Cretaceous deposits
[   ] Upper and Middle Chalk (195–460 m)
[||||||] Lower Chalk (50–75 m)
[:::::] Upper Greensand (30–50 m)

~90~ Contour on Chalk and Upper Greensand potentiometric surface in October 1973 (m above sea level)
----- Course of intermittent stream
♂ Spring
⊞ Borehole licensed to abstract > 0.05 Mm³ a⁻¹
▨ Urban area

(a)

Fig. 2.61 (a) Hydrogeological map of the north-east of Salisbury Plain showing the intermittent and perennial sections of the River Bourne. Position A is the site of the photograph shown in (b) located at Collingbourne Kingston and looking north in December 2002. The river bed is covered by lesser water parsnip *Berula erecta*, a plant that proliferates in still to medium flows of base-rich water (British Geological Survey 1978 and Haslam *et al.* 1975). *Sources:* British Geological Survey (1978) *1 : 100 000 Hydrogeological Map of the South West Chilterns and the Berkshire and Marlborough Downs.* Natural Environment Research Council.; Haslam, S., Sinker, C. and Wolseley, P. (1975) British water plants. *Field Studies* **4**, 243–351.

parameters, such as precipitation, groundwater levels, stream flow, evapotranspiration and water use. Therefore, a monitoring programme should aim to provide the data essential to generate a conceptual and quantitative

understanding of the status of a transboundary aquifer system.

Since the year 2000, UNESCO's Internationally Shared Aquifer Resources Management programme (ISARM, www.isarm.org) has been

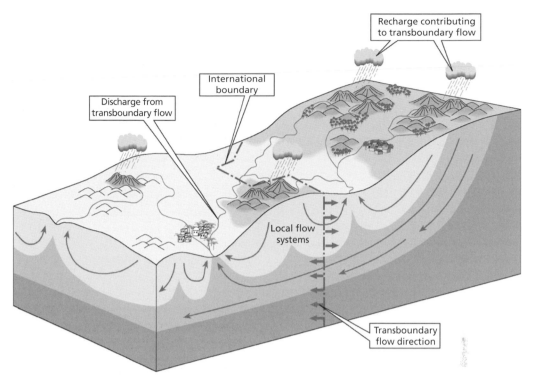

Fig. 2.62 Schematic illustration of a transboundary aquifer (Puri *et al.* 2001). (*Source*: Puri, S., Appelgren, B., Arnold, G. *et al.* (2001) *Internationally Shared (Transboundary) Aquifer Resources Management: Their Significance and Sustainable Management.* UNESCO, Paris.)

participating in the establishment of a groundwater database and the presentation of a detailed map of transboundary aquifers (see Plate 2.3). According to UNESCO (2006), the inventory comprised 273 shared aquifers: 68 on the American continent, 38 in Africa, 65 in Eastern Europe, 90 in Western Europe and 12 in Asia. As an example, Box 2.13 describes the transboundary Guarani Aquifer, a very large hydrogeological system that

---

Box 2.13   The transboundary Guarani Aquifer System

The Guarani Aquifer underlies areas of Brazil, Uruguay, Paraguay and Argentina (Fig. 2.63) and outcrops in dense populated areas such as São Paulo in Brazil. The aquifer contains water of very good quality and is exploited for urban, industrial and irrigation supplies and for thermal, mineral and tourist purposes. The Guarani Aquifer is one of the most important underground fresh water reservoirs in the world and comprises a very large hydrogeological system that underlies an area of about 1 100 000 km^2 mainly in the Paraná River Basin of Brazil (with about 62% of its known area), Paraguay, Uruguay and Argentina. It has an average thickness of about 250 m, varying from <50 to >600 m (Fili *et al.* 1998; Araújo *et al.* 1999), and reaches depths of over 1000 m (Fig. 2.63). The total

(*Continued*)

Box 2.13   (Continued)

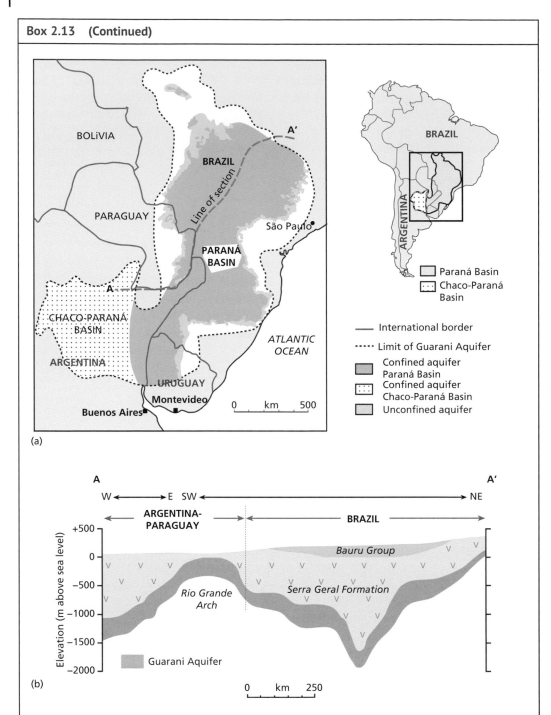

(a)

(b)

Fig. 2.63   Map of (a) outcrop and (b) cross-section of the Guarani Aquifer System (Campos 2000 and Puri *et al.* 2001). (*Sources:* Campos, H. (2000) *Mapa Hidrogeológico del Acuífero Guaraní.* Proceedings of the First Joint World Congress on Groundwater, Fortaleza, Brasil.; Puri, S., Appelgren, B., Arnold, G. *et al.* (2001) *Internationally Shared (Transboundary) Aquifer Resources Management: Their Significance and Sustainable Management.* UNESCO, Paris.)

Box 2.13 (Continued)

volume of freshwater contained in storage is estimated to be around 30 000 km^3, equivalent to 100 years of cumulative flow in the Paraná River. The aquifer extends across a number of international political boundaries, as well as those of many individual states of Brazil and provinces of Argentina, which are federal countries with groundwater resources essentially under state- or provincial-level jurisdiction (Foster *et al.* 2009).

The Guarani Aquifer System (SAG, from the Spanish and Portuguese abbreviation) comprises a sequence of mainly weakly cemented Triassic-Jurassic sandstone beds formed by continental aeolian, fluvial and lacustrine deposition on a regional Permo-Triassic erosion surface, and is overlain by Cretaceous basalt flows which are almost equally as extensive, exceeding 1000 m thickness in some areas. The stratigraphic equivalence of the sandstone beds was only recognized in the 1990s following drilling of oil exploration wells, after which the associated aquifer system was named 'Guarani' after the indigenous population of the area. The aquifer occurs in three main 'hydrogeological domains' delimited by two geological structures that control the aquifer thickness and depth, and also influence regional groundwater flow: the Ponta Grossa Arch (in the north of Paraná State, Brazil), which forces groundwater to flow from east to west in São Paulo State, Brazil; and the Asunción-Rio Grande Arch, which divides the portion south of the Ponta Grossa Arch into two semi-independent sedimentary basins, the Central Paraná and the south-western Chaco-Lower Paraná. The SAG is also affected by many tectonic structures and crossed by numerous volcanic dykes, but despite these important discontinuities at a local scale it is considered to be a

continuous groundwater body across the entire region (Foster *et al.* 2009).

Recharge of the SAG occurs by direct infiltration of excess rainfall and streamflow along the length of the unconfined aquifer outcrop area (Fig. 2.63) and in adjacent zones with a limited thickness of well-fractured basalt. Recharge also occurs via 'windows' in the basalt from overlying local groundwater bodies in Tertiary sedimentary formations. The high average rainfall across most of the SAG recharge area of 1000–2000 mm a^{-1} results in potentially elevated rates of aquifer recharge (300–400 and 500–600 mm^{-1} in the northern and southern regions, respectively). Although some recharge may be 'rejected' because of inadequate infiltration capacity or high water-tables, most of this potential recharge infiltrates to form local flow cells which discharge nearby as baseflow to rivers crossing the SAG outcrop. In these areas, groundwater hydraulic gradients are up to 3–5 m km^{-1} and actual flow velocities are over 5 m a^{-1}. However, there are substantial differences in detail between the recharge areas on the 'north-western flank' of the main basin (Paraguay to Mato Grosso do Sul-Brazil) and on the 'north-eastern flank' (Santa Catarina to São Paulo State in Brazil), where reduced formation thickness and steeper dip result in a much narrower outcrop area, creating a smaller zone where recharge through the basalt cover is favoured. Estimation of the overall current rate of SAG recharge is not straightforward because of uncertainties in the spatial variation of average potential recharge rates, the proportion of SAG outcrop area that permits recharge, and the extent of recharge in areas covered by basalt. It is known that the total SAG recharge area is only a minor proportion of the known aquifer extension and recharge

(Continued)

Box 2.13 (Continued)

is estimated to be in the range 45–55 km³ a⁻¹, equivalent to less than 0.2% of the estimated freshwater storage. Thus, the SAG is considered a totally 'storage-dominated' groundwater system (Foster *et al.* 2009).

Evidence from groundwater potentiometric levels (Foster *et al.* 2009) indicates that some regional flow occurs from the main recharge areas into the deeper structural basins, with subsequent southward flow parallel to the general axis of the Paraná catchment. Towards the centre of the structural basins, SAG groundwater becomes progressively more confined by an increasing thickness of overlying basalts and exhibits artesian overflowing conditions in deep boreholes over extensive areas. The SAG has a relatively high hydraulic conductivity of 5–10 m day⁻¹ and an estimated mean transmissivity of about 300 m² day⁻¹ (range 50–1200 m² day⁻¹), although the flat terrain and low hydraulic gradients in the confined aquifer (about 0.1–0.3 m km⁻¹) imply very low groundwater flow velocities (<0.5 m a⁻¹). Groundwater numerical modelling suggests that active groundwater flow into the deep confined aquifer is very limited, probably equivalent to 10–15 mm a⁻¹ of vertical infiltration in the recharge area, equivalent to about 1–2% of annual rainfall. With increasing depth and confinement, the groundwater temperature increases substantially as a result of normal geothermal gradients, such that it forms a low-enthalpy hydrothermal resource with temperatures widely exceeding 40°C and locally reaching 60°C (Foster *et al.* 2009).

Some natural discharge from the regional flow regime must occur but is not yet quantified due to difficulty in detecting and

Table 2.4 Hydrochemical and environmental isotope evolution of groundwaters of the Guarani Aquifer System down-gradient westwards from Ribeirão Petro (São Paulo), Brazil (Sracek and Hirata 2002).

		Down-gradient boreholes	
Parameter (units)	Outcrop boreholes	30 km	150 km
T (°C)	24	26	42
pH (–)	6.5	8.5	9.5
Ca^{2+} (mg L⁻¹)	30	20	2
Na^+ (mg L⁻¹)	1	5	90
HCO_3^- (mg L⁻¹)	15	75	160
Cl^- (mg L⁻¹)	1	2	10
F^- (mg L⁻¹)	<0.1	0.2	>1.0
SiO_2 (mg L⁻¹)	15	20	30
$\delta^{18}O$ (‰)	−7.3	−8.0	−9.8
δ^2H (‰)	−49	−50	−67
$\delta^{13}C$ (‰)	−19.0	−11.1	−6.3

(*Source:* Modified from Sracek, O. and Hirata, R. (2002) Geochemical and stable isotopic evolution of the Guarani Aquifer System in the state of São Paulo, Brazil. *Hydrogeology Journal* **10**, 643–655.)

Box 2.13 (Continued)

measuring small groundwater upwelling in areas with large river flows. Evidence exists in the small springs with a chemical composition similar to that of confined SAG groundwater in areas with volcanic dykes. Other potential discharge zones of favourable geological structure, groundwater potentiometric levels and reduced basalt thickness include sections of the Paraná River (along the Paraguay frontier) and the Uruguay River (in Rio Grande do Sul and Santa Catarina States, Brazil) and the Esteros de Ibera (Argentina) and Neembacu (Paraguay) wetlands. An extensive study of environmental isotope composition (^{3}H, δ^{18}O, δ^2H, δ^{13}C and ^{14}C) of SAG groundwaters (Sracek and Hirata 2002) has shown that groundwater in aquifer recharge areas generally has δ^{18}O and δ^2H values matching those of modern rainfall (δ^{18}O > −7.5‰) (Table 2.4). Moreover, the presence of ^{3}H up to 3 TU and ^{14}C activity close to 100 pmc confirm the presence of recently recharged water, including below 'windows' in areas with thicker basalts. The rapid decline of ^{14}C activity along groundwater flow paths towards the highly confined SAG is commensurate with extremely slow circulation, with most deep boreholes recording ^{14}C activity below detection limit (probably water recharged more than 35 ka). Subsequently, using both ^{14}C and the long-lived cosmogenic radio-isotope ^{81}Kr (half-life 229 000 years), Aggarwal *et al.* (2015) confirmed the extremely slow flow rates (0.3–1.3 m a^{-1}; average 0.7 m a^{-1}) in the SAG with groundwater ages of 728 ± 69 and 834 ± 91 ka at two sites at 560 km and 250 km from the outcrop, respectively, in São Paulo State, Brazil. However, these groundwater ages are not compatible with darcian calculations, suggesting that there

may be older water mixtures coming from pre-SAG formations (Hirata and Foster 2020). In addition, the δ^{18}O content of groundwater in some confined SAG areas (for example in São Paulo State, Brazil) appears anomalous given the more negative stable isotope composition (δ^{18}O of −8.0 to −9.8%) compared to modern rainfall and probably reflects palaeogroundwater recharged under cooler climatic conditions, although the same phenomenon is not found in the SAG further south.

Natural groundwater quality in the SAG is generally very good with low mineralization rates in most areas. A hydrochemical evolution is observed as recharge waters from outcrop areas flow slowly into the deeper confined aquifer (Table 2.4) accompanied by dissolution of carbonates leading to more positive δ^{13}C values, ion exchange processes (notably Na$^+$ replacing Ca^{2+} in solution), rising pH from 6.8 to 9.5 and also marked temperature increases. The hydrochemical and isotopic data show that mostly saline aquitard formations underlying parts of the SAG contribute to observed salinity and increases in trace elements, especially F (up to 13.3 mg L^{-1}) and more locally As, although this contribution is not significant in terms of associated groundwater flow volumes. There are also more marked and general down-dip increases in groundwater salinity in the extreme south-west of the SAG in Argentina, which effectively mark the limit of the potentially useful aquifer system. There have also been some concerns that the deep confined groundwater might locally contain significant levels of the soluble uranium isotopes, radium and radon gas (Foster *et al.* 2009).

Potential threats to the naturally excellent groundwater quality of the SAG include urbanization and the disposal of domestic

(Continued)

Box 2.13 (Continued)

urban wastewaters, industrial development and the potentially inadequate storage and handling and disposal of hazardous chemicals and liquid and solid effluents, and intensification of agricultural crop cultivation and forestry. The degree of groundwater pollution vulnerability of the SAG varies with water-table depth, the degree of consolidation of the sandstone units and the extent of fracturing of the overlying basalts. As a result of these factors, only parts of the SAG exhibit significant vulnerability to surface-derived groundwater pollution in the main recharge area comprising the aquifer outcrop and adjacent areas where the basalts are highly-fractured or 'windows' through the basalt exist. At a distance from outcrop below the basalt cover, the relatively old groundwater age indicates minimal pollution vulnerability, except perhaps from low levels of any highly persistent and mobile groundwater contaminants in the very long term (Foster *et al.* 2009).

For the future, the SAG is becoming increasingly important for potable water supply to many towns with populations of 50 000–250 000 and in supporting the numerous industrial applications and potential agro-industrial processes that are seen as essential to the economic growth of the Mercosur region. Currently, more than 1.1 $km^3 a^{-1}$ of groundwater are abstracted mainly in São Paulo State, with an economic value of about US\$ 600 million a^{-1}. The exploitation in São Paulo State arises not only because of its high water demand and capacity to drill deep water-wells (some to more than 1000 m and costing more than US\$ 1 million) but also because here the SAG has a high transmissivity (up to 1200 m^2 day^{-1}), resulting in large water-well production (Hirata and Foster 2020). Additionally, the SAG also represents a major low-enthalpy geothermal resource (often with overflowing artesian head) of very extensive distribution, with potential for future expansion of spa facilities in north-western Uruguay, neighbouring parts of Argentina and further north in the Iguazu international tourist area.

In considering current groundwater exploitation, Foster *et al.* (2009) concluded that most actual and potential groundwater resource management and protection needs of the SAG fundamentally do not have an 'international transboundary character', albeit that there exist some local 'transboundary hotspots' both between nations and between individual states of Brazil that share the aquifer. Furthermore, Puri *et al.* (2001) concluded that future exploitation of groundwater resources reliant on the SAG will partly depend on international and federal cooperation in sharing the benefits of advances in scientific understanding and positive groundwater management experiences of the Guarani Aquifer.

underlies four countries in South America and which is of increasing importance in the region for the supply of drinking water to many urban areas.

Political tensions can be exacerbated over access to groundwater such that unequal rights of use of aquifers can potentially lead to conflict (Bergkamp and Cross 2006), although Amery and Wolf (2004) recognized the importance of water resources in peace negotiations. Transboundary groundwater problems in the Middle East are an example of the socio-political impacts of groundwater depletion. In the case of Israel and Palestine, the coastal plain aquifer extends from Carmel (near Haifa) in the north to the Palestinian Gaza Strip in the south. According to Kandel (2003), serious conflict exists between Israel and Palestine over this aquifer, which is both directly and indirectly related to the conflict over land.

2.16 Submarine groundwater discharge

Hydrogeologists and coastal oceanographers now recognize that a significant proportion of freshwater input to the ocean occurs as submarine groundwater discharge (SGD) together with a significant contribution of dissolved solutes (nutrients, carbon and metals), probably exceeding the input of these materials by rivers (Moore 2010). To illustrate, an integrated tracer study (Moore *et al.* 2008) concluded that the total flux of SGD to the Atlantic Ocean was similar to the river flux to this ocean. Thus, SGD must be considered a primary component in global ocean budgets and models of these constituents. A worldwide compilation of observed SGD by Taniguchi *et al.* (2002) showed that groundwater discharge as seeps and springs from the land to the ocean occurs in many environments along the world's continental margins and that SGD has a significant influence on the environmental condition of many near-shore marine environments.

Defining the term 'SGD' has presented a dilemma for hydrologists and oceanographers. As explained by Taniguchi *et al.* (2002) oceanographers tend to view any subsurface fluid fluxes as groundwater, whereas hydrologists have conventionally termed groundwater as only that water originating from an aquifer, which does not include re-circulated seawater (Moore 1999). This discrepancy in definitions has resulted in confusion when comparing SGD results from hydrologic models and oceanographic mass balances. Adopting the definition that SGD represents the total direct discharge of subsurface fluids across the sea-floor (the land–ocean interface) then, with reference to Fig. 2.64, the rate of SGD is defined as the sum of the submarine fresh groundwater discharge (SFGD) driven by a hydraulic head and the re-circulated saline groundwater (RSGD), the latter comprising a number of components due to processes of wave set-up, and tidally driven oscillation and convection (either density- or thermally driven). Thus, the total net SGD includes both net fresh groundwater and re-circulated seawater discharge components. The marine-induced forces that result in flow into and out of the seabed have been described as creating subterranean estuaries, characterized by biogeochemical reactions that influence the transfer of nutrients to the coastal zone in a manner similar to that of surface estuaries (Moore 2010).

Coastal aquifers may consist of complicated arrays of confined, semi-confined and unconfined systems comprised of a complex assemblage of coastal sediments subject to fresh and salt water mixing. The measurement of SGD in such a dynamic environment is challenging. In a review of multiple methods used to quantify SGD, Burnett *et al.* (2006) identified seepage meters and piezometers, environmental tracer methods (thermal gradients, radon, radium isotopes and methane, and artificial tracers such as fluorescein dye saturated with SF_6) and hydrologic water balance and modelling methods as being applicable. The choice of technique depends on practical considerations (cost, availability of equipment, etc.). For many situations, seepage meters, the only device that measures seepage directly, can work well. These devices provide a flux at a specific time and location from a limited amount of seabed (generally about $0.25\,m^2$). Seepage meters range in cost from almost nothing for a simple bag-operated meter too expensive for those equipped with more sophisticated

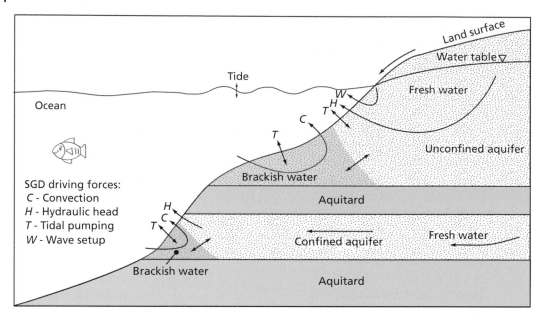

Fig. 2.64 Schematic diagram of the processes associated with submarine groundwater discharge (SGD). Three different units of measurement of SGD include (1) volume per unit time; (2) volume per unit time per unit length of shoreline; and (3) volume per unit time per unit area (i.e. specific discharge or darcy flux) (Taniguchi *et al.* 2002). (*Source:* Taniguchi, M., Burnett, W.C., Cable, J.E. and Turner, J.V. (2002) Investigation of submarine groundwater discharge. *Hydrological Processes* **16**, 2115–2129. © 2002, John Wiley & Sons.)

measurement devices. The use of natural geochemical tracers, for example radon (Burnett *et al.* 2001) involves the use of more costly equipment and requires personnel with special training and experience. The main advantage of the tracer approach is that the water column tends to integrate the signal and, as a result, smaller scale variations, which may be unimportant for larger scale studies, are smoothed out. Therefore, this approach may be optimal in environments where especially large spatial variation is expected (e.g. fractured rock aquifers). In addition to the spatial integration, tracers integrate the water flux over the time-scale of the isotope and the water residence time of the study area. The use of multiple tracers is recommended where possible, for example, the simultaneous measurement of radon and radium isotopes can be used to constrain the mixing loss of radon.

Simple water balance calculations have been shown to provide a useful estimate of the fresh groundwater discharge (see Box 2.14). Dual-density groundwater modelling can also be undertaken either as simple steady-state (annual average flux) or non-steady state (requires real-time boundary conditions) methods. Unfortunately, model results usually do not compare well with seepage meter and tracer measurements due to problems encountered in the proper scaling, both in time and space, and in parameterizing dispersion processes. Apparent inconsistencies between modelling and direct measurement approaches often arise because different components of SGD (fresh and salt water) are being evaluated or because the models do not include transient terrestrial (e.g. recharge cycles) or marine processes (tidal pumping, wave set-up, etc.) that drive part or all of the SGD. Geochemical

Box 2.14 Submarine groundwater discharge on the west coast of Ireland

Submarine and intertidal groundwater discharge is a principal source of freshwater entering Irish coastal waters between the major west coast estuaries of the Corrib and the Shannon. Discharges of groundwater around and off this coastal karst area are known from caves and groundwater resurgences, including those off the coast of the Burren in Counties Clare and Galway (Fig. 2.65). The Burren is an extensive area of Carboniferous limestone plateaux. Three levels of karst have been identified in the east of the Burren in the Gort Lowlands: a shallow depth karst at 15–25 m below ground level (m bgl); an intermediate level at 40–50 m bgl; and a deep level at 70–80 m bgl in the Kinvara area in Galway Bay (OPW 1997). The shallower two levels appear to transmit rapid groundwater flow at times of high water levels but are susceptible to saline intrusion at low water levels. The deepest karst limestone underlying Kinvara is considered to be a palaeo-karst containing older groundwater. The karst underlying Kinvara drains quickly through large conduits, some only active during wet periods.

Dye tracer experiments indicate that Kinvara Bay is the focal point for a large part of the underground drainage from the Gort-Kinvara lowland area, with groundwater taking several paths to the sea from the sites of tracer input (Fig. 2.66). Groundwater discharge is clearly visible at low tide at several sites around the head of Kinvara Bay (Plate 2.4). Even at high tide the inner waters of the bay show low salinities and water can be observed bubbling up, a phenomenon that continues during long spells of dry weather (Cave and Henry 2011). A small lake at Caherglassaun, about 5 km inland (Plate 2.5) rises and falls by a few centimetres over a tidal cycle with a lag of about 2 h, while remaining fresh, but a municipal well between this lake and the sea, about 2 km inland, occasionally pumps salt water when long dry periods coincide with spring tides (Cave and Henry 2011). Further evidence based on flooding frequency-duration curves and water level recession characteristics of ephemeral lakes known as turloughs (Plate 2.6) also show that there is very rapid flow through groundwater catchments draining the Burren, such that

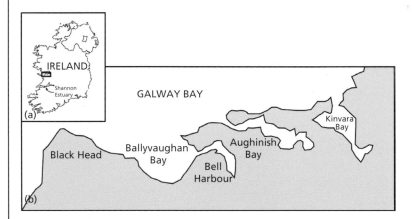

Fig. 2.65 Location maps of: (a) Ireland; and (b) bay areas with submarine groundwater discharge draining the north of the Burren and the Gort-Kinvara area (Cave and Henry 2011). (*Source:* Cave, R.R. and Henry, T. (2011) Intertidal and submarine groundwater discharge on the west coast of Ireland. *Estuarine, Coastal and Shelf Science* **92**, 415–423. © 2011, Elsevier.)

(Continued)

Box 2.14   (Continued)

storage of water in the turloughs only occurs at times of heavy and persistent rainfall, when precipitation exceeds the catchment drainage capacity (Tynan *et al.* 2007).

Measurements of groundwater discharges in the intertidal zone of Kinvara Bay during the autumn and winter of 2006 using the velocity–area method gave intertidal groundwater discharges of 3.5–5.9 m^3 s^{-1} at low tide from sites between Dunguaire Castle and Kinvara Pier, albeit at the lower end of the expected range in discharge (Cave and Henry 2011). Moreover, groundwater sampled at the coast contained relatively high concentrations of dissolved nutrients,

potentially representing a significant source of nutrients to these coastal waters which potentially, if unchecked, could lead to the development of eutrophic conditions. In their study, Cave and Henry (2011) measured the combined concentration of nitrite and nitrate in water sampled at the Dunguaire outfall at the head of Kinvara Bay as having an average concentration over the winters 2005/06 and 2006/07 of 3.2 mg L^{-1}, leading to the conclusion that the loads of fixed nitrogen entering Kinvara Bay by submarine groundwater discharge can approach the same magnitudes as those entering Galway Bay by the River Corrib (Table 2.5).

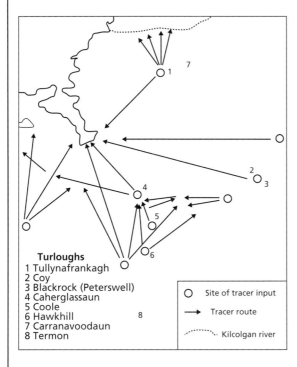

Turloughs
1 Tullynafrankagh
2 Coy
3 Blackrock (Peterswell)
4 Caherglassaun
5 Coole
6 Hawkhill
7 Carranavoodaun
8 Termon

○	Site of tracer input
→	Tracer route
⋯⋯	Kilcolgan river

Fig. 2.66   Routes of groundwater tracer inputs in the vicinity of Kinvara Bay. Numbers 1–8 show the location of turloughs (ephemeral lakes controlled by fluctuating groundwater levels in the Carboniferous limestone karst aquifer), with locations 1–6 representing tracer input locations (Cave and Henry 2011). (*Source:* Cave, R.R. and Henry, T. (2011) Intertidal and submarine groundwater discharge on the west coast of Ireland. *Estuarine, Coastal and Shelf Science* **92**, 415–423. © 2011, Elsevier.)

Box 2.14 (Continued)

Table 2.5 Winter nutrient loading (nitrite and nitrate as total oxidized nitrogen, TON) to Kinvara Bay from submarine groundwater discharge (SGD), compared with loading from the River Corrib (Cave and Henry 2011).

Location	Date	Mean flow ($m^3 s^{-1}$)	Mean TON ($mg L^{-1}$)	Mean TON flux ($kg day^{-1}$)
River Corrib	Nov 2006	162	1.0	13 960
	Dec 2006	306	1.0	26 434
	Jan 2007	269	1.0	23 240
	Feb 2007	189	1.0	16 358
Kinvara Bay Min. SGD Low TON	Nov 2006	14	1.4	1666
	Dec 2006	23	1.4	2809
	Jan 2007	31	1.4	3732
	Feb 2007	17	1.4	2084
Kinvara Bay Min. SGD High TON	Nov 2006	14	3.2	3809
	Dec 2006	23	3.2	6421
	Jan 2007	31	3.2	8530
	Feb 2007	17	3.2	4763
Kinvara Bay Max. SGD Low TON	Nov 2006	41	1.4	4979
	Dec 2006	61	1.4	7376
	Jan 2007	96	1.4	11 563
	Feb 2007	35	1.4	4180
Kinvara Bay Max. SGD High TON	Nov 2006	41	3.2	11 380
	Dec 2006	61	3.2	16 859
	Jan 2007	96	3.2	26 431
	Feb 2007	35	3.2	9554

(*Source:* Modified from Cave, R.R. and Henry, T. (2011) Intertidal and submarine groundwater discharge on the west coast of Ireland. *Estuarine, Coastal and Shelf Science* **92**, 415–423.)

tracers and seepage meters measure total flow, very often a combination of fresh groundwater and seawater and driven by a combination of oceanic and terrestrial forces. Water balance calculations and most models evaluate just the fresh groundwater flow driven by terrestrial hydraulic heads (Burnett *et al.* 2006).

Freshened groundwater has been detected tens to hundreds of kilometres offshore of coastlines worldwide below continental shelves (Post *et al.* 2013; Gustafson *et al.* 2019) and simulations indicate that this could be common across a range of geologic systems (Michael *et al.* 2016). Mechanisms for emplacing offshore groundwater include glacial processes that drove water into exposed continental shelves during sea-level low stands and active connections to onshore hydrologic systems (Gustafson *et al.* 2019; Fig. 5.5). The exposure of the continental shelves reached its most recent extent during the Last Glacial Maximum (LGM) from 26 500 to about 19 000 years ago.

Groundwater systems were slow to adapt to the reconfiguration of hydrological conditions at the end of the LGM, such that remnants of meteoric groundwater are likely to be found offshore (Post *et al.* 2013). The global volume of brackish water (TDS < 10 g L^{-1}) stored in offshore meteoric groundwater reserves for the total length of passive continental margins (105 000 km) was estimated by Post *et al.* (2013) to be 5×10^5 km^3. Adopting a threshold TDS concentration of <1 g L^{-1} yielded a volume of 3×10^5 km^3. These figures could vary up or down by a factor of about two owing to uncertainty in the value of sediment porosity used in the estimations (Post *et al.* 2013).

The widespread confirmation of the scale of offshore fresh and brackish groundwater reserves could potentially provide opportunities for the relief of water scarcity in densely populated coastal regions (Post *et al.* 2013) and be exploited for uses such as drinking water, agriculture and oil recovery (Yu and Michael 2019). However, Yu and Michael (2019) cautioned that coastal aquifers may be vulnerable to offshore pumping activities. Numerical simulations of variable-density groundwater flow and salt transport in coastal aquifers with different geologic structure subject to offshore pumping showed that offshore pumping could potentially reduce both onshore groundwater availability and SGD and cause widespread land subsidence (Yu and Michael 2019).

As an example of the ecological importance of SGD to near-shore coastal waters, Biscayne Bay, Florida is a coastal barrier island lagoon that relies on significant quantities of freshwater to sustain its estuarine ecosystem. During the twentieth century, field observations suggested that Biscayne Bay changed from a system largely controlled by widespread and continuous SGD from the highly permeable Pliocene-Pleistocene limestone aquifer (Biscayne aquifer) and overland sheet-flow, to one controlled by episodic discharge of surface water at the mouths of canals (Langevin 2003). Kohout and Kolipinski (1967) demonstrated

the ecological importance of SGD by showing that near-shore biological zonation in the shallow Biscayne Bay estuary was directly related to upward seepage of fresh groundwater (Fig. 2.67).

Throughout much of the area of southeastern Florida, a complex network of levees, canals and control structures is used to manage water resources (see Box 10.5). Beginning in the early 1900s, canals were constructed to lower the water table, increase the available land for agriculture and provide flood protection. By the 1950s, excessive drainage had lowered the water table by 1–3 m and caused salt water intrusion, thus endangering the freshwater resources and dependent ecosystems of the Biscayne aquifer. In an effort to reverse and prevent saltwater intrusion, control structures were built within the canals near Biscayne Bay to raise inland water levels. On the western side of the coastal control structures, water levels can be 1 m higher than the tidal water level east of the structures. Current ecosystem restoration efforts in southern Florida (see Box 10.5) are examining alternative water management scenarios that could further change the quantity and timing of freshwater delivery to the bay, although there are concerns that these proposed modifications could adversely affect bay salinities (Langevin 2003).

2.17 Groundwater resources of the world

To promote understanding of the world's groundwater systems, the collaborative Worldwide Hydrogeological Mapping and Assessment Programme (WHYMAP) was established in 1999 with the aim of collecting, compiling and visualizing hydrogeological information at a global scale (for more information see www.whymap.org). The basic concept underlying this mapping is the assumption that large-sized territories can be identified having an overall

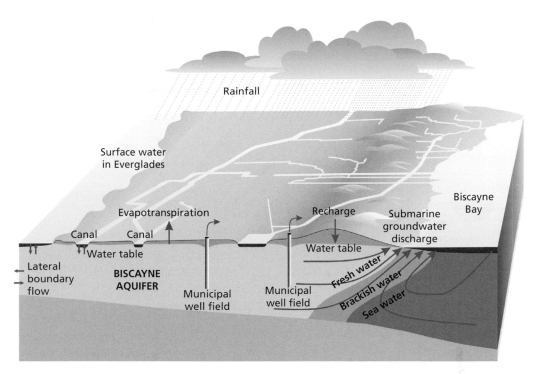

Fig. 2.67 Conceptual hydrological model of submarine groundwater discharge to Biscayne Bay, Florida (Langevin 2003). (*Source:* Langevin, C.D. (2003) Simulation of submarine ground water discharge to a marine estuary: Biscayne Bay, Florida. *Ground Water* **41**, 758–771. © 2003, John Wiley & Sons.)

groundwater setting, primarily based on the spatial extent of rock formations with similar hydraulic properties. The *Groundwater Resources Map of the World* at the scale of 1:25 000 000 presented in Plate 2.7 shows the aerial extent of characteristic groundwater environments. Blue colour shading in this map is used to indicate large and mainly uniform groundwater basins (aquifers and aquifer systems usually in large sedimentary basins that offer good conditions for groundwater exploitation). Green colour shading represents areas with complex hydrogeological structure (with highly productive aquifers in heterogeneous folded or faulted regions in close vicinity to non-aquifers), and brown colour shading symbolizes regions with limited groundwater resources in local and shallow aquifers. Within the three main hydrogeological units, up to five different categories are defined according to their modelled recharge rates from less than 2

to over $300 \, mm \, a^{-1}$. Dark colours (dark blue, green and brown) represent areas with very high recharge rates, while light blue, green and brown colours outline regions with very low recharge potential, with the latter category vulnerable to groundwater mining. Groundwater recharge rates refer to the historic baseline climate period 1961–1990 calculated from simulations using the WaterGAP Global Hydrology Model (Döll *et al.* 2003) as discussed next.

The example of the *Groundwater Resources Map of the World* shown in Plate 2.7 shows the areal extent of the named principal large aquifer systems that contain the largest storage of groundwater on Earth. Of note are those areas containing thick accumulations of sedimentary material deposited in various environments during different geological periods. Such large aquifer systems include the High Plains-Ogallala Aquifer in North America, the Guarani Aquifer System in South America, the Paris Basin in France, the North China Plain Aquifer System and the Great Artesian Basin of Australia. In some arid regions, these reserves constitute very important sources of groundwater, in spite of the lack of recent recharge, for example, the Nubian Aquifer System of North Africa and the Arabian Aquifer System. Altogether, 40 large aquifer systems are distinguished in Plate 2.7. None of the aquifer systems can be regarded as homogeneous, since they all include sub-regions or zones that have characteristics different from the dominant features of the regional system.

Another feature evident from the *Groundwater Resources Map of the World* (Plate 2..7) is that surface water catchments of river basins rarely coincide with underground hydrogeological structures. As a consequence, integrated water resources management units have to be chosen carefully to pay due attention to the complementary surface water and groundwater resources, with most of the large groundwater units categorized as shared transboundary aquifers (see Section 2.15).

Groundwater recharge is the major limiting factor for the sustainable use of groundwater resources and is, therefore, of considerable interest in the assessment of global groundwater resources. Long-term average diffuse groundwater recharge for the historic baseline climate period 1961–1990 has been computed using the WaterGAP Global Hydrology Model (WGHM) developed by Döll *et al.* (2003). With a spatial resolution of 0.5° latitude by 0.5° longitude, WGHM first computes total runoff based on a time series of monthly climate variables, as well as soil and land cover characteristics. Groundwater recharge is then calculated as a fraction of total runoff using data on relief, soil texture, geology and permafrost/glaciers. For semi-arid and arid areas, the model has been calibrated against estimates of groundwater recharge derived from chloride and isotope data (Döll *et al.* 2003).

Using two global datasets of gridded observed precipitation, Döll and Fiedler (2008) gave groundwater recharge values ranging from 0 to 960 mm a^{-1}, with the highest values occurring in the humid tropics. Values of over 300 mm a^{-1} are computed for parts of north-western Europe and the Alps. Europe is the continent with the smallest fraction of regions with groundwater recharge below 20 mm a^{-1}. Low recharge values occur in the dry sub-tropics and, mainly as a result of permafrost, in Arctic regions. For all land areas of the Earth, excluding Antarctica, groundwater recharge and, thus, renewable groundwater resources are estimated to be 12 666 km^3 a^{-1}, while the continental values range from 404 km^3 a^{-1} for Australia and Oceania to 4131 km^3 a^{-1} for South America (Table 2.6) (Döll and Fiedler 2008).

To estimate the renewable groundwater resource available for human consumption, Plate 2.8 shows the global distribution of groundwater recharge divided by the population of specific national and, in the case of 11 large countries (Argentina, Australia, Brazil, Canada, China, India, Kazakhstan, Mexico, Mongolia, Russia and the United States), at sub-national scales as presented by Döll and Fiedler (2008). Renewable groundwater resources are shown to range from 8 m^3 capita^{-1} a^{-1} for Egypt to greater than 1×10^6 m^3 capita^{-1} a^{-1} for the Falkland Islands, with a global average for the year 2000 of 2091 m^3 capita^{-1} a^{-1}. All countries in Northern Africa and the Near East, except Libya, have average per capita groundwater resources of less than 500 m^3 a^{-1}. Humid countries with high population densities, for example The Netherlands, Vietnam, Japan and Germany, demonstrate per capita groundwater resources below 1000 m^3 a^{-1}. In the United States, lowest values occur in the Southwest, while in Mexico the northern areas and the densely populated states in the central area show the smallest per capita groundwater resources. In Brazil, there is an evident difference between the water-rich and population-poor Amazon Basin and the rest of the country. In Argentina, the semi-arid southern states with low population densities have high per capita

Table 2.6 Long-term average continental renewable groundwater resources (total and per capita) and total renewable water resources as computed by the WaterGAP Global Hydrology Model for the historic baseline climate period 1961–1990 (Döll and Fiedler 2008).

Continent	Total renewable water resources A ($km^3 a^{-1}$)	Total renewable groundwater resources B ($km^3 a^{-1}$)	B/A (%)	Per capita renewable groundwater resources[e] ($m^3 capita^{-1} a^{-1}$)
Africa	4065	2072	51	2604
Asia[a,b]	13 168	3247	25	873
Australia and Oceania	1272	404	32	14 578
Europe[a]	3104	1191	38	1740
North/ Central America[c]	6493	1621	25	3336
South America	11 310	4131	37	11 949
Total land area[d]	39 414	12 666	32	2091

[a] Eurasia is subdivided into Europe and Asia along the Ural Mountains, with Turkey assigned to Asia.
[b] Including the whole island of New Guinea.
[c] Including Greenland.
[d] Excluding Antarctica.
[e] Population data based on CIESIN GPWv3 for the year 2000.
(*Sources:* Döll, P. and Fiedler, K. (2008) Global-scale modelling of groundwater recharge. *Hydrology and Earth System Sciences* **12**, 863–885.)

groundwater resources. In Russia, Mongolia, Australia and Canada, population density also dominates the spatial pattern. Of the large countries, India has the lowest per capita groundwater resources, with an average 273 $m^3 a^{-1}$, with most federal states below 250 $m^3 a^{-1}$. The average value for China is 490 $m^3 capita^{-1} a^{-1}$, but the densely populated northern states as well as the semi-arid north-west of the country show per capita groundwater resources below 250 $m^3 a^{-1}$.

Further information for the reader interested in the worldwide distribution of groundwater resources is hosted by the International Groundwater Resources Assessment Centre (IGRAC; see www.un-igrac.org) and is comprehensively described in the International Hydrological Programme-VI publication edited by Zektser and Everett (2004). A geographic synopsis is also provided by Margat and van der Gun (2013).

2.18 Hydrogeological environments of the United Kingdom

The occurrence of groundwater and the extent and distribution of aquifers and aquitards in a region are determined by the lithology, stratigraphy and structure of the geological strata present. The lithology refers to the general characteristics of the geological strata in terms of mineral composition and texture of the formations present. The stratigraphy describes the character of the rocks and their sequence in time, as well as the relationship between various deposits in different localities. Structural features, such as folds and faults, determine the geometric properties of the formations that are produced by deformation and fracturing after deposition or crystallization. In unconsolidated strata, the lithology and stratigraphy comprise the most important controls.

In any hydrogeological investigation, a clear understanding of the geology of an area is essential if the identification of aquifers and aquitards and the mechanisms of groundwater flow are to be properly understood. To illustrate the range of geological conditions in which groundwater resources can occur, the following description is based on the wide variation in groundwater occurrence in the United Kingdom.

The United Kingdom is fortunate in the variety of its rocks, structures and natural resources. As shown in Fig. 2.68, the principal groundwater resources are located in the Midlands and south-east of England, where the Cretaceous Chalk, Permo-Triassic sandstones

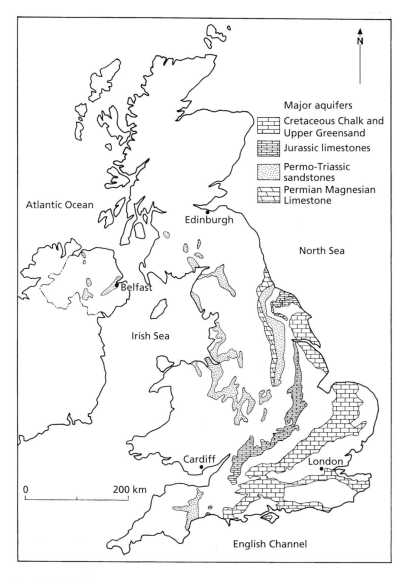

Fig. 2.68 Hydrogeological map of the United Kingdom showing the location of major aquifers. (*Source:* Modified from Colour-in geology map of the UK and Ireland, http://www.bgs.ac.uk/research/groundwater/ datainfo/levels/ngla.html)

and the Jurassic limestones contribute one-third of abstracted water supplies, of which half is reliant on the Chalk aquifer. Unlike many other countries, these important water supplies are dependent on fissure flow in making the limestones and sandstones permeable. Aquifers also exist in older rocks such as the Carboniferous limestones and in more recent formations such as the Pleistocene sands and gravels, but these aquifers are not of such regional significance. Although less important than surface water sources, the Precambrian and Palaeozoic rocks in the remoter areas of Ireland, Scotland and Wales have sufficient storage to be of local importance for domestic supplies and in supporting baseflows to minor rivers.

The hydrogeology of England, Wales, Scotland and Northern Ireland is presented in a number of maps produced by the Institute of Geological Sciences (1977) and the British Geological Survey (1986, 1988, 1989, 1994), with the hydrogeology of Scotland documented in reports by Robins (1990a, b) and Ó Dochartaigh *et al.* (2015) and for Wales by Robins and Davies (2015). In addition, Maps B3 Edinburgh and B4 London from the International Hydrogeological Map of Europe series cover the British Isles (UNESCO 1976, 1980). In the following sections, the hydrogeological environments of the British Isles are described in terms of the three major rock types (sedimentary, metamorphic and igneous) and their associated aquifer properties. To help in locating rock types, Fig. 2.69 shows the geological map of Britain and Ireland and a geological timescale is provided in Appendix 3.

2.18.1 Sedimentary rocks

The extensive sedimentary rocks range from unconsolidated Quaternary deposits to ancient, highly indurated Late Precambrian sandstones and siltstones. The most prolific aquifers are associated with the Mesozoic sandstones and limestones. The groundwater resources of the principal Mesozoic aquifers in England and Wales derived from the infiltration of rainfall amount to $7309\,km^3\,a^{-1}$ (Table 2.7) of which 63% is from the Cretaceous Chalk and 20% from the Permo-Triassic sandstones.

Recent coastal dune sands and raised beach deposits are restricted in distribution but provide limited supplies to individual farm and domestic users. Riverine alluvium occurs along many valley bottoms and includes alluvial fans, deltas, lake and estuarine deposits. Alluvial deposits include fine grained sands, silts and clays with the presence of lenses of gravels and cobbles. Together with glacial sands and gravels, these superficial deposits are of significance as locally important aquifers in the hard rock areas of Ireland, Scotland and Wales (Fig. 2.70). In Scotland, glacial sands and gravels form terraced and gently sloping hillocky ground with the groundwater potential dependent on the extent and thickness of the saturated material. Borehole yields of up to $10^{-2}\,m^3\,s^{-1}$ can typically be achieved in these deposits.

The fine, largely unconsolidated Pleistocene shelly marine sands and silts of eastern England form the regionally important Crag aquifer. The Crag can attain a thickness of 80 m representing a single water-bearing unit with overlying glacial sands and can yield supplies of $10^{-2}\,m^3\,s^{-1}$. Elsewhere in south-east England, Tertiary strata form a variable series of clays, marls and sands ranging in thickness from 30 to 300 m. The Eocene London Clay, up to 150 m thick, is an important confining unit in the London Basin. The underlying Lower London Tertiaries include clays, fine sands and pebble beds and where permeable sands rest on the underlying Chalk, these Basal Sands are generally in hydraulic continuity and can yield small supplies.

The Cretaceous Period resulted in the transgression of shallow, warm tropical seas and the deposition of the Chalk Formation, the most important source of groundwater in the south

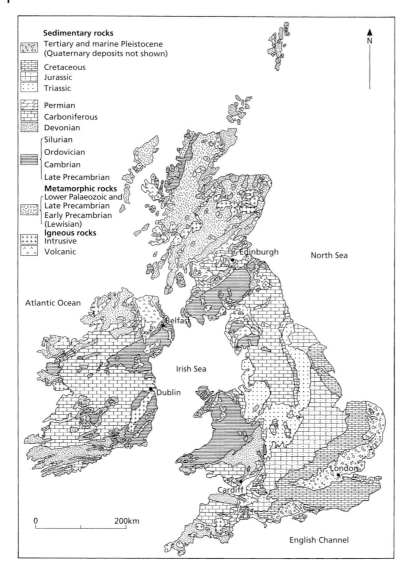

Fig. 2.69 Geological map of Britain and Ireland (Geological Museum 1978). (*Source:* Geological Museum (1978) *Britain Before Man*. HMSO for the Institute of Geological Sciences, London. © 1978, Natural History Museum.)

and east of England (Fig. 2.71) (Downing *et al.* 1993). The Chalk is a pure, white, microporous limestone made up of minute calcareous shells and shell fragments of plankton together with bands of harder nodular chalk and flints, marly in the lower part. In total, the Chalk is up to 500 m in thickness. The intrinsic permeability of the Chalk matrix is low, such that good yields of typically 10^{-1} m^3 s^{-1} depend on the

intersection of fissures and fractures, solutionally developed along bedding plains and joints. The permeability is best developed in the upper 80–100 m in the zone of greatest secondary permeability development. Below this level, the fissures are infrequent and closed by the overburden pressure and the groundwater becomes increasingly saline. In Northern Ireland, a hard microporous and fissured Chalk is found with

Table 2.7 Groundwater resources of the principal Mesozoic aquifers in England and Wales (in $km^3\ a^{-1}$). Abstraction data are for the year 1977 (Downing *et al.* 1993).

Aquifer	Infiltration (I)	Abstraction (A)	Balance	A/I
Cretaceous Chalk	4631	1255	3376	0.27
Cretaceous Lower Greensand	275	86	189	0.31
Jurassic Lincolnshire Limestone	86	43	43	0.50
Middle Jurassic limestones	627	65	562	0.10
Permo-Triassic sandstones	1443	587	856	0.41
Permian Magnesian Limestone	247	41	206	0.17
Total	7309	2077	5232	0.28

(*Source:* Downing, R.A., Price, M. and Jones, G.P. (1993) The making of an aquifer. In: *The Hydrogeology of the Chalk of North-West Europe* (eds R.A. Downing, M. Price and G.P. Jones). Clarendon Press, Oxford, pp. 1–13.)

Fig. 2.70 Site of a groundwater source developed in alluvial deposits in the Rheidol Valley, west Wales. In general, the well-sorted fluvial and glaciofluvial sands and gravels reach a thickness of 30 m and are exploited for locally important groundwater supplies. At this site, Lovesgrove, transmissivities are 0.05–0.07 $m^2\ s^{-1}$ (4000–6000 $m^2\ day^{-1}$), specific yield about 5% and sustainable yield 0.05 $m^3\ s^{-1}$. The number 1 borehole is positioned below the top of the concrete chamber shown at left and reaches a depth of 30 m through a sequence of river gravels (Hiscock and Paci 2000). (*Source:* Hiscock, K.M. and Paci, A. (2000) Groundwater resources in the Quaternary deposits and Lower Palaeozoic bedrock of the Rheidol catchment, West Wales. In: *Groundwater in the Celtic Regions: Studies in Hard Rock and Quaternary Hydrogeology* (eds N.S. Robins and B.D.R. Misstear). Geological Society, London, Special Publications **182**, pp. 141–155. DOI: 10.1144/GSL.SP.2000.182.01.14.)

Fig. 2.71 Cretaceous Chalk outcrop at Hunstanton in north-west Norfolk. Unusually at this location, the hard, well-fissured lower Chalk passes into the highly fossiliferous, iron-rich, red-stained Chalk known as the Red Chalk (now Hunstanton Formation). Below is the Lower Cretaceous Carstone Formation of medium to coarse, pebbly, glauconitic, quartz sand stained brown by a limonitic cement.

recrystallized calcite partly infilling pore spaces. The Chalk attains a maximum thickness of only 150 m and is largely covered by Tertiary basalt lavas. Recharge via the lavas supports numerous springs at the base of the outcrop along the Antrim coast. Borehole yields from the Chalk beneath the lavas are typically less than 10^{-3} m^3 s^{-1} and the number of boreholes is few.

Beneath the Chalk, the Lower Cretaceous glauconitic and ferruginous sands and sandstones of the Upper and Lower Greensand Formations and the alternating sequence of sandstones and clays of the Hastings Beds occur and form locally important aquifers in southern England.

The Jurassic Period also resulted in the formation of important limestone aquifer units, namely the Corallian and Lincolnshire Limestone aquifers. The Corallian is well-developed in Yorkshire where well-jointed oolitic limestones and grits are found up to 110 m thick. Southwards the formation thins to 20 m and is replaced by clay before reappearing in the Cotswolds, up to 40 m thick, giving yields of 10^{-3}–10^{-2} m^3 s^{-1}. The Great Oolite and Inferior Oolite limestones of Central England and the Cotswolds are part of a variable group of limestones, clays and sands, up to 60 m thick that can yield copious supplies. The Inferior Oolite Lincolnshire Limestone aquifer is partly karstic in nature with rapid groundwater flow through conduits (Figs 2.72 and 2.73). The highly developed secondary permeability supports yields of 10^{-1} m^3 s^{-1}. Beneath the limestones, the Upper Lias sands form locally important aquifers such as the Midford Sands close to Bath in the west of England.

The Permo-Triassic sandstones, marls and conglomerates comprise an extensive sequence up to

Fig. 2.72 The Lincolnshire Limestone at Clipsham Quarry, Lincolnshire. In this exposure, which is approximately 12 m high, solutional weathering of bedding surfaces and vertical joints imparts a high secondary permeability. The pale centre in the block at the top right-hand side is an example of the heart stone of unoxidized limestone. Above is the grey-green clay of the Rutland Formation that acts as an aquitard above the limestone aquifer.

Fig. 2.73 An exposed joint surface on a block of Lincolnshire Limestone showing a scallop-like feature caused by groundwater flow. The dark area to the right is a large vertical conduit formed by solutional weathering of a joint normal to the exposed face.

Fig. 2.74 Exposure of Triassic Sherwood Sandstone (Otter Sandstone Formation) at Ladram Bay, South Devon. The Otter Sandstone Formation comprises predominantly fine- to medium-grained red-brown, micaceous, variably cemented, ferruginous sands and sandstones, with occasional thin silt and conglomerate lenses. Frequent, well-developed bedding planes show a gentle 2–4° east-south-east dip, with, as shown in this cliff face, cross-bedding also typically present. Cementation along bedding planes exerts a small but potentially significant control in promoting horizontal groundwater movement (Walton 1981). Two prominent fissure openings along cemented bedding planes are visible to the left, in the lower one-third and upper two-thirds, of the face shown.

600 m thick and form deep sedimentary basins in the Midlands and North West of England and a smaller basin typically over 100 m thick in South West England (Fig. 2.74). The red sandstones originated in a desert environment and much of the fine to medium grained cross-bedded sandstones are soft, compact rock that is only weakly cemented. Groundwater can flow through the intergranular matrix but the presence of fractures enhances the permeability considerably giving good yields of up to 10^{-1} m^3 s^{-1} of good quality. Above the Triassic sandstones, the Mercia Mudstone confines the underlying aquifer, although local supplies are possible from minor sandstone intercalations. Up to 300 m of Triassic sandstone occur in the Lagan Valley and around Newtownards in Northern Ireland, where it is intruded by many basalt dykes and sills with yields of up to 10^{-2} m^3 s^{-1} obtained from a fine to medium grained sandstone. Smaller isolated Permian sands and sandstones are found in north-west England and south-west Scotland. At the base of the Permian, and overlain by red marls, the Magnesian Limestone forms a sequence of massive dolomitic and reef limestones that are important for water supply in the north-east of England where typical yields range up to 10^{-2} m^3 s^{-1}.

The Carboniferous strata include the massive, well-fissured karstic limestones that give large supplies of up to 10^{-1} m^3 s^{-1} from springs in the Mendip Hills (see Section 2.6.2), South Wales and, to a lesser extent, Northern Ireland. Later, rhythmic sequences of massive grits, sandstone, limestone, shale and coals produce minor supplies from fissured horizons in sandstone and limestone and form aquifers of local significance, particularly from the Yoredale Series in northern England and in the Midland Valley of Scotland.

The Old Red Sandstone is the principal aquifer unit of the Devonian Period and includes sandstones, marls and conglomerates that yield small supplies from sandstones in the Welsh borders. In Scotland, the Upper Old Red Sandstone is of much greater significance and around Fife and the southern flank of the Moray Firth consists of fine to medium grained sandstones, subordinate mudstones and conglomerates with good intergranular permeability yielding supplies of up to 10^{-2} m^3 s^{-1}. Widespread outcrops of fine to medium grained Lower and Middle Old Red Sandstones, in places flaggy, with siltstones, mudstones and conglomerates as well as interbedded lavas give borehole yields ranging from 10^{-3} m^3 s^{-1} in the Borders to 10^{-2} m^3 s^{-1} in Ayrshire and parts of Strathmore. In Northern Ireland, the principal Devonian lithology is conglomerate with some sandstone, subordinate mudstone and volcanic rocks but these

Fig. 2.75 Silurian shales at outcrop in the Rheidol Valley, west Wales showing the grey mudstones. Secondary permeability is developed in the weathered upper horizon from which shallow well supplies can be obtained as well as from occasional fracture openings (as seen at lower left) that intersect drilled boreholes.

are indurated and poorly jointed with small borehole yields of 10^{-3} m^3 s^{-1} where secondary permeability is present.

Silurian, Ordovician and other Lower Palaeozoic and Late Precambrian sedimentary facies predominate in southern Scotland, the north-east and north-west of Ireland and Wales and consist of great thicknesses of highly indurated and tectonically deformed shales, mudstones, slates and some limestones and sandstones (Fig. 2.75). Some groundwater may occur in shallow cracks and joints that produce a sub-surface permeable zone in which perched water tables may occur that support occasional springs and shallow boreholes providing small yields.

2.18.2   Metamorphic rocks

The Lower Palaeozoic and Precambrian crystalline basement rocks of the Highlands and Islands of Scotland and north-west Ireland offer little potential for groundwater storage and flow other than in cracks and joints that may be associated with tectonic features or near-surface weathering. Available groundwater can support a yield up to 10^{-3} m^3 s^{-1}.

2.18.3   Igneous rocks

Groundwater flow in well-indurated igneous intrusive and extrusive rocks occurs in shallow cracks and joints opened by weathering. Yields of occasional springs are small except where

tectonic influences have enhanced the secondary porosity (isolated yields of up to 10^{-3} m^3 s^{-1} are typical). Tertiary basalts up to 800 m thick occur in the Antrim and Lough Neagh areas of Northern Ireland. The basalts have some primary permeability in weathered zones, but the principal, secondary permeability is developed in joints and fissures which provide sustainable borehole yields of only 10^{-4} to 10^{-3} m^3 s^{-1}.

Further reading

Albu, M., Banks, D. and Nash, H. (1997) *Mineral and Thermal Groundwater Resources.* Chapman & Hall, London.

Cripps, J.C., Bell, F.G. and Culshaw, M.G. (eds) (1986) Groundwater in Engineering Geology: Proceedings of the 21st Annual Conference of the Engineering Group of the Geological Society. The Geological Society of London.

Domenico, P.A. and Schwartz, F.W. (1998) *Physical and Chemical Hydrogeology* (2nd edn). John Wiley & Sons, Inc., New York.

Downing, R.A., Price, M. and Jones, G.P. (1993) *The Hydrogeology of the Chalk of North-West Europe.* Clarendon Press, Oxford.

Jacob, C.E. (1950) Flow of groundwater. In: *Engineering Hydraulics* (ed. H. Rouse). John Wiley & Sons, Inc., New York.

Kresic, N. and Stevanovic, Z. (ed.) (2010) *Groundwater Hydrology of Springs. Engineering,*

Theory, Management and Sustainability.
Butterworth-Heinemann, Burlington, MA.

Margat, J. and van der Gun, J. (2013) *Groundwater Around the World: A Geographic Synopsis.* CRC Press, Boca Raton.

Robins, N.S. and Misstear, B.D.R. (2000) *Groundwater in the Celtic Regions: Studies in Hard Rock and Quaternary Hydrogeology.* Geological Society, London, Special Publications **182**.

Stevanovic, Z. (ed.) (2015) *Karst Aquifers – Characterization and Engineering.* Professional Practice in Earth Sciences, Springer International Publishing, Switzerland.

Tóth, J. (2009) *Gravitational Systems of Groundwater Flow: Theory, Evaluation, Utilization.* Cambridge University Press, Cambridge.

References

Aggarwal, P.K., Matsumoto, T., Sturchio, N.C. *et al.* (2015) Continental degassing of ^{4}He by surficial discharge of deep groundwater. *Nature Geoscience* **8**, 35–39, doi: 10.1038/NGEO2302.

Amery, H.A. and Wolf, A.T. (2004) *Water in the Middle East: A Geography of Peace.* University of Texas Press, Austin.

Andrews, J.N., Burgess, W.G., Edmunds, W.M. *et al.* (1982) The thermal springs of Bath. *Nature* **298**, 339–343.

Araújo, L.M., Franca, A.B. and Potter, P.E. (1999) Hydrogeology of the Mercosul aquifer system in the Paraná and Chaco-Paraná Basins, South America, and comparison with the Navajo-Nugget aquifer system, USA. *Hydrogeology Journal* **7**, 317–336.

Atkinson, T.C. (1977) Diffuse flow and conduit flow in limestone terrain in the Mendip Hills, Somerset (Great Britain). *Journal of Hydrology* **35**, 93–110.

Atkinson, T.C. and Davison, R.M. (2002) Is the water still hot? Sustainability and the thermal springs at Bath, England. In: *Sustainable Groundwater Development* (eds K.M. Hiscock, M.O. Rivett and R.M. Davison). Geological

Society, London, Special Publications **193**, pp. 15–40.

Back, W., Rosenshein, J.S. and Seaber, P.R. (eds) (1988) *Hydrogeology. The Geology of North America*, Vol. **O-2**. The Geological Society of North America, Boulder, Colorado.

Bauer-Gottwein, P., Gondwe, B.R.N., Charvet, G. *et al.* (2011) Review: The Yucatán Peninsula karst aquifer, Mexico. *Hydrogeology Journal* **19**, 507–524.

Bell, F.G., Cripps, J.C. and Culshaw, M.G. (1986) A review of the engineering behaviour of soils and rocks with respect to groundwater. In: *Groundwater in Engineering Geology* (eds J.C. Cripps, F.G. Bell and M.G. Culshaw). Geological Society, London, Engineering Geology Special Publication **3**, pp. 1–23.

Bentley, H.W., Phillips, F.M., Davis, S.N. *et al.* (1986) Chlorine 36 dating of very old groundwater: the Great Artesian Basin, Australia. *Water Resources Research* **22**, 1991–2001.

Bergkamp, G. and Cross, K. (2006) *Groundwater and Ecosystem Services: Towards Their Sustainable Use.* Proceedings on the International Symposium on Groundwater Sustainability (ISGWAS), Alicante, Spain, 177–193.

Bord, J. and Bord, C. (1985) *Sacred Waters: Holy Wells and Water Lore in Britain and Ireland.* Granada Publishing Ltd, London.

British Geological Survey (1978) *1 : 100 000 Hydrogeological Map of the South West Chilterns and the Berkshire and Marlborough Downs.* Natural Environment Research Council.

British Geological Survey (1986) *1 : 125,000 Hydrogeological Map of South Wales.* British Geological Survey, Keyworth, Nottingham.

British Geological Survey (1988) *1 : 625 000 Hydrogeological Map of Scotland.* Natural Environment Research Council.

British Geological Survey (1989) *1 : 100,000 Hydrogeological Map of Clwyd and the Cheshire Basin.* British Geological Survey, Keyworth, Nottingham.

British Geological Survey (1994) *1 : 250 000 Hydrogeological Map of Northern Ireland.* Natural Environment Research Council.

British Geological Survey (2020) *Sinkholes (or dolines).* https://www.bgs.ac.uk/caves/sinkholes/home.html (accessed 18 May, 2020).

Bryan, K. (1919) Classification of springs. *Journal of Geology* **27**, 552–561.

Burgess, W.G., Edmunds, W.M., Andrews, J.N. et al. (1980) *Investigation of the Geothermal Potential of the UK. The Hydrogeology and Hydrochemistry of the Thermal Water in the Bath-Bristol Basin.* Institute of Geological Sciences, London.

Bürgmann, R., Rosen, P.A. and Fielding, E.J. (2000). Synthetic aperture radar interferometry to measure Earth's surface topography and its deformation. *Annual Review of Earth and Planetary Sciences* **28**, 169–209.

Burnett, W.C., Aggarwal, P.K., Aureli, A. et al. (2006) Quantifying submarine groundwater discharge in the coastal zone via multiple methods. *Science of the Total Environment* **367**, 498–543.

Burnett, W.C., Kim, G. and Lane-Smith, D. (2001) A continuous radon monitor for assessment of radon in coastal ocean waters. *Journal of Radioanalytical and Nuclear Chemistry* **249**, 167–172.

Campos, H. (2000) *Mapa Hidrogeológico del Acuífero Guaraní.* Proceedings of the First Joint World Congress on Groundwater, Fortaleza, Brasil.

Cave, R.R. and Henry, T. (2011) Intertidal and submarine groundwater discharge on the west coast of Ireland. *Estuarine, Coastal and Shelf Science* **92**, 415–423.

Cedergren, H.R. (1967) *Seepage, Drainage and Flow Nets.* John Wiley and Sons, New York.

Christiansen, E.A., Acton, D.F., Long, R.J. et al. (1981) *Fort Qu'Appelle Geolog. The Valleys – Past and Present.* Interpretive Report No. 2. The Saskatchewan Research Council, Canada.

Cooper, A.H. (1998) Subsidence hazards caused by the dissolution of Permian gypsum in England: Geology, investigation and remediation. In: *Geohazards in Engineering Geology* (eds J.G. Maund and M. Eddleston). The Geological Society, London, Engineering Geology Special Publication **15**, pp. 265–275.

Cooper, A.H., Odling, N.E, Murphy, P.J. et al. (2013) The role of sulfate-rich springs and groundwater in the formation of sinkholes over gypsum in Eastern England. In: *Sinkholes and the Engineering and Environmental Impacts of Karst: Proceedings of the Thirteenth Multidisciplinary Conference, May 6–10* (eds L. Land, D.H. Doctor and J.B. Stephenson). National Cave and Karst Research Institute, Carlsbad, New Mexico, pp. 141–150.

Daly, D., Dassargues, A., Drew, D. et al. (2002) Main concepts of the "European approach" to karst-groundwater-vulnerability assessment and mapping. *Hydrogeology Journal* **10**, 340–345.

Davis, S.N. and De Wiest, R.J.M. (1966) *Hydrogeology.* John Wiley & Sons, Inc., New York.

Downing, R.A. (1993) Groundwater resources, their development and management in the UK: an historical perspective. *Quarterly Journal of Engineering Geology* **26**, 335–358.

Downing, R.A., Price, M. and Jones, G.P. (1993) The making of an aquifer. In: *The Hydrogeology of the Chalk of North-West Europe* (eds R.A. Downing, M. Price and G.P. Jones). Clarendon Press, Oxford, pp. 1–13.

Döll, P. and Fiedler, K. (2008) Global-scale modelling of groundwater recharge. *Hydrology and Earth System Sciences* **12**, 863–885.

Döll, P., Kaspar, F. and Lehner, B. (2003) A global hydrological model for deriving water availability indicators: model tuning and validation. *Journal of Hydrology* **270**, 105–134.

Drew, D. (1990) The hydrology of the Burren, County Clare. *Irish Geography* **23**, 69–89.

Dublyansky, Y.V. (1995) Speleogenetic history of the Hungarian hydrothermal karst. *Environmental Geology* **25**, 24–35.

Edmunds, W.M., Darling, W.G., Purtschert, R. and Corcho Alvarado, J.A. (2014) Noble gas, CFC and other geochemical evidence for the age and origin of the Bath thermal waters, UK. *Applied Geochemistry* **40**, 155–163.

Erhardt, I., Ötvös, V., Erőss, A. *et al.* (2017) Hydraulic evaluation of the hypogenic karst area in Budapest (Hungary). *Hydrogeology Journal* **25**, 1871–1891.

Famiglietti, J.S., Lo, M., Ho, S.L. *et al.* (2011) Satellites measure recent rates of groundwater depletion in California's Central Valley. *Geophysical Research Letters* **38**, L03403, doi: 10.1029/2010GL046442.

Faunt, C.C. (ed.) (2009) Groundwater availability of the Central Valley Aquifer, California. *United States Geological Survey Professional Paper* **1766**.

Fetter, C.W. (2001) *Applied Hydrogeology* (4th edn). Pearson Higher Education, Upper Saddle River, New Jersey.

Fili, M., Da Rosa Filho, E.F., Auge, M. *et al.* (1998) El acuífero Guaraní. Un recurso compartido por Argentina, Brasil, Paraguay y Uruguay (América del Sur). Hidrología Subterránea. *Boletin Geologico y Minero* **109**, 389–394.

Ford, D.C. and Williams, P.W. (1989) *Karst Geomorphology and Hydrology*. Unwin Hyman Ltd., London.

Ford, D. and Williams, P. (2007) *Karst Geomorphology and Hydrology*. John Wiley and Sons, Ltd., Chichester.

Foster, S., Hirata, R., Vidal, A. *et al.* (2009) *The Guarani Aquifer Initiative – Towards Realistic Groundwater Management in a Transboundary Context*. GW Mate Case Profile Collection No.9. World Bank, Washington DC, USA.

Freeze, R.A. and Cherry, J.A. (1979) *Groundwater*. Prentice-Hall, Inc., Englewood Cliffs, New Jersey.

Freeze, R.A. and Witherspoon, P.A. (1967) Theoretical analysis of regional groundwater flow. 2. Effect of water-table configuration and subsurface permeability variation. *Water Resources Research* **3**, 623–634.

Gale, J.E. (1982) Assessing permeability characteristics of fractured rock. *Geological Society of America, Special Paper* **189**, 163–181, Boulder CO.

Gallois, R.W. (2006) The geology of the hot springs at Bath Spa, Somerset. *Geoscience in South-West England* **11**, 168–173.

Galloway, D.L. and Burbey, T.J. (2011) Review: Regional land subsidence accompanying groundwater extraction. *Hydrogeology Journal* **19**, 1459–1486.

Geological Museum (1978) *Britain Before Man*. HMSO for the Institute of Geological Sciences, London.

Gleeson, T. and Manning, A.H. (2008) Regional groundwater flow in mountainous terrain: three-dimensional simulations of topographic and hydrologic controls. *Water Resources Research* **44**, W10403, doi: 10.1029/2008WR006848.

Gleeson, T., Marklund, L., Smith, L. and Manning, A.H. (2011) Classifying the water table at regional to continental scales. *Geophysical Research Letters* **38**, L05401, doi: 10.1029/2010GL046427.

Goldscheider, N., Mádl-Szőnyi, J., Erőss, A. and Schill, E. (2010) Review: Thermal water resources in carbonate rock aquifers. *Hydrogeology Journal* **18**, 1303–1318.

Guo, X., Jiang, G., Gong, X. *et al.* (2015) Recharge processes on typical karst slopes implied by isotopic and hydrochemical indexes in Xiaoyan Cave, Guilin, China. *Journal of Hydrology* **530**, 612–622.

Gustafson, C., Key, K. and Evans, R.L. (2019) Aquifer systems extending far offshore on the U.S. Atlantic margin. *Scientific Reports* **9**, 8709, doi: 10.1038/s41598-019-44611-7.

Gutiérrez, F., Parise, M., De Waele, J. and Jourde, H. (2014) A review on natural and human induced geohazards and impacts in karst. *Earth-Science Reviews* **138**, 61–88.

Habermehl, M.A. (1980) The Great Artesian Basin, Australia. *BMR Journal of Australian Geology and Geophysics* **5**, 9–38.

Habermehl, M.A. (2020) Review: The evolving understanding of the Great Artesian Basin (Australia), from discovery to current hydrogeological interpretations. *Hydrogeology Journal* **28**, 13–36.

Habermehl, M.A. and Lau, J.E. (1997) *1 : 2 500 000 Map of the Hydrogeology of the Great Artesian Basin, Australia*. Australian Geological Survey Organization, Canberra, A.C.T.

Haghshenas Haghighi, M. and Motagh, M. (2019) Ground surface response to continuous compaction of aquifer system in Tehran, Iran: results from a long-term multi-sector InSAR analysis. *Remote Sensing of Environment* **221**, 534–550.

Haitjema, H.M. and Mitchell-Bruker, S. (2005) Are water tables a subdued replica of the topography? *Ground Water* **43**, 781–786.

Harris, C.R. (1981) Oases in the desert: the mound springs of northern South Australia. *Proceedings of the Royal Geographical Society of Australasia, South Australian Branch* **81**, 26–39.

Haslam, S., Sinker, C. and Wolseley, P. (1975) British water plants. *Field Studies* **4**, 243–351.

Hirata, R. and Foster, S. (2020) The Guarani Aquifer System – from regional reserves to local use. *Quarterly Journal of Engineering Geology and Hydrogeology*, doi: 10.1144/qjegh2020-091.

Hiscock, K.M. and Paci, A. (2000) Groundwater resources in the Quaternary deposits and Lower Palaeozoic bedrock of the Rheidol catchment, West Wales. In: *Groundwater in the Celtic Regions: Studies in Hard Rock and Quaternary Hydrogeology* (eds N.S. Robins and B.D.R. Misstear). Geological Society, London, Special Publications **182**, pp. 141–155.

Huggett, R.J. (2011) *Fundamentals of Geomorphology* (3rd edn). Routledge, Abingdon, Oxon.

Institute of Geological Sciences (1977) *1 : 625 000 Hydrogeological Map of England and Wales*. Natural Environment Research Council.

Jacob, C.E. (1940) On the flow of water in an elastic artesian aquifer. *Transactions of the American Geophysical Union* **22**, 574–586.

Kandel, R. (2003) *Water from Heaven: The story from the Big Bang to the Rise of Civilisation, and Beyond*. Columbia University Press, New York.

Karst Working Group (2000) *The Karst of Ireland*. Karst Working Group, Geological Survey of Ireland, Dublin.

Khan, S. and Rushton, K.R. (1996) Reappraisal of flow to tile drains I. Steady state response. *Journal of Hydrology* **183**, 351–366.

Kirkaldy, J.F. (1954) *General Principles of Geology*. Hutchinson's Scientific and Technical Publications, London, pp. 284–286.

Knill, J.L. and Jones, K.S. (1968) Ground-water conditions in Greater Tehran. *Quarterly Journal of Engineering Geology* **1**, 181–194.

Kogovsek, J and Petric, M. (2014) Solute transport processes in a karst vadose zone characterized by long-term tracer tests (the cave system of Postojnska Jama, Slovenia). *Journal of Hydrology* **519**, 1205–1213.

Kohout, F.A. and Kolipinski, M.C. (1967) Biological zonation related to groundwater discharge along the shore of Biscayne Bay, Miami, Florida. *Estuaries* **83**, 488–499.

Krumbein, W.C. and Monk, G.D. (1943) Permeability as a function of the size parameters of unconsolidated sand. *Transactions of the American Institution of Mining and Metallurgy Engineers* **151**, 153–163.

Langevin, C.D. (2003) Simulation of submarine ground water discharge to a marine estuary: Biscayne Bay, Florida. *Ground Water* **41**, 758–771.

Love, A.J., Herczeg, A.L., Sampson, L. *et al.* (2000) Sources of chloride and implications for ^{36}Cl dating of old groundwater, southwestern Great Artesian Basin, Australia. *Water Resources Research* **36**, 1561–1574.

Lucas, H.C. and Robinson, V.K. (1995) Modelling of rising groundwater levels in the Chalk aquifer of the London Basin. *Quarterly Journal of Engineering Geology* **28**, S51–S62.

Mahmoudpour, M., Khamehchiyan, M., Nikudel, M.R. and Ghassemi, M.R. (2016) Numerical simulation and prediction of regional land subsidence caused by groundwater exploitation in the southwest plain of Tehran, Iran. *Engineering Geology* **201**, 6–28.

Margat, J. and van der Gun, J. (2013) *Groundwater Around the World: A Geographic Synopsis*. CRC Press, Boca Raton.

Marsh, T.J. and Davies, P.A. (1983) The decline and partial recovery of groundwater levels below London. *Proceedings of the Institution of Civil Engineers, Part 1* **74**, 263–276.

Marsily, G. (1986) *Quantitative Hydrogeology: Groundwater Hydrology for Engineers.* Academic Press, San Diego, California.

Meinzer, O.E. (1923) The occurrence of groundwater in the United States with a discussion of principles. *United States Geological Survey Water Supply Paper* **489**.

Meinzer, O.E. and Hard, H.H. (1925) The artesian water supply of the Dakota sandstone in North Dakota, with special reference to the Edgeley Quadrangle. *United States Geological Survey Water Supply Paper* **520-E**, 73–95.

Michael, H.A., Scott, K.C., Koneshloo *et al.* (2016) Geologic influence on groundwater salinity drives large seawater circulation through the continental shelf. *Geophysical Research Letters* **43**, 10,782–10,791.

Moore, W.S. (1999) The subterranean estuary: a reaction zone of ground water and sea water. *Marine Chemistry* **65**, 111–125.

Moore, W.S. (2010) The effect of submarine groundwater discharge on the ocean. *Annual Review of Marine Science* **2**, 59–88.

Moore, W.S., Sarmiento, J.L. and Key, R.M. (2008) Submarine groundwater discharge revealed by [228]Ra distribution in the upper Atlantic Ocean. *Nature Geoscience* **1**, 309–311.

Ó Dochartaigh, B.É., MacDonald, A.M., Fitzsimons, V. and Ward, R. (2015) *Scotland's aquifers and groundwater bodies.* British Geological Survey Open Report, OR/15/028, 76pp.

OPW (1997) *An investigation of the flooding problems in the Gort Ardrahan Area of South Galway.* Final Report, vols 1–12. Office of Public Works, Ireland.

Panagopoulos, G. and Lambrakis, N. (2006) The contribution of time series analysis to the study of the hydrodynamic characteristics of the karst systems: Application on two typical karst aquifers of Greece (Trifilia, Almyros Crete). *Journal of Hydrology* **329**, 368–376.

Parise, M., Ravbar, N., Živanovic, V. *et al.* (2015) Hazards in karst and managing water resources quality. In: *Karst Aquifers – Characterization and Engineering* (ed. Z. Stevanovic).

Professional Practice in Earth Sciences, Springer International Publishing, Switzerland, pp. 601–687.

Peyraube, N., Lastennet, R. and Denis, A. (2012) Geochemical evolution of groundwater in the unsaturated zone of a karstic massif, using the P_{CO2}—SIc relationship. *Journal of Hydrology* **430–431**, 13–24.

Poland, J.F. (1972) Subsidence and its control. *American Association of Petroleum Geologists, Memoir* **18**, 50–71.

Post, V.E.A., Groen, J., Kooi, H. *et al.* (2013) Offshore fresh groundwater reserves as a global phenomenon. *Nature* **504**, 71–78.

Powell, O., Silcock, J. and Fensham, R. (2015) Oases to oblivion: the rapid demise of springs in the south-eastern Great Artesian Basin, Australia. *Groundwater* **53**, 171–178.

Prescott, J.R. and Habermehl, M.A. (2008) Luminescence dating of spring mound deposits in the southwestern Great Artesian Basin, northern South Australia. *Australian Journal of Earth Sciences* **55**, 167–181.

Puri, S., Appelgren, B., Arnold, G. *et al.* (2001) *Internationally Shared (Transboundary) Aquifer Resources Management: Their Significance and Sustainable Management.* UNESCO, Paris.

Rahn, P.H. (1996) *Engineering Geology: An Environmental Approach* (2nd edn). Prentice Hall PTR, Upper Saddle River, New Jersey.

Ransley, T.R., Radke, B.M., Feitz, A.J. *et al.* 2015. *Hydrogeological Atlas of the Great Artesian Basin.* Geoscience Australia, Canberra, 134 pp. http://dx.doi.org/10.11636/9781925124668.

Ravilious, K. (2018) Tehran's drastic sinking exposed. *Nature* **564**, 17–18.

Robins, N.S. (1990a) *Hydrogeology of Scotland.* HMSO for the British Geological Survey, London.

Robins, N.S. (1990b) Groundwater chemistry of the main aquifers in Scotland. *British Geological Survey Report* **18/2**.

Robins, N.S. and Davies, J. (2015) *Hydrogeology of Wales.* British Geological Survey, Keyworth, Nottingham.

Rushton, K.R. (2003) *Groundwater Hydrology: Conceptual and Computational Models.* John Wiley & Sons, Ltd., Chichester.

Rushton, K.R. and Tomlinson, L.M. (1999) Total catchment conditions in relation to the Lincolnshire Limestone in South Lincolnshire. *Quarterly Journal of Engineering Geology* **32**, 233–246.

Sharp, J.M. (1988) Alluvial aquifers along major rivers. In: *Hydrogeology. The Geology of North America*, Vol. **O-2** (eds W. Back, J.S. Rosenshein and P.R. Seaber). The Geological Society of North America, Boulder, Colorado, pp. 273–282.

Shepherd, R.G. (1989) Correlations of permeability and grain size. *Ground Water* **27**, 633–638.

Sherlock, R.L. (1962) *British Regional Geology: London and Thames Valley* (3rd edn Reprint with Minor Additions). HMSO, London.

Simpson, B., Blower, T., Craig, R.N. and Wilkinson, W.B. (1989) *The Engineering Implications of Rising Groundwater Levels in the Deep Aquifer Beneath London.* Construction Industry Research & Information Association, London, Special Publication **69**.

Smith, D.I., Atkinson, T.C. and Drew, D.P. (1976) The hydrology of limestone terrains. In: *The Science of Speleology* (eds T.D. Ford and C.H.D. Cullingford). Academic Press, London.

Snow, D.T. (1969) Anisotropic permeability of fractured media. *Water Resources Research* **5**, 1273–1289.

Sracek, O. and Hirata, R. (2002) Geochemical and stable isotopic evolution of the Guarani Aquifer System in the state of São Paulo, Brazil. *Hydrogeology Journal* **10**, 643–655.

Taniguchi, M., Burnett, W.C., Cable, J.E. and Turner, J.V. (2002) Investigation of submarine groundwater discharge. *Hydrological Processes* **16**, 2115–2129.

Tobin, B.W., Springer, A.E., Kreamer, D.K. and Schenk, E. (2018) Review: The distribution, flow, and quality of Grand Canyon Springs, Arizona (USA). *Hydrogeology Journal* **26**, 721–732.

Tóth, J. (1963) A theoretical analysis of groundwater flow in small drainage basins. *Journal of Geophysical Research* **68**, 4795–4812.

Tynan, S., Gill, M. and Johnston, P. (2007) *Water Framework Directive: Development of a methodology for the characterisation of a karst groundwater body with particular emphasis on the linkage with associated ecosystems such as turlough ecosystems (2002-W-DS-8-M1).* Final report to the Environmental Protection Agency, Ireland. Department of Civil, Structural and Environmental Engineering, Trinity College, Dublin.

UNESCO (1976) *1 : 1 500 000 International Hydrogeological Map of Europe. Map B4 London.* Bundesanstalt für Geowissenschaften und Rohstoffe and UNESCO, Hannover.

UNESCO (1980) *1: 1 500 000 International Hydrogeological Map of Europe. Map B3 Edinburgh.* Bundesanstalt für Geowissenschaften und Rohstoffe and UNESCO, Hannover.

UNESCO (2006) *Groundwater Resources of the World – Transboundary Aquifer Systems. Special Edition for the 4th World Water Forum, Mexico City*, March 2006. World-wide Hydrogeological Mapping and Assessment Programme (WHYMAP), BGR, Hannover and UNESCO, Paris.

Waltham, T., Bell, F. and Culshaw, M. (2005) *Sinkholes and Subsidence: Karst and Cavernous Rocks in Engineering and Construction.* Praxis Publishing Ltd, Chichester.

Walton, N.R.G. (1981) *A Detailed Hydrogeochemical Study of Groundwaters from the Triassic Sandstone Aquifer of South-West England.* Institute of Geological Sciences, Report No. 81/5. HMSO, London.

White, W.B. and White, E.L. (eds) (1989) *Karst Hydrology: Concepts from the Mammoth Cave Area.* Springer Science+Business Media, New York.

Williams, M. (2017) *Subterranean Norwich: The Grain of the City.* Lasse Press, Norwich.

Williamson, A.K., Prudic, D.E. and Swain, L.A. (1989) Ground-water flow in the Central

Valley, California. *United States Geological Survey Professional Paper* **1401-D**.

Wilson, G. and Grace, H. (1942) The settlement of London due to the under drainage of the London Clay. *Journal of the Institution of Civil Engineers* **12**, 100–127.

Witherspoon, P.A., Wang, J.S.Y., Iwal, K. and Gale, J.E. (1980) Validity of cubic law for fluid flow in a deformable rock fracture. *Water Resources Research* **16**, 1016–1024.

Xiao, H., Li, H. and Tang, Y. (2018) Assessing the effects of rainfall, groundwater downward leakage, and groundwater head differences on the development of cover-collapse and cover-suffosion sinkholes in central Florida (USA). *Science of the Total Environment* **644**, 274–286.

Yu, X. and Michael, H.A. (2019) Offshore pumping impacts onshore groundwater resources and land subsidence. *Geophysical Research Letters* **46**, 2553–2562.

Zektser, I.S. and Everett, L.G. (eds) (2004) *Groundwater Resources of the World and Their Use*. International Hydrological Programme-VI, Series on Groundwater No. 6. UNESCO, Paris.

3

Groundwater and geological processes

3.1 Introduction

This chapter discusses the response of groundwater systems to geological processes such as plate tectonics, glaciation and sea-level fluctuations. Although topography-driven groundwater flow, as introduced in Chapter 2, is in many areas of the world the dominant mode of groundwater circulation, consideration of the additional impact of geological processes on groundwater flow is often important to fully appreciate the dynamics of groundwater systems on relatively long timescales (millennia). For example, only by considering the hydrogeological impacts of glaciation and associated sea-level low stand, can the occurrence of fresh water underneath the sea-floor along much of the continental shelf be understood. In the description of regional groundwater flow systems, which typically reach circulation depths of several kilometres, flow driven by the variable density of groundwater and flow induced by stresses imposed on pore fluids play an important role and often form the sole driver of fluid flow in the absence of topographically driven flow. The variable density of groundwater and anomalous pore pressures can be caused by a suite of geological processes. Thus, to understand fluid flow patterns in relatively deep hydrogeological systems requires an understanding of how the relevant geological processes cause fluid flow. It is the task of the hydrogeologist to evaluate, for a given geological setting, which processes can be expected to be dominant in driving fluid flow in the subsurface.

The way groundwater and other geofluids, such as hydrocarbons, respond to geological forcing often has a direct feedback on how geological processes proceed. This feedback mechanism is, for example, important when the influence of fluids on seismicity should be considered. In this chapter, however, discussion will be limited to a more or less one-way description of the primary impact of geological processes on fluid movement in the Earth's crust, while a consideration of feedback mechanisms would be the topic of a more specialized discussion than covered here.

This chapter demonstrates the societal relevance of understanding the role of geological processes in driving groundwater flow. Topical issues such as the safe sequestration of CO_2 in the subsurface, or the design of nuclear waste repositories in deep aquifers, require understanding of the interplay between groundwater and geological processes, knowledge of which are of paramount importance.

3.2 Geological processes driving fluid flow

Groundwater recharge is a hydrogeological process leading to topography-driven groundwater flow. The rates of groundwater recharge will vary with fluctuating climatic conditions, but there are also several geological processes

Hydrogeology: Principles and Practice, Third Edition. Kevin M. Hiscock and Victor F. Bense.
© 2021 John Wiley & Sons Ltd. Published 2021 by John Wiley & Sons Ltd.
Companion website: www.wiley.com/go/hiscock/hydrogeology3e

that modify water table gradients on geological timescales, such as the build-up and retreat of ice-sheets, and mountain building and erosion. There are two further groups of mechanisms unrelated to topography-driven groundwater flow by which hydraulic head gradients can arise in groundwater flow systems, which are often relevant when flow on a regional scale and on geological timescales is considered. First, contrasts in fluid densities resulting from geological processes can lead to hydraulic gradients. Density contrasts strong enough to cause significant hydraulic head gradients will result from developing temperature gradients and/or contrasts in total dissolved solids content (salinity). Density-driven flow is important in coastal areas, in deep sedimentary basins where highly mineralized, and therefore often dense, fluids occur and in hydrothermal systems. Second, stresses imposed on pore fluids as a result of geological processes can lead to fluid movement. Examples of such processes discussed in this chapter are sediment compaction in subsiding sedimentary basins, or underneath developing ice-sheets, and mountain building and erosion.

3.3 Topography-driven flow in the context of geological processes

In topography-driven groundwater flow systems, hydraulic gradients in an aquifer required for groundwater flow (see Chapter 2) are caused by the topography of the water table near the surface. Through fluid movement, the potential energy as reflected by hydraulic gradients dissipates and water table gradients will become smaller. Therefore, to sustain fluid flow, energy needs to be supplied to the system which, in topography-driven flow systems, occurs through groundwater recharge. Hence, the flow of groundwater from recharge areas to groundwater discharge areas will only continue when groundwater recharge is sufficient to maintain the driving water table gradient. If not, the water

table gradient will be reduced over time until groundwater movement halts. This is exemplified by arid areas which experience little groundwater recharge and have deep water tables. Thus, the rate of topography-driven flow not only will vary with climatic conditions affecting recharge but also with possible changes in topography over geological time. Topography can change with erosion and mountain building and the development of ice-sheets. Such changes in driving topographic gradients are usually accompanied by other geological processes causing density gradients and/ or changes in pore water pressure, which provide additional mechanisms to generate fluid flow. Before discussing examples of groundwater systems affected by changing configurations of topography, groundwater flow driven by pressure anomalies and density-driven flow are first introduced in some detail.

3.4 Compaction-driven fluid flow

Sedimentary basins (e.g. the North Sea Basin and Gulf of Mexico Basin) contain a sediment pile with a thickness of the order of kilometres. During development of such basins, sediments will become increasingly deeply buried underneath younger sediments. Whilst this happens, mechanical compaction takes place which will result in a loss of porosity. When pore volume is reduced, fluids residing in the pores will be expelled because water has a negligible compressibility to accommodate strain. This process will be accomplished by a rise in pore fluid pressures. Thus, the process of compaction drives fluid flow with the energy provided by the increasing weight of the overburden. Investigation of measurements of porosity versus depth for different types of sediments illustrates the process of compaction in more detail. Two observations are apparent from an example of such data depicted in Fig. 3.1. First, porosity is most rapidly lost in the upper few hundred metres of the subsurface. Second,

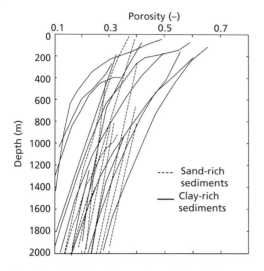

Fig. 3.1 Variation of porosity with depth in predominantly sand-rich sediments and clay-rich sediments (Bense and Person 2006). (*Source:* Based on Bense, V.F. and Person, M.A. (2006) Fault zones as conduit-barrier systems to fluid flow in siliciclastic sedimentary aquifers. *Water Resources Research* **42**, W05421, doi: 10.1029/2005WR004480.)

clay-rich sediments have a larger porosity (water-content) than sand-rich sediments at the surface. However, for clay-rich sediments, porosity is lost more rapidly with increasing depth than for sand-rich sediments. The difference in compressibility of clay and sands can be understood from the ability of platy clay-minerals to hold water molecules in between them.

As outlined by Terzaghi in his classic work from the 1920s (Terzaghi 1925), mechanical compaction of a pore network requires an increase of the effective stress (σ_e) which is equal to (see Section 2.9.1):

$$\sigma_e = \sigma_t - p \qquad (eq.3.1)$$

To accomplish a rise in effective stress the total weight of the overburden has to increase (σ_t) or the fluid pressure (p) needs to decrease. The latter will be accomplished by the drainage of pore fluids, whilst the former occurs during increasing sediment burial. When effective stress rises, a larger proportion of the weight of the overburden (total stress) has to be carried

by the sediment framework. If the sediment is mechanically compressible, an increase in effective stress will result in compaction and a reduction in porosity. The ease by which sediments compress as a result of a mechanical load (compressibility) is larger for relatively porous sediments such as clays and shales, while lower porosity sands are less easily compressible (Table 2.2).

Porosity (n) variations with depth, as shown in Fig. 3.1, are often described (e.g. Bethke and Corbet 1988) using a simple relationship taking into account effective stress, compressibility (β, Pa^{-1}) and assuming a porosity at the surface (n_o), of the form:

$$n = n_o \cdot e^{-\sigma_e \beta} \qquad (eq.3.2)$$

In eq. 3.2, porosity, n, exponentially declines with increasing effective stress. If a linear correlation between effective stress and depth is assumed, eq. 3.2 can be used to calculate porosity-depth curves as shown as the dashed lines in Fig. 3.1, for various values of compressibility and n_o (see Box 3.1).

When the vertical connectivity of the pore network is efficient, rises in pore pressure as a result of porosity loss will quickly even out and *hydrostatic conditions* will be met. Under hydrostatic conditions, the pore pressure in the subsurface increases linearly with depth representing only the weight of the overlying fluids hydraulically connected through the pore network. The hydrostatic pressure (p_h) with increasing depth (z) is thus simply given by:

$$p_h = z\rho_f g \qquad (eq.3.3)$$

in which g is the acceleration due to gravity and ρ_f is the water density. For purely fresh water ($\rho_f = 1000\ kg\ m^{-3}$), the hydrostatic gradient is about $10^4\ Pa\ m^{-1}$. One kilometre of water thus exerts a pressure of about 10 MPa. Likewise, the total vertical stress, σ_T, often called the lithostatic pressure, is equal to the pressure exerted by the overburden of sediment only, which is equal to:

$$\sigma_T = z\rho_s g \qquad (eq.3.4)$$

Box 3.1 Determining porosity and sediment compressibility from drill core and well data

To illustrate the basic concepts of porosity and how porosity loss can be related to sediment compressibility, consider a dry, water-free drill-core which comes from a depth of 324 m as retrieved from a borehole penetrating a sandstone. The circular drill-core has a length of 10 cm and a diameter of 50 mm. This core is submerged in a tank containing water which has dimensions ($l \times w \times h$) of $0.1 \times 0.1 \times 0.2$ m. The water level in this tank initially rises by an unspecified amount. Four hours after the core was submerged, the water level has dropped by 3.1 mm and the water level does not drop any further.

It can now be assumed that initially, the rock only contained air in its pore space which took time to be filled with water. Therefore, the volume of water filling the pores can be calculated as follows: 0.0031 m (the depth of water used to fill the pores in metres) $\times 0.1 \times 0.1$ (the surface area over which the drop in water level occurred in the tank) $= 3.1 \times 10^{-5}$ m^3.

Now, to calculate the porosity of the rock in the drill-core, the volume of the drill-core is found from the following: $\pi \times \text{radius}^2 \times h = \pi \times 0.025^2 \times 0.1$ m$^3 = 1.96 \times 10^{-4}$ m^3; and the porosity of the rock at a depth of 324 m follows from the following: $n = V_{\text{pores}}/V_{\text{drill-core}} = 3.1 \times 10^{-5}/1.96 \times 10^{-4} = 0.158$.

From a rock outcrop of the same geological formation from which the drill core was obtained, a similar determination yielded a porosity of 0.27. From this additional information, the compressibility of the sandstone can be derived using eq. 3.2:

$$n = n_o \cdot e^{-\sigma_e \beta} \qquad \text{(eq.1)}$$

and by assuming that the effective stress (eq. 3.1), σ_e, follows a simple linear relationship with depth. Therefore, σ_e can be replaced with depth, z, and as a result the compressibility, β, takes on units of length rather than pressure:

$$n = n_o \cdot e^{-z\beta} \qquad \text{(eq.2)}$$

β is calculated using the data obtained above using $n_o = 0.27$ (porosity of 27% at the surface), and $n = 0.158$ at $z = 324$ m. This calculation yields:

$$n(324\,\text{m}) = n_0(0\,\text{m}) \times \exp(-\beta \times 324)$$
$$n(324\,\text{m})/n_0(0\,\text{m}) = \exp(-\beta \times 324)$$
$$\ln(0.158/0.27) = -\beta \times 324$$
$$\beta = -\ln(0.158/0.27)/324$$
$$\beta = \sim 1.7 \times 10^{-3}\,\text{m}^{-1}$$

The value for β found this way can now be used to calculate a complete porosity-depth profile for this sandstone formation using eq. 2. Table 3.1 shows these calculated values for Well #1.

Table 3.1 Example porosity-depth data for three wells. Porosity values for Well #1 were calculated from field data as outlined in the text. Note that values of porosity, n, are given in %.

Depth (m)	Well #1	Well #2	Well #3
0	27.0	55.0	15.0
50	24.9	45.2	14.2
100	22.9	37.1	13.5
150	21.1	30.5	13.8
200	19.4	25.0	14.1
250	17.8	20.5	19.5
300	16.4	16.9	19.9
350	15.1	13.8	18.3
400	13.9	11.4	13.8
450	12.8	9.3	11.3
500	11.8	7.7	8.8
550	10.9	6.3	8.3
600	10.0	5.2	7.9
650	9.2	4.2	7.5
700	8.5	3.5	7.1
750	7.8	2.9	6.7

(Continued)

Box 3.1 (Continued)

A plot of the porosity values for Well #1 is shown in Fig. 3.2. Data for two additional wells (Well #2 and Well #3 in Table 3.1) are also plotted. The sediments found in Well #2 have a significantly higher compressibility of $4.04 \times 10^{-3}\ m^{-1}$, which can be confirmed by following the procedure outlined above. The porosity-depth profile for Well #3 shows an increase of porosity with depth over the interval roughly between 100–300 m that

cannot be explained from compaction as described by eq. 1. Over this depth, additional processes must be at work, such as water–rock interaction that leads to porosity enhancements which may result from dissolution. Another possible explanation for an interval of enhanced porosity to occur in borehole data such as these is if the borehole is intersected by a zone of fracturing over this depth interval.

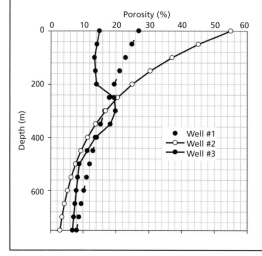

Fig. 3.2 A plot of the porosity-depth data presented in Table 3.1.

in which ρ_s is the density of the rock matrix. A typical density for sediments is 2700 kg m^{-3}. Hence, the lithostatic gradient under such conditions is 2.7 times larger than the hydrostatic gradient (Fig. 3.3a).

In many areas of active sedimentation, pore pressures in deeper aquifers (>1 km) are higher than would be expected from hydrostatic conditions (Fig. 3.3b). Elevated pressure above the hydrostatic gradient is often referred to in the literature as 'overpressure', 'anomalous pressure', 'excess pressure' or 'geopressure'. To drill in overpressured strata can present a hazard and result in a 'blow-out' when pressurized fluids start to escape through the borehole. The occurrence of overpressure is an indication that the high pressures developing during compaction do not dissipate efficiently. This is

typically the case in low permeability environments. Although any out-of-equilibrium pressures disappear given sufficient time, the rate of dissipation is in competition with the rate of overpressure generation due to ongoing mechanical loading (sedimentation rate). In mathematical terms, the net effect of the generation of hydraulic gradients (and thus fluid fluxes) via sedimentary loading can be incorporated into a one-dimensional version of the groundwater flow equation by describing flow in the depth (z) direction via a source term, Q_c, as follows:

$$K\frac{\partial^2 h}{\partial z^2} = S_s\frac{\partial h}{\partial t} + Q_c \qquad \text{(eq.3.5)}$$

Kooi (1999) showed that the magnitude of the source term (Q_c) relates to the subsidence

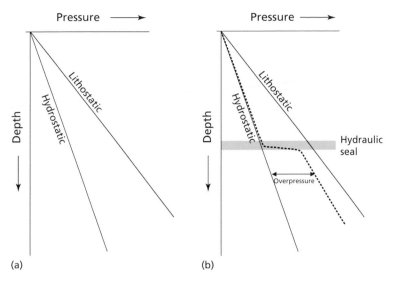

Fig. 3.3 Fluid pressure in a pore network will follow a hydrostatic trend when the vertical connectivity of the pore network is efficient. In (a), lithostatic pressures reflect the pressure exerted by the rock matrix. In (b), when the vertical equilibration of fluid pressures is hampered during basin subsidence, for example by the presence of a hydraulic seal such as a low-permeability shale, so-called 'overpressures' can develop at depth. A field example of pressure-depth profiles showing overpressure can be found in Box 3.2 (Domenico and Schwartz 1998). (*Source:* Domenico, P.A. and Schwartz, F.W. (1998) *Physical and Chemical Hydrogeology* (2nd edn). John Wiley & Sons, Inc., New York. © 1998, John Wiley & Sons.)

rate (v_z), the rate of sedimentation (v_{sed}), the density of the sediment settling at the sea bed (ρ_{gr}) and its porosity (n_{sed}) and the compressibilities of sediment (α) and the fluid residing in the pores (β), as given by:

$$Q_c = \alpha(\rho_{gr} - \rho_w)(1 - n_{sed})gv_{sed} - n\beta\rho_w gv_z$$

$$(eq.3.6)$$

Equation 3.6 shows that in more rapidly subsiding basins, as well as in more easily compressible sediments, the hydrodynamic effects of compaction will be larger.

While the concept of compaction-driven flow is introduced above using the example of flow driven by an increasingly large overburden, other geological processes can also lead to the compression of pore space and result in fluid movement. For example, the increasing overburden weight exerted by a developing ice-sheet or glacier can induce fluid flow in the aquifers underlying the ice. Also, tectonic stresses can cause a rise in fluid pressure (Ge and Garven 1994). The rates and magnitude with

which the load is applied (e.g. rate of either sedimentation, ice-sheet build up or tectonic movement) will primarily control the impact of these different processes on fluid pressure.

It is important to realise that compaction-driven flow is a transient phenomenon related to the geological process of basin formation leading to the burial of fluid-bearing sediments. In more mature basins, the effects of compaction-driven flow can be expected to be less important. Compaction-driven flow is a common phenomenon in actively subsiding sedimentary basins. In such basins, topography-driven flow can be relatively less dominant during development of the basin while in more mature sedimentary basins, in which compaction has ceased because the sediments are strengthened by diagenesis and therefore less compressible, topography-driven flow is dominant (Fig. 3.4). An example of a basin within the latter category is the Bath Basin (Box 2.11), while the Gulf of Mexico Basin exemplifies an active basin in which

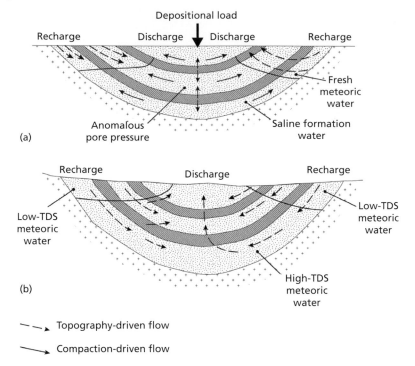

--- ➤ Topography-driven flow

——➤ Compaction-driven flow

Fig. 3.4 (a) Groundwater circulation pattern in an actively subsiding sedimentary basin where topography-driven flow from the flanks competes with compaction-driven flow originating mainly from the depositional centre of the basin. TDS = total dissolved solids. (b) A more geologically mature basin in which deposition has halted and topography-driven flow is dominant (Domenico and Schwartz 1998). (*Source:* Domenico, P.A. and Schwartz, F.W. (1998) *Physical and Chemical Hydrogeology* (2nd edn). John Wiley & Sons, Inc., New York. © 1998, John Wiley & Sons.)

both topography-driven and compaction-driven flow play an important role. Further analysis of the interplay between compaction-driven and topography-driven flow can be found in the case study (Box 2.12) describing overpressure generation in the Gulf of Mexico by employing a two-dimensional numerical model.

3.5 Variable-density driven fluid flow

3.5.1 Salinity gradients leading to variable-density flow

Density gradients in groundwater will cause buoyancy effects and the associated hydraulic gradients will drive groundwater flow. The classic setting in which to consider variable-density flow is that of coastal areas where saline water from the sea interfaces with fresh water originating from meteoric recharge over the land surface. Section 8.4.6 discusses the practical issues of the potential for saline intrusion when fresh water is exploited for important water supply needs. Furthermore, a certain group of organic contaminants can have properties that are less dense or denser than water such that their density needs to be taken into account when investigating their behaviour in groundwater flow systems (see Section 8.3.3).

Variable-density flow can be described mathematically by inclusion of buoyancy terms in the groundwater flow equation. Contrasts in fluid density differences can arise by variations in temperature, pressure and solute

Box 3.2   Compaction- and topography-driven fluid flow in the Gulf of Mexico Basin

The Gulf of Mexico Basin is a classic example of an actively subsiding sedimentary basin (Fig. 3.5). The Gulf of Mexico Basin contains economically important hydrocarbon reserves as well as groundwater aquifers exploited for water supply purposes. The hydrodynamical situation in the Gulf of Mexico is similar to other basins across the world such as the North Sea Basin, Amazon Delta, Niger Delta and Persian Gulf.

Offshore in the Gulf of Mexico, sediments of sand and clay accumulate and the increasing overburden is expected to lead to compaction and compaction-driven flow. However, onshore, on the flanks of the basin,

meteoric recharge occurs and topography-driven flow will dominate. Interesting hydrodynamic systems develop where topography-driven and compaction-driven flows interact. The interaction between topography-driven and compaction-driven flow is apparent when pressure-depth profiles taken within the basin are inspected (Fig. 3.6). Profiles 5 and 6 are taken near-shore and onshore in the Gulf of Mexico Basin. These profiles display hydrostatic conditions (Section 3.3) down to a depth of ~4 km, below which pressures are much higher than hydrostatic (overpressure). The transition between hydrostatic and overpressured conditions occurs at much

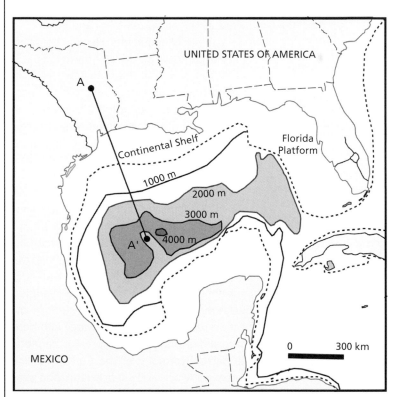

Fig. 3.5 Outline of the shape and size of the Gulf of Mexico sedimentary basin and its approximate bathymetry. The location of cross-section A–A′ is indicated, for which Harrison and Summa (1991) carried out a numerical model to simulate the hydrodynamic history of the basin (Harrison and Summa 1991). (*Source*: Adapted from Harrison, W.J. and Summa, L.L. (1991) Paleohydrology of the Gulf of Mexico Basin. *American Journal of Science* **291**, 109–176.)

(Continued)

Box 3.2 (Continued)

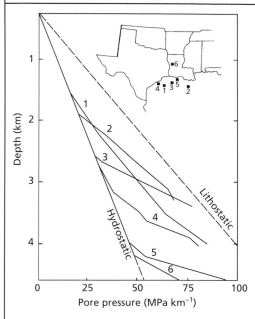

Fig. 3.6 Pore pressure variation with depth as observed in six wells in the Gulf of Mexico sedimentary basin. Locations of these wells are indicated on the accompanying map inset. The occurrence of overpressure at depth is observed in all wells, but nearer to the centre of the basin, overpressure occurs at increasingly shallower depth (Harrison and Summa 1991). (*Source:* Adapted from Harrison, W.J. and Summa, L.L. (1991) Paleohydrology of the Gulf of Mexico Basin. *American Journal of Science* **291**, 109–176.)

shallower depths further offshore (Profiles 1–4). Model simulations constrained by geological and pressure data from the Gulf of Mexico show the relation between the occurrence of overpressure and the geological build-up of the basin (Fig. 3.7).

Figure 3.7 illustrates the two-dimensional distribution of overpressure in the deeper parts of the basin. The overpressures at depth lead to an upward flow of fluids away from the basin centre. The model results show that the rates of fluid expulsion, taking into account the pressure gradients and sediment permeabilities, are of the order of 6.5 cm a^{-1} (Fig. 3.7a). At depths of up to ~2 km in the onshore and near-shore part of the basin, overpressure is absent and topography-driven groundwater flow is in the direction of the basin centre. Rates of shallow groundwater flow are an order of magnitude higher than those deeper in the basin at up to ~70 cm a^{-1}. At least in part, the observed patterns can be understood when aspects of the basin stratigraphy are considered (Fig. 3.7b). As a result of the geological history of the basin, the deeper basin sediments are more clay-rich than shallow sediments which are dominated by sand. A relatively high clay-content increases the compressibility and lowers the permeability of sediments. Hence, the dominance of compaction-driven flow in the basin centre, together with low fluid flow rates can, at least in part, be understood from the presence of clay-rich sediments. Likewise, the more rapid topography-driven groundwater flow circulation rates at shallow depths are facilitated by the presence of permeable sediments.

However, when analysed in detail, and next to sediment compaction, additional geological processes could explain the current overpressures presently observed in the Gulf of Mexico Basin. These are, for example, the role of thermal expansion of pore waters in causing overpressure and the de-hydration of clay minerals such as smectite during compaction releasing interlayer water into pore space and raising pore pressure. Furthermore, the palaeohydrology of the Gulf of Mexico system in which subsidence rates and sea-level will have fluctuated over the past millions of years can have an important impact on present pore pressure distributions.

Box 3.2 (Continued)

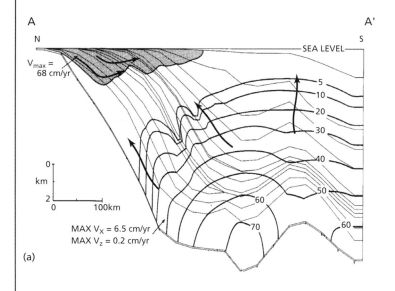

(a)

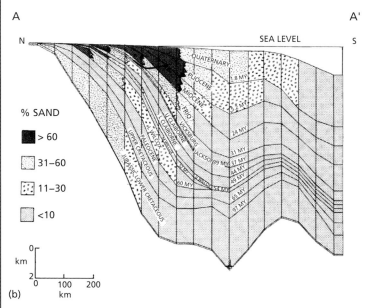

(b)

Fig. 3.7 (a) The current distribution of overpressure in the Gulf of Mexico Basin as simulated by Harrison and Summa (1991) with approximate direction of fluid flow directions indicated by arrows. The shaded area represents that part of the basin where fluids are driven by topographic gradients, while in the rest of the basin compaction-driven flow predominates. (b) The geological build-up of the basin shows that the largest overpressures develop in the more easily compressible lithologies which are more clay-rich, such as shale, and in the deepest parts of the basin (Harrison and Summa 1991). (*Source:* Adapted from Harrison, W. J. and Summa, L.L. (1991) Paleohydrology of the Gulf of Mexico Basin. *American Journal of Science* **291**, 109–176.)

Harrison and Summa (1991) presented a set of two-dimensional numerical models considering the palaeohydrology of the Gulf of Mexico Basin during the Tertiary and Quaternary. These models illustrate how the depth of meteoric groundwater circulation and the intensity of fluid expulsion from the basin centre have varied over time. However, the outcomes of the Harrison and Summa (1991) model largely support the concept that compaction-driven flow is the main driver of deeper fluid flow regimes in the Gulf of Mexico Basin.

concentration (see Appendix 2). Where variable-density driven flow develops in the absence of topographic gradients, it is commonly referred to as occurring through *free-convection*. Convection of groundwater caused by topography of the water table would then formally be called *forced-convection* which, in practice, is hardly used as a term. Free-convection will only occur when the permeability of the aquifer, the density contrast between the seawater and the groundwater in the aquifer and the thickness of the aquifer are large enough. These parameters can be combined to calculate a theoretical stability criterion, a Rayleigh number, Ra, for the onset of free-convection due to solute concentration effects (e.g. Schincariol and Schwartz 1990) and is given as follows:

$$Ra = \frac{gkL\left(\rho_s - \rho_f\right)}{\mu D} \qquad \text{(eq.3.7)}$$

in which μ is the dynamic viscosity, ρ_s is seawater density, ρ_f is density of water in the aquifer, k is intrinsic permeability, D is solute diffusivity coefficient and L is thickness of the aquifer. If Ra is larger than $\sim$10, free-convection of saline groundwater into fresh groundwater underneath is likely to occur.

3.5.2 Hydrothermal systems driven by variable-density flow

Where temperature gradients are large enough to lead to sufficient fluid density gradients, convective fluid flow patterns can result, which are often referred to as thermal convection. In freely convecting systems, areas of downwelling and upwelling develop. In a hydrothermal system driven by free-convection, hot fluids with a relatively low density rise to the surface and cool. Then, their increasing density counters the buoyancy and results in sinking.

Again, a Rayleigh number can be formulated to determine under which conditions thermal convection will start to occur. In a simplified system, in which the upper and the lower boundaries of a horizontal aquifer with thickness, L, are kept at a fixed temperature (denoted as T_u and T_l, respectively), the Rayleigh number for the onset of thermal convection is given as follows:

$$Ra = \frac{\alpha_f \rho_w^2 c_w gkL(T_l - T_u)}{\mu \kappa} \qquad \text{(eq.3.8)}$$

in which α_f is the thermal expansivity, k is aquifer permeability, κ is thermal conductivity, ρ_w is fluid density, μ is fluid dynamic viscosity and c_w is the fluid volumetric heat capacity. The reference temperature chosen to evaluate temperature-dependent properties in eq. 3.8 is typically taken as $(T_u + T_l)/2$. Free convection occurs when Ra exceeds a critical value. Commonly cited is a critical value of $\sim$$4\pi^2$. Using $Ra = 4\pi^2$ and common values for the other parameters in eq. 3.8, Fig. 3.8 shows the critical thermal gradients as a function of the thickness of the aquifer and for three different values of permeability. This suggests that thermal gradients mostly need to exceed that of normal

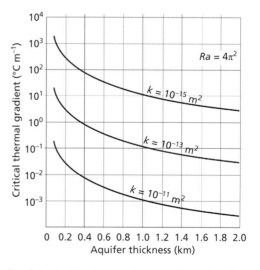

Fig. 3.8 In this diagram, by assuming a Rayleigh number (Ra) for the onset of hydrothermal convection (eq. 3.8), the critical thermal gradient can be calculated which, if exceeded, leads to groundwater convection. However, as shown, this critical thermal gradient is also a function of aquifer thickness and permeability (k).

Box 3.3   Hot springs along permeable fault zones in the Great Basin, Western USA

The Great Basin in the Western USA is formed by a number of extensional basins bounded by normal fault systems. The Great Basin is characterized by a relatively small crustal thickness (~20 km) intermediate between oceanic and continental crust, reminiscent of the geological history of the area as a former site of oceanic crust formation before being incorporated into the American continent during the Mesozoic. Cenozoic uplift and associated crustal extension created extensive normal-fault systems and new local basin formation. In combination with the high geothermal heat flow in the area, many of these fault systems are the location of hydro-geothermal systems. In this area, world-class ore deposits formed by gold-silver accumulations along fault zones are associated with the past circulation of hydrothermal water along faults (e.g. Person *et al.* 2007a).

One of the most closely studied spring systems associated with active faulting in the Great Basin is that just north of Borax Lake in northern Nevada (Fig. 3.9). In this area, no boreholes are available for the collection of hydraulic head and geological data, or to take samples of the deeper groundwater. However, the geochemistry and temperature of the spring waters can be measured by relatively simple means. Fairley and Hinds (2004) used a handheld digital thermometer to measure the water temperature of 175 springs and also the temperature of near-surface materials close to these springs. Geostatistical techniques were then used to properly interpolate these data to obtain an image of the near-surface temperature distribution in the area of the springs (Fig. 3.10).

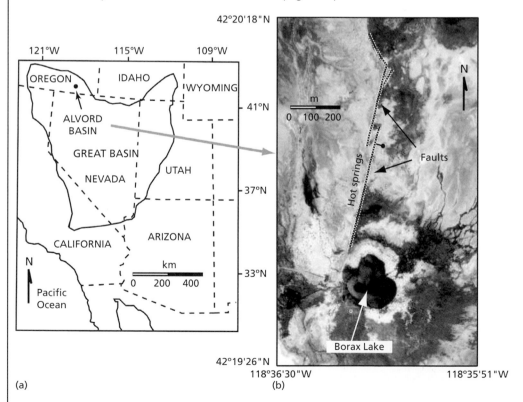

(a)   (b)

Fig. 3.9 Locations of (a) the Great Basin; and (b) Borax Lake (Fairley and Hinds 2004). *Source:* Adapted from Fairley, J. & Hinds, J. (2004). Field observation of fluid circulation patterns in a normal fault system. *Geophysical Research Letters* **31**, doi: 10.1029/2004GL020812. Reproduced with permission of John Wiley & Sons.

(Continued)

Box 3.3 (Continued)

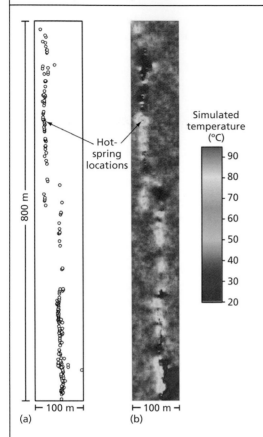

(a)   (b)

Fig. 3.10 (a) Location and (b) interpolated (simulated) near-surface temperatures of springs emerging in the Borax Lake area. Clearly, the springs align along the trace of a set of linear fault segments (Fairley and Hinds 2004). *Source:* Adapted from Fairley, J. & Hinds, J. (2004). Field observation of fluid circulation patterns in a normal fault system. *Geophysical Research Letters* **31**, doi: 10.1029/2004GL020812. Reproduced with permission of John Wiley & Sons.

Several patterns in the thermal image (Fig. 3.10b) are striking. The hot springs line up along what are interpreted to be two fault traces. These might be part of the same fault structure at depth showing a so-called "fault relay" structure at the surface where one fault strand ends and the fault displacement is transferred to a second fault trace. Also, along the fault trace the spring and ground temperatures are highly variable. This would indicate the presence of preferential flow paths in the fault zone with a relatively high permeability along which groundwater can rise more rapidly. Since along these flow paths, cooling will occur where the ambient rock is colder than the upwelling groundwater, higher surface temperatures would indicate a higher flow rate. The temperatures of some springs are near boiling (up to 94°C), whilst oxygen-isotope data indicate that recharge to the springs is meteoric. Sr-isotope data suggest circulation depths of up to 2–3 km depth (Anderson and Fairley 2008). To reconcile these data, it seems likely that thermal convection is the primary mechanism to circulate groundwater from and towards the surface accompanied by a component of more shallow topography-driven flow as seen, for example, in groundwater recharge from Borax Lake flowing into the fault zone.

As a first-pass quantitative analysis of the data, Fairley and Hinds (2004) used a simple one-dimensional heat flow model to estimate the possible variability in vertical groundwater fluxes reflected by the observed surface temperature distribution. This primary analysis yields insight into the distribution of fault permeability which then provides the basis of more elaborate numerical models that incorporate an explicit description of flow driven by thermal convection (Fairley 2009). In this model (Fig. 3.11), a constant temperature (250°C) boundary at 400 m depth is simulated from which groundwater rises upward at a rate controlled by a stochastic permeability structure that has statistical properties equivalent to those observed in the field. The latter model produces a variability of groundwater discharge rates as well as surface temperatures, the statistics of which can then be compared with field observations.

Box 3.3   (Continued)

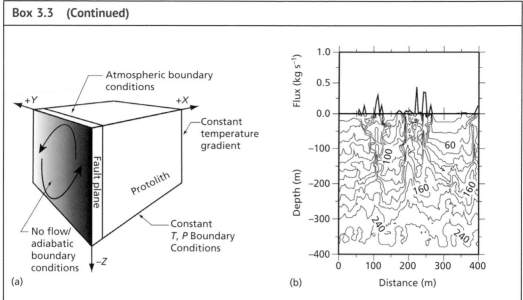

(a)   (b)

Fig. 3.11   (a) Set-up and boundary conditions of a numerical model describing thermal convection along a fault zone. (b) Temperature distribution and surface groundwater fluxes as calculated in the fault plane for one model of a set of model realizations (Fairley 2009). (*Source:* Fairley, J. (2009) Modeling fluid flow in a heterogeneous, fault-controlled hydrothermal system. *Geofluids* **9**, 154–166. © 2009, John Wiley & Sons.)

The groundwater circulation along faults near Borax Lake is a prime example of a system driven by thermal convection. The research at this location illustrates how relatively simple field observations, such as near-surface groundwater temperatures, in combination with conceptualized numerical models can provide a rather detailed insight into the permeability structure of complex hydrogeological systems such as fault zones.

geothermal conditions (e.g. $25°C\ km^{-1}$) substantially for free-convection to occur as a result of density differences caused by thermal effects. Permeable rock with a good vertical continuity (in the direction of the geothermal gradient) increases the likelihood of free-convection occurring. It is not surprising, therefore, that many permeable fault zones have been identified as loci of thermal convection of groundwater resulting in the emergence of hot springs at the surface (see Box 3.3 and discussion of the hydrothermal system in the Great Basin, USA). To an extent, many of these systems are also controlled by high crustal heat flow required to maintain the required temperature gradients.

3.6 Regional groundwater flow systems driven predominantly by variable-density flow

In the following, two regional-scale systems are considered in which variable-density flow plays the central role in controlling the hydrogeological evolution of the system. First, the longer-term shifts in the distribution of saline and fresh water in coastal areas as a result of sea-level fluctuations over the course of millennia are discussed. Second, salinity contrasts are important when considering where brines (highly saline fluids) in deep aquifers occur, often far away from coastal areas in mature sedimentary basins. The impacts not only on fluid flow of this type of groundwater situation needs to be addressed in specific cases, for example when the impacts of glaciations on continental-scale groundwater flow is considered but also in the design of nuclear waste repositories, or in consideration of the long-term safety of CO_2 sequestration in deep aquifers.

3.6.1 Fluctuating sea-level and its impact on the distribution of groundwater salinity in coastal areas

Seawater can either intrude into fresh water aquifers via tongues or via coastal flooding after which seawater can sink into underlying aquifers by free-convection. When the coastline is at a stable position, conditions such as depicted in Fig. 3.12a will generally apply.

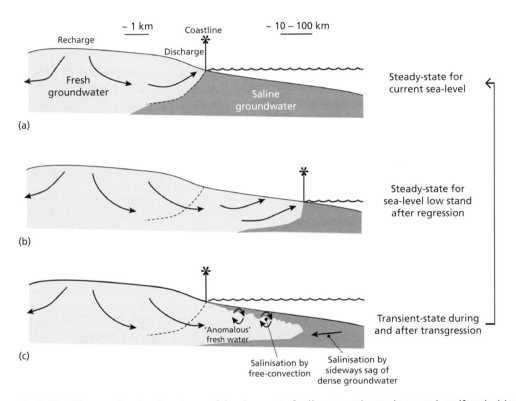

Fig. 3.12 Diagram showing the stages of development of saline groundwater in coastal aquifers. In (a) under steady-state conditions and considering a stable position of the coastline, seawater can be assumed to be present below the seafloor. The extent of saline water sag as a tongue at the base of the coastal aquifer is restricted by the coastward flow of fresh groundwater derived from meteoric recharge. In (b) during sea level low stand after marine regression, the saline groundwater formerly below the sea floor is displaced by fresh groundwater which leads to a new steady-state situation. In (c) during and after transgression, unstable transient conditions prevail for a period. Salinization processes in the coastal zone exposed to meteoric recharge during sea-level low stand are complex. Both free-convection of sea water sinking into less dense fresh water in the sub-seafloor aquifer as well as the sideways migration of the saline-fresh water interface inland will lead to an eventual, complete resalinization of offshore fresh groundwater as in (a). However, model simulations illustrate these processes can take many millennia before such a renewed steady-state is reached (Plate 3.1). Note the difference in length scale that should be assumed between the onshore and offshore domain shown in these sketches.

Tongues of saline water will be present in aquifers landward from the coastline underlying the fresh-water system fed by meteoric recharge. For a stable average position of the coastline (not considering tidal fluctuations), the extent of the landward ingress of seawater will depend on the density contrast between the fresh and the saline water in combination with the magnitude of the topography-driven seaward flow of fresh water which is controlled by the rates of meteoric recharge in the coastal zone.

Locally, over decades, groundwater over-abstraction from the fresh water parts of coastal aquifers has led to the upconing and inward migration of the salt-fresh interface in many coastal areas (see Section 8.4.4). On a global scale, the geological process of sea-level fluctuation, resulting from climate variability (primarily glaciation) in combination with crustal isostatic effects has, in the geological past, led to dramatic shifts in the position of the saline-fresh water interface. Our understanding of the dynamics of these changes and, therefore, of the impacts of future sea-level rise to the availability on fresh water resources in coastal areas, has improved considerably over past decades (e.g. Oude Essink *et al.* 2010).

In many submarine aquifer systems underneath continental shelves flanking the global oceans, pore waters have been found with salinities considerably lower than that of seawater (Fig. 3.13). For example, on the continental shelf off the coast of New England, drilling for oil and gas in the 1960s and 1970s showed the presence of water with a salinity of only 30–70% of that of seawater hundreds of metres below the seafloor (Person *et al.* 2003 and references therein). If only equilibrium conditions were considered, the occurrences of relatively fresh water underneath the seafloor cannot be readily understood. It was proposed (e.g. Kohout *et al.* 1977) that these relatively fresh water bodies underneath the present seafloor offshore of New England reflect hydrogeological conditions related to sea-level low-stand during glacial maxima in the Late Pleistocene.

Environmental isotopes provide further evidence for this view. Groundwater age determinations show that the anomalously fresh waters on the continental shelf are commonly ~7–20 ka, strengthening the hypothesis that these aquifers were recharged during the last global sea-level low-stand. Furthermore, ^{18}O and ^{2}H abundances in these groundwaters often indicate that recharge must have taken place during relatively cool climatic conditions as would be found during ice-house conditions (e.g. Morrissey *et al.* 2010). Similar 'anomalous' bodies of fresh water have since been found in European coastal aquifers (e.g. summarized in Edmunds and Milne 2001).

The hydrogeological mechanism to explain fresh water underneath the sea floor hundreds of kilometres offshore is as follows. After global sea-level dropped by ~120 m during the Last Glacial Maximum (LGM) about 18 000 years ago, the continental shelves were largely exposed. During this period, for example along most of the Atlantic Ocean, the coastline had moved seaward by 400 km or more. Thus, the aquifers on the continental shelves were at that time exposed to direct precipitation (Fig. 3.12b). Where these conditions persisted for long enough in combination with hydrogeological conditions favouring groundwater recharge (high effective rainfall and transmissive aquifers), significant volumes of fresh water could have displaced the saline water in these aquifers during this period, pushing the salt-fresh interface seaward. During eventual climate warming, continental ice-sheets thawed which resulted in a sea-level rise and the transgression of the sea. When transgression of the sea over the continental shelf occurred during the Holocene, seawater would have started to invade the fresh water bodies which developed during the sea-level low stand. This can happen from above where seawater will start to overlie fresh water in the aquifers underneath the seafloor. As a result of the density differences, seawater will potentially start to sink into the aquifer through free-convection, while at the same time, the

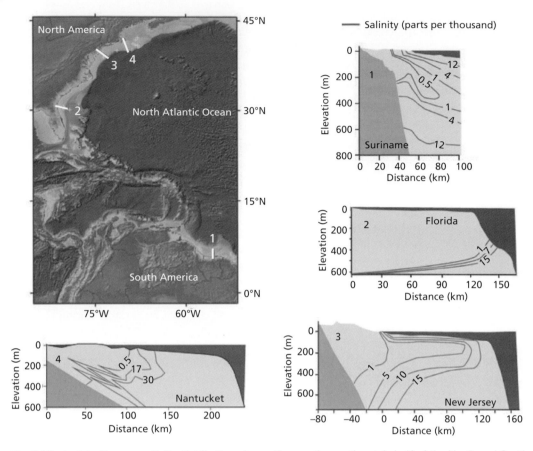

Fig. 3.13 In this diagram, salinity distributions in aquifers on the continental shelf of the North and South Atlantic Ocean illustrate the presence of off-shore fresh to brackish groundwater. By invoking the conceptual model shown in Fig. 3.12, it can be hypothesized that these bodies of 'anomalous' fresh water were emplaced during past sea-level low stand and reflect transient hydrogeological conditions on the continental shelf (Cohen *et al.* 2010). (*Source:* Cohen, D., Person, M., Wang, P. *et al.* (2010). Origin and extent of fresh paleowaters on the Atlantic continental shelf, USA. *Ground Water* **48**, 143–158. https://doi.org/10.1111/j.1745-6584. 2009.00627. Reproduced with permission of John Wiley & Sons.

salt-fresh interface at depth will migrate landward. However, it appears that the process of re-salinization has still not yet been completed. As a result of which, the present condition is not in equilibrium, and 'anomalous' water with a salinity much lower than that of seawater occurs on the continental shelf (Fig. 3.12c).

In order to get a more quantitative understanding of the process described above and, for example, to assess what type of conditions would favour the preservation of fresh water on continental shelves, there are several approaches available. As outlined earlier, a main process to salinize aquifers on

continental shelves would be free-convection, where seawater is transgressing over an aquifer containing fresh groundwater. Where free-convection does not occur, the fresh groundwater underlying the sea floor will still become saline over time but only via the much slower process of molecular diffusion. Another approach constitutes the use of numerical models (e.g. Kooi and Groen 2001; Post and Kooi 2003; Cohen *et al.* 2010) which allow investigation of a larger range of parameters controlling the timescales at which relatively fresh water in aquifers on the continental shelves might be preserved after transgression.

Examples of such parameters are the rate of sea-level rise, and the impact of the presence of aquitards near the sea-bed which control different modes of seawater intrusion that can be expected during transgression (Plate 3.1).

Analysis applying the type of methodology described previously, together with field observations, yields estimates of the amount of water with relatively low-salinity which can potentially be used for drinking water purposes after relatively minor treatment, as compared to desalination using seawater. These yield, for example Person *et al.* (2007b) and Cohen *et al.* (2010), that on the continental shelf offshore of New England a relatively fresh groundwater resource of the order of $\sim 1.7 \times 10^3$ km^3 is present. To put this into perspective, for a current water use of $\sim 4.5 \times 10^6$ m^3 day^{-1} for a city the size of New York, this resource could provide ~ 1035 years of fresh water after relatively minor treatment compared to complete desalination of seawater.

3.6.2 Brines in continental aquifers

The regional-scale occurrence of saline water is not constrained to coastal areas. It is a common phenomenon that solute concentrations in aquifers increase with depth as a result of which brines occur deep in sedimentary basins. Some continental sedimentary basins contain groundwater even at relatively shallow depths with a density much larger than that of seawater. Such a type of groundwater is usually called brine. Well studied examples of sedimentary basins containing brines at relative near-surface depths include the inter-cratonic Michigan, Illinois and Williston Basins in North America. The salinity of the pore waters in these basins is up to five to six times the salinity of seawater ($200\,\text{g}\,\text{L}^{-1}$ versus $35\,\text{g}\,\text{L}^{-1}$; Fig. 3.14). Where evaporite deposits such as halite and gypsum are present in sedimentary sequences, the dissolution of these could explain the extreme salinity of the pore waters. However, the Illinois Basin, for example, is not

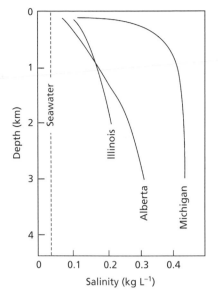

Fig. 3.14 Graph showing depth versus salinity relationships for three sedimentary basins: the Illinois, Alberta and Michigan Basins in North America. For reference, sea water salinity is indicated by the dashed line (Bense and Person 2008). (*Source:* Bense, V.F. and Person, M.A. (2008) Transient hydrodynamics in inter-cratonic sedimentary basins during glacial cycles. *Journal of Geophysical Research* **113**, F04005. © 2008, John Wiley & Sons.)

known to contain such deposits. Therefore, it has been proposed that these brines represent seawater which was partially evaporated before it was enclosed in the pore network as pore fluids (e.g. Hanor and McIntosh 2007). In most of the continental basins in North America, these brines are hosted by Devonian and Silurian sedimentary rocks; hence, the brines are believed to be of around the same age implying these fluids have been largely immobile since those ancient geological times.

As the salinity (and thus density) is increasing with depth (Fig. 3.14), free convection is not likely to occur. Topographic gradients could potentially drive fresh meteoric groundwater into these saline aquifers to dilute and displace the brines. However, in many basins, hydrogeological forcing by water-table gradients is not strong enough to flush out the brines. Therefore, it seems a good assumption that these deep brines, mostly due to their density, are more or less stagnant with very little fluid movement even on geological time scales. It is for this reason that deep aquifers containing these brines have been considered relatively safe localities for the sequestration of CO_2 or even the long-term storage of nuclear waste.

3.7 Regional groundwater flow systems driven predominantly by shifting topography and stress changes

Undulating water tables are often an important driving mechanism for regional fluid flow. When sufficient recharge occurs, the water table will mimic surface topography. However, over geological time, these topographic gradients can change. Such changes occurring on geological timescales are now believed to play a fundamental role in explaining the characteristics of regional- to continental-scale fluid flow systems.

There are two main geological processes that will result in a significant shift in topographic gradients over time which are discussed here. These are orogeny and subsequent erosion and the waxing and waning of continental ice-sheets. A well-founded understanding of the long-term development of fluid flow systems on a continental scale as a result of plate tectonic processes and glaciation is needed to correctly interpret, and potentially predict, the location of hydrocarbon accumulations, fresh water bodies in otherwise saline aquifers, and the presence of ore bodies.

3.7.1 Mountain building and erosion

Orogens are formed by the collision of continental plates forced by continental drift caused by mantle convection. Orogeny results in a vertical thickening of the crust, while laterally crustal shortening occurs. Consequently, mountain chains are built from rock originally forming in sedimentary basins in which compaction-driven flow and thermal convection dominate (Fig. 3.15a). The ensuing mountain belt is subsequently eroded through weathering processes. This cycle happens on geological timescales of tens of Ma. The development of strong topographic gradients in a mountain chain can be expected to drive an episodic mobilization of deep crustal fluids especially in combination with increased pore fluid pressures that will result from the tectonic compression during mountain building (Fig. 3.15b). Both the occurrence of lead-zinc ore deposits (often referred to as Mississippi-Valley-type (MVT) ore-deposits), as well as that of hydrocarbon accumulations at basin margins, have been explained as being the result of the expulsion of originally deep-seated fluids with a high temperature, and often a high salinity to the basin margins during and for a while after orogeny. These continental-scale migration pathways ceased to exist with erosion of the topography, resulting in a more local character of groundwater flow systems, and the waning of tectonic compression (Fig. 3.15c). Emplacement of these reactive, high-salinity fluids that are believed to have migrated laterally over hundreds of kilometres (literally at a continental scale), into a shallower environment that contained sulphur from organic sources meant that the metals contained in these brines precipitated out as sulphides such as galena (PbS) and sphalerite (ZnS). Deposits of this type found in the United States and in Carboniferous basins in Ireland now form a substantial part of the world's lead and zinc reserves. Where oil has migrated from depth to an environment near to the surface (Fig. 3.15b), bio-degradation and oxidation have strongly increased the viscosity of the original buoyant oil and a tar-like substance remains. The latter type of deposit is exemplified by the now economically viable tar sand deposits found at the eastern end of the Western Canada Sedimentary Basin (e.g. Head *et al.* 2003).

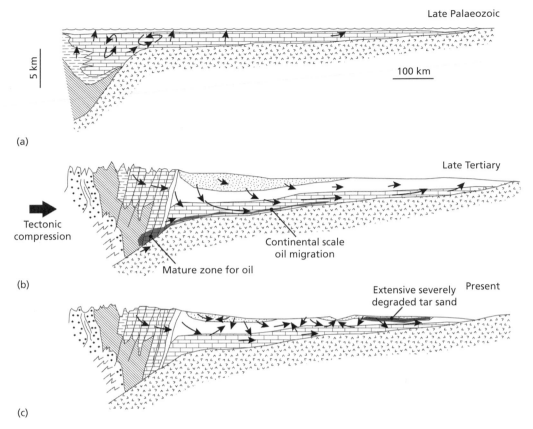

Fig. 3.15 Diagram showing the schematic evolution of the Western Canada Sedimentary Basin as an example of how tectonic compression and mountain building lead to episodic continental-scale fluid migration. (a) During the late Palaeozoic, flow is driven by compaction and thermal convection in a thick limestone aquifer. (b) During the Late Tertiary, uplift and mountain building result in topography-driven flow becoming the dominant driving fluid flow. The extra 'push' provided by tectonic compression means that continental-scale fluid migration pathways are established. These processes result in the long-range migration of oil produced at depth in the mountain belt where the temperature and pressure conditions favour the production of oil. (c) At present, only relatively local-scale topography-driven flow systems exist as a result of erosion and the absence of ongoing tectonic compression. However, the tar sand deposits found near the surface are interpreted as being emplaced during Late Tertiary times when continental-scale transport of hydrocarbons occurred. The occurrence of hydrocarbons near the surface leaves them prone to oxidation and biodegradation (Garven 1995). (*Source:* Modified from Garven, G. (1995) Continental-scale groundwater flow and geologic processes. *Annual Review of Earth and Planetary Sciences* **23**, 89–117.)

3.7.2 Impact of glaciations on regional hydrogeology

Ice-house conditions have prevailed across areas of the Northern Hemisphere for most of the Pleistocene (~2.5 Ma until ~12 ka), followed by the warm interglacial period, the Holocene, experienced at present. During the Late Pleistocene, ice-house conditions prevailed for roughly 50% of the time alternating with warm periods on a timescale of ~20–50 ka. The last glacial period lasted from ~90–12 ka and the period of maximum ice built-up during this period is commonly referred to as the Last Glacial Maximum (LGM) at around 18 ka. During the LGM, ice-sheets reached down to 40° N across the North American continent (Laurentian ice-sheet), and covered most of Scandinavia in Europe (Eurasian ice-sheet).

There is ample evidence that past glaciations resulted in a complete reorganization of regional hydrology and hydrogeology as compared to interglacial periods (Box 3.4). Many important aquifer systems at northern latitudes are now thought to have received significant volumes of fresh water recharge during glacial periods from underneath ice-sheets. Analogous to the situation on the continental shelves discussed above (Section 3.6.1), evidence from geochemistry and environmental isotopes suggests that groundwater recharge rates in many glaciated areas were significantly elevated during glaciations, which led to the emplacement of pockets of fresh water into, for example, the otherwise very saline groundwater aquifers of the intercratonic basins in the Canadian shield such as the Michigan, Illinois and Williston Basins (see Section 3.6.2). Although, appreciable groundwater recharge might have occurred from underneath ice sheets, in the pro-glacial area ahead of the ice sheet snout, permafrost conditions are likely to have prevailed and prevented groundwater recharge across those areas. The latter effect is, again, confirmed by geochemical and isotope data (see Jiráková *et al.* 2011).

Box 3.4   Reversal of regional groundwater flow patterns forced by ice-sheet waxing and waning in the Williston Basin, Western Canada

The Williston Basin is a thick (up to 5 km) sedimentary basin of Phanerozoic origin in south-western Canada and flanking states in the USA (Fig. 3.16). Since basin subsidence stopped at least 500 Ma, it is safe to assume that compaction-driven flow does not play a role in analysing data from the basin. Hence, the present groundwater flow in this trans-boundary aquifer system can be considered to be primarily topography-driven with recharge occurring over the topographic highs along the southern end of the basin such as the Black Hills in South Dakota and similar highs in Montana and Wyoming. Discharge of groundwater occurs mainly from Devonian limestone aquifers outcropping in Manitoba, Canada. Thus, the basin's hydrogeological architecture suggests a regional groundwater flow pattern from south-west to north-east (Fig. 3.17a). In the deepest aquifers centrally in the basin, brines occur (see Section 3.6.2), which currently emerge in springs in the main discharge area of the basin in Canada.

For the Williston Basin, a very extensive dataset is available of groundwater chemistry and stable isotopes obtained via samples taken in water supply wells, test wells, oil wells and springs. The collation of these data (Grasby *et al.* 2000) reveals that roughly three types of groundwaters with very distinct geochemical signatures are present in the system (Fig. 3.18). These are (1) recent groundwater derived from present-day recharge (low-salinity, $\delta^{18}O \sim -12‰$); (2) brine (high-salinity; $\delta^{18}O \sim +3‰$); and (3) low-salinity groundwater ($\delta^{18}O \sim -24‰$). The third type of groundwater is interpreted as derived from Pleistocene meltwater because of the low-salinity in combination with an isotope signature suggesting recharge temperatures substantially lower than those of today's climatic conditions.

The designation of three different types of groundwater found in the Williston Basin allows identification of the primary hydrodynamic mechanisms within the basin. Figure 3.18 shows that mixing between the three principal water types occurs in particular between the Pleistocene meteoric water and the deep basinal brines. Spatial

Box 3.4   (Continued)

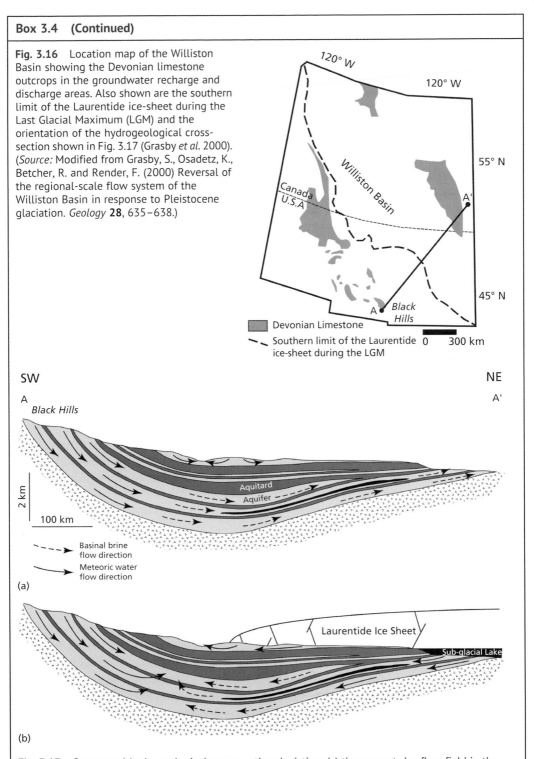

Fig. 3.16 Location map of the Williston Basin showing the Devonian limestone outcrops in the groundwater recharge and discharge areas. Also shown are the southern limit of the Laurentide ice-sheet during the Last Glacial Maximum (LGM) and the orientation of the hydrogeological cross-section shown in Fig. 3.17 (Grasby *et al.* 2000). (*Source:* Modified from Grasby, S., Osadetz, K., Betcher, R. and Render, F. (2000) Reversal of the regional-scale flow system of the Williston Basin in response to Pleistocene glaciation. *Geology* **28**, 635–638.)

Fig. 3.17 Conceptual hydrogeological cross-section depicting: (a) the present-day flow field in the Williston Basin; and (b) the Late Glacial Maximum with reversal of the regional groundwater flow system as inferred by Grasby *et al.* (2000) mainly from geochemical data (Fig. 3.18). (Grasby, S., Osadetz, K., Betcher, R. and Render, F. (2000) Reversal of the regional-scale flow system of the Williston Basin in response to Pleistocene glaciation. *Geology* **28**, 635–638.)

Box 3.4 (Continued)

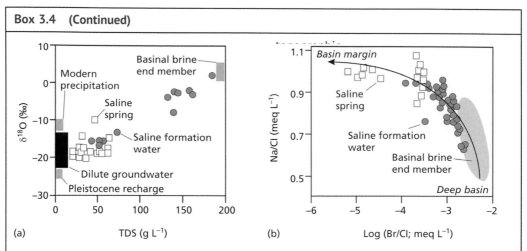

Fig. 3.18 (a) Plot of groundwater oxygen-isotope ratios and salinity (total dissolved solids, TDS) interpreted to show that regional-scale mixing of glacial meltwater and basinal brines has occurred in the current groundwater discharge area of the Williston Basin. (b) The mixing of groundwater in the current discharge area is further illustrated by Na : Cl and Br : Cl ratios (Grasby *et al.* 2000). (*Source:* Modified from Grasby, S., Osadetz, K., Betcher, R. and Render, F. (2000) Reversal of the regional-scale flow system of the Williston Basin in response to Pleistocene glaciation. *Geology* **28**, 635–638.)

identification of the zone of mixing between Pleistocene water and brines shows that this occurs within a distinct zone along the eastern flank of the basin. Supported by additional geomorphological evidence and patterns of the degree of biodegradation of crude oil in the basement, which indicate the influx of oxygenated water, the present-day distribution of Late Pleistocene water in the Williston Basin, and its mixing relation to the deeper brines, is interpreted by Grasby *et al.* (2000) to strongly suggest that during the Late Pleistocene a significant westward influx of groundwater occurred on the east flank of the basin. For this groundwater recharge regime to have existed in the past, however, requires a complete reversal of present-day hydraulic gradients since current

gradients and the occurrence of springs identify the eastern part of the Williston Basin as an area of groundwater discharge, rather than one of recharge. Therefore, Grasby *et al.* (2000) proposed that the southern limit of the Laurentide ice sheet covering the north-eastern half of the Williston Basin must have had the hydraulic capacity to accomplish such regional-scale reversal of groundwater flow directions, a situation which is depicted in Fig. 3.17(b). In this scenario, substantial recharge of the aquifers in the current discharge area occurs and a zone of intense hydrodynamic mixing is established, where the deep basinal brines are mobilized by the 'push' of the inflowing fresh water from underneath the ice-sheet, possibly from a sub-glacial lake.

During glacial periods, in glaciated areas, the thickness and the topography of the ice-sheet surface in combination with the availability of sub-glacial water are the main parameters controlling groundwater recharge and discharge

conditions on a regional scale. This is in strong contrast with the hydrogeological situation during temperate climatic conditions when groundwater recharge rates are determined by surface topography-controlled water table gradients in combination with the availability of water for recharge, constrained mostly by meteorological conditions. Appreciation of these long-term shifts in recharge conditions is important because it will lead to the realization that fresh water abstracted from aquifers that were recharged under ice-house conditions might not be rapidly replenished under present-day climatic conditions.

To understand hydrogeological conditions during glaciation, it is of paramount importance to evaluate what is controlling the volumes and rates of groundwater recharge from underneath an ice-sheet. Ice-sheets can either have an appreciable amount of liquid water at their base (wet-based or temperate ice sheets), or be mostly frozen to the ground (dry-based or polar ice sheets). Whether an ice-sheet is either wet- or dry-based depends on a number of factors controlling the temperature at the ice-bed. Liquid water at the base of an ice-sheet can come directly from in situ melting of ice, or be present indirectly from the routing towards the bed of surface melt water or of melt water from within the ice-sheet (englacially). The latter mechanism requires the existence of crevasses and/or other types of fracture systems in the ice providing an efficient conduit between the surface and the ice-sheet base (Fig. 3.19). The heat required for melting at the base of the ice sheet can come from mechanical friction (drag) between the ice and the substratum, in combination with a supply of geothermal heat from the rocks underneath the ice sheet, because the ice sheet will act partially as a thermal blanket. Hence, surface temperature conditions, ice thickness and ice flow dynamics will in this way largely control the possibility of liquid water occurring at the ice sheet base. Not all water present underneath a wet-based ice sheet will infiltrate. Water can either flow along the interface of the ice-sheet and the substratum, or it can infiltrate the sub-surface into aquifers present in the bedrock (Fig. 3.19).

The permeability of the rocks underlying the ice-sheet will to a large extent control the possibility of sub-glacial drainage via the subsurface as groundwater recharge. The presence of permafrost in the underlying strata, which would reduce permeability to almost zero, would prohibit infiltration of sub-glacial water. Permafrost will usually often be present in the pro-glacial area over which the ice-sheet expands. However, underneath wet-based ice-sheets, temperatures are near the melting point, and this will cause the permafrost underneath most of the ice-sheet to thaw over time, except for the zone near the ice-sheet snout.

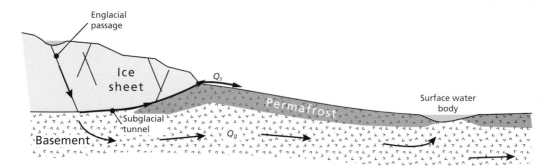

Fig. 3.19 Diagram showing the hydrogeology of ice-sheet margins. Englacial passages in ice sheets can deliver liquid water to the ice-sheet base where it can potentially recharge sub-glacial aquifers. In such a case, groundwater flow (Q_g) can occur from underneath the ice sheet to areas of groundwater discharge in the proglacial area, where discontinuities exist in the otherwise relatively impermeable permafrost. Examples of discontinuities include sufficiently insulating surface-water bodies such as lakes or large rivers. Otherwise, sub-glacial water will be routed along the ice-sheet bed via sub-glacial tunnels and exit at the toe of the ice sheet (Q_s).

As will be clear from the above, the evaluation of the potential for and rates of groundwater recharge from underneath ice sheets requires the consideration of a suite of processes and environmental conditions. Many of these are unknown for the LGM and idealized model representations of regional to continental-scale hydrogeological conditions (see Bense and Person 2008; Lemieux *et al.* 2008) take these processes into account to illustrate how highly elevated rates of groundwater recharge will have existed underneath continental ice sheets during past glaciations, as well as providing a sense of its spatial and temporal variability. With the insights of such idealized model results, current distributions of 'anomalous' fresh water occurrences recharged during glacial times can be better understood and the water resource they represent more efficiently managed.

While the relevance of understanding the impact of glaciations as a geological process to regional groundwater of relatively shallow depth (several hundreds of metres) is evident, the hydrogeology of much deeper aquifers is also affected by glaciation. As a result of the locally increasing weight of the ice-sheet while accreting at the start of a glacial period, underlying strata are mechanically loaded in an analogous way to sediments in subsiding sedimentary basins. Hence, the increase in total stress caused by the thickening of an ice sheet can lead to compaction and the associated generation of overpressure. The development of hydraulic head gradients via this mechanism can result in groundwater movement in very deep aquifers which are otherwise not directly impacted by the imposition of high hydraulic heads at the surface by the ice sheet (Plate 3.2). When the period of glaciation ends, the opposite effect occurs, generating sub-hydrostatic pressure regimes (under-pressure), in an analogous way to the impacts of erosion on fluid pressures. The generated over- and under-pressures caused by ice-sheet loading and unloading will in part cancel each other out. However, in between the phases of ice-sheet build up and retreat, pressures will start to dissipate so that a resultant anomalous pressure field will remain after glaciations in the deeper aquifers. The dissipation of anomalous pore pressures will take millennia, particularly in relatively thick sequences of low-permeability rocks. Numerical simulations (Bense and Person 2008) have shown that it is likely that present pore pressure distributions in deep aquifers underneath areas formerly covered by ice-sheets still reflect the hydrogeological conditions experienced during the ice-house period that ended ~12 ka.

The long-term stability of fluids in aquifers at several kilometres depth is of importance for consideration of the safety of the sequestration of CO_2 or the burial of nuclear waste in such locations. The potential hydrogeological impact of future glaciations on deep aquifer systems is one of the main concerns in the safety assessment of nuclear waste repositories at high-latitudes (e.g. Canada and Sweden) which require groundwater movements ideally to remain minimal for at least the coming few millennia. An improved understanding of the effects past glaciations have had on regional hydrogeological conditions will greatly aid the evaluation of whether such conditions can be expected with future changes of climate and possibly renewed glaciation.

3.8 Coupling and relative importance of processes driving fluid flow

This chapter demonstrates that on time scales of thousands to millions of years, geological processes such as subsidence in sedimentary basins, mountain building, sea-level fluctuation and glaciation can have profound impacts on the rates and direction of groundwater movement. In particular, when regional-scale systems are considered, patterns of groundwater flow and the distribution of fresh and saline groundwater will often be difficult to interpret if such processes are not taken into account. However, not only the current geological forcing of groundwater systems needs to be considered (e.g. present-day climatic conditions) but also past conditions which still have a lasting signature on the system. This is because geological processes often operate on timescales of millennia and longer, and groundwater systems are slow to respond to changes in boundary conditions. Many current hydrogeological patterns represent a transient state in which groundwater fluxes are reflecting the adjustment to present conditions from a past change in boundary conditions.

It is not just the common transient state of deep groundwater flow systems which complicates the interpretation of groundwater flow patterns. Another difficulty in the analysis of groundwater flow systems for geological processes is that these can operate simultaneously as a result of which it becomes difficult to unravel the relative importance of each and to judge which simplifications are justified in the analysis of hydrogeological data. One example of the interference of processes is that the build-up of continental ice-sheets and their subsequent waning can have comparable impacts on regional fluid flow as mountain building and erosion. This has led to discussion in the literature (e.g. Neuzil 2012) as to whether under-pressures observed in thick shale units in the Williston Basin can be attributed to deglaciation or erosional unloading, or a combination of the two processes.

The field examples and mechanisms presented in this chapter show that the impact of geological processes on deep and long-term groundwater flow is profound. Such processes have implications for humanity's informed use and exploitation of deep aquifer systems either as sources of drinking water (e.g. the Nubian aquifer in north-east Africa (Box 10.1) and off-shore groundwater) or as loci to store substances unwanted at the surface (such as excess CO_2 or nuclear waste).

Further reading

Ingebritsen, S.E., Sanford, W.E. and Neuzil, C.E. (2006) *Groundwater in Geologic Processes* (2nd edn). Cambridge University Press, New York.

Person, M., Raffensperger, J., Ge, S. and Garven, G. (1996) Basin-scale hydrogeological modelling. *Reviews of Geophysics* **34**, 61–87.

Post, V.E.A., Groen, J., Kooi, H., Person, M., Ge, S. and Edmunds, W.M. (2013) Offshore fresh groundwater reserves as a global phenomenon. *Nature* **504**, 71–78.

References

Anderson, T.R. and Fairley, J.P. (2008) Relating permeability to the structural setting of a fault-controlled hydrothermal system in southeast Oregon, USA. *Journal of Geophysical Research* **113**, B05402, doi: 10.1029/2007/JB004962.

Bense, V.F. and Person, M.A. (2006) Fault zones as conduit-barrier systems to fluid flow in siliciclastic sedimentary aquifers. *Water Resources Research* **42**, W05421, doi: 10.1029/2005WR004480.

Bense, V.F. and Person, M.A. (2008) Transient hydrodynamics in inter-cratonic sedimentary basins during glacial cycles. *Journal of Geophysical Research* **113**, F04005.

Bethke, C.M. and Corbet, T. (1988) Linear and non-linear solutions for one-dimensional compaction flow in sedimentary basins. *Water Resources Research* **24**, 461–467.

Cohen, D., Person, M., Wang, P. *et al.* (2010) Origin and extent of fresh paleowaters on the Atlantic continental shelf, USA. *Ground Water* **48**, 143–158.

Domenico, P.A. and Schwartz, F.W. (1998) *Physical and Chemical Hydrogeology* (2nd edn). John Wiley & Sons, Inc., New York.

Edmunds, W.M. and Milne, C.J. (eds) (2001) *Palaeowaters of Coastal Europe: Evolution of Groundwater Since the Late Pleistocene*. Geological Society, London. Special Publications **189**.

Fairley, J. (2009) Modeling fluid flow in a heterogeneous, fault-controlled hydrothermal system. *Geofluids* **9**, 154–166.

Fairley, J. and Hinds, J. (2004) Field observation of fluid circulation patterns in a normal fault system. *Geophysical Research Letters* **31**, L19502, doi: 10.1029/2004GL020812.

Garven, G. (1995) Continental-scale groundwater flow and geologic processes. *Annual Review of Earth and Planetary Sciences* **23**, 89–117.

Ge, S. and Garven, G. (1994) A theoretical model for thrust-induced deep groundwater expulsion with application to the Canadian Rocky Mountains. *Journal of Geophysical Research* **99**, 13851–13868.

Grasby, S., Osadetz, K., Betcher, R. and Render, F. (2000) Reversal of the regional-scale flow system of the Williston Basin in response to Pleistocene glaciation. *Geology* **28**, 635–638.

Hanor, J.S. and McIntosh, J.C. (2007) Diverse origins and timing of formation of basinal brines in the Gulf of Mexico sedimentary basin. *Geofluids* **7**, 227–237.

Harrison, W.J. and Summa, L.L. (1991) Paleohydrology of the Gulf of Mexico Basin. *American Journal of Science* **291**, 109–176.

Head, I.M., Jones, D.M. and Larter, S.R. (2003) Biological activity in the deep subsurface and the origin of heavy oil. *Nature* **426**, 344–352.

Jiráková, H., Huneau, F., Celle-Jeanton, H. *et al.* (2011) Insights into palaeorecharge conditions for European deep aquifers. *Hydrogeology Journal* **19**, 1545–1562.

Kohout, F.A., Hathaway, J.C., Folger, D.W. *et al.* (1977) Fresh ground water stored in aquifers under the continental shelf: implications from a deep test, Nantucket Sound, Massachusetts. *Water Resources Bulletin* **13**, 373–386.

Kooi, H. (1999) Competition between topography- and compaction-driven flow in a confined aquifer: some analytical results. *Hydrogeology Journal* **7**, 245–250.

Kooi, H. and Groen, J. (2001) Offshore continuation of coastal groundwater systems; predictions using sharp-interface approximations and variable-density flow modeling. *Journal of Hydrology* **246**, 19–35.

Lemieux, J.-M., Sudicky, E., Peltier, W. and Tarasov, L. (2008) Dynamics of groundwater recharge and seepage over the Canadian landscape during the Wisconsinian glaciation. *Journal of Geophysical Research* **113**, F01011, doi: 10.1029/2007JF000838.

Morrissey, S.K., Clark, J.F., Bennett, M. *et al.* (2010) Groundwater reorganization in the Floridan aquifer following Holocene sea-level rise. *Nature Geoscience* **3**, 683–687.

Neuzil, C.E. (2012) Hydromechanical effects of continental glaciation on groundwater systems. *Geofluids* **12**, 22–37.

Oude Essink, G.H.P., van Baaren, E.S. and de Louw, P.G.B. (2010) Effects of climate change on coastal groundwater systems: a modeling study in the Netherlands. *Water Resources Research* **46**, doi: 10.1029/2009WR008719.

Person, M., Dugan, B., Swenson, J.B. *et al.* (2003) Pleistocene hydrogeology of the Atlantic continental shelf, New England. *Geological Society of America Bulletin* **115**, 1324–1343.

Person, M., Mulch, A., Teyssier, C. and Gao, Y. (2007a) Isotope transport and exchange within metamorphic core complexes. *American Journal of Science* **307**, 555–589.

Person, M.A., McIntosh, J., Bense, V.F. and Remenda, V. (2007b) Pleistocene hydrology of North America: the role of icesheets in reorganizing groundwater flow systems. *Reviews of Geophysics* **45**, RG3007, doi: 10.1029/2006RG000206.

Post, V. and Kooi, H. (2003) On rates of salinization by free convection in high-permeability sediments; insights from numerical modelling and application to the Dutch coastal area. *Hydrogeology Journal* **11**, 549–559.

Schincariol, R.A. and Schwartz, F.W. (1990) An experimental investigation of variable density flow and mixing in homogeneous and heterogeneous media. *Water Resources Research* **26**, 2317–2329, doi: 10.1029/90WR01161.

Terzaghi, K. (1925) *Erdbaummechanic*. Franz Deuticke, Vienna.

4

Chemical hydrogeology

4.1 Introduction

The study of groundwater chemistry, or hydrochemistry, is useful in hydrogeology in a number of ways. Interpretation of the distribution of hydrochemical parameters in groundwater can help in the understanding of hydrogeological conditions and can also aid decisions relating to the quality of water intended for drinking water or as a process water. Hydrochemical processes are also significant in attenuating groundwater contaminants. In this chapter, the major hydrochemical processes of importance in groundwater are introduced. Interpretation techniques for combining data and defining hydrochemical types are also discussed as part of an integrated approach to understanding groundwater flow mechanisms.

4.2 Properties of water

The chemical structure of water is illustrated in Fig. 4.1, which shows one oxygen atom bonded asymmetrically to two hydrogen ions with a bond angle of 105°. The shape results from the geometry of the electron orbits involved in the bonding. Oxygen has a much higher electronegativity (a measure of the tendency of an atom to attract an additional electron) than hydrogen and pulls the bonding electrons towards itself and away from the hydrogen atom. The oxygen thus carries a partial negative charge (usually expressed as $\delta-$) and each

hydrogen a partial positive charge ($\delta+$), creating a dipole, or electrical charges of equal magnitude and opposite sign a small distance apart. As a consequence, the opposite charges of water molecules attract each other to form clusters of molecules, through a type of interaction known as hydrogen bonding. The size of the clusters increases with decreasing temperature reaching a maximum at 4 °C. When water is cooled from 4 to 0 °C the size of the clusters creates a more open structure and the water becomes less dense, with further expansion on freezing. Hence, ice has a lower density than liquid water. Values for water density, viscosity, vapour pressure and surface tension over a temperature range of 0–100 °C are given in Appendix 2.

As illustrated in Table 4.1, water is not simply H_2O, but rather a mixture of six molecules depending on the hydrogen and oxygen isotopes that combine to form the water molecule. Eighteen combinations are possible, the most common of which is $^1H_2^{16}O$. Pure water contains hydrogen and oxygen in ionic form as well as in the combined molecular form. The ions are formed when water dissociates as follows:

$$H_2O \Leftrightarrow H^+ + OH^- \qquad \text{(eq. 4.1)}$$

The H^+ ion is normally in the form H_3O^+ (the hydronium ion) and in rock–water interactions, the transfer of protons (H^+ ions) between the liquid and solid phases is known as proton transfer, an important consideration in carbonate chemistry.

Hydrogeology: Principles and Practice, Third Edition. Kevin M. Hiscock and Victor F. Bense.
© 2021 John Wiley & Sons Ltd. Published 2021 by John Wiley & Sons Ltd.
Companion website: www.wiley.com/go/hiscock/hydrogeology3e

Table 4.1 The relative abundance of hydrogen and oxygen isotopes in the water molecule.

Isotope		Relative abundance	Type
^1H	Proteum	99.984	Stable
^2H	Deuterium	0.016	Stable
^3H	Tritium	$0–10^{-15}$	Radioactive[a]
^{16}O	Oxygen	99.76	Stable
^{17}O	Oxygen	0.04	Stable
^{18}O	Oxygen	0.20	Stable

[a] Half-life = 12.3 years.

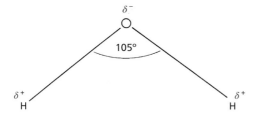

Fig. 4.1 The structure of the water molecule showing the dipole created by the partial negative charge on the oxygen atom and partial positive charge on each hydrogen atom.

The polarity of the water molecule makes water an effective solvent for ions; the water molecules are attracted to the ions by electrostatic forces to form a cluster with either oxygen or hydrogen oriented towards the ions as shown in Fig. 4.2. This phenomenon is known as hydration and acts to stabilize the solution. Polar solvents such as water easily dissolve crystalline solids like sodium chloride (NaCl) and break down the ionic crystal into a solution of separately charged ions:

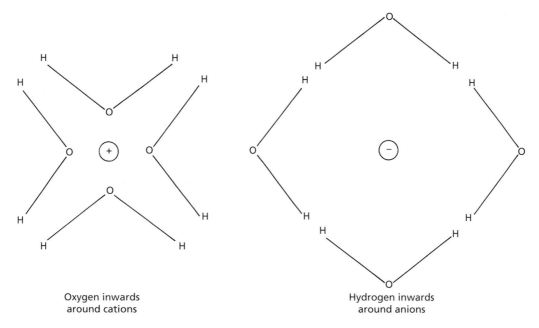

Oxygen inwards
around cations

Hydrogen inwards
around anions

Fig. 4.2 Water molecules surrounding a positively charged cation and a negatively charged anion to form hydrated ions.

$$Na^+Cl^- \overset{H_2O}{\Leftrightarrow} Na^+_{(aq)} + Cl^-_{(aq)} \quad \text{(eq. 4.2)}$$

Positively charged atoms like sodium are known as cations, while negatively charged ions like chloride are called anions. When writing chemical equations, the sum of charges on one side of the equation must balance the sum of charges on the other side. On the left-hand side of eq. 4.2, NaCl is an electrically neutral compound, while on the right-hand side, the aqueous sodium and chloride ions each carry a single but opposite charge so that the charges cancel, or balance, each other.

The degree of hydration increases with increasing electrical charge of the dissociated ion and also with decreasing ionic radius. Because cations are generally smaller than anions, cations are usually more strongly hydrated. The effect of hydration is to increase the size of an ion and so reduce its mobility and affect the rates of chemical reaction. In addition, the complexation of cations, particularly the transition metals such as iron, copper, zinc, cobalt and chromium, where the ions chemically bond with water molecules, leads to the creation of complex ions of fixed composition and great chemical stability.

4.3 Chemical composition of groundwater

The chemical and biochemical interactions between groundwater and the geological materials of soils and rocks provide a wide variety of dissolved inorganic and organic constituents. Other important considerations include the varying composition of rainfall and atmospheric dry deposition over groundwater recharge areas, the modification of atmospheric inputs by evapotranspiration, differential uptake by biological processes in the soil zone and mixing with seawater in coastal areas. As shown in Table 4.2, and in common with freshwaters in the

Table 4.2 Chemical composition of groundwater divided into major and minor ions, trace constituents and dissolved gases (Freeze and Cherry).

Major ions ($>5\,\text{mg L}^{-1}$)	
Bicarbonate	Sodium
Chloride	Calcium
Sulphate	Magnesium
	Silicon

Minor ions (0.01–$10.0\,\text{mg L}^{-1}$)	
Nitrate	Potassium
Carbonate	Strontium
Fluoride	Iron
	Boron

Trace constituents ($<0.1\,\text{mg L}^{-1}$)	
Aluminium	Molybdenum
Arsenic	Nickel
Barium	Phosphate
Bromide	Platinum
Cadmium	Radium
Caesium	Selenium
Chromium	Silver
Cobalt	Thorium
Copper	Tin
Gold	Titanium
Iodide	Uranium
Lead	Vanadium
Lithium	Zinc
Manganese	

Dissolved gases (trace – $10\,\text{mg L}^{-1}$)	
Nitrogen	Methane
Oxygen	Hydrogen sulphide
Carbon dioxide	Nitrous oxide

(*Source:* Freeze, R.A. and Cherry, J.A. (1979) *Groundwater*. Prentice-Hall, Inc., Englewood Cliffs, New Jersey. © 1979, Pearson Education.)

terrestrial aquatic environment, the principal dissolved components of groundwater are the six major ions sodium (Na^+), calcium (Ca^{2+}), magnesium (Mg^{2+}), chloride (Cl^-), bicarbonate (HCO_3^-) and sulphate (SO_4^{2-}). These cations and anions normally comprise over 90% of the total dissolved solids (TDS) content, regardless of whether the water is dilute rainwater or has a salinity greater than seawater (typical analyses of rainwater and seawater are given in Appendix 5). Minor ions include potassium (K^+), dissolved iron (Fe^{2+}), strontium (Sr^{2+}) and fluoride (F^-), while aqueous solutions commonly also contain amounts of trace elements and metal species. The introduction of contaminants into groundwater from human activities can result in some normally minor ions reaching concentrations equivalent to major ions. An example is nitrate, as excessive application of nitrogenous fertilizers can raise nitrate concentrations in soil water and groundwater to levels in excess of $50 \, mg \, L^{-1}$ (see Box 4.1 for a discussion of the different concentration units used in hydrochemistry).

Box 4.1   Concentration units used in hydrochemistry

Concentration is a measure of the relative amount of *solute* (the dissolved inorganic or organic constituent) to the *solvent* (water). A list of atomic weights is supplied in Appendix 4.

There are various types of concentration unit as follows:

Molarity (M): number of moles of solute dissolved in 1 L of solution ($mol \, L^{-1}$). For example, if we have 10 g of potassium nitrate (molar mass of KNO_3 = 101 g per mole), then this is (10 g)/(101 g mol^{-1}) = 0.10 moles of KNO_3. If we place this in a flask and add water until the total volume = 1 L, we would then have a 0.1 molar solution. Molarity is usually denoted by a capital M, for example a 0.10 M solution. It is important to recognize that molarity is moles of solute per litre of solution, not per litre of solvent, and that molarity changes slightly with temperature because the volume of a solution changes with temperature.

Molality (m): the number of moles of solute dissolved in 1 kg of solvent ($mol \, kg^{-1}$). Notice that, compared with molarity, molality uses mass rather than volume and uses solvent instead of solution. Unlike molarity, molality is independent of temperature because mass does not change with temperature. If we were to place 10 g of KNO_3 (0.10 moles) in a flask

and then add one kilogram of water, we would have a 0.50 molal solution. Molality is usually denoted with a small m, for example a 0.10 m solution.

Mass concentration: this unit of concentration is often used to express the concentration of very dilute solutions in units of parts per million (ppm) or, more commonly, $mg \, L^{-1}$. Since the amount of solute relative to the amount of solvent is typically very small, the density of the solution is approximately the same as the density of the solvent. For this reason, parts per million may be expressed in the following two ways:

$$ppm = mg \text{ of solute/L of solution}$$
$$ppm = mg \text{ of solute/kg solution}$$

Chemical equivalence: the concept of chemical equivalence takes into account ionic charge and is useful when investigating the proportions in which substances react. This aspect of chemistry is called *stoichiometry*.

Equivalents per litre ($eq \, L^{-1}$) = number of moles of solute multiplied by the valence of the solute in 1 L of solution. From this, it follows that $meq \, L^{-1} = mg \, L^{-1} \times$ (valence/atomic weight).

As an example of the application of chemical equivalence, take the effect of ion

---

Box 4.1 (Continued)

exchange when a fresh groundwater in contact with a rock is able to exchange a chemically equivalent amount of calcium (a divalent cation with atomic weight = 40 g) with sodium (a monovalent cation with atomic weight = 23 g) contained within the aquifer. If the groundwater has an initial calcium concentration of 125 mg L^{-1} and sodium concentration = 12 mg L^{-1}, what will be the new groundwater sodium concentration if all the calcium were exchanged with the clay material?

Initial calcium concentration

$$= 125 \, \text{mg L}^{-1} \times (2/40) = 6.25 \, \text{meq L}^{-1}$$

Initial sodium concentration

$$= 125 \, \text{mg L}^{-1} \times (1/23) = 0.52 \, \text{meq L}^{-1}$$

New sodium concentration after ion exchange

$$= 6.25 + 0.52 = 6.77 \, \text{meq L}^{-1}$$

$$= 6.77 \times (23/1) = 155.71 \, \text{mg L}^{-1}$$

Therefore, the extra sodium contributed to the groundwater by ion exchange is $\sim$144 mg L^{-1}.

Organic compounds are usually present in groundwater at very low concentrations of less than $0.1 \, \text{mg L}^{-1}$ as a result of oxidation of organic matter to carbon dioxide during infiltration through the soil zone (Section 4.7). In environments rich in organic carbon such as river floodplains and wetlands, biogeochemical processes can generate anaerobic groundwater conditions and the production of dissolved gases such as nitrogen (N_2), hydrogen sulphide (H_2S) and methane (CH_4). Other dissolved gases include oxygen (O_2) and carbon dioxide (CO_2) mostly of an atmospheric source, and nitrous oxide (N_2O) from biogeochemical processes in soils and groundwater. Radon (^{222}Rn) gas, a decay product of uranium (U) and thorium (Th), is common in groundwater and can accumulate to undesirable concentrations in unventilated homes, mines and caves. Uranium is present in crustal rocks (e.g. in the mineral uranite, UO_2), silicates (e.g. in the mineral zircon, $ZrSiO_4$) and phosphates (e.g. in the mineral apatite, $Ca_5(PO_4)_3(OH, F, Cl)$) and is common not only in granitic rocks but also in other rock types, sediments and soil. Radon and its decay products such as polonium (^{218}Po and ^{216}Po) and ultimately isotopes of lead (Pb) are harmful when inhaled by humans.

Minor concentrations of the inert gases argon (Ar), helium (He), krypton (Kr) and xenon (Xe) are found dissolved in groundwater, and these can provide useful information on the age and temperature of groundwater recharge and therefore help in the interpretation of hydrochemical and hydrogeological conditions in aquifers (Section 4.5).

The degree of salinization of groundwater expressed as the TDS content is a widely used method for categorizing groundwaters (Table 4.3). In the absence of any specialist analytical equipment for measuring individual dissolved components, a simple determination of TDS by weighing the solid inorganic and organic residue remaining after evaporating a measured volume of filtered sample to dryness, can help determine the hydrochemical characteristics of a regional aquifer. Equally, a measurement of the electrical conductivity (EC) of a solution will also give a relative indication of

Table 4.3 Simple classification of groundwater based on TDS content (Freeze and Cherry 1979).

Category	TDS (mg L^{-1})
Freshwater	0–1000
Brackish water	1000–10 000
Saline water (seawater)	10 000–100 000 (35 000)
Brine water	>100 000

Note: TDS > 2000–3000 mg L^{-1} is too salty to drink. (*Source:* Freeze, R.A. and Cherry, J.A. (1979) *Groundwater.* Prentice-Hall, Inc., Englewood Cliffs, New Jersey. © 1979, Pearson Education.)

the amount of dissolved salts made possible by the fact that groundwater is an electrolytic solution with the dissolved components present in ionic form. For any investigation, it is possible to relate the TDS value to EC (usually expressed in units of micro-siemens centimetre^{-1}) as follows:

$$TDS(mg\ L^{-1}) = k_e \times EC(\mu S\ cm^{-1})$$
$$(eq.\ 4.3)$$

where the correlation factor, k_e, is typically between 0.5 and 0.8 and can be determined for each field investigation. The EC for fresh groundwater is of the order of 100s $\mu S\ cm^{-1}$, while rainwater is of the order of 10s $\mu S\ cm^{-1}$ and brines 100 000s $\mu S\ cm^{-1}$. Given that ionic activity, and therefore EC, increases with temperature at a rate of about a 2% per °C, measurements are usually normalized to a specific temperature of 25 °C and recorded as SEC$_{25}$.

To demonstrate the influence of geological materials and the hydrogeological environment on the range of chemical composition of groundwater, Box 4.2 describes the characteristics of thermal and mineral waters at spa locations in Great Britain. Also, the distribution of hydrochemical parameters in groundwater can assist decisions relating to the quality of water intended for drinking water or as a process water. For example, Box 4.3 describes the influence of hydrochemistry on the history of malting and brewing in Europe.

4.4 Sequence of hydrochemical evolution of groundwater

In a series of three landmark papers, and based on nearly ten thousand chemical analyses of natural waters, Chebotarev (1955) put forward the concept that the salinity distribution of groundwaters obeys a definite hydrological

---

Box 4.2 Hydrochemical characteristics of British spa waters

The term 'spa' is derived from the town of Spa, now in Belgium, which in turn is derived from a Walloon word *espa* meaning a fountain (Mather 2013). The word 'spa' is now used in a generic sense to mean a mineral spring considered to have health-giving properties or a place or resort with such a spring. All spas in Britain (Fig. 4.3) developed originally as a result of the characteristics of local groundwater sourced from springs and wells. The compilation of hydrochemical characteristics of spas shown in Table 4.4 are compositionally diverse, ranging from potable waters to brines. The term 'saline' is used here in its medical sense, to mean water containing sodium chloride and/or a salt or salts of magnesium or another alkali metal (Mather 2013), in contrast to the definition of brackish and saline shown in Table 4.3 that is based on total solids content.

The waters exploited by the spas are derived from a wide range of geological formations ranging in age from the Precambrian of the Malvern Hills to the Lower Cretaceous of South East England. The rock type is an important influence on not only the composition of the spa waters but also the hydrogeological environment. As explained in Section 4.4, the hydrogeological environment controls the length of the groundwater flowpath and processes such as mixing, the time for rock–water interaction and the redox conditions encountered. Measurement of the environmental isotope composition of British spa waters demonstrate that they are meteoric in origin and up to 10 000 years old, potentially mixed with recharge water from the Pleistocene and perhaps small volumes of even older water (Mather 2013).

The spa waters presented in Table 4.4 are divided into three groups based on the physical hydrogeological environment in which they are found. Group I spas are located in natural discharge zones at the end of long flowpaths with long groundwater residence times. Discharges are commonly associated

Box 4.2   (Continued)

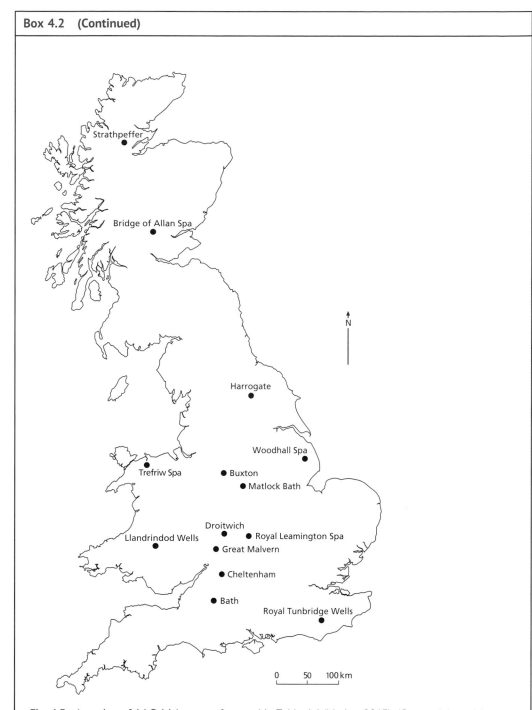

Fig. 4.3   Location of 14 British spas referenced in Table 4.4 (Mather 2013). (*Source:* Adapted from Mather, J.D. (2013) Britain's spa heritage: a hydrogeological appraisal. In: *A History of Geology and Medicine* (eds C.J. Duffin, R.T.J. Moody and C. Gardner-Thorpe). Geological Society, London, Special Publications 375, pp. 243–260.)

(*Continued*)

Table 4.4 Major ion chemistry of British spa waters. Location numbers are shown in Fig. 4.3. Group numbers I, II and III are referred to in the text. Except for pH, concentrations are in mg L⁻¹. The analyses are taken from various sources as compiled by Mather (2013). In most cases, the concentration of iron (Fe) is likely to be total iron rather than dissolved iron. Blank spaces indicate that no concentration is recorded. Concentrations of total dissolved solids (TDS) are given for comparative purposes and are an estimate based on a simple summation of the major ion species.

Location	1	2	3	4	5	6	7	8	9	10	11	12	13	14
Spa	Royal Tunbridge Wells	Bath, King's Spring	Cheltenham, Pittville	Great Malvern, Pewtriss Spring	Llandrindod Wells, Chalybeate Spring	Droitwich	Royal Leamington Spa	Matlock Bath, New Bath	Buxton, St Anne's Well	Woodhall Spa, Shaft	Trefriw Spa	Harrogate, Old Sulphur	Bridge of Allan Spa	Strathpeffer, Strong
Geology of water source	Lower Cretaceous sandstones and ironstones	Carboniferous limestone	Jurassic Lower Lias mudstone	Precambrian schists and gneisses	Silurian mudstone	Triassic Mercia Mudstone evaporites	Triassic Mercia Mudstone over Sherwood Sandstone	Lower Carboniferous limestones and evaporites	Lower Carboniferous limestone	Upper Jurassic sandstone	Upper Ordovician slates	Carboniferous sandstones and shales	Lower Devonian volcanics	Lower Devonian sandstones and shales
Date of analysis	1956	1979	1995	2006	1991	1958	1983	1969	2011	1911	1953	ca. 1930	1928	ca.1930
Type	Mild chalybeate	Thermal/Weak saline	Saline	Potable	Saline/Sulfur	Brine	Saline	Thermal (Potable)	Thermal (Potable)	Saline	Strong chalybeate	Saline/Sulfur	Saline	Sulfur
Group	II	I	II	II	I	II	II	I	I	III	III	I	III	II
Na^+	20	183	3080	15	1580	120 690	6000	30	24	7897	7	5243	2005	32
K^+	8	17	15	1	12			1	1	28		73	10	13
Ca^{2+}	27	382	26	35	362	1368	520	105	55	544	188	428	1414	302
Mg^{2+}	6	53	42	19	38	132	550	32	19	151	66	185	21	90
Fe	37	0.9	<0.02		1.5		0.7		0	7	516	Trace	0.04	
Cl^-	41	287	1930	39	3180	185 550	7300	57	37	13 582	4	9088	5423	Trace
SO_4^{2-}	65	1032	599	35	<0.5	4491	2200	150	13	67	2267	3	225	924
HS^-		<0.01			Present					Trace		99		9
HCO_3^-	321	192	922	123	113	26	120	271	248		0	1222	106	259
NO_3^-		<1	11	8				0.4	<0.1			3	3	
NH_4^+					0.85							3		
TDS	525	2147	6625	275	5288	312 257	16 691	646	397	22 276	3048	16 252	9207	1620
pH		6.65	8.8	8.0	7.27		7.0	7.35	7.4		ca. 3.5	6.87		

(*Source*: Mather, J.D. (2013) Britain's spa heritage: a hydrogeological appraisal. In: *A History of Geology and Medicine* (eds C.J. Duffin, R.T.J. Moody and C. Gardner-Thorpe), Geological Society, London, Special Publications 375, pp. 243–260.)

Box 4.2   (Continued)

with faults along which flowpaths merge as groundwater travels under the hydraulic gradient to the surface. Three of the best known spas in Britain (Bath, Buxton and Harrogate) are all within this group and are associated with deeply sourced groundwater. At the thermal springs in Bath (see Box 2.12), meteoric water circulates to a depth of between 2.7 and 4.3 km in fissured Carboniferous limestone before discharging at 46.5 °C at the King's Spring with an age of about 10 000 years. At Harrogate, meteoric-derived formational brines are the source of the salinity, with dilution and bacterial reduction of sulphate by shallow groundwaters (Bottrell *et al.* 1996) accounting for a wide range of spring compositions and the presence of sulphide. In comparison to the springs at Bath that are weakly mineralized due to the relatively hard limestones, the other Group I waters originating in interbedded mudstones and sandstones have much higher concentrations of dissolved rock-derived minerals. At Llandrindod Wells in mid-Wales, reducing hydrochemical conditions cause the replacement of sulphate and nitrate by sulphide (HS^-) and ammonium (NH_4^+), respectively. In addition, the concentrations of many trace elements at Llandrindod Wells are elevated, for example bromine (as bromide), and are indicative of long groundwater residence times (Mather 2013).

Most of the British spas waters shown in Table 4.4 are within Group II in which the hydrogeological environment is not the only factor in the development of the spa, given that similar conditions occur widely. The Group II spas embrace many different chemical types including ferruginous, saline and sulphurous sources as well as those characterized by brines. For example, the saline waters of Cheltenham Spa are derived from the Jurassic Lower Lias clays that crop out over an extensive area. However, Cheltenham only grew into a fashionable Georgian spa by virtue of the activities of local entrepreneurs (Mather

2013). Royal Tunbridge Wells and Great Malvern have little to do with the quality of local groundwaters. Discharges of iron-rich (ferruginous or chalybeate) waters, such as at Royal Tunbridge Wells, occur throughout the Weald in South East England. Flowpaths are short and both discharges and water quality vary widely from year to year, with a high risk of contamination, particularly during periods of low flow (Mather 2013). The weakly mineralized groundwaters at Great Malvern are not especially distinguishable from public water supplies sourced widely. Droitwich and Royal Leamington Spas both rely on the dissolution of halite within the Triassic Mercia Mudstone Group for their salinity, a condition that is widely found in the Midlands and North West England. Under natural conditions, the brine is almost static unless it is disturbed by, for example, the erosion of a river valley, as at Droitwich, or leakage into underlying sandstones, as at Royal Leamington Spa (Mather 2013).

A limited number of British spas (Group III in Table 4.4) have resulted from unplanned intervention of the hydrogeological environment where the undisturbed flow is relatively static. The intervention typically creates a hydraulic gradient and a surface discharge that would otherwise not have occurred or have been too dispersed to be noticed. For example, at Woodhall Spa in eastern England, the sinking of an exploration shaft to prospect for coal eventually led to a discharge of mineralized water at the surface, which would have been otherwise undetectable (Mather 2013). Bromide concentrations are above $25\,mg\,L^{-1}$ and iodide is as high as $6\,mg\,L^{-1}$ at Woodhall Spa (Edmunds *et al.* 1969). The strongly ferruginous spa at Trefriw Spa in North Wales arises from a cave that is either a trial level or an actual pyrite mine and has a total iron concentration of 516 mg/L and an aluminium concentration of $129\,mg\,L^{-1}$ (Edmunds *et al.* 1969).

Box 4.3   Influence of groundwater chemistry on the history of malting and brewing in Europe

The brewing of cereal-based ales is known from Ancient Egyptian times, and today, in the British Isles and parts of western, central and eastern Europe, a wide variety of fermented barley beverages are produced, generally known as beer or ale. Much of the variety derives from the geology of the water supplies, with the hydrochemistry of the water used in the brewing process exerting a strong control over the flavour, alcohol content, colour, head retention and clarity of the beer (Cribb 2005). Historically, groundwater sources have been preferred to surface water sources for three main reasons: the natural constituents of groundwater can provide the right quality of water; the quality and temperature tend to be more consistent; and normally there are fewer pollution problems (Lloyd 1986). In general, anions in groundwater exert flavour effects, whereas cations affect the brewing process. Efficient mashing, in which fermentable sugars are leached from the cereals, is crucial and typically needs a pH within the range of 5.2–5.8 (Maltman 2019).

Four cations are particularly important, of which calcium is the most significant. Calcium stabilizes the enzyme α-amylase and helps the breakdown of starch from the barley malt. It also precipitates phosphate and so increases the acidity of the wort, which influences the strength and character of the fermentation process. Magnesium promotes a sour to bitter taste, but retards phosphate precipitation, which in turn stops the required drop in pH level. Sodium and potassium in small amounts give a salty taste. Of the anions, sulphate, carbonate and chloride are important (Cribb 2005). With increased knowledge of the biochemistry of malting and wort production, it is apparent that calcium and carbonate (or bicarbonate) are significant factors, given their influence on pH (Hough 1985).

Large quantities of water are used in the brewing process, including for cooling purposes, with around 300 L of water required to produce 1 L of beer. The concentration of large populations in urban conurbations during the Industrial Revolution enabled centralized brewing to become commercially viable, and it soon became evident that certain urban areas with a reliable water supply, for example London and Edinburgh, consistently delivered a more desirable product (Maltman 2019). To illustrate, hydrochemical analyses of brewing waters in famous beer-producing towns and cities in the 'beer belt' of Europe are given in Table 4.5.

Burton-upon-Trent in the English Midlands is regarded as the 'home of British brewing' with beer first brewed in a local abbey in the sixth century. The presence in groundwater of high sulphate concentrations derived from Triassic gypsiferous marls allows the full extraction of bitter oils from the hops, not only as an important taste factor but also acts as a preservative, such that Burton ales were exported for the British army in the nineteenth century as India Pale Ale (Lloyd 1986; Cribb 2005). The hydrogeology in the River Trent valley at Burton is unique. The river has eroded the Triassic Mercia Mudstone Group that overlies the regional Triassic sandstone aquifer, depositing a sequence of alluvial sands and gravels. The Mercia Mudstone Group contains gypsum beds and thin, persistent sandstones rich in sulphate as a result of gypsum dissolution. Groundwater abstracted from the alluvial deposits contains not only bicarbonate but also sulphate derived from mixing with groundwater from the adjacent Mercia Mudstone Group. The mixed waters have traditionally provided the brewing waters at Burton (Lloyd 1986). Because of the variable mixing of groundwaters at the alluvial margin, the chemistry varies. Hence, careful well management is necessary, given the considerable water requirements for production from both the sands and gravels and underlying Triassic sandstone aquifers (Lloyd 1986).

Similar, suitable hydrogeological conditions exist at other early brewery towns.

Box 4.3   (Continued)

Table 4.5   Major ion analyses of waters used for brewing (concentrations in mg L^{-1}) (Lloyd 1986; Cribb 2005).

Location	Ca^{2+}	Mg^{2+}	Na$^+$	SO$_4{}^{2-}$	Cl$^-$	HCO$_3{}^-$	TDS
Burton-upon-Trent, Midlands, England	283	90	29	725	54	171	1401
Birmingham, Midlands, England	148	48	28	240	77	260	750
Manchester, North West England	97	33	36	112	66	250	557
Tadcaster, North East England	184	71	44	405	87	262	930
London, South East England	109	5	29	69	37	128	428
Wrexham, North Wales	22	3	20	33	25	36	140
Edinburgh, Scotland	140	36	92	231	60	210	800
Dublin, Ireland	100	16	0	17	17	150	305
Dortmund, Germany	260	23	69	283	106	270	1011
Munich, Germany	80	19	1	5	1	164	273
Plzeň, Czech Republic	7	1	3	6	5	9	31

(*Sources:* Lloyd, J.W. (1986) Hydrogeology and beer. *Proceedings of the Geologists Association* **97**, 213–219.; Cribb, S.J. (2005) Geology of beer. In: *Encyclopedia of Geology* (eds R.C. Selley, L.R.M. Cocks and I.R. Plimer). Elsevier Ltd., Oxford, pp. 78–81.) Note that values of TDS content for Manchester and Tadcaster are estimated from EC values (857 and 1430 μS cm^{-1}, respectively) using eq. 4.3 and a correlation factor, k_e, of 0.65.

For example, Tadcaster in Yorkshire in North East England is underlain by Upper Permian marl containing lenses of gypsum and anhydrite overlying the dolomitic Magnesian Limestone aquifer, and with a water yield enhanced by several permeable faults (Maltman 2019). The English Midlands and North West England, centred on Permo-Triassic sandstones below Birmingham, Liverpool (Fig. 4.4) and Manchester, include evaporitic halite deposits and are associated with the production of a sweet, yet still bitter, mild ale due to the presence of sodium, sulphate and chloride.

In London and Jutland in Denmark, where the water is dominated by carbonate-rich sources derived from the Cretaceous Chalk aquifer, and in Dublin and Cork, where carbonate-dominated water is sourced from Carboniferous limestones, the higher pH of these groundwaters and lower sulphate values result in a sweet, dark beer or stout, flavoured with roasted malt. Roasted malts contain phosphates that are released on mashing as phosphoric acid, which assist in reducing the pH to an acceptable range (Maltman 2019). In London, this drink was known as porter due its popularity in the Billingsgate and Covent Garden markets (Cribb 2005). Perhaps the most extreme case of using roasted malts is in Dublin, most famously at the Guinness brewery. The brewery obtains water from Blessington Reservoir in the Wicklow Mountains to the west of Dublin that is fed with hard water from limestone gravels (Maltman 2019).

In continental Europe, the beer from Dortmund, Germany, is traditionally supplied by water from the Carboniferous Coal Measures. Groundwater from the Coal Measures contains high carbonate content and enough sulphate to produce a clean, dry beer that is well hopped. In Munich, water drawn from more recent Pleistocene fluvial and glacio-fluvial sands and gravels derived from the nearby calcareous rocks of the Bavarian Alps, has a high bicarbonate content, similar in composition to Dublin water, and produces a sweet, full, dark and brown beer, similar to stouts and porters, again with the use of roasted malts.

(Continued)

Box 4.3   (Continued)

Fig. 4.4   The former Robert Cain Brewery, Stanhope Street, Liverpool, situated on the Permo-Triassic sandstone aquifer of North West England (see Box 4.9 for a description of the hydrochemical water types found in the Lower Mersey Basin). The arrangement of the external buildings in the form of a vertical tower identifies the brewery as an example of a tower brewery, common in the late Victorian period. The purpose of a tower brewery is to allow the multi-stage brewery flow process to continue by gravity, rather than lifting or pumping the brew liquor between stages. As here, the highest floor would have been the water tower. Cain's Mersey Brewery started around 1875 and the elaborate, five-storey front section, with its red brick and Welsh Ruabon terracotta facade, was built in 1896–1902 (Pearson and Anderson 2010). (*Source:* Dave Ellison/Alamy Stock Photo.)

In contrast, in the Czech Republic, soft water from older, Palaeozoic strata low in calcium and sulphate produces alcoholically strong, lightly hopped and texturally thin beer. Such beer was initially developed in Plzeň (Pilsen) located partly on Stephanian (Upper Carboniferous) lacustrine, feldspathic sandstones and partly on the weakly fissured Upper Proterozoic metamorphic rocks from which groundwater was drawn. The finished product, fermented beer, was stored in caves at a cool temperature in a process known as 'lagering'. The water from České Budějovice (Budweis), 150 km to the south-east of Plzeň, is even softer and drawn from a 300-m deep artesian well in fractured, older Proterozoic gneisses and migmatites. The original Budweiser beer is still produced in this locality (Maltman 2019).

The approach to brewing lagers with waters of low ionic strength spread rapidly in the late nineteenth century throughout northern Europe in regions with similar waters, including in North Wales, the Netherlands, Germany and Denmark. In Scandinavia, water is derived from poorly mineralized Precambrian strata, and beers of a lager-style, often flavoured with fruit, for example juniper berries in Finland, are produced (Cribb 2005).

It is concluded that hydrogeological conditions and groundwater chemistry have had a considerable influence on the location and continuance of the brewing industry both in terms of supply and taste. Today, with knowledge of the ionic composition of natural waters and the biochemistry of malting and wort production, it is possible to mimic the waters of the historic centres such as Burton-upon-Trent or elsewhere. Now modern breweries worldwide can use mains water that is deionised and then reconstituted to a given composition for a particular brew (Hough 1985; Cribb 2005).

and geochemical law which can be formulated as the cycle of metamorphism of natural waters in the crust of weathering. Chebotarev (1955) recognized that the distribution of groundwaters with different hydrochemical *facies* depended on rock–water interaction in relation to hydrogeological environment, with groundwaters evolving from bicarbonate waters at outcrop to saline waters at depth in the Earth's crust.

In a later paper, Hanshaw and Back (1979) described the chemistry of groundwater as a result of the intimate relationship between mineralogy and flow regime because these determine the occurrence, sequence, rates and progress of reactions. With reference to carbonate aquifers, Hanshaw and Back (1979) presented the conceptual model shown in Fig. 4.5 to depict the changes in groundwater chemistry from the time of formation of a carbonate aquifer through to the development of the aquifer system. When carbonate sediments first emerge from the marine environment, they undergo flushing of seawater by freshwater during which

time the salinity decreases and the hydrochemical facies becomes dominated by Ca-HCO$_3$. At this time, the carbonate sediments are selectively dissolved, recrystallized, cemented and perhaps dolomitized to form the rock aquifer. Gradually, as recharge moves downgradient (R → D in Fig. 4.5), Mg^{2+} increases due to dissolution of dolomite and high-magnesium calcite while Ca^{2+} remains relatively constant. With this chemical evolution, SO$_4^{2-}$ increases as gypsum dissolves and HCO$_3^-$ remains relatively constant. For coastal situations or where extensive accumulations of evaporite minerals occur, highly saline waters or brines result (R → M/B in Fig. 4.5).

In interpreting groundwater chemistry and identifying hydrochemical processes, it is useful to adopt the concept of hydrochemical facies or water type introduced by Chebotarev (1955). A hydrochemical facies is a distinct zone of groundwater that can be described as having cation and anion concentrations within definite limits. A pictorial representation of the typical changes in

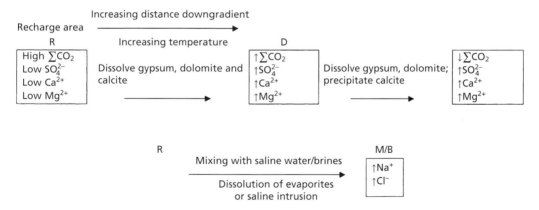

Fig. 4.5 Schematic model showing the evolution of the chemical character of groundwater in carbonate aquifers. In areas of recharge (R), the high concentrations of CO$_2$ and low dissolved solids content cause solution of calcite, dolomite and gypsum. As the concentrations of ions increase and their ratios change downgradient (D), groundwater becomes saturated with respect to calcite which begins to precipitate. Dedolomitization (dissolution of dolomite to form calcite with a crystalline structure similar to dolomite) occurs in response to gypsum solution with calcite precipitation. Where extensive accumulations of evaporite minerals occur, their dissolution results in highly saline brines (B). Another common pathway is caused by mixing with seawater (M) that has intruded the deeper parts of coastal aquifers (Hanshaw and Back 1979). (*Source:* Hanshaw, B.B. and Back, W. (1979) Major geochemical processes in the evolution of carbonate-aquifer systems. *Journal of Hydrology* **43**, 287–312.)

hydrochemical facies along a groundwater flowpath is shown in Fig. 4.6. Dilute rainwater with a Na–Cl water type and containing CO_2 enters the soil zone whereupon further CO_2, formed from the decay of organic matter, dissolves in the infiltrating water. Where relevant, the application of agricultural chemicals such as fertilizers add further Na^+, Cl^- and K^+, NO_3^-, PO_4^{3-}. Within the soil and unsaturated zone, the dissolved CO_2 produces a weakly acidic solution of carbonic acid, H_2CO_3, which itself dissociates and promotes the dissolution of calcium and magnesium carbonates giving a Ca–Mg–HCO_3 water type.

Away from the reservoir of oxygen in the soil and unsaturated zone, the groundwater becomes increasingly anoxic below the water table with progressive reduction of oxygen, nitrate and sulphate linked to bacterial respiration and mineralization of organic matter. Under increasing reducing conditions, Fe and Mn become mobilized and then later precipitated as metal sulphides (Section 4.9). In the presence of disseminated clay material within the aquifer, ion exchange causes Ca^{2+} to be replaced by Na^+ in solution and the water evolves to a Na–HCO_3 water type. In the deeper, confined section of the aquifer, mixing

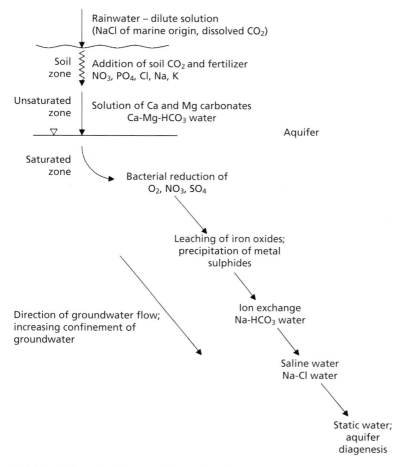

Fig. 4.6 Schematic diagram of the evolution of groundwater along a flowpath from recharge area to confined section showing important hydrochemical processes that affect the chemical composition of groundwater. The processes shown are not wholly sequential in that ion exchange can occur without substantial redox reactions (Section 4.9).

with saline water occurs to produce a Na–Cl water type before a region of static water and aquifer diagenesis is reached. Either part or all of this classic sequence of hydrochemical change is identified in a number of aquifers including the Great Artesian Basin of Australia (see Box 2.11), the Floridan aquifer system (see Box 4.4) and the Chalk aquifer of the London Basin (Ineson and Downing 1963; Mühlherr *et al.* 1998).

Box 4.4   Hydrochemical evolution in the Floridan aquifer system

The Floridan aquifer system occurs in the south-east of the United States (Fig. 4.7) and is one of the most productive aquifers in the world. The aquifer system is a vertically continuous sequence of Tertiary carbonate rocks of generally high permeability. Limestones and dolomites are the principal rock types, although in south-western and north-eastern Georgia and in South Carolina, the limestones grade into lime-rich sands and clays. The Floridan aquifer is composed primarily of calcite and dolomite with minor gypsum, apatite, glauconite, quartz, clay minerals and trace amounts of metallic oxides and sulphides.

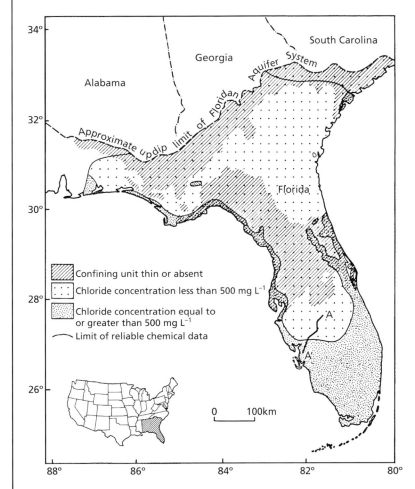

Fig. 4.7 Map of the extent of the Upper Floridan aquifer showing the relation between the spatial distribution of chloride concentrations and areas of confined-unconfined groundwater conditions (Sprinkle 1989). (*Source:* Sprinkle, C.L. (1989). Geochemistry of the Florida Aquifer System in Florida and in parts of Georgia, South Carolina, and Alabama. *United States Geological Survey Professional Paper* 1403-I, 105 pp.)

(*Continued*)

Box 4.4 (Continued)

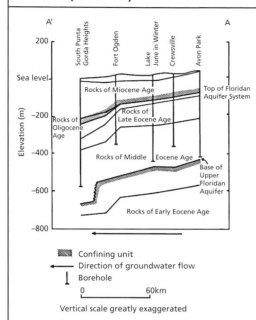

Fig. 4.8 Hydrogeological section of part of the Upper Floridan aquifer. The line of section A-A′ from Highlands County to Charlotte County is shown in Fig. 4.7 (Sprinkle 1989). (*Source:* Sprinkle, C.L. (1989). Geochemistry of the Florida Aquifer System in Florida and in parts of Georgia, South Carolina, and Alabama. *United States Geological Survey Professional Paper* 1403-I, 105 pp.)

The aquifer system generally consists of an Upper and Lower Floridan aquifer separated by a less permeable confining unit having highly variable hydraulic properties (Fig. 4.8). The Upper Floridan aquifer is present throughout, but the Lower Floridan is absent in most of northern Florida and Georgia. Recharge occurs primarily in outcrop areas of Alabama, Georgia and north-central Florida. Most discharge is to rivers and springs with only a small fraction (<5%) discharged directly into the sea. Where the system is unconfined, recharge is rapid and groundwater circulation and discharge rates are high, and secondary permeability is developed by mineral dissolution. Where confining units are thick, the carbonate chemistry of the groundwater evolves in a closed-system (Section 4.7) and the development of secondary permeability and flushing of residual saline water within the aquifer system is slow (Sprinkle 1989).

In an extensive study of the hydrochemistry of the Floridan aquifer system, Sprinkle (1989) identified the following major hydrochemical processes:

1) dissolution of aquifer minerals towards equilibrium;
2) mixing of groundwater with seawater, recharge or leakage;
3) sulphate reduction; and
4) cation exchange between water and rock minerals.

A sequence of hydrochemical evolution is observed starting with calcite dissolution in recharge areas that produces a $Ca-HCO_3$-dominated water type with a TDS concentration of generally less than 250 mg L^{-1}. Down-gradient, dissolution of dolomite leads to a $Ca-Mg-HCO_3$ hydrochemical facies. Where gypsum is abundant, sulphate becomes the predominant anion. In coastal areas, as shown in Fig. 4.7, seawater increases the TDS concentrations and the hydrochemical facies changes to $Na-Cl$. Leakage from underlying or adjacent sand aquifers in south-central Georgia enters the Floridan aquifer and lowers TDS concentrations but does not change the hydrochemical facies. In the western panhandle of Florida, cation exchange leads to the development of a $Na-HCO_3$ water type.

The hydrogeological and hydrochemical sections shown in Figs. 4.8 and 4.9 illustrate the main features of the hydrochemical environment in south Florida. The confining unit of the Upper Floridan aquifer is thick along this section, and there are evident changes in the concentrations of the major ions in the general direction of groundwater flow. Chemical stratification of the Upper Floridan aquifer may be indicated by the fact that concentrations of Ca^{2+}, Mg^{2+} and SO_4^{2-} in water from the Lake June in Winter (LJIW) well are greater than in water from the downgradient Fort Ogden well, which is about 100 m shallower than the LJIW well. However, the Fort Ogden well is nearer the coast and Na^+ and Cl^- concentrations increase by about three times compared with concentrations at the LJIW well.

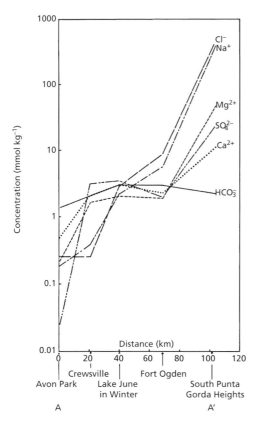

Fig. 4.9 Hydrochemical section along line A-A' (see Fig. 4.7) showing the variation in the concentrations of major ions downgradient in the direction of groundwater flow (Sprinkle 1989). (*Source:* Sprinkle, C.L. (1989). Geochemistry of the Florida Aquifer System in Florida and in parts of Georgia, South Carolina, and Alabama. *United States Geological Survey Professional Paper* 1403-I, 105 pp.)

4.5 Groundwater sampling and graphical presentation of hydrochemical data

The aim of field sampling is to collect a raw water sample that is representative of the hydrochemical conditions in the aquifer. To obtain a groundwater sample for analysis, some type of sampling device is required, for example a bailer, depth sampler, gas lift sampler or pump (inertial, suction or submersible) in either an open hole or borehole, or from purpose-designed piezometers or nested piezometers. Care must be taken to first flush the water standing in the well or borehole by removing up to three or four times the well or borehole volume prior to sampling. Abstraction boreholes that are regularly pumped provide a sample that is representative of a large volume of aquifer, whereas depth-specific sampling from a multi-level system, or from a section of borehole column isolated by inflatable packers, provides discrete samples that are representative of a small volume of aquifer. For a comparison of the performances of various depth sampling methods, the reader is referred to Price and Williams (1993) and Lerner and Teutsch (1995).

Groundwater samples can be conveniently collected in new or cleaned screw-cap, high-density polyethylene bottles. When filling sampling bottles, the bottle should first be rinsed two to three times with the water sample and, for samples containing suspended sediment or particles, filtered using a 0.45 μm membrane filter. A 0.1 μm filter is recommended where trace metals are to be analysed. Typically, several filled sample bottles are collected at one site, with one sample preserved to stabilize the dissolved metals in solution. The conventional method of preservation prior to analysis is to add a few drops of concentrated nitric or acetic acid after the sample has been collected and filtered to lower the pH to 2. A test using pH paper can be applied to indicate if the pH after acidification is adequate. Once collected, samples should be stored in a cool box and shipped as soon as possible, ideally the same day, to the laboratory for storage at 4 °C in order to limit bacterial activity and degradation of nutrient species (NO_3^-, SO_4^{2-} and PO_4^{3-}). Freezing of samples should be avoided as this can lead to precipitation of some elements.

While at a sampling location, and as part of the sampling methodology, the well-head chemistry should be measured, again with the aim of collecting data representative of the hydrochemical conditions in the aquifer.

Parameters such as temperature, pH, redox potential (E*h*) and dissolved oxygen (DO) content will all change once the groundwater sample is exposed to ambient conditions at the ground surface, and during storage and laboratory analysis. On-site measurement of EC and alkalinity should also be conducted. To limit the exsolution of gases from the groundwater, for example the loss of CO_2 which will cause the pH value of the sample to increase, and in order to prevent mixing of atmospheric oxygen with the sample and so affecting the E*h* and DO values, measurements should be made on a flowing sample within an isolation cell or flow cell (Fig. 4.10). The flow cell is designed with a plastic tube leading from the sampling tap into the base of the cell and an overflow tube at the top of the cell. The lid of the flow cell has access holes into which the various measurement probes can be inserted. A typical time to allow for the E*h* and DO electrodes to stabilize in the flow cell prior to recording the final values is between 20 and 30 minutes. For further information on hydrochemical parameter measurement and sample collection, including sampling of springs and pore waters, the reader is referred to Lloyd and Heathcote (1985) and Appelo and Postma (2005).

Before attempting any hydrochemical interpretation, it is first necessary to check the quality of laboratory chemical analyses and that the condition of electroneutrality has been met whereby the sum of the equivalent weight of cations equates to the sum of the equivalent weight of anions. This check is commonly carried out by calculating the ionic balance error of the major ions where

$$
\text{Ion balance error}(\%)
$$
$$
= \left\{ \left(\sum \text{cations} - \sum \text{anions} \right) / \right.
$$
$$
\left. \left(\sum \text{cations} + \sum \text{anions} \right) \right\} \times 100
$$

(eq. 4.4)

An ionic balance error of less than 5% should be achievable with modern analytical equipment and certainly less than 10%. Larger errors are unacceptable and suggest that one or more analyses are in error.

Various methods have been developed for the visual inspection of hydrochemical data in order to look for discernible patterns and trends. By grouping chemical analyses, it

Fig. 4.10 Flow cell with inserted measurement probes for monitoring the well-head chemistry of a pumped groundwater sample.

becomes possible to identify hydrochemical facies and begin to understand the hydrogeological processes that influence the groundwater chemistry. The simplest methods include plotting distribution diagrams, bar charts, pie charts, radial diagrams and pattern diagrams (as presented by Stiff 1951). Although these are easy to construct, they are not convenient for graphical presentation of large numbers of analyses and for this reason, other techniques are used including Schoeller (named after Schoeller 1962), trilinear (Piper 1944) and Durov (Durov 1948) diagrams.

The distribution diagram shown in Fig. 4.11 represents concentrations of Na^+ and Cl^- in groundwaters of the Milk River aquifer system located in the southern part of the Western Canadian Sedimentary Basin. The Milk River aquifer is an artesian aquifer, one of several sandstone units developed within a Tertiary–Cretaceous section comprising mainly shale and mudstones. The aquifer crops out in the southern part where recharge occurs, but in the northern part is covered by up to 400 m of younger rocks. Groundwater flow is generally northward with a significant component of upward leakage through confining beds. Concentrations of Na^+ and Cl^- are characterized by marked spatial variability and, as shown in Fig. 4.11, are lowest in the south where most of the freshwater recharge occurs. An important feature is the marked northward tongue of fresh groundwater (delimited by the 100 mg L^{-1} Cl$^-$ contour) that coincides with a well-developed zone of aquifer permeability. This hydrochemical pattern is interpreted as a broad zone of mixing that forms as meteoric recharge water flushes pre-existing more saline formation water (Schwartz and Muehlenbachs 1979).

The Schoeller diagram visualizes concentrations as meq L^{-1} (see Box 4.1 for a discussion of concentration units). The example shown in Fig. 4.12 is for the chemistry of groundwater in crystalline rocks in which the near-surface groundwater is typically recharged by rain and snowmelt. The rock-forming minerals in

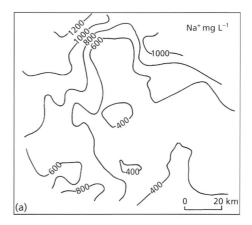

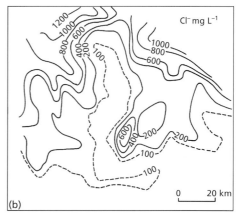

Fig. 4.11 Distribution diagrams of (a) Na$^+$ and (b) Cl$^-$ concentrations in groundwaters from the Milk River aquifer system, Alberta (Schwartz and Muehlenbachs 1979). (*Source:* Schwartz, F.W. and Muehlenbachs, K. (1979) Isotope and ion geochemistry of groundwaters in the Milk River Aquifer, Alberta. *Water Resources Research* **15**, 259–268. © 1979, John Wiley & Sons.)

crystalline rocks are silicates, and their chief cations are Ca^{2+}, Mg^{2+}, Na^+ and K^+. For North America, the principal weathering agent is carbonic acid, chiefly from CO_2 dissolved in the soil zone. Under these conditions, the water typically attains a Ca– or Na–HCO_3 composition. Variants of this general water type are due to major differences in the composition of the aquifer rock (Trainer 1988).

To illustrate simple methods for presenting hydrochemical data, Fig. 4.13 shows bar charts, pie charts, a radial diagram and pattern

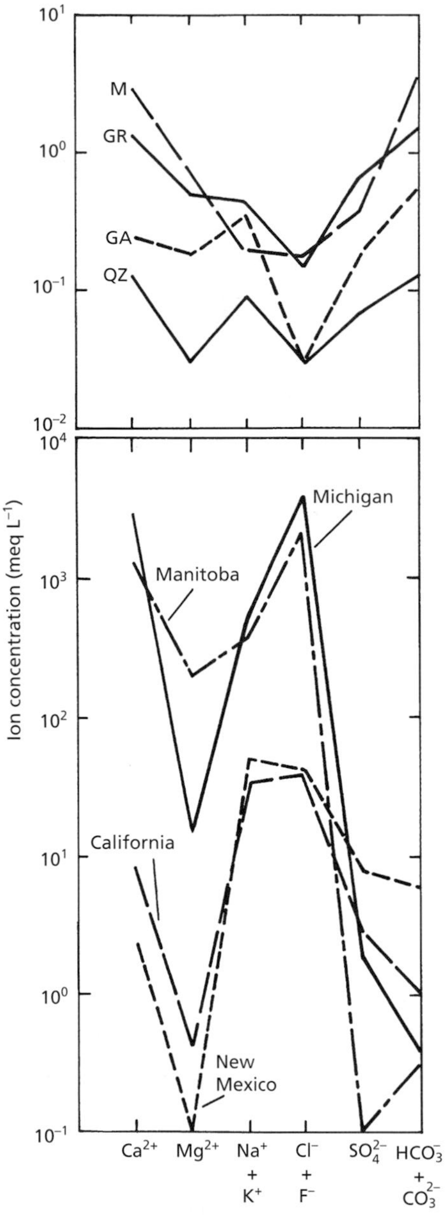

Fig. 4.12 Schoeller (semi-logarithmic) diagram illustrating near-surface groundwater chemistry in crystalline rocks. Symbols on plot indicate rock type (QZ, quartzite; GA, gabbro; GR, granite; M, marble). Analyses shown are for Houghton County, Michigan; Thompson, Manitoba; New Mexico (mixture of native and injected surface waters) and California (Trainer 1988). (*Source:* Trainer, F.W. (1988) Plutonic and metamorphic rocks. In: *Hydrogeology. The Geology of North America, Vol. O-2*, (eds W. Back, J.S. Rosenshein and P.R. Seaber). The Geological Society of North America, Boulder, Colorado, pp. 367–380.)

diagrams for the major ion analyses given in Table 4.6. These analyses are for the Milligan Canyon area, south-west Montana, in the eastern region of the Northern Rocky Mountain Province. The area is a broad synclinal basin with folded carbonate rocks of Palaeozoic and Mesozoic age on the southern rim, and volcanic breccia and andesitic lava overlying older, deformed rocks on the northern rim. The basin is infilled with unconsolidated alluvial and aeolian deposits and Tertiary sediments of siltstones, limestones and sandstones that contain deposits of gypsum ($CaSO_4.2H_2O$) and anhydrite ($CaSO_4$). A number of Upper Cretaceous and early Tertiary igneous intrusives are also present. Groundwater flow is predominantly from west to east but with a contribution of upward groundwater flow from the Madison limestone aquifer underlying the basin. High groundwater yields are obtained from the Tertiary basin.

The same major ion analyses for the Milligan Canyon area are also presented as trilinear and Durov diagrams in Figs. 4.14 and 4.15. With these methods of graphical presentation, the concentrations of individual samples are plotted as percentages of the total cation and/or anion concentrations, such that samples with very different total ionic concentrations can occupy the same position in the diagrams. Also, with the trilinear diagram (Fig. 4.14), samples that plot on a straight line within the central diamond field represent mixing of groundwaters between two end-member solutions, for example freshwater and saline water. Further hydrochemical interpretations can be obtained from the Durov diagram (Fig. 4.15). Lines from the central square field can be extended to the adjacent scaled rectangles to allow for representation in terms of two further parameters.

The next step in the hydrochemical interpretation after plotting the chemical data in a variety of ways is to identify the hydrochemical facies present and to prepare maps and cross-sections to show the regional

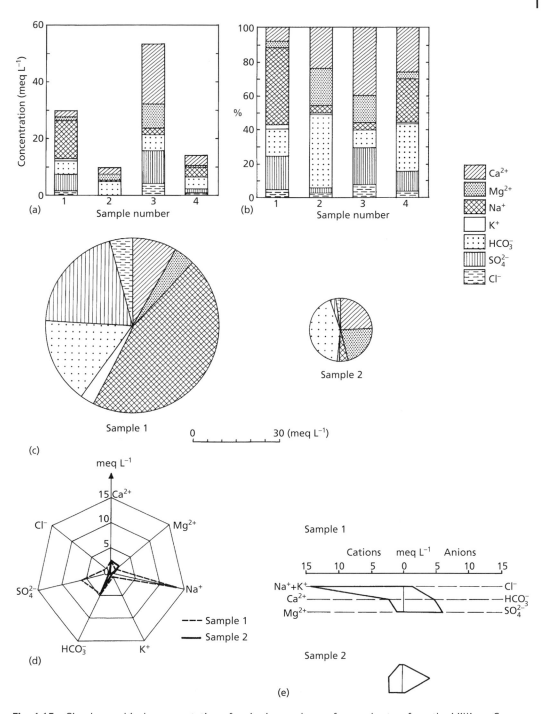

Fig. 4.13 Simple graphical representation of major ion analyses of groundwaters from the Milligan Canyon area, south-west Montana, given in Table 4.6. A stacked bar chart is shown in (a) with major ion concentration values given in meq L^{-1}. The same information is shown in (b) but with the individual major ions shown as a percentage of the total ion concentration for each sample. The two pie charts in (c) represent samples 1 and 2 with the radii scaled to the total ion concentration of each sample. Samples 1 and 2 are also shown in the radial diagram (d) and the pattern, or Stiff, diagram (e).

Table 4.6 Major ion analyses of four groundwater samples from the Milligan Canyon area, south-west Montana, used to plot Figs. 4.13–4.15 (Krothe and Bergeron 1981).

Water type	Sample number	Ca^{2+} (mg L^{-1})	Mg^{2+} (mg L^{-1})	Na^+ (mg L^{-1})	K^+ (mg L^{-1})	HCO_3^- (mg L^{-1})	SO_4^{2-} (mg L^{-1})	Cl^- (mg L^{-1})
Ca–HCO₃	2	47	25.2	10	3.1	251	29	7
Ca–SO₄	3	426	103	56	5.4	328	1114	143
Na–HCO₃	4	73	6.4	83	4.5	236	158	17
Na–SO₄	1	49	13.7	312	29.2	286	566	43

Water type	Sample number	Ca^{2+} (meq L^{-1})	Mg^{2+} (meq L^{-1})	Na^+ (meq L^{-1})	K^+ (meq L^{-1})	HCO_3^- (meq L^{-1})	SO_4^{2-} (meq L^{-1})	Cl^- (meq L^{-1})
Ca–HCO₃	2	2.35	2.10	0.43	0.08	4.11	0.30	0.20
Ca–SO₄	3	21.30	8.58	2.43	0.14	5.38	11.60	4.09
Na–HCO₃	4	3.65	0.53	3.61	0.12	3.87	1.65	0.49
Na–SO₄	1	2.45	1.14	13.57	0.75	4.69	5.90	1.23

(*Source:* Data from Krothe, N.C. and Bergeron, M.P. (1981) Hydrochemical facies in a Tertiary basin in the Milligan Canyon area, Southwest Montana. *Ground Water* **19**, 392–399.)

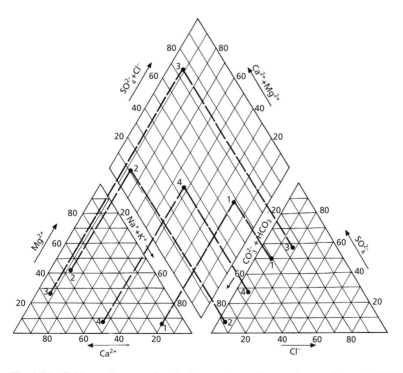

Fig. 4.14 Trilinear diagram of major ion analyses of groundwaters from the Milligan Canyon area, south-west Montana, given in Table 4.6. The individual cation and anion concentration values are expressed as percentages of the total cations and total anions and then plotted within the two triangular fields at the lower left and lower right of the diagram. The two points representing each sample are then projected to the central diamond field and the point of intersection found.

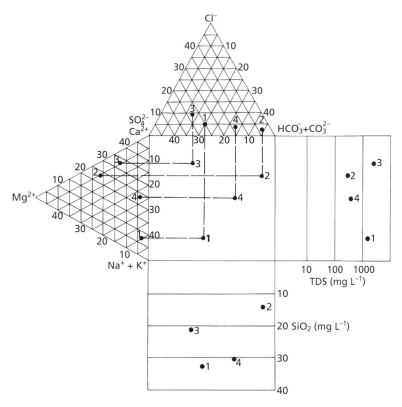

Fig. 4.15 Durov diagram of major ion analyses of groundwaters from the Milligan Canyon area, south-west Montana, given in Table 4.6 with additional SiO_2 and TDS data from Krothe and Bergeron (1981). The method of plotting is similar to that for the trilinear diagram shown in Fig. 4.14, but with the additional projection of points from the central square to the two adjacent scaled rectangles. (*Source:* Data from Krothe, N.C. and Bergeron, M.P. (1981) Hydrochemical facies in a Tertiary basin in the Milligan Canyon area, Southwest Montana. *Ground Water* **19**, 392–399.)

distribution of water types. In the example of groundwaters in the Milligan Canyon area, Fig. 4.16 shows the overall hydrochemical and hydrogeological interpretation. In Fig. 4.16a, the central diamond field of the trilinear diagram is shown, and the samples grouped depending on the hydrochemical facies present (Ca–HCO_3, Ca–SO_4, Na–HCO_3 and Na–SO_4). When the spatial distribution of the hydrochemical facies is plotted in Fig. 4.16b and compared with the regional geology and groundwater flow direction obtained from a potentiometric map, it is clear that an elongate groundwater body with a distinct Ca–SO_4 water type is found in the centre of the basin. The

hydrochemical interpretation given by Krothe and Bergeron (1981) is that recharge occurring in the structurally high areas around the rim of the basin, and formed by older carbonate rocks, results in groundwater of a Ca–HCO_3 character. As groundwater flows through the Tertiary deposits, the recharge water undergoes a change in chemical character with increasing residence time and possible solution of anhydrite and/or gypsum resulting in the Ca–SO_4 water type and high concentrations of SO_4^{2-} and TDS. Groundwaters with high Na^+ concentrations are localized in the north-east and east-central portions of the basin at the point of groundwater discharge out of the basin

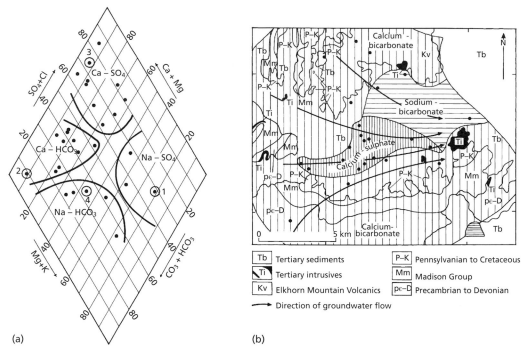

Fig. 4.16 (a) Identification using a trilinear diagram and (b) mapping of the distribution of hydrochemical facies in the Milligan Canyon area, south-west Montana. The four numbered points refer to the major ion analyses given in Table 4.6 (Krothe and Bergeron 1981). (*Source:* Krothe, N.C. and Bergeron, M.P. (1981) Hydrochemical facies in a Tertiary basin in the Milligan Canyon area, Southwest Montana. *Ground Water* **19**, 392–399. © 1981, John Wiley & Sons.)

where the HCO_3^- ion is again dominant. A combination of three factors accounts for these chemical changes, including: sulphate reduction; mixing between Ca–HCO3 and Ca–SO4 waters at the exit from the basin; and possible ion exchange between Ca^{2+} and Na^+ associated with montmorillonite clay in the Tertiary sediments and the weathering of the Elkhorn Mountain Volcanics that are rich in Na-plagioclases.

It might be concluded from Fig. 4.16 that sharp boundaries exist between adjacent water types. In reality, the groundwater chemistry is evolving along a flowpath, and this can be illustrated by constructing a hydrochemical section such as the example shown for the Floridan aquifer system described in Box 4.4

(Fig. 4.9). Another interpretation technique for understanding regional hydrochemistry is to prepare a series of X–Y plots and dilution diagrams on either linear or semi-logarithmic paper that can demonstrate hydrochemical processes such as simple mixing, ion exchange and chemical reactions. Examples are shown in Fig. 4.17.

All the graphical techniques described here have been applied principally in regional hydrochemical studies. For contaminated groundwater investigations, these techniques are not always appropriate, except for the simpler diagrams such as bar charts, due to the wide spatial variation in concentrations of contaminant species between background and contaminated groundwaters. In this case,

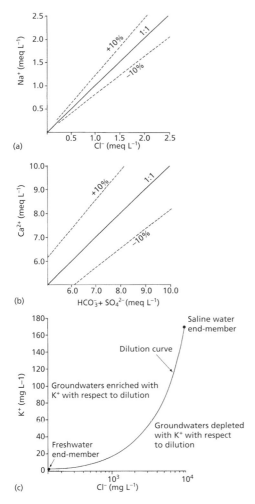

(a)

(b)

(c)

Fig. 4.17 Graphical methods for exploring hydrochemical processes. The X–Y plots in (a) and (b) indicate simple mixing within confidence limits of ±10% between groundwaters that plot close to the lines with a 1:1 ratio. Processes such as ion exchange and sulphate reduction would cause samples to plot above the line in (a) and deviate from the line in (b). The semi-logarithmic plot of a dilution diagram in (c) shows a line representing mixing between fresh and saline end-member groundwaters. Points plotting above and below the dilution line represent enrichment and depletion, respectively, of the ionic concentration with respect to the conservative chloride ion.

plotting the contaminant concentrations as pie charts on a site map can give a visual indication of the 'hot spots' of contamination. An example is shown in Fig. 4.18.

4.6 Concept of chemical equilibrium

Hydrochemical processes in groundwater can be viewed as proceeding slowly towards chemical equilibrium, a concept that is common to aqueous chemistry. Shifts in a system's equilibrium can be qualitatively described by Le Chatelier's Principle that states that, if a system at equilibrium is perturbed, the system will react in such a way as to minimize the imposed change. For example, consider groundwater flowing through a limestone aquifer composed of calcite:

$$CaCO_3 + H_2CO_3 \Leftrightarrow Ca^{2+} + 2HCO_3^{2-}$$

(eq. 4.5)

The chemical reaction described by eq. 4.5 will proceed to the right (mineral dissolution) or to the left (mineral precipitation) until equilibrium is reached. Looking at this in another way, any increase in Ca^{2+} or HCO_3^- in solution would be lessened by a tendency for the reaction to shift to the left, precipitating calcite. The concept of chemical equilibrium establishes boundary conditions towards which chemical processes will proceed and can be discussed from either a kinetic or an energetic viewpoint. The kinetic approach is described first.

4.6.1 Kinetic approach to chemical equilibrium

The equilibrium relationship is often called the law of mass action which describes the equilibrium chemical mass activities of a reversible reaction. The rate of reaction is proportional to the effective concentration of the reacting substances. For the reaction,

$$aA + bB \Leftrightarrow cC + dD \qquad (eq. 4.6)$$

the law of mass action expresses the relation between the reactants and the products when the chemical reaction is at equilibrium such that

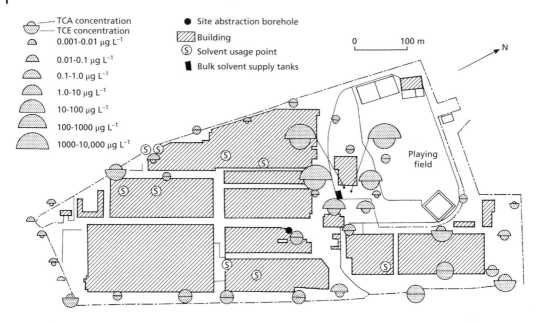

Fig. 4.18 Site plan showing points of solvent use and soil gas concentrations of TCE (trichloroethene) and TCA (1,1,1-trichloroethane) at an industrial site in the English Midlands (Bishop *et al.* 1990). (*Source:* Bishop, P.K., Burston, M.W., Lerner, D.N. and Eastwood, P.R. (1990) Soil gas surveying of chlorinated solvents in relation to groundwater pollution studies. *Quarterly Journal of Engineering Geology* **23**, 255–265.)

$$K = \frac{[C]^c [D]^d}{[A]^a [B]^b} \qquad \text{(eq. 4.7)}$$

where the square brackets indicate the thermodynamically effective concentration, or activity (Box 4.5).

Equation 4.7 is a statement of chemical equilibrium, where K is the thermodynamic equilibrium constant (or stability constant). Values of K depend on temperature with solute concentrations expressed in terms of activities. An equilibrium constant greater than unity suggests that

---

Box 4.5 Active concentration

The active concentration or *activity* of an ion is an important consideration not only in concentrated and complex solutions such as seawater but also for groundwaters and surface waters that contain dissolved ions from many sources. Ions in a concentrated solution are sufficiently close to one another for electrostatic interactions to occur. These interactions reduce the effective concentration of ions available to participate in chemical reactions and, if two salts share a common ion,

they mutually reduce each other's solubility and exhibit the *common ion effect*. In order to predict accurately chemical reactions in a concentrated solution, it is necessary to account for the reduction in concentration as follows:

$$a = \gamma m \qquad \text{(eq. 1)}$$

where a is the solute activity (dimensionless), γ is the constant of proportionality known as the activity coefficient (kg mol^{-1}) and m is the

Box 4.5	(Continued)

molality. In most cases, it is convenient to visualize the activities of aqueous species as modified molalities in order to take account of the influence on the concentration of a given solute species of other ions in solution. The activity coefficient of an ion is a function of the ionic strength, I, of a solution given by

$$I = \frac{1}{2}\sum_i c_i z_i^2 \qquad \text{(eq. 2)}$$

where c_i is the concentration of ion, i, in mol L^{-1}, z_i, is the charge of ion, i and Σ represents the sum of all ions in the solution. As a measure of the concentration of a complex electrolyte solution, ionic strength is better than a simple sum of molar concentrations since it accounts for the effect of the charge of multivalent ions. Freshwaters typically have ionic strengths between 10^{-3} and 10^{-4} mol L^{-1}, whereas seawater has a fairly constant ionic strength of 0.7 mol L^{-1}.

For dilute solutions such as rainwater, γ is about equal to unity. Activity coefficients can be calculated by the extended Debye–Hückel equation, examples of which for a number of charged and uncharged species are shown graphically in Fig. 4.19. In most practical applications involving dilute or fresh groundwaters, it is adequate to assume that the activity of a dissolved species is equal to the concentration, although measured concentrations of any chemical species should strictly be converted to activities before comparison with thermodynamic data.

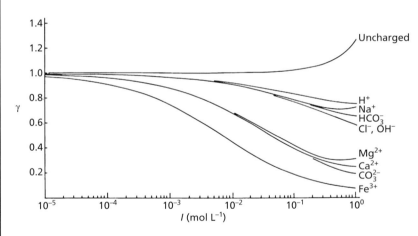

Fig. 4.19 Activity coefficient, γ, as a function of ionic strength for common ionic constituents in groundwater.

equilibrium lies to the right-hand side of the equation describing the chemical reaction and that the forward reaction is favoured.

4.6.2 Energetic approach to chemical equilibrium

In this approach to chemical equilibrium, the most stable composition of a mixture of reactants is the composition having the lowest energy. This more rigorous thermodynamic treatment, compared with the kinetic approach, involves enthalpy, the heat content, H, at constant pressure and entropy, S, a measure of the disorder of a system. The change in enthalpy (ΔH, measured in J mol^{-1}) in a reaction is a direct measure of the energy emitted or adsorbed. The change in entropy in most reactions (ΔS, measured in $J\,mol^{-1}\,K^{-1}$) proceeds to increase disorder, for example by splitting a compound into its constituent ions. For a reversible process, the change in entropy is equal to the amount of heat taken up by a reaction divided by the absolute temperature, T, at which the heat is absorbed.

The total energy released, or the energy change in going from reactants to products, is termed the Gibbs free energy, G (measured in $kJ\,mol^{-1}$). If energy is released, in which case, the products have lower free energy than the reactants, G is considered negative. The change in Gibbs free energy is defined as follows:

$$\Delta G = \Delta H - T\Delta S \qquad \text{(eq. 4.8)}$$

By convention, elements in their standard state ($25\,°C$ and 1 atmosphere pressure) are assigned enthalpy and free energy values of zero. Standard state thermodynamic data, indicated by the superscript $°$, and tabulated as values of standard free energies, enthalpies and entropies, are given in most geochemistry and aqueous chemistry textbooks, for example, Krauskopf and Bird (1995) and Stumm and Morgan (1981). Values of $\Delta G°$ for different reactions can be calculated by simple arithmetic combination of the tabulated values. Any reaction with a negative $\Delta G°$ value will, in theory, proceed spontaneously (the chemical equivalent of water flowing down a hydraulic gradient), releasing energy. The reverse reaction requires an input of energy. For example, consider the reaction of aqueous carbon dioxide (H_2CO_3) with calcite (eq. 4.5). Relevant data for this reaction at standard state are $\Delta G°\ (H_2CO_3) = -623.1\,kJ\,mol^{-1}$, $\Delta G°\ (CaCO_3) = -1128.8\,kJ\,mol^{-1}$, $\Delta G°\ (Ca^{2+}) = -553.6\,kJ\,mol^{-1}$ and $\Delta G°\left(HCO_3^-\right) = -586.8\,kJ\,mol^{-1}$. Therefore, the change in standard free energy for this reaction, $\Delta G° = \sum \Delta G°_{products} - \sum \Delta G°_{reactants} = -1727.2 - (-1751.9) = 24.7\,kJ\,mol^{-1}$.

Since an energetically favoured reaction proceeds from reactants to products, the relationship between ΔG and the equilibrium constant, K, for a reaction is given by

$$\Delta G = -RT\log_e K \qquad \text{(eq. 4.9)}$$

where R is the universal gas constant relating pressure, volume and temperature for an ideal gas ($8.314\,J\,mol^{-1}\,K^{-1}$).

Hence, one useful application of the energetic approach to chemical equilibrium is the use of thermodynamic data to derive equilibrium constants, K, using eq. 4.9. Now, for the reaction given in eq. 4.5, and using eq. 4.9 at standard conditions ($T = 298\,K$), then $\log_e K = -\Delta G°/RT = -(24.7 \times 10^3)/(8.314 \times 298) = -9.97$. Hence, K, the thermodynamic equilibrium constant for the reaction of dissolved carbon dioxide with calcite at $25\,°C$ is equal to 4.68×10^{-5} or, expressed as the negative logarithm to base 10 of K (pK) = 4.33 and $K = 10^{-4.33}$.

4.7 Carbonate chemistry of groundwater

Acids and bases exert significant control over the chemical composition of water (Box 4.6). The most important acid–base system with respect to the hydrochemistry of most natural waters is the carbonate system. The fate of many types of contaminants, for example metal species, can depend on rock–water interactions involving groundwater and carbonate minerals. Later, in Section 5.5.2, the interpretation of groundwater ages based on the carbon-14 dating method will require knowledge of carbonate chemistry and how the water has interacted with carbonate minerals in an aquifer.

The fundamental control on the reaction rates in a carbonate system is the effective concentration of dissolved CO_2 contained in water. The proportion of CO_2 in the atmosphere is about 0.03%, but this increases in the soil zone due to the production of CO_2 during the decay of organic matter, such that the amount of CO_2 increases to several per cent of the soil atmosphere. As groundwater infiltrates the soil zone and recharges the aquifer, reactions can occur

---

Box 4.6   Acid–base reactions

Acids and bases are important chemical compounds that exert particular control over reactions in water. Acids are commonly considered as compounds that dissociate to yield hydrogen ions (protons) in water:

$$HCl_{(aq)} \Rightarrow H^+_{(aq)} + Cl^-_{(aq)} \qquad \text{(eq. 1)}$$

Bases (or alkalis) can be considered as those substances which yield hydroxide (OH^-) ions in aqueous solutions:

$$NaOH_{(aq)} \Rightarrow Na^+_{(aq)} + OH^-_{(aq)} \qquad \text{(eq. 2)}$$

Acids and bases react to neutralize each other, producing a dissolved salt plus water:

$$HCl_{(aq)} + NaOH_{(aq)}$$
$$\Rightarrow Cl^-_{(aq)} + Na^+_{(aq)} + H_2O_{(l)} \qquad \text{(eq. 3)}$$

Hydrochloric acid (HCl) and sodium hydroxide (NaOH) are recognized, respectively, as strong acids and bases that dissociate completely in solution to form ions. Weak acids and bases dissociate only partly.

The acidity of aqueous solutions is often described in terms of the pH scale. The pH of a solution is defined as follows:

$$pH = -\log_{10}[H^+] \qquad \text{(eq. 4)}$$

Water undergoes dissociation into two ionic species as follows:

$$H_2O = H^+ + OH^- \qquad \text{(eq. 5)}$$

In reality, H^+ cannot exist, and H_3O^+ (hydronium) is formed by the interaction of water and H^+. However, it is convenient to use H^+ in chemical equations. From the law of mass action, the equilibrium constant for this dissociation is

$$K_{H_2O} = \frac{[H^+][OH^-]}{[H_2O]} = 10^{-14}$$
$$\text{(eq. 6)}$$

For water that is neutral, there are exactly the same concentrations (10^{-7}) of H^+ and OH^- ions such that pH = $-\log_{10}[10^{-7}]$ = 7. If pH < 7, there are more H^+ ions than OH^- ions and the solution is acidic. If pH > 7, there are more OH^- ions than H^+ ions and the solution is basic. It is important to notice that pH is a logarithmic scale and so it is not appropriate to average pH values of solutions. Instead, it is better to average H^+ concentrations.

Acid–base pairs commonly present in groundwater are those associated with carbonic acid and water itself. Boric, orthophosphoric and humic acids are minor constituents of groundwater but are relatively unimportant in controlling acid–base chemistry. Many aquifers of sedimentary origin contain significant amounts of solid carbonate such as calcite ($CaCO_3$), a fairly strong base that contributes a carbonate ion, thus rendering the solution more alkaline, and dolomite ($CaMg(CO_3)_2$) which participate in equilibrium reactions involving carbonic acid. All acid–base reactions encountered in natural aqueous chemistry are fast such that acid–base systems are always in equilibrium in solution.

---

with carbonate minerals, typically calcite and dolomite, which are present.

CO_2 dissolves in water forming small quantities of weak carbonic acid, as follows:

$$CO_{2_{(g)}} + H_2O \Leftrightarrow H_2CO_3 \qquad \text{(eq. 4.10)}$$

From the law of mass action, the equilibrium constant for this reaction is

$$K_{CO_2} = \frac{[H_2CO_3]}{[H_2O]\left[CO_{2_{(g)}}\right]} \qquad \text{(eq. 4.11)}$$

According to Henry's Law, in dilute solutions, the partial pressure of a dissolved gas, expressed in atmospheres, is equal to its molality (or activity for dilute solutions). Also, given that the activity of water is unity except for very saline solutions, then eq. 4.11 becomes:

$$K_{CO_2} = \frac{[H_2CO_3]}{P_{CO_2}} \quad \text{(eq. 4.12)}$$

Carbonic acid is polyprotic (i.e. it has more than one H^+ ion) and dissociates in two steps:

$$H_2CO_3 \Leftrightarrow H^+ + HCO_3^- \quad \text{(eq. 4.13)}$$

$$HCO_3^- = H^+ + CO_3^{2-} \quad \text{(eq. 4.14)}$$

From the law of mass action, dissociation constants can be expressed as follows:

$$K_{H_2CO_3} = \frac{[H^+][HCO_3^-]}{[H_2CO_3]} \quad \text{(eq. 4.15)}$$

$$K_{HCO_3^-} = \frac{[H^+][CO_3^{2-}]}{HCO_3^-} \quad \text{(eq. 4.16)}$$

Using a mass balance expression for the dissolved inorganic carbon (DIC) in the acid and its dissociated anionic species, expressed in terms of molality, then:

$$DIC = (H_2CO_3) + (HCO_3^-) + (CO_3^{2-}) \quad \text{(eq. 4.17)}$$

Rearranging eqs. 4.13–4.17, and taking an arbitrary value of unity (one) for DIC, equations for the relative concentration of H_2CO_3, HCO_3^- and CO_3^{2-} as a function of pH are obtained as

shown graphically in Fig. 4.20. It can be seen from Fig. 4.20 that over most of the normal pH range of groundwater (6–9), HCO_3^- is the dominant carbonate species, and this explains why HCO_3^- is one of the major dissolved inorganic species in groundwater.

To calculate actual concentrations of inorganic carbon species in groundwater, first consider the dissolution of calcite by carbonic acid (eq. 4.5). With reference to eq. 4.10, if the partial pressure of carbon dioxide (P_{CO2}) in the infiltrating groundwater increases, then reaction 4.5 proceeds further to the right to achieve equilibrium. Now, at 25 °C, substitution of eqs. 4.12, 4.15 and 4.16 into the equation expressing the equilibrium constant for the dissociation of calcite:

$$K_{calcite} = [Ca^{2+}][CO_3^{2-}] \quad \text{(eq. 4.18)}$$

yields:

$$[H^+] = 10^{-4.9}\{[Ca^{2+}]P_{CO_2}\}^{\frac{1}{2}} \quad \text{(eq. 4.19)}$$

To obtain the solubility of calcite for a specified P_{CO2}, an equation of electroneutrality is required for the condition $\Sigma z m_c = \Sigma z m_a$ for calcite dissolution in pure water, where z is the ionic valence and m_c and m_a are,

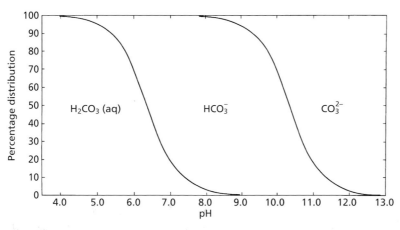

Fig. 4.20 Percentage distribution of inorganic carbon species in water as a function of pH at 25 °C and 1 atmosphere pressure.

respectively, the molalities of the cation and anion species involved:

$$2(Ca^{2+}) + (H^+) = (HCO_3^-)$$
$$+ 2(CO_3^{2-}) + (OH^-)$$
$$(eq. 4.20)$$

The concentrations of H^+ and OH^- are negligible compared with the other terms in eq. 4.20 with respect to the groundwater P_{CO2} values and by combining eqs. 4.19 and 4.20 with eqs. 4.12, 4.15 and 4.16 gives a polynomial expression in terms of two of the variables and the activities. For a specified P_{CO2} value, iterative solutions by computer can be obtained. The results of these calculations for equilibrium calcite dissolution in water for a condition with no limit on the supply of carbon dioxide are shown graphically in Fig. 4.21.

It can be seen from Fig. 4.21, that the solubility of calcite is strongly dependent on the P_{CO2}

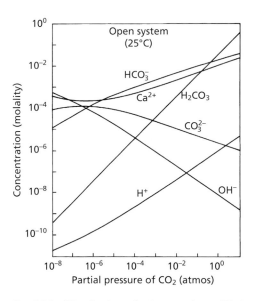

Fig. 4.21 Dissolved species in water in equilibrium with calcite as a function of P_{CO2} at 25 °C. Note the parallel lines at the top right side for Ca^{2+} and HCO_3^- (just 0.301 unit apart) that demonstrate that at high P_{CO2} values, such as found in groundwater, these are the major species formed by dissolving calcite (Guenther 1975). (*Source*: Guenther, W.B. (1975) *Chemical Equilibrium. A Practical Introduction for the Physical and Life Sciences.* Plenum Press, New York. © 1975, Springer Nature)

and that the equilibrium value of H^+ (i.e. pH) also varies strongly with P_{CO2}. Hence, the accurate calculation of the inorganic carbon species in groundwater requires the careful measurement of pH in the field with no loss of carbon dioxide by exsolution from the sample. In the groundwater environment, P_{CO2} is invariably $>10^{-4}$ atmospheres, and again, it is noticed that HCO_3^- rather than CO_3^{2-} is the dominant ionic species of DIC in groundwater. In Box 4.7, the carbonate chemistry of a limestone aquifer in the west of England is described to illustrate the changes in the distribution of carbonate species along a groundwater flowpath.

The chemical evolution of the carbonate system can be considered as occurring under 'open' or 'closed' conditions. As shown in Fig. 4.25, under open conditions of carbonate dissolution, a constant P_{CO2} is maintained, while under 'closed' conditions, carbonate dissolution occurs without replenishment of CO_2. Using the theory outlined previously, it is possible to model paths of chemical evolution for groundwater dissolving carbonate. Steps along the paths are computed by hypothetically dissolving small amounts of carbonate material (calcite or dolomite) for a given temperature and starting condition for P_{CO2} until the water becomes saturated (lines 2 and 3 at 15 °C in Fig. 4.26). Lower temperatures will shift the saturation line to higher solubilities, while higher temperatures will result in saturation at lower solubilities.

It is noticeable in Fig. 4.26 that for closed-system dissolution, the pH values at saturation are higher and the HCO_3^- concentrations lower. In reality, very small quantities of calcite and dolomite exert a strong influence on the carbonate chemistry of groundwater flowing through the soil zone such that open-systems typically dominate, resulting in a pH invariably between 7 and 8.

To illustrate carbonate dissolution pathways in a limestone aquifer, Fig. 4.27 shows the results of chemical modelling of recharge to the Chalk aquifer in north Norfolk, eastern

Box 4.7   Carbonate chemistry of the Jurassic limestones of the Cotswolds, England

The Jurassic limestones of the Cotswolds, England, form two major hydrogeological units in the Upper Thames catchment: (1) limestones in the Inferior Oolite Series; and (2) limestones in the Great Oolite Series. These aquifer units are separated by the Fullers Earth clay which, in this area, acts as an aquitard. The groundwater level contours shown in Fig. 4.22 indicate a regional groundwater flow direction to the south-east, although with many local variations superimposed upon it. In the upper reaches, river valleys are often eroded to expose the underlying Lias clays. The presence of these clays together with the Fullers Earth clay, intercalated clay and marl bands within the limestones, as well as the presence of several faults gives rise to numerous springs. Flow through the limestones is considered to be dominantly through a small number of fissures. At certain horizons, a component of intergranular movement may be significant, but on the whole, the limestone matrix has a very low intrinsic intergranular permeability due to its normally well-cemented and massive lithology (Morgan-Jones and Eggboro 1981).

A survey of the hydrochemistry of boreholes and springs between 1976 and 1979 is reported by Morgan-Jones and Eggboro (1981). The hydrochemical profiles shown in Fig. 4.23 are for the selected locations shown in Fig. 4.22. The dominant major ions in the limestone groundwaters are Ca^{2+} and HCO_3^- in the unconfined areas, but in areas where the limestones become increasingly confined beneath clays, the groundwater changes to a $Na–HCO_3$ type. Groundwaters from deep within the confined limestones contain a $Na–Cl$ component as a result of mixing with increasingly saline groundwater.

The degree of groundwater saturation with respect to calcite and dolomite is shown

diagrammatically in Fig. 4.24. A confidence limit of ±10% is shown to account for any errors in the chemical analyses or in the basic thermodynamic data used to compute the saturation indices (Box 4.8). Samples from locations in the unconfined limestones where high P_{CO2} values are maintained under open-system conditions are supersaturated with respect to calcite. The consistently high Ca^{2+}/Mg^{2+} ratios and low degrees of dolomite saturation in the samples analysed suggest that both limestone series contain predominantly low magnesium calcite.

Values of pH of about 7.5 were recorded for all unconfined groundwater samples. With the onset of confined conditions, pH values increase as a result of changes in HCO_3^- concentration and ion exchange reactions that lead to an increase in Na^+ and decrease in Ca^{2+} concentrations (Fig. 4.23). The removal of Ca^{2+} causes the groundwater to dissolve more carbonate material, but without the replenishment of CO_2 in the closed-system conditions that prevail, the pH begins to rise as H^+ ions are consumed in the calcite dissolution reaction (eq. 4.5). For example, the confined groundwater of sample 8 has a pH of 9.23 and Ca^{2+}, HCO_3^- and Na^+ concentrations of 4, 295 and 137 mg L^{-1}, respectively, compared with the unconfined sample 2 which has corresponding values of 7.31 (pH) and 100 (Ca^{2+}), 246 (HCO_3^-) and 6.9 (Na^+) mg L^{-1}.

Of the minor ions, fluoride shows considerable enrichment with increasing confinement of the groundwaters. The majority of spring samples have F^- concentrations of <100 μg L^{-1} (Fig. 4.23). Most limestones contain small amounts of fluorite (CaF_2), but this mineral has a very low solubility (K_{sp} at 25 °C = $10^{-10.4}$) and the majority of Ca-rich waters have low F^- values. Morgan-Jones and Eggboro (1981) considered that rainwater is the

Box 4.7 (Continued)

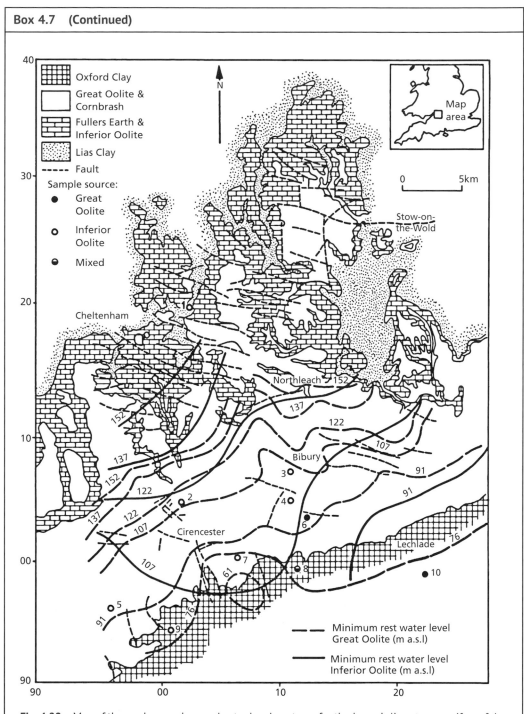

Fig. 4.22 Map of the geology and groundwater level contours for the Jurassic limestone aquifers of the Cotswolds. The sample points shown are from the survey of Morgan-Jones and Eggboro (1981) with the numbered samples used to construct the hydrochemical sections shown in Fig. 4.23. (*Source:* Morgan-Jones, M. and Eggboro, M.D. (1981) The hydrogeochemistry of the Jurassic limestones in Gloucestershire, England. *Quarterly Journal of Engineering Geology* **14**, 25–39.)

(*Continued*)

Box 4.7 (Continued)

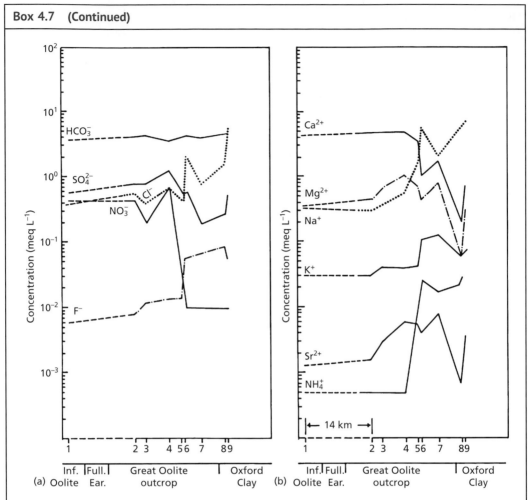

Fig. 4.23 Hydrochemical sections showing major and minor anion (a) and cation (b) data for the sample locations shown in Fig. 4.22 (Morgan-Jones and Eggboro 1981). (*Source:* Morgan-Jones, M. and Eggboro, M.D. (1981) The hydrogeochemistry of the Jurassic limestones in Gloucestershire, England. *Quarterly Journal of Engineering Geology* **14**, 25–39.)

principal source of F^- in the unconfined aquifer areas. A maximum value of 9.8 mg L^{-1} was recorded at sample location 10 in the confined Great Oolite aquifer. Here, the increase is related to the onset of ion exchange and to the availability and solubility of fluorite within the limestone. The equilibrium activity

of F^- is dependent on the activity of Ca^{2+} as defined by the equilibrium constant:

$$K_{CaF_2} = \frac{[Ca^{2+}][F^-]^2}{[CaF_2]} \qquad \text{(eq. 1)}$$

Hence, the decline in the activity of Ca^{2+} with increasing ion exchange allows the F^- activity to increase.

Box 4.7   (Continued)

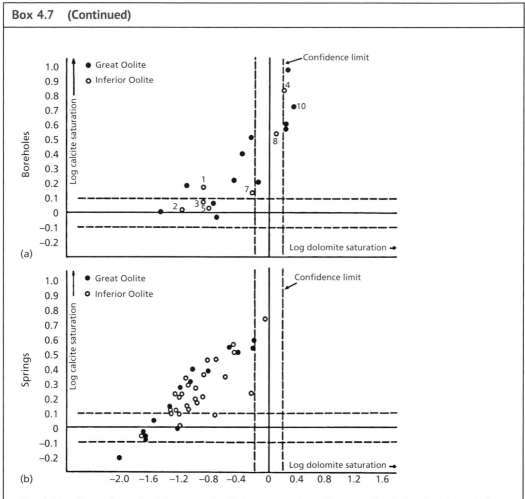

Fig. 4.24   Cross-plots of calcite saturation index versus dolomite saturation index for (a) borehole and (b) spring samples in the Jurassic limestone aquifers of the Cotswolds. The numbered samples shown are those used in Fig. 4.22 (Morgan-Jones and Eggboro 1981). (*Source:* Morgan-Jones, M. and Eggboro, M.D. (1981) The hydrogeochemistry of the Jurassic limestones in Gloucestershire, England. *Quarterly Journal of Engineering Geology* **14**, 25–39.)

England. In this area, the Chalk aquifer is covered by glacial deposits including outwash sands and gravel and two types of till deposits that are distinguished by the mixture of contained clay, sand and carbonate fractions. The evolution of the carbonate chemistry for this aquifer system can be modelled for both open- and closed-systems as demonstrated by Hiscock (1993). Two shallow well waters in the glacial deposits that are both

undersaturated with respect to calcite represent starting conditions for the two models. Using the computer program MIX2 (Plummer *et al.* 1975) small increments of carbonate are added to the groundwaters until saturation is reached. The initial and final chemical compositions of the well waters are given in Table 4.7 and the evolution paths followed in each case are shown in Fig. 4.27. The agreement between the modelled chemistry

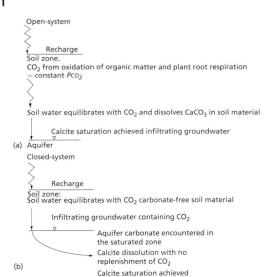

Open-system

Recharge
Soil zone.
CO₂ from oxidation of organic matter and plant root respiration
→ constant P_{CO_2}

Soil water equilibrates with CO_2 and dissolves $CaCO_3$ in soil material

Calcite saturation achieved infiltrating groundwater

(a) Aquifer
Closed-system

Recharge
Soil zone:
Soil water equilibrates with CO_2 carbonate-free soil material

Infiltrating groundwater containing CO_2

Aquifer carbonate encountered in
the saturated zone
Calcite dissolution with no
replenishment of CO_2
Calcite saturation achieved

(b)

Fig. 4.25 Schematic representation of the development of open- and closed-systems of calcite dissolution in soil-aquifer systems.

and the actual Chalk groundwaters sampled from beneath the glacial deposits demonstrates that calcite saturation is achieved under open-system conditions. The distribution of points about the phase boundary in Fig. 4.27 is, as explained by Langmuir (1971), the result of either variations in the soil P_{CO_2} at the time

of groundwater recharge or changes in groundwater chemistry that affect P_{CO_2}. For the Chalk aquifer in north Norfolk, the main reason is variations in soil P_{CO_2}. In areas of sandy till cover, recharge entering soils depleted in carbonate attain calcite saturation for lower values of soil P_{CO_2}, in the range $10^{-2.5}$–$10^{-2.0}$ atmospheres. In contrast, the carbonate-rich soils developed in areas of chalky, clay-rich till experience calcite saturation for higher soil P_{CO_2} values, in the range $10^{-2.1}$–$10^{-2.0}$ (Hiscock 1993).

Although a useful framework for considering the chemical evolution of the carbonate system, several factors have not been considered including seasonal variation in soil temperature and P_{CO_2}; processes such as adsorption, cation exchange and gas diffusion and dispersion that influence the concentrations of Ca^{2+} and P_{CO_2}; and the process of incongruent dissolution whereby the dissolution is not stoichiometric, with one of the dissolution products being a mineral phase sharing a common ionic component with the dissolving phase.

The above treatment of calcite dissolution assumed independent dissolution of calcite and dolomite. However, if both minerals occur

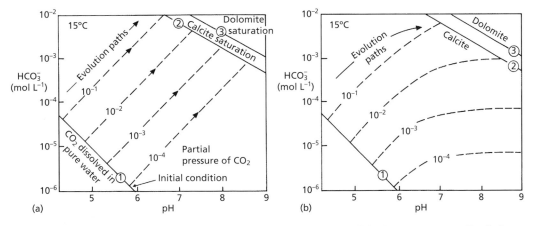

Fig. 4.26 Chemical evolution paths for water dissolving calcite at 15 °C for (a) open-system dissolution and (b) closed-system dissolution. Line 1 represents the initial condition for the water containing dissolved CO_2; line 2 represents calcite saturation and line 3 represents dolomite saturation if dolomite is dissolved under similar conditions (Freeze and Cherry 1979). (*Source:* Freeze, R.A. and Cherry, J.A. (1979) *Groundwater.* Prentice-Hall, Inc., Englewood Cliffs, New Jersey.)

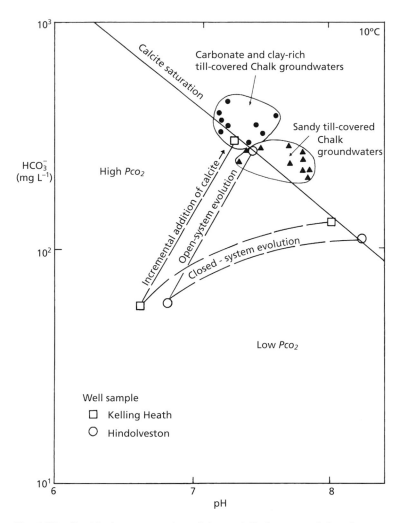

Fig. 4.27 Graphical representation of the modelled open- and closed-system carbonate dissolution paths given in Table 4.7. Starting conditions are represented by shallow well samples contained in glacial till deposits. The two fields showing Chalk groundwaters represent saturated conditions with respect to calcite with the differences explained by the greater dissolution of carbonate material at higher P_{CO2} values in soils developed on the carbonate and clay-rich till deposits (Hiscock 1993). (*Source:* Hiscock, K.M. (1993) The influence of pre-Devensian glacial deposits on the hydrogeochemistry of the Chalk aquifer system of north Norfolk UK. *Journal of Hydrology* **144**, 335–369. © 1993, Elsevier.)

in a hydrogeological system, they may both dissolve simultaneously or sequentially leading to different equilibrium relations compared to those shown in Fig. 4.26. In this situation, a comparison of equilibrium constants for calcite and dolomite for the particular groundwater temperature is necessary to define which mineral is dissolving incongruently.

For example, considering the thermodynamic data shown in Table 4.8, at about $20\,°C$, $K_{calcite} = K_{dolomite}^{1/2}$. Under these conditions, since the solubility product (Box 4.8) of calcite is equal to $[Ca^{2+}][CO_3^{2-}]$ and for dolomite is equal to $[Ca^{2+}][Mg^{2+}][CO_3^{2-}][CO_3^{2-}]$, if groundwater saturated with dolomite flows into a zone that contains calcite, no calcite

Table 4.7 Chemical models for the formation of Chalk groundwaters beneath glacial deposits in north Norfolk, eastern England, under the processes of open- and closed-system dissolution of calcite (Hiscock 1993).

Well sample name	CaCO$_3$ added (×10^{-3} mol L^{-1})	Ca^{2+} (mg L^{-1})	HCO$_3^-$ (mgL^{-1})	pH	P_{CO2} (atmos)	$\Omega_{calcite}$
Open-system conditions						
Kelling Heath	0	10.0	56.7	6.62	1.94	−2.47
	2.00	90.2	299.5	7.30	1.94	−0.03
Hindolveston	0	9.4	59.2	6.81	2.12	−2.09
	1.75	79.5	271.5	7.44	2.12	0.02
Closed-system conditions						
Kelling Heath	0	10.0	56.7	6.62	1.94	−2.47
	0.70	36.0	130.5	8.02	3.00	−0.05
Hindolveston	0	9.4	59.2	6.81	2.12	−2.09
	0.44	28.4	111.6	8.24	3.29	0.02

(*Source:* Hiscock, K.M. (1993) The influence of pre-Devensian glacial deposits on the hydrogeochemistry of the Chalk aquifer system of north Norfolk UK. *Journal of Hydrology* **144**, 335–369. © 1993, Elsevier.)

Table 4.8 Thermodynamic equilibrium constants for calcite, dolomite and major aqueous carbonate species in pure water for a temperature range of 0–30 °C and 1 atmosphere total pressure. Note that pK = −log$_{10}K$ Thermodynamic data from Langmuir (1971); Plummer and Busenberg (1982).

Temperature (°C)	pK_{CO_2}	p$K_{H_2CO_3}$	p$K_{HCO_3^-}$	p$K_{calcite}$	p$K_{dolomite}$	$K_{calcite}/(K_{dolomite})^{1/2}$
0	1.11	6.58	10.63	8.38	16.18	0.51
5	1.19	6.52	10.56	8.39	16.39	0.63
10	1.27	6.47	10.49	8.41	16.57	0.75
15	1.34	6.42	10.43	8.43	16.74	0.87
20	1.41	6.38	10.38	8.45	16.88	0.97
25	1.47	6.35	10.33	8.48	17.00	1.05
30	1.52	6.33	10.29	8.51	17.11	1.11

(*Sources:* Langmuir, D. (1971) The geochemistry of some carbonate groundwaters in Central Pennsylvania. *Geochimica et Cosmochimica Acta* **35**, 1023–1045.; Plummer, L.N. and Busenberg, E. (1982) The solubilities of calcite, aragonite and vaterite in CO_2-H_2O solutions between 0 and 90°C, and an evaluation of the aqueous model for the system $CaCO_3$-CO_2-H_2O. *Geochimica et Cosmochimica Acta* **46**, 1011–1040.)

dissolution will occur because the water is already saturated with respect to calcite. At temperatures lower than 20 °C, $K_{calcite} < K_{dolomite}^{1/2}$ and if groundwater dissolves dolomite to equilibrium the water becomes supersaturated with respect to calcite which can then precipitate. In a system where the rate of dolomite dissolution is equal to the rate of calcite precipitation, this is the condition of incongruent dissolution of dolomite. At temperatures higher than 20 °C, $K_{calcite} > K_{dolomite}^{1/2}$ and if dolomite saturation is achieved with the groundwater then entering a region containing calcite, calcite dissolution will occur leading to an increase in Ca^{2+} and CO_3^{2-} concentrations. The water will now be supersaturated with

---

Box 4.8 Solubility product and saturation index

The dynamic equilibrium between a mineral and its saturated solution when no further dissolution occurs is quantified by the thermodynamic equilibrium constant. For example, the reaction:

$$CaCO_{3(calcite)} \Leftrightarrow Ca^{2+}_{(aq)} + CO^{2-}_{3(aq)}$$

$$(eq. 1)$$

is quantified by the equilibrium constant, K, found from

$$K_{calcite} = \frac{[Ca^{2+}][CO_3^{2-}]}{[CaCO_3]} \qquad (eq. 2)$$

Since $CaCO_3$ is a solid crystal of calcite, its activity is effectively constant and by convention is assigned a value of one or unity.

The equilibrium constant for a reaction between a solid and its saturated solution is known as the solubility product, K_{sp}. Solubility products have been calculated for many minerals, usually using pure water under standard conditions of 25 °C and 1 atmosphere pressure and are tabulated in many textbooks, for example data for major components in groundwater are given by Appelo and Postma (2005). The solubility product for calcite (eq. 2) is $10^{-8.48}$ and eq. 2 now becomes:

$$K_{sp} = \frac{[Ca^{2+}][CO_3^{2-}]}{1}$$
$$= [Ca^{2+}][CO_3^{2-}] = 10^{-8.48} \ mol^2L^{-2}$$

$$(eq. 3)$$

The solubility product can be used to calculate the solubility (mol L^{-1}) of a mineral in pure water. The case for calcite is straightforward since each mole of $CaCO_3$ that dissolves produces one mole of Ca^{2+} and one mole of CO_3^{2-}. Thus, the calcite solubility = $[Ca^{2+}] = [CO_3^{2-}]$ and therefore calcite solubility = $(10^{-8.48})^{1/2} = 10^{-4.24} = 5.75 \times 10^{-5}$ mol L^{-1}.

The state of saturation of a mineral in aqueous solution can be expressed using a saturation index, where

$$\Omega = \frac{IAP}{K_{sp}} \qquad (eq. 4)$$

in which IAP is the ion activity product of the ions in solution obtained from analysis. A Ω value of 1 indicates that mineral saturation (equilibrium) has been reached. Values greater than one represent oversaturation or supersaturation and the mineral is likely to be precipitated from solution. Values less than one indicate undersaturation and further mineral dissolution can occur. An alternative to eq. 4 is to define $\Omega = \log_{10}(IAP/K_{sp})$ in which case the value of Ω is zero at equilibrium with positive values indicating supersaturation and negative values undersaturation.

By calculating saturation indices, it is possible to determine from hydrochemical data the equilibrium condition of groundwater with respect to a given mineral. For example, a groundwater from the unconfined Chalk in Croydon, South London, gave the following results: temperature = 12 °C, pH = 7.06, Ca^{2+} concentration = 121.6 mg L^{-1} ($10^{-2.52}$ mol L^{-1}) and HCO_3^- = 217 mg L^{-1} ($10^{-2.45}$ mol L^{-1}). By making the assumption that the concentrations are equal to activities for this dilute groundwater sample, the first step in calculating a calcite saturation index is to find the CO_3^{2-} concentration. The dissociation of HCO_3^- can be expressed as follows:

$$HCO_3^- = H^+ + CO_3^{2-} \qquad (eq. 5)$$

for which the approximate equilibrium constant at 10 °C (Table 4.8) is

$$K_{HCO_3^-} = \frac{[H^+][CO_3^{2-}]}{HCO_3^-} = 10^{-10.49}$$

$$(eq. 6)$$

(Continued)

Box 4.8 (Continued)

Rearranging eq. 6 for the unknown CO_3^{2-} concentration and substituting the measured values for H^+ and HCO_3^- gives

$$[CO_3^{2-}] = \frac{K_{HCO_3^-}[HCO_3^-]}{[H^+]}$$

$$= \frac{10^{-10.49} \times 10^{-2.45}}{10^{-7.06}} = 10^{-5.88}$$

(eq. 7)

Now, using the result of eq. 7, the calcite saturation index is found from

$$\Omega_{calcite} = \log_{10} \frac{[Ca^{2+}][CO_3^{2-}]}{K_{sp}}$$

(eq. 8)

If K_{sp} for calcite = $10^{-8.41}$ at 10 °C, then

$$\Omega_{calcite} = \log_{10} \frac{[10^{-2.52}][10^{-5.88}]}{10^{-8.41}}$$

$$= \log_{10} 10^{0.01} = 0.01$$

(eq. 9)

Hence, the Chalk groundwater is marginally supersaturated and, given the assumptions used in the calculation of $\Omega_{calcite}$, can be regarded as at equilibrium. A more accurate calculation using the chemical program WATEQ (Truesdell and Jones 1973) that accounts for the chemical activity and speciation of the sample as well as the actual sample temperature of 12.0 °C gives a calcite saturation index of −0.12, again indicating equilibrium with respect to calcite for practical purposes (Mühlherr *et al.* 1998).

---

respect to dolomite and dolomite precipitation, although sluggish, will occur, to achieve a condition of incongruent dissolution of calcite. In cases where groundwater first dissolves calcite to equilibrium and then encounters dolomite, dolomite dissolves regardless of the temperature because the water must acquire appreciable Mg^{2+} before dolomite equilibrium is achieved. However, as the water becomes supersaturated with respect to calcite due to the influx of Ca^{2+} and CO_3^{2-} ions from dolomite dissolution, calcite precipitates and the dolomite dissolution becomes incongruent (Freeze and Cherry 1979).

Over long periods of time, incongruent calcite and dolomite dissolution may exert an important influence on the chemical evolution of the groundwater and on the mineralogical evolution, or diagenesis, of the aquifer rock; for example, dolomitization of calcareous sediments. Dedolomitization, the process whereby a dolomite-bearing rock is converted to a calcite-bearing rock has been reported, for example in the case of the Floridan aquifer system described in Box 4.4. Concentrations of Mg^{2+}

in the Upper Floridan aquifer generally range from 1 to 1000 mg L^{-1}, with the highest concentrations occurring where the aquifer contains seawater. Where the aquifer contains freshwater, Mg^{2+} concentrations generally increase in downgradient directions because of dedolomitization of the aquifer, although data are insufficient to prove that formation of dolomite limits Mg^{2+} concentrations in the Upper Floridan aquifer (Sprinkle 1989).

4.8 Adsorption and ion exchange

Adsorption and ion exchange reactions in aquifers can significantly influence the natural groundwater chemistry and are an important consideration in predicting the migration of contaminants such as heavy metals and polar organic chemicals (see Section 8.3.2 for further discussion). Major ion exchange reactions affect not only the exchanging ions but also other species, especially via dissolution and precipitation reactions. The attenuation of

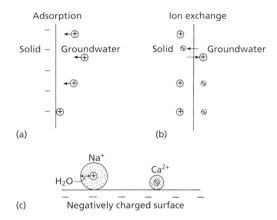

Fig. 4.28 Pictorial representation of (a) adsorption and (b) cation exchange reactions in groundwater. In (c) the divalent Ca^{2+} ion with a smaller hydrated radius is more strongly adsorbed by the negatively charged surface (e.g. clay) than is the monovalent Na^+ ion with a larger hydrated radius.

some pollutants, for example NH_4^+, is mainly by the process of ion exchange (Carlyle *et al.* 2004). Ion exchange reactions can also lead to changes in the hydraulic conductivity of natural materials (Zhang and Norton 2002).

Ionic species present in groundwater can react with solid surfaces. As shown in Fig. 4.28a, adsorption occurs when a positively charged ion in solution is attracted to and retained by a solid with a negatively charged surface. Depending on the point of zero charge (PCZ) of the rock-forming mineral, where PCZ is the pH at which the mineral has zero charge (at pH values less than the PCZ the mineral has a net positive charge, at values greater than the PCZ a net negative charge), different minerals will attract anions or cations to their surfaces depending on the pH of the solution. Clay minerals have negative surface charges in all but the most acidic solutions and therefore attract cations to their surfaces to neutralize the negative charge. The surface charge for oxides and hydroxides of Fe can be either negative or positive in the pH range of most groundwaters (PCZ values for haematite and goethite are between 5–9 and 7.3–7.8, respectively; Krauskopf and Bird 1995), giving the potential

for anion adsorption (for example, arsenic oxyanions) at low pH, and release at high pH.

In general, the degree of adsorption increases with an increasing surface area and with decreasing grain size. Hence, clays are typically most reactive since they have a small grain size and therefore have a large surface area on which sorption reactions can occur. In addition, clays tend to be strong adsorbers since they have an excess of negative charge at the surface due to crystal lattice defects on to which cations can adsorb. The adsorption may be weak, essentially a physical process caused by van der Waals' force, or strong, if chemical bonding occurs. Divalent cations are usually more strongly adsorbed than monovalent ions as a result of their greater charge density, a consequence of valence and smaller hydraulic radius (Fig. 4.28c). The adsorptive capacity of specific soils or sediments is usually determined experimentally by batch or column experiments.

Ion exchange occurs when ions within the mineral lattice of a solid are replaced by ions in the aqueous solution (Fig. 4.28b). Ion exchange sites are found primarily on clays, soil organic matter and metal oxides and hydroxides which all have a measurable cation exchange capacity (CEC). In cation exchange, the divalent ions are more strongly bonded to a solid surface such that the divalent ions tend to replace monovalent ions. The amounts and types of cations exchanged are the result of the interaction of the concentration of cations in solution and the energy of adsorption of the cations at the exchange surface. The monovalent ions have a smaller energy of adsorption and are therefore more likely to remain in solution. As a result of a larger energy of adsorption, divalent ions are more abundant as exchangeable cations. Ca^{2+} is typically more abundant as an exchangeable cation than is Mg^{2+}, K^+ or Na^+. The energy absorption sequence is $Ca^{2+} > Mg^{2+} > K^+ > Na^+$ and this provides a general ordering of cation exchangeability for common ions in groundwater.

Values of CEC are found experimentally with laboratory results reported in terms of meq $(100\,g)^{-1}$. CEC is commonly determined by extraction of the cations from soils or aquifer materials with a solution containing a known cation, normally NH_4^+. Ammonium acetate (CH_3COONH_4), the salt of a weak acid and weak base, is usually used for this purpose, its pH being adjusted to the value most suited to the investigation. A review of methods for determining exchangeable cations is provided by Talibudeen (1981). The surface area and CEC of various clays and Fe and Al oxy-hydroxides are given in Table 4.9.

An example of cation exchange occurring in groundwater is found in the Jurassic Lincolnshire Limestone aquifer in eastern England. The Lincolnshire Limestone aquifer comprises 10–30 m of oolitic limestone with a variable content of finely disseminated iron minerals and dispersed clay and organic matter acting as reactive exchange sites for Ca^{2+} and Na^+. The aquifer is confined down-dip by thick marine clays. A hydrochemical survey (Fig. 4.29) showed that cation exchange occurs at around 12 km from the aquifer outcrop and that within a further 12 km downgradient the Ca^{2+} decreases to a minimum of $<4\,mg\,L^{-1}$ as a result of exchange with Na^+ (Edmunds and Walton 1983). The lack of cation exchange

closer to the aquifer outcrop is explained by the exhaustion of the limited cation exchange capacity of the limestone. The concentrations of Sr^{2+} and, to a lesser extent, Mg^{2+} continue

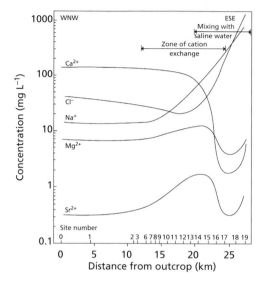

Fig. 4.29 Hydrogeochemical trends in the Lincolnshire Limestone for Ca^{2+}, Mg^{2+}, Na^+, Sr^{2+} and Cl^-. The trend lines are for 1979 and illustrate the effect of cation exchange between Ca^{2+} and Na^+ and the onset of mixing with saline water in the deeper aquifer (Edmunds and Walton 1983). (*Source:* Edmunds, W.M. and Walton, N.G.R. (1983) The Lincolnshire Limestone – hydrogeochemical evolution over a ten-year period. *Journal of Hydrology* **61**, 201–211. © 1983, Elsevier.)

Table 4.9 Surface area and cation exchange capacity (CEC) values for clays and Fe and Al oxy-hydroxides (Talibudeen 1981 and Drever 1988).

	Surface area $(m^2\ g^{-1})$	CEC $(meq\ 100\ g^{-1})$
Fe and Al oxy-hydroxides (pH ~ 8.0)	25–42	0.5–1
Smectite	750–800	60–150
Vermiculite	750–800	120–200
Bentonite	750	100
Illite	90–130	10–40
Kaolinite	10–20	1–10
Chlorite	–	<10

(*Sources:* Talibudeen, O. (1981) Cation exchange in soils. In: *The Chemistry of Soil Processes* (eds D.J. Greenland and M. H.B. Hayes). Wiley, Chichester, pp. 115–177.; Drever, J.I. (1988) *The Geochemistry of Natural Waters* (2nd edn). Prentice-Hall, Inc., Englewood Cliffs, New Jersey.)

to increase for around 22 km from outcrop as a result of incongruent dissolution. The removal of Ca^{2+} by cation exchange causes calcite dissolution to occur to restore carbonate equilibrium. However, once the Ca^{2+}/Sr^{2+} and Ca^{2+}/Mg^{2+} equivalents ratios fall below a certain critical level ($\sim$20:1 and 1:1, respectively) cation exchange reactions become dominant and both Sr^{2+} and Mg^{2+} concentrations begin to decrease (Edmunds and Walton 1983).

Cation exchange reactions are a feature of saline water intrusion in coastal areas. Freshwater in coastal areas is typically dominated by Ca^{2+} and HCO_3^- ions from the dissolution of calcite such that cation exchangers present in the aquifer have mostly Ca^{2+} adsorbed on their surfaces. In seawater, Na^+ and Cl^- are the dominant ions and aquifer materials in contact with seawater will have Na^+ attached to the exchange surfaces. When seawater intrudes a coastal freshwater aquifer, the following cation exchange reaction can occur

$$Na^+ + \frac{1}{2}Ca - X \rightarrow Na - X + \frac{1}{2}Ca^{2+}$$

(eq. 4.21)

where X indicates the exchange material. As the exchanger takes up Na^+, Ca^{2+} is released, and the hydrochemical water type evolves from Na–Cl to Ca–Cl. The reverse reaction can occur when freshwater flushes a saline aquifer:

$$\frac{1}{2}Ca^{2+} + Na - X \rightarrow \frac{1}{2}Ca - X + Na^+$$

(eq. 4.22)

where Ca^{2+} is taken up from water in return for Na^+ resulting in a Na–HCO_3 water type. An example of this reaction is given in Box 4.9 for the Lower Mersey Basin Permo-Triassic sandstone aquifer of north-west England.

Box 4.9   Cation exchange in the Lower Mersey Basin Permo-Triassic sandstone aquifer, England

The Lower Mersey Basin Permo-Triassic sandstone aquifer of north-west England demonstrates the effect of very long-term natural flushing of a saline aquifer. The aquifer comprises two main units: the Permian Collyhurst Sandstone Formation and the Triassic Sherwood Sandstone Formation that dip southwards at about 5° and are up to 500 m thick. To the south, the aquifer unit is overlain by the Triassic Mercia Mudstone Group, a formation which contains evaporites. Underlying the aquifer is a Permian sequence, the upper formations of which are of low permeability that rest unconformably on Carboniferous mudstones. The sequence is extensively faulted with throws frequently in excess of 100 m. Overlying the older formations are highly heterogeneous, vertically variable Quaternary deposits dominated by glacial till. Pumping of the aquifer system has caused a decline in water levels such that much of the sandstone aquifer is no longer confined by the till.

Typical compositions of the sandstones are quartz 60–70%, feldspar 3–6%, lithic clasts 8%, calcite 0–10% and clays (including smectite) <15%. Haematite imparts a red colour to most of the sequence. The sandstones contain thin mudstone beds often less than 10 cm in thickness. Cation exchange capacities are of the order of 1 meq $(100\,g)^{-1}$. Underlying the fresh groundwaters present in the area are saline groundwaters attaining a Cl^- concentration of up to $100\,g\,L^{-1}$, which appear to be derived from dissolution of evaporites in the overlying Mercia Mudstone Group (Tellam 1995). This saline water is present within 50 m of ground level immediately up-flow of the Warburton Fault Block and along the Mersey Valley, but the

(Continued)

Box 4.9 (Continued)

freshwater-saline water interface is found deeper both to the north and south (Tellam *et al.* 1986).

A hydrochemical survey of around 180 boreholes across the Lower Mersey Basin was conducted in the period 1979–1980, and the results presented by Tellam (1994) in which five water types were identified (Table 4.10). Salinities ranged from 100 mg L^{-1} up to brackish water concentrations. Ion proportions varied widely, with $Ca-HCO_3$, $Ca-SO_4$ and $Na-HCO_3$ being dominant water types in various locations (Fig. 4.30). The large storativity of the aquifer means that the groundwater chemistry does not substantially change seasonally. The spatial distribution of water types is shown in Fig. 4.31. In general, the hydrochemical distribution of water types correlates with the broad pattern of groundwater flow in the aquifer. Type 1

Table 4.10 Hydrochemical water types for the Lower Mersey Basin Permo-Triassic sandstone aquifer as determined from borehole water chemistry.

Water type	General characteristics	Occurrence	Interpretation
1A [1B]	High NO_3, SO_4, Cl, tritium Low HCO_3, Ca, Mg, pH SIC<<0	Predominantly where aquifer outcrops	Samples dominated by recent recharge, affected by agricultural and industrial activity
	[As for 1A but SIC ~ 0]	[As above]	[Slightly older water, or water which has been induced to flow through carbonate-containing parts of the aquifer near outcrop]
2	NO_3 often low, but not always Variable SO_4 and Cl pH ~ 7 SIC ~ 0	Predominantly under glacial till	Mixing in borehole of water recharged locally through Quaternary deposits and water recharged at outcrop. Through-Quaternary recharge occurs because of leakage induced by abstraction. Approximately a mixture of Types 1 and 4, though modified during travel through the Quaternary deposits
4	NO_3 < d.l., low SO_4, low Cl High HCO_3, Ca, Mg pH ~ 8 SIC ~ 0; SID ~ 0	Predominantly under glacial till	Samples dominated by pre-industrially recharged water with low NO_3, SO_4, Cl. Has encountered enough mineral carbonate to become saturated with respect to calcite and dolomite

Box 4.9 (Continued)

Table 4.10 (Continued)

Water type	General characteristics	Occurrence	Interpretation
5	NO_3 < d.l., low SO_4, low to very high Cl Very high HCO_3 pH 8 to 9 Very low to low Ca, Mg Na/Cl > to >> 1 SIC ~ 0; SID ~ 0	Below glacial till, adjacent to saline groundwater	Type 4 water, but in a part of the aquifer previously occupied by saline groundwaters. The Na/Cl ratio indicates ion exchange, the exchangers releasing Na sorbed when the saline groundwater was present. As Ca and Mg are taken up, more carbonate is dissolved, and pH and HCO_3 rise. Cl concentration depends on whether there is any saline water left in the system
3	Generally low and variable concentrations of major determinants. SIC < 0	Collyhurst Sandstone Formation and in an isolated fault block in west of the area	Post-industrial recharge, but with limited pollutants. Limited carbonate in aquifer results in carbonate undersaturation, even for waters with considerable residence time
Saline groundwater S1	Cl to 5 g/l, high I, SO_4 < expected from Mersey Estuary High Ca/Cl, low Na/Cl, SIC ~ 0	Bordering the Mersey Estuary (average Cl concentration ~ 5 g/l)	Intrusion from Mersey Estuary (Carlyle *et al.* 2004), accompanied by various reactions including SO_4 reduction, CO_2 degassing and ion exchange
Saline groundwater S2	Cl to 105 g/l Low Br/Cl compared with S1 Low SO_4/Cl compared with S1 Very light $\delta^{18}O$ and δ^2H	Below fresh groundwater in Mersey Valley inland from Mersey Estuary and adjacent areas; interface with freshwater up to 250 m below ground level	Brines resulting from dissolution of evaporites in the Mercia Mudstone Group, and subsequent migration. Upper part of the saline groundwater was diluted by freshwater recharged under climatic conditions significantly cooler than at present

Notes: SIC saturation index for calcite, *SID* saturation index for dolomite [$\log_{10}$IAP/K]; *d.l.* detection limit.
(*Source:* Courtesy of J.H. Tellam.)

water is found in areas of recent groundwater recharge, in contrast to older fresh groundwaters (Types 4 and 5) and saline groundwater (Type S2) located in areas of the aquifer with low hydraulic gradients. The effect of modern groundwater abstractions is reflected by saline intrusion (Type S1) and recently induced recharge through the Quaternary deposits (Type 2).

Example analyses of water Types 4 and 5 are shown in Table 4.11. Type 5 water that occurs in the eastern confined areas and in

(Continued)

Box 4.9 (Continued)

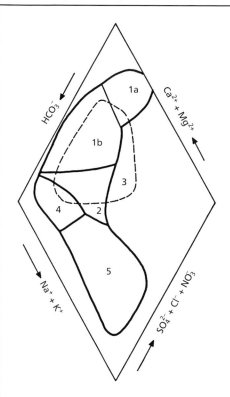

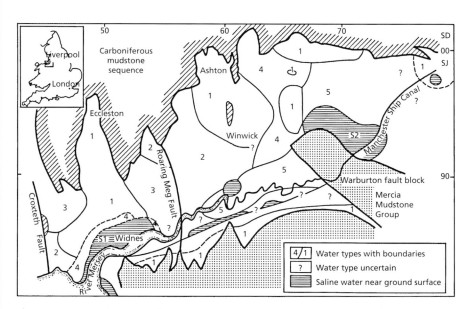

Fig. 4.30 The central area of a Piper diagram showing the distribution of hydrochemical *water* types in the Lower Mersey Basin Permo-Triassic sandstone aquifer (Tellam 1994). (*Source:* Tellam, J.H. (1994) The groundwater chemistry of the Lower Mersey Basin Permo-Triassic Sandstone Aquifer system, UK: 1980 and pre-industrialisation-urbanisation. *Journal of Hydrology* **161**, 287–325. © 1994, Elsevier.)

Fig. 4.31 Spatial distribution of hydrochemical water types defined in Fig. 4.30 for the Lower Mersey Basin. Saline groundwater underlies much of the fresh groundwater in the Permo-Triassic sandstone aquifer (Tellam 1994). (*Source:* Tellam, J.H. (1994) The groundwater chemistry of the Lower Mersey Basin Permo-Triassic Sandstone Aquifer system, UK: 1980 and pre-industrialisation-urbanisation. *Journal of Hydrology* **161**, 287–325. © 1994, Elsevier.)

Box 4.9 (Continued)

the Mersey Valley area are very similar to Type 4 except it has a high Na^+/Cl^- equivalents ratios (Table 4.11) and a range of Cl^- concentrations from <10 to 1000 mg L^{-1}. In both areas, the saline water interface is at a high level and groundwater flow rates are low. The hydrochemistry of Type 5 water is interpreted as having experienced Na^+ release by cation exchange following invasion by Type 4 water into regions originally occupied by saline groundwater. Where

flushing has been less complete, higher Cl^- concentrations occur. The Na^+ release is accompanied by Ca^{2+} uptake by the aquifer as described by eq. 4.22. An extreme example is provided by sample 123 in Table 4.11, where the water contains greater than 1000 mg L^{-1} Cl^-, 2 mg L^{-1} Ca^{2+} and 2 mg L^{-1} Mg^{2+}. The removal of Ca^{2+} promotes calcite dissolution with HCO_3^- concentrations able to reach 500–600 mg L^{-1} (Tellam 1994).

Table 4.11 Example hydrochemical analyses of Types 4 and 5 groundwaters from the Lower Mersey Basin Permo-Triassic sandstone aquifer (Tellam 1994).

Water type	Site (depth in (m); P, pumped)	Na^+ (mg L^{-1})	K^+ (mg L^{-1})	Ca^{2+} (mg L^{-1})	Mg^{2+} (mg L^{-1})	Cl^- (mg L^{-1})	SO_4^{2+} (mg L^{-1})	HCO_3^- (mg L^{-1})	NO_3^- (mg L^{-1})	Na^+/Cl^-	pH	SIC^a
4	20 (100)	15	4.5	48	32	20	2	322	0.0	1.14	7.7	+0.1
	88 (P)	46	4.0	108	35	24	17	540	1.1	2.92	7.8	+0.8
	104 (180)	17	4.0	83	48	28	8	513	0.4	0.92	7.3^b	+0.3
5	109 (47)	192	1.2	3.7	1.7	25	9	487	0.4	11.69	6.8	−1.7
	123 (95)	947	2.7	2.3	1.7	1196	76	454	0.4	1.20	8.8	−0.1
	140 (P)	170	3.7	37	14	102	8	438	0.0	2.54	7.9	−0.2

^{a} SIC = saturation index for calcite [$\log_{10}$IAP/K].
^{b} Laboratory measured pH value.
(*Source:* Modified from Tellam, J.H. (1994) The groundwater chemistry of the Lower Mersey Basin Permo-Triassic Sandstone Aquifer system, UK: 1980 and pre-industrialisation-urbanisation. *Journal of Hydrology* **161**, 287–325.)

The chemical reactions that occur during freshwater and saline water displacements in aquifers can be identified from a consideration of conservative mixing of fresh and saline water end member solutions and comparing with individual water analyses. For conservative mixing:

$$c_{i,mix} = f_{saline} \times c_{i,saline} + (1 - f_{saline})c_{i,fresh}$$

$$\text{(eq. 4.23)}$$

where c_i is the concentration of ion i; $_{mix}$, $_{fresh}$ and $_{saline}$ indicate the conservative mixture and end-member fresh and saline waters; and f_{saline} is the fraction of saline water. Any change in the sample composition as a result of reactions, for example cation exchange, other than by simple mixing ($c_{i,react}$) is then simply found from

$$c_{i,react} = c_{i,sample} - c_{i,mix} \qquad \text{(eq. 4.24)}$$

Table 4.12 Selected hydrochemical data for the Lincolnshire Limestone aquifer to illustrate cation exchange (Edmunds and Walton 1983).

	Na^+ (mg L^{-1})	Ca^{2+} (mg L^{-1})	Cl^- (mg L^{-1})	Na^+/Cl^-
Freshwater (Sample 0, Ropsley)	14	135	42	0.51
Mixed water (Sample 14, Pepper Hill)	280	17	114	3.74
Saline water (Sample 19, Deeping St. Nicholas)	920	9.5	1100	1.27

(*Source:* Modified from Edmunds, W.M. and Walton, N.G.R. (1983) The Lincolnshire Limestone – hydrogeochemical evolution over a ten-year period. *Journal of Hydrology* **61**, 201–211.)

As an example calculation, the data shown in Table 4.12 are for samples 0, 14 and 19 of Fig. 4.29 representing fresh, mixed and saline groundwaters present in the Lincolnshire Limestone aquifer. To calculate how much Na^+ has been added to the mixed groundwater sample by cation exchange, eq. 4.23 can be re-written as follows:

$$c_{Na,mix} = f_{saline} \times c_{Na,saline} + (1 - f_{saline})c_{Na,fresh}$$
(eq. 4.25)

Now, using the mixed and saline Cl^- concentration values (samples 14 and 19) to indicate the fraction of the saline water ($f_{saline} = 114/1100$), and assuming a freshwater end-member Cl^- concentration value equal to zero, then substituting the values from Table 4.12 into eq. 4.23:

$$c_{Na,mix} = (114/1100)920 + (1 - 114/1100)14$$
$$= 107.9 \text{ mg } L^{-1}$$

Similarly, eq. 4.24 can be re-written to calculate the amount of Na^+ involved in the cation exchange reaction:

$$c_{Na,react} = c_{Na,sample} - c_{Na,mix} \quad \text{(eq. 4.26)}$$

Now, using the result for $c_{Na,mix}$ found above

$$c_{Na,react} = 280 - 107.9 = 172.1 \text{ mg } L^{-1}$$
$$(\text{or } 7.5 \text{ meq } L^{-1})$$

A similar calculation for Ca^{2+} removed from the mixed groundwater sample results in

$$c_{Ca,mix} = (114/1100)9.5 + (1 - 114/1100)135$$
$$= 122.0 \text{ mg } L^{-1}$$

and

$$c_{Ca,react} = 17 - 122.0 = -105.0 \text{ mg } L^{-1}$$
$$(\text{or } 5.3 \text{ meq } L^{-1})$$

4.9 Redox chemistry

Reactions involving a change in oxidation state are referred to as oxidation–reduction or redox reactions. Redox reactions have a controlling influence on the solubility and transport of some minor elements in groundwater such as Fe and Mn and also on redox sensitive species such as NO_3^- and SO_4^{2-}. The extent to which redox reactions occur in groundwater systems is therefore significant with respect to many practical problems, for example issues of groundwater quality for drinking water, the attenuation of landfill leachate plumes and the remediation of sites contaminated by organic pollutants (McMahon and Chapelle 2008). The major redox sensitive components of groundwaters and aquifers are O_2, $NO_3^-/N_2/NH_4^+$, SO_4^{2-}/HS^-, $Mn(II)/Mn(IV)$ and $Fe(II)/Fe(III)$. Redox sensitive trace elements include As, Se, U and Cr in addition to Fe and Mn. The toxic effects of these elements differ greatly for various redox species, for example $Cr(III)/Cr(VI)$, in which hexavalent chromium and its compounds are toxic, and so it is important that the behaviour of these elements can be predicted on the basis of the groundwater redox conditions.

The evolution of redox processes in groundwater is dependent on the source and distribution of electron donors and acceptors in an aquifer, relative rates of redox reaction and groundwater flow, aquifer confinement, position in the flow system and groundwater mixing. Redox gradients are largely vertical in recharge areas of unconfined aquifers dominated by natural sources of electron donors, whereas lateral gradients predominate in confined aquifers. Electron-donor limitations can result in the persistence of oxic groundwater conditions over distances on many kilometres and groundwater residence times of several thousand years in some aquifers. Where electron donors are abundant, redox conditions can evolve from oxygen reducing to methanogenic over much shorter flow distances and residence times (McMahon *et al.* 2011).

The most common electron donor supporting microbial populations in groundwater systems is organic carbon. Aquifer materials deposited in sedimentary environments commonly contain particulate organic carbon and can be present in groundwater recharge areas, in discharge areas or at intermediate points depending on the hydrogeological setting. Dissolved organic carbon (DOC) can enter aquifers by processes such as water percolating through the unsaturated zone mobilized from surface sources, typically derived from decaying plant material, or by diffusion into aquifers from adjacent confining layers (McMahon *et al.* 2011).

Of the commonly available electron acceptors, oxygen, nitrate, sulphate and carbon dioxide are present in the atmosphere and in the unsaturated zone of many environments, are soluble in near-neutral pH groundwater conditions, and often enter the groundwater system through recharge processes. Manganese and iron, the other commonly available electron acceptors, primarily exist in the solid phase of rocks and minerals and are less likely to be present in groundwater recharge. Iron oxyhydroxide minerals, such as goethite, and amorphous manganese and iron solid phases can occur as coatings on other minerals. Iron within the structure of minerals such as

smectite also can serve as an electron acceptor. The minerals gypsum and anhydrite can serve as subsurface sources of sulphate, particularly in carbonate-rock aquifers. Sulphate that has diffused from confining layers can also act as an electron acceptor for redox processes in marine sedimentary aquifers. Subsurface carbon dioxide can be produced by dissolution of carbonate minerals and by microbial respiration in many types of aquifers (McMahon *et al.* 2011).

During redox reactions, electrons are transferred between dissolved, gaseous or solid constituents and result in changes in the oxidation states of the reactants and products. The oxidation state (or oxidation number) represents the hypothetical charge that an atom would have if the ion or molecule were to dissociate. The oxidation states that can be achieved by the most important multi-oxidation state elements that occur in groundwater are listed in Table 4.13. By definition, oxidation is the loss of electrons and reduction is the gain of electrons. Every oxidation is accompanied by a reduction and *vice versa*, so that an electron balance is always maintained (Freeze and Cherry 1979).

For every redox half-reaction, the following form of an equation can be written, where n is the number of electrons transferred:

$$\text{oxidized state} + ne^- = \text{reduced state}$$
$$(\text{eq. } 4.27)$$

As an example, the redox reaction for the oxidation of Fe can be expressed by two half-reactions:

$$\frac{1}{2}O_2 + 2H^+ + 2e^- = H_2O \quad (\text{reduction})$$
$$(\text{eq. } 4.28)$$

$$2Fe^{2+} = 2Fe^{3+} + 2e^- \quad (\text{oxidation})$$
$$(\text{eq. } 4.29)$$

The complete redox reaction for the oxidation of Fe is found from the addition of eqs 4.28 and 4.29 and expresses the net effect of the electron transfer with the absence of free electrons, thus

Table 4.13 Examples of oxidation states for various compounds that occur in groundwater. The oxidation state of free elements, whether in atomic or molecular form, is zero. Other rules for assigning oxidation states include the following: the oxidation state of an element in simple ionic form is equal to the charge on the ion; the sum of oxidation states is zero for molecules; and for ion pairs or complexes it is equal to the formal charge on the species (Freeze and Cherry 1979). (Freeze, R.A. and Cherry, J.A. (1979) *Groundwater*. Prentice-Hall, Inc., Englewood Cliffs, New Jersey. © 1979, Pearson Education.)

Carbon compounds		Sulphur compounds		Nitrogen compounds		Iron compounds	
Substance	C state	Substance	S state	Substance	N state	Substance	Fe state
HCO_3^-	+IV	S	0	N_2	0	Fe	0
CO_3^{2-}	+IV	H_2S	-II	SCN^-	+II	FeO	+II
CO_2	+IV	HS^-	-II	N_2O	-III	$Fe(OH)_2$	+II
CH_2O	0	FeS_2	-I	NH_4^+	+III	$FeCO_3$	+II
$C_6H_{12}O_6$	0	FeS	-II	NO_2^-	+V	FeO_3	+III
CH_4	-IV	SO_3^{2-}	+IV	NO_3^-	-III	$Fe(OH)_3$	+III
CH_3OH	-II	SO_4^{2-}	+VI	HCN	-I	FeOOH	+III

$$\frac{1}{2}O_2 + 2Fe^{2+} + 2H^+ = 2Fe^{3+} + H_2O$$

(eq. 4.30)

By expressing redox reactions as half-reactions, the concept of pe is used to describe the relative electron activity, where

$$pe = -\log_{10}[e^-] \qquad \text{(eq. 4.31)}$$

pe is a dimensionless quantity and is a measure of the oxidizing or reducing tendency of the solution, where pe and pH are functions of the free energy involved in the transfer of 1 mole of electrons or protons, respectively, during a redox reaction.

For the general half-reaction:

$$\text{oxidants} + ne^- = \text{reductants}$$

(eq. 4.32)

then, from the law of mass action:

$$K = \frac{[\text{reductants}]}{[\text{oxidants}][e^-]^n} \qquad \text{(eq. 4.33)}$$

A numerical value for such an equilibrium constant can be computed using Gibbs free energy data for conditions at 25 °C and 1 atmosphere pressure. By convention, the equilibrium constant for a half-reaction is always

expressed in the reduction form. The oxidized forms and electrons are written on the left and the reduced products on the right.

Rearrangement of eq. 4.33 gives the electron activity $[e^-]$ for a half-reaction as follows:

$$[e^-] = \left\{\frac{[\text{reductants}]}{[\text{oxidants}]K}\right\}^{\frac{1}{n}} \qquad \text{(eq. 4.34)}$$

Rewriting eq. 4.34 by taking the negative logarithm of both sides yields:

$$-\log_{10}[e^-] = pe =$$
$$\frac{1}{n}\left\{\log_{10}K - \log_{10}\frac{[\text{reductants}]}{[\text{oxidants}]}\right\}$$

(eq. 4.35)

When a half-reaction is written in terms of a single electron transfer, or $n = 1$, the $\log_{10}K$ term is written as $pe°$ such that:

$$pe = pe° - \log_{10}\frac{[\text{reductants}]}{[\text{oxidants}]}$$

(eq. 4.36)

Tabulations of thermodynamic data for redox reactions are commonly expressed as $pe°$ values. A set of reduction reactions of importance in groundwater, together with

Table 4.14 Table of reduction reactions of importance in groundwater (Champ *et al.* 1979).

Reaction	$pe° = log_{10}K$
(1) $\frac{1}{4}O_{2(g)} + H^+ + e^- = \frac{1}{2}H_2O$	+20.75
(2) $\frac{1}{5}NO_3^- + \frac{6}{5}H^+ + e^- = \frac{1}{10}N_{2(g)} + \frac{3}{5}H_2O$	+21.05
(3) $\frac{1}{2}MnO_{2(s)} + 2H^+ + e^- = \frac{1}{2}Mn^{2+} + H_2O$	+20.8
(4) $\frac{1}{8}NO_3^- + \frac{5}{4}H^+ + e^- = \frac{1}{8}NH_4^+ + \frac{3}{8}H_2O$	+14.9
(5) $Fe(OH)_{3(s)} + 3H^+ + e^- = Fe^{2+} + 3H_2O$	+17.1
(6) $\frac{1}{8}SO_4^{2-} + \frac{9}{8}H^+ + e^- = \frac{1}{8}HS^- + \frac{1}{2}H_2O$	+4.25
(7) $\frac{1}{8}CO_{2(g)} + H^+ + e^- = \frac{1}{8}CH_{4(g)} + \frac{1}{4}H_2O$	+2.87
(8) $\frac{1}{6}N_{2(g)} + \frac{4}{3}H^+ + e^- = \frac{1}{3}NH_4^+$	+4.68
(9) $\frac{1}{4}CO_{2(g)} + H^+ + e^- = \frac{1}{4}CH_2O + \frac{1}{4}H_2O$	−1.20

(*Source*: Adapted from Champ, D.R., Gulens, J. and Jackson, R.E. (1979) Oxidation-reduction sequences in ground water flow systems. *Canadian Journal of Earth Sciences* **16**, 12–23. © 1979, Canadian Science Publishing.)

their respective $pe°$ values, is listed in Table 4.14. The reactions are listed on the basis of decreasing oxidizing ability, such that species associated with a reaction of more positive $pe°$ act as electron acceptors or oxidizing agents in the oxidation of species associated with reactions of significantly more negative $pe°$.

As an example of this law of mass action approach to redox reactions, the following equilibrium constants can be written for the two half-reactions describing the oxidation of Fe^{2+} to Fe^{3+} by free oxygen (eqs. 4.28 and 4.29):

$$K = \frac{1}{P_{O_2}^{1/4}[H^+][e^-]} = 10^{20.75}$$

$$\text{(eq. 4.37)}$$

$$K = \frac{[Fe^{2+}]}{[Fe^{3+}][e^-]} = 10^{13.05} \quad \text{(eq. 4.38)}$$

Rewriting eqs. 4.37 and 4.38 in logarithmic form produces:

$$pe = 20.75 + \frac{1}{4}\log_{10}P_{O_2} - pH$$

$$\text{(eq. 4.39)}$$

$$pe = 13.05 + \log_{10}\left(\frac{[Fe^{3+}]}{[Fe^{2+}]}\right)$$

$$\text{(eq. 4.40)}$$

If the complete redox reaction (eq. 4.30) is at equilibrium, and if the concentrations of Fe^{2+} and Fe^{3+}, P_{O2} and pH are known, then the pe obtained from both these relations (eqs. 4.39 and 4.40) is the same.

In groundwater systems, there is an interdependency of pe and pH which can be conveniently represented as pe-pH stability diagrams. Methods for the construction of pe-pH diagrams are presented by Stumm and Morgan (1981). As an example, a Fe stability diagram is shown in Fig. 4.32. The equilibrium equations required to construct this type of diagram provide boundary conditions towards which a redox system is proceeding. In practical terms, stability diagrams can be used to predict the likely dissolved ion or mineral phase that may be present in a groundwater for a measured pe-pH condition.

As an alternative to pe, the redox condition for equilibrium processes can be expressed in terms of Eh. Eh is commonly referred to as the (platinum electrode) redox potential and is defined as the energy gained in the transfer of 1 mole of electrons from an oxidant to H_2. Eh is defined by the Nernst equation:

$$Eh(\text{volts}) = Eh° + \frac{2.303RT}{nF}\log_{10}\left(\frac{[\text{oxidants}]}{[\text{reductants}]}\right)$$

$$\text{(eq. 4.41)}$$

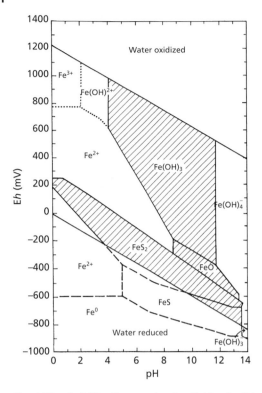

Fig. 4.32 Stability diagram showing fields of solid and dissolved forms of Fe as a function of Eh and pH at 25 °C and 1 atmosphere pressure. The diagram represents a system containing activities of total sulphur species of 96 mg L^{-1} as SO_4^{2-}, total carbon dioxide species of 61 mg L^{-1} as HCO_3^-, and dissolved Fe of 56 μg L^{-1}. Solids indicated by the shaded areas would be thermodynamically stable in their designated domains. Boundaries between solute species are not sensitive to specific dissolved Fe activity, but the domains of solid species will increase in area if more dissolved Fe is present. The boundaries for sulphides and elemental Fe extend below the water stability boundary and show the conditions required for thermodynamic stability of Fe^0. Under the conditions specified, siderite (FeCO$_3$) saturation is not reached. Therefore, FeCO$_3$ is not a stable phase and does not have a stability domain in the diagram (Hem 1985). (*Source:* Adapted from Hem, J.D. (1985). Study and interpretation of the chemical characteristics of natural water (3rd edn). *United States Geological Survey Water Supply Paper* 2254, 263 pp.)

are readily available, for example in Krauskopf and Bird (1995).

The equation relating E$h°$ to the thermodynamic equilibrium constant is

$$Eh° = \frac{RT}{nF} \log_e K \qquad \text{(eq. 4.42)}$$

Finally, the relationship between Eh and pe is

$$Eh = \frac{2.303RT}{nF} pe \qquad \text{(eq. 4.43)}$$

which at 25 °C becomes

$$Eh = \frac{0.059}{n} pe \qquad \text{(eq. 4.44)}$$

The Nernst equation (eq. 4.41) assumes that the species participating in the redox reactions are at equilibrium and that the redox reactions are reversible, both in solution and at the electrode-solution interface. Since the redox reactions involving most of the dissolved species in groundwater are not reversible, such correlations between Eh, pe and pH can be limited in practice. Instead, and as proposed by Champ *et al.* (1979), it is valuable to consider redox processes from a qualitative point of view in which overall changes in redox conditions in an aquifer are described. As part of the concept of redox sequences, Champ *et al.* (1979) suggested that three redox zones exist in aquifer systems: the oxygen-nitrate, iron-manganese and sulphide zones (Fig. 4.33).

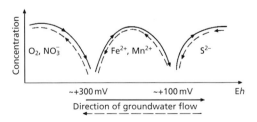

Fig. 4.33 Variation in the concentration of the major dissolved species affected by redox processes within a groundwater flow system (Champ *et al.* 1979). (*Source:* Champ, D.R., Gulens, J. and Jackson, R. E. (1979) Oxidation-reduction sequences in ground water flow systems. *Canadian Journal of Earth Sciences* **16**, 12–23. © 1979, Canadian Science Publishing.)

where F is the Faraday constant (9.65 × 10^4 C mol^{-1}), R the gas constant, and T the absolute temperature (degrees K). E$h°$ is a standard or reference condition at which all substances involved in the redox reaction are at a hypothetical unit activity. Tabulated values of E$h°$

Two general types of hydrochemical systems are recognized by Champ *et al.* (1979): closed and open oxidant systems. In the closed-system, the groundwater initially contains dissolved oxidized species such as O_2, NO_3^-, SO_4^{2-} and CO_2 and also excess reduced DOC. After entering the groundwater flow system via the recharge area, the groundwater is then closed to the input of further oxidants or oxidized species. In the open-system, excess dissolved oxygen is present which may react with reduced species such as HS^- and NH_4^+, for example in situations where landfill leachate is in contact with groundwater. Champ *et al.* (1979) recognized that in closed-systems containing excess reducing agent (DOC) and open-systems containing an excess of dissolved oxygen, sequences of redox reactions can be identified as summarized in Tables 4.15 and 4.16, respectively. An example of the

Table 4.15 Sequence of redox processes in a closed system. In this example, the simplest carbohydrate, CH_2O, represents the dissolved organic carbon that acts as a reducing agent to reduce the various oxidized species initially present in a recharging groundwater. For a confined aquifer containing excess DOC and some solid phase Mn(IV) and Fe(III), it is predicted, on the basis of decreasing negative values of free energy change, that the oxidized species will be reduced in the sequence O_2, NO_3^-, Mn(IV), Fe(III), SO_4^{2-}, HCO_3^- and N_2. As the reactions proceed, and in the absence of other chemical reactions such as ion exchange, the equations show that the sum of dissolved inorganic carbon species (H_2CO_3, HCO_3^-, CO_3^{2-} and complexes) rises as DOC is consumed. The pH of the groundwater may also increase depending on the relative importance of the Fe and Mn reduction processes (Champ *et al.* 1979).

Reaction	Equation
Aerobic respiration	$CH_2O + O_2 = CO_2 + H_2O$
Denitrification	$CH_2O + \frac{4}{5}NO_3^- + \frac{4}{5}H^+ = CO_2 + \frac{2}{5}N_2 + \frac{7}{5}H_2O$
Mn(IV) reduction	$CH_2O + 2MnO_2 + 4H^+ = 2Mn^{2+} + 3H_2O + CO_2$
Fe(III) reduction	$CH_2O + 8H^+ + 4Fe(OH)_3 = 4Fe^{2+} + 11H_2O + CO_2$
Sulphate reduction	$CH_2O + \frac{1}{2}SO_4^{2-} + \frac{1}{2}H^+ = \frac{1}{2}HS^- + H_2O + CO_2$
Methane fermentation	$CH_2O + \frac{1}{2}CO_2 = \frac{1}{2}CH_4 + CO_2$
Nitrogen fixation	$CH_2O + H_2O + \frac{2}{3}N_2 + \frac{4}{3}H^+ = \frac{4}{3}NH_4^+ + CO_2$

(*Source:* Champ, D.R., Gulens, J. and Jackson, R.E. (1979) Oxidation-reduction sequences in ground water flow systems. *Canadian Journal of Earth Sciences* **16**, 12–23. © 1979, Canadian Science Publishing.)

Table 4.16 Sequence of redox processes in an open-system in which an excess of dissolved oxygen reacts with reduced species. Under these conditions, the reduced species will be oxidized in the sequence: dissolved organic carbon (CH_2O), HS^-, Fe^{2+}, NH_4^+ and Mn^{2+}. As each of these reactions proceeds, the Eh will become more positive, while the pH of the groundwater should decrease (Champ *et al.* 1979).

Reaction	Equation
Aerobic respiration	$O_2 + CH_2O = CO_2 + H_2O$
Sulphide oxidation	$O_2 + \frac{1}{2}HS^- = \frac{1}{2}SO_4^{2-} + \frac{1}{2}H^+$
Fe(II) oxidation	$O_2 + 4Fe^{2+} + 10H_2O = 4Fe(OH)_3 + 8H^+$
Nitrification	$O_2 + \frac{1}{2}NH_4^+ = \frac{1}{2}NO_3^- + H^+ + \frac{1}{2}H_2O$
Mn(II) oxidation	$O_2 + 2Mn^{2+} + 2H_2O = 2MnO_2 + 4H^+$

(*Source:* Champ, D.R., Gulens, J. and Jackson, R.E. (1979) Oxidation-reduction sequences in ground water flow systems. *Canadian Journal of Earth Sciences* **16**, 12–23. © 1979, Canadian Science Publishing.)

identification of a groundwater redox sequence is provided in Box 4.10, and further discussion of the specific redox reaction of microbially mediated denitrification is presented in Box 4.11.

4.10 Groundwater in crystalline rocks

Weathering processes participate in controlling the hydrogeochemical cycles of many elements. In soluble carbonate and evaporite deposits, solution processes are rapid and, generally, congruent but in lithologies composed of silicates and quartz, the solution processes are very slow and incongruent. In many sedimentary sandstone aquifers, the solution of traces of carbonate present either as cement or detrital grains may predominate over any chemical contribution from the silicate minerals. Crystalline rocks of igneous or metamorphic origin, on the other hand, contain appreciable amounts of quartz and aluminosilicate minerals such as feldspars and micas. Although the dissolution of most silicate minerals results in very dilute solutions, the weathering of silicate minerals is estimated to contribute about 45% to the total dissolved load of the world's rivers, underlining the significant role of these processes in the overall chemical denudation on the Earth's surface (Stumm and Wollast 1990).

Upper crustal rocks have an average composition similar to the rock granodiorite (Table 4.17) that is composed of the framework silicates plagioclase feldspar, potassium

---

Box 4.10   Redox processes in the Lincolnshire Limestone aquifer, England

The Jurassic Lincolnshire Limestone aquifer in eastern England provides an example of the zonation of redox conditions in an aquifer system. In this fissured, oolitic limestone aquifer (Fig. 4.34), an eastward decline in NO_3^- concentration accompanies progressive removal of dissolved oxygen and lowering of the redox potential (Eh) from an initial value of about +400 mV. Over a distance of 10 km, the groundwater NO_3^- concentration declines from a maximum of 25 mg L^{-1} as N at outcrop to less than 5 mg L^{-1} as N when confined. Four cored boreholes were drilled along a groundwater flow line and the material recovered examined for organic carbon content and the presence of denitrifying bacteria. The results showed that the DOC content of the pore water ranged between 13 and 28 mg L^{-1} and that of the mobile groundwater between 1.6 and 3.4 mg L^{-1}. Denitrifying bacteria were cultured from samples scraped from fissure walls, but not from samples incubated with pore water. Thus, it appears that the source of DOC supporting denitrification is contained in the limestone matrix, and that the very small pore size of the matrix restricts denitrification to short distances from fissure walls (Lawrence and Foster 1986).

Further into the confined aquifer, at about 12 km from outcrop, the Eh falls to less than +100 mV and the groundwater environment remains anaerobic. In this region, SO_4^{2-} reduction is noticeable by the presence of H_2S and lowered SO_4^{2-} concentrations, although the process is sluggish with SO_4^{2-} concentrations persisting for at least 5 km beyond the decline in redox potential. According to Edmunds and Walton (1983), Fe is initially quite soluble in this reducing environment (Fe^{2+} < 0.5 mg L^{-1}), but once sulphide is produced by SO_4^{2-} reduction (see reaction equation in Table 4.15) much of the Fe^{2+} is removed as ferrous sulphide (FeS) as follows:

$$Fe^{2+} + HS^- = FeS_{(s)} + H^+ \quad (eq.\ 1)$$

Box 4.10   (Continued)

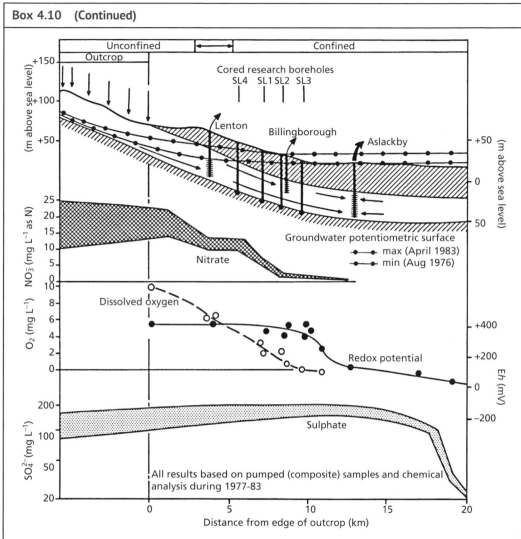

Fig. 4.34 Downgradient hydrogeological and hydrochemical cross-section of the southern Lincolnshire Limestone showing the sequential changes in the redox sensitive species, O_2, NO_3^- and SO_4^{2-} (Lawrence and Foster 1986). (*Source:* Lawrence, A.R. and Foster, S.S.D. (1986) Denitrification in a limestone aquifer in relation to the security of low-nitrate groundwater supplies. *Journal of the Institution of Water Engineers and Scientists* **40**, 159–172.)

Box 4.11   Microbially mediated denitrification

The sequences of redox reactions shown in Tables 4.15 and 4.16 are thermodynamically favoured, but their reaction rates are slow in the absence of catalysts. In natural waters, the most important electron-transfer mechanism is catalysis associated with microbially produced enzymes. As shown in Fig. 4.35, the most stable nitrogen species within the E*h*-pH range encountered in the majority of groundwaters is erroneously predicted to be gaseous nitrogen (N_2). The observed departure from equilibrium is explained by the

(Continued)

Box 4.11 (Continued)

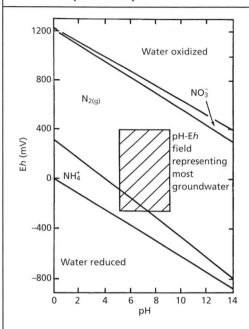

Fig. 4.35 Stability diagram showing fields of dissolved and gaseous forms of N as a function of E*h* and pH at 25 °C and 1 atmosphere pressure. The diagram represents a system containing an activity of total N species of 14 mg L^{-1} (Stumm and Morgan 1981). (*Source: Stumm, W. and Morgan, J.J. (1981) Aquatic Chemistry: An Introduction Emphasizing Chemical Equilibria in Natural Waters* (2nd edn). Wiley, New York. © 1981, John Wiley & Sons.)

(C, H, O, N, P, S), minor amounts of minerals (K, Na, Mg, Ca, Fe) and trace amounts of certain metals (Mn, Zn, Cu, Co and Mo). On the basis of average cellular composition, the favourable ratio of C:N:P:S is about 100:20:4:1 (Spector 1956). For energy generation, electron donors (DOC, H$_2$S, NH$_3$, Fe^{2+}) and electron acceptors (DO, NO$_3^-$, Fe(III), Mn(IV), SO$_4^{2-}$, CO$_2$) are required. It is normally considered that most groundwaters should be capable of supplying the very low concentrations of minerals and trace metals required by microbes, as well as the electron donors, particularly DOC (Box 4.10).

From a survey of one hundred groundwaters, Thurman (1985) reported a median DOC content of 0.7 mg L^{-1} for sandstone, limestone and sand and gravel aquifers and from a global synthesis of 9404 groundwater DOC concentrations, with a dataset dominated by countries in the low and mid-latitudes, McDonough *et al.* (2020) calculated global mean, median and standard deviation of groundwater DOC concentrations of 3.8, 1.2 and 14.8 mg C L^{-1}, respectively, with most groundwaters (84.1%) falling within the range 0–5 mg C L^{-1}. McDonough *et al.* (2020) found that groundwater age and depth appear to control groundwater DOC with major groundwater basins in the United States containing significantly lower DOC concentrations than local and shallow aquifers and those with complex hydrogeological structures. Also, significantly higher DOC concentrations were identified in aquifers with <100 mm a^{-1} of recharge compared to those with high and very high recharge rates (100–300 mm a^{-1} and >300 mm a^{-1}, respectively), which could indicate a dilution effect. Hence, based on these reported values, DOC concentrations in groundwater should meet microbial requirements which have been reported to be less than 0.1 mg L^{-1} (Zobell

catalysing effect of bacteria in accelerating the biological reduction of NO$_3^-$ at lower redox potentials (Hiscock *et al.* 1991).

The viability of micro-organisms in groundwater is dependent on two important factors that limit enzymatic function and cell growth, namely temperature and nutrient availability. At low temperatures, microbial activity decreases markedly but is measurable between 0 and 5 °C. For the process of denitrification, a general doubling is observed with every 10 °C increase in temperature (Gauntlett and Craft 1979). The nutrients required for biosynthesis include those elements required in large amounts

Box 4.11 (Continued)

and Grant 1942). In aquifer situations where the availability of DOC is limited, then other electron donors such as reduced sulphur species become important.

Denitrification is observed to proceed at reduced oxygen levels via a number of microbially mediated steps, the end product of which is normally gaseous nitrogen (N_2) (Korom 1992). The denitrification process requires a suitable electron donor or donors to complete the dissimilatory reduction of NO_3^- to N_2, the most likely of which are organic carbon (heterotrophic denitrification) and reduced Fe and S species (autotrophic denitrification). The stoichiometry of denitrification reactions can be expressed by the following simple equations which describe a generalized progression of reactions with depth below the water table:

1) Heterotrophic denitrification in which an arbitrary organic compound (CH_2O) is oxidized:

$$5CH_2O + 4NO_3^- + 4H^+$$
$$= 5CO_2 + 2N_{2(g)} + 7H_2O$$

(eq. 1)

In this reaction, 1.1 g of C is required to reduce 1.0 g of N.

2) Autotrophic denitrification by reduced iron in which ferrous iron is oxidized:

$$5Fe^{2+} + NO_3^- + 12H_2O$$
$$= 5Fe(OH)_3 + 1/2N_{2(g)} + 9H^+$$

(eq. 2)

In this reaction, 20.0 g of Fe^{2+} is required to reduce 1.0 g of N.

3) Autotrophic denitrification by reduced sulphur in which pyrite is oxidized:

$$5FeS_{2(s)} + 14NO_3^- + 4H^+$$
$$= 5Fe^{2+} + 10SO_4^{2-} + 7N_{2(g)} + 2H_2O$$

(eq. 3)

In this reaction, 1.6 g of S is required to reduce 1.0 g of N.

In some highly reducing, carbon-rich environments, NO_3^- may be converted to NH_4^+ by dissimilatory nitrate reduction to ammonium (DNRA). DNRA may be important in some marine sediments, but in less-reducing

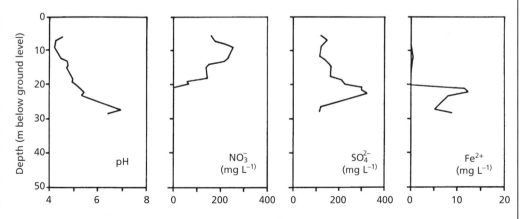

Fig. 4.36 Depth profiles of groundwater pH, NO_3^-, SO_4^{2-} and Fe for multilevel observation well installation NP40 located near to Vierlingsbeek, south-east Netherlands (van Beek 2000). (*Source:* Adapted from van Beek, C.G.E.M. (2000) Redox processes active in denitrification. In: *Redox: Fundamentals, Processes and Applications* (eds J. Schüring, H.D. Schulz, W.R. Fischer *et al.*). Springer-Verlag, Berlin, pp. 152–160.)

(Continued)

Box 4.11 (Continued)

groundwater environments, denitrification is generally favoured over DNRA (Korom 1992).

Although denitrification by organic matter is thermodynamically favourable, the reactivity of organic matter in denitrification is much lower than the reactivity of pyrite. To demonstrate, Fig. 4.36 shows depth profiles for pH and the redox sensitive species NO_3^-, SO_4^{2-} and Fe^{2+} for a multilevel observation well installed near to the Vierlingsbeek wellfield in the south-east of the Netherlands. The unconfined aquifer at this site is composed of unconsolidated fluvial sands containing no calcite and low amounts of organic matter (0.1–2%) and pyrite (<0.01–0.2%), the amounts of which decrease with depth (Fig. 4.37). An aquitard composed of fine-grained cemented deposits exists at 30 m below ground level (m bgl) and the water table varies between 2 and 4 m bgl. Very intensive cattle farming in the vicinity of the wellfield with the application of liquid manure has provided a source of high NO_3^- concentrations in the shallow groundwater (van Beek 2000).

At Vierlingsbeek, at 10 m bgl, Fig. 4.36 shows that measured NO_3^- concentrations are above 200 mg L^{-1}, but that below 21 m bgl NO_3^- is absent. The decrease in NO_3^- coincides with an increase in SO_4^{2-} and Fe^{2+} between 20 and 21 m bgl that can be explained by autotrophic denitrification in which pyrite is oxidized (eq. 3). In this reaction, and in the absence of carbonate to act as a buffer, protons are consumed, and a steady rise in pH is observed below 20 m bgl.

The weight percentage of pyrite in the fluvial sands is about a factor of 10 lower than the content of organic matter (Fig. 4.37) and it would be expected that the reduction capacity of organic matter per unit weight of solid material would be far greater than that of pyrite. Hence, the measured hydrochemical profiles demonstrate the higher reactivity of pyrite compared with organic

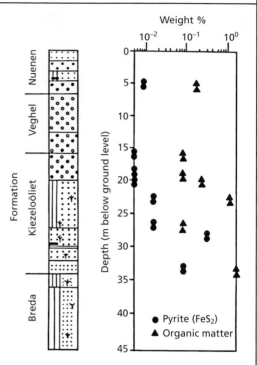

Fig. 4.37 Depth profile of pyrite and organic matter content in the unconsolidated fluvial sands aquifer recorded at multi-level observation well NP1 located near to Vierlingsbeek, south-east Netherlands (van Beek 2000). (*Source:* Adapted from van Beek, C.G.E.M. (2000) Redox processes active in denitrification. In: *Redox: Fundamentals, Processes and Applications* (eds J. Schüring, H.D. Schulz, W.R. Fischer *et al.*). Springer-Verlag, Berlin, pp. 152–160.)

matter in the denitrification process at this site (van Beek 2000).

This example illustrates how, with consumption of the source of electron donor, in this case pyrite, the denitrification front will migrate downwards such that the aquifer will gradually lose the ability to attenuate NO_3^-. For a similar hydrogeological situation to Vierlingsbeek, and also for denitrification in the presence of reduced sulphur with the oxidation of pyrite, Robertson *et al.* (1996) measured a downward rate of movement of a denitrification front in silt-rich sediments of 1 mm a^{-1}.

Table 4.17 Estimated average mineralogical composition of the upper continental crust (volume %). The composition of the exposed crust differs from the upper continental crustal estimate primarily in the presence of volcanic glass (Nesbitt and Young 1984).

	Upper continental crust	Exposed continental crust surface
Plagioclase feldspar	39.9	34.9
Potassium feldspar	12.9	11.3
Quartz	23.2	20.3
Volcanic glass	0.0	12.5
Amphibole	2.1	1.8
Biotite mica	8.7	7.6
Muscovite mica	5.0	4.4
Chlorite	2.2	1.9
Pyroxene	1.4	1.2
Olivine	0.2	0.2
Oxides	1.6	1.4
Others	3.0	2.6

(*Source:* Modified from Nesbitt, H.W. and Young, G.M. (1984) Prediction of some weathering trends of plutonic and volcanic rocks based on thermodynamic and kinetic considerations. *Geochimica et Cosmochimica Acta* **48**, 1523–1534.)

feldspar and quartz, with plagioclase feldspar being the most abundant. Depending on the nature of the parent rocks, various secondary minerals such as gibbsite, kaolinite, smectite and illite are formed as reaction products. In all cases, water and carbonic acid (H_2CO_3), which is the source of H^+, are the main reactants. The net result of the reactions is the release of cations (Ca^{2+}, Mg^{2+}, K^+, Na^+) and the production of alkalinity via HCO_3^-.

The two following reactions (eqs 4.45 and 4.46) provide examples of important silicate weathering processes. First, taking Ca-rich plagioclase feldspar (anorthite), the incongruent weathering reaction resulting in the aluminosilicate residue kaolinite is written (Andrews *et al.* 2004):

$$CaAl_2Si_2O_{8(s)} + 2H_2CO_{3(aq)} + H_2O$$
$$\rightarrow Ca^{2+}_{(aq)} + 2HCO_{3(aq)}^- + Al_2Si_2O_5(OH)_{4(s)}$$
$$\text{(eq. 4.45)}$$

In this weathering reaction, H^+ ions dissociated from H_2CO_3 hydrate the silicate surface

and naturally buffer the infiltrating soil water or groundwater. The ionic bonds between Ca^{2+} and the SiO_4 tetrahedra are easily broken, releasing Ca^{2+} into solution resulting in a Ca-HCO_3 water type.

Second, for the Na-rich plagioclase feldspar (albite), the incongruent reaction producing kaolinite and releasing Na^+ and HCO_3^- ions is

$$2NaAlSi_3O_{8(s)} + 9H_2O + 2H_2CO_{3(aq)}$$
$$\rightarrow Al_2Si_2O_5(OH)_{4(s)} + 2Na^+_{(aq)}$$
$$+ 2HCO_{3(aq)}^- + 4Si(OH)_{4(aq)}$$
$$\text{(eq. 4.46)}$$

Further examples of weathering reactions for some common primary minerals are listed in Table 4.18. When ferrous iron (Fe^{2+}) is present in the lattice, as in the case of biotite mica, oxygen consumption may become an important factor affecting the rate of dissolution that results in Fe-oxide as an insoluble weathering product. For example, in the

Table 4.18 Reactions for incongruent dissolution of some aluminosilicate minerals (solid phases are underlined) (Freeze and Cherry 1979).

Gibbsite-kaolinite	$\underline{Al_2O_3.3H_2O} + 2Si(OH)_4 = \underline{Al_2Si_2O_5(OH)_4} + 5H_2O$
Na-montmorillonite-kaolinite	$\underline{Na_{0.33}Al_{2.33}Si_{3.67}O_{10}(OH)_2} + \frac{1}{3}H^+ + \frac{23}{6}H_2O$ $= \frac{7}{6}\underline{Al_2Si_2O_5(OH)_4} + \frac{1}{3}Na^+ + \frac{4}{3}Si(OH)_4$
Ca-montmorillonite-kaolinite	$\underline{Ca_{0.33}Al_{4.67}Si_{7.33}O_{20}(OH)_4} + \frac{2}{3}H^+ + \frac{23}{2}H_2O$ $= \frac{7}{3}\underline{Al_2SiO_2O_5(OH)_4} + \frac{1}{3}Ca^{2+} + \frac{8}{3}Si(OH)_4$
Illite-kaolinite	$\underline{K_{0.6}Mg_{0.25}Al_{2.30}Si_{3.5}O_{10}(OH)_2} + \frac{11}{10}H^+ + \frac{63}{60}H_2O = \frac{23}{30}\underline{Al_2Si_2O_5(OH)_4}$ $+ \frac{3}{5}K^+ + \frac{1}{4}Mg^{2+} + \frac{6}{5}Si(OH)_4$
Biotite-kaolinite	$\underline{KMg_3AlSi_3O_{10}(OH)_2} + 7H^+ + \frac{1}{2}H_2O = \frac{1}{2}\underline{Al_2Si_2O_5(OH)_4}$ $+ K^+ + 3Mg^{2+} + 2Si(OH)_4$
Albite-kaolinite	$\underline{NaAlSi_3O_8} + H^+ + \frac{9}{2}H_2O = \frac{1}{2}\underline{Al_2Si_2O_5(OH)_4} + Na^+ + 2Si(OH)_4$
Albite-Na-montmorillonite	$\underline{NaAlSi_3O_8} + \frac{6}{7}H^+ + \frac{20}{7}H_2O = \frac{3}{7}\underline{Na_{0.33}Al_{2.33}Si_{3.67}O_{10}(OH)_2} + \frac{6}{7}Na^+ + \frac{10}{7}Si(OH)_4$
Microcline-kaolinite	$\underline{KAlSi_3O_8} + H^+ + \frac{9}{2}H_2O = \frac{1}{2}\underline{Al_2Si_2O_5(OH)_4} + K^+ + 2Si(OH)_4$
Anorthite-kaolinite	$\underline{CaAl_2Si_2O_8} + 2H^+ + H_2O = \underline{Al_2Si_2O_5(OH)_4} + Ca^{2+}$
Andesine-kaolinite	$\underline{Na_{0.5}Ca_{0.5}Al_{1.5}Si_{2.5}O_8} + \frac{3}{2}H^+ + \frac{11}{4}H_2O = \frac{3}{4}\underline{Al_2Si_2O_5(OH)_4}$ $+ \frac{1}{2}Na^+ + \frac{1}{2}Ca^{2+} + Si(OH)_4$

(*Source:* Freeze, R.A. and Cherry, J.A. (1979) *Groundwater*. Prentice-Hall, Inc., Englewood Cliffs, New Jersey. © 1979, Pearson Education.)

following equation, biotite weathers to gibbsite and goethite:

$$KMgFe_2AlSi_3O_{10}(OH)_{2(s)} + \frac{1}{2}O_{2(aq)} + 3CO_{2(aq)}$$
$$+ 11H_2O \rightarrow Al(OH)_{3(s)} + 2Fe(OH)_{3(s)} + K^+_{(aq)}$$
$$+ Mg^{2+}_{(aq)} + 3HCO^-_{3(aq)} + 3Si(OH)_{4(aq)}$$
$$(eq. 4.47)$$

Differences in solution rates between the silicate minerals lead to their successive disappearance as weathering proceeds. This kinetic control on the distribution of primary silicates is known as the Goldich weathering sequence (Fig. 4.38). As shown in Fig. 4.38, olivine and Ca-plagioclase are the most easily weathered

minerals, while quartz is the most resistant to weathering.

The leaching of different weathering products depends not just on the rate of mineral weathering but also on the hydrological conditions. Montmorillonite ($Na_{0.5}Al_{1.5}Mg_{0.5}Si_4$ $O_{10}(OH)_2$) is formed preferentially in relatively dry climates, where the flushing rate in the soil is low, and its formation is favoured when rapidly dissolving material such as volcanic rock is available. In contrast, gibbsite ($Al(OH)_3.3H_2O$) forms typically in tropical areas with intense rainfall and under well-drained conditions. Here, gibbsite and other Al-hydroxides may form a thick weathering residue of bauxite.

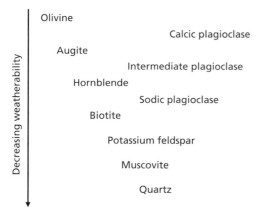

Fig. 4.38 The Goldich weathering sequence based on observations of the sequence of disappearance of primary silicate minerals in soils (Goldich 1938). (*Source:* Modified from Goldich, S.S. (1938) A study in rock-weathering. *Journal of Geology* **46**, 17–58.)

Because of differences in residence times, water in areas where montmorillonite is forming is high in dissolved ions, while in areas with high rainfall with the formation of gibbsite, the dissolved ion concentrations are low (Appelo and Postma 2005).

Now, by taking a thermodynamic equilibrium approach, it is possible to gain further insight into some of the more specific results of groundwater interactions with feldspars and clays. For example, considering the albite ($NaAlSi_3O_8$) dissolution reaction given in Table 4.18, then from the law of mass action:

$$K_{\text{albite–kaolinite}} = \frac{[Na^+][Si(OH)_4]^2}{[H^+]}$$
(eq. 4.48)

where $K_{\text{albite-kaolinite}}$ is the equilibrium constant. In this approach, the activities of the mineral phases and water are taken as unity. Expressing eq. 4.48 in logarithmic form gives:

$$\log_{10}K_{\text{albite–kaolinite}} = \log_{10}[Na^+] \\ + 2\log_{10}[Si(OH)_4] - pH$$
(eq. 4.49)

or

$$\log_{10}K_{\text{albite–kaolinite}} = \log_{10}\left(\frac{[Na^+]}{[H^+]}\right) \\ + 2\log_{10}[Si(OH)_4]$$
(eq. 4.50)

which indicates that the equilibrium condition for the albite-kaolinite reaction can be expressed in terms of pH and activities of Na^+ and $Si(OH)_4$. Equilibrium relations such as this are the basis for the construction of stability diagrams, examples of which are shown in Fig. 4.39. These types of diagrams represent minerals with ideal chemical compositions, which may not accurately represent real systems, but, nevertheless, are useful in the interpretation of chemical data from hydrogeological systems. In igneous terrain, nearly all groundwaters within several hundred metres of the ground surface plot in the kaolinite fields of Fig. 4.39. A small percentage of samples plot in the montmorillonite fields and hardly any occur in the gibbsite, mica or feldspar fields or exceed the solubility limit of amorphous silica. This observation suggests that alteration of feldspars and micas to kaolinite is a common process in groundwater flow systems in igneous rocks (Freeze and Cherry 1979).

The chemical composition of groundwaters in crystalline rocks is characterized by very low major ion concentrations. Normally, without exception, HCO_3^- is the dominant anion (see eqs. 4.45 and 4.46) with silicon present in major concentrations relative to the cations. In the pH range that includes nearly all groundwaters (pH = 6–9), the dominant dissolved silicon released by weathering is the extremely stable and non-ionic monosilicic acid, $Si(OH)_4$.

$Si(OH)_4$ is usually expressed as SiO_2 in water analyses. Representative chemical analyses of groundwaters from granitic massifs are shown in Table 4.19 in which the SiO_2 concentration ranges from 1 to 85 mg L^{-1}. For comparison, Haines and Lloyd (1985) present data for the major British sedimentary limestone and

Table 4.19 Mean chemical composition of groundwaters from European and African granitic massifs as presented by Tardy (1971).

Location	Number of samples	pH	HCO$_3^-$ (mg L^{-1})	Cl$^-$ (mg L^{-1})	SO$_4^{2-}$ (mg L^{-1})	SiO$_2$ (mg L^{-1})	Na$^+$ (mg L^{-1})	K$^+$ (mg L^{-1})	Ca^{2+} (mg L^{-1})	Mg^{2+} (mg L^{-1})
Norway	28	5.4	4.9	5.0	4.6	3.0	2.6	0.4	1.7	0.6
Vosges	51	6.1	15.9	3.4	10.9	11.5	3.3	1.2	5.8	2.4
Brittany	7	6.5	13.4	16.2	3.9	15.0	13.3	1.3	4.4	2.6
Central Massif	10	7.7	12.2	2.6	3.7	15.1	4.2	1.2	4.6	1.3
Alrance Spring F	77	5.9	6.9	<3	1.15	5.9	2.3	0.6	1.0	0.4
Alrance Spring A	47	6.0	8.1	<3	1.1	11.5	2.6	0.6	0.7	0.3
Corsica	25	6.7	40.3	22.0	8.6	13.2	16.5	1.4	8.1	4.0
Sahara	8	6.9	30.4	4.0	20	9	30	1.8	40	-
Senegal	7	7.1	43.9	4.2	0.8	46.2	8.4	2.2	8.3	3.7
Chad	2	7.9	54.4	<3	1.4	85	15.7	3.4	8.0	2.5
Ivory Coast (dry season)	54	5.5	6.1	<3	0.4	10.8	0.8	1.0	1.0	0.10
Ivory Coast (wet season)	59	5.5	6.1	<3	0.5	8.0	0.2	0.6	<1	<0.1
Malagasy (High Plateaux)	2	5.7	6.1	1	0.7	10.6	0.95	0.62	0.40	0.12

(*Source*: Tardy, Y. (1971) Characterization of the principal weathering types by the geochemistry of waters from some European and African crystalline massifs. *Chemical Geology* **7**, 253–271. © 1971, Elsevier.)

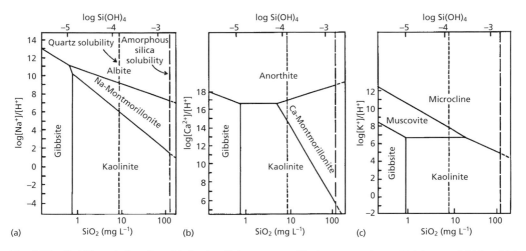

Fig. 4.39 Stability relations for gibbsite, kaolinite, montmorillonite, muscovite and feldspar at 25 °C and 1 bar pressure as functions of pH and the activities of Na^+, Ca^{2+}, K^+ and $Si(OH)_4$. (a) Gibbsite, $Al_2O_3.H_2O$; kaolinite, $Al_2Si_2O_5(OH_4)$; Na-montmorillonite, $Na_{0.33}Al_{2.33}Si_{3.67}O_{10}(CH)_2$; and albite, $NaAlSi_3O_8$. (b) Gibbsite; kaolinite; Ca-montmorillonite, $Ca_{0.33}Al_{4.67}Si_{7.33}O_{20}(OH_4)$; and anorthite, $CaAl_2Si_2O_8$. (c) Gibbsite; kaolinite; muscovite, $KAl_2(AlSi_3O_{10})(OH)_2$; and microcline (feldspar), $KAlSi_3O_8$ (Tardy 1971). (*Source:* Tardy, Y. (1971) Characterization of the principal weathering types by the geochemistry of waters from some European and African crystalline massifs. *Chemical Geology* **7**, 253–271. © 1971, Elsevier.)

sandstone aquifers that demonstrate a groundwater SiO_2 concentration in the range 6–27 mg L^{-1}, with the higher values associated with weathering of Mg-smectites found in carbonate aquifers.

In crystalline rocks, the groundwater concentrations of the anions Cl^- and SO_4^{2-} usually occur in only minor or trace amounts and typically have an atmospheric source or occur as trace impurities in rocks and minerals. Saline groundwaters can occur in granitic rocks, with the example of the Carnmenellis Granite in Cornwall, south-west England illustrated in Box 4.12.

Box 4.12 Hydrogeochemical characteristics of the Carnmenellis Granite, Cornwall, England

The Carnmenellis Granite and its aureole in south-west England contain the only recorded thermal groundwaters in British granites and occur as springs in tin mines. Most of the groundwaters are saline with a maximum mineralization of 19 310 mg L^{-1} (Edmunds *et al.* 1984). The Carnmenellis Granite forms a near-circular outcrop of the Cornubian batholith (Fig. 4.40) which was intruded about 290 Ma into Devonian argillaceous sediments. The rock is highly fractured.

Beneath a weathered zone of variable thickness, the granite is characterized by secondary permeability. Most groundwater flow and storage occurs in open horizontal fractures and is shallow in depth (commonly above 50 m) and localized.

The granite is composed of coarse- and fine-grained porphyritic muscovite-biotite granite, the former being more common, and has undergone a long history of alteration. The granite is enriched in volatile

(Continued)

Box 4.12 (Continued)

elements (B, Cl, F, Li) compared with many other granite terrains. There is extensive hydrothermal mineralization of Variscan age which has produced economic vein deposits of Sn, Cu, Pb and Zn. The principal mineral lodes occur in a mineralized belt north of the Carnmenellis Granite (Edmunds *et al.* 1985).

Saline groundwaters are encountered in four accessible mines (Fig. 4.40 for locations) in the granite or its thermal aureole, as well as in several disused mines, all at the northern margin. The saline waters generally issue from cross-courses with discharges between 1 and $10 \, L \, s^{-1}$ at depths between 200 and 700 m below surface. The discharge temperatures of up to 52 °C are typically in excess of the average regional thermal gradients of $30 °C \, km^{-1}$ in the granite and $50 °C \, km^{-1}$ in the aureole. As proposed by Edmunds *et al.* (1985) and shown schematically in Fig. 4.41, the temperature anomaly implies that ancient, warmer saline fluids are upwelling by convective circulation and mixing with recent, fresh, shallow groundwaters. The driving force for the current circulation system is the hydraulic sink created by the former mining operations. Also, the existence of old, flooded mine workings locally increases the secondary porosity of the rocks.

Chemical analyses of the four mine waters and two shallow groundwater samples are given in Table 4.20. The fresh, shallow groundwater is generally of good quality, has a low TDS content and may be acidic (pH < 5.5). The most important features of the hydrochemistry of the Carnmenellis Granite, in addition to the high Cl^- concentrations, are the depletion of Na^+ relative to Cl^-, the enhanced Ca^{2+} levels and especially the significantly enriched Li^+, with Li^+ values as high as $125 \, mg \, L^{-1}$. The unusual chemistry combined with stable isotope data demonstrates a meteoric origin for all the groundwaters that excludes seawater as the source of the salinity.

In their hydrogeochemical interpretation, Edmunds *et al.* (1985) concluded that hydrolysis of biotite $(K_2(Mg,Fe)_4(Fe,Al, Li)_2[Si_6Al_2O_{20}](OH)_2(F,Cl)_2)$ could account for the increase in Cl^-, Li^+, K^+ and other species in the groundwaters. The Li^+ and Cl^- are considered as conservative products of biotite alteration, unlikely to be assimilated by reaction products. The fact that the Li/Cl ratio is only 1:100 compared to 1:2 for unaltered granite suggests that Cl^- release is a relatively easy and rapid process that accounts for the observed salinities over very long timescales, while Li^+ and Mg^{2+} and other ions are only released upon a breakdown of the trioctahedral biotite structure. The depletion of Mg^{2+} in most of the saline groundwaters is consistent with the observed biotite alteration to chlorite.

A second important silicate weathering process affecting the hydrogeochemistry is the acid hydrolysis of plagioclase feldspars (eqs. 4.45 and 4.46) that contributes the principal sources of Na^+ and Ca^{2+} to the groundwaters. Both biotite and plagioclase alteration by weathering produces significant amounts of silica as well as kaolinite and other clay minerals; in some areas, kaolinization along fractures reduces the available secondary porosity. Concentrations of SiO_2 (Table 4.20) in these groundwaters represent saturation or supersaturation with respect to chalcedony, and this tendency for silica deposition is argued to rule out fluid inclusions acting as a potential source of salinity (Edmunds *et al.* 1985).

Box 4.12 (Continued)

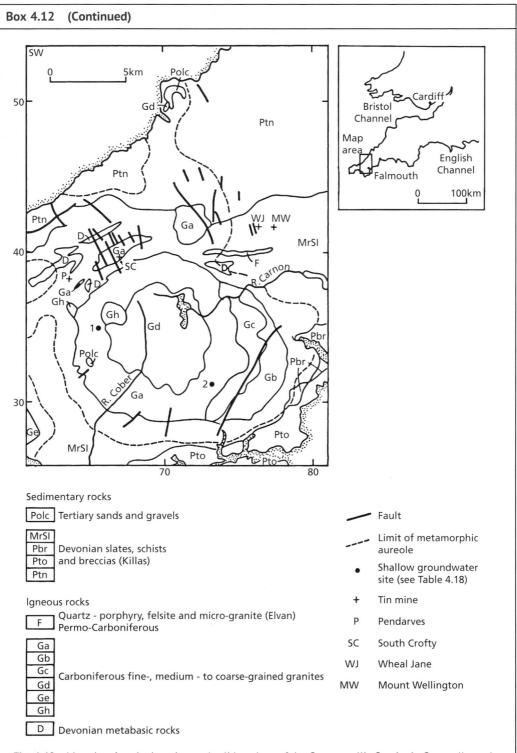

Sedimentary rocks

| Polc | Tertiary sands and gravels |

| MrSl |
| Pbr | Devonian slates, schists |
| Pto | and breccias (Killas) |
| Ptn |

Igneous rocks

| F | Quartz - porphyry, felsite and micro-granite (Elvan) Permo-Carboniferous |

| Ga |
| Gb |
| Gc |
| Gd | Carboniferous fine-, medium - to coarse-grained granites |
| Ge |
| Gh |

| D | Devonian metabasic rocks |

—— Fault

-- -- Limit of metamorphic aureole

• Shallow groundwater site (see Table 4.18)

+ Tin mine

P Pendarves

SC South Crofty

WJ Wheal Jane

MW Mount Wellington

Fig. 4.40 Map showing the location and solid geology of the Carnmenellis Granite in Cornwall, south-west England. The sites of tin mines are shown as follows: P, Pendarves Mine; SC, South Crofty Mine; WJ, Wheal Jane Mine; and MW, Mount Wellington Mine (British Geological Survey 1990). (*Source:* British Geological Survey (1990) *1:50 000 Map of The Carnmenellis Granite. Hydrogeological, Hydrogeochemical and Geothermal Characteristics.* Natural Environment Research Council. © 1990, British Geological Survey.)

(Continued)

Box 4.12   (Continued)

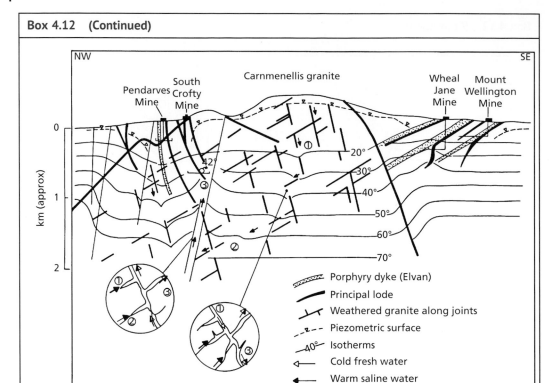

Fig. 4.41   Conceptual cross-section through the Carnmenellis Granite showing the location of tin mines and saline groundwaters. Isotherms in the granite and its aureole show convective distortion due to groundwater circulation. The suggested circulated route of groundwater flow is shown as follows: (1) Rapid percolation of recent meteoric water with vertical drainage enhanced by mining. Local distortion of isotherms by groundwater flow; (2) Ancient meteoric water with high salinity derived from granite-water reactions and stored in the fracture system; and (3) Circulation of mixed groundwater which discharges as springs in the mines. Local enhancement of thermal gradient by upward groundwater flow (Edmunds *et al.* 1985). (*Source:* Edmunds, W.M., Kay, R.L.F. and McCartney, R.A. (1985) Origin of saline groundwaters in the Carnmenellis Granite (Cornwall, England): natural processes and reaction during hot dry rock reservoir circulation. *Chemical Geology* **49**, 287–301. © 1985, Elsevier.)

Table 4.20   Representative chemical analyses of saline and thermal groundwaters from four tin mines together with two fresh shallow groundwater sites in the Carnmenellis Granite or its aureole. Locations are shown in Fig. 4.40. δ^2H and δ^{18}O are measured in per mil relative to Standard Mean Ocean Water (SMOW; Section 5.2) (Edmunds *et al.* 1985).

Site	Mount Wellington Mine	Pendarves Mine	Wheal Jane Mine	South Crofty Mine	Shallow groundwater	
					1	2
Depth (m)	240	260	300	690	30	41
Flow rate (L s^{-1})	15	0.5	10	3.5	–	–
Temperature (°C)	21.6	21.4	39.5	41.5	10.8	10.5
pH	5.6	6.9	6.4	6.5	4.92	–

Box 4.12   (Continued)

Table 4.20   (Continued)

Site	Mount Wellington Mine	Pendarves Mine	Wheal Jane Mine	South Crofty Mine	Shallow groundwater 1	2
Na (mg L^{-1})	125	29	1250	4300	14	12
K	12	3.1	72.0	180	2.8	4.0
Li	3.55	0.06	26.0	125	<0.01	<0.01
Ca	93	18	835	2470	7	13
Mg	11.9	5.0	22.0	73.0	2.7	2.4
Sr	1.43	0.06	12.8	40.0	<0.06	0.23
HCO$_3$ (mg L^{-1})	9	67	21	68	4	9
NO$_3$	11.2	8.3	10	<0.2	2.1	26.0
SO$_4$	275	38	148	145	15	17
Cl	287	32	3300	11 500	25	21
F	0.29	2.90	3.30	2.70	<0.1	0.11
Br	0.9	<0.3	–	43.0	–	–
B	0.80	<0.01	3.3	11.0	–	–
SiO$_2$	19.2	34.2	28.4	34.2		
Fe (mg L^{-1})	43.0	0.62	22.4	4.75	0.280	0.014
Mn	2.90	0.30	4.00	4.50	0.031	0.008
Cu	0.024	0.002	0.005	0.023	0.004	0.027
Ni	0.134	0.007	0.027	0.190	0.002	0.0015
Total mineralization (mg L^{-1})	885	230	5747	19 002	73	105
δ^2H (‰ SMOW)	−35	−38	−31	−29		
δ^{18}O (‰ SMOW)	−5.4	−5.2	−5.7	−5.2		

(*Source:* Edmunds, W.M., Kay, R.L.F. and McCartney, R.A. (1985) Origin of saline groundwaters in the Carnmenellis Granite (Cornwall, England): natural processes and reaction during hot dry rock reservoir circulation. *Chemical Geology* **49**, 287–301. © 1985, Elsevier.)

4.11   Geochemical modelling

Geochemical speciation codes enable comprehensive calculation of chemical activities, speciation and mineral saturation indices, as well as enabling the modelling of hydrochemical processes. Details of available models, their application and limitations are given by Appelo and Postma (2005). Two of the most widely used model codes are NETPATH and PHREEQC.

The interactive geochemical model NETPATH is used to interpret net geochemical mass–balance reactions between initial and final groundwaters along a hydrologic flow-path (Plummer *et al.* 1994). Alternatively, NET-PATH can compute the mixing proportions of two to five initial waters and net geochemical reactions that account for the observed composition of a final water. The program utilizes previously defined chemical and isotopic data for waters from a hydrochemical system. For a set of mineral and (or) gas phases hypothesized to be the reactive phases in the system, NET-PATH calculates the mass transfers in every possible combination of the selected phases that accounts for the observed changes in the selected chemical and (or) isotopic compositions observed along the flowpath. The calculations are of use in interpreting geochemical reactions, mixing proportions, evaporation and (or) dilution of waters, and mineral mass transfer in the chemical and isotopic evolution of natural and environmental waters. Rayleigh distillation calculations are applied to each mass–balance model that satisfies the constraints to predict carbon, sulphur, nitrogen and strontium isotopic compositions at the end point, including radiocarbon dating.

NETPATH uses a mass–balance approach (chemical and isotopic), rather than thermodynamic equilibrium and is best suited for regional, confined flow systems where changes between sampling points can be observed. NETPATH is particularly useful in flow systems in a geochemically complicated system, such as sulphate reduction and methanogenesis (Clark and Fritz 1997). A Microsoft Windows® version (NETPATH-WIN) is presented by El-Kadi *et al.* (2011). An application of NETPATH-WIN in modelling the relative importance of biotite, plagioclase and amphibole weathering and dissolution of secondary carbonate minerals along a flowpath in the hydrochemical evolution of groundwater in the Plateaux Region of Togo is presented by Akpataku *et al.* (2019).

The geochemsical modelling code PHREEQC is designed to perform a wide variety of aqueous geochemical calculations. PHREEQC implements several types of aqueous models: two ion-association aqueous models (including WATEQ4F; Ball and Nordstrom 1991), a Pitzer specific-ion-interaction aqueous model and the Specific Ion Interaction Theory (SIT) aqueous model. Using any of these aqueous models, PHREEQC has capabilities for (1) speciation and saturation-index calculations; (2) batch-reaction and one-dimensional transport calculations with reversible and irreversible reactions that include aqueous, mineral, gas, solid-solution, surface-complexation and ion-exchange equilibria; and (3) inverse modelling to compute sets of mineral and gas mole transfers that account for differences in composition between waters within specified compositional uncertainty limits (Parkhurst and Appelo 2013). An application of PHREEQC to investigate mixing between groundwaters in shallow Quaternary and Permo-Triassic aquifers with elevated fluoride and a deep Cambrian-Ordovician carbonate aquifer system in the central Shaanxi Province, China is presented by Dou *et al.* (2020).

Further reading

Chapelle, F.H. (1993) *Ground-Water Microbiology and Geochemistry.* Wiley, New York.

Domenico, P.A. and Schwartz, F.W. (1998) *Physical and Chemical Hydrogeology* (2nd edn). Wiley, New York.

Edmunds, W.M. and Shand, P. (eds) (2008). Groundwater baseline quality. In: *Natural Groundwater Quality* (eds W.M. Edmunds and P. Shand). Blackwell Publishing, Oxford.

Fetter, C.W. (2001) *Applied Hydrogeology* (4th edn). Pearson Higher Education, Upper Saddle River, New Jersey.

Mazor, E. (2004) *Chemical and Isotopic Groundwater Hydrology: The Applied Approach* (3rd edn). Marcel Dekker, Inc., New York.

Stumm, W. and Morgan, J.J. (1996) *Aquatic Chemistry: Chemical Equilibria and Rates in Natural Waters* (3rd edn). Wiley, New York.

References

Akpataku, K.V., Rai, S.P., Gnazou, M.D.-T. *et al.* (2019) Hydrochemical and isotopic characterisation of groundwater in the southeastern part of the Plateaux Region, Togo. *Hydrological Sciences Journal* **64**, 983–1000.

Andrews, J.E., Brimblecombe, P., Jickells, T.D. *et al.* (2004) *An Introduction to Environmental Chemistry* (2nd edn). Blackwell Science, Oxford.

Appelo, C.A.J. and Postma, D. (2005) *Geochemistry, Groundwater and Pollution* (2nd edn). A.A. Balkema, Leiden, The Netherlands.

Ball, J.W. and Nordstrom, D.K. (1991). *User's manual for WATEQ4F, with revised thermodynamic data base and test cases for calculating speciation of major, trace, and redox elements in natural waters.* Open-File Report 91–183. United States Geological Survey, Menlo Park, California, 189 pp. https://pubs.usgs.gov/of/1991/0183/report.pdf.

Bishop, P.K., Burston, M.W., Lerner, D.N. and Eastwood, P.R. (1990) Soil gas surveying of chlorinated solvents in relation to groundwater pollution studies. *Quarterly Journal of Engineering Geology* **23**, 255–265.

Bottrell, S.H., Raiswell, R. and Leosson, M.A. (1996) The influence of sulphur redox reactions and mixing on the chemistry of shallow groundwaters: the Harrogate mineral waters. *Journal of the Geological Society* **153**, 231–242.

British Geological Survey (1990) *1:50 000 Map of The Carnmenellis Granite. Hydrogeological, Hydrogeochemical and Geothermal Characteristics.* Natural Environment Research Council.

Carlyle, H.F., Tellam, J.H. and Parker, K.E. (2004) The use of laboratory-determined ion exchange parameters in the predictive modelling of field-scale major cation migration in groundwater over a 40-year period. *Journal of Contaminant Hydrology* **68**, 55–81.

Champ, D.R., Gulens, J. and Jackson, R.E. (1979) Oxidation-reduction sequences in ground water flow systems. *Canadian Journal of Earth Sciences* **16**, 12–23.

Chebotarev, I.I. (1955) Metamorphism of natural water in the crust of weathering. *Geochimica et Cosmochimica Acta* **8**, 22–48, 137–170, 198–212.

Clark, I.D. and Fritz, P. (1997) *Environmental Isotopes in Hydrogeology.* Lewis Publishers, Boca Raton, Florida.

Cribb, S.J. (2005) Geology of beer. In: *Encyclopedia of Geology* (eds R.C. Selley, L.R.M. Cocks and I.R. Plimer). Elsevier Ltd., Oxford, pp. 78–81.

Dou, Y., Howard, K., Yang, L. *et al.* (2020) Hydrochemical relation between shallow groundwater with elevated fluoride and groundwater in underlying carbonates. *Groundwater* **58**, 1000–1011. DOI: 10.1111/gwat.12985.

Drever, J.I. (1988) *The Geochemistry of Natural Waters* (2nd edn). Prentice-Hall, Inc., Englewood Cliffs, New Jersey.

Durov, S.A. (1948) Natural waters and graphic representation of their composition. *Doklady Akademii Nauk SSSR* **59**, 87–90.

Edmunds, W.M., Andrews, J.N., Burgess, W.G. *et al.* (1984) The evolution of saline and thermal groundwaters in the Carnmenellis granite. *Mineralogical Magazine* **48**, 407–424.

Edmunds, W.M., Kay, R.L.F. and McCartney, R.A. (1985) Origin of saline groundwaters in the Carnmenellis Granite (Cornwall, England): natural processes and reaction during hot dry rock reservoir circulation. *Chemical Geology* **49**, 287–301.

Edmunds, W.M., Taylor, B.J. and Downing, R.A. (1969) Mineral and thermal waters of the United Kingdom. In: *Report of the Twenty-Third Session of the International Geological Congress Czechoslovakia, 1968. Proceedings of Symposium II, Mineral and Thermal Waters of the World, A – Europe.* Academia, Prague, pp. 139–158.

Edmunds, W.M. and Walton, N.G.R. (1983) The Lincolnshire Limestone – hydrogeochemical evolution over a ten-year period. *Journal of Hydrology* **61**, 201–211.

El-Kadi, A.I., Plummer, L.N. and Aggarwal, P. (2011) NETPATH-WIN: An interactive user version of the mass-balance model, NETPATH. *Ground Water* **49**, 593–599.

Freeze, R.A. and Cherry, J.A. (1979) *Groundwater.* Prentice-Hall, Inc., Englewood Cliffs, New Jersey.

Gauntlett, R.B. and Craft, D.G. (1979) *Biological Removal of Nitrate from River Water.* Report TR 98. Water Research Centre, Medmenham, Buckinghamshire.

Goldich, S.S. (1938) A study in rock-weathering. *Journal of Geology* **46**, 17–58.

Guenther, W.B. (1975) *Chemical Equilibrium. A Practical Introduction for the Physical and Life Sciences.* Plenum Press, New York.

Haines, T.S. and Lloyd, J.W. (1985) Controls on silica in groundwater environments in the United Kingdom. *Journal of Hydrology* **81**, 277–295.

Hanshaw, B.B. and Back, W. (1979) Major geochemical processes in the evolution of carbonate-aquifer systems. *Journal of Hydrology* **43**, 287–312.

Hem, J.D. (1985). Study and interpretation of the chemical characteristics of natural water (3rd edn). *United States Geological Survey Water Supply Paper* 2254, 263 pp.

Hiscock, K.M. (1993) The influence of pre-Devensian glacial deposits on the hydrogeochemistry of the Chalk aquifer system of north Norfolk UK. *Journal of Hydrology* **144**, 335–369.

Hiscock, K.M., Lloyd, J.W. and Lerner, D.N. (1991) Review of natural and artificial denitrification of groundwater. *Water Research* **25**, 1099–1111.

Hough, J.S. (1985) *The Biotechnology of Malting and Brewing.* Cambridge University Press, Cambridge.

Ineson, J. and Downing, R.A. (1963) Changes in the chemistry of groundwaters of the Chalk passing beneath argillaceous strata. *Bulletin of the Geological Survey Great Britain* **20**, 176–192.

Korom, S.F. (1992) Natural denitrification in the saturated zone: a review. *Water Resources Research* **28**, 1657–1668.

Krauskopf, K.B. and Bird, D.K. (1995) *Introduction to Geochemistry* (3rd edn). McGraw-Hill, Inc., New York.

Krothe, N.C. and Bergeron, M.P. (1981) Hydrochemical facies in a Tertiary basin in the Milligan Canyon area, Southwest Montana. *Ground Water* **19**, 392–399.

Langmuir, D. (1971) The geochemistry of some carbonate groundwaters in Central Pennsylvania. *Geochimica et Cosmochimica Acta* **35**, 1023–1045.

Lawrence, A.R. and Foster, S.S.D. (1986) Denitrification in a limestone aquifer in relation to the security of low-nitrate groundwater supplies. *Journal of the Institution of Water Engineers and Scientists* **40**, 159–172.

Lerner, D.N. and Teutsch, G. (1995) Recommendations for level-determined sampling in wells. *Journal of Hydrology* **171**, 355–377.

Lloyd, J.W. (1986) Hydrogeology and beer. *Proceedings of the Geologists Association* **97**, 213–219.

Lloyd, J.W. and Heathcote, J.A. (1985) *Natural Inorganic Hydrochemistry in Relation to Groundwater: An Introduction.* Clarendon Press, Oxford.

Maltman, A. (2019) Wine, whisky and beer – and geology? *Geoscientist* **29**, 10–15.

Mather, J.D. (2013) Britain's spa heritage: a hydrogeological appraisal. In: *A History of Geology and Medicine* (eds C.J. Duffin, R.T.J. Moody and C. Gardner-Thorpe). Geological Society, London, Special Publications 375, pp. 243–260.

McDonough, L.K., Santos, I.R., Andersen, M.S. *et al.* (2020) Changes in global groundwater organic carbon driven by climate change and urbanization. *Nature Communications* **11**, 1279: DOI: 10.1038/s41467-020-14946-1.

McMahon, P.B. and Chapelle, F.H. (2008) Redox processes and water quality of selected Principal Aquifer systems. *Ground Water* **46**, 259–271.

McMahon, P.B., Chapelle, F.H. and Bradley, P.M. (2011) Evolution of redox processes in groundwater. In: *Aquatic Redox Chemistry* (eds P.G. Tratnyek, T.J. Grundl and S.B. Haderlein).

ACS Symposium Series 1071, ACS Publications, Washington, DC, pp. 581–597. DOI:10.1021/bk-2011-1071.ch026.

Morgan-Jones, M. and Eggboro, M.D. (1981) The hydrogeochemistry of the Jurassic limestones in Gloucestershire, England. *Quarterly Journal of Engineering Geology* **14**, 25–39.

Mühlherr, I.H., Hiscock, K.M., Dennis, P.F. and Feast, N.A. (1998) Changes in groundwater chemistry due to rising groundwater levels in the London Basin between 1963 and 1994. In: *Groundwater Pollution, Aquifer Recharge and Vulnerability* (ed. N.S. Robins). Geological Society, London, Special Publications 130, pp. 47–62.

Nesbitt, H.W. and Young, G.M. (1984) Prediction of some weathering trends of plutonic and volcanic rocks based on thermodynamic and kinetic considerations. *Geochimica et Cosmochimica Acta* **48**, 1523–1534.

Parkhurst, D.L. and Appelo, C.A.J. (2013). *Description of input and examples for PHREEQC version 3–A computer program for speciation, batch-reaction, one-dimensional transport, and inverse geochemical calculations.* United States Geological Survey Techniques and Methods, Book 6, Chapter A43. United States Geological Survey, Denver, Colorado, 497 pp.

Pearson, L. and Anderson, R. (2010) *Gazetteer of Operating pre-1940 Breweries in England.* Brewery History Society, Longfield, Kent.

Piper, A.M. (1944) A graphic procedure in the geochemical interpretation of water analyses. *Transactions of the American Geophysical Union* **25**, 914–923.

Plummer, L.N. and Busenberg, E. (1982) The solubilities of calcite, aragonite and vaterite in CO_2-H_2O solutions between 0 and 90°C, and an evaluation of the aqueous model for the system $CaCO_3$-CO_2-H_2O. *Geochimica et Cosmochimica Acta* **46**, 1011–1040.

Plummer, L.N., Parkhurst, D. and Kosier, D.R. (1975). MIX2, a computer program for modeling chemical reactions in natural water. *United States Geological Survey Report* **61–75**, 130 pp.

Plummer, L.N., Prestemon, E.C. and Parkhurst, D.L. (1994). *An interactive code (NETPATH) for modeling net geochemical reactions along a flow path version 2.0.* Water-Resources Investigations Report 94-4169. United States Geological Survey, Reston, Virginia, 130 pp. https://pubs.er.usgs.gov/publication/wri944169.

Price, M. and Williams, A. (1993) *Journal of the Institution of Water and Environmental Management* **7**, 651–659.

Robertson, W.D., Russell, B.M. and Cherry, J.A. (1996) Attenuation of nitrate in aquitard sediments of southern Ontario. *Journal of Hydrology* **180**, 267–281.

Schoeller, H. (1962) *Les Eaux Souterraines.* Maison et Cie, Paris.

Schwartz, F.W. and Muehlenbachs, K. (1979) Isotope and ion geochemistry of groundwaters in the Milk River Aquifer, Alberta. *Water Resources Research* **15**, 259–268.

Spector, W.S. (1956) *Handbook of Biological Data.* Saunders, Philadelphia, Pennsylvania.

Sprinkle, C.L. (1989). Geochemistry of the Florida Aquifer System in Florida and in parts of Georgia, South Carolina, and Alabama. *United States Geological Survey Professional Paper* 1403-I, 105 pp.

Stiff, H.A. (1951) The interpretation of chemical water analysis by means of patterns. *Journal of Petroleum Technology* **3**, 15–17.

Stumm, W. and Morgan, J.J. (1981) *Aquatic Chemistry: An Introduction Emphasizing Chemical Equilibria in Natural Waters* (2nd edn). Wiley, New York.

Stumm, W. and Wollast, R. (1990) Coordination chemistry of weathering: kinetics of the surface-controlled dissolution of oxide minerals. *Reviews of Geophysics* **28**, 53–69.

Talibudeen, O. (1981) Cation exchange in soils. In: *The Chemistry of Soil Processes* (eds D.J. Greenland and M.H.B. Hayes). Wiley, Chichester, pp. 115–177.

Tardy, Y. (1971) Characterization of the principal weathering types by the geochemistry of waters from some European and African crystalline massifs. *Chemical Geology* **7**, 253–271.

Tellam, J.H. (1994) The groundwater chemistry of the Lower Mersey Basin Permo-Triassic Sandstone Aquifer system, UK: 1980 and pre-industrialisation-urbanisation. *Journal of Hydrology* **161**, 287–325.

Tellam, J.H. (1995) Hydrochemistry of the saline groundwaters of the lower Mersey Basin Permo-Triassic sandstone aquifer, UK. *Journal of Hydrology* **165**, 45–84.

Tellam, J.H., Lloyd, J.W. and Walters, M. (1986) The morphology of a saline groundwater body: its investigation, description and possible explanation. *Journal of Hydrology* **83**, 1–21.

Thurman, E.M. (1985) *Organic Geochemistry of Natural Waters*. Nijhoff-Junk, Dordrecht.

Trainer, F.W. (1988) Plutonic and metamorphic rocks. In: *Hydrogeology. The Geology of North America, Vol. O-2*, (eds W. Back, J.S. Rosenshein and P.R. Seaber). The Geological Society of North America, Boulder, Colorado, pp. 367–380.

Truesdell, A.H. and Jones, B.R. (1973). WATEQ, a computer program for calculating chemical equilibria of natural waters. United States Geological Survey, National Technical Information Service, PB-220 464, 77 pp.

van Beek, C.G.E.M. (2000) Redox processes active in denitrification. In: *Redox: Fundamentals, Processes and Applications* (eds J. Schüring, H. D. Schulz, W.R. Fischer *et al.*). Springer-Verlag, Berlin, pp. 152–160.

Zhang, X.C. and Norton, L.D. (2002) Effect of exchangeable Mg on saturated hydraulic conductivity, disaggregation and clay dispersion of disturbed soils. *Journal of Hydrology* **260**, 194–205.

Zobell, C.E. and Grant, C.W. (1942) Bacterial activity in dilute nutrient solutions. *Science* **96**, 189.

5

Environmental isotope hydrogeology

5.1   Introduction

The stable and radioactive isotopes of the common elements of oxygen, hydrogen, carbon, sulphur and nitrogen have a wide range of applications in hydrogeology. The stable isotopes of water (^{16}O, ^{18}O, ^{1}H, ^{2}H) can be used as tracers of the origin of groundwater recharge and, together with noble (inert) gases (Ne, Ar, Kr, Xe), used to provide information on aquifer evolution. Nitrogen and sulphur stable isotopes have applications in contaminant hydrogeology in the identification of pollution sources and their fate in the groundwater environment. The radioactive isotopes of water, ^{3}H (tritium), carbon (^{14}C) and chlorine (^{36}Cl) are useful in providing estimates of aquifer residence times that can assist in managing groundwater sources.

The following sections describe, with examples, the basis for the application of environmental isotopes in groundwater investigations with emphasis given to groundwater source identification and age dating using the stable and radioactive isotopes of water, ^{14}C and ^{36}Cl. In discussing the origin of groundwater recharge, a section is also included to demonstrate the application of noble gas concentration data in reconstructing present and past groundwater recharge temperatures. The combined interpretation of environmental isotope and noble gas data enables reconstruction of palaeoenvironmental conditions and gives insight into the history of aquifer evolution.

Finally, a further example of the application of a multi-tracer approach using noble gases, stable isotopes and major ions is presented to demonstrate the upward migration of deep crustal fluids into a shallow aquifer environment.

5.2   Stable isotope chemistry and nomenclature

Modern double inlet, double collector mass spectrometers are capable of detecting small changes in relative isotopic abundances with the results expressed using the δ notation. In general, the δ notation, normally expressed in parts per thousand (per mil or ‰) with respect to a known standard, is written as follows:

$$\delta = \frac{R_{sample} - R_{standard}}{R_{standard}} \times 1000 \qquad (eq.5.1)$$

where R_{sample} and $R_{standard}$ are the isotopic ratios (for example $^{18}O/^{16}O$ and $^{2}H/^{1}H$) of the sample and standard, respectively. With this notation, an increasing value of δ means an increasing proportion of the rare, heavy isotope. In this case, the sample is said to have a heavier, more positive or enriched isotope composition compared with another, isotopically lighter sample. For water, the accepted international standard is V_{SMOW} (Vienna Standard Mean Ocean Water) with values of $\delta^{18}O$ and $\delta^{2}H$ equal to zero. Measurements of $\delta^{18}O$

Hydrogeology: Principles and Practice, Third Edition. Kevin M. Hiscock and Victor F. Bense.
© 2021 John Wiley & Sons Ltd. Published 2021 by John Wiley & Sons Ltd.
Companion website: www.wiley.com/go/hiscock/hydrogeology3e

and δ^2H can usually be determined to an accuracy of better than ±0.2 and $\pm2‰$, respectively.

Groundwater isotope data can help quantify molecular movements and chemical reactions in the interpretation of groundwater recharge, quality, storage, flow and discharge (Jasechko 2019). The mechanisms of isotope separation that lead to enrichment and depletion of isotopic ratios between phases or species can be divided into three types (Krauskopf and Bird 1995):

1) mechanisms depending on physical properties, for example evaporation or precipitation;
2) exchange reactions resulting in isotopic equilibrium between two or more substances; and
3) separation depending on reaction rate.

An example of mechanism (1) is the evaporation of water which leads to the concentration of the light isotopes ^{16}O and ^1H in the vapour phase and the heavy isotopes in the liquid phase. This is because water molecules containing the light isotopes move more rapidly and thus have a higher vapour pressure. Most samples of freshwater have negative values of $\delta^{18}O$ (ranging down to $-60‰$) in that the light ^{16}O isotope is concentrated in the vapour evaporating from the sea surface. Oxygen in air has a high positive isotopic signature of $+23.5‰$.

As an example of mechanism (2), if CO_2 containing only ^{16}O is mixed with water containing only ^{18}O, exchange will occur according to the following reaction, until equilibrium is reached among the four species:

$$\frac{1}{2}C^{16}O_2 + H_2{}^{18}O \Leftrightarrow \frac{1}{2}C^{18}O_2 + H_2{}^{16}O$$

$$(eq.5.2)$$

Although the bond strengths in the two compounds are different, at equilibrium, the ratio $^{18}O/^{16}O$ will be nearly the same in the CO_2 and H_2O.

The variable separation of isotopes depending on reaction rates (mechanism 3) is particularly associated with reactions catalysed by bacterial activity. For example, in the bacterial reduction of SO_4^{2-}, the production of sulphide (S^{2-}, HS^- and H_2S) is faster for the light isotope, ^{32}S, than for the heavy isotope, ^{34}S, such that the light isotope becomes concentrated in the sulphide species and the heavy isotope enriched in residual SO_4^{2-}.

Regardless of mechanism, the extent of isotope separation between two phases A and B can be represented by a fractionation factor, α, where

$$\alpha_{A-B} = \frac{R_A}{R_B} \qquad (eq.5.3)$$

where R_A is the ratio of concentrations of heavy to light isotope in phase A ($^{18}O/^{16}O$ in liquid water, for example) and R_B is the same ratio in phase B ($^{18}O/^{16}O$ in water vapour). If equilibrium is established between liquid water and vapour at 25°C, the value of α is about 1.0092. Similar fractionation factors, very slightly greater or less than one, are obtained for other examples of isotope separation, and it is for this reason that the descriptive δ notation is adopted.

The relationship between δ and α is given by the expression:

$$\alpha_{A-B} = \frac{R_A}{R_B} = \frac{1000 + \delta_A}{1000 + \delta_B} \qquad (eq.5.4)$$

For example, in the condensation of water vapour (phase v) to liquid (phase l), eq. 5.4 becomes:

$$\alpha_v^l = \frac{1000 + \delta^{18}O_l}{1000 + \delta^{18}O_v} \qquad (eq.5.5)$$

and if $\delta^{18}O_l = -5‰$ and $\delta^{18}O_v = -14‰$, then the fractionation factor can be calculated:

$$\alpha_v^l = \frac{(1000 + (-5))}{(1000 + (-14))} = 1.0092 \quad (eq.5.6)$$

5.3 Stable isotopes of water

The relative abundances of hydrogen and oxygen isotopes found naturally in the water molecule are given in Table 4.1. Meteoric water shows a wide range of $\delta^{18}O$ and δ^2H values reflecting the extent of isotope fractionation during successive cycles of evaporation and condensation of water originally evaporated from the sea. When condensation occurs to form precipitation, the isotopic concentration changes according to a Rayleigh distillation process for which the isotopic ratio, R, in a diminishing reservoir of reactant is a function of its initial ratio, R_o, the remaining reservoir fraction, f, and the fractionation factor, α, such that $R = R_o f^{(\alpha - 1)}$. The $^2H/^1H$ fractionation is proportional to, and about eight times as large as, the $^{18}O/^{16}O$ fractionation. Both fractionations change proportionally as temperature changes. Craig (1961) showed that δ values for meteoric water samples of global distribution, for the most part, define a straight line on a cross-plot of δ^2H against $\delta^{18}O$, represented by the approximate equation, known as the World Meteoric Water Line (WMWL):

$$\delta^2H = 8\delta^{18}O + 10 \qquad (eq.5.7)$$

In general, samples with $\delta^{18}O$ and δ^2H lighter than -22 and $-160‰$, respectively, represent snow and ice from high latitudes, while tropical samples show very small depletions relative to ocean water. This distribution is expected for an atmospheric Rayleigh process as vapour is removed from tropospheric air that is moving poleward. Linear correlations with coefficients only slightly different to eq. 5.7 are obtained from studies of local precipitation. For example, in the British Isles, a 20-year monthly dataset from 1982 to 2001 for Wallingford, a station in central, southern England gave the following regression: $\delta^2H = 7.00\delta^{18}O + 0.98$, with a slope a little less than the WMWL but consistent with those from other long-term stations in north-west Europe (Darling and Talbot 2003).

In addition to latitudinal and, similarly, altitudinal effects of temperature, the location of a site in relation to the proximity of the evaporating water mass is also important. As water vapour moves inland across continental areas, and as the process of condensation and evaporation is repeated many times, rain or snow becomes increasingly isotopically depleted. This continental 'rain-out' effect of the heavy isotopes has been shown for Europe by Rozanski (1985) as shown in Fig. 5.1 and is also apparent in the stable isotope composition of recent groundwaters measured for the British Isles (Fig. 5.2).

Since both condensation and isotope separation are temperature-dependent, the isotope composition of meteoric water displays a strong seasonal variation at a given location. In interpreting groundwater isotopic compositions, individual recharge events are mixed in the region of water table fluctuation such that isotopic variations over short timescales become obscured. Thus, it is possible to use a weighted mean isotopic composition to represent the isotopic signature of the seasonal recharge. In Fig. 5.3, a one-year record of monthly precipitation amount and composite $\delta^{18}O$ values is shown for a rain gauge situated in north Norfolk, eastern England. The data show the general trend of isotopically enriched precipitation in the warmer summer months and isotopically depleted precipitation in the colder winter months. The winter rainfall provides a representative volume-weighted mean isotopic composition for groundwater recharge of -7.20 and $-47.6‰$ for $\delta^{18}O$ and δ^2H, respectively.

Recognition of the effects of season, latitude, altitude and continentality is the basis for using $\delta^{18}O$ and δ^2H isotope values as non-reactive, naturally occurring tracers to identify the climatic and palaeogeographic conditions of groundwater recharge. For example, modern recharge waters in the Chalk aquifer of Norfolk, with a $\delta^{18}O$ composition of about $-7.0‰$, are found in the major unconfined river valleys (compare with the isotopic composition of rainfall; Fig. 5.3). Palaeogroundwaters with a $\delta^{18}O$ composition of $<-7.5‰$ are found trapped below extensive

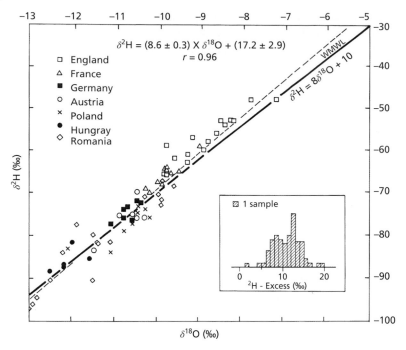

Fig. 5.1 The δ^2H versus δ^{18}O relationship for Western and Central European palaeowaters (groundwaters having a radiocarbon age of between approximately 15 and 30 ka). Old European groundwaters are depleted in deuterium (^{2}H) by about 12‰ compared with modern recharge waters (as shown in the frequency histogram of deuterium excess values, *d*, where $d = \delta^2$H $- 8\delta^{18}$O). The continental gradient in deuterium is very similar to modern meteoric water as indicated by the essentially identical comparison with the WMWL and strongly suggests a constant atmospheric circulation regime over Europe during the past 35 ka years (Rozanski 1985.). (*Source:* Rozanski, K. (1985) Deuterium and oxygen-18 in European groundwaters – links to atmospheric circulation in the past. *Chemical Geology (Isotope Geoscience Section)* **52**, 349–363. © 1985, Elsevier.)

low-permeability glacial till deposits in the interfluvial areas. The existence of the palaeogroundwater, and an isotopic shift of greater than 0.5‰ between the modern water and palaeogroundwater, is evidence for groundwater recharge during the late Pleistocene when the mean surface air temperature is estimated to have been at least 1.7°C cooler than at present. This estimate is based on the slope of the best-fit line of δ^{18}O data for global precipitation in the temperature range 0–20°C and equal to 0.58‰°C^{-1}, in close agreement with theoretical predictions based on the Rayleigh condensation model (Rozanski *et al.* 1993). In the Norfolk Chalk aquifer, the isotopic data confirm the conceptual hydrogeological model (Fig. 5.4) by which most active

groundwater recharge and flow is restricted to relatively limited areas where the overlying glacial till deposits are thin or absent (Hiscock *et al.* 1996).

Other examples of the application of the stable isotopes of water in the interpretation of palaeogroundwaters include the identification of freshwaters at depth in European coastal aquifers (Fig. 5.5) and the existence of fossil freshwater bodies in arid and semi-arid areas (Fig. 5.6). A compilation and synthesis of stable isotope data (δ^{18}O, δ^2H) for palaeogroundwaters in the British Isles is given by Darling *et al.* (1997). Evidence from the major British sandstone and limestone aquifers show that atmospheric circulation patterns over Britain have probably remained the same since the late

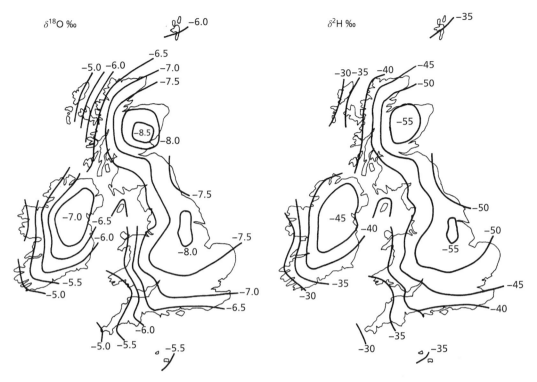

Fig. 5.2 Contour maps of $\delta^{18}O$ and δ^2H in recent (i.e. within the Holocene, 0–10 ka) groundwaters of the British Isles. The maps show similar features with relatively large variations in isotopic composition covering ranges of almost 4‰ in $\delta^{18}O$ and 30‰ in δ^2H. The areal isotopic composition is controlled mainly by the predominant source of rainfall from the south-west of the British Isles with some topographic variation noticeable over the Highlands of Scotland and the Pennines of England, where isotopic depletion occurs between the west and east of the country due to the orographic patterns of rainfall distribution (Darling *et al.* 2003). (*Source:* Adapted from Darling, W.G., Bath, A.H. and Talbot, J.C. (2003) The O and H stable isotopic composition of fresh waters in the British Isles. 2. Surface waters and groundwater. *Hydrology and Earth System Sciences* **7**, 183–195.)

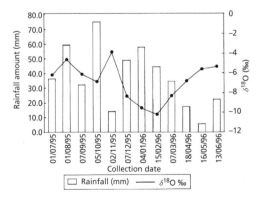

Fig. 5.3 Record of monthly precipitation amount and composite $\delta^{18}O$ values for a rain gauge situated at Salle, Norfolk (NGR TG 6126 3243), eastern England. The established Local Meteoric Water Line (LMWL) for precipitation in north Norfolk is $\delta^2H = 6.48\delta^{18}O - 0.62$. The volume-weighted mean winter rainfall (recharge) values are $\delta^2H = -47.6‰$ and $\delta^{18}O = -7.20‰$ (George 1998). (*Source:* Adapted from George, M.A. (1998) *High precision stable isotope imaging of groundwater flow dynamics in the Chalk aquifer systems of Cambridgeshire and Norfolk.* PhD Thesis, University of East Anglia, Norwich, 164–165.)

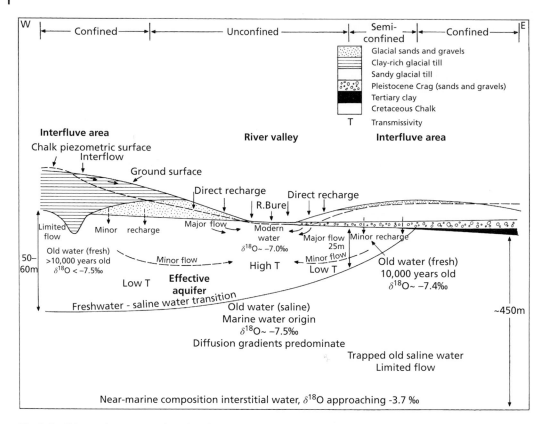

Fig. 5.4 Schematic cross-section showing a conceptual model of the extent and nature of the effective Chalk aquifer of north Norfolk, eastern England (Hiscock *et al.* 1996). (*Source:* Hiscock, K.M., Dennis, P.F., Saynor, P.R. and Thomas, M.O. (1996) Hydrochemical and stable isotope evidence for the extent and nature of the effective Chalk aquifer of north Norfolk, UK. *Journal of Hydrology* **180**, 79–107. © 1996, Elsevier.)

Pleistocene. However, additional ^{14}C groundwater age data highlight a hiatus in recharge occurrence under periglacial conditions between the late Pleistocene and early Holocene at about the time of the Last Glacial Maximum (LGM).

In deep aquifers where temperatures can exceed 50–100°C, the ^{18}O content of groundwater emerging as hot springs can be significantly altered by chemical interactions with the host rock. Measurements in such areas fall close to a horizontal line, indicating that the hot water contains an excess of ^{18}O over the meteoric water of the same region, but with approximately the same ^{2}H content (Fig. 5.7). This suggests that the infiltrating water has exchanged some of its ^{16}O for ^{18}O from the

silicate minerals of the host rock. A similar exchange of hydrogen isotopes is insignificant because most minerals only contain a small amount of this element (Krauskopf and Bird 1995). Further discussion of the isotopic composition of thermal waters can be found in Albu *et al.* (1997).

Another isotopic effect is observed for water evaporating from shallow soil or surface water bodies. Under natural conditions, the surface water becomes enriched in the heavy isotopes as evaporation occurs and provides a means for identifying surface water inputs to groundwater, and also for separating stream hydrographs into components of event (rainfall) and pre-event (soil) water (Section 6.7.1).

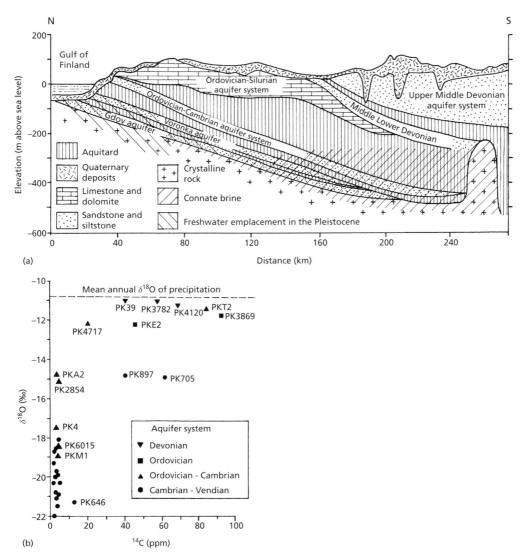

Fig. 5.5 (a) Hydrogeological cross-section of bedrock aquifers of Estonia showing the Voronka and Gdov aquifers of the Cambrian-Vendian system. (b) Distribution of $\delta^{18}O$ values of groundwater in Estonian aquifers grouped according to age as indicated by ^{14}C concentrations. Samples PK705 and PK897 are affected by mixing with infiltrated modern water. The groundwater in the Cambrian-Vendian system has a strongly depleted stable isotope composition and low radiocarbon content and represents periglacial or possibly sub-glacial recharge of meltwater through tunnel valleys from the Fennoscandian ice sheet during the late Weichselian (Devensian) ice age (Vaikmäe *et al.* 2001). (*Source:* Adapted from Vaikmäe, R., Vallner, L., Loosli, H.H. *et al.* (2001) Palaeogroundwater of glacial origin in the Cambrian-Vendian aquifer of northern Estonia. In: *Palaeowaters in Coastal Europe: Evolution of Groundwater Since the Late Pleistocene* (eds W.M. Edmunds and C.J. Milne). Geological Society, London, Special Publications **189**, pp. 17–27.)

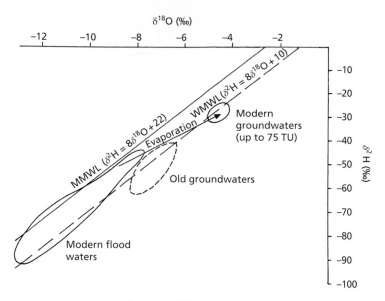

Fig. 5.6 Cross-plot of δ^2H versus $\delta^{18}O$ for groundwaters in southern Jordan. Old groundwaters with ^{14}C ages of 5–30 ka plot as a distinct group compared with modern, tritiated groundwaters that have a flood water source. The floodwaters plot close to the Mediterranean Meteoric Water Line (MMWL) while the modern (evaporated) and old groundwaters relate more closely to the WMWL. The fact that the old groundwaters conform to the WMWL suggests that their recharge was associated with storms tracking through the area from either the Atlantic or the Indian Oceans during the late Pleistocene and early Holocene. The limited occurrence of groundwaters showing a modern isotope composition also suggests that current recharge in this arid region is limited (Lloyd and Heathcote 1985). (*Source:* Lloyd, J.W. and Heathcote, J.A. (1985) *Natural Inorganic Hydrochemistry in Relation to Groundwater: An Introduction.* Clarendon Press, Oxford. © 1985, Oxford University Press.)

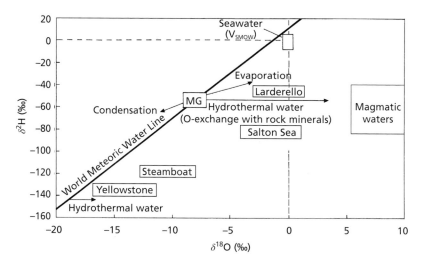

Fig. 5.7 Diagram illustrating the possible isotopic shifts from the WMWL on a plot of δ^2H versus $\delta^{18}O$. The arrows indicate isotope separation effects away from a typical groundwater of modern meteoric origin (MG). The diagram includes the plotting positions of the range of isotopic composition of hydrothermal waters (Larderello, Salton Sea, Steamboat Springs and Yellowstone), illustrating a shift to the right (enrichment of ^{18}O) relative to the meteoric water line (Albu *et al.* 1997). (*Source:* Albu, M., Banks, D. and Nash, H. (1997) *Mineral and Thermal Groundwater Resources.* Chapman & Hall, London. © 1997, Springer Nature.)

5.4   Stable isotopes of nitrogen and sulphur

The application of nitrogen and sulphur stable isotopes in hydrogeology is based on the premise that different sources of nitrogen and sulphur have different isotopic signatures that can be used to identify the origins of nitrate and sulphate in groundwater. This section describes the scientific basis for using the nitrogen and sulphur isotopic composition of dissolved nitrate and sulphate in order to distinguish possible sources and their fate in groundwater. For the purpose of illustration, the example of the southern Lincolnshire Limestone, a dual-porosity carbonate aquifer (Figs. 2.72 and 2.73), is used to describe the sources and behaviour of nitrate and sulphate in a regional aquifer system with particular emphasis on denitrification and pyrite oxidation processes.

5.4.1   Nitrogen stable isotopes

Values of the stable isotopes of nitrogen species in groundwater vary relative to the sources of nitrogen (Kendall 1998). A summary of the ranges of $^{15}N/^{14}N$ ratios (or $\delta^{15}N$ in per mil relative to the atmospheric nitrogen standard of 0‰) for commonly occurring sources of nitrogen inputs is given in Fig. 5.8. Nitrogen in rainfall is generally isotopically depleted, with dissolved ammonia, nitrate and organic nitrogen all depleted in ^{15}N relative to the atmospheric nitrogen standard. If nitrate is formed by photochemical oxidation of ammonia (Feth 1966), then isotopically depleted values

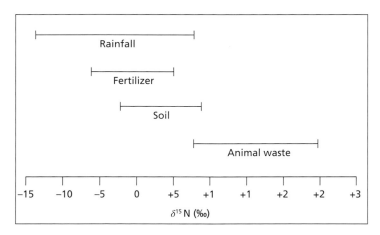

Fig. 5.8   Summary of the ranges of $\delta^{15}N$ values of nitrate for commonly occurring sources of nitrogen inputs to groundwater. The bars shown encompass the majority of reported analyses (Kendall and Krabbenhoft 1995). (Sources: Heaton, T.H.E. (1986) Isotopic studies of nitrogen pollution in the hydrosphere and atmosphere: a review. *Chemical Geology (Isotope Geoscience Section)* 59, 87–102, doi: 10.1016/0168-9622(86)90059-X.; Létolle, R. (1980) Nitrogen-15 in the natural environment. In: *Handbook of Environmental Isotope Geochemistry. Volume 1: The Terrestrial Environment, A* (eds P. Fritz and J.C. Fontes eds). Elsevier, Amsterdam, pp. 407–433.; Hübner, H. (1986) Isotope effects of nitrogen in the soil and biosphere. In: *Handbook of Environmental Isotope Geochemistry, Volume 2 The Terrestrial Environment, B* (eds P. Fritz and J.C. Fontes). Elsevier, Amsterdam, pp. 361–426.; Mariotti, A., Landreau, A. and Simon, B. (1988) ^{15}N isotope biogeochemistry and natural denitrification process in groundwater: application to the chalk aquifer of northern France. *Geochimica et Cosmochimica Acta* **52**, 1869–1878, doi: 10.1016/0016-7037(88)90010-5.; Kendall, C. and Krabbenhoft, D.P. (1995) Applications of isotopes to tracing sources of solutes and water in shallow systems. *Proceedings of the International Symposium on Groundwater Management* (ed. R.J. Charbeneau). American Society of Civil Engineers, San Antonio, TX, pp. 390–395.; Spalding, R.F., Exner, M.E., Martin, G.E. and Snow, D.D. (1993) Effects of sludge disposal on groundwater nitrate concentrations. *Journal of Hydrology* **142**, 213–228, doi: 10.1016/0022-1694(93)90011-W.)

are expected because of the equilibrium fractionation between ammonium and ammonia. Reported values for rainfall typically range from -10 to $+7‰$ (Hübner 1986; Fig. 5.8).

Nitrogen fertilizers are usually produced by quantitative industrial fixation of atmospheric nitrogen involving little or no fractionation (Heaton 1986). Since, by definition, the nitrogen isotopic composition of atmospheric nitrogen is 0‰, then it follows that most nitrogenous fertilizers should have an isotopic composition close to 0‰, although in reality there are differences in the $\delta^{15}N$ value of nitrogen fertilizers according to fertilizer type (Hübner 1986) with a range of -7 to $+5‰$.

The isotopic composition of nitrate produced by the mineralization of soil organic nitrogen (see Fig. 8.38) depends on the environmental conditions during the mineralization process as catalysed by aerobic bacteria. Fractionation of nitrogen isotopes during the mineralization of natural soil nitrogen produces more positive $\delta^{15}N\text{-}NO_3^-$ values than for synthetic fertilizers but the two ranges tend to overlap at around $+4‰$ (Fig. 5.8) in a range from -3 to $+8‰$.

As with soil nitrogen, the isotopic composition of nitrified sewage and animal waste (urea $(CO(NH_2)_2)$ and manure) is controlled by fractionation processes during the transformation to nitrate. If the transformation to ammonia occurs under alkaline conditions, then ammonia may be lost to the atmosphere. The ammonia that is lost is isotopically depleted because of equilibrium fractionation between ammonium and ammonia as well as unidirectional fractionation (Heaton 1986). Alkaline conditions may be created by the hydrolysis of urea or be naturally occurring within the soils in the recharge area. Since isotopically depleted ammonia is lost to the atmosphere, the nitrate formed from nitrification of the remaining ammonium is isotopically enriched relative to the initial animal nitrogen. The nitrogen isotopic composition of animal manure is slightly more enriched than that of plants (Létolle 1980) and so nitrate formed from animal waste is generally significantly heavier than for the

plant–soil system, with values typically between $+7$ and $+25‰$ (Fig. 5.8). The isotopic signature associated with animal waste varies according to the environmental conditions under which it is converted to nitrate, for example concentrated as a point source by leakage from septic tanks or broken sewers, or from farmyards around feedlots, or as a diffuse source by the spreading of manure to farmland.

Denitrification as a natural sink for nitrate is an important process in groundwater with the potential to attenuate nitrate contamination of aquifers as discussed in Box 4.11. Stable isotopes of nitrogen may be used to identify denitrification within the groundwater environment. Since denitrification involves the breakage of chemical bonds it induces a kinetic isotope fractionation effect. Because the lighter isotope is more reactive, the products of denitrification become enriched in ^{14}N relative to the initial nitrate composition while the remaining nitrate becomes isotopically enriched. Thus, denitrification may be inferred within a system by monitoring the isotopic composition of nitrate. The process has been observed in culture experiments with bacteria and in soil incubation experiments as well as in groundwater (Mariotti et al. 1988; Spalding et al. 1993).

If the denitrification reaction is treated as a single-step, unidirectional reaction, the Rayleigh equation relates the isotopic enrichment with the nitrate concentration value by an enrichment factor, as follows:

$$\delta_s = \delta s_o + \varepsilon \ln\left(\frac{C}{C_o}\right) \qquad (eq.5.8)$$

where δ_s is the isotopic composition of $NO_3^-\text{–N}$ ($\delta^{15}N$ in ‰), δs_o is the initial isotopic composition of NO_3^-, ε is the isotopic enrichment factor (‰), and $\ln(C/C_o)$ is the fraction of NO_3^- remaining.

As the nitrate concentration decreases, the isotopic composition of the remaining nitrate should exponentially increase. If δ_s is plotted against $\ln(C/C_o)$, then the values should fit a straight line with ε being equal to the gradient of the line. If the reduction in nitrate concentration was purely as a result of dilution with nitrate-free water, there would be no variation in the isotopic composition as nitrate concentrations decrease. A full derivation of the fractionation equation (eq. 5.8) is given by Mariotti *et al.* (1981).

As an illustration of the application of nitrogen stable isotopes to the study of regional hydrochemical processes, nitrogen isotope ratios of dissolved nitrate in groundwaters were measured at 16 locations in the southern Lincolnshire Limestone by Feast (1995), with the data presented in Hiscock *et al.* (2011). Values ranged from −7.1 to +14.3‰, with a mean of +7.6‰ and a standard deviation (1σ) of 3.2‰ (Fig. 5.9a). At high nitrate concentrations, the $\delta^{15}N$ values ranged from +3.2 to +7.9‰. As nitrate concentrations decrease, the mean $\delta^{15}N$ values increase such that, at low nitrate concentrations ($<5\,\mathrm{mg\,L^{-1}}$), the

$\delta^{15}N$ composition is between +7.5 and +15‰, except for two samples that did not fit this trend with isotopically depleted signatures of +1.1‰ and −7.1‰. Generally, $\delta^{15}N$ values increased from west to east (Fig. 5.9a). Samples taken from the unconfined limestone outcrop ranged between 3.2 and 7.9‰, whereas further east in the confined aquifer, the $\delta^{15}N$ values increased to +14.3‰. The two anomalously depleted $\delta^{15}N$ values were taken from the most eastern part of the confined aquifer.

The modelled field data yielded an enrichment factor value of −1.8‰ (Hiscock *et al.* 2011), considerably lower than values of about +5.0‰ obtained from other groundwater denitrification studies (Mariotti *et al.* 1988; Spalding and Parrott 1994). Comparison with these studies showed that the reason for the low isotopic enrichment factor in the Lincolnshire Limestone is the absence of enriched isotope values at very low nitrate concentrations. Hiscock *et al.* (2011) considered that factors other than

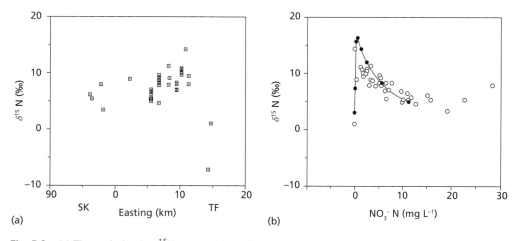

Fig. 5.9 (a) The variation in $\delta^{15}N$ composition of groundwater nitrate from west to east, downdip in the southern Lincolnshire Limestone aquifer. $\delta^{15}N$ values general increase from west to east until easting TF14 whereupon the isotopic composition becomes much more depleted. (b) Isotope evolution model applied to measured NO_3^- and $\delta^{15}N$ values for the southern Lincolnshire Limestone aquifer. The results of a denitrification–pore water interaction model are shown in which equal volumes of a low-nitrate concentration pore water with a depleted isotopic composition (NO_3^- = 0.2 mg L^{-1}; $\delta^{15}N$ = −5‰) are mixed with a high-nitrate water that has been variably denitrified (initial conditions: NO_3^- = 22.6 mg L^{-1}; $\delta^{15}N$ = 5‰; ε = −5‰) (Hiscock *et al.* 2011). (*Source:* Hiscock, K.M., Iqbal, T., Feast, N.A. and Dennis, P.F. (2011) Isotope and reactive transport modelling of denitrification in the Lincolnshire Limestone aquifer, eastern England. *Quarterly Journal of Engineering Geology and Hydrogeology* **44**, 93–108, doi: 10.1144/1470-9236/08-110.)

denitrification must have affected the isotopic fractionation and assumed that the mixing of modern water with old nitrate-free water had occurred.

A mixing model relating the concentrations and isotopic values of the mixing waters is described by the following equation:

$$\delta_m C_m = \delta_A C_A + \delta_B C_B \qquad (eq.5.9)$$

where δ denotes the δ^{15}N values of nitrate of the mixed component, m, and end components, A and B; and C denotes nitrate concentration values. The best model that matched the field data assumed the mixing of denitrified water with isotopically depleted (old, nitrate-free) water present in the confined aquifer. In this model, only a small quantity of isotopically depleted water is needed to lower the highly enriched values expected from near-complete denitrification and so lower the degree of isotopic enrichment at low concentrations (Fig. 5.9b). As the concentration of the apparent, isotopically depleted component is small, the effect is noticeable only at low nitrate concentrations, where it becomes very significant owing to the large difference in δ^{15}N values.

Different model results may be obtained by varying the concentration and isotopic composition of the isotopically depleted end member, using simultaneous denitrification and mixing, or by assuming that the level of mixing increases down-gradient as the proportion of ancient water increases. However, the basic principle of isotopic depletion of low-nitrate groundwaters by mixing with pore water nitrogen is the same. For this mechanism to be plausible, evidence is needed that the limestone pore water contains low concentrations of isotopically depleted (pre-intensive agriculture) nitrate. Evidence is provided by the two sites in the confined zone that yielded low quantities of nitrate with isotopically depleted signatures, which cannot be the product of denitrification of arable nitrate as this would leave these two samples relatively enriched. For this nitrate to remain, the water must have been

inadequately flushed from the pore space of the limestone by recent groundwater circulation. Tritium data (Downing *et al.* 1977), ^{14}C data (Bishop and Lloyd 1991) and noble gas data (Wilson *et al.* 1990) support this view.

The isotopic enrichment of ^{15}N in nitrate due to isotopic fractionation during the denitrification process means that the δ^{15}N signature alone does not unambiguously identify the source of nitrogen atoms. Usually, if the values of δ^{15}N–NO$_3^-$ in groundwater are in the range of +7 to +25‰, the source of nitrate can be associated with nitrogen in sewage or animal manure, but values in this range could also result from isotopic enrichment associated with denitrification of soil-derived nitrate. To overcome this potential ambiguity, δ^{18}O–NO$_3^-$ is increasingly used as an additional tracer for groundwater nitrate contamination. Several studies have confirmed that δ^{18}O–NO$_3^-$ is useful for determining nitrate reactions. As an example, Fukada *et al.* (2004) applied this dual-isotope method to the identification of sewage-derived contamination of the urban Triassic Sherwood sandstone aquifer in Nottingham in the English Midlands. In this study, depth sample measurements at one multi-level piezometer (Old Basford) gave δ^{15}N–NO$_3^-$ in the range +9.2 to +11.4‰ and δ^{18}O–NO$_3^-$ in the range +8.2 to +10.9‰, together with nitrate concentrations from 31.7 to 66.7 mg L^{-1}, interpreted as evidence for nitrification of sewage-derived inputs. In contrast, at a second multi-level piezometer (the Meadows), isotopically enriched samples (δ^{15}N–NO$_3^-$ in the range +24.3 to +42.2‰ and δ^{18}O–NO$_3^-$ in the range +20.5 to +29.4‰) provided evidence for denitrification, although the compositional range of δ^{15}N–NO$_3^-$ does not identify the nitrogen source without corroborating, microbiological data. For the Meadows location, a cross-plot of δ^{15}N–NO$_3^-$ versus δ^{18}O–NO$_3^-$ gave an enrichment of the ^{15}N isotope relative to the ^{18}O isotope by a factor of 1.9, within the range of 1.3–2.1 reported for denitrification in other studies (Fukada *et al.* 2003).

5.4.2 Sulphur stable isotopes

Sulphur stable isotopes can be used to understand the controls on the distribution of sulphur species in the hydrogeological environment revealing information on sulphate sources and hydrochemical processes. The range of sulphur isotope ratios ($^{34}S/^{32}S$ or $\delta^{34}S$ in per mil relative to the reference standard (Vienna-CDT) based on artificially prepared silver sulphide with a value of −0.3‰) for various sources of sulphur are shown in Fig. 5.10. The majority of natural samples fall between −40 and +40‰ (Nielsen 1979). Most meteorites, basic sills and volcanogenic sources of sulphur have isotopic compositions close to 0‰. The sulphur isotope composition of sulphate in seawater is remarkably uniform at around +20‰. Over geological time, the sulphur isotope composition of sulphate has varied greatly such that palaeoseawater sulphate is likely to possess a different sulphur isotope signature to modern seawater (Hoefs 2015).

Sulphate in rainwater may be derived from industrial emissions of SO_2, H_2S produced by anaerobic bacteria in soils and lakes, sea-spray sulphate, or oxidation of dimethylsulphide and similar compounds. Rainwater sulphate commonly has sulphur isotope values between 0 and +20‰, with coastal sites generally exhibiting marine sulphate values close to +20‰ and urban area sulphate having a more anthropogenic signature between 0 and +10‰ (Nriagu *et al.* 1991). Sulphur isotope values of sulphur in coals may vary widely but values for the more commonly burned low-sulphur coals typically lie in a narrow range, for example +4.6 to +7.3‰ (Smith and Batts 1974).

Sulphate in fertilizers is commonly derived from the sulphuric acid used in the chemical treatment of phosphorites. Chesterikoff *et al.* (1981) quoted a mean value of +8‰ and Moncaster *et al.* (1992) measured a value of +8.5‰.

No significant fractionation of evaporite deposits occurs during dissolution (Nriagu *et al.* 1991) and only very minor fractionation occurs during precipitation of gypsum from sulphate solutions (Thode and Monster 1965). Thus, specific evaporite beds will yield sulphate with an isotopic composition characteristic of the formation being dissolved, with sulphur isotope values generally in the range +8 to +32‰.

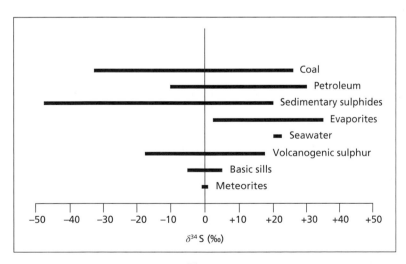

Fig. 5.10 Sulphur isotope values ($\delta^{34}S$) for various sources of sulphur (Thode 1991). (*Source:* Adapted from Thode, H.G. (1991) Sulphur isotopes in nature and the environment: an overview. In: *Stable Isotopes: Natural and Anthropogenic Sulphur in the Environment* (eds H.R. Krouse and V.A. Grinenko). John Wiley & Sons, Chichester, SCOPE **43**, pp. 1–26.)

Sedimentary sulphides typically have sulphur isotope values lighter than the standard value because they are formed from the bacterial reduction of sulphate, which favours the lighter isotope. The oxidation of reduced sulphur species may proceed by inorganic reaction or be bacterially mediated. The chemical oxidation of reduced sulphur species is probably not fractionating (Ault and Kulp 1959), causing the sulphate formed to retain the isotopic signature of the reduced precursor. The bacterially mediated oxidation of pyrite is reported to yield sulphate that is isotopically slightly depleted, the same or slightly enriched compared to the reduced sulphur. Generally, it has been shown that sulphate derived from oxidation of reduced sulphides possesses a sulphur isotope signature similar to the starting material (Strebel *et al.* 1990). Although the range of world values illustrated in Fig. 5.10 is large, the range observed for a particular region is likely to be much smaller. The world mean for sedimentary sulphides has been calculated to be $-12‰$ (Holser and Kaplan 1966).

In a similar way to bacterially mediated denitrification, the reduction of dissolved sulphate can result in a very large kinetic isotopic fractionation. This fractionation causes the reduced sulphur product phase (for example, pyrite or H_2S) to be depleted in the heavier isotope, while the $\delta^{34}S$ value of the remaining sulphate increases. This process has been demonstrated conclusively by Rye *et al.* (1981) in the groundwaters of Florida, where a correlation was found between $\delta^{34}S$ values of dissolved sulphate and dissolved sulphide concentrations. Rye *et al.* (1981) showed that the oldest waters had achieved isotopic equilibrium between sulphide and sulphate, indicating a very slow rate of sulphate reduction.

Fast rates of sulphate reduction are accompanied by negligible isotope effects. Isotope fractionation effects also depend on the sulphate concentration, being small when the concentration is low. Sulphate reduction is complicated by the numerous bond breakages and formations involved, meaning that simple one-step fractionation models are inadequate (Krouse 1980).

As an illustration of the application of sulphur stable isotopes to the study of regional hydrochemical processes, 50 sulphur isotope ratios of dissolved sulphate in groundwaters were measured at 29 locations in the southern Lincolnshire Limestone, with the data presented by Feast (1995). Values ranged from -14.6 to $+3.3‰$ with a mean of $-8.6‰$ and a standard deviation (1σ) of $3.4‰$. No apparent relationship was observed between the dissolved sulphate concentration and sulphur isotope ratio, but a clear trend of decreasing $\delta^{34}S$ values was observed from west to east in the downgradient direction of groundwater flow (Fig. 5.11a). Samples from the unconfined outcrop recorded $\delta^{34}S$ values between -2.9 and $+3.3‰$, becoming increasingly negative downdip reaching $-13‰$. The two most easterly samples in the confined aquifer had more enriched $\delta^{34}S$ values than expected and have probably undergone sulphate reduction leading to an enrichment in ^{34}S in the remaining sulphate, and one sample had a much lighter $\delta^{34}S$ than all other samples ($\delta^{34}S = -14.6‰$). The correlation coefficient ($r^2 = 0.82$) for the remaining samples (Fig. 5.11a) suggested that there is a significant relationship between the location of boreholes in the Lincolnshire Limestone and the $\delta^{34}S$ values.

In order to model the observed sulphur stable isotope data, then either an additional, isotopically depleted component must be added, or an isotopically enriched sulphur must be removed to explain the observed decrease in $\delta^{34}S$ values of sulphate downdip. The only source with an isotope composition depleted enough to explain the observed variations is sedimentary sulphide deposits with $\delta^{34}S$ values that commonly encompass the world mean of $-12‰$. Sulphide deposits are common within the Lincolnshire Limestone and in the overlying Middle to Upper Jurassic mudstone deposits. Oxidation of this pyrite imparts the variation in colour of the limestone. Close to the fissures, the limestone is often brown, where pyrite has been oxidized, whereas

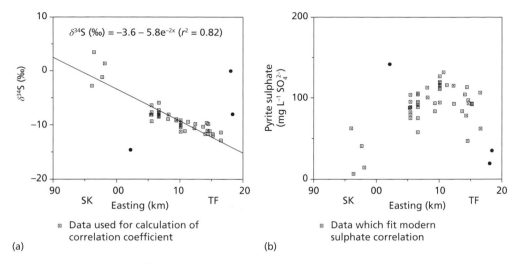

Fig. 5.11 (a) Variation in $\delta^{34}S$ composition of dissolved sulphate from west to east, downdip in the southern Lincolnshire Limestone aquifer. (b) Isotope evolution model applied to measured SO_4^{2-} and $\delta^{34}S$ values to calculate the change in the concentration of dissolved SO_4^{2-} derived from pyrite oxidation from west to east downdip in the southern Lincolnshire Limestone aquifer (Feast 1995). (*Source:* Feast, N.A. (1995) *Application of nitrogen and sulphur isotope hydrochemistry in groundwater studies.* PhD Thesis, University of East Anglia, Norwich, 269 pp.)

deeper in the limestone matrix diagenesis has been restricted and the unaffected pyrite imparts a blue colour to the rock.

Feast (1995) applied a simple two-member mixing model (eq. 5.9) with δ now the $\delta^{34}S$ value of sulphate and C the sulphate concentration of the mixed component, m, and with end-member components of modern recharge (A) ($\delta_A = +5‰$) and pyrite sulphate (B) ($\delta_B = -15‰$). Since δ_m, C_m, δ_A and δ_B are known, it is possible to calculate the quantity of sulphate contributed to the total amount of sulphate by the two respective inputs. This is achieved by substitution of $C_B = C_m - C_A$ in eq. 5.9 and rearranging to give

$$n = \frac{C_A}{C_m} = \frac{\delta_m - \delta_B}{\delta_A - \delta_B} \qquad (eq.5.10)$$

where n is the proportion of total sulphate derived from modern recharge.

The results of isotope modelling yielding the concentration of pyrite-derived sulphate is shown in Fig. 5.11(b). In reality, the actual isotopic values for sulphate will depend on the balance of different inputs in the recharge zone, the $\delta^{34}S$ value of the sulphide being oxidized (which will be controlled by fractionation effects during sulphide formation) and any isotope effects during sulphide oxidation, although these are likely to be small. In the model, the amount of pyrite-derived sulphate increases downdip to a peak of 140 mg L^{-1} and then declines further east to around 50 mg L^{-1} (Fig. 5.11b). Two samples do not fit the modelled trend of decreasing amounts of recharge sulphate downdip in the aquifer. This is because they have unusually enriched $\delta^{34}S$ values, which lead to an unrealistically high component of modern sulphate. Groundwater at these two sites have probably undergone sulphate reduction leading to an enrichment in ^{34}S in the remaining sulphate. The two sites are at the most easterly position in the confined aquifer where there is limited potential for groundwater circulation and long groundwater residence times (Bishop and Lloyd 1991) that favour sulphate reduction to occur. In addition, H$_2$S has been detected in boreholes within this region of the aquifer as a result of sulphate reduction (Edmunds and Walton 1983).

The major oxidizing agents of pyrite in natural environments are Fe^{3+} and O_2. Dissolved iron concentrations in Lincolnshire Limestone groundwaters are low and since solid phase ferric iron does not oxidize pyrite, then oxidation is likely restricted to zones where oxygen is present (see Box 8.5, eq. 1). Pyrite oxidation occurs in the limestone closest to fissure surfaces, giving rise to the buff/grey colouration of the limestone, with oxidation likely to be controlled by diffusion of oxygen from the highly oxic fissure water into the anaerobic porewaters. Sulphate produced from oxidation may similarly be released by diffusion from the high sulphate concentrations in the porewaters towards the lower concentrations in the fissure waters.

The sulphur isotope modelling showed that the amount of pyrite-derived sulphate increases steadily from an average of $30\,mg\,L^{-1}$ for the outcrop waters to approximately $130\,mg\,L^{-1}$, coincident with the eastern margin of the denitrification zone (Fig. 4.34). Thus, it is possible that some of the pyrite oxidized in this zone is accompanying the autotrophic reduction of nitrate (see Box 4.11, eq. 3), but most of the pyrite oxidation must be caused by the presence of dissolved oxygen in the groundwaters and the entry of air into the unconfined zone with seasonal fluctuations in the water table (Feast 1995). Although it is not possible to say with exact certainty whether organic carbon (heterotrophic denitrification; Box 4.11, eq. 1) or reduced sulphur is implicated in the denitrification process, it is not inconceivable that both species are acting as electron donors. Similar conclusions to the study by Feast (1995) were obtained by Bottrell *et al.* (2000) and Moncaster *et al.* (2000) in their study of the central Lincolnshire Limestone aquifer. For example, Moncaster *et al.* (2000) also considered that reduction of nitrate with pyrite as the electron donor was occurring, but that the mechanism responsible for denitrification is ultimately limited by pyrite availability near fissure surfaces where the reaction takes place.

5.5 Age dating of groundwater

A simple definition of groundwater age is the interval of time that has elapsed since groundwater at a location in a flow regime entered the sub-surface (Bethke and Johnson 2008), with groundwater age a function of distance from the recharge area. In contrast, mean residence time is the volume of water in a groundwater system divided by the volumetric recharge (or discharge) rate (see Section 1.5), which gives an average turnover time for the system (Suckow 2014).

Given that groundwater velocities are typically small and variable, a wide range of residence times are encountered in natural systems from a few days in karst aquifers to millennia in unfractured mudstones. Fossil groundwater is groundwater that was recharged by precipitation more than approximately 12,000 years ago, prior to the start of the Holocene epoch, whereas modern groundwater is often defined as being less than about 50 years old (Jasechko *et al.* 2017).

The exploitation of groundwater resources at a rate in excess of the time to replenish the aquifer storage will risk mining the groundwater. Therefore, knowledge of the age of groundwater is useful in aquifer management, although as commented by Ferguson *et al.* (2020), groundwater age does not provide a direct measure of whether groundwater resources can be developed sustainably. Pumping young groundwater does not guarantee sustainability and pumping old water does not guarantee non-sustainability. However, groundwater age measurements can provide valuable insights into how groundwater systems function under natural and perturbed (pumping) conditions (Ferguson *et al.* 2020).

Ages are typically derived from interpretation of various isotope tracers and may differ from the actual age of the water due to mixing and transport processes that occur within groundwater systems (Suckow 2014), as well as the different flow paths over the screened interval of wells used for sampling (Zinn and

Konikow 2007). Qualitative indicators of the age of a groundwater body include whether the groundwater is chemically oxidizing (aerobic, modern water) or reducing (anoxic, older water) in chemical character. Quantitative measures of the age of groundwater use radio-isotopes as a dating method. To demonstrate, the next section defines the law of radioactive decay. The following sections then present applications of the ^{14}C and tritium dating techniques together with an introduction to the more advanced methods of ^{36}Cl dating and $^{3}H/^{4}He$ dating. In addition, reference to ^{81}Kr dating is presented in Box 2.13.

5.5.1 Law of radioactive decay

The activity of a radioisotope at a given time can be calculated using the basic radioactive decay law:

$$\frac{A}{A_o} = 2^{-t/t_{1/2}} \qquad (eq.5.11)$$

where A_o is the radioactivity at time $t = 0$, A is the measured radioactivity at time t, and $t_{1/2}$ is the half-life of the radionuclide found from

$$t_{\frac{1}{2}} = \frac{\log_e 2}{\lambda} \qquad (eq.5.12)$$

where λ is the decay constant of the radionuclide.

Hence, with knowledge of A/A_o, the fraction or percentage of a radionuclide remaining at time t and the decay constant λ, then it is possible to calculate the apparent age of the groundwater. The age is considered an apparent age due to interpretation difficulties that arise from the general problem of mixing of groundwater bodies with different ages and, in the case of ^{14}C dating, from reactions between groundwater and aquifer carbonate material. To overcome these problems, corrections are required to the apparent groundwater age in order to obtain a corrected age.

5.5.2 ^{14}C dating

The radioisotope ^{14}C is produced by cosmic ray bombardment of nitrogen in the upper atmosphere. For every 10^{12} atoms of the stable isotopes of carbon (^{12}C and ^{13}C) in the atmosphere and oceans, there is an abundance of one atom of ^{14}C. ^{14}C decays back to nitrogen together with the emission of a β particle. The half-life of ^{14}C is measured as 5730 years and provides a useful dating tool in the age range up to 40,000 years for the most accurate determinations. Using the radioactive decay law (eq. 5.11) and substituting $t_{1/2} = 5730$ years, the dating equation becomes

$$t = -8267 \log_e \frac{A}{A_o} \qquad (eq.5.13)$$

The ^{14}C activity measured in the laboratory by accelerator mass spectrometry is given in terms of per cent modern carbon (pmc), with a counting statistics error of about ± 0.6 pmc. Calculated groundwater ages are quoted in years Before Present (pre-1950) using the notation, 'a' (for example, 1 ka indicates 1000 years Before Present).

Allowance must be made in the interpretation of the apparent age provided by measurement of the ^{14}C activity of the sample for reaction with sources of inorganic carbon encountered along the groundwater flowpath. The two principal sources are

1) 'biogenic' carbon with a source in the CO_2 of the atmosphere and also respired by the decay of organic carbon in the soil zone, which is usually assumed to contain 100 pmc; and
2) 'lithic' carbon with a source in the soil and rock carbonate and containing 'dead' carbon in which all the ^{14}C has decayed away (0 pmc).

The distribution of biogenic and lithic sources of carbon in groundwater can be described by the solution of calcite by weak carbonic acid (eq. 4.5) as follows:

$\text{CaCO}_3(\text{lithic carbon}) + \text{H}_2\text{CO}_3(\text{biogenic carbon})$
$\rightarrow \text{Ca}^{2+} + 2\text{HCO}_3^- (\text{sample carbon})$

$$(\text{eq.}5.14)$$

If this reaction predominates, then the stoichiometry of eq. 5.14 predicts that the ^{14}C activity of the groundwater bicarbonate will be 50% of the modern activity. In other words, because of dilution of the sample with 'dead' carbon from soil and rock carbonate then, without correction, the apparent age of the groundwater will appear older than it actually is. To correct for this effect, one method is to use the $^{13}\text{C}/^{12}\text{C}$ ratio ($\delta^{13}\text{C}$) of the groundwater bicarbonate as a chemical tracer of the ^{14}C activity. The interpretation of the ^{14}C data from the sample $\delta^{13}\text{C}$ value is based on the fact that the principal sources of carbon (lithic and biogenic) contributing to the carbonate system in the water have different $\delta^{13}\text{C}$ values.

Analyses of $\delta^{13}\text{C}$ values are relatively easy to obtain by conversion of the sample carbonate to CO_2 followed by measurement of the isotope ratio on a mass spectrometer. Results are quoted relative to the standard Pee Dee Belemnite (PDB), a rock unit from the Cretaceous Period, with an accuracy of about $\pm0.2‰$. The CO_2 associated with organic carbon in temperate soils has a $\delta^{13}\text{C}$ value of about $-26‰$ for Calvin photosynthetic cycle plants ($-12‰$ for Hatch–Slack cycle plants in hot, arid climates). Values for limestone rock are usually between 0 and $+2‰$ (Schoelle and Arthur 1980).

The result of calcite dissolution (eq. 5.14) is that the groundwater would be expected to have a $\delta^{13}\text{C}$ value of about $-12‰$. Away from the soil zone, $\delta^{13}\text{C}$ values are less negative than those near to the aquifer outcrop as a result of the continuous precipitation and dissolution of carbonate as the water flows through the aquifer. The isotopic composition of the precipitated carbonate mineral differs from that of the carbonate species in solution. As a result there is isotopic fractionation between the aqueous phase and solid carbonate-containing mineral phase.

To account for the two sources of inorganic and organic carbon in the groundwater, let Q equal the proportion of biogenic carbon in the sample. To determine the proportions of lithic and biogenic carbon, then the following relationship can be used with the $\delta^{13}\text{C}$ data:

$$\delta^{13}\text{C}_{sample} = \delta^{13}\text{C}_{biogenic}(Q) + \delta^{13}\text{C}_{lithic}(1-Q)$$

$$(\text{eq.}5.15)$$

or rearranged:

$$Q = \frac{\delta^{13}\text{C}_{sample} - \delta^{13}\text{C}_{lithic}}{\delta^{13}\text{C}_{biogenic} - \delta^{13}\text{C}_{lithic}} \qquad (\text{eq.}5.16)$$

Using the calculated value of Q, the corrected groundwater age, t_c, can be found from the dating equation (eq. 5.13) by applying the fraction of biogenic carbon to the initial ^{14}C activity, A_o, such that

$$t_c = -8267 \log_e \frac{A}{A_o Q} \qquad (\text{eq.}5.17)$$

or separating terms:

$$t_c = -8267 \log_e \frac{A}{A_o} + 8267 \log_e Q$$

$$(\text{eq.}5.18)$$

As an example, the following information was obtained for a groundwater sample from Waltham Abbey PS (sample 12, Table 5.1) in the Chalk aquifer of the London Basin:

$$A_o = {}^{14}\text{C}_{atmospheric} = 100 \text{ pmc};$$
$$A = {}^{14}\text{C}_{sample} = 13 \text{ pmc};$$
$$\delta^{13}\text{C}_{sample} = -6.3‰;$$
$$\delta^{13}\text{C}_{lithic} = +2‰;$$
$$\delta^{13}\text{C}_{biogenic} = -26‰.$$

From eq. 5.16, the fraction of biogenic carbon, $Q = (-6.3 - (+2))/(-26 - (+2)) = 0.2964$ and substitution in eq. 5.18 gives the following corrected groundwater age:

$$t_c = -8267 \log_e(13/100) + 8267 \log_e(0.2964)$$
$$= 6813.5 \text{ a}$$

or in practical terms, about 7 ka.

Table 5.1 Tritium, carbon-14 and stable isotope measurements for Chalk groundwaters in the London Basin. The locations of numbered sampling sites are shown in Fig. 5.12(a) (Smith *et al.* 1976).

Site	Tritium (TU)	^{14}C age (ka)	^{14}C (pmc)	δ^{13}C (‰)	δ^{18}O (‰)	δ^2H (‰)
1. Burnham PS	57	modern	60.8	−13.2	−7.1	−46
2. Duffield House	13	3.9	31.4	−11.9	−7.1	−45
3. Iver PS	19	9.1	6.8	−3.5	−7.4	−48
4. West Drayton	0	22	1.0	−1.8	−7.9	−50
5. Crown Cork	0	19	1.1	−0.8	−7.7	−50
6. Callard and Bowser	0	>20	0.8*	−0.5	−7.9	−51
7. Polak's Frutal	3	20	1.1	−1.1	−7.8	−50
8. Southall AEC	1	>20	0.5*	−0.9	−7.8	−51
9. Morganite Carbon	4	15	4.1	−4.5	−7.4	−48
10. White City Stadium	0	14	2.2	−0.9	−7.9	−51
11. Broxbourne PS	37	0.8	49.1	−13.0	−7.2	−47
12. Waltham Abbey PS	3	7.0	13.0	−6.3	−7.1	−46
13. Hadley Road PS	2	20	1.7	−3.3	−7.2	−48
14. Hoe Lane PS	2	8.2	14.9	−9.1	−7.1	−46
15. Chingford Mill PS	12	4.3	21.5	−7.9	−7.2	−46
16. Initial Services	53	modern	41.4	−8.3	−6.6	−44
17. Berrygrove	45	modern	58.5	−13.1		
18. Kodak	1	11	3.8	−1.7	−7.8	−47
19. New Barnet	1	17	3.3	−4.9	−7.8	−48
20. Schwepps	0	>25	0.7*	−1.5		
21. Kentish Town	1	>25	1.2*	−4.0		
22. Bouverie House	1	>25	0.6*	−1.7		

* Samples near the limit of detection. In these cases, values given represent the lowest possible value.
PS – Pumping station
Source: Smith, D.B., Downing, R.A., Monkhouse, R.A. *et al.* (1976) The age of groundwater in the Chalk aquifer of the London Basin. *Water Resources Research* **12**, 392–404. © 1976, John Wiley & Sons.

Further corrected groundwater ages for the Chalk aquifer of the London Basin are given in Table 5.1 and the results shown as age contours in Fig. 5.12(c). The oldest groundwaters are found below the Eocene London Clay in the confined aquifer at the centre of the basin. Modern groundwaters are present in the recharge area on the northern limb of the basin. The groundwater becomes progressively older towards the centre of the basin with flow induced by pumping of the aquifer (compare with the map of the Chalk potentiometric surface shown in Box 2.4). The upper limit for the age measurement is about 25 ka with groundwater at the centre of the basin known to be older than this limit for the ^{14}C dating method (Smith *et al.* 1976).

In situations where a negative corrected groundwater age is obtained, then the sample has become swamped by the modern atmospheric ^{14}C content which is in excess of 100 pmc. The extra ^{14}C was contributed by the

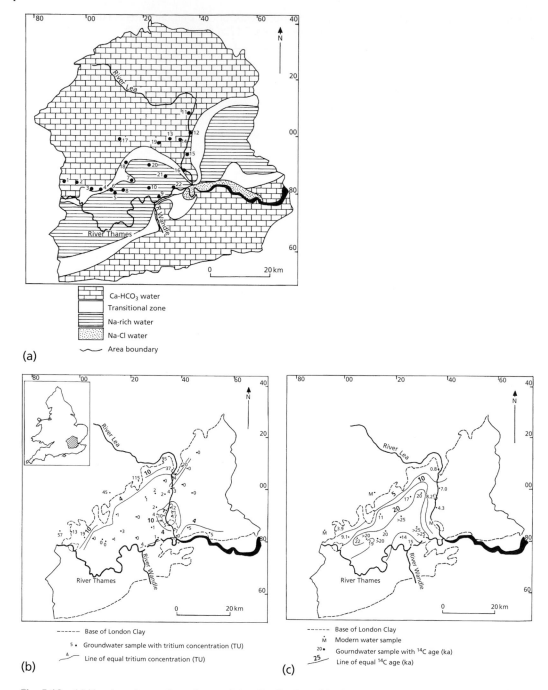

Fig. 5.12 (a) Numbered sampling sites and the distribution of basic hydrochemical water types. (b) Tritium concentrations (in TU). (c) ^{14}C ages (in ka) of Chalk groundwaters in the central part of the London Basin for the data presented in Table 5.1 (Smith *et al.* 1976). (*Source:* Smith, D.B., Downing, R.A., Monkhouse, R.A. *et al.* (1976) The age of groundwater in the Chalk aquifer of the London Basin. *Water Resources Research* **12**, 392–404, doi: 10.1029/WR012i003p00392. © 1976, John Wiley & Sons.)

detonation of thermonuclear devices in the 1950s and early 1960s. In such cases, the groundwater age can be assumed to be modern. This effect and other inherent shortcomings, for example contamination with atmospheric ^{14}C during field sampling and laboratory errors in the measurement of δ^{13}C and ^{14}C, mean that any groundwater sample with an age of up to 0.5 ka can be regarded as modern.

More advanced approaches to correcting ^{14}C ages that account for carbon isotopic exchange between soil gas, dissolved carbonate species and mineral carbonate in the unsaturated and saturated zones under open- and closed-system conditions (Section 4.7) are included in Mook (1980). However, the extra effort expended may not be justified especially if mixing between two groundwaters is suspected. For example, in the case of the Chalk aquifer discussed above, diffusive mixing of old water contained in the pores of the rock matrix with modern water moving through the fissured component will dilute the ^{14}C content of the pumped groundwater sample, thus increasing the apparent groundwater age. Hence, even after correction for isotopic exchange with mineral carbonate, the calculated age may not be the true age.

5.5.3 ^{36}Cl dating

Chlorine-36 (^{36}Cl), with a half-life of 301 000 ± 4000 years, is produced primarily in the atmosphere via cosmic ray bombardment of ^{40}Ar. ^{36}Cl is potentially an ideal tracer for age dating on timescales of up to 1.5 Ma in large groundwater systems, provided that sources and sinks of ^{36}Cl and Cl can be accounted for. The advantage of using ^{36}Cl is that the Cl anion behaves conservatively and, in the absence of Cl-bearing minerals such as halite, it is neither added nor removed from solution via rock-water interactions and moves at approximately the same velocity as the groundwater.

Although ^{36}Cl and Cl concentrations in groundwater may be modified after recharge

by mixing between aquifers or by diffusion from adjacent aquitards, these problems can be overcome by incorporating supplementary chemical and isotopic data to account for these added contributions. Cosmogenic production of ^{36}Cl in the near-surface environment by interaction of cosmic rays with minerals in surface rocks and soils, and nucleogenic production via neutrons generated within the aquifer matrix through decay of U and Th can be reasonably estimated and are relatively small compared with atmospheric production. The most difficult parameter to estimate in the age determination is the initial ^{36}Cl/Cl ratio at the time of recharge. A further issue is the secular variation of ^{36}Cl production over long timescales (Love *et al.* 2000).

For ^{36}Cl determinations, about 20 mg of Cl is precipitated as AgCl and analysed by accelerator mass spectrometry. A decrease in the measured ^{36}Cl/Cl ratio along a groundwater hydraulic gradient represents ^{36}Cl decay, and after correcting for different sources of Cl, ^{36}Cl ages can be estimated. In an ideal situation, if Cl and ^{36}Cl are solely derived from atmospheric sources with no internal sources or sinks, except for ^{36}Cl decay and nucleogenic production, and if the initial ^{36}Cl/Cl ratio (R_o) and the secular equilibrium ^{36}Cl/Cl ratio (R_{se}) can be estimated, then groundwater age estimates can be determined by the following equation (Bentley *et al.* 1986):

$$t = -\frac{1}{\lambda_{36}} \log_e \frac{R - R_{se}}{R_o - R_{se}} \qquad \text{(eq.5.19)}$$

where λ_{36} is the decay constant for ^{36}Cl, and R is the ^{36}Cl/Cl measured in groundwater.

Love *et al.* (2000) presented two modifications of eq. 5.19 to allow for the addition of Cl via leakage or diffusion from an adjacent aquitard. In their study, Love *et al.* (2000) applied the ^{36}Cl dating technique to groundwaters of the south-west flow system of the Great Artesian Basin (Box 2.11) in north-east South Australia. The main aquifer system comprises Cretaceous gravels, sands and silts of the

Jurassic-Cretaceous aquifer, with mean flow velocities, calculated from the rate of decrease of absolute ^{36}Cl concentrations, of 0.24 ± 0.03 m a^{-1}. Calculated ^{36}Cl ages of the confined aquifers, although complicated by addition of Cl via diffusion from the overlying aquitard, range from 200 to 600 ka (Table 5.2; Love *et al.* 2000). Groundwater trends, from the eastern margin of the Basin in north-east Queensland towards the centre of the Basin in South Australia, show a decrease in the $^{36}Cl/Cl$ ratio in the direction of the hydraulic gradient from approximately 100×10^{-15} to about 10×10^{-15} over a distance of 1000 km indicating ages approaching 1.1 Ma near the end of the flowpath (Bentley *et al.* 1986).

5.5.4 Tritium dating

Tritium (^3H or T), the radioisotope of hydrogen, has a relative abundance of about $0–10^{-15}\%$. The unit of measurement is the tritium unit (TU) defined as 1 atom of tritium occurring in 10^{18} atoms of H and equal to 3.19 pCi L^{-1} or 0.118 Bq L^{-1}. The half-life of tritium is 12.38 years, ^3H decaying to ^3He with the emission of a β particle. In a similar way to ^{14}C production, tritium is produced naturally mainly in the upper atmosphere by interaction of cosmic ray-produced neutrons with nitrogen. After oxidation to $^1H^3HO$, tritium becomes part of the hydrological cycle. Analysis of tritium requires distillation followed by electrolytic enrichment of the tritium content. The enriched sample is converted to ethane and gas scintillation techniques are used to measure the tritium content. Combined field and laboratory errors yield an accuracy of ± 2 TU or better.

Natural levels of tritium in precipitation are estimated to be between 0.5 and 20 TU. The tritium concentrations of four long-term precipitation records are shown in Fig. 5.13. The records illustrate the effect of atmospheric testing of thermonuclear devices between 1952, prior to which there were no measurements of natural tritium levels in the Earth's

atmosphere, and the test ban treaty of 1963 when tritium concentrations, as a result of nuclear fusion reactions, reached a peak of over 2000 TU in the northern hemisphere. A clear seasonal variation is also evident with measured tritium concentrations less in the winter compared with the summer when tritium is rained out of the atmosphere. In the southern hemisphere (and at coastal sites in general), the greater influence of the oceans leads to a greater dilution with water vapour resulting in lower tritium concentrations overall. Currently, atmospheric background levels in the northern hemisphere are between about 5 and 30 TU and in the southern hemisphere between 2 and 10 TU (IAEA/WMO 1998).

The application of the dating equation (eq. 5.11) with tritium data to obtain groundwater ages is problematic, given the variation in the initial activity, A_o. Ambiguity arises in knowing whether the input concentration relates to the time before or after the 1963 bomb peak. Nevertheless, tritium concentrations provide a relative dating tool with the presence of tritium concentrations above background concentrations indicating the existence of modern, post 1952 water. Again, as with ^{14}C dating, mixing between different water types with different recharge histories complicates the interpretation. Some authors (for example, Downing *et al.* 1977) have attempted to use the amount of tritium in groundwater to correct the ^{14}C content of a sample for the effects of dilution with modern water.

Although the bomb peak has decayed away, making tritium less useful as a dating tool, earlier studies in the 1970s and 1980s were able to relate large quantities of tritium in a sample to groundwater recharge in the 1960s. For example, in Fig. 5.12b, modern water with a tritium concentration up to 57 TU (Table 5.1) was sampled in the recharge area of the Chalk outcrop on the northern rim of the London Basin. Away from the Chalk outcrop, tritium values declined rapidly below the confining Eocene London Clay. Another area of interest is the River Lea valley

Table 5.2 Environmental isotope data and modelled ^{36}Cl ages for groundwaters from the southwest margin of the Great Artesian Basin (26°30′–28°30′S; 133°30′–136°30′E) (Love et al. 2000).

Unit number	Sample name	Distance along transect (km)	δ^2H (‰ V$_{SMOW}$)	δ^{18}O (‰ V$_{SMOW}$)	δ^{13}C (‰ PDB)	^{14}C (pmc)	^{36}Cl/Cl × 10^{-15}	^{36}Cl × 10^6 (atoms L^{-1})	^{36}Cl age (ka) (1)
Northern Transect (NW to SE)									
574400015	Lambina Homestead	47	−46.3	−5.84	−11.0	6.6 ± 1.2	129 ± 8	1888 ± 140	
574400003	Warrungadinna	61	−42.6	−5.97	−10.6	11 ± 1.3	132 ± 10	2520 ± 235	
574400004	Lambina Soak	71	−49.7	−6.33	−11.7	2.4 ± 1.2	108 ± 7	2026 ± 160	
584300025	Marys Well 3	129	−47.6	−6.71	−11.1		102 ± 8	1162 ± 110	58–94
584300026	Murdarinna 2	145	−51.3	−6.66	−11.7	2.5 ± 2.5	109 ± 10	1186 ± 125	47–190
594300017	Midway Bore	154	−45.7	−6.56			89 ± 4	934 ± 60	160–290
594200001	Oodnadatta Town Bore 1	203	−48.2	−6.71	−11.6	2.0 ± 2.0	52 ± 5	594 ± 65	380–490
604200021	Watson Creek 2	246	−42.7	−6.35	−11.0		25 ± 3	439 ± 60	550–790
614200004	Duckhole 2 (24701)	261	−49.5	−6.61	−11.0	1.7 ± 1.2	31 ± 4	444 ± 65	540–670
574300005	Appatinna Bore (2)	96	−35.7	−4.85	−10.4	49.1 ± 5.1	134 ± 13	780 ± 85	
Southern Transect (W to E)									
574100007	C.B. Bore	6	−33.0	−3.81	−8.9	41.2 ± 4.5	115 ± 8	1030 ± 85	
574100014	Ross Bore	33	−39.4	−5.02			54 ± 6	1478 ± 175	
574100049	Evelyn Downs Homestead	43	−40.0	−5.16	−10.9		54 ± 5	1296 ± 140	
584100053	Woodys Bore Windmill	67	−38.5	−4.89			47 ± 5	1914 ± 250	460
584100050	Robyns Bore 2	80	−39.4	−5.10	−11.4	1.8 ± 1.2	41 ± 4	1691 ± 210	30–530
584100011	Ricky Bore 2	89	−41.4	−5.47			37 ± 3	1295 ± 85	110–580
594100003	Paulines Bore	95	−43.3	−5.29	−11.9	3.3 ± 1.2	40 ± 4	1149 ± 130	200–540
594100017	Nicks Bore	114	−42.8	−5.73			46 ± 5	1135 ± 145	200–470
594100013	Leos Bore	136	−42.0	−5.68			53 ± 5	1012 ± 115	250–400
594100006	Fergys Bore	141	−44.1	−6.07	−11.5		61 ± 1.2	969 ± 210	270–340
604100037	Lagoon Hill Drill Hole 15 (New Peake)	183	−46.4	−6.56	−10.7	3.8 ± 1.2	29 ± 3	461 ± 60	600–710

Note: pmc = % modern carbon; (1) age range calculated using eq. 5.19 and two variants to allow for the addition of Cl via leakage or diffusion from an adjacent aquitard; (2) data for the unconfined Appatinna Bore have been used to determine the initial ^{36}Cl/Cl ratio.
Source: Love, A.J., Herczeg, A.L., Sampson, L. *et al.* (2000) Sources of chloride and implications for ^{36}Cl dating of old groundwater, southwestern Great Artesian Basin, Australia. *Water Resources Research* **36**, 1561–1574. © 2000, John Wiley & Sons.

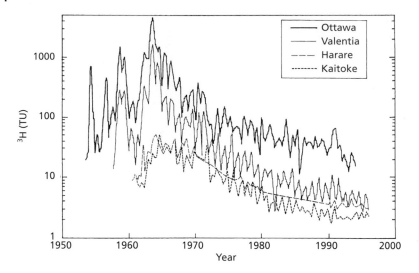

Fig. 5.13 Variations in mean monthly tritium concentrations in precipitation since 1953 at four IAEA stations (IAEA/WMO 1998): Ottawa, Canada (northern hemisphere, continental); Valencia, Ireland (northern hemisphere, marine); Harare, Zimbabwe (southern hemisphere, continental); and Kaitoke, New Zealand (southern hemisphere, marine). (*Source:* IAEA/WMO (1998) Global network of isotopes in precipitation. The GNIP Database. http://isohis.iaea.org/ (accessed 7 June 2013).)

where a tongue of water containing measurable tritium (53 TU maximum) extends south from the Chalk outcrop. This area coincides with the region of the North London Artificial Recharge Scheme (Box 10.2) and it appears likely that the tritium anomaly is due to artificial recharge of treated mains water containing a component of modern water.

Another application for tritium is providing a tracer for modern pollutant inputs to aquifers such as agricultural nitrate and landfill leachate. For example, sequential profiling of tritium in porewater contained in the unsaturated zone of the English Chalk has helped in the understanding of the transport properties of this dual-porosity aquifer. As shown in Fig. 5.14a for a site in west Norfolk, sequential re-drilling and profiling of tritium concentrations in the Chalk matrix revealed the preservation of the tritium bomb peak. The results are interpreted as indicating slow downward migration of water through the low-permeability, saturated

Chalk matrix. The tritium appears to migrate with a 'piston-like' displacement with only limited dispersion of the peak concentrations. Detailed analysis of the profiles in terms of their mass balance and peak movement revealed the possibility of rapid water movement, or 'by-pass' flow. It is estimated that up to 15% of the total water movement occurs through the numerous fissures that comprise the secondary porosity of the Chalk once the fluid pressure in the unsaturated zone (Section 6.4.1) increases to the range −5 to 0 kPa (Gardner *et al.* 1991). At other locations, a 'forward tailing' and broadening of the tritium peaks are observed in sequential tritium profiles providing evidence of greater dispersion (Parker *et al.* 1991).

Laboratory measurements have shown that tritiated water and NO_3^- have similar diffusion coefficients and so similar concentration profiles might be expected for NO_3^- (Fig. 5.14b). Dispersion of NO_3^- is suggested where there

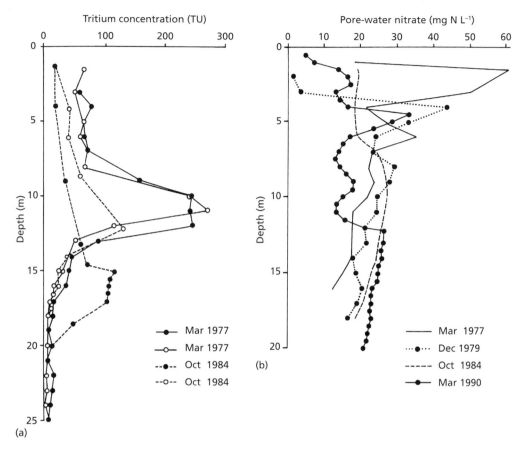

Fig. 5.14 Sequential profiles of (a) tritium and (b) NO$_3^-$ in the unsaturated zone of the Chalk aquifer beneath arable land in west Norfolk, Eastern England (Parker *et al.* 1991). (*Source:* Parker, J.M., Young, C.P. and Chilton, P. J. (1991) Rural and agricultural pollution of groundwater. In: *Applied Groundwater Hydrology: A British Perspective* (eds R.A. Downing and W.B. Wilkinson). Clarendon Press, Oxford, pp. 149–163. © 1991, Oxford University Press.)

has been less downward movement than would be expected if piston flow were occurring. An apparent flattening of the original NO$_3^-$ peaks leads to higher concentrations than expected in the deeper part of the profiles. The rate of solute movement in the Chalk implied by the tritium and NO$_3^-$ profiles is between 0.4 and 1.1 m a^{-1}. The actual rate of movement at a specific site will depend on many factors including the physical properties of the Chalk, temporal variations in rainfall and antecedent moisture content, and seasonal variations in solute input (Parker *et al.* 1991).

5.5.5 ^{3}H/^{3}He dating

The limitations of using tritium alone to estimate groundwater residence times that arise from its short radioactive half-life and the decay of the bomb peak, can be addressed with the additional measurement of ^{3}He. As tritium in groundwater enters the unsaturated zone and is isolated from the atmosphere below the water table, dissolved ^{3}He concentrations from the decay of tritium will increase as the groundwater becomes older. Although groundwaters contain helium from several sources other than tritium decay, including atmospheric and radiogenic ³He and ⁴He, determination of both tritium and tritiogenic ³He (³He*) can be used as a dating tool (Solomon *et al.* 1993). The tritium-helium (³H/³He) age is defined as:

$$t_{\,^3H/^3He} = \frac{1}{\lambda_T} \log_e \left(\frac{^3He^*}{^3H} + 1 \right) \quad (eq.5.20)$$

where λ_T is the tritium decay constant. Analytical techniques for tritium and ³He determinations are such that under ideal conditions, ages between 0 and about 30 years can be determined with typical analytical uncertainties of <10% (Solomon *et al.* 1993). Further confidence in deriving an apparent age is obtained when multiple techniques are applied to the same sample, for example combining measurements of ³H/³He and chlorofluorocarbons (CFCs) (Section 7.3.3). Applications for ³H/³He dating are found in the accurate dating

of groundwater residence times of a few tens of years (Plummer *et al.* 2001), in identifying groundwater flowpaths (Beyerle *et al.* (1999) and in tracing sources and rates of contaminant movement (Shapiro *et al.* 1999).

As an example, Beyerle *et al.* (1999) applied ³H/³He dating at a bank filtration site at Linsental in north-east Switzerland. The Linsental aquifer is in the lower reach of the River Töss valley and has a maximum thickness of 25 m and an average width of 200 m. The aquifer is composed of heterogeneous Quaternary gravels reworked by fluvial erosion. The lower aquifer boundary is formed by a low permeability molasse bedrock. Hydraulically, the Linsental aquifer is predominantly influenced by groundwater abstraction from two pumping stations, Sennschür and Obere Au (PS1 and PS2, respectively, in Fig. 5.15), and infiltration from the River Töss.

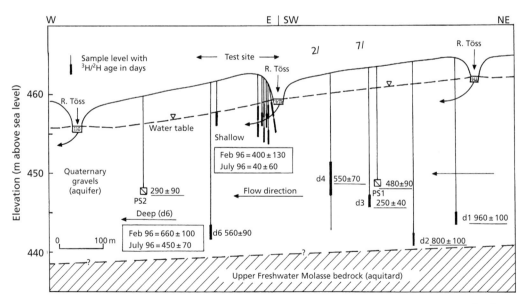

Fig. 5.15 Hydrogeological section of the Linsental aquifer, north-east Switzerland, indicating the ³H/³He ages (in days) of samples from boreholes and two pumping stations, Sennschür (PS1) and Obere Au (PS2), for various sampling dates. The ages of samples collected from shallow and deep (point d6) sections of the aquifer in February and July 1996 are also shown. In February, older groundwater dominated in the shallow aquifer during a time of decreased river water infiltration. In July, heavy rain and flood events were common and locally infiltrated river water with a younger age dominated in the shallow aquifer (Beyerle *et al.* 1999). (*Source:* Beyerle, U., Aeschbach-Hertig, W., Hofer, M. *et al.* (1999) Infiltration of river water to a shallow aquifer investigated with ³H/³He, noble gases and CFCs. *Journal of Hydrology* **220**, 169–185. © 1999, Elsevier.)

Groundwater and river water samples were collected between 1995 and 1997 and analysed for their tritium and helium contents and the resulting ^{3}H/^{3}He ages interpreted along a hydrogeological section. As shown in Fig. 5.15, the ^{3}H/^{3}He groundwater ages from the deep part of the aquifer are similar (about 600 days), whereas water ages of the shallow aquifer depend on the time of sampling. As river infiltration is reduced at times of low river stage, only older groundwater remains and the ^{3}H/^{3}He ages close to the river become comparable to the age of groundwater in the deeper aquifer. The ages of groundwater abstracted from the two pumping stations are about 1 year old, much higher than initially considered. An estimated mixing ratio of 50% between the younger infiltrated river water and older deep groundwater explains the age of the water abstracted from the production boreholes. Hence, the ^{3}H/^{3}He ages have provided useful information on the recharge dynamics and residence times of the bank filtration system (Beyerle *et al.* 1999).

5.6 Noble gases

The temperatures at which recharge water is equilibrated with air in the soil zone can be determined from the noble (inert) gas contents of groundwater. The concentrations of Ne, Ar, Kr and Xe dissolved during equilibration of recharge water with air are controlled by their solubility relationship with temperature, generally decreasing in concentration with increasing temperature. In temperate latitudes, the potential effects of solar insolation at the ground surface are minor and the recharge temperature generally reflects the mean annual temperature of the soil zone. Any increase in groundwater temperature as the water moves from the soil zone to greater depths in the aquifer will not result in exsolution of noble gases since the increase in hydrostatic pressure maintains the groundwater undersaturated with respect to the gases. It is

therefore valid to derive recharge temperatures from noble gas contents (Andrews and Lee 1979). The quantity of a dissolved noble gas, for example Ar, is given by

$$[Ar] = s_T P_{Ar} \quad (cm^3\ STP\ cm^{-3}\ H_2O)$$

(eq.5.21)

where s_T is the solubility of Ar at 1 atmosphere pressure and the temperature of recharge, T; and P_{Ar} is the partial pressure of Ar in the atmosphere. Similar relationships apply to the other noble gases.

Noble gas concentrations of groundwater commonly exceed those calculated for thermodynamic solubility equilibrium with air in the unsaturated zone (eq. 5.21). This additional component, termed 'excess air', is most likely the result of fluctuations in the water table trapping and partially or entirely dissolving small bubbles under increased hydrostatic pressure or surface tension. Corrections for the excess air component are possible, normalized to dissolved Ne contents (Elliot *et al.* 1999). Some of this excess air may be lost by gas exchange across the water table. The rates of gas exchange of the individual noble gases decrease with molecular weight such that fractionation of the residual excess air occurs with significant gas loss. Typically, noble gas recharge temperatures are determined by an iterative procedure. Initially, unfractionated air is subtracted successively from the measured concentrations, and the remaining noble gas concentrations are converted into temperatures on the basis of solubility data and the atmospheric pressure at the elevation of the water table. This procedure is repeated until optimum agreement among the four calculated noble gas temperatures is achieved (Stute *et al.* 1995).

When combined with stable isotope data and ^{14}C ages, noble gas recharge temperatures can further elucidate the history of aquifer evolution and provide a proxy indicator of palaeoenvironmental conditions. Noble gas data are presented by Dennis *et al.* (1997) and Elliot *et al.* (1999) for the Chalk aquifer of the London

Basin and provide evidence for Late Pleistocene (cold stage interstadial) groundwaters recharged at temperatures 5–7°C cooler than present in confined zones north of the River Thames (Table 5.3). Recharge conditions at this time were probably controlled by the occurrence of areas of unfrozen ground within the summer permafrost (talik zones) and so the noble gas recharge temperature may be more representative of the mean summer air temperature. Mean annual air temperatures spanning the frozen winter period were probably lower than the noble gas recharge temperatures.

As shown in Fig. 5.16, similar mean recharge temperatures during the last glacial period of about 5°C cooler than modern Holocene waters are reported for the East Midlands Triassic sandstone aquifer in the United Kingdom (Andrews and Lee 1979) and Devonian sandstones in the semi-arid Piaui Province in north-east Brazil (Stute et al. 1995). Hence, terrestrial records of palaeogroundwaters provide convincing evidence for surface temperature cooling at both high and low latitudes during the last (Devensian or Weichselian) ice age.

Dissolved helium has been extensively used for estimating groundwater residence times and in tracing the origin of crustal fluids. Deep circulating groundwater typically has a radiogenic source of ^{4}He from in situ α-decay of U-Th series elements (Aggarwal et al. 2015). However, modern groundwater (>10 years) can also exhibit high concentrations of helium which originate in previously trapped reservoirs in the subsurface. Helium enrichment in palaeogroundwater is reported in the literature at up to six orders of magnitude higher than expected due to solubility equilibrium with the atmosphere.

Terrigenic ^{3,4}He is also significant in deep circulating fluids. The ^{3}He/^{4}He isotope ratio is often used to distinguish between helium

Table 5.3 Isotope and noble gas recharge temperatures for Chalk groundwaters in the London Basin along a N–S transect from the outcrop on the North Downs to the confined centre of the Basin in the City of London (Elliot et al. 1999).

Sample location	Distance N along transect (km)	δ^2H (1)	δ^{18}O (1)	δ^{13}C (2)	^{14}C (3)	RT °C	±1s.e.
1. Paynes u	0	−51	−7.4	−13.6	66.2	12.5	0.7
2. Philips 2 u	1.7	−48	−7.2	−13.4		11.4	0.9
3. BXL Plastics c	3.7			−3.6	4.3	9.0	0.5
4. Modeluxe c	6.3	−50	−7.4	−3.8	4.5	9.2	0.8
5. Sunlight c	10.9	−49		−8.7	14.6	10.2	0.6
6. Unigate c	13.4	−47	−7.2	−12.8	12.6	10.9	1.0
7. Harrods c	15.1	−51	−7.4	−4.8	17.1	9.5	0.8
8. Buchanan House c	16.3	−52	−7.3	−3.5		8.7	0.7
11. Unilever c	16.3	−53	−7.6	−2.5	1.0	6.8	0.6
10. Sainsburys c	16.3	−51	−7.4	−1.7		7.1	0.5
9. Dorset House c	17.4	−53	−7.8	−2.4	0.8	7.1	0.5
13. Kentish Town c	20.0	−53	−7.8	−2.6	0.8	5.4	0.6
12. Hornsey Road c	21.1	−54	−7.7	−2.6	1.4	5.8	0.5

Notes: u = unconfined; c = confined; RT = noble gas recharge temperature; 1 s.e. = 1 standard error; (1) ±1‰ and ±0.1‰, respectively, relative to V$_{SMOW}$; (2) ±0.1‰ relative to PDB; (3) pmc, % modern carbon.
Source: Adapted from Elliot, T., Andrews, J.N. and Edmunds, W.M. (1999) Hydrochemical trends, palaeorecharge and groundwater ages in the fissured Chalk aquifer of the London and Berkshire Basins, UK. Applied Geochemistry **14**, 333–363.

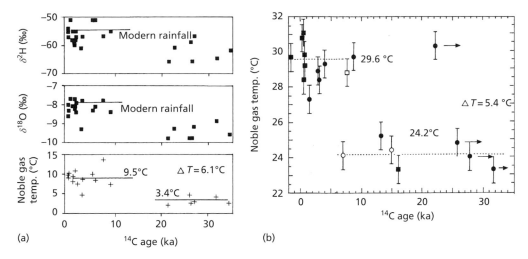

Fig. 5.16 Noble gas palaeotemperature records for (a) the East Midlands Triassic sandstone aquifer, United Kingdom (53.5°N, 1°W) (Edmunds 2001) (*Source:* Edmunds, W.M. (2001) Palaeowaters in European coastal aquifers – the goals and main conclusions of the PALAEAUX project. In: *Palaeowaters in Coastal Europe: Evolution of Groundwater since the Late Pleistocene* (eds W.M. Edmunds and C.J. Milne). Geological Society, London, Special Publications **189**, pp. 1–16. © 2001, Geological Society of London.), and (b) Devonian sandstones of the Cabeças (circles) and Serra Grande (squares) aquifers of Piaui Province, Brazil (7°S, 41.5°W) (Stute *et al.* 1995). (*Source:* Stute, M., Forster, M., Frischkorn, H. *et al.* (1995) Cooling of tropical Brazil (5°C) during the last glacial maximum. *Science* **269**, 379–383. © 1995, American Association for the Advancement of Science.) Noble gas recharge temperatures are plotted as a function of ^{14}C age. Relative to modern waters, the climate of both tropical and mid-latitudes was between 5 and 6°C cooler during the last ice age. At mid-latitudes, the mean annual temperatures may have been cooler, given that recharge was probably restricted to talik zones within the summer permafrost. The ^{14}C data for the East Midlands aquifer demonstrate that groundwater recharge occurred during the Holocene as well as during the Devensian (the oldest waters are beyond the limits of radiocarbon dating). An age gap is evident between 10 and 20 ka when the recharge areas were frozen at the time of the LGM. In lowland Brazil, the arrows shown indicate those samples with a ^{14}C content below the detection limit and the open symbols represent samples obtained by Verhagen *et al.* (1991). The cooling of surface temperatures in Brazil by 5.4°C during the period between 10 and 35 ka is consistent with a lowering of the snow line altitudes in the tropical mountain ranges of about 1000 m for the LGM.

of crustal or mantle origin. However, due to groundwater typically containing helium from several different sources, helium age dating is usually only used as a technique for estimating groundwater age, rather than a precise quantitative tool (Gumm *et al.* 2016). However, Torgersen (2010) suggested that it is now possible to determine uncertainty limits associated with crustal fluxes of helium using the

increased number of published data sets, thus improving the accuracy of [4]He age estimates.

An example of the application of a multi-tracer approach using noble gases, stable isotopes and major ions to provide hydrochemical evidence for the upward migration of deep crustal fluids along the Bornheim fault in the Lower Rhine Embayment in Germany is presented in Box 5.1.

Box 5.1   Fluid flow along faults in the Lower Rhine Embayment, Germany

The role of faults as barriers to groundwater movement has been widely described in the literature but much less documented is the ability of faults to simultaneously act as conduits for

sub-vertical flow along a fault (Bense and Person 2006). The application of noble gas tracers has been particularly useful in characterizing fluid flow associated with faults as conduit-

Box 5.1 (Continued)

barrier structures in both deep and shallow settings, and dissolved noble gases and stable isotopes are particularly useful in studies of aquifer systems that contain pre-Holocene 'palaeo' groundwater or fluids that have a deep crustal origin (Gumm *et al.* 2016).

In the unconsolidated sedimentary aquifers of the Lower Rhine Embayment (LRE), Germany, large hydraulic gradients exist across many faults, which suggest that they act as barriers to groundwater flow (Bense *et al.* 2008). The LRE forms part of the Roer Valley Rift System, which is the southward extension of the North Sea Basin and part of a Cenozoic mega-rift system that crosses western and central Europe (Ziegler 1994). The German region of the LRE covers an area of 3800 km^2 and contains approximately 1300 m of Oligocene to Pleistocene unconsolidated siliciclastic sediments that form a highly complex multi-layered aquifer sequence within six tectonic blocks. Each tectonic block consists of 5–15 different aquifers to a depth of 200 m and the Erft block has as many as 21 separate aquifer units to a depth of approximately 400 m. The LRE is intersected by numerous northwest–southeast striking fault zones which have a significant impact on regional groundwater flow patterns. The LRE primarily consists of Oligocene, Miocene and Pliocene marine sediments and the more recent Pleistocene sediments are mainly fluvial deposits derived from the Rhenish Massif to the south (Schäfer *et al.* 2005). Highly porous loess soils have formed due to the accumulation of silts, sands and clays, and these overlay the Pleistocene deposits (Kemna 2008).

The LRE contains two commercially important lignite seams. The main lignite seam has a thickness of up to 100 m and the second, upper seam has a thickness of up to 40 m (Hager 1993). The Miocene lignite deposits of the LRE form one of the largest reserves

in Europe with an estimated 55,000 × 10^6 t (Hager 1993) with an overburden thickness up to 300 m. The geological structures that underlie the Oligocene aquifers below a depth of 400–800 m are less well known. The oldest geological sequences identified from borehole records are Palaeozoic (Geluk *et al.* 1994).

A programme of groundwater sampling from observation boreholes in the LRE for analysis of nobles gases (He, Ne, Ar, Kr and Xe), stable isotopes (δ^2H and δ^{18}O) and major ions undertaken by Gumm *et al.* (2016) showed that groundwater in shallow unconsolidated sedimentary aquifers close to the Bornheim fault has relatively low δ^2H and δ^{18}O values in comparison to regional modern groundwater recharge, and ^{4}He concentrations up to 1.7×10^{-4} cm^3 (STP) g^{-1}, approximately four orders of magnitude higher than expected due to solubility equilibrium with the atmosphere. A conceptual hydrogeochemical model presented by Gumm *et al.* (2016) (Fig. 5.17) illustrates the observed occurrence of palaeo groundwater with a deep crustal component at very shallow depth in the Brühl region. *In situ* ^{4}He production derived from α-decay of U-Th series elements in the subsurface as well as a significant external flux of helium are necessary to account for the observed helium concentrations. Hydrochemical analysis indicates that Na$^+$ and Cl$^-$ are dominant ions in the palaeo groundwater and that the hydrochemical evolution involved very high PCO$_2$ conditions and a significant subsurface source of H$^+$ ions.

Conduit flow associated with the Bornheim fault is possibly linked to historic groundwater abstraction in the region. At the location of the Bornheim fault, no thermal anomaly is observed, suggesting that either the flux of upwelling fluid along the fault zone is not vigorous enough to create such a thermal anomaly, or that the plume of fluid observed was emplaced during an episode of enhanced

Box 5.1   (Continued)

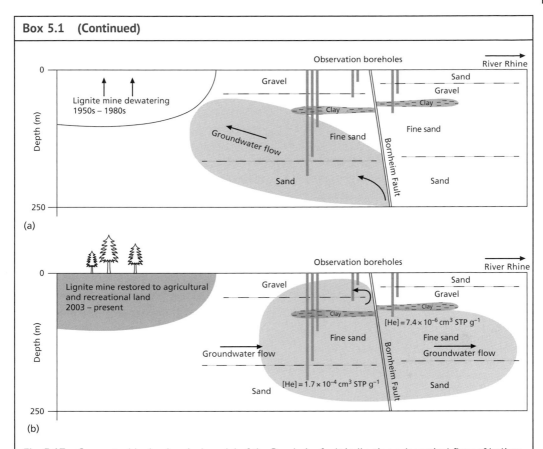

Fig. 5.17 Conceptual hydrochemical model of the Bornheim fault indicating sub-vertical flow of helium-enriched fluid from depth along the fault during (a) periods of lignite mine dewatering and (b) subsequent dispersal following mine restoration (represented by the grey shaded area). The cross-section illustrates the groundwater flow dynamics in the shallow aquifer system of the Brühl region. The deeper (>250 m) aquifer system is currently influenced by dewatering of the Hambach mine further to the north-west but is effectively isolated from the shallow aquifer system by a very thick confining layer of clay (Gumm *et al.* 2016). (*Source:* Gumm, L.P., Bense, V.F., Dennis, P.F. *et al.* (2016) Dissolved noble gases and stable isotopes as tracers of preferential fluid flow along faults in the Lower Rhine Embayment, Germany. *Hydrogeology Journal* **24**, 99–108.)

fluid flow caused by groundwater abstraction. A shallow source of helium from U-Th rich lignite deposits could explain the decoupling of helium and heat, but excess helium is absent from groundwater sampled at all other observation boreholes situated near lignite seams, and additional data such as hydrochemistry, water isotopes and ^{14}C-dating all support the idea of deep fluid migration from depth to shallow aquifer units (Gumm *et al.* 2016). Although fluid exchange between the deep basal aquifer system and the upper aquifer layers is generally impeded by confining clay layers and lignite, the geochemical data suggest that deep circulating fluids penetrate shallow aquifers in the locality of fault zones, implying that fluid flow occurs along preferential sub-vertical faults within the fault core material in the LRE. The geochemical and temporal hydraulic head data suggest that the groundwater originates at depth and flows upwards. However, large hydraulic-head gradients observed across many faults suggest that they also act as barriers to lateral groundwater flow.

Box 5.1 **(Continued)**

Gumm *et al.* (2016) hypothesized that the emplacement of deep fluids at shallow depth occurred during the active dewatering of nearby lignite mines and that conduit flow from depth is either reduced or not occurring at present along the Bornheim fault. The emplaced anomalous fluids also appear to be slow to disperse, perhaps due to the small hydraulic head gradients that currently exist

in the shallow aquifers of the Brühl region (Fig. 5.17). Overall, the geochemical data substantiate a conduit-barrier model of fault-zone hydrogeology in unconsolidated sedimentary deposits, as well as corroborating the concept that faults in unconsolidated aquifer systems can create hydraulic connectivity between deep and shallow aquifers (Gumm *et al.* 2016).

Further reading

Attendorn, H.-G. and Bowen, R.N.C. (1997) *Radioactive and Stable Isotope Geology.* Chapman & Hall, London.

Clark, I.D. and Fritz, P. (1997) *Environmental Isotopes in Hydrogeology.* Lewis Publishers, Boca Raton, FL, doi: 10.1016/S0962-8924(97)01152-5.

Faure, G. and Mensing, T.M. (2005) *Isotopes: Principles and Applications.* John Wiley & Sons, Inc., Hoboken, NJ.

Fritz, P. and Fontes, J.C. (1980) *Handbook of Environmental Isotope Chemistry, Vol. 1: The Terestrial Envionment, A.* Elsevier, Amsterdam.

Kendall, C. and McDonnell, J.J. (eds) (1998) *Isotope Tracers in Catchment Hydrology.* Elsevier, Amsterdam.

Mazor, E. (2004) *Chemical and Isotopic Groundwater Hydrology: The Applied Approach* (3rd edn). Marcel Dekker, Inc., New York, doi: 10.1097/01.id.0000148554.83439.00.

White, W.M. (2015) *Isotope Geochemistry.* John Wiley & Sons, Chichester.

References

Aggarwal, P.K., Matsumoto, T., Sturchio, N.C. *et al.* (2015) Continental degassing of ^{4}He by surficial discharge of deep groundwater. *Nature Geoscience* **8**, 35–39, doi: 10.1038/NGEO2302.

Albu, M., Banks, D. and Nash, H. (1997) *Mineral and Thermal Groundwater Resources.* Chapman & Hall, London.

Andrews, J.N. and Lee, D.J. (1979) Inert gases in groundwater from the Bunter sandstone of England as indicators of age and palaeoclimatic trends. *Journal of Hydrology* **41**, 233–252, doi: 10.1016/0022-1694(79)90064-7.

Ault, W.U. and Kulp, J.L. (1959) Isotopic geochemistry of sulphur. *Geochimica et Cosmochimica Acta* **16**, 201–235, doi: 10.1016/0016-7037(59)90112-7.

Bense, V.F. and Person, M.A. (2006) Faults as conduit-barrier systems to fluid flow in siliciclastic sedimentary aquifers. *Water Resources Research* **42**, W05421, doi: 10.1029/2005WR004480.

Bense, V.F., Person, M.A., Chaudhary, K. *et al.* (2008) Thermal anomalies indicate preferential flow along faults in unconsolidated sedimentary aquifers. *Geophysical Research Letters* **35**, L24406, doi: 10.1029/2008GL036017.

Bentley, H.W., Phillips, F.M., Davis, S.N. *et al.* (1986) Chlorine 36 dating of very old groundwater: The Great Artesian Basin, Australia. *Water Resources Research* **22**, 1991–2001, doi: 10.1029/WR022i013p01991.

Bethke, A.M. and Johnson, T.M. (2008) Groundwater age and groundwater age dating. *Annual Review of Earth and Planetary Sciences* **36**, 121–152, doi: 10.1146/annurev.earth.36.031207.124210.

Beyerle, U., Aeschbach-Hertig, W., Hofer, M. *et al.* (1999) Infiltration of river water to a shallow aquifer investigated with ^{3}H/^{3}He, noble gases and CFCs. *Journal of Hydrology* **220**, 169–185, doi: 10.1016/S0022-1694(99)00069-4.

Bishop, P.K. and Lloyd, J.W. (1991) Use of ^{14}C modelling to determine vulnerability and pollution of a carbonate aquifer: the Lincolnshire Limestone, eastern England. *Applied Geochemistry* **6**, 319–331, doi: 10.1016/0883-2927(91)90008-D.

Bottrell, S.H., Moncaster, S.J., Tellam, J.H. *et al.* (2000) Controls on bacterial sulfate reduction in a dual porosity aquifer system: the Lincolnshire Limestone aquifer, England. *Chemical Geology* **169**, 461–470, doi: 10.1016/S0009-2541(00)00222-9.

Chesterikoff, A., Lécolle, P., Létolle, R. and Carbonnel. J.P. (1981) Sulfur and oxygen isotopes as tracers of the origin of sulfate in Lake Créitel (southeast of Paris, France). *Journal of Hydrology* **54**, 141–150, doi: 10.1016/0022-1694(81)90156-6.

Craig, H. (1961) Isotopic variations in meteoric water. *Science* **133**, 1702–1703, doi: 10.1126/science.133.3465.1702.

Darling, W.G. and Talbot, J.C. (2003) The O and H stable isotopic composition of fresh waters in the British Isles. 1. Rainfall. *Hydrology and Earth System Sciences* **7**, 163–181, doi: 10.5194/hess-7-163-2003.

Darling, W.G., Edmunds, W.M. and Smedley, P.L. (1997) Isotopic evidence for palaeowaters in the British Isles. *Applied Geochemistry* **12**, 813–829, doi: 10.1016/S0883-2927(97)00038-3.

Darling, W.G., Bath, A.H. and Talbot, J.C. (2003) The O and H stable isotopic composition of fresh waters in the British Isles. 2. Surface waters and groundwater. *Hydrology and Earth System Sciences* **7**, 183–195, doi: 10.5194/hess-7-183-2003.

Dennis, F., Andrews, J.N., Parker, A. *et al.* (1997) Isotopic and noble gas study of Chalk groundwater in the London Basin, England.

Applied Geochemistry **12**, 763–773, doi: 10.1016/S0883-2927(97)00046-2.

Downing, R.A., Smith, D.B., Pearson, F.J. *et al.* (1977) The age of groundwater in the Lincolnshire Limestone, England and its relevance to the flow mechanism. *Journal of Hydrology* **33**, 201–216, doi: 10.1016/0022-1694(77)90035-X.

Edmunds, W.M. (2001) Palaeowaters in European coastal aquifers – the goals and main conclusions of the PALAEAUX project. In: *Palaeowaters in Coastal Europe: Evolution of Groundwater since the Late Pleistocene* (eds W. M. Edmunds and C.J. Milne). Geological Society, London, Special Publications **189**, pp. 1–16.

Edmunds, W.M. and Walton, N.R.G. 1983. The Lincolnshire Limestone: hydrogeochemical evolution over a ten-year period. *Journal of Hydrology* **61**, 201–211, doi: 10.1016/0022-1694(83)90248-2.

Elliot, T., Andrews, J.N. and Edmunds, W.M. (1999) Hydrochemical trends, palaeorecharge and groundwater ages in the fissured Chalk aquifer of the London and Berkshire Basins, UK. *Applied Geochemistry* **14**, 333–363, doi: 10.1016/S0883-2927(98)00060-2.

Feast, N.A. (1995) *Application of nitrogen and sulphur isotope hydrochemistry in groundwater studies.* PhD Thesis, University of East Anglia, Norwich, 269 pp.

Feth, J.H. (1966) Nitrogen compounds in natural water—a review. *Water Resources Research* **2**, 41–58, doi: 10.1029/WR002i001p00041.

Fukada, T., Hiscock, K.M., Dennis, P.F. and Grischek, T. (2003) A dual isotope approach to identify denitrification in groundwater at a river-bank infiltration site. *Water Research* **37**, 3070–3078, doi: 10.1016/S0043-1354(03)00176-3.

Fukada, T., Hiscock, K.M. and Dennis, P.F. (2004) A dual-isotope approach to the nitrogen hydrochemistry of an urban aquifer. *Applied Geochemistry* **19**, 709–719, doi: 10.1016/j.apgeochem.2003.11.001.

Ferguson, G., Cuthbert, M.O., Befus, K. *et al.* (2020) Rethinking groundwater age. *Nature Geoscience* 13, 592–594. doi: 10.1038/s41561-020-0629-7.

Gardner, C.M.K., Bell, J.P., Cooper, J.D. et al. (1991) Groundwater recharge and water movement in the unsaturated zone. In: *Applied Groundwater Hydrology: A British Perspective* (eds R.A. Downing and W.B. Wilkinson). Clarendon Press, Oxford, pp. 54–76.

Geluk, M.C., Duin, E.J.T., Dusar, M. *et al.* (1994) Stratigraphy and tectonics of the Roer Valley Graben. *Geologie en Mijnbouw* 73, 129–141.

George, M.A. (1998) *High precision stable isotope imaging of groundwater flow dynamics in the Chalk aquifer systems of Cambridgeshire and Norfolk*. PhD Thesis, University of East Anglia, Norwich, 164–165.

Gumm, L.P., Bense, V.F., Dennis, P.F. *et al.* (2016) Dissolved noble gases and stable isotopes as tracers of preferential fluid flow along faults in the Lower Rhine Embayment, Germany. *Hydrogeology Journal* 24, 99–108, doi: 10.1007/s10040-015-1321-7.

Hager, H. (1993) The origin of the Tertiary lignite deposits in the Lower Rhine region, Germany. *International Journal of Coal Geology* 23, 251–262, doi: 10.1016/0166-5162(93)90051-B.

Heaton, T.H.E. (1986) Isotopic studies of nitrogen pollution in the hydrosphere and atmosphere: a review. *Chemical Geology (Isotope Geoscience Section)* 59, 87–102, doi: 10.1016/0168-9622(86)90059-X.

Hiscock, K.M., Dennis, P.F., Saynor, P.R. and Thomas, M.O. (1996) Hydrochemical and stable isotope evidence for the extent and nature of the effective Chalk aquifer of north Norfolk, UK. *Journal of Hydrology* 180, 79–107, doi: 10.1016/0022-1694(95)02895-1.

Hiscock, K.M., Iqbal, T., Feast, N.A. and Dennis, P.F. (2011) Isotope and reactive transport modelling of denitrification in the Lincolnshire Limestone aquifer, eastern England. *Quarterly Journal of Engineering Geology and Hydrogeology* 44, 93–108, doi: 10.1144/1470-9236/08-110.

Hoefs, J. (2015) *Stable Isotope Geochemistry* (7th edn). Springer International Publishing, Cham, Switzerland.

Holser, W.T. and Kaplan, I.R. (1966) Isotope geochemistry of sedimentary sulfates. *Chemical Geology* 1, 93–135, doi: 10.1016/0009-2541(66)90011-8.

Hübner, H. (1986) Isotope effects of nitrogen in the soil and biosphere. In: *Handbook of Environmental Isotope Geochemistry, Volume 2 The Terrestrial Environment, B* (eds P. Fritz and J.C. Fontes). Elsevier, Amsterdam, pp. 361–426.

IAEA/WMO (1998) Global network of isotopes in precipitation. The GNIP Database. http://isohis.iaea.org/ (accessed 7 June 2013).

Jasechko, S. (2019) Global isotope hydrogeology–review. *Reviews of Geophysics* 57, 835–965. doi: 10.1029/2018RG000627.

Jasechko, S., Perrone, D., Befus, K.M. *et al.* (2017) Global aquifers dominated by fossil groundwaters but wells vulnerable to modern contamination. *Nature Geoscience* 10, 425–430. doi: 10.1038/NGEO2943.

Kemna, H.A. (2008) A revised stratigraphy for the Pliocene and Lower Pleistocene deposits of the Lower Rhine Embayment. *Netherlands Journal of Geosciences* 87, 91–105, doi: 10.1017/S0016774600024069.

Kendall, C. (1998) Tracing nitrogen sources and cycling in catchments. In: *Isotope-Tracers in Catchment Hydrology* (eds C. Kendall and J.J. McDonnel). Elsevier, Amsterdam, pp. 534–569.

Kendall, C. and Krabbenhoft, D.P. (1995) Applications of isotopes to tracing sources of solutes and water in shallow systems. *Proceedings of the International Symposium on Groundwater Management* (ed. R.J. Charbeneau). American Society of Civil Engineers, San Antonio, TX, pp. 390–395.

Krauskopf, K.B. and Bird, D.K. (1995) *Introduction to Geochemistry* (3rd edn). McGraw-Hill, Inc., New York.

Krouse, H.R. (1980) Sulphur isotopes in our environment. In: *Handbook of Environmental Isotope Geochemistry, Volume 1 The Terrestrial Environment, A* (eds P. Fritz and J.C. Fontes). Elsevier, Amsterdam, pp. 435–472.

Létolle, R. (1980) Nitrogen-15 in the natural environment. In: *Handbook of Environmental Isotope Geochemistry. Volume 1: The Terrestrial Environment, A* (eds P. Fritz and J.C. Fontes eds). Elsevier, Amsterdam, pp. 407–433.

Lloyd, J.W. and Heathcote, J.A. (1985) *Natural Inorganic Hydrochemistry in Relation to Groundwater: An Introduction.* Clarendon Press, Oxford.

Love, A.J., Herczeg, A.L., Sampson, L. *et al.* (2000) Sources of chloride and implications for ^{36}Cl dating of old groundwater, southwestern Great Artesian Basin, Australia. *Water Resources Research* **36**, 1561–1574, doi: 10.1029/2000WR900019.

Mariotti, A., Germon, J.C., Hubert, P. *et al.* (1981) Experimental determination of nitrogen kinetic isotope fractionation: some principles; illustration for the denitrification and nitrification processes. *Plant and Soil* **62**, 413–430, doi: 10.1007/BF02374138.

Mariotti, A., Landreau, A. and Simon, B. (1988) ^{15}N isotope biogeochemistry and natural denitrification process in groundwater: application to the chalk aquifer of northern France. *Geochimica et Cosmochimica Acta* **52**, 1869–1878, doi: 10.1016/0016-7037(88)90010-5.

Moncaster, S.J., Bottrell, S.H., Tellam, J.H. and Lloyd, J.W. (1992) Sulphur isotope ratios as tracers of natural and anthropogenic sulphur in the Lincolnshire Limestone aquifer, eastern England. In: *Water-Rock Interaction* (eds Y. Kharaka and A.S. Maest). A.A.Balkema, Rotterdam, pp. 813–816.

Moncaster, S.J., Bottrell, S.H., Tellam, J.H. *et al.* (2000) Migration and attenuation of agrochemical pollutants: insights from isotopic analysis of groundwater sulphate. *Journal of Contaminant Hydrology* **43**, 147–163, doi: 10.1016/S0169-7722(99)00104-7.

Mook, W.G. (1980) Carbon-14 in hydrogeological studies. In: *Handbook of Environmental Isotope Geochemistry, Volume 1 The Terrestrial Environment, A* (eds P. Fritz and J.C. Fontes). Elsevier Scientific Publishing Company, Amsterdam, pp. 49–74.

Nielsen, H. (1979) Sulphur isotopes. In: *Lectures in Isotope Geology* (eds E. Jäger and J.C. Hunziker). Springer-Verlag, Berlin, pp. 283–312.

Nriagu, J.O., Rees, C.E., Mekhtiyeva, V.L. *et al.* (1991) Hydrosphere. In: *Stable Isotopes: Natural and Anthropogenic Sulphur in the Environment* (eds H.R. Krouse and V.A. Grinenko). John Wiley & Sons, Chichester, SCOPE **43**, pp. 177–265.

Parker, J.M., Young, C.P. and Chilton, P.J. (1991) Rural and agricultural pollution of groundwater. In: *Applied Groundwater Hydrology: A British Perspective* (eds R.A. Downing and W.B. Wilkinson). Clarendon Press, Oxford, pp. 149–163.

Plummer, L.N., Busenberg, E., Böhlke, J.K. *et al.* (2001) Groundwater residence times in Shenandoah National Park, Blue Ridge Mountains, Virginia, USA: a multi-tracer approach. *Chemical Geology* **179**, 93–111, doi: 10.1016/S0009-2541(01)00317-5.

Rozanski, K. (1985) Deuterium and oxygen-18 in European groundwaters – links to atmospheric circulation in the past. *Chemical Geology (Isotope Geoscience Section)* **52**, 349–363, doi: 10.1016/0168-9622(85)90045-4.

Rozanski, K., Araguás-Araguás, L. and Gonfiantini, R. (1993) Isotopic patterns in modern global precipitation. *Geophysical Monograph of the American Geophysical Union* **78**, 1–36.

Rye, R.O., Back, W., Hanshaw, B.B. *et al.* (1981) The origin and isotopic composition of dissolved sulphide in groundwater from the carbonate aquifers of Florida and Texas. *Geochimica et Cosmochimica Acta*

45, 1941–1950, doi: 10.1016/0016-7037(81) 90024-7.

Schäfer, A., Utescher, T., Klett, M. and Valdivia-Manchego, M. (2005) The Cenozoic Lower Rhine Basin:, sedimentation, and cyclic stratigraphy. *International Journal of Earth Sciences* **94**, 621–639, doi: 10.1007/s00531-005-0499-7.

Schoelle, P.A. and Arthur, M.A. (1980) Carbon isotope fluctuations in Cretaceous pelagic limestones: potential stratigraphic and petroleum exploration tool. *American Association of Petroleum Geologists Bulletin* **64**, 67–87.

Shapiro, S.D., LeBlanc, D., Schlosser, P. and Ludin, A. (1999) Characterizing a sewage plume using the ^{3}H-^{3}He dating technique. *Ground Water* **37**, 861–878, doi: 10.1111/j.1745-6584.1999.tb01185.x.

Smith, D.B., Downing, R.A., Monkhouse, R.A. *et al.* (1976) The age of groundwater in the Chalk aquifer of the London Basin. *Water Resources Research* **12**, 392–404, doi: 10.1029/WR012i003p00392.

Smith, J.W. and Batts, B.D. (1974) The distribution and isotopic composition of sulphur in coal. *Geochimica et Cosmochimica Acta* **38**, 121–131, doi: 10.1016/0016-7037(74) 90198-7.

Solomon, D.K., Schiff, S.L., Poreda, R.J. and Clarke, W.B. (1993) A validation of the ^{3}H/^{3}He method for determining groundwater recharge. *Water Resources Research* **29**, 2951–2962, doi: 10.1029/93WR00968.

Spalding, R.F. and Parrott, J.D. (1994) Shallow groundwater denitrification. *Science of the Total Environment* **141**, 17–25, doi: 10.1016/0048-9697(94)90014-0.

Spalding, R.F., Exner, M.E., Martin, G.E. and Snow, D.D. (1993) Effects of sludge disposal on groundwater nitrate concentrations. *Journal of Hydrology* **142**, 213–228, doi: 10.1016/0022-1694(93)90011-W.

Strebel, O., Böttcher, J. and Fritz, P. (1990) Use of isotope fractionation of sulphate-sulphur and sulfate oxygen to assess bacterial desulphurication in a sandy aquifer. *Journal of Hydrology* **121**, 155–172, doi: 10.1016/0022-1694(90)90230-U.

Stute, M., Forster, M., Frischkorn, H. *et al.* (1995) Cooling of tropical Brazil (5°C) during the last glacial maximum. *Science* **269**, 379–383, doi: 10.1126/science.269.5222.379.

Suckow, A. (2014) The age of groundwater: definitions, models and why we do not need the term. *Applied Geochemistry* **50**, 222–230, doi: 10.1016/j.apgeochem.2014.04.016.

Thode, H.G. (1991) Sulphur isotopes in nature and the environment: an overview. In: *Stable Isotopes: Natural and Anthropogenic Sulphur in the Environment* (eds H.R. Krouse and V.A. Grinenko). John Wiley & Sons, Chichester, SCOPE **43**, pp. 1–26.

Thode, H.G. and Monster, J. (1965) Sulphur-isotope geochemistry of petroleum evaporites and ancient seas. In: *Fluids in the Subsurface Environments* (eds A. Young and J.E. Galley). American Association of Petroleum Geologists, Tulsa, OK. Memoir 4, 367–377.

Torgersen, T. (2010) Continental degassing flux of ^{4}He and its variability. *Geochemistry, Geophysics, Geosystems* **11**, Q06002. doi: 10.1029/2009GC002930.

Vaikmäe, R., Vallner, L., Loosli, H.H. *et al.* (2001) Palaeogroundwater of glacial origin in the Cambrian-Vendian aquifer of northern Estonia. In: *Palaeowaters in Coastal Europe: Evolution of Groundwater Since the Late Pleistocene* (eds W.M. Edmunds and C.J. Milne). Geological Society, London, Special Publications **189**, pp. 17–27.

Verhagen, B.T., Geyh, M.A., Fröhlich, K. and Wirth, K. (1991) *Isotope hydrological methods for the quantitative evaluation of ground water resources in arid and semiarid areas.* Research Report of the Federal Ministry of Economic Cooperation of the Federal Republic of Germany, Bonn.

Wilson, G.B., Andrews, J.N. and Bath, A.H. (1990)
Dissolved gas evidence for denitrification in the
Lincolnshire Limestone groundwaters, Eastern
England. *Journal of Hydrology* **113**, 51–60, doi:
10.1016/0022-1694(90)90166-U.

Ziegler, P.A. (1994) Cenozoic rift system
of Western and Central Europe: an
overview. *Geologie en Mijnbouw* **73**,
99–127.

Zinn, B.A. and Konikow, L.F. (2007)
Potential effects of regional pumpage on
groundwater age distribution. *Water Resources
Research* **43**, W06418. doi: 10.1029/
2006WR004865.

6

Groundwater and catchment processes

6.1 Introduction

This chapter and the next introduce basic field methods and techniques used in hydrological and hydrogeological investigations. Taking an integrated, whole river basin approach, groundwater is a central component of the hydrological cycle supporting terrestrial aquatic environments such as rivers and wetlands. To be able to assess this interaction and to begin to manage catchment water resources for various functions such as water supply and ecosystem support, it is useful to adopt the concept of a catchment water balance in which all known inflows and outflows are considered. The measurements needed to complete a water balance, such as precipitation, evapotranspiration, streamflow and groundwater recharge are presented in this chapter. In addition, methods and models to estimate river flow depletion as a result of groundwater abstraction are presented to demonstrate the potential harmful impacts of groundwater over-abstraction on surface water systems at catchment and global scales.

6.2 Water balance equation

The basic raw resource within a catchment is precipitation (P) and, as shown in Fig. 6.1, precipitation is either lost to evaporation and transpiration (usually combined and referred to as evapotranspiration, ET) or routed through the hydrological pathways of overland flow and interflow to give surface water runoff (S_R) and groundwater discharge (G_R). The groundwater discharge component is supplied by groundwater recharge and includes natural discharge to springs and rivers (the river baseflow, Q_G) and artificial abstractions (Q_A). Depending on the catchment geology, the groundwater catchment may or may not coincide with the surface catchment area such that additional components of cross-formational groundwater flow (Q_U) may need to be considered.

Now, assuming that the surface water and groundwater catchments coincide, the following catchment water balance equation can be written:

$$P = E + S_R + G_R \pm \Delta S \qquad \text{(eq. 6.1)}$$

On a short timescale of weeks to months, eq. 6.1 is balanced by changes in the water held in soil and groundwater storage (ΔS) as represented by changes in soil moisture content and groundwater levels. Over longer timescales of several years, changes in storage balance out to zero and, expressing $G_R = Q_G + Q_A$, eq. 6.1 becomes:

$$P - E = S_R + Q_G + Q_A \qquad \text{(eq. 6.2)}$$

In other words, the difference between precipitation and evapotranspiration, or effective precipitation, supports surface runoff, groundwater discharge as river baseflow and borehole abstractions. The total flow in a river is calculated as the sum of S_R and Q_G. Methods for

Hydrogeology: Principles and Practice, Third Edition. Kevin M. Hiscock and Victor F. Bense.
© 2021 John Wiley & Sons Ltd. Published 2021 by John Wiley & Sons Ltd.
Companion website: www.wiley.com/go/hiscock/hydrogeology3e

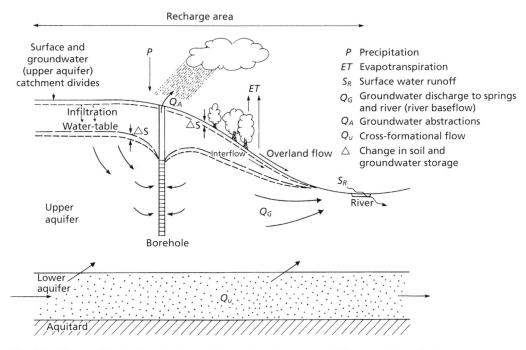

Fig. 6.1 Diagram illustrating the inputs, flowpaths and outputs within a river basin that comprise a catchment water balance. The total groundwater discharge, G_R, is the sum of $Q_G + Q_A + Q_u$.

calculating the amount of effective precipitation (hydrological excess) are discussed in Section 6.5. Clearly, if eq. 6.2 is to be balanced, any increase in the amount of groundwater abstraction (Q_A) will be at the expense of river baseflow (Q_G) and so potentially causing environmental impacts. As a general rule, an example measure of the quantity of water required to maintain acceptable minimum river flows is the Q95 low flow statistic. This means that during the baseflow recession period, when river flows are dominated by the groundwater input (Q_G), it is undesirable for the river flow to fall below the long-term average of flow that is equalled or exceeded for 95% of the time.

For sustainable groundwater development, and in order to meet conflicting environmental (Q_G) and socio-economic (Q_A) demands, it is highly desirable that these demands do not exceed the rate of groundwater recharge (equated to G_R). Three conditions can be defined in which (1) $Q_A < G_R - Q_G$ and further groundwater resources are available for exploitation;

(2) $Q_A = G_R - Q_G$ in which case the safe yield has been achieved; and (3) $Q_A > G_R - Q_G$ when the groundwater resources are over-exploited or mined. In the case of (3), groundwater in support of abstractions is taken from aquifer storage with potential long-term impacts on groundwater levels.

Gleeson *et al.* (2012) introduced the term 'groundwater footprint', defined as the aquifer area required to sustain groundwater use and groundwater-dependent ecosystems of a region of interest, such as an aquifer, catchment or community, as a method for assessing the impact of groundwater consumption on natural groundwater stores and flows. Using the previous nomenclature, the groundwater footprint, *GF*, is equal to $A_A(Q_A/(G_R - Q_G))$, where A_A is aquifer area with dimensions $[L^2]$ and Q_A, Q_G and G_R are given as area-averaged annual values with dimensions $[L\,T^{-1}]$. It is instructive to compare the ratio of the groundwater footprint, *GF*, to the aquifer area, A_A as an indicator of groundwater stress. As shown in Table 6.1

Table 6.1 Properties of aquifers with the largest groundwater footprints (Gleeson *et al.* 2012).

Aquifer	Country	GF (10⁶ km²)	A_A (10⁶ km²)	GF/A_A
Upper Ganges	India, Pakistan	26.1 ± 7.5	0.48	54.2 ± 15.6
North Arabian	Saudi Arabia	17.3 ± 4.7	0.36	48.3 ± 13.5
South Arabian	Saudi Arabia	9.5 ± 3.6	0.25	38.5 ± 14.7
Persian	Iran	8.4 ± 3.7	0.42	19.7 ± 8.6
South Caspian	Iran	5.9 ± 2.0	0.06	98.3 ± 32.6
Western Mexico	Mexico	5.5 ± 2.0	0.21	26.6 ± 9.4
High Plains	USA	4.5 ± 1.2	0.50	9.0 ± 2.4
Lower Indus	India, Pakistan	4.2 ± 1.5	0.23	18.4 ± 6.5
Nile Delta	Egypt	3.1 ± 0.8	0.10	31.7 ± 7.9
Danube Basin	Hungary, Austria, Romania	2.4 ± 0.8	0.32	7.4 ± 2.6
Central Mexico	Mexico	1.8 ± 0.5	0.20	9.1 ± 2.6
North China Plain	China	1.8 ± 0.6	0.23	7.9 ± 2.8
Northern China	China	1.4 ± 0.6	0.31	4.5 ± 1.8
North Africa	Algeria, Tunisia, Libya	0.9 ± 0.3	0.36	2.6 ± 0.9
Central Valley	USA	0.4 ± 0.2	0.07	6.4 ± 2.4
Other aquifers		38.6 ± 10.8	34.17	1.1 ± 0.3
All aquifers		131.8 ± 24.9	38.27	3.5 ± 0.7

Notes: The values of groundwater footprint (*GF*) and *GF*/A_A are the mean and standard deviation of 10 000 Monte Carlo realizations based on the independent estimates of groundwater recharge and abstraction given by Wada *et al.* (2012). Note that only the 15 aquifers with the largest *GF* are listed individually. The remaining 768 'Other aquifers' are included in 'All aquifers'. *GF*/A_A is calculated before rounding the *GF* to one decimal place. A_A is aquifer area. (*Source:* Gleeson, T., Wada, Y., Bierkens, M.F.P. and van Beek, L.P.H. (2012) Water balance of global aquifers revealed by groundwater footprint. *Nature* **488**, 197–200. © 2012, Springer Nature.)

and Plate 6.1, groundwater footprints of aquifers that are important in irrigated agriculture are often significantly larger than their geographic areas, for example the High Plains, North Arabian, North China Plain, Persian, Upper Ganges and Western Mexico aquifers. Values of *GF*/A_A >1 indicate unsustainable groundwater consumption that could affect groundwater availability and groundwater-dependent ecosystems. Values of *GF*/A_A $\gg 1$ would suggest unsustainable groundwater mining, for example of fossil groundwater recharged under past climatic conditions.

The size of the global groundwater footprint shown in Table 6.1 is $(131.8 \pm 24.9) \times 10^6$ km², or 3.5 ± 0.7 times the actual area of hydrologically active aquifers. The global groundwater footprint is dominated by a handful of

countries including China, India, Iran, Mexico, Pakistan, Saudi Arabia and the United States. However, in an analysis of regional-scale, hydrologically active aquifers, Gleeson *et al.* (2012) showed that the majority of aquifers in the world have groundwater footprints of less than 10^6 km² and that 80% of aquifers have values of *GF*/A_A <1, suggesting that groundwater depletion globally is not ubiquitous.

6.3 Precipitation and evapotranspiration

Near-surface hydrological processes such as precipitation, evapotranspiration and infiltration have a profound influence on streamflow generation and groundwater recharge.

Precipitation, namely in the form of rainfall, provides the raw input of water to a catchment, but its availability for supporting river flows, replenishing aquifer storage and supporting water supplies depends on catchment conditions such as soil type, geology, climate and land use that affect catchment runoff properties. Evapotranspiration is a major component of the hydrological cycle and its accurate estimation is required in applications such as catchment water balance calculations, water resources planning, and irrigation scheduling.

6.3.1 Precipitation measurement

Precipitation falls mainly as rain but may also occur as hail, sleet, snow, fog or dew. The design of the standard rain gauge used in the United Kingdom is shown in Fig. 6.2 with a cylinder diameter of 5 inches (127 mm). In the United States and Canada, standard rain gauges have diameters of 8 and 9 inches (203 and 229 mm), respectively. Standard rain gauges are read daily, for example at 09:00 h in the United Kingdom. In exposed locations, the rain catch of the gauge is affected by high

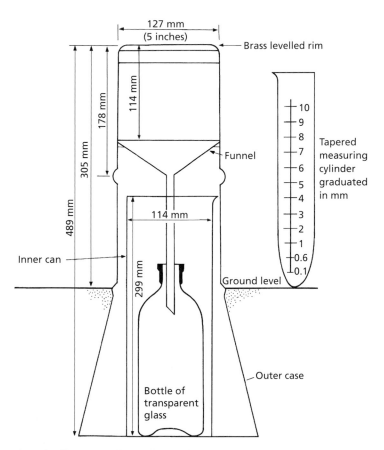

Fig. 6.2 The copper 5-inch standard rain gauge used by the UK Meteorological Office. The rain gauge consists of a 5-inch (127-mm) diameter funnel with a sharp rim, the spout of the funnel being inserted into a glass collecting jar. The jar is in an inner copper can and the two are contained in the main body of the gauge, the lower part of which is sunk into the ground with the rim 12 inches (305 mm) above the surrounding short grass or gravel, this height being chosen so that no rain splashes from the surroundings into the funnel. The funnel has a narrow spout in order to reduce evaporation loss. Normally, the gauge is sited such that its distance from any obstructions (trees, houses, etc.) is at least four times the height of the obstruction.

winds, and it is generally accepted that more accurate results will be obtained from a rain gauge set with its rim at ground level. Although more expensive, with a ground-level installation, it is necessary to house the gauge in a pit and to surround it with an anti-splash grid. The measurement of snowfall is also possible with a standard rain gauge, although subject to error due to turbulence around the rim of the gauge. The snow is caught and melted, and the equivalent amount of water recorded.

Recording gauges (or autographic gauges) are able to automatically measure or weigh precipitation and are useful at remote, rarely visited sites. An example is the tipping-bucket rain gauge in which the number of times a small bucket of known volume fills and tips is recorded. Each tip activates a reed switch which sends an electrical pulse to a logger. The tilting siphon autographic gauge has a chart pen that floats up as rainfall fills a chamber, which tips when full, thus returning the pen to the bottom of the chart. Recording gauges are more expensive and more prone to error (for example, very low rainfall amounts are not recorded) but have the advantage of measuring rainfall intensity as well as rainfall total.

The number of rain gauges required to give a reliable estimate of catchment rainfall increases where rainfall gradients are marked. A minimum density of 1 gauge per 25 km^2 is recommended, considering that large thunderstorm systems may only cover an area of about 20 km^2. In hilly terrain, where orographic effects may cause large and consistent rainfall variations over short distances, higher rain gauge densities are necessary in the first years of measurement (Table 6.2). In tropical areas, there is large spatial variation in daily rainfall, but only a small gradient in annual totals. In such areas, the rain gauge densities in Table 6.2 will be excessive and higher priority should be given to obtaining homogeneous records of long duration at a few reliable sites.

To be able to assess a representative value of rainfall over large areas, it is necessary to

Table 6.2 Density of rain gauges required in a hill area.

Catchment area (km^2)	Number of gauges
4	6
20	10
80	20
160	30

employ a method of averaging the individual gauge measurements. The simplest method is to take the arithmetic mean of the amounts recorded for all rain gauges in an area. If the distribution of points is uniform and rainfall gradients are small, then this method gives acceptable results. Another method is to use Thiessen polygons to define the zone of influence of each rain gauge by drawing lines between adjacent gauges, bisecting the lines with perpendiculars, and then joining the perpendiculars to form a boundary enclosing the gauge (Fig. 6.3a). It is assumed that the area inside the boundary has received an average rainfall amount equal to the enclosed gauge. A variation of this technique is to draw the perpendiculars by bisecting the height difference between adjacent gauges, although this altitude-corrected approach does not produce a greatly different result. As shown in Fig. 6.3b, a further method is to draw contours of equal rainfall depth, or isohyets, with the areas between successive isohyets measured and assigned an average rainfall amount.

A map showing rainfall isohyets for the United Kingdom is given in Fig. 6.4 and illustrates the strong west-east gradient in mean annual rainfall as a result of the prevailing, rain-bearing westerly air flow and orographic and rain shadow effects. A yearly compendium of rainfall, river flows, groundwater levels and river water quality data was formerly published by the Natural Environment Research Council in the series entitled *Hydrological Data UK Yearbook*. Yearbooks were published for 1981–1995 but are now

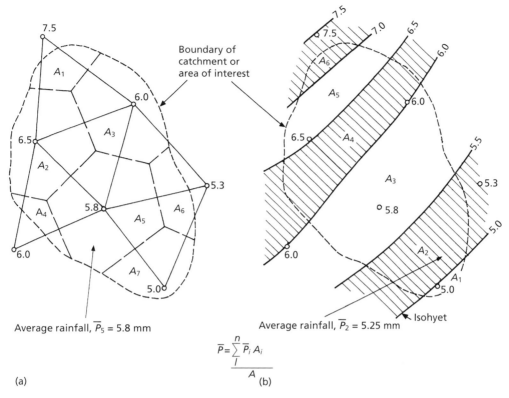

Average rainfall, $\overline{P}_5$ = 5.8 mm

Average rainfall, $\overline{P}_2$ = 5.25 mm

$$\overline{P} = \frac{\sum_{l}^{n} \overline{P}_i \, A_i}{A}$$

(a)

(b)

Fig. 6.3 Calculation of area-weighted average rainfall amount using the methods of (a) Thiessen polygons and (b) isohyets.

superseded by the National Water Archive which is accessed via the Internet at www.ceh.ac.uk/data/NWA.htm.

6.3.2 Evapotranspiration measurement and estimation

In the hydrological cycle, evaporation of water occurs from free water surfaces such as lakes, reservoirs and rivers with the rate dependent on factors such as water temperature and the temperature and humidity of the layer of air above the water surface. Wind speed is also a factor in determining the rate at which vapour is carried away from the water surface. Evaporation can be measured directly using an evaporation tank or pan. Tanks are buried in the ground, whereas pans are set above ground on a small plinth. The British standard tank has dimensions of 1.83 m × 1.83 m and is 0.61 m deep. The United States National Weather Service (NWS) Class A pan, with a diameter of 1.22 m and a depth of 0.25 m, is made of unpainted, galvanized metal and is generally accepted as an international standard. Evaporation tanks tend to give a more accurate measurement of true evaporation, unlike pans that are affected by higher losses of vapour caused by heating of their exposed walls and shallow water. Pan coefficients (correction factors) can be applied to correct measurements to 'true evaporation'. Pan coefficients for the British standard tank and US NWS Class A pan range between 0.93–1.07 and 0.60–0.80, respectively (WMO 1994).

For vegetated surfaces, water is lost by evaporation from bare soil and also by transpiration through the leaf stomata of plants. The term 'evapotranspiration' is used to describe the

A more process-based approach to calculating *PE*, following Penman's method and extended by experimental work, is given by the Penman–Monteith formula (eq. 6.5) that incorporates canopy stomatal and aerodynamic resistance effects, to calculate evapotranspiration rate in mm day^{-1} as follows:

$$ET = \frac{\Delta H + \frac{\rho c_p (e_{sat} + e_{act})}{r_{sfc}}}{\lambda \left\{ \Delta + \frac{\gamma (r_{sfc} + r_{aero})}{r_{aero}} \right\}} \qquad \text{(eq. 6.5)}$$

where further to the symbols applied in eq. 6.3: ρ = density of water; λ = latent heat of vaporization; c_p = specific heat capacity of water; r_{sfc} = surface resistance; and r_{aero} = aerodynamic roughness.

Practical guidance on the application of the Penman–Monteith formula is presented by the Food and Agriculture Organization (FAO) of the United Nations which details a procedure for calculating reference and crop evapotranspiration from meteorological data and crop coefficients (FAO Irrigation and Drainage Paper 56; Allen *et al.* 1998). The procedure, which is intended to provide guidance in computing crop water requirements for both irrigated and rain-fed agriculture, and for computing water consumption by agricultural and natural vegetation, is termed the 'K_c-ET_o' approach. In this approach, the effect of climate on crop water requirements is given by the reference evapotranspiration, ET_o, and the effect of the crop by the crop coefficient, K_c. The approach defines the reference crop as a hypothetical crop with an assumed height of 0.12 m, with a surface resistance of 70 s m^{-1} and an albedo of 0.23, closely resembling the evaporation from an extensive surface of green grass of uniform height, actively growing and adequately watered.

The FAO Penman–Monteith method uses standard climatic data that can be easily measured or derived from commonly measured data. All calculation procedures are standardized according to the available weather data and time scale of computation. In the

'K_c-ET_o' approach, differences in the crop canopy and aerodynamic resistance relative to the reference crop are accounted for within the crop coefficient. The K_c coefficient serves as an aggregation of the physical and physiological differences between crops. Two calculation methods to derive crop evapotranspiration from ET_o are presented. The first approach integrates the relationships between evapotranspiration of the crop and the reference surface into a single K_c coefficient. In the second approach, K_c is split into two factors that separately describe the evaporation (K_e) and transpiration (K_{cb}) components. The selection of the K_c approach depends on the purpose of the calculation and the time step for which the calculations are to be executed. Additional procedures are presented to adjust crop coefficients to account for deviations from standard conditions, such as water and salinity stress, low plant density, environmental factors and management practices. Most of the computations, namely all those required for the reference evapotranspiration and the single crop coefficient approach, can be performed using a spreadsheet, or similar computing procedure, and applying the various tables of values given by Allen *et al.* (1998).

The Penman–Monteith formula (eq. 6.5) is also used as the basis for the national computerized system, MORECS, the UK Meteorological Office Rainfall and Evaporation Calculation System (Thompson *et al.* 1981). MORECS provides an areal-based estimate of evapotranspiration which supplements the Penman approach with simulation of soil–water flux and a consideration of local vegetation cover. The model is based on a two-layer soil and provides estimates of areal precipitation, *P*, potential evapotranspiration, *PE*, actual evapotranspiration, *AE*, soil moisture deficit, *SMD*, and hydrologically effective rainfall (*P–AE–SMD*) for 40 × 40 km grid squares on a weekly basis.

The MORECS operational service for predicting soil moisture has been provided since 1978 but has now reached its limit in providing accurate and real-time soil moisture and runoff information for Numerical Weather Prediction (NWP) models due to its limited resolution of a 40-km grid and daily updates (Smith *et al.* 2006). In the late 1990s, the UK Meteorological Office Surface Exchanges Scheme (MOSES) was developed to calculate the fluxes of water and energy at the land surface and in the root zone of the soil at a range of timescales from seasonal to daily. MOSES has been enhanced by including a surface runoff generation component based on a probability-distributed moisture store scheme (PDM) developed at the UK Centre for Ecology and Hydrology (Essery and Clark 2003). The soil moisture predicted by the new MOSES-PDM model has been integrated in the UK Meteorological Office's Nimrod system (Smith *et al.* 2006), operationally producing a 5-km grid product across the United Kingdom. Although MOSES is used for both climate modelling and numerical weather prediction, the model still needs further validation and improvement with respect to applications (Kong *et al.* 2011), such that higher spatial resolution models such as JULES are expected to operate in the future to improve local scale forecasts. JULES is the UK Joint Land Environment Simulator model that has been developed from MOSES to improve the land phase representation of the land surface energy and water budget. JULES has been applied at a 1-km^2 scale for the United Kingdom and further development of the model should improve the reliability of the soil moisture estimates. In the longer term, it is likely that models like JULES will provide progressively improving model estimates of soil moisture and thermal behaviour where direct measurements are not available (Garcia Gonzalez *et al.* 2012).

A disadvantage of the Penman and Penman–Monteith formulae (eqs 6.3 and 6.5) is that the main weather variables and other variables used in the equations are not always available in records from standard weather stations worldwide, which routinely measure air temperature, T, relative humidity, RH, solar radiation, R_s (or, more frequently, bright sunshine hours, n, from which R_s is indirectly estimated) and wind speed, u. Furthermore, the complexity of calculations increases given that each of the parameters can be expressed by a variety of units. In an attempt to simplify the Penman formula, Valiantzas (2006) presented an algebraic formula equivalent in accuracy to the Penman formula using the variables T (°C found as $(T_{max} + T_{min})/2$), R_s (MJ m^{-2} day^{-1} where 1 MJ m^{-2} day^{-1} = 11.57 W m^{-2}), RH (%), u (m s^{-1}) and site elevation, z (m). The only additional parameter appearing in the simplified formula is the extraterrestrial radiation, R_A, which can also be calculated using a simplified expression. Valiantzas (2006) presented the following formula for estimating reference crop (grass) evapotranspiration, ET:

$$ET \approx 0.051(1-\alpha)R_S\sqrt{T+9.5} - 2.4\left(\frac{R_S}{R_A}\right)^2$$
$$+ 0.048(T+20)\left(1-\frac{RH}{100}\right)(0.5+0.536u)$$
$$+ 0.000127z$$

$$(\text{eq. 6.6})$$

with the values of R_A (MJ m^2 day^{-1}) and N, the length of daylight hours, calculated from the following empirical equations (Valiantzas 2006):

$$N \approx 4\varphi \sin(0.53i - 1.65) + 12 \qquad (\text{eq. 6.7})$$

$$R_A \approx 3N \sin(0.131N - 0.95\varphi) \quad \text{for}$$
$$|\varphi| > \frac{23.5\pi}{180}$$

$$(\text{eq. 6.8a})$$

$$R_A \approx 118N^{0.2} \sin(0.131N - 0.2\varphi) \quad \text{for}$$
$$|\varphi| < \frac{23.5\pi}{180}$$

$$(\text{eq. 6.8b})$$

where φ is latitude of the site (radians), positive for the northern hemisphere, negative for the southern, and i is rank of the month (first month is January). Equation 6.8a is valid for the temperate zone, whereas eq. 6.8b is valid for the tropical zone. The applied value for albedo, α, used in eq. 6.6 is equal to 0.25. For places where wind speed data are not available or reliable, then a value of u equal to $2\,\text{m s}^{-1}$ can be substituted (the global average wind speed). The approximate formula (eq. 6.6) results in relatively good estimates of evapotranspiration compared with the Penman formula, although tending to overestimate ET.

In comparison to the above methods, satellite remote sensing (SRS) is the only method capable of providing global coverage of environmental variables at economically affordable cost. A major challenge in the use of SRS for monitoring ET at regional and global scales is that the phase change of water molecules produces neither emission nor absorption of an electromagnetic signal. Hence, ET is not directly quantifiable from satellite observations. Complex, deterministic models based on Soil-Vegetation-Atmosphere Transfer (SVAT) modules are employed to compute the different components of the energy budget. An example is the MODIS (Moderate Resolution Imaging Spectroradiometer sensor onboard the Aqua and Terra satellites) ET algorithm based on the Penman–Monteith equation. The algorithm incorporates surface stomatal resistance and vegetation information derived from MODIS land products to estimate daily and potential ET. The MODIS global ET, latent heat flux and potential ET datasets have a 0.5 km spatial resolution for a 109.03×10^6 km^2 global vegetated land area at 8-day, monthly and annual intervals accessible in ArcMap via the MODIS toolbox. For more details of the MODIS Global Evapotranspiration Product (MOD16), the reader is referred to Running *et al.* (2017).

6.4 Soil water and infiltration

Understanding soil water distribution, storage and movement is important in hydrology in predicting when flooding will occur and also in irrigation scheduling. In hydrogeology, understanding infiltration of water in the unsaturated zone is a necessary prerequisite to quantifying groundwater recharge to the water table. The branch of hydrology dealing with soil water and infiltration is studied in detail by soil physicists and suggested further reading in this topic is provided at the end of this chapter.

6.4.1 Soil moisture content and soil water potential

Water contained in the soil zone is held as a thin film of water adsorbed to soil grains and also as capillary water occupying the smaller pore spaces (Fig. 6.5). The main forces responsible for holding water in the soil are those of capillarity, adsorption and osmosis. Adsorption is mainly due to electrostatic forces in which the polar water molecules are attracted to the charged surfaces of soil particles. Osmosis is often ignored but acts to retain water in the soil as a result of osmotic pressure due to solutes in the soil water. This occurs particularly where there is a difference in solute concentration across a permeable membrane such as the surface of a plant root, making water less available to plants, especially in saline soils, and is of importance when considering irrigation water quality (Section 8.2.2).

Capillary forces result from surface tension at the interface between the soil air and soil water. Molecules in the liquid are attracted

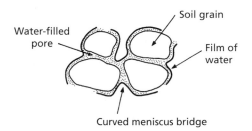

Fig. 6.5 Sketch of the occurrence of water within unsaturated material showing both soil grains coated with a film of adsorbed water and soil pores filled with capillary water.

more to each other than to the water vapour molecules in the air, resulting in a tendency for the liquid surface to contract. This effect creates a greater fluid pressure on the concave (air) side of the interface than the convex (water) side such that a negative pressure head, indicated by $-\psi$ (in centimetres or metres head of water), develops relative to atmospheric pressure. The smaller the neck of the pore space, the smaller the radius of curvature and the more negative the pressure head. In soil physics, the negative pressure is often termed the suction head or tension head and describes the suction required to obtain water from unsaturated material such as soils and rocks of the unsaturated zone. With increasing moisture content, the larger pore spaces become saturated and the radius of curvature of the menisci increases creating a more positive pressure head (i.e. the suction head is reduced). Close to the water table, the pore space is fully saturated, but the pressure head is still negative as a result of water being drawn up above the water table by the capillary effect (Fig. 2.20). Hence, the measurement of soil moisture content and pressure head and the understanding of their inter-relationship are important in the understanding of soil water movement.

The measurement of soil moisture content, θ, can be undertaken in the laboratory by gravimetric determination or in the field with a neutron probe. The gravimetric method requires a known volume, V_t, of soil to be removed and the total mass, m_t, found. The sample is then oven-dried at 105°C until the final mass of the dried sample, m_s, is constant. The difference in mass, $m_t - m_s$, before and after drying is equal to the mass of water originally contained in the sample, m_w. The water content of the soil by mass is then equal to:

$$\theta_m = \frac{m_w}{m_s} \qquad \text{(eq. 6.9)}$$

By calculating the bulk density of the soil sample, ρ_b, from m_s/V_t, it is then possible to express the water content by volume, as:

$$\theta = \frac{V_w}{V_t} = \theta_m \frac{\rho_b}{\rho_w} \qquad \text{(eq. 6.10)}$$

where ρ_w is the density of water.

A disadvantage of the gravimetric determination is that the sample is destroyed providing only a one-off measurement. An alternative method is to use a neutron probe to give a direct measurement of soil moisture content. As shown in Fig. 6.6a, a radioactive source (for example Am-Be) is lowered into an augered hole and the impedance of emitted, fast neutrons, caused by impact with hydrogen nuclei contained in the soil water, is

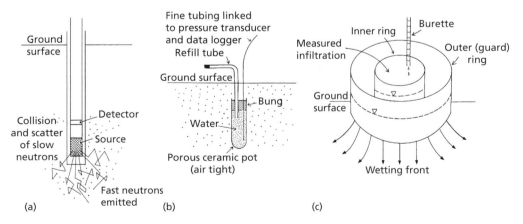

Fig. 6.6 Methods for the measurement of the physical properties of soils: (a) a neutron probe for measurement of soil moisture content; (b) a porous pot tensiometer for measurement of pressure head (suction head or tension head); and (c) a double-ring infiltrometer for measurement of infiltration rate.

determined from the scatter of slow neurons measured by a detector. By calibrating the density of scattered slow neutrons with measurements of soil moisture content made by gravimetric determination, it is then possible to make repeated measurements of soil moisture content at various depths in the access tube. The neutron probe is calibrated for each soil type investigated, with each field site having its own relationship of probe reading versus soil moisture content, and used to make regular measurements of vertical profiles of water content.

Newer techniques for measurement of soil water content that do not require destruction of the sample or the use of hazardous radioactive material have been developed that take advantage of the high dielectric constant of water to quantify the volumetric water content. The most common instruments use the capacitance technique or time domain reflectometry (TDR) method, although the precision and accuracy of these instruments vary in that their calibration may change with soil type, electrical conductivity (for example, high salt content) and temperature (Topp and Davis 1985).

Instruments that use the capacitance technique rely on the high dielectric constant of water (80) relative to mineral soil (3–5) and air (1), because water in the soil influences the propagation of an electrical signal through the soil medium. Empirical relationships between soil water content and the change in electrical signal are used to estimate θ. Instruments available for the measurement of θ include the TDR, which determines the apparent soil dielectric constant, K_a, by measuring the time required for an electromagnetic pulse to travel and up and down a pair of metal transmission lines (wave guides) of fixed length, and the frequency domain reflectometer (FDR) which propagates an electromagnetic pulse along two parallel rods and measures the oscillating frequency of the pulse as affected by the capacitance of the soil in order to predict K_a (Czarnomski *et al.* 2005). A field example of a soil moisture probe that uses the capacitance technique is shown in Fig. 6.7.

The field measurement of soil water pressure (the amount of suction) is made with a tensiometer. One design is shown in Fig. 6.6b and consists of a ceramic pot filled with deionized

Fig. 6.7 Field soil moisture monitoring station with an installed capacitance (FDR) soil moisture probe (next to flag at right) and associated solar-powered, radio-controlled telemetry system (fixed to pole at right).

water. The pressure within the pot equalizes with the fluid pressure in the surrounding soil and is measured via a hydraulic link to a manometer or pressure transducer and data logger. Porous pots operate best in relatively wet conditions up to suctions of about 0.85 bar. In drier conditions, there is a danger of air entering the pot and affecting pressure transmission. Porous pots also work under positive fluid pressure, when a vertical bank of tensiometers can be used to monitor the position of the water table surface (Jones 1997). To measure suctions above the limit of a tensiometer (about 0.85 bar), an electrical resistance block is used. A porous block of gypsum, nylon or fibreglass mesh containing two electrodes is situated in the ground in close contact with the soil and hydraulic equilibrium established between the block and soil. The water content of the block is calibrated against the electrical resistance between the two electrodes to provide a means of measurement of the soil moisture content. Disadvantages of the electrical resistance block are that it has a long response time for water to seep through the block and

reach hydraulic equilibrium and measurements can be affected by dissolved salts in the soil water. The technique is less reliable at suctions of <4 m but does have the advantage of operating at high suctions.

The relationship between soil moisture content and pressure head is conveniently displayed by the presentation of soil characteristic curves or retention curves. Also, given that the degree of interconnection between saturated pore spaces will affect the hydraulic conductivity, K, of the soil, characteristic curves can also be constructed to show hydraulic conductivity as a function of pressure head. Single-value characteristic curves for hypothetical uniform sand, silty sand and silty clay soils are shown in Fig. 6.8 and demonstrate the different behaviour of the soil types with decreasing pressure head. When the soil is completely saturated, the maximum volumetric moisture content, θ, is equal to the soil porosity such that eq. 6.10 equates to eq. 2.1. As the soil dries, the uniform sand with its larger diameter pore space is quickly drained and both the water content and hydraulic conductivity decrease

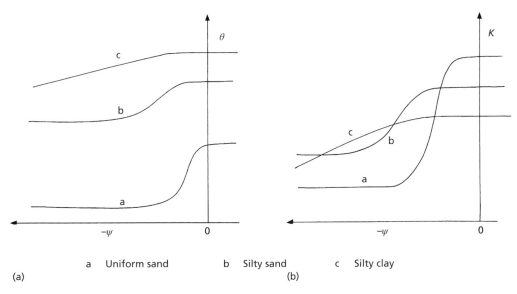

| a | Uniform sand | b | Silty sand | c | Silty clay |

(a)   (b)

Fig. 6.8 Single-value characteristic curves showing the relationship of (a) volumetric moisture content, θ, and (b) hydraulic conductivity, K, against pressure head, ψ, for three hypothetical soils. Values of pressure head less than zero indicate unsaturated soil conditions (Freeze and Cherry 1979). (*Source:* Freeze, R.A. and Cherry, J.A. (1979) *Groundwater*. Prentice-Hall, Inc., Englewood Cliffs, New Jersey. © 1979, Pearson Education)

rapidly. On the other hand, the larger clay content of the silty sand and silty clay creates a larger porosity as a result of the smaller grain size distribution. Because of the smaller pore diameter, the water is held in the silty soils for longer until a greater suction creating an increasingly negative pressure head is able to drain the pore water. As a result of the finer soil texture, the silty clay retains both a higher moisture content and hydraulic conductivity compared with the uniform sand under dry soil conditions. The slope of the line representing the soil moisture curve, $d\theta/d\psi$, is referred to as the specific moisture capacity, C, and is a measure of the storage behaviour of the unsaturated zone.

The process of drying a soil causes the soil grains to compact and alters the soil structure through shrinkage and air entrapment. On wetting, the soil swells as water is added. Consequently, the resulting characteristic curves for the drying and wetting phases are different and a hysteretic effect is observed as shown in Fig. 6.9. Further important causes of hysteresis are the 'ink-bottle' effect and the 'contact angle' effect. As explained by Ward and Robinson (2000), the 'ink bottle' effect results from the fact that a larger suction is necessary to enable air to enter the narrow pore neck, and hence drain the pore, than is necessary during wetting. The 'contact angle' effect results from the fact that the contact angle of fluid interfaces on the soil particles tends to be greater when the interface is advancing during wetting than when it is receding during drying, such that a given water content tends to be associated with a greater suction in drying than in wetting.

At the start of the drying process, the soil is tension-saturated, a condition analogous to a capillary fringe above a water table, until the air-entry or bubbling pressure is reached, at which point the soil begins to drain. For soils that are only partially dried, then wetted, and vice versa, the soil characteristic curves follow the scanning curves shown in Fig. 6.9.

From the above explanation of characteristic curves, it is clear that moisture content, specific moisture capacity and hydraulic conductivity in soils and unsaturated rocks and sediments are themselves a function of pressure head. Expressed in mathematical notation, $\theta = \theta(\psi)$, $C = C(\psi)$ and $K = K(\psi)$. It also follows that $K = K(\theta)$. As a result, it is recognized that flow in the unsaturated zone will vary according to the value of hydraulic conductivity at a given soil moisture content. In general, and by considering one-dimensional flow, the quantity of discharge, Q, across a sectional area, A, is found from Darcy's Law applied to the unsaturated zone:

$$Q = -AK(\psi)\frac{dh}{dx} \qquad \text{(eq. 6.11)}$$

where h is the fluid potential or soil water potential. As discussed in Section 2.7, fluid potential is the work done in moving a unit mass of fluid from the standard state to a point in a flow system. Ignoring osmotic potential, the total soil water potential at a given point comprises the sum of the gravitational (or elevation) potential, ψ_g, and pressure potential, ψ_p, as follows:

$$\Phi = \psi_g + \psi_p \qquad \text{(eq. 6.12)}$$

By applying gravitational acceleration, eq. 6.12 is identical to eq. 2.22, emphasizing that the gradient of potential energy for

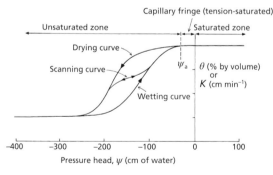

Fig. 6.9 Soil characteristic curves showing the variation of volumetric moisture content, θ, and hydraulic conductivity, K, with pressure head, ψ. The unequal variation of the drying and wetting curves is caused by the effects on soil structure of the drying (soil shrinkage) and wetting (soil swelling) processes. ψ_a is the air-entry or bubbling pressure.

Table 6.3 Values of soil water potential ($\Phi = \psi_g + \psi_p$) relative to a ground surface datum and volumetric moisture content (θ) for a representative clay soil measured on two dates in August and used to plot Fig. 6.10. Calculated evaporation and drainage losses during this 7-day period are given in Table 6.4.

Gravitational potential, ψ_g (cm)	Pressure potential, ψ_p (cm of water)		Soil water potential, Φ (cm of water)		Volumetric moisture content, θ	
	1 August	8 August	1 August	8 August	1 August	8 August
−5	−2493	−3399	−2498	−3404	0.356	0.349
−15	−1124	−1834	−1139	−1849	0.375	0.363
−25	−521	−1011	−546	−1036	0.394	0.378
−35	−197	−618	−232	−653	0.418	0.390
−45	−122	−608	−167	−653	0.428	0.390
−55	−219	−598	−274	−653	0.413	0.390
−70	−388	−645	−458	−715	0.399	0.387
−90	−694	−946	−784	−1036	0.385	0.378
−110	−833	−1029	−943	−1139	0.380	0.375

subsurface water is continuous throughout the full depth of the unsaturated and saturated zones. In studies of the unsaturated zone, it is common to use the ground surface as the datum level for soil water potential values. As shown in Table 6.3, gravitational potential declines uniformly with depth below the ground surface and is negative when referred to a ground surface value of zero.

6.4.2 Calculation of drainage and evaporation losses

From the above explanation of soil water potential, it follows that water will move from a point where the total potential energy is high to one where it is lower. Hence, by plotting a profile of soil water potential it is possible to identify the direction of water movement in the unsaturated zone. As shown in Fig. 6.10b, there is one level in the unsaturated zone where there is no potential gradient and, there-fore, no vertical soil water movement. At this level, known as the zero flux plane (ZFP), the soil profile is divided into a zone with an upward flux of water above the ZFP and a zone of downward flux below this level. Such a

divergent ZFP initially develops at the soil sur-face, as a result of evaporation exceeding rain-fall, and moves downwards into the soil during warm weather as the profile dries out, stabiliz-ing at a depth, typically between 1 and 6 m below ground level, depending on climate and soil conditions and the depth of the water table (Wellings and Bell 1982). If the dry period is followed by a wet period, then a convergent ZFP develops at the ground surface, moving down the soil profile until it reaches the origi-nal, divergent ZFP, at which point both ZFPs disappear and downward drainage of the soil water can then take place throughout the soil profile (Ward and Robinson 2000).

By combining measurements of soil moisture content and soil water potential, and adopting a soil water balance approach, it is possible to use these data to quantify both the amounts of deep drainage downwards to the water table and also upward flux to the ground surface due to evapotranspiration. This approach is demon-strated in Table 6.4 using the data contained in Table 6.3 and plotted in Fig. 6.10. In this exam-ple, the profiles of volumetric moisture content show that between the two measurements dates (1 and 8 August), the soil has become

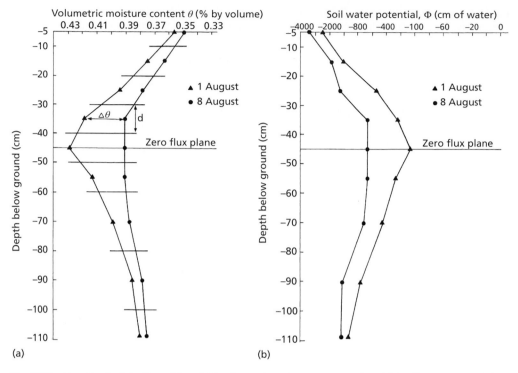

Fig. 6.10 Hypothetical depth profiles of (a) volumetric moisture content, θ, and (b) soil water potential, Φ, for a representative clay soil measured on two dates in August. Notice the ZFP developed at −45 cm relative to ground surface.

Table 6.4 Evaporation and drainage losses during a 7-day period for a representative clay soil calculated using the soil water data given in Table 6.3 and with reference to the ZFP identified in Fig. 6.10.

Depth range (cm)	d (mm)	$\Delta\theta$	$\Delta\theta \times d$ (mm)	Flux OUT = Flux IN + [$\Delta\theta \times d$] (mm)
Evaporation losses				
40–45	50	0.036	1.80	1.80 = 0 + 1.80
30–40	100	0.028	2.85	4.65 = 1.80 + 2.85
20–30	100	0.016	1.67	6.32 = 4.65 + 1.67
10–20	100	0.012	1.19	7.51 = 6.32 + 1.19
0–10	100	0.007	0.71	8.22 = 7.51 + 0.71
				Total evaporation losses = 8.22 mm
Drainage losses				
45–50	50	0.035	1.75	1.75 = 0 + 1.75
50–60	100	0.023	2.37	4.12 = 1.75 + 2.37
60–80	200	0.012	2.38	6.50 = 4.12 + 2.38
80–100	200	0.007	1.44	7.94 = 6.50 + 1.44
100–120	200	0.005	0.94	8.88 = 7.94 + 0.94
				Total drainage losses = 8.88 mm

drier (Fig. 6.10a) and that a ZFP, identified by the profiles of soil water potential, has descended to a depth of 45 cm below ground level (Fig. 6.10b). By dividing the profiles of volumetric moisture content into convenient depth intervals, d, an approximate amount of water lost during the 7-day period across a single depth interval is $d \times \Delta\theta$. By starting at the level of the ZFP, the total amount of water draining or evaporating from the soil can be calculated by accumulating the individual amounts of water lost from each separate depth interval. In the example calculation shown in Table 6.4, by the end of the 7-day period, the total evaporation and drainage losses are approximately 8.22 and 8.88 mm, respectively, expressed as a depth of water for a unit area of soil surface.

6.4.3 Infiltration theory and measurement

Water in the soil zone is generally replenished by precipitation or surface runoff at the ground surface. The process by which water enters the soil is known as infiltration and can be defined as the entry into the soil of water made available at the ground surface, together with the associated flow away from the ground surface within the unsaturated zone (Freeze and Cherry 1979). The infiltration rate of a soil will depend on the initial soil moisture condition and the physical properties of the soil. For sandy, open-textured soils infiltration will be rapid, whereas for fine-textured, low-permeability clay soil infiltration is limited. Soil structure is also important with the type of vegetated surface, type of cultivation, whether compacted by machinery or poached by cattle, as well as location with respect to proximity to a stream or on a hillside, all affecting the soil's ability to accept infiltrating water.

The paper by Horton (1933) is celebrated in the hydrological literature in presenting the original concept that storm runoff is primarily a result of overland flow generated by an excess of rainfall over the infiltration capacity of the soil. Hortonian overland flow, or infiltration excess flow, occurs when the soil is saturated from above by infiltration of rainwater. Rain that falls in excess of the infiltration rate will runoff as overland flow. The Horton model of infiltration is most likely to apply to conditions of compact soil or sparse vegetation cover, especially in arid and semi-arid environments where a crust develops at or near the surface.

The Horton overland flow model provides a method of calculating infiltration using the Horton equation:

$$f_t = f_c + (f_0 - f_c)e^{-kt} \qquad \text{(eq. 6.13)}$$

where f_t is the infiltration rate at time, t; f_0 is the initial infiltration rate or maximum infiltration rate; f_c is a constant or equilibrium infiltration rate after the soil has been saturated or the minimum infiltration rate; and k is a decay constant specific to the soil. The total volume of infiltration, F, after time, t, is found from:

$$F_t = f_c t + \frac{(f_0 - f_c)}{k}\left(1 - e^{-kt}\right) \qquad \text{(eq. 6.14)}$$

The Horton equation assumes that infiltration starts at a constant rate, f_0, and decreases exponentially with time. The model assumes that the rainfall intensity exceeds the infiltration rate. The empirical relationship assumes an infinite water supply at the surface with saturation conditions at the soil surface, such that there must be ponding at the surface and a reduction in infiltration rate with time. After some time when the soil saturation level reaches a certain value, the rate of infiltration will level off as represented graphically by the Horton curve (Fig. 6.11). The Horton curve shows that infiltration rate, f_t, declines rapidly at the start of a rainfall event as the soil pores fill with water, eventually reaching a constant value once the soil has become completely saturated at which point the soil is said to have reached its infiltration capacity, f_c. This constant infiltration rate is equivalent to the saturated hydraulic conductivity of the soil at field capacity (the volumetric moisture content of

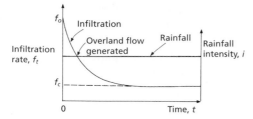

Fig. 6.11 Schematic diagram of the Horton curve showing a decrease in infiltration rate, f_t (in mm minute^{-1}) with time during a rainfall event with intensity, i (in mm minute^{-1}). As the rain continues the soil becomes saturated and the soil reaches its infiltration capacity, f_c, equivalent to the saturated hydraulic conductivity. At the point where the rainfall intensity equals the infiltration rate, ponding of water occurs at the ground surface and overland flow is generated.

the soil after the saturated soil has drained to an equilibrium under the influence of gravity). As also shown in Fig. 6.11, if the rainfall event continues at an intensity greater than the ability of the soil to accept infiltration, then surface ponding of rainfall will occur potentially leading to the generation of overland flow and the possibility of sheet or gully erosion.

A challenge to the *status quo* concept of infiltration excess overland flow was introduced by Hewlett and Hibbert (1967). Hewlett and Hibbert (1967) presented the concept of saturation overland flow, or saturation excess flow, that occurs at locations where the soil becomes saturated as a result of a rise in the groundwater table from below. The Hewlett model of overland flow generation is typical of conditions in stream corridors where the water table is close to the ground surface. As Fig. 6.12 shows, as a precipitation event progresses, the catchment area contributing to the discharge at the catchment outlet extends progressively upstream and uphill from the outlet. When precipitation ceases, the contributing area shrinks again. The actual dynamic development of the contributing area depends on the distribution and intensity of precipitation and antecedent soil moisture conditions and is referred to as the variable source area (VSA)

concept. The VSA concept is suited to the description of the infiltration process for vegetated, humid-temperate river basins and is the basis of many topographically based rainfall–runoff models (McDonnell 2009).

The field measurement of infiltration rate is made using an infiltrometer ring. In the example of a double infiltrometer (Fig. 6.6c), at the start of the experiment water is added to the inner ring, either from a graduated burette, volumetric flask or constant head device, followed by further, measured additions of water to restore the water to a constant level at regular time intervals. The infiltration rate for each time interval is then calculated and measurements continued until a constant rate is achieved. The outer ring is also flooded with water, with the water draining from this ring intended to prevent lateral seepage from beneath the inner ring which, if it occurred, would lead to an over-estimation of infiltration rate. Other complicating factors, such as soil heterogeneity, soil swelling and shrinkage and soil aggregation, which limit the interpretation of field measurements based on simple soil physical theory, are discussed by Youngs (1991).

The classical description and analysis of the infiltration process and the uptake of water in dry soils is contained in work by Philip (1969) and Rubin (1966), respectively, with the analytical solution to one-dimensional, vertical infiltration obtained by differentiation of the Philip infiltration equation to give, for large values of time, t:

$$f_t = \frac{1}{2} S t^{-1/2} + K_s \qquad \text{(eq. 6.15)}$$

where f_t is infiltration rate, S is sorptivity and K_s is saturated hydraulic conductivity. The first term on the right-hand side describes the temporal pattern of the absorption of water resulting from the sudden application of water at the surface of a homogeneous soil at time $t = 0$. The second term describes the steady conduction of flow under a potential gradient between the flooded surface and the drier soil below. For

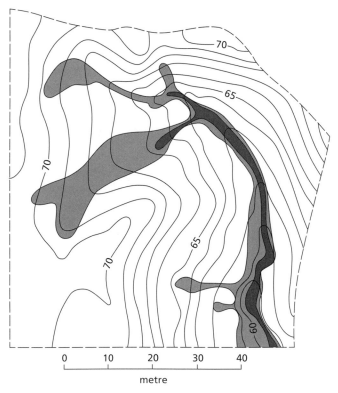

0   10   20   30   40

metre

Contour interval 1 m

Fig. 6.12 Map showing expansion of saturated areas during a single rainstorm demonstrating the variable source area concept (Hewlett model) of runoff generation. The solid black area shows the area of ground saturation at the beginning of the rain event and the light shaded area the expanded saturated area at the end of the event, representing the area over which the water table has risen to the ground surface.

further discussion of the theory of infiltration, the reader is referred to Smith (2002).

The classical explanation of infiltration envisages a front of infiltrating water moving down through the soil profile leading to a gradual increase in soil moisture content and pressure head. An inverted water table is conceived to propagate downwards leading to a rise of the true water table in response to the moisture infiltrating from above. At this point, recharge of the groundwater below the water table occurs. The chance of infiltration becoming recharge is greater if a number of conditions are met, namely: low-intensity rainfall of long duration; a shallow water table; wet antecedent moisture conditions; and for soils whose characteristic curves exhibit a high hydraulic conductivity, K, and low-specific moisture capacity, C, the ability to maintain a high moisture content over a wide range of values of pressure head, ψ. Although the calculation of the amount of recharge from infiltration measurements is possible, other methods that are less time-intensive and more applicable at a catchment scale are available (Section 6.5).

In reality, the infiltration process is affected by soil heterogeneities. Compared with the classical description of a homogeneous infiltration front moving downwards through the soil profile by a piston-flow type displacement, in reality macropores creating high permeability pathways, such as drying cracks, rootlet channels and worms burrows (Fig. 6.13a), can short-circuit the flow of water both vertically and

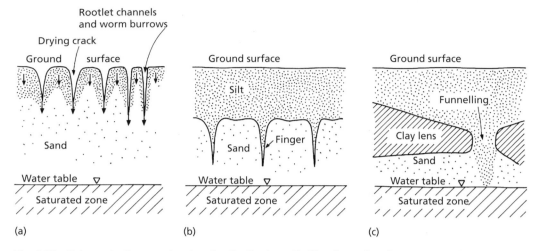

Fig. 6.13 Schematic diagram showing the distribution of infiltrating soil moisture content (densely stippled area) and nature of preferential flow of water in the vadose zone above a water table caused by (a) macropores; (b) instability due to a permeability contrast creating fingering of flow; and (c) a stratified soil profile producing funnelling.

horizontally leading to preferential flow (Beven and Germann 1982). Equally, a layered heterogeneity with a lower permeability layer overlying a higher permeability layer can cause the flow to concentrate in downward-reaching fingers (Fig. 6.13b) as a result of flow instability at the boundary between the two layers (Hillel and Baker 1988). A further form of preferential flow is reported by Kung (1990a,b) who observed funnelling in the vadose zone of the Central Sand Area of Wisconsin. Funnelled flow occurs where coarse sand layers or densely packed fine layers present in an interbedded soil profile behave like the wall of a funnel which can concentrate an initially unsaturated flow into irregularly spaced columns (Fig. 6.13c). Instead of penetrating through the funnel walls, water will flow laterally on top of these layers, eventually becoming a more concentrated columnar flow after passing the lower edges of these layers.

Apart from affecting the infiltration mechanism in the unsaturated zone, soil heterogeneities such as macropores, fingering and funnelling are also significant in transporting dissolved solutes or contaminants. Hence, soil heterogeneity can increase the risk of groundwater pollution, but the resulting uneven distribution of contaminant mass in the soil profile makes the location of such groundwater pollution difficult to predict.

6.5 Recharge estimation

Understanding the balance between recharge input to and discharge output from aquifers forms the basis of groundwater resources management. Groundwater recharge is the amount of surface water that reaches the permanent water table either by direct contact in the riparian zone or by downward percolation through the unsaturated zone, and is largely controlled by climate, soils, geology, topography, hydrology, vegetation and land use. Groundwater recharge is the quantity which, in the long term, is available for both abstraction and supporting the baseflow component of rivers (Rushton and Ward 1979). The calculation of effective rainfall, the amount of rainfall remaining after evapotranspiration, and the partitioning of this hydrological excess water between surface water and groundwater is an important consideration in a catchment water balance (Section 6.2).

Although recharge is one of the most important components in groundwater resources management, it is one of the least constrained, largely because recharge rates vary greatly in space and time and are difficult to measure directly (Moeck *et al.* 2018). At the catchment scale, recharge cannot be measured experimentally, with direct measurement of recharge only possible at plot scale, for example using lysimeters that permit direct measurement of evapotranspiration, soil moisture and seepage through the unsaturated zone. Direct measurements typically result in a relatively small number of measurements (Scanlon *et al.* 2002; von Freyberg *et al.* 2015). From a literature search, Moeck *et al.* (2020) compiled a global dataset of >5000 plot-scale groundwater recharge values for application in the study of recharge processes. In order to estimate groundwater recharge at catchment to global scales, then large-scale numerical models are needed to compensate for the lack of observations and fill spatial gaps (Döll and Zhang 2010; Wada *et al.* 2016; de Graaf *et al.* 2017).

The paper by Lerner *et al.* (1990) discusses in detail methods for estimating direct recharge to the water table and indirect recharge, where recharge occurs via fractures and fissures in hard rock or limestone terrain, as localized infiltration below water-filled surface depressions, or as lateral runoff to an aquifer at the edge of a confining layer. Later papers reviewing methods for groundwater recharge estimation include Gee and Hillel (1988), Petheram *et al.* (2002), Scanlon *et al.* (2006) and Kim and Jackson (2012), and for evapotranspiration processes, particularly in relation to remote sensing and uncertainty analysis, by Kalma *et al.* (2008), Glenn *et al.* (2011) and Doble and Crosbie (2017). In their review, Doble and Crosbie (2017) considered methods used to incorporate recharge and evapotranspiration as boundary conditions and as outputs from catchment-scale groundwater models, with an emphasis on shallow groundwater.

The available methods for calculating direct recharge, four of which are discussed in the following sections, can be classified as: direct measurement using lysimeters over areas up to $100\,\text{m}^2$ (for further details, see Kitching *et al.* 1980; Kitching and Shearer 1982); Darcian approaches to calculate flow in the unsaturated zone above the water table (eq. 6.11); borehole and stream hydrograph analysis; empirical methods that simplify recharge as a function of rainfall amount; soil water budget methods either at a field (Section 6.4.2) or catchment scale; and application of environmental or applied tracers to image the saturated movement of water in the unsaturated zone (Fig. 5.14a).

6.5.1 Borehole hydrograph method

The borehole hydrograph method, in conjunction with stream hydrograph separation, provides a convenient means of calculating the partitioning of effective rainfall between surface water runoff and groundwater discharge during a recharge season. The method is based on the premise that a rise in groundwater level in an unconfined aquifer is due to recharge water arriving at the water table. No assumptions are made pertaining to the mechanism by which water travels through the unsaturated zone. Hence, the presence of preferential pathways within the unsaturated zone does not restrict application of this method. The measurement of water level in an observation borehole represents an area of at least several square metres, and so the borehole hydrograph method is considered to be an integrated approach and less a point measurement compared with methods based on data from the unsaturated zone (Healy and Cook 2002). Fluctuations in borehole hydrographs represent changes in aquifer storage and, as shown in Fig. 6.14, multiplication of the amplitude of water level change, Δh, by the aquifer storage coefficient provides a value for the net recharge. The total recharge is equal to the addition of net recharge and groundwater

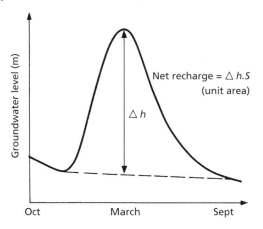

Fig. 6.14 Borehole hydrograph record representing the change in aquifer storage during a single recharge period. Over a long time interval (season or annual) the borehole hydrograph method produces an estimate of change in subsurface recharge, termed the net recharge. The net recharge is equal to the product of the amplitude of groundwater level rise (Δh) and the aquifer storage coefficient, S, or specific yield, S_y, for an unconfined aquifer.

outflows (baseflow, found by hydrograph separation (Section 6.7.1) and spring flow). The method is best applied to shallow water tables that display a sharp rise and fall in water level and is useful in the preparation of a preliminary catchment water balance or in support of regional groundwater flow modelling. The method is limited by the need for a good distribution of observation boreholes in the catchment of interest to allow for differences in elevation, geology, land surface slope, vegetation and other factors (Healy and Cook 2002).

6.5.2 Soil moisture budget method

The conventional method of estimating recharge using a soil moisture budgeting approach is based on the studies of Penman and Grindley (Penman 1948, 1949; Grindley 1967, 1969). The method is conceptually simple. Water is held in a soil moisture store, precipitation adds to the store and evapotranspiration depletes it. When full, the conceptual quantity of soil moisture deficit (*SMD*), a

measure of the amount of water required to return the soil to field capacity, is zero and surplus precipitation (the hydrological excess, *HXS*) is routed to surface water and groundwater as recharge. The most difficult aspect is to calculate actual evapotranspiration (*AE*). In general, the potential evapotranspiration (*PE*) is first defined as the maximum rate of evapotranspiration under prevailing meteorological conditions over short-rooted vegetation with a limitless water supply. A budgeting procedure is used to convert *PE* to *AE* with the degree to which potential and actual evapotranspiration rates diverge being controlled by a root constant (*RC*), a function of soil and vegetation characteristics and a measure of readily available water within the root range. Representative values of *RC* expressed as an equivalent rainfall are given in Table 6.5.

The extent to which *PE* and *AE* diverge is a matter of debate, with various models having been proposed to represent the reduction in plant transpiration with decreasing soil moisture content. An example of a drying curve is shown in Fig. 6.15, illustrating the decline in *AE* as plant wilting occurs until finally die-off is reached. The complexity of the wilting process is simply represented as a single-step function in Fig. 6.15 with the *AE* equal to the full *PE* until a *SMD* equal to *RC* + 0.33*RC* is reached, at which point the *AE* decreases drastically to one-tenth of the *PE* rate. No further decline in *AE* is shown until die-off occurs. By adopting a daily, weekly or monthly budgeting approach, it is possible to account for *AE* by following the development of *SMD* during the drying phase. An example of the method is shown in Table 6.6 with the calculation started on 1 April when it is assumed that the *SMD* = 0 and the soil is at field capacity. The effect of introducing the step function, shown in Fig. 6.15, is to reduce the rate at which the *SMD* develops with a cap effectively reached for values above a *SMD* = 100 mm. No further reduction in *SMD* is expected at the point of die-off. At the end of the accounting procedure, and in

Table 6.5 Monthly root constant (*RC*) values for the Penman–Grindley method. Values in mm (Grindley 1969).

Month	Crop type														
	1	2	3	4	5	6	7	8	9	10	11	12	13	14
Jan and Feb	25	25	25	25	25	25	25	25	25	25	56	76	13	203
Mar	56	56	56	25	25	56	25	25	25	25	56	76	13	203
Apr	76	76	76	76	56	56	56	25	25	25	56	76	13	203
May	97	97	97	56	56	56	56	56	25	25	56	76	13	203
Jun and Jul	140	140	140	76	76	25	56	56	56	25	56	76	13	203
Aug	140	140	25	97	97	25	25	56	56	25	56	76	13	203
Sept	140	25	25	97	25	25	25	25	56	25	56	76	13	203
Oct	25	25	25	97	25	25	25	25	56	25	56	76	13	203
Nov and Dec	25	25	25	25	25	25	25	25	25	25	56	76	13	203

Notes: Valid for England and Wales (temperate, maritime climate). Crop types are: 1 cereals, Sept. harvest; 2 cereals, Aug. harvest; 3 cereals, July harvest; 4 potatoes, Sept. harvest; 5 potatoes, May harvest; 6 vegetables, May harvest; 7 vegetables, July harvest; 8 vegetables, Aug. harvest; 9 vegetables, Oct. harvest; 10 bare fallow; 11 temporary grass; 12 permanent grass; 13 rough grazing; 14 woodland; and 15 riparian (not shown) since *RC* effectively infinite. (*Source:* Based on Grindley, J. (1969) *The Calculation of Actual Evaporation and Soil Moisture Deficits Over Specified Catchment Areas.* Hydrological Memoir 38. Meteorological Office, Bracknell.)

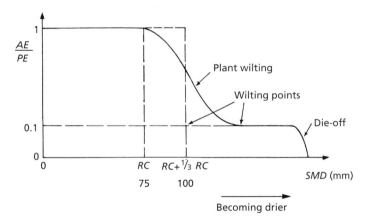

Fig. 6.15 Drying curve for a vegetated soil of short-rooted grass with a root constant (*RC*) of 75 mm showing the decline in the ratio of actual to potential evapotranspiration (*AE/PE*) for increasing values of soil moisture deficit (*SMD*).

the absence of surface runoff, the cumulative hydrological excess that is predicted to occur for those times when the *SMD* = 0 provides an estimate of groundwater recharge for the period of interest (291 mm in the example shown in Table 6.6).

According to the conventional soil water budgeting method, recharge cannot occur when a *SMD* exists. A critique by Rushton and Ward (1979) showed that recharge amount calculated using the Penman–Grindley method often under-predicts the amount calculated by other

Table 6.6 Example of the Penman–Grindley soil water budget method to calculate values of actual evapotranspiration (*AE*), soil moisture deficit (*SMD*) and recharge (hydrological excess, *HXS*) using precipitation (*P*) and potential evapotranspiration (*PE*) data. The calculation assumes that the permeable soil is covered by a short-rooted grass with a root constant (*RC*) of 75 mm and that the soil is at field capacity on 1 April (*SMD* = 0).

Month	*P* (mm)	*PE* (mm)	*P – PE* (mm)	*AE* (mm)	Δ*SMD* (mm)	*SMD* (mm)	*HXS* (mm)
Apr	20	48	−28	48	28	28	0
May	12	56	−44	56	44	72	0
Jun	24	72	−48	(24 + [100 − 72] + 0.1 [20]) = 54	(28 + 2) = 30	102	0
Jul	9	68	−59	(9 + 0.1[59]) = 15	6	108	0
Aug	31	42	−11	(31 + 0.1[11]) = 32	1	109	0
Sep	60	28	32	28	−32	77	0
Oct	75	20	55	20	−55	22	0
Nov	106	10	96	10	−22	0	74
Dec	94	5	89	5	0	0	89
Jan	69	5	64	5	0	0	64
Feb	40	18	22	18	0	0	22
Mar	72	30	42	30	0	0	42
Apr	18	50	−32	50	32	32	0
							Total recharge = 291 mm

methods based on lysimeters, tracers and hydrograph analysis. In particular, the Penman–Grindley method appears to underestimate summer and early autumn recharge. From a sensitivity analysis of recharge calculations for an area of Chalk aquifer in Lincolnshire, England, Rushton and Ward (1979) found that the time-step of the accounting procedure, the estimate of *PE* (Section 6.3.2), choice of root constant, the functional relationship between *AE* and *PE* and the date of harvesting produced an error in calculated recharge by up to 15%. To account for an underestimate in their recharge calculation compared with known outflows and borehole hydrograph records, Rushton and Ward (1979) permitted a direct component of recharge, conceptualized as a bypass flow component via Chalk fissures, equivalent to 15% of actual precipitation in excess of 5 mm plus 15% of the effective precipitation, with the remainder of the recharge

calculated with the conventional Penman–Grindley method. However, this method must not be seen as giving the correct daily recharge. On a monthly basis, and as input to a regional groundwater flow model, the monthly distribution of recharge calculated this way is regarded as acceptable.

6.5.3 Chloride budget method

Soil water budgeting methods as described above were developed for temperate climates and therefore have less validity in semi-arid and arid zones, where these methods normally underestimate recharge, often giving zero values. An alternative, geochemical method is to use a conservative tracer species such as chloride to estimate the amount of recharge and, in favourable circumstances, the recharge history. An example of the application of the chloride budget method is given in Box 6.1.

Box 6.1   Application of the chloride budget method to the Quaternary sand aquifer of Senegal, West Africa

The water balance of the coastal Quaternary sand aquifer in the West African Sahel region of Senegal is sensitive to short- and long-term climatic change, with groundwater resources in many areas dependent on recharge during former wet periods (Edmunds and Gaye 1994). Hence, it is important to be able to determine the recharge amount in this region in order to quantify the available groundwater resource. For this purpose, the chloride budget method was employed and included collection of well waters and porewaters from the unsaturated zone recovered from hand-augered material down to depths of 35 m (Fig. 6.16). From analysis of the well water and extracted porewater, and assuming both negligible surface runoff in the sandy terrain and a chloride source from atmospheric sources only, the direct recharge, R_d, is estimated from:

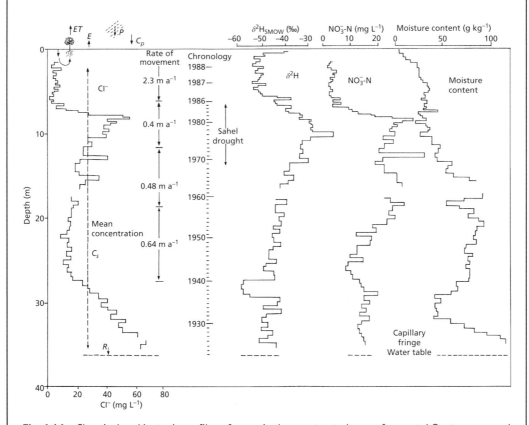

Fig. 6.16   Chemical and isotopic profiles of water in the unsaturated zone of a coastal Quaternary sand aquifer from a site west of Louga in Senegal, West Africa, interpreted by the chloride budget method to provide a 60-year recharge chronology. The symbols C_p, C_s, E, ET, P and R represent mean chloride concentration in rainfall, mean chloride concentration in well water or interstitial porewater, evaporation, evapotranspiration, precipitation and recharge, respectively (Edmunds 1991). (*Source:* Adapted from Edmunds, M. (1991) Groundwater recharge in the West African Sahel. NERC News April 1991, 8–10. © 1991, Natural Environment Research Council.)

(Continued)

Box 6.1 (Continued)

$$R_d = P \frac{C_p}{C_s} \qquad \text{(eq. 1)}$$

where P is the mean annual precipitation amount, C_p is the mean chloride concentration in rainfall and C_s is the mean chloride concentration in the well water or interstitial porewater. The ratio C_p/C_s represents the evaporative concentration of chloride in porewater or shallow groundwater and so measures the effect of evapotranspirative loss of precipitation. For example, using representative data for Senegal, where P = 443.2 mm, C_p = 1.1 mg L^{-1} and C_s = 11.2 mg L^{-1}, then:

$$R_d = 443.2 \frac{1.14}{11.23} = 45 \text{ mma}^{-1}$$
$$\text{(eq. 2)}$$

Seven unsaturated zone profiles from a small, 1 km^2 area to the west of Louga in Senegal gave a mean C_s value of 82 mg L^{-1} corresponding to a long-term recharge rate of 13 mm a^{-1}. When considered on a regional scale, this long-term recharge represents a sufficient volume to supply the present domestic water needs of the traditional villages, even during a drought when the recharge rate may be halved (Edmunds 1991).

The recharge history can also be calculated if the moisture content profile of the unsaturated zone is known. The rate of downward water movement, $\bar{v}$, can be calculated from:

$$\bar{v} = \frac{i}{\theta} \qquad \text{(eq. 3)}$$

where i is the recharge or infiltration rate and θ is the moisture content. For the recharge rate calculated earlier (eq. 2), and for a moisture content of 0.02 measured in 1988, a calculated downward water movement is 0.045/0.02 = 2.25 m a^{-1}. Thus, for this year, the soil water is estimated to infiltrate a depth of 2.25 m. By inspecting the chloride value in each sampling interval, it is then possible to calculate, from dividing the sampling interval by the rate of downward water movement, the residence time of the water in the interval. Hence, it is possible to build-up a recharge chronology for the unsaturated zone profile.

The oscillations in chloride concentration, shown in Fig. 6.16, indicate that the recharge rate has not been constant during the constructed, 60-year recharge chronology for Senegal. The period of Sahel Drought (1968–1986) is clearly visible as a zone of higher chloride concentrations. The drought is also emphasized by the deuterium stable isotope data (δ^2H in ‰) which show that water from this period is enriched in the heavier isotope as a result of greater evaporation. The water quality data for nitrate show values often in excess of 10 mg L^{-1} as N, although these high values are unrelated to surface pollution and instead arise from natural fixation of nitrogen by plants and micro-organisms (Fig. 8.38) with subsequent concentration by evaporation.

6.5.4 Temperature profile methods

Temperature versus depth profiles (TD profiles) can yield useful information about groundwater recharge conditions because groundwater flow impacts the distribution of heat in aquifers. If only heat conduction is considered in an homogeneous aquifer the temperature distribution in the aquifer will increase with depth, starting from the average annual surface temperature, at a rate (the thermal gradient) governed by the magnitude of geothermal heat flow at the base of the aquifer in combination with the thermal conductivity of the aquifer. This relationship is summarized by Fourier's Law of heat conduction which can be stated for heat flow in the depth direction (z) as follows:

$$q_G = -\kappa_e \frac{\partial T}{\partial z} \qquad \text{(eq. 6.16)}$$

in which q_G is the heat flow density, κ_e is the effective thermal conductivity of the subsurface, and T is temperature. The quantity $\partial T/\partial z$ is the thermal gradient and is observed in boreholes by taking measurements of temperature variations with depth. Thermal gradients in the upper parts (the upper kilometres) of the crust are generally of the order of several degrees per km (Fig. 6.17a). Additional heat transport by groundwater flow resulting in heat advection can, if vigorous enough, alter the shape of the temperature-depth profile. In areas of groundwater recharge, the profile becomes concave downward whilst in areas of discharge the opposite effect can be observed (Fig. 6.17b).

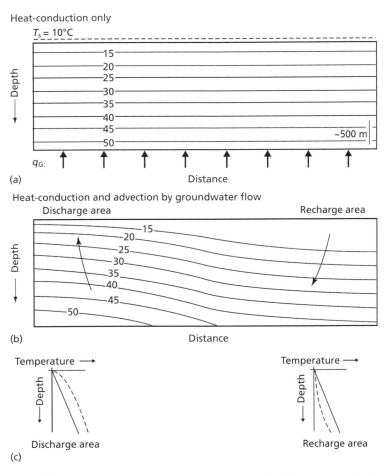

Fig. 6.17 In (a) isotherms in an idealized cross-section through the subsurface where heat flow is homogeneous and temperatures increase linearly with depth at a rate governed by the heat flow density (q_G) at the base of the domain and the effective thermal conductivity of the rock. The annual average surface temperature (T_s) controls the absolute values of the isotherms as the upper boundary condition to the system. In (b) when heat flow is not purely conductive, groundwater flow is vigorous enough to perturb the distribution of heat as shown. In (c) when heat advection by groundwater flow is contributing to heat flow in the subsurface, temperature-depth profiles in groundwater discharge and recharge areas (dashed lines) become clearly discernible from a heat-conduction only profile (solid line).

Detailed vertical temperature profiles across aquitards can be analysed using analytical solutions to the applicable equations to enable groundwater flow rates to be calculated. One well-known example of such a solution employs the use of type curves developed by Bredehoeft and Papadopoulos (1965). How this approach is used with temperature data from boreholes is illustrated in Box 6.2.

Box 6.2 Application of temperature-depth profiles to calculate vertical specific discharge across an aquitard

The temperature-depth profiles shown in Fig. 6.18 have been observed in two boreholes in the discharge and recharge areas of a groundwater flow basin. The borehole descriptions for the two wells are very similar. At both locations, reasonably uniform sand is found to a depth of 80 m (porosity, $n = 0.25$). From 80 to 110 m sandy clay is described ($n = 0.64$) below which sand is again found ($n = 0.25$) to at least 200 m depth. The hydraulic conductivity of these hydrogeological units is unknown. Table 6.7 lists the temperatures measured in each well. The depths between 80 and 110 m are across the sandy clay unit. The regional geothermal heat flux in the area is known to be 50 mW m^{-2}.

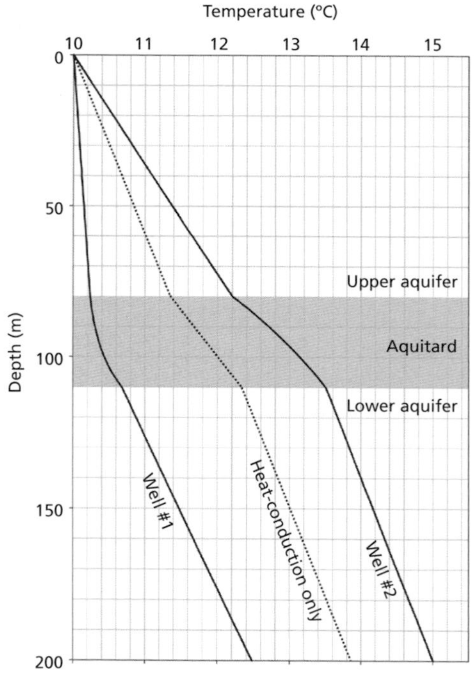

In the classic paper by Bredehoeft and Papadopoulos (1965), an analytical solution technique to the advection-diffusion equation of heat transfer is presented, using type curves to interpret temperature-depth profiles resulting from vertical down- or up-flow across a semi-confining aquifer or aquitard (Fig. 6.19). In short, the method uses a determination of the dimensionless Peclet number by matching normalized temperature data with type curves as shown in Fig. 6.20. From the Peclet number, a fluid flux across the unit can be calculated. The Peclet number (N_{Pe}) is the dimensionless ratio between conductive and advective heat transport given by:

$$N_{Pe} = \frac{\rho_w c_w q_z L}{\kappa_e} \quad \text{(eq. 1)}$$

in which q_z (m s^{-1}) is vertical specific discharge, L (m) is the thickness of the aquitard, κ_e (W m$^{-1\circ}$C^{-1}) is the effective thermal conductivity of the aquitard, ρ_w is groundwater density (=1000 kg m^{-3}) and c_w is volumetric heat capacity (=4190 J kg^{-1} $^\circ$C^{-1}). κ_e is calculated from:

Fig. 6.18 Plot of temperature-depth data presented in Table 6.7 for Well #1 and Well #2. The dashed line shows the calculated temperature-depth profile for heat-conduction-only conditions as discussed in the text.

Box 6.2   (Continued)

Table 6.7   Observed temperature-depth profiles in Well #1 and Well #2. Over the aquitard (depth interval 80–110 m) analytical parameters are calculated and plotted in Fig. 6.20 on the type curves given by Bredehoeft and Papadopoulos (1965). The symbols T_z, T_0 and T_L are denoted in Fig. 6.19.

Depth (m)	Well #1 T (°C)	Well #2 T (°C)	Well #1 $(T_z - T_0)/(T_L - T_0)$	Well #1 $(T_z - T_0)/(T_L - T_0)$	z/L
0	10.00	10.00			
10	10.02	10.28			
20	10.05	10.55			
30	10.07	10.83			
40	10.10	11.10			
50	10.12	11.38			
60	10.15	11.65			
70	10.17	11.93			
80	10.19	12.20	0.000	0.000	0.000
82	10.20	12.31	0.023	0.083	0.067
84	10.22	12.42	0.050	0.164	0.133
86	10.23	12.52	0.080	0.242	0.200
88	10.25	12.62	0.114	0.318	0.267
90	10.26	12.71	0.154	0.390	0.333
92	10.28	12.80	0.198	0.461	0.400
94	10.30	12.89	0.249	0.529	0.467
96	10.33	12.98	0.306	0.595	0.533
98	10.36	13.06	0.371	0.659	0.600
100	10.39	13.14	0.445	0.721	0.667
102	10.43	13.22	0.530	0.780	0.733
104	10.47	13.29	0.626	0.838	0.800
106	10.52	13.37	0.735	0.894	0.867
108	10.58	13.44	0.859	0.948	0.933
110	10.64	13.50	1.000	1.000	1.000
120	10.80	13.67			
130	10.97	13.84			
140	11.14	14.00			
150	11.30	14.17			
160	11.47	14.34			
170	11.64	14.50			
180	11.80	14.67			
190	11.97	14.84			
200	12.14	15.00			

(*Source:* Modified from Bredehoeft, J.D. and Papadopoulos, I.S. (1965) Rates of vertical groundwater movement estimated from the Earth's thermal profile. *Water Resources Research* **1**, 325–328.).

(*Continued*)

Box 6.2   (Continued)

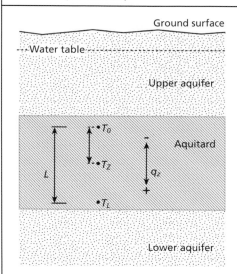

Ground surface

Water table

Upper aquifer

Aquitard

Lower aquifer

Fig. 6.19 The parameters needed to apply the analytical solution technique to the advection-diffusion equation of heat transfer as presented by Bredehoeft and Papadopoulos (1965). (*Source:* Bredehoeft, J.D. and Papadopoulos, I.S. (1965) Rates of vertical groundwater movement estimated from the Earth's thermal profile. *Water Resources Research* **1**, 325–328. © 1965, American Geophysical Union.)

$$\kappa_e = n\kappa_w + (1-n)\kappa_s \qquad \text{(eq. 2)}$$

in which the thermal conductivity of water, $\kappa_w = 0.11$ W m$^{-1\circ}$C^{-1}, and that of the solid fraction, $\kappa_s = 3.97$ W m^{-1} °C^{-1}. Following eq. 2 for the example of the sand aquifer, $\kappa_{\text{aquifer}} = (0.25 \times 0.11) + (1 - 0.25) \times 3.97 = 3.0$ W m^{-1} °C^{-1} and for the sandy clay aquitard, $\kappa_{\text{aquitard}} = (0.64 \times 0.11) + (1 - 0.64) \times 3.97 = 1.5$ W m^{-1} °C^{-1}.

First, by employing Fourier's Law of heat conduction (eq. 6.16) and by applying boundary conditions of surface temperature ($T_s = 10$°C) and basal heat flow ($q_b = 50 \times 10^{-3}$ W m^{-2}), a steady-state conduction-only temperature can be calculated for this aquifer–aquitard system as follows. In the upper aquifer, temperature will rise per metre increase in depth by $50 \times 10^{-3}/3.0 = 0.0166$°C m^{-1}. If the average surface temperature is 10°C,

the temperature at the base of the upper aquifer (at 80 m depth) is then found to be $(80 \times 0.0166) + 10 = 11.33$°C. In a similar way, the thermal gradient in the aquitard unit is double that in the aquifers and the temperature at the base of the aquitard (thickness = 30 m) will be 1°C higher than at its top so that at the aquitard base the temperature will be: $(30 \times 0.033) + 11.33 = 12.33$°C. The temperature-depth profile in the lower aquifer can now easily be constructed by assuming the same thermal gradient as in the upper aquifer. The resulting temperature-depth profile is shown in Fig. 6.18 as the dashed line. As can be seen, the heat conduction-only profile lies in between the two profiles measured in the field. Where the temperature-depth profile is cooler than the conduction-only profile one might suspect that this is caused by down-welling of relatively cool water from shallow depth and, vice versa, the relatively warm profile is indicative of upward heat transport by groundwater discharge.

Second, the magnitude of discharge and recharge across the aquitard can be quantified by looking at the curvature of the temperature profile across the aquitard. By normalization between 0 and 1 of the temperature data by calculating the quantity $(T_z - T_0)/(T_L - T_0)$ for each depth, and plotting these values against the normalized depth given by z/L (see Fig. 6.19 for what these symbols denote). Following this procedure for the data provided in Table 6.7 yields the diagram in Fig. 6.20 in which the data points are plotted together with the type-curves of the same quantities for different values for N_{Pe} (eq. 1). From this, it can be seen that the temperature-depth profile over the aquitard in Well #1 is consistent with $N_{Pe} = 2$, while for Well# 2 $N_{Pe} = -0.5$ is found. Now, using eq. 1, it follows that for Well #1, $q_z = 2.4 \times 10^{-8}$ m s^{-1} (downward flow) and for Well #2, $q_z = -6.0 \times 10^{-8}$ m s^{-1} (upward flow).

Box 6.2   (Continued)

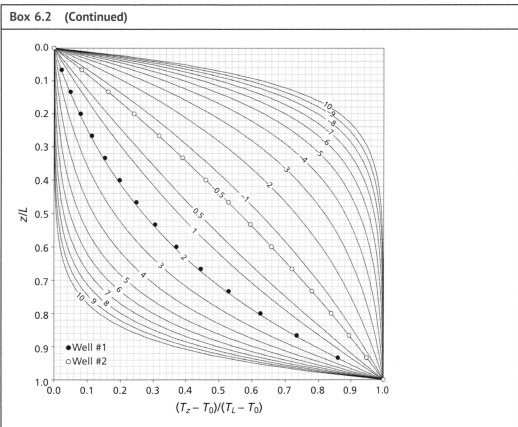

Fig. 6.20   Normalized temperatures from Well #1 and Well #2 plotted against normalized distance over which the temperature curvature is considered. The type curves for different values of N_{Pe} as presented by Bredehoeft and Papadopoulos (1965) allow an estimation of vertical specific discharge. (*Source:* Bredehoeft, J.D. and Papadopoulos, I.S. (1965) Rates of vertical groundwater movement estimated from the Earth's thermal profile. *Water Resources Research* **1**, 325–328. © 1965, American Geophysical Union.)

An alternative or complementary methodology for detecting groundwater recharge and discharge rates using temperature-depth profiles exploits the dependency on vertical groundwater flow rates of the penetration depth and phase shift of the seasonal wave of heat which is generated by the seasonal fluctuation of surface temperature. This seasonal temperature variation is usually detectable down to ∼20 m depth. For an analysis using such data, repeated measurements of shallow temperature profiles throughout the season are required. An example of such data and their quantitative analysis can be found in Taniguchi (1993).

Not only in a warming climate but also under the influence of certain land-use changes such as urbanization, the impact of rising annual average ground surface temperatures can be seen in temperature-depth profiles below the seasonal zone. The depth to which this climatic disturbance can be detected depends on the magnitude and nature (gradual versus instant), subsurface thermal properties and the time since surface warming was initiated. In addition, vertical groundwater flow is an important process, controlling the depth penetration of longer-term surface temperature fluctuations. Temperature-depth measurements that are repeated over time-spans of decades illustrate this beautifully as they document the rates of temperature changes at different depths in detail, and vertical groundwater flow rates can be derived from such datasets using numerical models of coupled groundwater flow and transient heat flow (Fig. 6.21; Bense and Kurylyk 2017; Kurylyk *et al.* 2019). See Section 6.8.1 for further examples on the use of temperature for the delineation of groundwater flow in near-surface environments (e.g. surface water–groundwater interaction).

6.6 Stream gauging techniques

The recording of streamflow data is fundamental to water resources studies and management, flood studies and water quality management. In hydrogeology, the importance of river flow data extends to groundwater resources, with aquifer recharge deduced from the balance of a number of measurements including baseflow, which is assessed directly from streamflow gauging. Flow data are also necessary for the derivation and application of operating rules for surface reservoirs, groundwater resources and river regulation (Section 10.2).

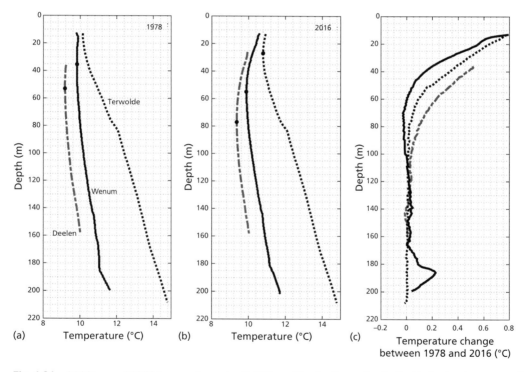

Fig. 6.21 (a) Measured 1978 temperature–depth (TD) profiles for three sites in the Netherlands representing recharge (Deelen), intermediate (Wenum) and discharge (Terwolde) conditions (Bense and Kurylyk 2017). (*Source:* Data from Bense, V.F. and Kurylyk, B.L. (2017) Tracking the subsurface signal of decadal climate warming to quantify vertical groundwater flow rates. *Geophysical Research Letters* **44**, 12,244–12,253. doi: 10.1002/2017GL076015). Black dots show inflection points in the TD profile (see text). (b) Data from the same boreholes reprofiled in 2016 with inflection points that have migrated downwards. Inflection points appear earlier and migrate more deeply in recharge zones (Deelen) than in discharge zones (Terwolde). (c) Thermal difference between (a) and (b) (Kurylyk *et al.* 2019). (*Source:* Kurylyk, B.L., Irvine, D.J. and Bense, V.F. (2019) Theory, tools, and multidisciplinary applications for tracing groundwater fluxes from temperature profiles. *WIREs Water* **6**, e1329. doi: 10.1002/wat2.1329.)

Historic flow data are used in setting minimum residual flow in a river in order to support the aquatic ecology or provide sufficient dilution to achieve water quality standards. A number of simple and advanced techniques are employed to measure or estimate river flows (discharge), with the main techniques described in the following sections. General guidelines for the selection of methods of discharge measurement (velocity-area, slope-area, dilution, ultrasonic, electromagnetic, weirs and flumes) are included in BSI (1998).

6.6.1 Velocity area methods

6.6.1.1 Surface floats

This velocity-area method is particularly useful when conditions, for example during a flood, make it dangerous for other discharge measurement procedures. The method requires the choice of a length of river reach sufficient to allow accurate timing of a float released in the middle of the channel and far enough upstream to attain ambient velocity before entering the reach. By measuring the distance of the reach and the time taken for the float to travel the length of the reach, the water velocity can be calculated by dividing the length by the time. The procedure is repeated a number of times to obtain the average maximum surface velocity, converted to mean velocity using coefficients (Table 6.8). By measuring the flow area upstream and downstream of the reach and taking the average value, the mean flow area for the reach is obtained. The river discharge is then found by multiplying the mean velocity by the mean flow area.

6.6.1.2 Current metering

Current metering of stream discharge is another velocity-area method and commonly employs one of two types of current meter (cup type or propeller type) to obtain point measurements of velocity (Fig. 6.22). At each measurement point, the meter is allowed to run for about 60 seconds and the number of

Table 6.8 Coefficients by which the maximum surface velocity of a river should be multiplied to give the mean velocity in the measuring reach.

Average depth in reach (m)	Coefficient
0.3	0.66
0.6	0.68
0.9	0.70
1.2	0.72
1.5	0.74
1.8	0.76
2.7	0.77
3.7	0.78
4.6	0.79
6.1 and above	0.80

revolutions made by the cup or propeller is obtained from a counter. The velocities can subsequently be calculated given a calibration equation linking velocity and count rate. Another type of measuring device is the electromagnetic current meter which measures the voltage resulting from the motion of a conductor (water flow velocity) through a magnetic field. The magnetic field is produced by a coil in the sensor, and the voltage is detected by electrodes on the surface of the sensor. The sensor has no moving parts and the meter has the advantage of giving a direct reading of velocity. Cleaning with clean water and mild soap is recommended to remove dirt and non-conductive grease and oil from the sensor's electrodes and surface. For a comparison of the performance of different current meters used for stream gauging, the reader is referred to Fulford *et al.* (1993).

Field methods for obtaining values of stream discharge across a section are described in detail by Rantz (1982). The main objective is the systematic measurement of point velocities across the river channel. For shallow rivers, wading techniques can be employed, while for deeper sections, the meter is suspended from a cableway, boat or bridge. The procedure

Cup type current meter

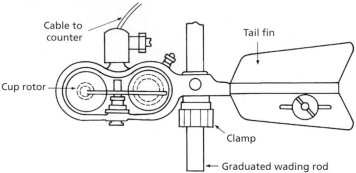

Propeller type current meter

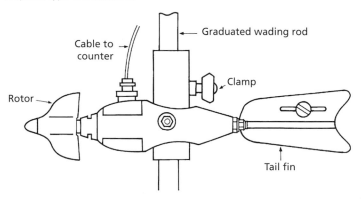

Fig. 6.22 Design of two types of current meter for measuring stream discharge. The cup type has an impeller consisting of six small cups which rotate on a horizontal wheel. The propeller type has a rotor as an impeller. In both types, the rate of rotation of the impeller is recorded by an electrically operated counter and is converted to a velocity using a calibration equation (Brassington 1998). (*Source:* Brassington, R. (1998) *Field Hydrogeology* (2nd edn). John Wiley & Sons, Ltd., Chichester. © 1998, John Wiley & Sons.)

for stream gauging using wading rods is as follows:

1) Choose a straight, uniform channel so that the flow is parallel to the banks.
2) Set-up a tag line and measuring tape across the channel perpendicular to the line of the bank and secure.
3) Measure water depth, d, and current meter count rate at a depth of $0.4d$ from the bed of the river, if $d \leq 0.75$ m, or $0.2d$ and $0.8d$ if $d \geq 0.75$ m, at 20 equal intervals across the section. Provided that the velocity profile is logarithmic, the point velocity at $0.4d$ or $(0.2d + 0.8d)/2$ represents the mean velocity

for the vertical. When taking velocity readings it is necessary to stand 0.5 m downstream and to one side of the meter, to hold the wading rod vertically, and ensure that the meter is aligned perpendicular to the section.

4) Convert count rate values to velocities using a calibration equation.
5) Assuming that the average velocity at a vertical is representative of the area bounded by the mid-points between adjacent verticals, calculate the discharge (velocity × area) for each segment. The discharges for each segment are then summed to obtain the total discharge of the section (Fig. 6.23).

Current meter measurement point

Sectional area for measurement vertical

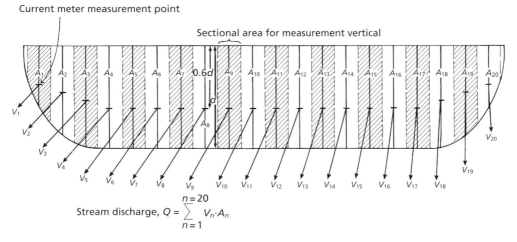

$$\text{Stream discharge, } Q = \sum_{n=1}^{n=20} V_n \cdot A_n$$

Fig. 6.23 Standard sectional method for calculating stream discharge using results from current metering. The cross-section is divided into 20 sections (or about 5% of the total width and containing less than 10% of the total discharge) and the current meter set in the middle of each section (mid-section method) or at either side and the velocities averaged (mean section method).

6.6.1.3 Acoustic Doppler current profiler

The acoustic Doppler current profiler (ADCP) is a velocity-area method using a boat, launch or towed platform (Fig. 6.24) to traverse a river to perform a single measurement of discharge normally for plotting on a rating curve (Box 6.3, Fig. 6.25). As it passes across the measuring section, velocity is measured by the Doppler principle, and the area is measured by tracking the bed to provide river depth and boat position. In the Doppler principle, the reflection of sound waves (back scattering) from moving sediment particles or air bubbles in the flow causes an apparent change in frequency. The frequency difference (Doppler shift) between the transmitted and reflected sound waves is a measure of the relative velocity of flow in both magnitude and direction. In design, an ADCP is basically a cylinder with a transducer head on the end being a ring of three or four transducers with their faces angled to the horizontal and at right angles to each other. When an ADCP is used to determine discharge, a series of acoustic pulses (pings) are transmitted that measure the flow (water pings) and boat velocity (bottom-tracking pings). A group of these pings are

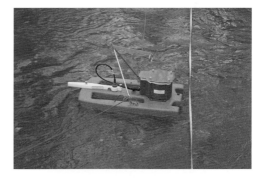

Fig. 6.24 Details of a StreamPro ADCP assembly on a catamaran with tow rope connection.

referred to as an ensemble, analogous to a vertical in the conventional current metering method. An ADCP measurement in a river that is say 5 m deep may make up to 18 averaged velocity measurements in depth cells spaced at 0.25 m in the vertical which are then used to compute a discharge for each cell. Given that velocity information cannot be collected near the water surface or near the stream bed, representing typically 7–10% of the total depth, the velocity cells are used to produce a vertical velocity curve which estimates discharge

Box 6.3 Rating curve adjustment

A rating curve allows the discharge of a river to be found from measurements of stage alone (Fig. 6.25), with observations of water level made with a recorder positioned over a stilling well linked to the stream in which water turbulence is reduced. A rating curve can be established at a gauging station with a fixed gauging structure (a weir or flume) or in a straight, uniform stream section (the control or rated section) that does not contain a gauging structure by taking a series of discharge measurements at different levels of flow. Thus, the successful operation of a gauging station involves the production of a reliable, accurate and continuous record of stage level.

The array of discharge and stage measurements usually lies on a curve which is approximately parabolic (Fig. 6.25) and, at most gauging stations, the zero stage does not correspond to zero flow. If the measurement points do not describe a single smooth curve,

then the control section governing the stage–discharge relationship has some variation in its nature, for example, a change in the slope of the river banks, the alternate scouring and deposition of loose bed material or growth of vegetation, the backing-up of flow due to the entry of a tributary stream or the operation of a sluice gate downstream, or if the gauging structure is drowned-out at higher flows. Another break in the curve at high stages can often be related to the normal bankfull level, above which the stage–discharge relationship could be very different from the within-banks curve due to the different hydraulics of floodplain flow. Changes in the bed profile in the control section will necessitate recalibration of the rating curve and it soon becomes apparent that maintaining an up-to-date stage–discharge relationship is a continuous activity. In certain rivers, it may be advisable to establish rating curves for different seasons of the year.

Discharges depend on both the stage and slope of the water surface. Typically, the slope is not the same for the rising and falling stages as a non-steady flow, such as a flood wave (Fig. 6.26), passes a gauging station. Discharge measurements made on the rising stage will be greater than those measured at the same stage level on the falling stage, and thereby demonstrating hysteresis resulting in a looped rating curve. When the river sustains a steady flow at a particular point, an average of the two discharges may be taken, otherwise an adjustment to the measured discharge value is required before including in the rating curve. In the latter case, the discharge measurements may require correction for both channel storage and water surface slope effects, especially when measuring higher discharges during flood events, in order to obtain a steady-state rating curve.

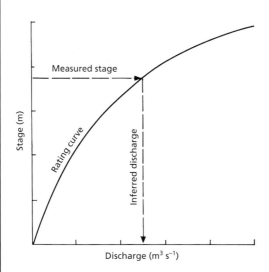

Fig. 6.25 A rating curve to convert measurements of depth of flow (stage or water level) to stream discharge.

Box 6.3 (Continued)

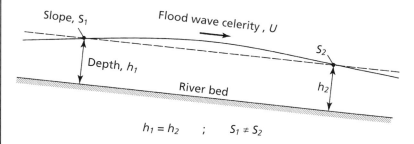

$$h_1 = h_2 \quad ; \quad S_1 \neq S_2$$

Fig. 6.26 Surface water slope variation during the passage of a flood wave. Slope $S_1 > S_2$.

Example of rating curve adjustment for channel storage

To correct for the effect of channel storage, suppose the gauge shows a rise in water level at the rate of 0.25 m hour^{-1} during a discharge measurement of 120 m^3 s^{-1} and that this rate of rise applies to a channel control section of 1000 m^2. If the average width of the river is 110 m, then the rate of change of storage in the reach, dS is given by:

$$dS = 1000 \times 110 \times 0.25$$
$$= 27\ 500 \text{ m}^3 \text{ hour}^{-1}$$
$$= 7.6 \text{ m}^3 \text{ s}^{-1}$$

Thus, the discharge measurement should be plotted on the rating curve as 112.4 m^3 s^{-1} (not 120 m^3 s^{-1}), since this is the discharge past the control section corresponding to the mean gauge height.

Example of rating curve adjustment for channel slope

The second reason for the looping of rating curves is the variation in surface slope that occurs as a flood wave moves along the channel. As shown in Fig. 6.26, representing a longitudinal section of a flood wave passing along a river channel, the rising stage is associated with the greater slope discharge (see the Manning formula, eq. 6.20) and so measurements made on the rising limb will plot to the right of the rating curve, and those

on the falling stage (lesser slope discharge) to the left. As a consequence, and depending on the peakiness of the flood wave, it is often the case that maximum river discharge occurs before the maximum stage is reached, since the influence of the steeper slope on velocity may outweigh the slight increase in cross-sectional area. It is usually necessary to correct the discharge measurements taken on either side of the flood wave to the theoretical steady-state condition. Based on the method presented by Wilson (1990), the following example explains the procedure to correct a measured discharge for the passage of a flood wave.

In this example, to calculate the discharge measurement that should be plotted on the rating curve, an actual river discharge measurement, Q_a, made during a flood indicated 3200 m^3 s^{-1}. During the measurement, which took 1.5 hours, the gauge height increased from 40.50 to 40.86 m. Stage readings taken 400 m upstream and 300 m downstream of the observation site differed by 110 mm. The river is 600 m wide with an average depth of 4.5 m at the time of measurement.

The cross-sectional area of the river, A = 600 × 4.5 = 2700 m^2, and the average water velocity = Q_a/A = 3200/2700 = 1.19 m s^{-1}.

Assuming that the flood wave celerity, U, can be found from:

$$U = 1.3 \frac{Q_a}{A} \qquad \text{(eq. 1)}$$

(Continued)

Box 6.3 (Continued)

Then, $U = 1.3 \times 1.19 = 1.55$ m s^{-1}

The rate of change in stage level of the river, dh/dt, equals $(40.86 - 40.50)/5400 = 6.67 \times 10^{-5}$ m s^{-1}.

The slope of the surface of the river, S, = $0.11/700 = 1.57 \times 10^{-4}$.

Now, the equation for corrected discharge, Q_c, for a rising river is:

$$Q_c = \frac{Q_a}{\sqrt{\left(1 + \frac{A\frac{dh}{dt}}{1.3Q_a S}\right)}} \qquad \text{(eq. 2)}$$

Then, for this example:

$$Q_c = \frac{3200}{\sqrt{\left(1 + \frac{2700 \times 6.67 \times 10^{-5}}{1.3 \times 3200 \times 1.57 \times 10^{-4}}\right)}}$$

$$Q_c = \frac{3200}{\sqrt{1.276}}$$

Hence, $Q_c = 2833$ m^3 s^{-1}.

Therefore, the corrected coordinates for plotting in the rating curve are stage 40.68 m and discharge 2833 m^3 s^{-1}.

---

within the unmeasured top and bottom sections of the profile. Further discussion on the ADCP method of stream flow measurement is provided by Herschy (2009).

6.6.2 Dilution gauging

This technique relies on the dilution of a tracer to measure stream discharge and is a useful technique in smaller, upland streams where flows and mixing of the tracer solution are rapid and the shallow and irregular bed form exclude current metering. The tracer is usually a fluorescent dye such as fluorescein or rhodamine WT (Table 7.4) that can be measured at trace concentrations (μg L^{-1}) using a fluorometer. The fact that low concentrations can be measured is an advantage in that only a few grams of harmless dye solution are required to obtain a measurement of tracer concentration above the background fluorescence. Salt can also be used but is less environmentally satisfactory since a much larger mass is required to detect chloride concentrations above background concentrations. Steam discharge measured by dilution gauging can be estimated to within 2% of current metering results provided that the tracer is fully mixed.

There are two principal dilution gauging techniques: steady-state (constant rate) and slug injection. With the steady-state method, a tracer solution of known concentration is run into the stream at a constant rate using a constant flow device such as a Mariotte bottle. By conservation of tracer mass:

$$cq = CQ \qquad \text{(eq. 6.17)}$$

where c is the initial tracer concentration, q is the rate of injection, C is the concentration in the stream at some downstream point, and Q is the unknown discharge. Thus, the discharge can be determined from a single sample taken at a point far enough downstream for full mixing to have occurred over the stream cross-section (Fig. 6.27). The degree of mixing must be checked first by taking several samples across the section. The method assumes steady discharge that does not vary along the length of the reach and also no losses of tracer, for example, in dead zones along the banks of the river.

In the slug injection method, a known volume of tracer solution, v, of known concentration, c, is injected instantaneously into the main flow of the stream. The concentration is measured in samples taken at frequent intervals from a point far enough downstream for

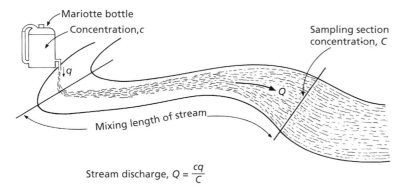

Stream discharge, $Q = \dfrac{cq}{C}$

Fig. 6.27 The steady-state or constant rate injection dilution gauging method of stream discharge measurement. In this example, tracer of known concentration, *c*, is dispensed from a Mariotte bottle at a constant rate, *q*. The sample concentration, *C*, is for a water sample taken where complete mixing of the tracer has occurred over the mixing length of the stream flowing with steady discharge, *Q*.

full mixing to have occurred. Sampling should begin before the tracer arrives at the sampling point and end after the cloud of tracer has passed. As shown in Fig. 6.28, results are plotted as a graph of concentration versus time and the area beneath the curve found, as follows:

$$\text{Area, } A = \int_0^\infty C \cdot dt \qquad \text{(eq. 6.18)}$$

The stream discharge is given by:

$$Q = \frac{cv}{A} \qquad \text{(eq. 6.19)}$$

As with the steady-state method, the slug injection method is based on the assumptions of complete mixing in the stream cross-section, no tracer losses and steady, uniform discharge.

6.6.3 Ultrasonic, electromagnetic and integrating float methods

These three methods rely on the velocity-area measurement approach, but permit automated monitoring. Ultrasonic gauging uses pulses of high-frequency ultrasound which are transmitted from both banks at an angle of 45° to the flow, one upstream and one downstream. The difference in time taken for the sound waves to travel in either direction and received by transducers is proportional to the average velocity of flow across the stream. Sampling can be at one or more depths but measurements can be affected by suspended sediment and other matter. The technique does not obstruct navigation, with the measurement section usually smoothed and lined to create a stable, rectangular cross-section.

Electromagnetic gauging depends on an electric cable buried in the stream bed. An applied electric current creates an electromotive force in the above flowing water which is proportional to the average velocity in the cross-section as measured by bankside probes. The technique is expensive and requires a mains electricity supply.

The integrated float technique, or bubble line method, uses bubbles of compressed air released at regular intervals from a pipe laid across the stream bed. Photographic monitoring reveals the amount of displacement of the bubbles by the flowing water, with the vertical pattern of displacement proportional to the stream velocity profile.

6.6.4 Slope-area method

It is possible to estimate the average velocity of flow through a channel using a friction equation provided that the roughness coefficient for the stretch of channel can be determined. This estimation method is particularly useful for reconnaissance surveys and for estimating

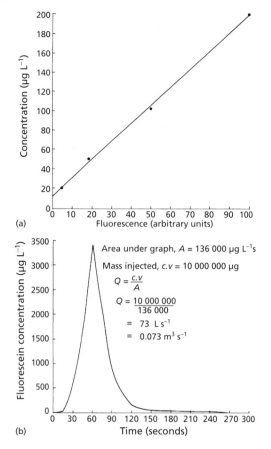

(a)

(b)

Fig. 6.28 The slug injection dilution gauging method of stream discharge measurement. The results shown were obtained using a slug injection of 10 g of the fluorescent dye fluorescein. The laboratory calibration curve used to obtain fluorescein concentrations in the steam water samples is shown in (a). The stream discharge is calculated by finding the area, A, under the concentration-time curve shown in (b) and equating the area to the mass of injected fluorescein (equal to the product of the concentration, c, and volume, v, of injected fluorescein).

flood flows after the peak discharge has subsided. The procedure is as follows:

1) Estimate the roughness coefficient (Manning's n) for the channel from a set of photographs of similar channels with known values of n or from a table of values (Table 6.9);

Table 6.9 Table of typical values of Manning's n for application in the estimation of stream discharge. (Wilson 1990).

Type of channel	n
Smooth timber	0.011
Cement-asbestos pipes, welded steel	0.012
Concrete-lined (high-quality formwork)	0.013
Brickwork well-laid and flush-jointed	0.014
Concrete and cast iron pipes	0.015
Rolled earth: brickwork in poor condition	0.018
Rough-dressed-stone paved, without sharp bends	0.021
Natural stream channel, flowing smoothly in clean conditions	0.030
Standard natural stream or river in stable condition	0.035
River with shallows and meanders and noticeable aquatic growth	0.045
River or stream with rocks and stones, shallow and weedy	0.060
Slow flowing meandering river with pools, slight rapids, very weedy and overgrown	0.100

(*Source:* Wilson, E.M. (1990) *Engineering Hydrology* (4th edn). Macmillan, London. © 1990, Macmillan Publishers.)

2) Measure the slope, S, of the water surface over a distance of approximately 200 m;
3) Survey the cross-section of the channel at a representative site to obtain the hydraulic radius, R, equal to the cross-sectional area of flow divided by the wetted perimeter;
4) Calculate the average velocity, v, in units of m s^{-1} using the Manning formula:

$$v = \frac{R^{2/3} S^{1/2}}{n} \qquad \text{(eq. 6.20)}$$

5) Calculate the stream discharge from $Q = vA$.

Another formula for application in estimating peak discharge is given by the Darcy–Weisbach equation for pipe flow:

$$v = \sqrt{\frac{8RgS}{f}} \qquad \text{(eq. 6.21)}$$

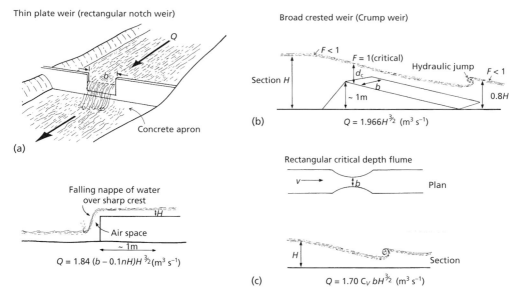

Fig. 6.29 Examples of: (a) thin plate weir (rectangular notch weir) and (b) broad crested weir (Crump weir); and (c) a rectangular critical depth flume for the measurement of stream discharge. In the formula given for the rectangular notch weir shown in (a), n is the number of side constrictions (=2 for a rectangular notch; =0 if notch across whole channel), b is the width of the notch and H is upstream head of water above the crest. In (b) the critical depth, d_c, is shown where the Froude number, F, is equal to 1 in the equation $F = v/(gd_c)^{1/2}$, where g is gravitational acceleration and v is average velocity. In (c) the term C_v is the approach velocity coefficient. The introduction of a shape factor is also possible for trapezoidal or U-shaped throats.

where f is the Darcy–Weisbach friction factor, g is gravitational acceleration and v, R and S are as defined for eq. 6.20. The friction factor, or flow resistance, is dependent on the flow geometry, the roughness height of the stream bed and the cross-sectional variation in roughness heights.

6.6.5 Weirs and flumes

A gauging station is a site on a river which has been selected, equipped and operated to provide the basic data from which systematic records of water level and stream discharge may be derived. Essentially, a gauging station consists of an artificial river cross-section (a weir) where a continuous record of stage (water level upstream of the weir crest) can be obtained and where a relation between the stage and discharge (the rating curve, see Box 6.3) can be determined.

Fixed gauging structures such as weirs and flumes (Fig. 6.29) are designed so that stream discharge is made to behave according to well-known hydraulic laws of the general form:

$$Q = KbH^a \qquad \text{(eq. 6.22)}$$

where H is the measured depth, or head, of water, K and a are coefficients reflecting the design of the structure, and b is the width of flow over the weir crest or in the throat (the constricted section) of a flume.

Many specialized weirs, such as V-notch, rectangular notch, compound and Crump weirs, provide accurate discharge data by observations of water level upstream of the weir. The same applies to flumes, where a stream is channelled through a geometrically, often trapezoidal shaped regular channel section. Flumes are designed so that the point of transition from sub-critical to critical flow, when a standing wave is formed accompanied by an increase

Fig. 6.30 Marham gauging station on the River Nar, Norfolk, England (NGR TF 723 119; catchment area 153.3 km^2; mean flow 1.15 m^3 s^{-1}). The gauging structure is a critical depth flume, 7.16 m wide. The stilling well is positioned behind the metal fence on the downstream, left wall of the flume. Prior to April 1982, the flume (7.47 m wide) contained a low flow notch at the centre. Weed growth can be a problem during summer if not cut regularly.

in velocity and a lowering in water level, occurs at a fixed location at the upstream end of the throat of the flume. Flumes are self-flushing and can be used in streams that carry a high sediment load, unlike a weir that can become silted-up. Generally, weirs and flumes are restricted in application to streams and small rivers (Fig. 6.30) since, for large flows and wide rivers, such structures become expensive to construct.

6.7 Hydrograph analysis

Rain falling on a catchment is classically considered to partition between overland flow, interflow and baseflow. These three components of total runoff are shown schematically in Fig. 6.31a and combine to generate the storm hydrograph shown in Fig. 6.31b. Overland flow is rarely observed on natural, vegetated surfaces but may occur where soils are compacted by vehicle movement or are completely saturated, for example at the bottom of a slope next to a stream channel. Interflow is water moving laterally within the soil zone in the direction of

the topographic slope and is potentially accelerated by flow through field drains. Together, overland flow and interflow represent the quickflow or surface runoff from a stream catchment. Baseflow is the component of total runoff contributed by groundwater discharge as springs or seepages and supports surface flows during dry periods when there is little or no rainfall. From the perspective of groundwater resources investigations, techniques of baseflow separation from the quickflow component are useful in contributing to an assessment of groundwater recharge.

6.7.1 Quickflow and baseflow separation

As shown by the storm hydrograph in Fig. 6.31b, and following passage of the flood peak, surface runoff declines along the recession limb until its contribution to total runoff may eventually disappear. During the storm, infiltration and percolation of water continue, resulting in an elevated groundwater table which enhances the rate of baseflow. With time, as the aquifer drains following the

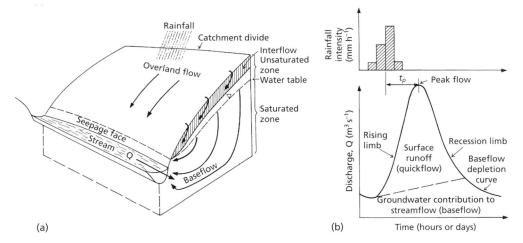

Fig. 6.31 Components of total catchment runoff contributing to streamflow. In (a) the directions of overland flow, interflow and baseflow are shown and in (b) the flood hydrograph from a rainfall event is shown. At the start of rainfall there is an initial period of interception and infiltration after which runoff reaches the stream and continues until a peak value occurs at time, t_p.

cessation of infiltration, the baseflow component also declines along its depletion curve. In catchments with a permeable geology, for example a limestone or sandstone, the baseflow component is a large fraction of the total surface runoff, but in clay-dominated or hard rock terrains the storm hydrograph shows a flashy response with a lower percentage of baseflow. These variations can be seen clearly when the record of streamflow is plotted as a flow duration curve (Fig. 6.32).

In practice, hydrograph separation can be conveniently achieved by plotting the streamflow data on semi-logarithmic graph paper, with the baseflow depletion curves identified as a series of straight lines, such as shown in Fig. 6.33. The variability encountered in the recession behaviour of individual segments represents different stages in the groundwater discharge and presents a problem in deriving a characteristic recession. One method, as demonstrated in Fig. 6.34, is to derive a master depletion curve and apply this to individual hydrograph peaks in order to separate surface runoff and baseflow discharge.

In general, the section of the hydrograph representing baseflow recession follows an exponential curve and the quantity at any time may be represented by:

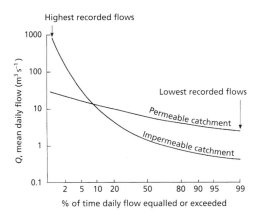

Fig. 6.32 Flow duration curves for permeable and impermeable catchments. Notice the higher peak flows generated in the impermeable catchment and the higher low flows in the permeable catchment. In permeable catchments, rainfall infiltration reduces the effects of flooding and long-term groundwater discharge as baseflow acts to lessen the impacts of droughts.

$$Q_t = Q_o e^{-at} \qquad \text{(eq. 6.23)}$$

where Q_o is discharge at the start of baseflow recession, Q_t is discharge at later time t, and a is an aquifer coefficient. Once the aquifer coefficient is found using eq. 6.23 for two known stream discharge values, the volume of water discharged from an aquifer in support

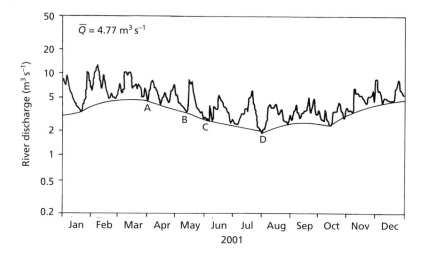

Fig. 6.33 Hydrograph of daily gauged flows during 2001 for Costessey Mill on the River Wensum, Norfolk, England (NGR TG 177 128; catchment area 570.9 km^2). The period of baseflow recession is identified by the straight lines AB, BC and CD.

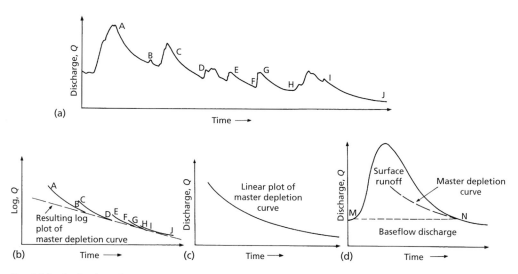

Fig. 6.34 Derivation of a master depletion curve by hydrograph analysis. First, examine the continuous stream discharge record and identify the individual sections of streamflow recession (a). Second, plot the individual sections of baseflow recession on semi-logarithmic graph paper (b). This is most easily achieved by moving tracing paper over the plots, keeping the axes parallel, until each log Q plot in successively increasing magnitude fits into the growing curve, extending it fractionally upwards. The tangential curve then established to the highest possible discharge is next converted back to linear graph paper and called the master depletion curve (c). The master depletion curve can now be applied to any particular storm hydrograph in which the depletion curves are matched at their lower ends and the point of divergence, N, marked to represent the point at which surface runoff has effectively finished (d). The line MN shown in (d) represents the base line of the hydrograph of surface runoff, the area below which can then be analysed to find the volume of baseflow discharge for a particular gauging station (Wilson 1990). (*Source:* Wilson, E.M. (1990) *Engineering Hydrology* (4th edn). Macmillan, London. © 1990, Macmillan Publishers.)

of baseflow during a recession period, t, can either be found by re-plotting the straight line baseflow depletion curve on linear graph paper and then finding the area under the curve, or by integrating eq. 6.23 to give:

$$\text{Volume of baseflow} = \frac{Q_o}{a}[1 - e^{-at}]$$

$$\text{(eq. 6.24)}$$

From the analysis of long-term records of baseflow, when variations in aquifer storage are assumed to be zero, it is possible to establish the volume of groundwater discharge (G_R in the water balance equation, eq. 6.1) and, in the absence of groundwater abstractions, equate this volume to the quantity of groundwater recharge. Further discussion of baseflow recession analysis and different ways of characterizing the baseflow recession rate is provided by Tallaksen (1995).

Hydrograph separation is also possible by hydrochemical means by adopting a two-component mixing model, or mass balance equation, where the pre-event water (baseflow) and event (quickflow) chemical compositions can be easily distinguished as follows:

$$C_T Q_T = C_P Q_P + C_E Q_E \qquad \text{(eq. 6.25)}$$

rearranging eq. 6.25 and recognizing that $Q_T = Q_P + Q_E$, then:

$$\frac{Q_P}{Q_T} = \frac{C_T - C_E}{C_P - C_E} \qquad \text{(eq. 6.26)}$$

where Q is discharge, C is tracer concentration and P, E and T represent the pre-event component, event component and total (peak) discharges.

To illustrate this method, the study by Durand *et al.* (1993) used dissolved silica and stable isotopes of oxygen ($\delta^{18}O$) to separate storm hydrographs in small, granitic mountainous catchments in south-east France. In one event that occurred in early autumn, the following data were obtained for a total stream discharge, Q_T, of 0.6 m^3 s^{-1}:

$\delta^{18}O$ of pre-event streamwater $= -7‰$

$\delta^{18}O$ of event water $= -9‰$
$\delta^{18}O$ of total discharge $= -8‰$

Substituting in eq. 6.26 gives:

$$\frac{Q_P}{Q_T} = \frac{-8 - (-9)}{-7 - (-9)} = \frac{1}{2} \qquad \text{(eq. 6.27)}$$

and it becomes apparent that $Q_P = Q_T/2$ is equal to 0.3 m^3 s^{-1} and $Q_E = (Q_T - Q_P)$ is also equal to 0.3 m^3 s^{-1}. In this example, the baseflow component calculated by hydrochemical means was a higher percentage of the total discharge than was interpreted by the graphical hydrograph separation method.

The hydrochemical separation technique was readily applied in the above example in that the pre-event and event stable isotope compositions were easily distinguishable as a result of evaporative enrichment of the heavier isotope (^{18}O) in groundwater stored in the peaty catchment soils. In general, the technique is best applied to small catchments of the order of 10 km^2. In larger catchments, variation in catchment geology may obscure the chemical signatures of individual components of baseflow and storm runoff.

6.7.2 Unit hydrograph theory

The concept of the unit hydrograph as proposed by Sherman (1932) is that given the constant physical characteristics of a catchment, for example its shape, size and slope, then the shape of hydrographs from storms of similar rainfall characteristics might be expected. The unit hydrograph is a typical hydrograph for a catchment. It is called a unit hydrograph because, for convenience, the runoff volume under the hydrograph is commonly adjusted to a unit (1 mm or 1 cm) equivalent depth over the catchment. Although the physical characteristics of a catchment remain relatively constant, the variable characteristics of storm events, such as rainfall duration, time-intensity pattern, areal distribution and amount, cause variations in the shape of the resulting hydrograph. The areal pattern of runoff can cause variations in hydrograph shape and so the unit hydrograph method is best applied to catchments small enough to minimize this factor. The limiting catchment size is determined by the required accuracy and regional climatic characteristics. Generally, unit hydrographs should not be used for catchments greater than 5000 km². Where convective rainfall predominates, the acceptable limit is much smaller.

Central to the unit hydrograph concept is the assumption that ordinates of flow are proportional to volume runoff for all storms of a given duration and that the time bases of all such hydrographs are equal. This assumption is not completely valid since, from the character of recession curves, the recession duration is a function of peak flow. Moreover, unit hydrographs for storms of the same duration but different magnitudes do not always agree. Peaks of unit hydrographs derived from very small events are commonly lower than those derived from larger storms. This observation may be because the smaller events contain less surface water and relatively more interflow and groundwater than the larger events or because channel flow time is longer at low flows.

Following from this introduction, the unit hydrograph can be defined as the hydrograph of 1 mm or 1 cm of direct runoff (quickflow) from a storm of specified duration. Once derived, then for a storm of the same duration but with a different amount of runoff, the hydrograph of direct runoff is assumed to have the same time base as the unit hydrograph and ordinates of flow approximately proportional to the volume of runoff. The duration assigned to a unit hydrograph should be the duration of rainfall producing significant runoff, determined by inspection of hourly rainfall. Methods for the derivation of unit hydrographs and their conversion to accommodate different storm durations are given in Box 6.4. Once derived, unit hydrographs can then be used to synthesize a storm hydrograph from a series of rainfall periods with varying intensity. Using the principle of superposition, the total hydrograph resulting from contiguous and/or isolated periods of uniform intensity effective rainfall is the sum of all the incremental hydrographs and estimated baseflow. For further discussion of unit hydrograph theory, the reader is directed to textbooks on engineering hydrology such as Linsley *et al.* (1982) and Wilson (1990).

Box 6.4 Derivation of unit hydrographs of various durations

The unit hydrograph is best derived from the hydrograph of a storm of reasonably uniform intensity, duration of desired length and relatively large runoff volume. As shown in Fig. 6.35a, the first step is to separate the baseflow from direct runoff. The volume of the direct runoff is then found (Table 6.10), and the ordinates of the direct runoff hydrograph are divided by the observed runoff depth. The adjusted ordinates produce the unit hydrograph (Fig. 6.35b).

To convert an existing unit hydrograph from one storm duration to another, then if two unit hydrographs of given duration of say t_r-hour, with one lagged by time t_r-hour with respect to the other, are added as shown

in Fig. 6.36, the result is the characteristic hydrograph for two units of rainfall excess and duration of $2t_r$-hour. Dividing the ordinates by two yields the $2t_r$-hour unit hydrograph.

hydrograph to a shorter or longer duration that is not a multiple of t_r can be achieved by application of the S-curve (summation-curve) method. The S-curve is the hydrograph that would result from an infinite series of unit runoff increments. Thus, each S-curve applies to a specific duration within which each unit of runoff is generated. The S-curve is constructed by adding together a series of unit hydrographs, each lagged by time, t_r with respect

Box 6.4 (Continued)

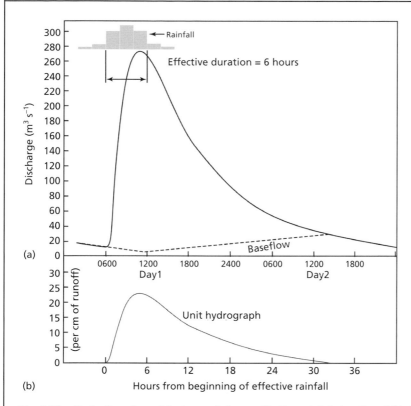

Fig. 6.35 Derivation of a unit hydrograph for an effective rainfall duration of 6 h. The ordinates of the unit hydrograph shown are obtained using the procedure shown in Table 6.10.

to the preceding one. If the time base of the unit hydrograph is t hours, then a continuous rainfall producing one unit of runoff every period would develop a constant outflow at the end of t hours. Hence, only t/t_r unit hydrographs need to be combined to produce an S-curve which should reach equilibrium at flow, Q_e (m^3 s^{-1}) found from:

$$Q_e = \frac{2.78A}{t_r} \qquad \text{(eq. 1)}$$

where A is the drainage area (km^2), the runoff is in centimetres and t_r is again the unit duration.

Table 6.11 illustrates the construction of an S-curve, starting with an initial unit hydrograph for which t_r = 3 hours. For the first 3 hours, the unit hydrograph and S-curve are

identical (columns 2 and 4). The S-curve additions (column 3) are the ordinates of the S-curve set ahead 3 h. Since an S-curve ordinate is the sum of all concurrent unit hydrograph ordinates, combining the S-curve additions with the initial unit hydrograph is equivalent to adding all previous unit hydrographs. The difference between two S-curves with initial points displaced by t_r'-hour gives a unit hydrograph for the new duration, t_r'-hour (columns 5 and 6). Since the S-curve represents runoff produced at a rate of one unit in t_r-hour, the runoff volume represented by this new hydrograph will be t_r'/t_r units. Thus, the ordinates of the unit hydrograph for t_r'-hour are computed by multiplying the S-curve differences by the ratio t_r/t_r' (column 7).

(Continued)

Box 6.4 (Continued)

Table 6.10 Derivation of the unit hydrograph ordinates shown in Fig. 6.35b for an effective rainfall duration of 6 h and a catchment drainage area of 104 km².

Day	Hour	Total flow ($m^3\ s^{-1}$)	Baseflow ($m^3\ s^{-1}$)	Direct runoff ($m^3\ s^{-1}$)	Unit hydrograph ordinate ($m^3\ s^{-1}$ for 1 cm of runoff)
1	0600	11	11	0	0.0
	0800	170	8	162	13.9
	1000	260	6	254	21.7
	1200	266	6	260	22.3
	1400	226	8	218	18.7
	1600	188	9	179	15.3
	1800	157	11	146	12.5
	2000	130	12	118	10.1
	2200	108	14	94	8.0
	2400	91	16	75	6.4
2	0200	76	17	59	5.1
	0400	64	19	45	3.9
	0600	54	21	33	2.8
	0800	46	22	24	2.1
	1000	38	24	14	1.2
	1200	32	26	6	0.5
	1400	27	27	0	0.0
				Total = 1687	

Note: Direct runoff (quickflow) = $(1687 \times 2\ \text{hours} \times 3600\ \text{s hour}^{-1})/104 \times 10^6\ \text{m}^2 = 0.1168\ \text{m or } 11.68\ \text{cm}.$

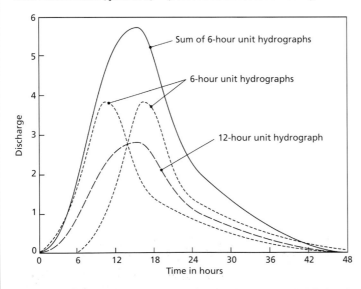

Fig. 6.36 Construction of a unit hydrograph for a longer rainfall duration (12 h) as an even multiple of a shorter rainfall duration (6 h).

Box 6.4 (Continued)

Table 6.11 Application of the S-curve method to derive a unit hydrograph for a shorter rainfall duration.

(1) Time (hour)	(2) 4-hour unit hydrograph	(3) S-curve additions	(4) S-curve columns (2) + (3)	(5) Lagged S-curve	(6) Columns (4) – (5)	(7) 3-hour unit hydrograph Column (6) × 4/3
0	0	–	0	–	0	0
1	5	–	5	–	5	7
2	35	–	35	–	35	47
3	65	–	65	0	65	87
4	90	0	90	5	85	113
5	105	5	110	35	75	100
6	92	35	127	65	62	83
7	78	65	143	90	53	71
8	67	90	157	110	47	63
9	57	110	167	127	40	53
10	48	127	175	143	32	43
11	40	143	183	157	26	35
12	33	157	190	167	23	31
13	26	167	193	175	18	24
14	22	175	197	183	14	19
15	16	183	199	190	9	12^a
16	12	190	202	193	9	12^a
17	8	193	201	197	4	5^a
18	5	197	202	199	3	4^a
19	2	199	201	202	0	0^a
20	0.5	202	202.5	201	1.5	2^a
21	0	201	201	202	−1	

^{a} Slight adjustment is required to the tail of the 3-h unit hydrograph, most easily done by eye.

6.8 Surface water – groundwater interaction

The link between groundwater and river flows is fundamental to conserving the riparian environment yet is one of the more difficult hydrogeological situations to predict. This difficulty is due to the complex nature of river–groundwater interactions and uncertainties in the nature of the hydraulic connection at a particular location. Hence, making decisions about the impacts of groundwater abstraction on rivers is technically challenging.

An important determining factor of the flux of water between a river and an aquifer is the degree of connection between the river and aquifer as controlled by the material properties of the river bed and river bank sediments and the extent to which the channel of the river intersects the saturated part of the aquifer. In general, and as shown in Fig. 6.37, there are three types of hydrogeological situation that lead to flow between an aquifer and a river.

The rate of change in river flow and the attenuation of short-term fluctuations in river flow, for example a flood event, can be strongly influenced by storage of water in the floodplain deposits surrounding a river. As the river stage and groundwater level increase, the extra water saturates the alluvial sediments and fills the available bank storage. When water levels recede, this bank storage is released and can have a short-term beneficial effect in

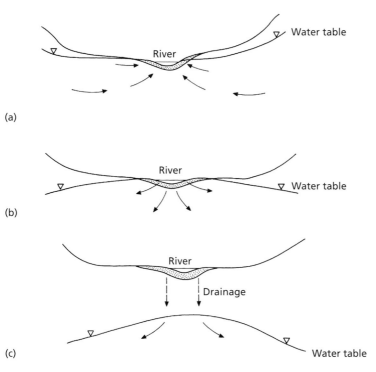

(a)

(b)

(c)

Fig. 6.37 Three hydrogeological situations that represent flow between an aquifer and a river. In (a) the water table in the aquifer is above the river stage and there is potential for flow from the aquifer to the gaining river (effluent condition) with the flux generally proportional to the difference between the elevations of the water table and river stage. In (b) the water table is below the river stage and the losing river (influent condition) potentially loses water to the aquifer with the flux also generally proportional to the difference between the elevations of the river stage and water table. In (c) a common situation is shown in which a partially penetrating river (where the saturated aquifer extends beneath the river) experiences a declining water table below the base of the river. In this situation, water will drain from the perched river under gravity with a unit head gradient. The unit head gradient creates a limiting infiltration rate such that river losses will not increase as the water table falls further. In each case shown, the nature of the river–aquifer interaction will also depend on the properties of the river bed sediments. Sediments with very low permeability can result in a significant resistance to flow (Kirk and Herbert 2002). (*Source:* Kirk, S. and Herbert, A.W. (2002) Assessing the impact of groundwater abstractions on river flows. In: *Sustainable Groundwater Development* (eds K.M. Hiscock, M.O. Rivett and R.M. Davison). Geological Society, London, pp. 211–233. Special Publications 193.)

alleviating the immediate impact of adjacent abstractions on the river flow.

In the long term, and usually within 1 or 2 years for boreholes a few hundred metres from a river, groundwater abstraction will deplete the river flow at a rate equal to the pumping rate. As shown in Fig. 6.38a, the river flow depletion consists of two components:

1) interception of flow that would otherwise reach the river; and
2) induced recharge from the river.

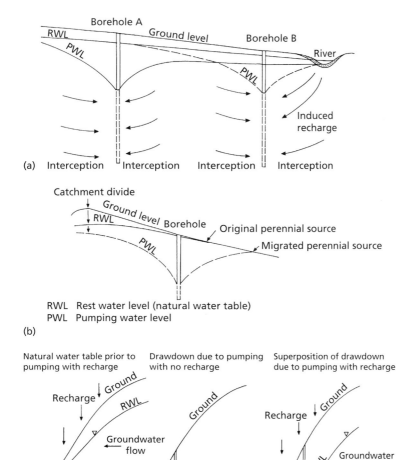

Fig. 6.38 Cross-sections to illustrate the impacts of groundwater abstraction on a groundwater fed river. In (a) river flow is depleted by the effects of interception and induced recharge and in (b) the perennial source of the river migrates downstream in response to a drawdown in the water table (Owen 1991). (*Source:* Owen, M. (1991) Relationship between groundwater abstraction and river flow. *Journal of the Institution of Water and Environmental Management* **5**, 697–702.) In (c) the principle of superposition is illustrated as applied to abstraction close to a river (Kirk and Herbert 2002). (*Source:* Kirk, S. and Herbert, A.W. (2002) Assessing the impact of groundwater abstractions on river flows. In: *Sustainable Groundwater Development* (eds K.M. Hiscock, M.O. Rivett and R.M. Davison). Geological Society, London, pp. 211–233. Special Publications 193.)

In general, the depletion of river flow caused by pumping increases with time and will increase more rapidly the closer the abstraction point is to the river. The degree of depletion is also dependent on the aquifer properties of transmissivity and storage coefficient.

A dramatic illustration of the impacts of groundwater abstraction in depleting river flows is the case of the River Colne valley, north of London (Fig. 6.39). In the first half of the twentieth century, the substantial growth of residential areas here and elsewhere on the outskirts of London was supplied by direct groundwater abstractions from the underlying Chalk aquifer. This rapid development had marked effects on certain rivers, particularly the Rivers Ver and Misbourne, tributaries of the River Colne, in the Chiltern Hills. In the River Ver catchment, about 75% of the average annual recharge to the Chalk is allocated to licensed groundwater abstraction which is now almost taken up (Owen 1991). Approximately half of the abstracted water is exported to supply areas outside of the catchment and is therefore effectively lost. Of the remainder used within the catchment, effluent returns via sewage treatment works are typically in the lower reaches of the River Colne Valley and therefore are unavailable for supporting river flows higher up in the catchment.

The effects of the large demand for water, especially from those boreholes situated towards the head of the River Colne Valley, has been to dry up those springs at the source of the perennial rivers (Fig. 6.38b shows the hydraulic mechanism). In the case of the River Ver, the upper 10 km section of originally perennial or intermittent river is now normally dry. Further downstream, the remaining perennial section experiences much reduced flows and this once typical Chalk stream has suffered substantial environmental degradation with major changes in riparian habitat, the loss of naturally sustained fisheries, the loss of water cress farming and reductions in the general amenity value of the river (Owen 1991).

This example illustrates the need for careful management of catchment water resources and the need to be able to predict potential environmental impacts. However, without very significant effort towards field investigation and numerical modelling (Section 7.5), it is often difficult to evaluate the impacts of abstractions on rivers with any degree of confidence. A not unusual limitation is the availability of accurate data, particularly for the physical properties of the river bed and river bank sediments. As discussed in Section 6.8.2, a further approach is to employ an analytical solution in the hydrogeological assessment of river flow depletion caused by groundwater abstraction.

6.8.1 Temperature-based methods of detection

Groundwater temperatures down to depths of $\sim$25 m are strongly affected by seasonal variations in surface temperature. This surface temperature fluctuation creates a temperature wave which propagates down into the subsurface. The penetration depth of this temperature wave is strongly controlled by rates of mainly vertical heat advection by shallow groundwater discharge or recharge. As a result, in strongly heterogeneous hydrogeological environments, such as river sediments, measurements of subsurface temperature at shallow depth (e.g. 1–2 m) can show a strong variability over short distances which can be related to contrasts in near-surface groundwater flow into and out of the river, exchanging water with the underlying aquifer. The principle of the use of shallow temperatures to delineate surface water–groundwater interaction is shown in Fig. 6.40a. The diagram in Fig. 6.40a makes it clear that the strongest lateral variability in temperature between zones of contrasting rates of groundwater flow will occur in early spring and early autumn. In early autumn, zones of groundwater discharge at a depth of, for example, a metre can be several degrees cooler than where discharge does not occur. Vice versa, these zones

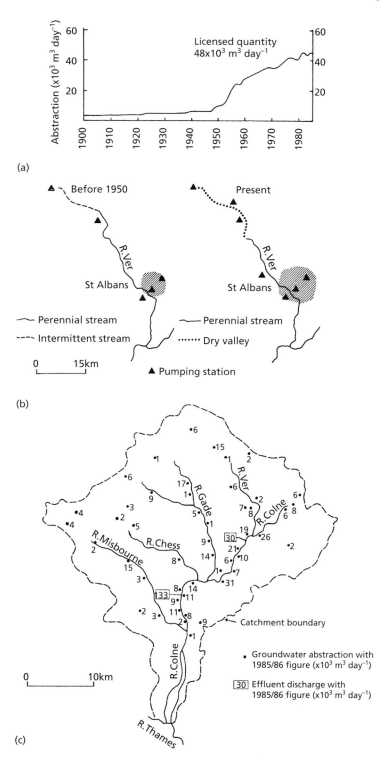

Fig. 6.39 The history of Chalk groundwater abstractions in the River Ver catchment, north London, and the impacts on river flow including (a) a graph of annual groundwater abstractions; (b) a sketch of the reduction in length of the perennial section of the River Ver; and (c) a location map of the River Ver tributary in the Colne catchment (Owen 1991). (*Source:* Owen, M. (1991) Relationship between groundwater abstraction and river flow. *Journal of the Institution of Water and Environmental Management* **5**, 697–702.)

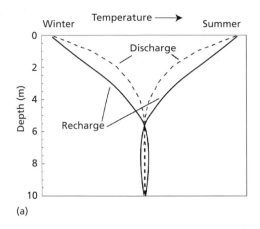

(a)

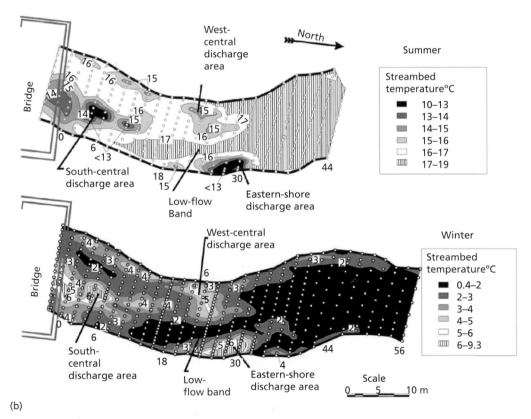

(b)

Fig. 6.40 (a) Qualitative sketch showing the difference in vertical temperature gradients between an area of groundwater discharge and recharge during winter and summer. Relatively small differences in temperature gradient near the surface as a result of advective heat transport through vertical groundwater flow will result in significant temperature anomalies in horizontal profiles of groundwater temperature (Bense and Kooi 2004). (*Source:* Bense, V.F. and Kooi, H. (2004) Temporal and spatial variations of shallow subsurface temperature as a record of lateral variations in groundwater flow. *Journal of Geophysical Research* **109**, B04103. doi: 10.1029/2003JB002782. © 2004, John Wiley & Sons.) (b) Patterns of shallow streambed temperature based on a large number of point measurements in the streambed of a river. Measurements were completed in summer (upper panel) and winter (lower panel). The pattern of temperature variability is similar between these two measurements but the zones of groundwater discharge (also independently identified with seepage meters) are warmer by several degrees in winter than other areas, but cooler in summer (Conant 2004). (*Source:* Conant, B. (2004) Delineating and quantifying ground-water discharge zones using streambed temperatures. *Ground Water* **42**, 243–257. © 2004, John Wiley & Sons.)

of groundwater discharge will be anomalously warm in early spring.

The impact of shallow groundwater flow on temperature distributions has been reported extensively in the literature, for example, by Conant (2004), as depicted in Fig. 6.40b. Conant (2004) also illustrated that the quantification of surface water–groundwater interaction with seasonal data, such as shown in Fig. 6.40b, often requires careful calibration. However, where data are collected more continuously, which is usually only feasible at selected point locations, results can yield accurate estimates of vertical groundwater flow (e.g. see Constantz *et al.* 2002; Constantz 2008).

Current efforts to use temperature as a tracer for hydrogeological processes in and near the streambed of rivers focus on applying relatively novel technologies such as Distributed Temperature Sensing using fibre-optic cables (see Selker *et al.* 2006) and thermal imaging from helicopters and aeroplanes (Tonolla *et al.* 2012) to detect thermal patterns at the streambed rather than from below (as in Fig. 6.40b). However, these efforts are hampered by other environmental factors that impact the detected temperatures in addition to groundwater flow, such as shading by vegetation, variations in water depth and anthropogenic factors.

6.8.2 Simulating river flow depletion

6.8.2.1 Analytical solutions

Analytical solutions simplify the aquifer properties, often assigning single representative values to the transmissivity and storage parameters. It is also implicit that the aquifer can be characterized by a single value of groundwater level away from the influence of the abstraction point. The effect of abstraction is to create a cone of depression around the borehole or well. Using analytical solutions, this drawdown is simplified by adopting assumptions of idealized radial flow to calculate the shape of the drawdown zone that can be superimposed on the distribution of head due to the natural behaviour of the system (Fig. 6.38c).

A discussion of the application of analytical solutions for the evaluation of stream flow depletion caused by pumping is presented in the classic papers by Jenkins (1968, 1970). A more recent review of analytical solutions is contained in Kirk and Herbert (2002). The method presented by Jenkins (1968) enables the estimation of the total depletion of stream flow as a function of time due to nearby abstraction with the following assumptions:

1) the aquifer is isotropic, homogeneous, semi-infinite in areal extent and bounded by an infinite, straight, fully penetrating stream;
2) that water is released instantaneously from aquifer storage;
3) that the borehole or well fully penetrates the aquifer;
4) the pumping rate is steady; and
5) the residual effects of previous pumping are negligible.

The model further assumes that the aquifer is confined or that, for an unconfined aquifer, the water table drawdown is negligible compared to the saturated aquifer thickness (in other words, the transmissivity remains constant). The temperature of the stream is assumed to be constant and equal to the temperature of the groundwater. A cross-section through the idealized conceptual model is shown in Fig. 6.41.

The mathematical solution to the problem, shown in Fig. 6.41, gives the rate of stream depletion as a proportion of the groundwater abstraction rate as follows:

$$\frac{q}{Q} = erfc\left(\frac{1}{2\tau}\right) \qquad \text{(eq. 6.28)}$$

where τ is a dimensionless length scale for the system given by:

$$\tau = \sqrt{\frac{tT}{L^2 S}} \qquad \text{(eq. 6.29)}$$

where T is the aquifer transmissivity, S is the aquifer storage coefficient (specific yield for unconfined aquifer approximations), L is the

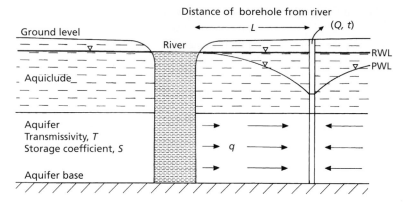

Fig. 6.41 Cross-section, drawn perpendicular to the river, showing the idealized conceptual model of river–aquifer interaction for a confined aquifer. Prior to pumping the initial potentiometric surface (shown as line RWL, the rest water level) is horizontal and equal to the constant water level of the stream. The drawdown due to pumping is shown as line PWL, the pumping water level.

perpendicular distance of the borehole or well to the line of the river, Q is the abstraction rate of the borehole or well, q is the rate of stream flow depletion, t is time and *erfc* is the complementary error function (see Appendix 8).

Similarly, the volume of stream depletion as a proportion of the groundwater abstraction rate is given by

$$\frac{v}{Qt} = 4i^2 erfc\left(\sqrt{\frac{1}{2\tau}}\right) \qquad \text{(eq. 6.30)}$$

where v is the volume of stream depletion during time, t, and i^2erfc is the second repeated integral of the error function.

Although the assumptions behind the solution of eqs 6.28 and 6.30 are an over-simplification of reality, analytical results can provide rough estimates of the local impacts of abstraction on river flow and the timescales over which flow depletion occurs. By neglecting river bed and river bank sediments and assuming full aquifer penetration, the impact of pumping on the stream flow is over-estimated and the time delay between abstraction starting and the impact of pumping on stream flow is underestimated. A further assumption that prior to pumping the initial potentiometric surface is horizontal and equal

to the constant water level of the stream (Fig. 6.41) is, in fact, not a major limitation (Kirk and Herbert 2002) but Wilson (1993, 1994) presented steady-state analytical solutions for two-dimensional, vertically integrated models of induced infiltration from surface water bodies for various combinations of aquifer geometry in the presence of ambient aquifer flow.

Wallace *et al.* (1990) extended the approach of Jenkins (1968) to show that pumping impacts may develop over several annual cycles. For cases where the stream depletion impacts develop over long timescales due to either the distance between the borehole and river, the type of aquifer properties or the possible role of river deposits, the maximum impact in later years may exceed the maximum depletion in the first year. Such delayed impacts are potentially an important catchment management consideration in order to avoid future low river flows.

An analytical solution for estimation of drawdown and stream depletion under conditions that are more representative of those in natural systems (that is, finite-width stream of shallow penetration adjoining an aquifer of limited lateral extent) is presented by Butler *et al.* (2001). The solution shows that the conventional assumption of a fully penetrating stream can lead to significant over-estimation of stream depletion (>100%) in many practical situations, depending on the value of the stream leakance parameter and the distance from the pumping well to the stream. An important assumption underlying this solution is that the penetration of the stream channel is negligible relative to aquifer thickness, although an approximate extension to the method provides reasonable results for the range of relative penetrations found in most natural systems (up to 85%) (Butler *et al.* 2001).

To assist in the practical application of analytical solutions, stream depletion caused by groundwater abstraction can be readily calculated using dimensionless type curves and tables. Jenkins (1968) presented a number of worked examples including computations of the rate of stream depletion for the pumping and following non-pumping periods, the volume of water induced by pumping and the effects (both rate and volume of stream depletion) of any selected pattern of intermittent pumping. An example calculation is given in Box 6.5.

6.8.2.2 Catchment resource modelling

Local-scale impacts of groundwater abstraction on river flows can be investigated with the above analytical solutions or by numerical models where data availability permits. A further quantitative method is to naturalize the river flow record to remove the effects of catchment abstractions and discharges. The derived (naturalized) and actual (historic) river flow records can then be conveniently presented as flow duration curves and compared to assess the environmental impacts of catchment water resources management.

The derivation of a naturalized flow duration curve and simulation of river–aquifer interaction can be computed using a simple numerical catchment resource model. For example, Hiscock *et al.* (2001) developed a resource model to assess the impacts of surface water and groundwater abstractions on lowland river flows in eastern England. The resource model simulated river flows from the basic components of baseflow (aquifer discharge) and surface flows and included the net effect of surface water and groundwater abstractions, thus enabling the naturalization of measured

---

Box 6.5   Computation of the rate and volume of stream depletion by boreholes and wells

The following example illustrates the application of an analytical solution to the problem of stream flow depletion caused by groundwater abstraction. The solution uses one of the type curves and tables presented by Jenkins (1968) to assist the calculation of the rate and volume of stream flow depletion.

The problem to be solved is as follows. A new borehole is to be drilled for supporting municipal water supply from an unconfined alluvial aquifer close to a stream. The alluvial aquifer has a transmissivity of 3×10^{-3} m^2 s^{-1} and a specific yield equal to 0.20. To protect the riparian habitat, the borehole should be located at a sufficient distance from the stream so that downstream of the new source stream flow depletion should not exceed a volume of 5000 m^3 during the dry season. The dry season is typically about 200 days in duration. The borehole is to be pumped continuously at a rate of 0.03 m^3 s^{-1} during the dry season only. During the wet season, recharge is sufficient to replace groundwater storage depleted by pumping in the previous dry season; hence, the residual effects on steam flow during the following non-pumping period can be neglected.

The problem requires us to find the minimum allowable distance between the

(Continued)

Box 6.5   (Continued)

borehole and stream using the following given information:

v, volume of stream flow depletion during time t_p = 5000 m³

t_p, total time of pumping = 200 days
Q, net pumping rate = 0.03 m³ s⁻¹
T, transmissivity = 3 × 10⁻³ m² s⁻¹
S_y, specific yield = 0.20
Qt, net volume of water pumped = (0.03 m³ s⁻¹) × (200 days) × (86 400 seconds per day) = 5.184 × 10⁵ m³

As a first step, compute v/Qt, the dimensionless ratio of the volume of stream depletion to volume of water pumped for the pumping period of interest which, in this example is equal to 5000/5.184 × 10⁵ = 0.01. Now, with this value of v/Qt, and using Curve B shown in Fig. 6.42, or the table of values given in Appendix 7, find the value of t/F, where F is the stream depletion factor

(arbitrarily defined by Jenkins (1968) as the time coordinate of the point where the volume of stream depletion, v, is equal to 28% of the volume of water pumped on a curve of v against t). If the system meets the assumptions of the idealized conceptual model shown in Fig. 6.41, then $F = L^2 S_y/T$, where L is the perpendicular distance from the abstraction borehole to the stream.

In this example, for v/Qt = 0.01, t/F = 0.12 (from Curve B), then:

$$t/F = 0.12 = tT/L^2 S_y$$

Rearranging and substituting values for T, S_y and t gives

$$L^2 = \left(200 \times 86\,400 \times 3 \times 10^{-3}\right)/(0.12 \times 0.20)$$

$$= 2.16 \times 10^6 \text{ m}^2$$

Hence, L, the required distance between the borehole and stream to avoid environmental impacts = 1470 m.

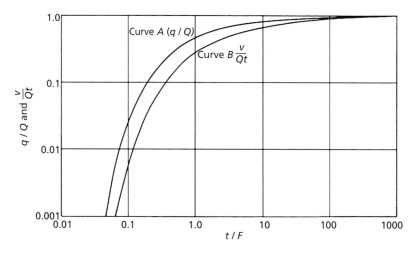

Fig. 6.42   Type curves to determine the rate and volume of stream depletion by boreholes and wells (Jenkins 1968). (*Source:* Jenkins, C.T. (1968) Computation of rate and volume of stream depletion by wells. In: *Techniques of Water-Resources Investigations of the United States Geological Survey, Chapter D1. Book 4, Hydrologic Analysis and Interpretation.* United States Department of the Interior, Washington, 17 pp.)

river flows. Calibration of the resource model was achieved against historic flows prior to naturalization of the river flow record. The main output from the model was the construction of flow duration curves at selected points in a river using predicted mean weekly flows.

The catchment resource model is an example of a lumped model that depends on summing all inputs to and outputs from defined river reaches within the catchment or sub-catchment (Fig. 6.43). The inflow to a single reach during a chosen time step is calculated as the sum of surface runoff, baseflow, abstractions and effluent returns distributed uniformly over the length of the reach. In each catchment, the underlying aquifer is represented as a single storage cell into which recharge is added and from which abstractions and baseflow are subtracted. The concept of catchment averaging greatly simplifies the modelling of aquifer behaviour which, in calculating net aquifer storage, apportions the effects of groundwater abstractions to predictions of baseflow output. By approximating

the aquifer area to an equivalent rectangular area, baseflow is related to storage within the cell by the empirical relationship, T/L^2S, controlled by the aquifer transmissivity, T, storage coefficient, S, and the distance, L, between the river and catchment boundary. As with river abstractions, groundwater abstractions are assumed to be distributed uniformly over the entire aquifer area.

Using their resource model, Hiscock *et al.* (2001) obtained reasonable results when comparing simulated and observed 95th percentile 7-day flows (the flows equalled or exceeded for 95% of the time) as the calibration target for lowland rivers in eastern England. As to be expected, predictive errors were caused by the simplification of adopting a single river reach (with no in-channel storage) and a single aquifer storage cell for the underlying Chalk aquifer, and also the adoption of a simple representation of runoff independent of antecedent soil moisture conditions.

With a catchment resource model, the naturalization of river flows is achieved by setting

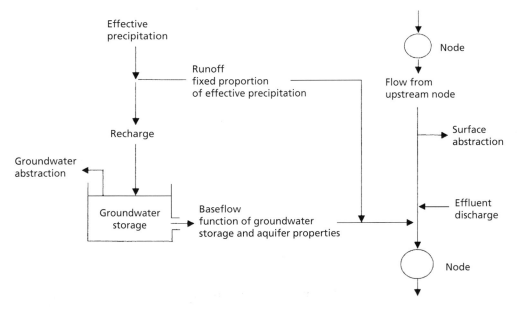

Fig. 6.43 Schematic diagram showing the type of lumped calculations performed for a single river reach within a catchment resource model (Hiscock *et al.* 2001). (*Source:* Hiscock, K.M., Lister, D.H., Boar, R.R. and Green, F.M.L. (2001) An integrated assessment of long-term changes in the hydrology of three lowland rivers in eastern England. *Journal of Environmental Management* **61**, 195–214.)

all net abstractions to zero. Then, the arithmetic difference between the historic net abstractions and nil net abstractions produces a series of net effects of abstractions and discharges. An example is shown in Fig. 6.44a for the River Wensum in Norfolk, presented as percentage depletions of mean weekly flows. The predicted maximum depletion in the mean weekly flows of 14% occurred in the drought of 1989–1992. In Fig. 6.44b, the modelled historic and

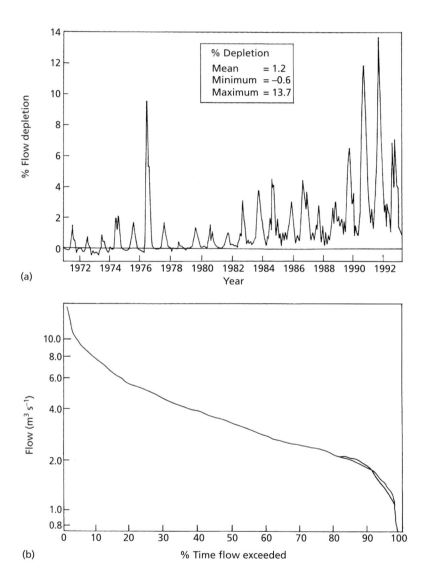

(a)

(b)

Fig. 6.44 Catchment resource model results illustrating the predicted effects of water resources management on flows in the River Wensum at Costessey Mill in Norfolk, eastern England. In (a) the percentage depletion of mean weekly flows are shown due to the net effect of surface water and groundwater abstractions and discharges for the period 1971–1992. A negative percentage depletion indicates a gain in flow. In (b) the 7-day flow duration curves are shown for both the naturalized and observed flows for the period 1971–1992. The upper curve shows natural flows; the lower curve observed flows (Hiscock *et al.* 2001). (*Source:* Hiscock, K.M., Lister, D.H., Boar, R.R. and Green, F.M.L. (2001) An integrated assessment of long-term changes in the hydrology of three lowland rivers in eastern England. *Journal of Environmental Management* **61**, 195–214.)

naturalized surface flow characteristics are shown as flow duration curves. Again, in the case of the River Wensum, the impacts of net abstractions are small, with the higher percentage flow depletion occurring when river flows are very low.

The catchment resource model was also used to predict the impact of an increase in spray irrigation to the total licensed amount from both surface water and groundwater sources (an important water use in the predominantly agricultural area of eastern England). For this scenario, the maximum potential depletion of the long-term (1971–1992) 95th percentile 7-day flow in the River Wensum is predicted to be 17%. Compared with flows during the drought of 1989–1992 when mean weekly flows in the river fell to $<0.8 \, \mathrm{m}^3 \, \mathrm{s}^{-1}$, the predicted flow depletion is approximately 30% of this mean. By highlighting spray irrigation, the catchment resource model illustrates the future challenge of managing water resources in eastern England with the additional factor of climate change (Section 10.4). Unless farmers adapt, the expected warmer and drier summers in this area are likely to lead to a greater demand for irrigation water in direct conflict with other water users and the need for water for environmental protection.

6.8.2.3 Global-scale surface water-groundwater modelling

In a study of the impact of global-scale exploitation of groundwater on streamflow, de Graaf *et al.* (2019) used a physically based model (GSGM) that simulates hydrological processes at 5 arc-minute resolution (approximately 10 km × 10 km at the equator) (see also Section 7.5). The model consists of the global hydrology and water resources model (PCR-GLOBWB) (Sutanudjaja *et al.* 2018) that is dynamically coupled via groundwater recharge and capillary rise, and by groundwater discharge and river infiltration to a two-layer global groundwater flow model based on MODFLOW (de Graaf *et al.* 2017) that

simulates lateral groundwater flow. The hydrological model runs at a daily time step and the groundwater model at a monthly time step. The hydrological model includes a water-use model that dynamically allocates sectoral water demand from irrigated agriculture, industries, households or livestock to abstraction of desalinated water, groundwater or surface water, based on the availability of these resources. To investigate the impact of past and future climate forcing (from 1960–2010 to 2011–2100, respectively), GSGM was run once with groundwater and surface water abstractions and once without (a natural run). Projected future water abstractions were assigned as a 'business-as-usual' scenario, in which industrial and domestic demands, as well as the extent of irrigated areas, remain unchanged after 2010, and where irrigation demands vary as a result of climate change only. For the climate scenarios, CMIP5 RCP8.5 was used as the greenhouse gas emissions scenario (the worst case), with the sensitivity of GSGM to climate input tested using the results of three global climate models (GCMs) selected for the wettest, average and driest model outcomes in terms of projected future global precipitation change.

GSGM model outcomes were considered in terms of when simulated streamflow reaches its environmental flow limit for the first time as a result of groundwater pumping when the threshold of environmentally critical streamflow is crossed for at least three consecutive months for two consecutive years. Environmentally critical streamflow is defined as the 90th percentile over 5 years (10% exceedance) of groundwater discharge. This measure of low flow highlights the dependence of ecosystem functions and services on streamflow when the contribution of groundwater discharge to streamflow is greatest. By comparing the 'natural' and human-influenced results, de Graaf *et al.* (2019) were able to exclude environmental flow limits that are reached as a result of climate-driven drought events alone.

The GSGM model results show that environmental flow limits caused by groundwater pumping have already been reached for a substantial number of catchments, estimated as approximately 15, 17 and 21% for the wettest, average and driest climate projections, respectively, and are likely to be reached for more than half of all catchments before the end of 2050. Globally, the estimated first times at which environmental flow limits will be reached peak at around 2030. Regions that have already reached their environmental flow limit are mainly located in the drier climates of the world, where discharge is small and irrigation depends more on groundwater. Hotspots of 'early limits', reached before 2010, are found for known groundwater depletion hotspots such as the High Plains aquifer and part of the Central Valley aquifer in North America, parts of Mexico, and the Upper Ganges and Indus basins. However, a considerable number of catchments where environmental flow limits have been reached are found outside of depletion hotspots, such as in the northeast USA and parts of Argentina. By 2050, new regions that are modelled to reach their environmental flow limit become apparent where the pressure on groundwater resources will increase due to projected drier climatic conditions that lead to an increase in irrigation water demand, such as in southern and central Europe and part of Africa.

de Graaf *et al.* (2019) also showed that only a small decline in groundwater level is needed to affect streamflow making model estimates uncertain for streams near a transition to reversed groundwater discharge. For many areas, however, groundwater pumping rates are high and environmental flow limits are known to be severely exceeded. The model results also highlight that, unlike surface water use, which immediately affects streamflow, the effect of groundwater pumping on streams can be substantially delayed by months to decades, with the serious implication that unsustainable groundwater abstractions are a 'ticking time bomb' for streamflow.

Further reading

Beven, K. (2012) *Rainfall-Runoff Modelling: The Primer* (2nd edn). John Wiley & Sons, Ltd, Chichester.

Healy, R.W. (2010) *Estimating Groundwater Recharge*. Cambridge University Press, Cambridge.

Hillel, D. (1982) *Introduction to Soil Physics*. Academic Press, Inc., Orlando, Florida.

Holden, J. (ed.) (2020) *Water Resources: An Integrated Approach* (2nd edn). Routledge, Abingdon.

Jury, W.A., Gardner, W.R. and Gardner, W.H. (1991) *Soil Physics* (5th edn). John Wiley & Sons, Inc., New York.

Shaw, E.M., Beven, K.J., Chappell, N.A. and Lamb, R. (2011) *Hydrology in Practice* (4th edn). Spon Press, Abingdon, Oxon.

Twort, A.C., Ratnayaka, D.D. and Brandt, M.J. (2000) *Water Supply* (5th edn). Butterworth-Heinemann, Oxford.

References

Allen, R.G., Luis, S., Pereira, L.S. *et al.* (1998) *Crop evapotranspiration: guidelines for computing crop water requirements*. FAO Irrigation and Drainage Paper 56. Food and Agriculture Organization of the United Nations, Rome.

Bense, V.F. and Kooi, H. (2004) Temporal and spatial variations of shallow subsurface temperature as a record of lateral variations in groundwater flow. *Journal of Geophysical Research* **109**, B04103. doi: 10.1029/2003JB002782.

Bense, V.F. and Kurylyk, B.L. (2017) Tracking the subsurface signal of decadal climate warming to quantify vertical groundwater flow rates. *Geophysical Research Letters* **44**, 12,244–12,253. doi: 10.1002/2017GL076015.

Beven, K. and Germann, P. (1982) Macropores and water flow in soil. *Water Resources Research* **18**, 1311–1325.

Blaney, H.F. and Criddle, W.D. (1962) Determining consumptive use and irrigation water requirements. *United States Department of Agriculture Technical Bulletin* 1275, 59p.

Brassington, R. (1998) *Field Hydrogeology* (2nd edn). John Wiley & Sons, Ltd., Chichester.

Bredehoeft, J.D. and Papadopoulos, I.S. (1965) Rates of vertical groundwater movement estimated from the Earth's thermal profile. *Water Resources Research* **1**, 325–328.

BSI (1998) *Measurement of Liquid Flow in Open Channels – General Guidelines for the Selection of Method.* BS ISO/TR 8363: 1997. British Standards Institution, London.

Butler, J.J., Zlotnik, V.A. and Tsou, M.-S. (2001) Drawdown and stream depletion produced by pumping in the vicinity of a partially penetrating stream. *Ground Water* **39**, 651–659.

Conant, B. (2004) Delineating and quantifying ground-water discharge zones using streambed temperatures. *Ground Water* **42**, 243–257.

Constantz, J. (2008) Heat as a tracer to determine streambed water exchanges. *Water Resources Research* **44**, W00D10. doi: 10.1029/ 2008WR006996.

Constantz, J., Stewart, A.E., Niswonger, R. and Sarma, L. (2002) Analysis of temperature profiles for investigating stream losses beneath ephemeral channels. *Water Resources Research* **38**, 1316. doi: 10.1029/2001WR001221.

Czarnomski, N.M., Moore, G.W., Pypker, T.G. *et al.* (2005) Precision and accuracy of three alternative instruments for measuring soil water content in two forest soils of the Pacific Northwest. *Canadian Journal of Forest Research* **35**, 1867–1876.

de Graaf, I.E.M., van Beek, R.L.P.H., Gleeson, T. *et al.* (2017) A global-scale two-layer transient groundwater model: Development and application to groundwater depletion. *Advances in Water Resources* **102**, 53–67.

de Graaf, I.E.M., Gleeson, T., van Beek, L.P.H. *et al.* (2019) Environmental flow limits to global groundwater pumping. *Nature* **574**, 90–94. doi: 10.1038/s41586-019-1594-4.

Doble, R.C. and Crosbie, R.S. (2017) Review: Current and emerging methods for catchment-scale modelling of recharge and evapotranspiration from shallow groundwater. *Hydrogeology Journal* **25**, 3–23.

Döll, P. and Zhang, J. (2010) Impact of climate change on freshwater ecosystems: A global-scale analysis of ecologically relevant river flow alterations. *Hydrology and Earth System Sciences* **14**, 783–799.

Durand, P., Neal, M. and Neal, C. (1993) Variations in stable oxygen isotope and solute concentrations in small submediterranean montane streams. *Journal of Hydrology* **144**, 283–290.

Edmunds, M. (1991) Groundwater recharge in the West African Sahel. NERC News April 1991, 8–10.

Edmunds, W.M. and Gaye, C.B. (1994) Estimating the spatial variability of groundwater recharge in the Sahel using chloride. *Journal of Hydrology* **156**, 47–59.

Essery, R. and Clark, D.B. (2003) Developments in the MOSES 2 land-surface model for PILPS 2e. *Global and Planetary Change* **38**, 161–164.

Freeze, R.A. and Cherry, J.A. (1979) *Groundwater.* Prentice-Hall, Inc., Englewood Cliffs, New Jersey.

Fulford, J.M., Thibodeaux, K.G. and Kaehrle, W. R. (1993) *Comparison of current meters used for stream gaging.* http://water.usgs.gov/osw/ pubs/CompCM.pdf (accessed 7 June, 2013).

Garcia Gonzalez, R., Verhoef, A., Luigi Vidale, P. and Braud, I. (2012) Incorporation of water vapor transfer in the JULES land surface model: Implications for key soil variables and land surface fluxes. *Water Resources Research* **48**, W05538 doi: 10.1029/2011WR011811.

Gee, G.W. and Hillel, D. (1988) Groundwater recharge in arid regions: Review and critique of estimation methods. *Hydrological Processes* **2**, 255–266.

Gleeson, T., Wada, Y., Bierkens, M.F.P. and van Beek, L.P.H. (2012) Water balance of global aquifers revealed by groundwater footprint. *Nature* **488**, 197–200.

Glenn, E.P., Doody, T.M., Guerschman, J.P. *et al.* (2011) Actual evapotranspiration estimation by ground and remote sensing methods: The

Australian experience. *Hydrological Processes* **25**, 4103–4116.

Grindley, J. (1967) The estimation of soil moisture deficits. *Meteorological Magazine* **96**, 97–108.

Grindley, J. (1969) *The Calculation of Actual Evaporation and Soil Moisture Deficits Over Specified Catchment Areas*. Hydrological Memoir 38. Meteorological Office, Bracknell.

Healy, R.W. and Cook, P.G. (2002) Using groundwater levels to estimate recharge. *Hydrogeology Journal* **10**, 91–109.

Hargreaves, G.H. and Samani, Z.A. (1982) Estimating potential evapotranspiration. *Proceedings of the American Society of Civil Engineers, Journal of the Irrigation and Drainage Division* **108**, 225–230.

Herschy, R.W. (2009) *Streamflow Measurement* (3rd edn). Taylor & Francis, Abingdon, Oxon.

Hewlett, J.D. and Hibbert, A.R. (1967) Factors affecting the response of small watersheds to precipitation in humid areas. In: *Forest Hydrology* (eds W.E. Sopper and H.W. Lull). Pergamon Press, New York, pp. 275–290.

Hillel, D. and Baker, R.S. (1988) A descriptive theory of fingering during infiltration into layered soils. *Soil Science* **146**, 51–56.

Hiscock, K.M., Lister, D.H., Boar, R.R. and Green, F.M.L. (2001) An integrated assessment of long-term changes in the hydrology of three lowland rivers in eastern England. *Journal of Environmental Management* **61**, 195–214.

Horton, R.E. (1933) The rôle of infiltration in the hydrologic cycle. *Transactions of the American Geophysical Union* **14**, 446–460.

Jenkins, C.T. (1968) Computation of rate and volume of stream depletion by wells. In: *Techniques of Water-Resources Investigations of the United States Geological Survey, Chapter D1. Book 4, Hydrologic Analysis and Interpretation*. United States Department of the Interior, Washington, 17 pp.

Jenkins, C.T. (1970) Techniques for computing rate and volume of stream depletion by wells. *Ground Water* **6**, 37–46.

Jones, J.A.A. (1997) *Global Hydrology: Processes, Resources and Environmental Management*. Addison Wesley Longman Ltd., Harlow, Essex.

Kalma, J.D., McVicar, T.R. and McCabe, M.F. (2008) Estimating land surface evaporation: A review of methods using remotely sensed surface temperature data. *Surveys in Geophysics* **29**, 421–469.

Kim, J.H. and Jackson, R.B. (2012) A global analysis of groundwater recharge for vegetation, climate, and soils. *Vadose Zone Journal*. doi: 10.2136/vzj2011.0021RA.

Kirk, S. and Herbert, A.W. (2002) Assessing the impact of groundwater abstractions on river flows. In: *Sustainable Groundwater Development* (eds K.M. Hiscock, M.O. Rivett and R.M. Davison). Geological Society, London, pp. 211–233. Special Publications 193.

Kitching, R. and Shearer, T.R. (1982) Construction and operation of a large undisturbed lysimeter to measure recharge to the chalk aquifer, England. *Journal of Hydrology* **58**, 267–277.

Kitching, R., Edmunds, W.M., Shearer, T.R. *et al.* (1980) Assessment of recharge to aquifers. *Hydrological Sciences Bulletin* **25**, 217–235.

Kong, X., Dorling, S. and Smith, R. (2011) Soil moisture modelling and validation at an agricultural site in Norfolk using the Met Office surface exchange scheme (MOSES). *Meteorological Applications* **18**, 18–27.

Kung, K.-J.S. (1990a) Preferential flow in a sandy vadose zone: 1. Field observation. *Geoderma* **46**, 51–58.

Kung, K.-J.S. (1990b) Preferential flow in a sandy vadose zone: 2. Mechanism and implications. *Geoderma* **46**, 59–71.

Kurylyk, B.L., Irvine, D.J. and Bense, V.F. (2019) Theory, tools, and multidisciplinary applications for tracing groundwater fluxes from temperature profiles. *WIREs Water* **6**, e1329. doi: 10.1002/wat2.1329.

Lerner, D.N., Issar, A.S. and Simmers, I. (1990) *Groundwater Recharge: A Guide to Understanding and Estimating Natural Recharge*. International Contributions to Hydrogeology **8**, http://www.iah.org/ downloads/pubfiles/IAHbook_ICH8.zip (accessed 7 June, 2013). Verlag Heinz Heise, Hannover.

Linsley, R.K., Kohler, M.A. and Paulhus, J.L.H. (1982) *Hydrology for Engineers* (3rd edn). McGraw-Hill, Inc., New York.

McDonnell, J.J. (2009) Classics in physical geography revisited. Hewlett, J.D. and Hibbert, A.R. 1967: Factors affecting the response of small watersheds to precipitation in humid areas. In Sopper, W.E. and Lull, H.W., editors, Forest hydrology, New York: Pergamon Press, 275–90. *Progress in Physical Geography* **33**, 288–293.

Moeck, C., von Freyberg, J. and Schirmer, M. (2018) Groundwater recharge predictions in contrasted climate: The effect of model complexity and calibration period on recharge rates. *Environmental Modelling and Software* **103**, 74–89.

Moeck, C., Grech-Cumbo, N., Podgorski, J. *et al.* (2020) A global-scale dataset of direct natural groundwater recharge rates: A review of variables, processes and relationships. *Science of the Total Environment* **717**, 137042.

Monteith, J.L. (1965) Evaporation and the environment. *Proceedings of the Symposium of the Society for Experimental Biology* **19**, 205–234.

Monteith, J.L. (1985) Evaporation from land surfaces: progress in analysis and prediction since 1948. In: *Advances in Evapotranspiration, Proceedings of the ASAE Conference on Evapotranspiration* (16–17 December 1985). American Society of Agricultural Engineers, St. Joseph, Michigan, pp. 4–12.

Owen, M. (1991) Relationship between groundwater abstraction and river flow. *Journal of the Institution of Water and Environmental Management* **5**, 697–702.

Penman, H.L. (1948) Natural evaporation from open water, bare soil and grass. *Proceedings of the Royal Society of London, Series A* **193**, 120–145.

Penman, H.L. (1949) The dependence of transpiration on weather and soil conditions. *Journal of Soil Science* **1**, 74–89.

Petheram, C., Walker, G., Grayson, R. *et al.* (2002) Towards a framework for predicting impacts of land-use on recharge: 1. A review of recharge studies in Australia. *Australian Journal of Soil Research* **40**, 397–417.

Philip, J.R. (1969) Theory of infiltration. In: *Advances in Hydroscience, Volume 5* (ed. V.T. Chow). Academic Press, New York, pp. 215–296.

Priestley, C.H.B. and Taylor, R.J. (1972) On the assessment of surface heat flux and evaporation using large-scale parameters. *Monthly Weather Review* **100**, 81–92.

Rantz, S.E. (1982) Measurement and computation of streamflow (2 vols). United States Geological Survey Water Supply Paper 2175.

Rubin, J. (1966) Theory of rainfall uptake by soils initially drier than their field capacity and its applications. *Water Resources Research* **2**, 739–749.

Running, S.W., Mu, Q., Zhao, M. and Moreno, A. (2017) *User's guide to MODIS global terrestrial evapotranspiration (ET) product (NASA MOD16A2/A3), Version 1.5*. NASA Earth Observing System, MODIS Land Algorithm. https://landweb.modaps.eosdis.nasa.gov/ QA_WWW/forPage/user_guide/ MOD16UsersGuide2016V1.52017May23.pdf (accessed 2 July, 2020).

Rushton, K.R. and Ward, C. (1979) The estimation of groundwater recharge. *Journal of Hydrology* **41**, 345–361.

Scanlon, B.R., Healy, R.W. and Cook, P.G. (2002) Choosing appropriate techniques for quantifying groundwater recharge. *Hydrogeology Journal* **10**, 18–39.

Scanlon, B.R., Keese, K.E., Flint, A.L. *et al.* (2006) Global synthesis of groundwater recharge in semiarid and arid regions. *Hydrological Processes* **20**, 3335–3370.

Selker, J., van de Giesen, N., Westhoff, M. *et al.* (2006) Fiber optics opens window on stream dynamics. *Geophysical Research Letters* **33**, L24401. doi: 10.1029/2006GL027979.

Sherman, L.K. (1932) Streamflow from rainfall by the unit-graph method. *Engineering News Record* **108**, 501–505.

Shuttleworth, W.J., Gash, J.H.C., Lloyd, C.R. *et al.* (1988) An integrated micrometeorological system for evaporation measurement.

Agricultural and Forestry Meteorology **43**, 295–317.

Smith, R.E. (2002) *Infiltration Theory for Hydrologic Applications*. Water Resources Monograph 15. American Geophysical Union, Washington, DC.

Smith, R.N.B., Blyth, E.M., Finch, J.W. *et al.* (2006) Soil state and surface hydrology diagnosis based on MOSES in the Met Office Nimrod nowcasting system. *Meteorological Applications* **13**, 89–109.

Sutanudjaja, E.H., van Beek, R., Wanders, N. *et al.* (2018) PCR-GLOBWB 2: A 5 arcminute global hydrological and water resources model. *Geoscientific Model Development* **11**, 2429–2453.

Tallaksen, L.M. (1995) A review of baseflow recession analysis. *Journal of Hydrology* **165**, 349–370.

Taniguchi, M. (1993) Evaluation of vertical groundwater fluxes and thermal properties of aquifers based on transient temperature-depth profiles. *Water Resources Research* **29**, 2021–2026. doi: 10.1029/93WR00541.

Thompson, N., Barrie, I.A. and Ayles, M. (1981) *The Meteorological Office Rainfall and Evaporation Calculation System: MORECS*. Hydrological Memorandum No. 45. Meteorological Office, Bracknell.

Thornthwaite, C.W. (1948) An approach towards a rational classification of climate. *Geographical Reviews* **38**, 55–94.

Tonolla, D., Wolter, C., Ruhtz, T. and Tockner, K. (2012) Linking fish assemblages and spatiotemporal thermal heterogeneity in a river-floodplain landscape using high-resolution airborne thermal infrared remote sensing and in-situ measurements. *Remote Sensing of Environment* **125**, 134–146.

Topp, G.C. and Davis, J.L. (1985) Measurement of soil water content using time-domain-reflectometry (TDR): A field evaluation. *Soil Science Society of America Journal* **49**, 19–24.

Valiantzas, J.D. (2006) Simplified versions for the Penman evaporation equation using routine weather data. *Journal of Hydrology* **331**, 690–702.

von Freyberg, J., Moeck, C. and Schirmer, M. (2015) Estimation of groundwater recharge and drought severity with varying model complexity. *Journal of Hydrology* **527**, 844–857.

Wada, Y., de Graaf, I.E.M. and van Beek, L.P.H. (2016) High-resolution modeling of human and climate impacts on global water resources. *Journal of Advances in Modeling Earth Systems* **8**, 735–763.

Wada, Y., van Beek, L.P.H. and Bierkens, M.F.P. (2012) Nonsustainable groundwater sustaining irrigation: A global assessment. *Water Resources Research* **48**, W00L06.

Wallace, R.B., Darama, Y. and Annable, M.D. (1990) Stream depletion by cyclic pumping of wells. *Water Resources Research* **26**, 1263–1270.

Ward, R.C. and Robinson, M. (2000) *Principles of Hydrology* (4th edn). McGraw-Hill Publishing Co., Maidenhead, Berkshire.

Wellings, S.R. and Bell, J.P. (1982) Physical controls of water movement in the unsaturated zone. *Quarterly Journal of Engineering Geology* **15**, 235–241.

Wilson, E.M. (1990) *Engineering Hydrology* (4th edn). Macmillan, London.

Wilson, J.L. (1993) Induced infiltration in aquifers with ambient flow. *Water Resources Research* **29**, 3503–3512.

Wilson, J.L. (1994) Correction to "induced infiltration in aquifers with ambient flow". *Water Resources Research* **30**, 1207.

WMO (1994) *Guide to Hydrological Practices* (5th edn). WMO Publication number 168. World Meteorological Office, Geneva.

Youngs, E.G. (1991) Infiltration measurements – a review. *Hydrological Processes* **5**, 309–320.

7

Groundwater investigation techniques

7.1 Introduction

Chapter 6 introduced the measurements needed to complete a catchment water balance including the calculation of groundwater recharge. In developing this approach, specifically the assessment of groundwater resources, this chapter describes the field methods that are applied in hydrogeological investigations. The measurement of groundwater levels and the presentation and interpretation of borehole hydrographs and groundwater level contour maps are first introduced. Next, field methods such as piezometer and pumping tests for the estimation of the aquifer properties of transmissivity and storativity are described. The application of tracer techniques in groundwater resources investigations is covered, together with an introduction to the application of downhole and surface geophysical techniques, remote sensing techniques and numerical groundwater flow and solute transport modelling in hydrogeology.

7.2 Measurement and interpretation of groundwater level data

The measurement and collection of groundwater level data are of fundamental importance in hydrogeology. Groundwater level data for an aquifer unit can be used for several purposes including plotting a hydrograph, determining the direction of groundwater flow by constructing a map of the potentiometric surface and in completing a flow net (Box 2.3). Values of hydraulic head are also essential in the process of designing and testing a numerical groundwater flow model for the purpose of making predictions of aquifer behaviour under future conditions.

7.2.1 Water-level measurement

Measurement of groundwater level in the field is undertaken using either a water level dipper or by use of a submersible pressure transducer positioned just below the lowest expected groundwater level. The required field measurement is the depth from a convenient measurement datum, for example the well top or borehole flange, to the position of the groundwater level. If the elevation of the measurement datum is known from levelling techniques (Pugh 1975), then the elevation of the groundwater level (or groundwater head, h) can be recorded as the height above or below a local base level, typically mean sea level.

A water level dipper for use in the field comprises a length of twin-core cable, graduated in centimetres and metres, wound on to a drum and with a pair of electrodes attached to the end (Fig. 7.1 and Fig. 7.2). When the electrodes touch the water surface, a circuit is completed which activates either a light or a buzzer or both. Water levels can be measured to a precision of ± 0.005 m (Brassington 1998).

Hydrogeology: Principles and Practice, Third Edition. Kevin M. Hiscock and Victor F. Bense.
© 2021 John Wiley & Sons Ltd. Published 2021 by John Wiley & Sons Ltd.
Companion website: www.wiley.com/go/hiscock/hydrogeology3e

General assembly

Detail of probe

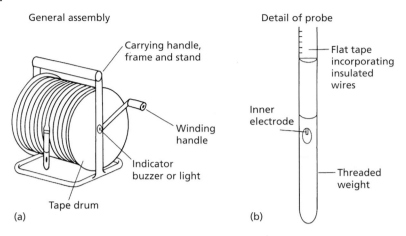

Fig. 7.1 Design of a water level dipper showing (a) the general assembly and (b) detail of the probe (Brassington 1998). (*Source:* Brassington, R. (1998) *Field Hydrogeology* (2nd edn). John Wiley & Sons, Ltd, Chichester. © 1998, John Wiley & Sons.)

Fig. 7.2 Water level dipper in use at a groundwater observation borehole.

A pressure transducer consists of a solid-state pressure sensor encapsulated in a stainless steel, submersible housing. A waterproof cable, moulded to the transducer, connects the water pressure sensor to a monitoring device such as a data logger from which a temporal record of groundwater level fluctuations can be obtained. When employing a pressure transducer and in order to convert to a value of groundwater head, h, the measurement of fluid pressure, P ($= \rho g \psi$), recorded at a depth

elevation, z, is converted to a value of groundwater head using the relationship given in eq. 2.22.

Groundwater levels are measured in either a well (Fig. 7.3a) or a purpose-built observation borehole (Fig. 7.3b). Wells are typically shallow in depth, lined with unmortared bricks and penetrate the top of the local water table in an unconfined aquifer. Observation boreholes can either be uncased (open) or cased, depending on the strength of the aquifer rock and

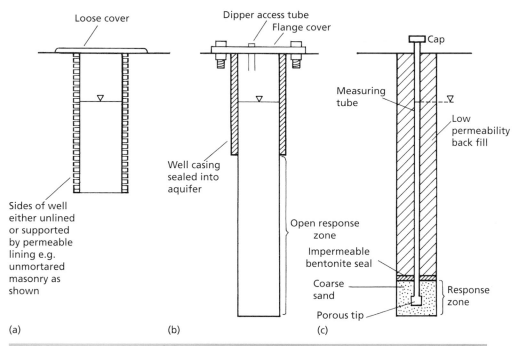

(d)

Fig. 7.3 General designs of (a) well, (b) observation borehole and (c) piezometer for the measurement of groundwater level (Brassington 1998). (*Source:* Brassington, R. (1998) *Field Hydrogeology* (2nd edn). John Wiley & Sons, Ltd, Chichester. © 1998, John Wiley & Sons.) In (d) a small percussion rig, particularly suited to site investigation or shallow exploration work, is based on a tripod that can be collapsed and towed behind a vehicle. The reciprocating action of the tool string, including the steel tube (shell) with its cutting shoe (seen resting at a shallow angle on the ground and attached to the cable) is achieved by direct operation of the cable winch.

record the groundwater level in unconfined aquifers (for which there is a water table) or confined aquifers (for which there exists a potentiometric surface). A special type of installation known as a piezometer (Fig. 7.3c) is designed to provide a measurement of the hydraulic head at a given depth in an aquifer. A bundle of piezometers nested in a single borehole installation can provide information on hydraulic heads at several depths in an aquifer from which the vertical component of groundwater flow, either downwards in a recharge area or upwards in a discharge zone, can be ascertained.

7.2.2 Well and borehole design and construction methods

Traditionally, boreholes are drilled with either an auger rig or, more usually, a percussion or rotary rig. Percussion drilling is used most often at shallow depths, while rotary methods predominate in the construction of deep boreholes and wells. Auger rigs vary from small-diameter manual augers for soil sampling to large truck- or crane-mounted augers used for drilling shafts that are more than a metre in diameter. The most common auger design is the screw auger, with a blade welded in a continuous spiral to a central solid shaft, usually supplied in 1-m sections. In continuous flight augering, drilling progresses by screwing the auger into the ground for one auger section and then withdrawing the auger. Another section is then added to the flight and the auger lowered to the bottom of the hole. The samples of material recovered by ordinary augering are held on the blades of the auger but are disturbed. Undisturbed core samples for core logging can be obtained by hollow-stem augering. The mechanics of the drilling are identical, but with the hollow-stem auger, the spiral blade is welded to a tube. As augering progresses, a core of sediment is forced into the tube or hollow stem to be recovered when the augers are withdrawn. Auger drilling is suitable for rapid formation sampling

at shallow depths, or even for drilling small observation boreholes provided that the formation is soft and cohesive. Augering is not possible in hard rocks, dry sands, or in gravels for which other methods are required. Below the water table, penetration by the auger may be impossible because of formation collapse.

Compared to auger drilling, percussion drilling is suited for shale, indurated sandstone and limestone, and even for slate, basalt, or granite, although progress is slow in these harder rocks. A heavy, solid-steel chisel-bit is suspended by a steel cable (cable-tool drilling) from a tripod or derrick, and dropped repeatedly. Once a certain amount of rock has been pulverized, a separate bit, known as the bailer and comprising a length of heavy-duty steel tube with a clack-valve set into its lower end, is lowered down the borehole to remove the debris, usually with the addition of water to loosen and collect the cuttings. A special tool is used for percussion drilling in soft, unstable formations such as clays and granular material. A steel tube or shell, either plain or with windows cut in the side to help sample removal, is employed with a cutting shoe at the bottom. The top end of the shell is open and is very similar in design and appearance to a normal bailer (Fig. 7.3d). Another difference between percussion drilling in unstable formations and hard rocks is that the former need support during drilling, and this means that temporary casing of sufficient diameter is needed to allow the string of permanent casing/screen to pass inside it on completion of the borehole.

Rotary drilling overcomes the problem of having to use temporary casing by using the hydrostatic pressure of circulating fluids to support the borehole wall. The use of drilling fluids enables boreholes to be drilled to much greater depths than with percussion rigs. Rotary rigs are divided into two types: direct circulation and reverse circulation, depending on the method used to circulate the drilling fluid. The drill string comprises lengths of heavy-duty steel tubing or drill pipe, with the drill bit assembly attached at the bottom. The design of the drill bit depends on the formation to be drilled. In soft formations, a simple drag bit equipped with hardened blades can be used. The commonest rotary bit is the tricone bit which has three hardened steel, toothed conical cutters which can rotate on bearings. The drilling fluid passes through ports which are placed to clean and cool the teeth as well as carry away the cuttings. The teeth on the cutters vary in size and number to suit the formation being drilled, with small, numerous teeth for hard formations and larger teeth for softer formations. The drilling fluid used can be clean water, if the formations are hard and competent, but the most common general-purpose drilling fluid for sediments is mud based on natural bentonite clay. A good bentonite mud will have a specific gravity of 1.2.

In direct circulation rotary drilling, the mud, which is a suspension of clay in water, is under the hydrostatic pressure of the column of mud in the borehole and is forced from suspension into the adjacent formations. The water leaves the clay behind as a layer or cake attached to the borehole wall. Bentonite-based mud has some important disadvantages for drilling water wells. The mud-cake can be securely keyed to the porous formation such that it is difficult to remove during well development. Alternatives to bentonite include organic polymers, foam and compressed air.

In reverse circulation rotary drilling, the drilling fluid is usually water which is pulled up the drill pipe by a centrifugal pump commonly aided by airlift. The reverse circulation system enables the drilling of large-diameter boreholes in loose formations and is based on the ability of the borehole walls to be supported by the hydrostatic pressure exerted by the water column and the positive flow of water from the borehole to the formations.

Further guidance on well and borehole design and construction methods are contained in useful textbooks by Driscoll (1986), Clark (1988), Misstear *et al.* (2017) and Glotfelty (2019).

7.2.3 Borehole hydrographs and barometric efficiency

Well or borehole hydrographs typically display data collected at monthly intervals, or at shorter (say 30-min) intervals if collected with a pressure transducer and provide a record of fluctuations in groundwater levels. As shown in Fig. 7.4, additional data can be shown on a hydrograph to indicate the position of measured monthly groundwater levels relative to the long-term average and to historic minima and maxima. Long-term records are invaluable. Climatic effects such as the frequency of wet and dry years can be identified (Box 10.7) as well as artificial effects, for example the over-exploitation of groundwater resources leading to a gradual decline in groundwater level (see Fig. 7.4b). Groundwater level drawdown data recorded during pumping tests (Section 7.3.2) can also be corrected for background trends in the regional potentiometric surface by reference to a hydrograph record unaffected by the pumping test.

Large fluctuations in water levels in wells and boreholes in confined aquifers can be caused by changes in atmospheric pressure. With increasing barometric pressure, water levels are noticed to decrease. This phenomenon relates to a change in the stress field applied to the aquifer (Jacob 1940). Using the principle of effective stress (Fig. 2.37) and considering the situation shown in Fig. 7.5a, the stress equilibrium at position X at the top of a confined aquifer is given by

$$\sigma_T + P_A = \sigma_e + P_w \qquad (\text{eq. 7.1})$$

where P_A is atmospheric pressure, σ_T is the total stress created by the weight of the overlying aquitard, σ_e is the effective stress acting on the aquifer material and P_w is the fluid pressure in the aquifer. The fluid pressure creates a pressure head, ψ, that is measured in the well penetrating the aquifer. At position Y in the well, the balance of pressures is

$$P_A + \gamma\psi = P_w \qquad (\text{eq. 7.2})$$

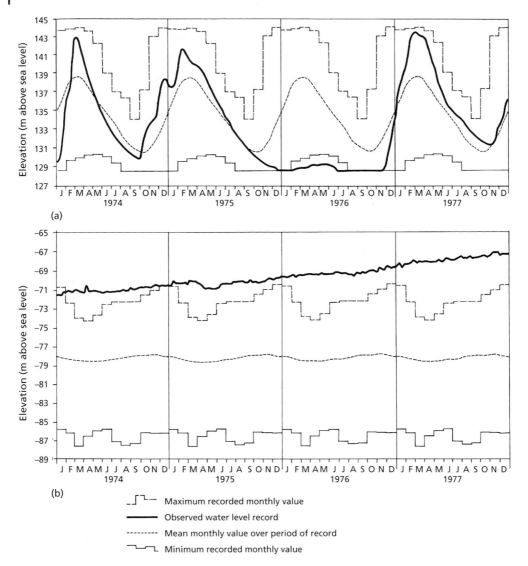

Fig. 7.4 Borehole hydrograph records (1974–1977) for the Chalk aquifer in southern England: (a) Rockley, Ogbourne St. Andrew (NGR 1655 7174; datum 146.6 m above sea level; records from 1933 to 1973 used to produce the maximum, minimum and average values); and (b) Trafalgar Square, London (NGR TQ 2996 8051; datum 12.6 m above sea level; records from 1953 to 1973 used to produce the maximum, minimum and average values). The record for Rockley shows the effect of drought in the dry year 1976 followed by the wet year 1977 when the groundwater level recovered. The record for Trafalgar Square shows the effect of rising groundwater levels in the London Basin following the reduction in abstraction rates after over-exploitation of the Chalk aquifer during the nineteenth and first half of the twentieth centuries (see Box 2.5). The rate of increase in groundwater level shown in this section of the borehole hydrograph is 1.2 m a^{-1}.

where γ is the specific weight of water. If, as shown in Fig. 7.5(b), the atmospheric pressure is increased by an amount dP_A, the change in the stress field at position X is given by

$$dP_A = d\sigma_e + dP_w \qquad \text{(eq. 7.3)}$$

Now, it can be seen that the change in dP_A is greater than the change in dP_w such that at

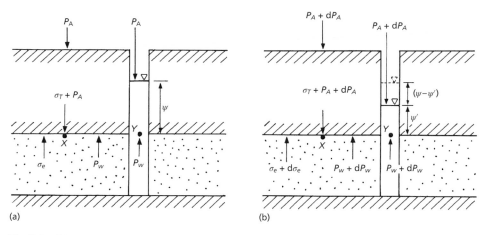

Fig. 7.5 Diagram showing the change in aquifer stress conditions caused by an increase in atmospheric pressure from P_A in (a) to $P_A + dP_A$ in (b) resulting in a decrease in groundwater level $(\psi - \psi')$ in a confined aquifer.

position Y in the well, the new balance of pressures is

$$P_A + dP_A + \gamma\psi' = P_w + dP_w \quad \text{(eq. 7.4)}$$

which, on substitution of eq. 7.2 in eq. 7.4, gives

$$dP_A - dP_w = \gamma(\psi - \psi') \quad \text{(eq. 7.5)}$$

Since $dP_A - dP_w$ is greater than zero, then $\psi - \psi'$ is also greater than zero, proving that an increase in atmospheric pressure leads to a decrease in water level (Fig. 7.6). In a horizontal, confined aquifer, the change in pressure head, $d\psi = \psi - \psi'$ in eq. 7.5, is equivalent to the change in hydraulic head, dh, and so provides a definition of barometric efficiency, B, expressed as follows:

$$B = \frac{\gamma dh}{dP_A} \quad \text{(eq. 7.6)}$$

The barometric efficiency of confined aquifers is usually in the range 0.20–0.75 (Todd 1980). Jacob (1940) further developed expressions relating barometric efficiency of a confined aquifer to aquifer and water properties, including the storage coefficient. In unconfined aquifers, atmospheric pressure changes are transmitted directly to the water table, both in the aquifer and in the well, such that the water level in an observation well does

not change. However, air bubbles trapped in pores below the water table are affected by pressure changes and can cause fluctuations similar to but smaller than observed in confined aquifers.

Fluctuations in water levels in wells and boreholes in confined aquifers can also be caused by the transient effect of external loading from passing trains, construction blasting, earthquakes (see example shown in Fig. 7.7) and terrestrial loading of water. Groundwater level fluctuations in response to distant earthquakes are generally associated with surface seismic waves, namely Rayleigh waves (Cooper *et al.* 1965), although small amplitude, early fluctuations have also been associated with S and Love waves (Brodsky *et al.* 2003). The commonly accepted mechanism is that seismic waves cause aquifers to expand and contract which in turn causes pore-pressure to oscillate. The degree to which the water level in a well fluctuates is determined by the dimensions of the well, the transmissivity, storage coefficient and porosity of the aquifer and also the type, period and amplitude of the seismic wave (Cooper *et al.* 1965).

Earthquake-induced permeability changes have been widely documented and can be categorized into two classes: horizontal

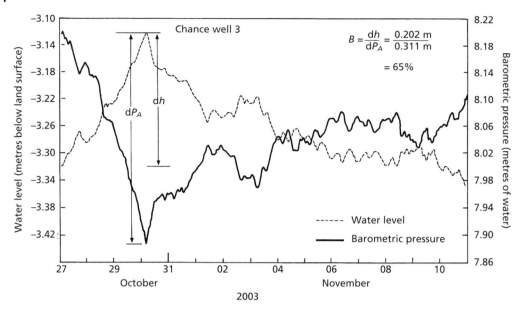

Fig. 7.6 Relationship between groundwater level fluctuation and atmospheric pressure in Chance Well 3 (CW-3), Long Valley Caldera, California (well depth 267 m; altitude 2155.90 m above sea level; latitude 37°38′49″, longitude 118°51′30″). At this site, water level and barometric pressure are monitored using a pressure transducer and barometer, respectively. These data are recorded on a data logger every 30 minutes and the data transmitted by satellite to the United States National Water Information System database. (*Source:* USGS Water Data for the Nation, The United States National Water Information System database. United States Department of the Interior.)

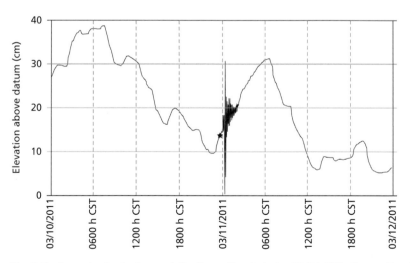

Fig. 7.7 Groundwater hydrograph for Bexar County Index Well J-17 in the confined karst limestone Edwards Aquifer, Texas showing the water level response to the maginitude-9.0 Sendai Earthquake that struck at 05 : 46 23.0 seconds UTC (or 2346 h CST as shown by the marked star) on 11 March 2011, approximately 100 km off the east coast of Japan. The earthquake, with its epicentre at longitude 38.322′ N, latitude 142.369′ E and depth of 24.4 km, is at a distance of approximately 10 000 km from the index well. Note that the vertical scale is with respect to a datum that is 203.34 m above mean sea level. (*Source:* Data from Edwards Aquifer Authority (http://www.edwardsaquifer.org)).

permeability changes in aquifers; and vertical permeability changes in confining layers (or aquitards) (Ingebritsen and Manga 2019; Zhang *et al.* 2019). Changes in the permeability of aquifers may have a large impact on groundwater flow and, thus, on groundwater supplies and solute transport, while vertical permeability changes in confining layers may compromise the safety of underground waste repositories (Zhang *et al.* 2019).

Conversely, changes in groundwater pressure caused by groundwater extraction (Foulger *et al.* 2018) or in response to heavy rainfall events can be a factor in the initiation of earthquakes and the possible outpouring of magma at the surface (Farquharson and Amelung 2020). The mechanical failure of rocks can occur when new fractures open, or existing faults slip, processes that can be promoted by pressure changes in groundwater. In the former case, increases in fluid pressure decrease the amount of stress needed to open new fractures. For the latter, faults can slip when the stresses acting parallel to the fault (shear stresses) overcome those perpendicular to the fault (normal stresses). Increasing fluid pressure in rock, for example in response to heavy rainfall, lowers normal stresses without changing shear stresses, thus promoting fault failure (Manga 2020). The minimum amount of stress observed to modulate earthquake activity is a few hundredths of a megapascal and possibly as little as a few thousandths, equivalent to a few tens of centimetres of water-table depth (Foulger *et al.* 2018).

7.2.3.1 Groundwater level fluctuations in the Bengal Basin Aquifer

Groundwater level fluctuations at coastal and inland sites in the fluvio-deltaic Bengal Aquifer System (BAS) in southern Bangladesh have been shown to respond dominantly to hydromechanical loading and unloading due to changes in terrestrial water storage above the aquifer surface by tidal and hydrological processes (heavy monsoon rainfall and deep surface water flooding) acting over periods

ranging from hours to months (Burgess *et al.* 2017). The poroelastic responses in the BAS are favoured by the low topographic relief that restricts gravitational flow, the low vertical hydraulic conductivity of the many silt-clay layers distributed throughout the aquifer, which further constrains the depth of groundwater circulation, and a pronounced sediment compressibility reflected in high-specific storage, S_s. Observations of groundwater levels responding to changes in near-surface water mass underlie the concept of 'geological weighing lysimeters'. The equation governing onedimensional, transient groundwater pressure ($\frac{\partial p}{\partial t}$) in permeable, elastically compressible sediments under purely vertical strain is

$$\frac{\partial p}{\partial t} = D\frac{\partial^2 p}{\partial z^2} + C\frac{\partial \sigma_T}{\partial t} \qquad (eq.7.7)$$

where the first term on the right-hand side describes flow-induced changes and the second term describes the effect of changes in vertical stress, or mass loading. D is the sediment hydraulic diffusivity (hydraulic conductivity divided by specific storage). C is the sediment loading efficiency, a dimensionless term with a value between 0 and 1 that depends on the distribution of surface load between the confined water and the solid matrix. σ_T is the total vertical stress applied as a mechanical load on the aquifer formation. Where the groundwater flow is negligible, as in the BAS, groundwater pressure responds instantaneously to addition or subtraction of mechanical load, as determined by

$$\frac{dp}{\partial t} = C\frac{\partial \sigma_T}{\partial t} \qquad (eq. 7.8)$$

Groundwater head response to mechanical loading potentially provides an empirical method for a direct measure of loading efficiency, C, the proportional change in groundwater pressure relative to change in total vertical stress, determined through its relationship with barometric efficiency, B, where

$$B + C = 1 \qquad (eq. 7.9)$$

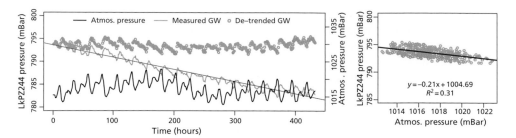

Fig. 7.8 Barometric efficiency, *B*, in the BAS determined from the inverse relationship between atmospheric pressure and groundwater pressure. Left: Groundwater hydrograph with time in hours from 14 to 31 January 2014 for piezometer LkPZ244 recorded adjacent to an open screened interval at 244 m in fine sand at Laksmipur, Bangladesh. Right: Barometric efficiency determined from the negative value of the linear regression slope (0.21) of the de-trended groundwater pressure versus atmospheric pressure (Burgess *et al.* 2017). (*Source:* Adapted from Burgess, W.G., Shamsudduha, M., Taylor, R.G. *et al.* (2017) Terrestrial water load and groundwater fluctuation in the Bengal Basin. *Nature Scientific Reports* **7**, 3872, doi: 10.1038/s41598-017-04159-w.)

From an analysis of the inverse relationship between atmospheric pressure and groundwater pressure during undisturbed periods of dry-season recession, when the trend of groundwater head decline can be determined and subtracted (see Fig. 7.8), Burgess *et al.* (2017) calculated a barometric efficiency, *B*, for the BAS of the order of 0.20 ± 0.03, giving values of loading efficiency, *C*, of 0.80 ± 0.03.

Applying the geological weighing lysimeter approach, the seasonal inundation of surface water, ΔTWS_m, can be estimated directly from the rise in freshwater-equivalent groundwater head recorded between the start (h_1) and end (h_2) of the rising limbs of the groundwater hydrograph and accounting for loading efficiency, *C*, where

$$\Delta TWS_m = (h_2 - h_1)/C \qquad \text{(eq. 7.10)}$$

Using estimates of *C* and groundwater hydrograph records for the BAS, Burgess *et al.* (2017) estimated the seasonal water inundation during the 2013 monsoon period to average an effective water depth of 0.90 m, compared with a value of 0.51 m derived from analysis of GRACE satellite data (Section 7.4) for the same year. The discrepancy between the two values arise potentially from the large contrast in observation scale and possible amplitude damping that occurs during the processing of GRACE data for which there is currently no ground-truth basis for a scaling correction (Burgess *et al.* 2017).

7.2.4 Construction of groundwater level contour maps

To be able to construct a map of the groundwater level and therefore depict the potentiometric surface and determine the direction of groundwater flow, a minimum of three observation points is required as shown in Fig. 7.9. The procedure is first to relate the field groundwater levels to a common datum (map datum or sea level for convenience) and then plot the positions of the observation points on a scale plan. Next, lines are drawn between three groundwater level measurements and divided into a number of short, equal lengths in proportion to the difference in elevation at each end of the line (in the example shown, each division on line AB and BC is 0.2 m, while on line AC each division is 0.1 m). The next step is to join points of equal elevation on each of the lines and then to select a contour interval which is appropriate to the overall variation in water levels in the mapped area (here 0.5 m). The same procedure is followed for other pairs of field observation points until one or two key contour lines can be mapped. At this point, the remaining contour lines can be drawn by interpolating between the field values.

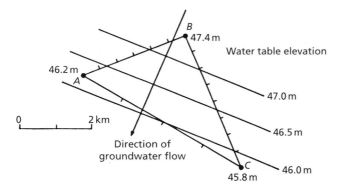

Fig. 7.9 Graphical construction method for determining the direction of groundwater flow from three groundwater level measurements.

Additional information that can be used in completing a potentiometric surface map is knowledge of the general topography of a region, and records of the elevations of springs known to discharge from an aquifer as well as the elevations of gaining streams and rivers (Fig. 6.37a) that flow over the aquifer outcrop since these points represent ground surface interception of the water table. For unconfined aquifers bordering the sea, it is usual to represent the coastline as a groundwater contour with an elevation equal to sea level (0 m). Similar assumptions can be made in respect of large surface water bodies at inland locations (Brassington 1998).

The direction of groundwater flow in an isotropic aquifer can be drawn at right angles to the contour lines on the potentiometric surface in the direction of decreasing hydraulic head. This assumes that the aquifer is an isotropic material (Section 2.4). In anisotropic material, for example fissured or fractured aquifers, the flow lines will be at an angle to the potentiometric contour lines (Box 2.3). An example of a completed potentiometric surface map for the Chalk aquifer in the London basin is shown in Box 2.4. Construction of potentiometric surface maps at times of low and high groundwater levels can be of assistance in calculating changes in the volume of water stored in an aquifer and in assessing the local effects of groundwater recharge and abstraction (Brassington 1998).

7.3 Field estimation of aquifer properties

7.3.1 Piezometer tests

Piezometer tests are small in scale and relatively cheap and easy to execute and provide useful site information, but are limited to providing values of hydraulic conductivity representative of only a small volume of ground in the immediate vicinity of the piezometer.

It is possible to determine the hydraulic conductivity of an aquifer by tests carried out in a single piezometer. Tests are carried out by causing a sudden change in the water level in a piezometer through the rapid introduction (slug test) or removal (bail test) of a known volume of water or, to create the same effect, by the sudden introduction or removal of a solid cylinder of known volume. Either way, the recovery of the water level with time subsequent to the sudden disturbance is monitored and the results interpreted.

For point piezometers that are open only for a short interval at their base (Fig. 7.10a), the interpretation of the water level versus time data commonly employs the Hvorslev (1951) method. Hvorslev (1951) found that the return of the water level to the original, static level occurs at an exponential rate, with the time taken dependent on the hydraulic conductivity of the porous material. Also, the recovery

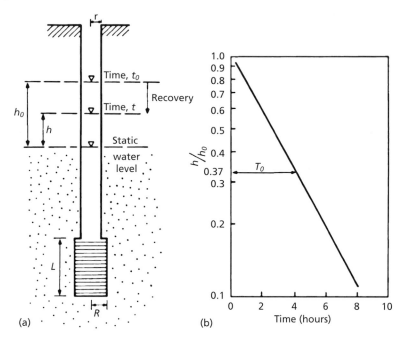

Fig. 7.10 Hvorslev piezometer test showing (a) the piezometer geometry and (b) the graphical method of analysis. T_o is the time lag or time taken for the water level to rise or fall to 37% of the initial change.

rate depends on the piezometer design; piezometers with a large area available for water to enter the response zone recover more rapidly than wells with a small open area. Now, if the height to which the water level rises above the static water level immediately at the start of a slug test is h_o and the height of the water level above the static water level is h after time, t, then a semi-logarithmic plot of the ratio h/h_o versus time should yield a straight line (Fig. 7.10b). In effect, using the ratio h/h_o normalizes the recovery between zero and one. If the length of the piezometer, L, is more than eight times the radius of the well screen, R, then the hydraulic conductivity, K, can be found from

$$K = \frac{r^2 \log_e(L/R)}{2LT_o}$$ (eq. 7.11)

where r is the radius of the well casing and T_o is the time lag or time taken for the water level to rise or fall to 37% of the initial change (Fig. 7.10b).

The Hvorslev method as presented here assumes a homogeneous, isotropic and infinite material and can be applied to unconfined conditions for most piezometer designs where the length is typically greater than the radius of the well screen. Hvorslev (1951) also presented formulae for anisotropic material and for a wide variety of piezometer geometries and aquifer conditions. For slug tests performed in fully or partially penetrating open boreholes or screened wells, the reader is referred to the method of Bouwer and Rice (1976) for unconfined aquifers and Bouwer (1989) for confined aquifers. The approach is similar to the Hvorslev method but involves using a set of curves to determine the radius of influence of the test.

7.3.2 Pumping tests

Pumping tests are generally of larger scale and duration compared with piezometer tests and are therefore more expensive, but can provide measurements of aquifer transmissivity and storativity that are representative of a large volume of the aquifer. In addition to measuring aquifer properties, pumping tests of wells and boreholes are also carried out to measure the variation of well performance with the discharge rate. Long-term pumping tests are invaluable in identifying boundary conditions, effectively describing the units of the aquifer providing water to the borehole being pumped. Pumping tests also provide a good opportunity to obtain information on water quality and its variation in time and perhaps with discharge rate.

When water is pumped from a well, the groundwater level in the well is lowered, creating a localized hydraulic gradient which causes water to flow to the well from the surrounding aquifer. The head in the aquifer is reduced and the effect spreads outwards from the well forming a cone of depression. The shape and growth of the cone of depression of the potentiometric surface depends on the pumping rate and on the hydraulic properties of the aquifer. Hence, by recording the changes in the position of the potentiometric surface in observation wells located around the pumping well it is possible to monitor the growth of the cone of depression and so determine the aquifer properties.

Different types of pumping test are undertaken with the most common being the step drawdown (variable discharge) and constant discharge tests. Step drawdown tests measure the well efficiency and the well performance. Constant discharge tests measure well performance and aquifer characteristics and help to identify the nature of the aquifer and its boundaries.

In a step drawdown test, the drawdown of water level below the pre-test level, s, in the pumped well is measured while the discharge rate, Q, is increased in steps. Observation boreholes are not required and analysis of the data provides a measure of the variation in specific capacity (Q/s) of the well with discharge rate, information that is invaluable in choosing the pump size and pump setting for the well in long-term production. Further discussion of the interpretation of step drawdown data is provided by Clark (1977) and Karami and Younger (2002).

The usual procedure for a constant discharge test is for water to be pumped at a constant rate from one well (the production well or pumped well) and the resulting change in the potentiometric surface to be monitored in one or more observation wells in close proximity to the pumped well (Figs. 7.11 and 7.12a). The constant discharge test programme has three parts: pre-test observations; pumping test; and observations during potentiometric recovery after the pumping has stopped. Prior to the start of the test, the initial water levels relative to a local datum must be measured and monitored for effects external to the pumping test, for example tidal fluctuations and barometric variations (Section 7.2.3), and the details of the site hydrogeology recorded, for example well depths and diameters, strata penetrated and the location of nearby streams that could act as recharge boundaries. From the start of the test, the pumping rate is monitored and paired values of drawdown and time are measured in the pumped and observation wells at specific time intervals that increase as the test progresses (Fig. 7.12b). Initially, the cone of depression expands rapidly, but the expansion slows logarithmically with time as the volume of aquifer contributing to the pumped well increases. Hence, the measurement time interval can also increase approximately logarithmically as the cone of depression grows. Eventually, a state of quasi- or actual equilibrium conditions may be reached, when the rate of recharge to the borehole catchment balances the rate of abstraction.

The last stage of the constant discharge test is the recovery phase after the pump has been switched off. On cessation of pumping, groundwater levels will recover to a static water level following a drawdown versus time curve that is approximately the converse of the drawdown curve (Fig. 7.12b). The groundwater levels should be measured from the time the pump is switched off with a similar, logarithmically increasing time internal as in the pumping test. In theory, the length of the recovery test is the same as the pumping test, but, in practice, the

Fig. 7.11 Monitoring groundwater drawdown in an observation borehole using a water level dipper and stop clock during a pumping test of the Luckington river support borehole in the southern Cotswolds, England.

recovery is monitored until the water level is within about 10 cm of the original static water level (Clark 1988).

For further details on the requirements of a pumping test, the reader is referred to the code of practice produced by the British Standards Institution (BSI 1992) and the procedures given by Walton (1987), Clark (1988) and Brassington (1998). One important consideration is the disposal of the discharged water. In tests of shallow aquifers, this water may infiltrate back into the aquifer and interfere with the test results. To avoid this problem, the pumped water should be abstracted to a point beyond the range of influence of the test.

The following sections provide a description of the common methods of pumping test analysis that are applied to confined aquifers and, where conditions allow, unconfined aquifers. For a fuller treatment of the various solutions and techniques for more complex aquifer conditions, the reader is directed to the specialist handbook by Kruseman and de Ridder (1990).

7.3.2.1 Thiem equilibrium method

Depending on whether equilibrium or transient conditions apply, the field data collected during a constant discharge test can be compared with theoretical equations (the Thiem and Theis equations, respectively) to determine the transmissivity of an aquifer. The Theis equation can also be applied to the transient data to determine aquifer storativity, which cannot be determined from equilibrium data.

For equilibrium conditions, and assuming radial, horizontal flow in a homogeneous and isotropic aquifer that is infinite in extent, the discharge, Q, for a well completely penetrating a confined aquifer can be expressed from a consideration of continuity as follows:

$$Q = Aq = 2\pi r b K \frac{dh}{dr} \qquad \text{(eq. 7.12)}$$

where A is the cross-sectional area of flow ($2\pi rb$), q is the specific discharge (darcy velocity) found from Darcy's Law (eq. 2.9), r is radial distance to the point of head measurement, b is

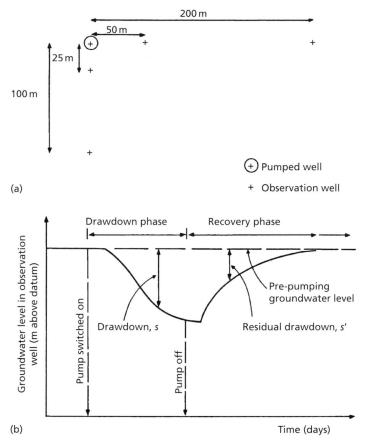

(a)

(b)

Fig. 7.12 (a) Suggested cruciform borehole array and (b) diagrammatic representation of the drawdown response in an observation well for a constant discharge test followed by a recovery test. By siting observation boreholes along two radii at right angles to each other, this enables the aquifer characteristics to be measured and an indication of aquifer geometry and anisotropy to be obtained (Toynton 1983). As a minimum, there should be at least one observation borehole in order to obtain reliable pumping test data. The recommended minimum period of a constant discharge test where observation wells are more than 100 m from the pumped well, or in an unconfined aquifer where response time is slow, is three days. A five-day test may be advised in situations where derogation effects on neighbouring wells need to be measured. Constant discharge tests involving more than one pumping well (a group test) are used to test wide areas of aquifer and require extended periods of pumping (Clark 1988). (*Source:* Clark, L. (1988) *The Field Guide to Water Wells and Boreholes.* Open University Press, Milton Keynes, Geological Society of London Professional Handbook Series. © 1988, John Wiley & Sons.)

aquifer thickness and K is the hydraulic conductivity. Rearranging and integrating eq. 7.12 for the boundary conditions at the well, $h = h_w$ and $r = r_w$, and at any given value of r and h (Fig. 7.13a), then

$$Q = 2\pi Kb \frac{(h - h_w)}{\log_e(r/r_w)} \qquad \text{(eq. 7.13)}$$

which shows that drawdown develops logarithmically with distance from the well. Equation 7.13 is known as the equilibrium, or Thiem, equation and enables the hydraulic conductivity or the transmissivity of a confined aquifer to be determined from a well being pumped at equilibrium, or steady-state,

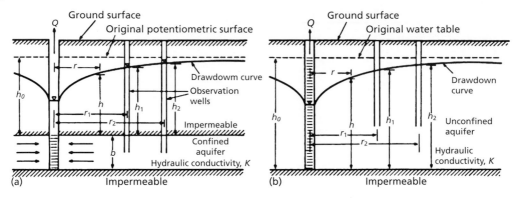

Fig. 7.13 Nomenclature and set-up of radial flow to (a) a well penetrating an extensive confined aquifer and (b) a well penetrating an unconfined aquifer.

conditions. Application of the Thiem equation requires the measurement of equilibrium groundwater heads (h_1 and h_2) at two observation wells at different distances (r_1 and r_2) from a well pumped at a constant rate. The transmissivity is then found from

$$T = Kb = \frac{Q}{2\pi(h_2 - h_1)} \log_e \frac{r_2}{r_1}$$

(eq. 7.14)

A similar equation for steady radial flow to a well in an unconfined aquifer can also be found for the set-up shown in Fig. 7.13(b). For a well that fully penetrates the aquifer, and from a consideration of continuity, the well discharge, Q, is

$$Q = 2\pi rKh \frac{dh}{dr}$$

(eq. 7.15)

which, upon integrating and converting to heads and radii at two observation wells and rearranged to solve for hydraulic conductivity, K, yields

$$K = \frac{Q}{\pi(h_2^2 - h_1^2)} \log_e \frac{r_2}{r_1}$$

(eq. 7.16)

This equation provides a reasonable estimate of K but fails to describe accurately the drawdown curve near to the well where the large vertical flow components contradict the Dupuit assumptions (Box 2.9). In practice, the drawdowns caused by pumping should be small (<5%) in relation to the saturated thickness

of the unconfined aquifer before eq. 7.16 is applied.

As an example of the application of the Thiem equation to find aquifer transmissivity, consider a well in a confined aquifer that is pumped at a rate of $2500 \ m^3 \ day^{-1}$ with the groundwater heads measured at two observation boreholes, A and B, at distances of 250 and 500 m, respectively, from the well. Once equilibrium conditions are established, the groundwater head measured at observation well A is 40.00 m and at observation well B is 43.95 m, both with reference to the horizontal top of the aquifer. Using this information, the aquifer transmissivity can be found from eq. 7.14 as follows:

$$T = \frac{2500}{2\pi(43.95 - 40.00)} \log_e \frac{500}{250} = 70 \ m^2 day^{-1}$$

(eq. 7.17)

7.3.2.2 Theis non-equilibrium method

Application of the Thiem equation is limited in that: it does not provide a value of the aquifer storage coefficient, S; it requires two observation wells in order to calculate transmissivity, T; and it generally requires a long period of pumping until steady-state conditions are achieved. These problems are overcome when the transient or non-equilibrium data are considered. In a major contribution to hydrogeology, Theis (1935) provided a solution to the following partial differential equation that describes unsteady, saturated, radial flow in a confined aquifer with transmissivity, T and storage coefficient, S:

$$\frac{\partial^2 h}{\partial r^2} + \frac{1}{r}\frac{\partial h}{\partial r} = \frac{S}{T}\frac{\partial h}{\partial t} \qquad \text{(eq. 7.18)}$$

By making an analogy with the theory of heat flow, and for the boundary conditions $h = h_o$ for $t = 0$ and $h \to h_o$ as $r \to \infty$ for $t \geq 0$, where h_o is the constant initial potentiometric surface (Fig. 7.13a), Theis derived an analytical solution to eq. 7.18, known as the non-equilibrium or Theis equation, written in terms of drawdown, s, as follows:

$$s = \frac{Q}{4\pi T} \int_u^\infty \frac{e^{-u}du}{u} \qquad \text{(eq. 7.19)}$$

where

$$u = \frac{r^2 S}{4Tt} \qquad \text{(eq. 7.20)}$$

For the specific definition of u given by eq. 7.20, the exponential integral in eq. 7.19 is known as the well function, $W(u)$, such that eq. 7.19 becomes:

$$s = \frac{Q}{4\pi T} W(u) \qquad \text{(eq. 7.21)}$$

A table of values relating $W(u)$ and u is provided in Appendix 6 and the graphical relationship of $W(u)$ versus $1/u$, known as the Theis curve, is given in Fig. 7.14.

The assumptions required by the Theis solution are

1) the aquifer is homogeneous, isotropic, of uniform thickness and of infinite areal extent;
2) the piezometric surface is horizontal prior to the start of pumping;
3) the well is pumped at a constant discharge rate;
4) the pumped well penetrates the entire aquifer, and flow is everywhere horizontal within the aquifer to the well;
5) the well diameter is infinitesimal so that storage within the well can be neglected; and
6) water removed from storage is discharged instantaneously with decline of groundwater head.

These assumptions are rarely met in practice, but the condition that the well is pumped at a constant rate should be checked during the field pumping test in order to limit calculation errors.

The Theis equations (eqs 7.20 and 7.21) can be used to predict the drawdown in hydraulic head in a confined aquifer at any distance, r, from a well at any time, t, after the start of pumping at a known rate, Q. For example, if a confined aquifer with a transmissivity of $500\ \text{m}^2\ \text{day}^{-1}$ and a storage coefficient of 6.4×10^{-4} is pumped at a constant rate of $2500\ \text{m}^3\ \text{day}^{-1}$, the drawdown after 10 days at an observation well located at a distance of $250\ \text{m}$ can be calculated as follows. First, a value of u is found from eq. 7.20:

$$u = \frac{250^2 \times 6.4 \times 10^{-4}}{4 \times 500 \times 10} = 2.0 \times 10^{-3}$$
$$\text{(eq. 7.22)}$$

and using the table in Appendix 6, the respective value of $W(u)$ is found to be 5.64. Substituting this value of $W(u)$ in eq. 7.21 gives the value of drawdown:

$$s = \frac{2500}{4 \times \pi \times 500} 5.64 = 2.24\ \text{m}$$
$$\text{(eq. 7.23)}$$

Conversely, the Theis equation enables determination of the aquifer transmissivity and

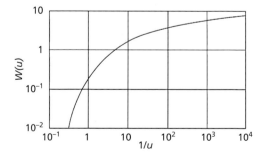

Fig. 7.14 The non-equilibrium type curve (Theis curve) for a fully confined aquifer.

storage coefficient by analysis of pumping test data. The Theis non-equilibrium method of analysis is based on a curve matching technique. An example of the interpretation of pumping test from a constant discharge test is given in Box 7.1.

7.3.2.3 Cooper–Jacob straight-line method

A modification of the Theis method of analysis was developed by Cooper and Jacob (1946) who noted that for small values of u ($u < 0.01$) at large values of time, t, the sum of the series beyond the term $\log_e u$ in the expansion of the well function, $W(u)$ (see Appendix 6) becomes negligible, so that drawdown, s, can be approximated as follows:

$$s = \frac{Q}{4\pi T}(-0.5772 - \log_e u) \quad \text{(eq. 7.24)}$$

By substituting eq. 7.20 for u in eq. 7.24 and noting that $\log_e u = 2.3\log_{10} u$, gives:

$$s = \frac{2.3Q}{4\pi T}\log_{10}\frac{2.25Tt}{r^2 S} \quad \text{(eq. 7.25)}$$

Since Q, r, T and S have constant values, a plot of drawdown, s, against the logarithm of time, t, should give a straight line. Furthermore, for two values of drawdown, s_1 and s_2, then:

$$s_2 - s_1 = \frac{2.3Q}{4\pi T}\left[\log_{10}\frac{2.25Tt_2}{r^2 S} - \log_{10}\frac{2.25Tt_1}{r^2 S}\right]$$
$$\text{(eq. 7.26)}$$

Therefore,

$$s_2 - s_1 = \frac{2.3Q}{4\pi T}\log_{10}\frac{t_2}{t_1} \quad \text{(eq. 7.27)}$$

and if $t_2 = 10t_1$ then:

$$s_2 - s_1 = \frac{2.3Q}{4\pi T} \quad \text{(eq. 7.28)}$$

Hence, from a semi-logarithmic plot of drawdown against time, the difference in drawdown over one log cycle of time on the straight-line portion of the curve will yield a value of transmissivity, T, using eq. 7.28. To find a value for the storage coefficient, S, it is necessary to

identify the intercept of the straight line plot with the time axis at $s = 0$, whereupon:

$$s = \frac{2.3Q}{4\pi T}\log_{10}\frac{2.25Tt_0}{r^2 S} = 0 \quad \text{(eq. 7.29)}$$

Therefore,

$$S = \frac{2.25Tt_0}{r^2} \quad \text{(eq. 7.30)}$$

To demonstrate the Cooper–Jacob method of analysis, Table 7.1 provides data for a pumping test in which a well in a confined aquifer is pumped at a rate of 0.01 m^3 s^{-1} and the drawdown in the potentiometric surface is recorded at an observation well situated 30 m away. The drawdown data are plotted in Fig. 7.15 and the difference in drawdown over one log cycle is found to be 1.75 m. Substitution of this value in eq. 7.28, taking care to work in a consistent set of units (here metres and days), gives

$$T = \frac{2.3 \times 0.01 \times 86400}{1.75 \times 4\pi} = 90 \text{ m}^2\text{day}^{-1}$$
$$\text{(eq. 7.31)}$$

The storage coefficient is found from eq. 7.30 with $t_o = 0.0022$ days where $s = 0$ m:

$$S = \frac{2.25 \times 90 \times 0.0022}{30^2} = 5 \times 10^{-4}$$
$$\text{(eq. 7.32)}$$

Table 7.1 Record of drawdown in an observation well situated 30 m from a well in a confined aquifer pumping at a rate of 0.01 m^3 s^{-1}.

Time since start of pumping (days)	Drawdown (m)
0.010	1.24
0.025	1.89
0.050	2.40
0.075	2.69
0.10	2.92
0.25	3.61
0.50	4.14
0.75	4.45
1.0	4.67
2.5	5.37

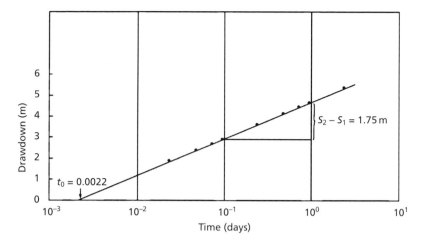

Fig. 7.15 Diagram showing the Cooper–Jacob semi-logarithmic plot of drawdown versus time for the data given in Table 7.1.

7.3.2.4 Recovery test method

At the end of a pumping test, when the pump is switched off, the water levels in the abstraction and observation wells begin to recover. As water levels recover, the residual drawdown, s', decreases (Fig. 7.12b). On average, the rate of recharge, Q, to the well during the recovery period is assumed to be equal to the mean pumping rate. Unlike the drawdown phase when the pumping rate is likely to vary (as seen in Box 7.1, Table 7.2), an advantage of monitoring the recovery phase is that the rate of recharge can be assumed to be constant and therefore satisfying one of the previous Theis solution assumptions.

The Theis method requires that pumping is continuous. Therefore, for the method to be applied to the recovery phase of a pumping test, a hypothetical situation must be conceptualized. If a well is pumped for a known period of time and then switched off, the following drawdown will be the same as if pumping had continued and a hypothetical recharge well with the same discharge were superimposed on the pumping well at the time the pump is switched off. From the principle of superimposition of drawdown (see next section), the residual drawdown, s', can be given as

$$s' = \frac{Q}{4\pi T}[W(u) - W(u')] \qquad \text{(eq. 7.33)}$$

where, for time t, measured since the start of pumping and time t', since the start of the recovery phase:

$$u = \frac{r^2 S}{4Tt} \quad \text{and} \quad u' = \frac{r^2 S}{4Tt'} \qquad \text{(eq. 7.34)}$$

For small values of u' and large values of t', the well functions can be approximated by the first two terms of eq. A6.1 so that eq. 7.33 becomes

$$s' = \frac{2.3Q}{4\pi T}\log_{10}\frac{t}{t'} \qquad \text{(eq. 7.35)}$$

Hence, a plot of residual drawdown, s', versus the logarithm of t/t' should provide a straight line. The gradient of the line equals 2.3 $Q/4\pi T$ so that $\Delta s'$, the change in residual drawdown over one log cycle of t/t', enables a value of transmissivity to be found from

$$T = \frac{2.3Q}{4\pi \Delta s'} \qquad \text{(eq. 7.36)}$$

It is not possible for a value of storage coefficient, S, to be determined by this recovery test method, although unlike the Theis and Cooper–Jacob methods, a reliable estimate of transmissivity can be obtained from measurements in either the pumping well or observation well. An example of the interpretation of recovery test data following a constant discharge pumping test is presented in Box 7.1.

Box 7.1   Interpretation of a constant discharge pumping test and recovery test

As an example of the interpretation of pumping test data, Table 7.2 gives the results of a constant discharge pumping test for a confined Chalk borehole site at Woolhampton in the south of England. A geological log for the abstraction borehole and a flow log, obtained using an impeller device, are shown in Fig. 7.16. The abstraction borehole was pumped at a rate of 6×10^3 m^3 day^{-1} and values of drawdown were recorded with time in an observation borehole located at a distance of 376 m. To obtain representative values of transmissivity and storativity, Fig. 7.17 shows the recorded values of

drawdown, s and time, t, plotted on log-log paper of the same scale as the type curve (Fig. 7.14). This field curve is then superimposed over the type curve, keeping the axes parallel and adjusting its position until the best match between field and types curves is achieved (Fig. 7.17). By selecting any point on the overlap as a match point, four values are obtained which define $W(u)$, $1/u$, s and t. For Woolhampton, the four match point values are: $W(u) = 1$; $1/u = 1$; $s = 1.5$ m; and $t = 2600$ seconds. Hence, rearranging eq. 7.21, and being

Table 7.2   Record of the constant discharge pumping test conducted at the Woolhampton Chalk borehole site, Berkshire.

Pumping test: Woolhampton River Regulation Borehole 56/106
Date: 19 April 1995

Observation borehole: 56/117	Depth: 74 m	Distance: 376 m
Location: Woolhampton, Berkshire	NGR: SU 574 662	
Reference level: 57.223 mOD (flange level)		
Initial water level: 56.673 mOD	Final water level: 55.953 mOD	

Time since start (seconds)	Depth to water level (m)	Drawdown (m)	Discharge (ML day^{-1})
0	0.55	0	0
30	0.56	0.01	6.07
60	0.56	0.01	6.02
90	0.56	0.01	5.94
120	0.565	0.015	5.96
150	0.565	0.015	6.00
180	0.565	0.015	5.94
240	0.57	0.02	6.02
300	0.57	0.02	5.97
360	0.57	0.02	6.01
480	0.575	0.025	6.00
600	0.58	0.03	5.94
720	0.585	0.035	5.90
900	0.595	0.045	6.02
1200	0.62	0.07	5.95
1500	0.65	0.10	5.95
1800	0.705	0.155	5.93
2400	0.815	0.265	5.89
3000	0.9	0.35	5.83
3600	1.02	0.47	5.91
4200	1.12	0.57	5.91
4800	1.20	0.65	5.91
5400	1.27	0.72	5.92

Box 7.1 (Continued)

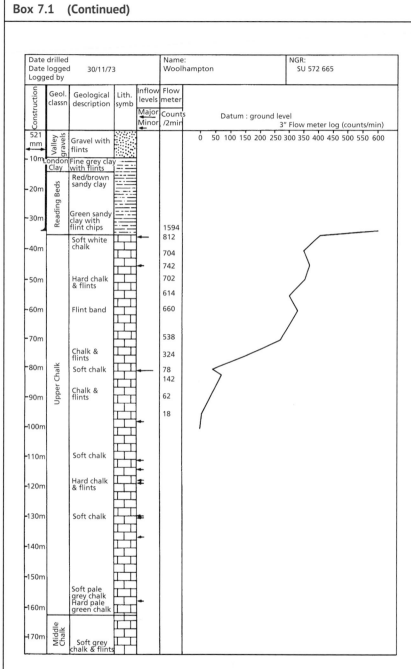

Fig. 7.16 Geological log and impeller flow log for the Woolhampton river regulation borehole, Berkshire, showing the confining beds of Tertiary strata overlying the Cretaceous Chalk aquifer. Notice the increase in flow in the upper section of the Chalk associated with fissured inflow horizons.

(Continued)

Box 7.1   (Continued)

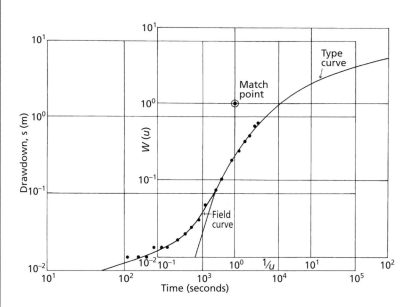

Fig. 7.17   Diagram showing the match point found from the overlay of a type curve (Theis curve, Fig. 7.14) to the field curve for the constant discharge pumping test data obtained for Woolhampton river regulation borehole (Table 7.2).

Table 7.3   Record of the recovery test conducted at the Woolhampton Chalk borehole site, Berkshire.

Pumping test: Woolhampton River Regulation Borehole 56/106

Date: 6 April 1999   NGR: SU 572 665

Reference level: 57.21 mOD (top of dip tube)

Time since start of test, t (seconds)	Time since pump stopped, t' (seconds)	t/t'	Residual drawdown, s' (m)
5460	60	91.0	3.275
5550	150	37.0	2.475
5580	180	31.0	2.455
5610	210	26.7	2.325
5640	240	23.5	2.235
5670	270	21.0	2.145
5700	300	19.0	2.025
5760	360	16.0	1.725
5820	420	13.9	1.715
5880	480	12.3	1.595
5940	540	11.0	1.500

Box 7.1 **(Continued)**

Table 7.3 (Continued)

Time since start of test, t (seconds)	Time since pump stopped, t′ (seconds)	t/t′	Residual drawdown, s′ (m)
6000	600	10.0	1.415
6120	720	8.5	1.305
6240	840	7.4	1.195
6360	960	6.6	1.065
6480	1080	6.0	0.985
6600	1200	5.5	0.845
6720	1320	5.1	0.755
6840	1440	4.8	0.675
6960	1560	4.5	0.595
7080	1680	4.2	0.525
7200	1800	4.0	0.475
7320	1920	3.8	0.425
7440	2040	3.6	0.375
7560	2160	3.5	0.275
7800	2400	3.3	0.205

careful to work in a consistent set of units (here in metres and seconds), a value for the Chalk aquifer transmissivity is found from

$$T = \frac{Q}{s4\pi} W(u) = \frac{6.0 \times 10^3}{1.5 \times 4 \times \pi \times 86400}$$
$$= 0.003684 \text{ m}^2\text{s}^{-1} \text{ or } 318 \text{ m}^2\text{day}^{-1}.$$
(eq. 1)

and using eq. 7.20, a value for the confined Chalk storage coefficient is found from

$$S = \frac{u4Tt}{r^2} = \frac{1 \times 4 \times 0.003684 \times 2600}{376^2}$$
$$= 2.7 \times 10^{-4}$$
(eq. 2)

As an example of the recovery test method, Table 7.3 lists the residual drawdown recorded following the Chalk borehole

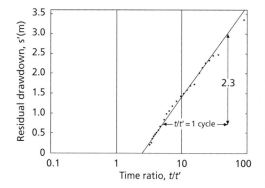

Fig. 7.18 Diagram showing the semi-logarithmic plot of residual drawdown, s′, versus the ratio t/t′ for the recovery test data obtained for Woolhampton river regulation borehole (Table 7.3). Time t is measured since the start of the drawdown phase and time t′ since the start of the recovery phase.

constant discharge pumping test. A plot of residual drawdown, s′, against the logarithm of t/t′ is shown in Fig. 7.18 and the value of Δs′ for one log cycle of t/t′ is found to equal

(Continued)

Box 7.1 (Continued)

2.3. Substitution of this value in eq. 7.36 gives, with attention to units in metres and seconds:

$$T = \frac{2.3 \times 6.0 \times 10^3}{4\pi \times 2.3 \times 86400}$$

$$= 0.005526 \ \mathrm{m \ s^{-1}} = 477 \ \mathrm{m^2 day^{-1}}$$

(eq. 3)

The calculated value of transmissivity from the recovery test is larger than the value obtained above from the constant discharge test ($318 \ \mathrm{m^2 \ day^{-1}}$) and is due to error introduced in the recovery test method in not having satisfactorily met the condition of a large recovery time, t', in order to satisfy a small value of u' as required by the approximation to $W(u)$ (eq. 7.24)

7.3.2.5 Principle of superposition of drawdown

As discussed in the previous section, the recovery test method relies on the principle of superposition of drawdown. As shown in Fig. 7.19, the drawdown at any point in the area of influence caused by the discharge of several wells is equal to the sum of the drawdowns caused by each well individually, thus:

$$s_t = s_1 + s_2 + s_3 + \cdots + s_n \quad \text{(eq. 7.37)}$$

where s_t is the drawdown at a given point and $s_1, s_2, s_3 \ldots s_n$ are the drawdowns at this point caused by the discharges of wells 1, 2, 3 ... n, respectively. Solutions to find the total drawdown can be found using the equilibrium (Thiem) or non-equilibrium (Theis) equations of well drawdown analysis and are of practical use in designing the layout of a well-field to minimize interference between well drawdowns or in designing an array of wells for the purpose of dewatering a ground excavation site.

For example, if two wells are 100 m apart in a confined aquifer ($T = 110 \ \mathrm{m^2 \ day^{-1}}$) and one well is pumped at a steady rate of 500 $\mathrm{m^3 \ day^{-1}}$ for a long period after which the drawdown in the other well is 0.5 m, what will be the steady-state drawdown at a point midway between the two wells if both wells are pumped at a rate of 500 $\mathrm{m^3 \ day^{-1}}$? First, considering one well pumping on its own, the drawdown at the mid-way position ($r_1 = 50 \ \mathrm{m}$) can be found from the Thiem equation (eq. 7.13). In this case, the head at the mid-way position (h_1) is equal to $h_o - s_1$, where h_o is the original

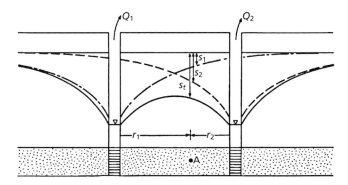

Fig. 7.19 Cross-section through a confined aquifer showing the cones of depression for two wells pumping at rates Q_1 and Q_2. From the principle of superposition of drawdown, at position A between the two pumping wells, the total drawdown, s_t, is given by the sum of the individual drawdowns s_1 and s_2 associated with Q_1 and Q_2, respectively.

piezometric surface prior to pumping and s_1 is the drawdown due to pumping. Therefore, $s_1 = h_o - h_1$. Similarly, $s_2 = h_o - h_2$ at $r_2 = 100$ m and eq. 7.13 becomes, expressed in terms of drawdown:

$$s_1 - s_2 = \frac{Q}{2\pi T} \log_e \frac{r_2}{r_1} \qquad \text{(eq. 7.38)}$$

In this example, with one well pumping on its own, s_1 is the unknown, $s_2 = 0.5$ m, $r_1 = 50$ m, $r_2 = 100$ m, $Q = 500$ m^3 day^{-1} and $T = 110$ m^2 day^{-1} giving:

$$s_1 = 0.5 + \frac{500}{2\pi 110} \log_e \frac{100}{50} = 1.0 \text{ m}$$

$$\text{(eq. 7.39)}$$

Now, considering both wells pumping simultaneously, and since the discharge rates for both are the same, then from the principle of superposition of drawdown, it can be determined that the total drawdown at the point mid-way between the two wells will be the sum of their individual effects, in other words 2.0 m.

7.3.2.6 Leaky, unconfined and bounded aquifer systems

The above solution methods for the non-equilibrium equation of radial flow apply to ideal, confined aquifers, but for leaky, unconfined and bounded aquifers variations of the curve matching technique must be applied. A summary of aquifer responses is provided here, but for a further treatment with worked examples of solution methods, including the case of partially penetrating wells, the reader is referred to Kruseman and de Ridder (1990). The determination of aquifer parameters from large-diameter dug well pumping tests is presented by Herbert and Kitching (1981). Once familiar with the various techniques for the analysis of pumping test data in different hydrogeological situations, it is then possible to use computer programs for the ease of estimating aquifer properties.

In the case of a leaky, or semi-confined aquifer, when water is pumped from the aquifer,

water is also drawn from the saturated portion of the overlying aquitard. By lowering the potentiometric head in the aquifer by pumping, a hydraulic gradient is created across the aquitard that enables groundwater to flow vertically downwards. From a consideration of Darcy's Law (eq. 2.7) and the sketch in Fig. 7.20b, the amount of downward flow is inversely proportional to the thickness of the aquitard (b') and directly proportional to both the hydraulic conductivity of the aquitard (K') and the difference between the water table in the upper aquifer unit and the potentiometric head in the lower aquifer unit. Compared with an ideal, confined aquifer (Fig. 7.20a), the effect of a leaky aquifer condition on the drawdown response measured at an observation well is to slow the rate of drawdown until a true steady-state situation is reached where the amount of water pumped is exactly balanced by the amount of recharge through the aquitard, assuming the water table remains constant (Fig. 7.20b). Methods of solution for the situation of steady-state and non-equilibrium conditions in a leaky aquifer with or without storage in the aquitard layer are provided by Hantush (1956) and Walton (1960).

Methods of pumping test analysis for confined aquifers can be applied to unconfined aquifers providing that the basic assumptions of the Theis solution are mostly satisfied. In general, if the drawdown is small in relation to the saturated aquifer thickness, then good approximations are possible. Where drawdowns are larger, the assumption that water released from storage is discharged instantaneously with a decline in head is frequently not met. As shown in Fig. 7.20(c), the drawdown response of unconfined aquifers typically resembles an S-curve with three distinct sections. At early time, following switching on the pump, water is released from storage due to compression of the aquifer matrix and expansion of the water in an analogous way to a confined aquifer (Section 2.9.1). A Theis type curve matched to this early data would give a value for storage coefficient comparable to a confined aquifer. As pumping continues and the water table is lowered, gravity drainage of water from the unsaturated zone in the developing cone of depression contributes delayed yield at a variable rate.

The pattern of drawdown in an unconfined aquifer depends on the vertical and horizontal hydraulic conductivity and the thickness of the aquifer. Once delayed yield begins, the drawdown curve appears to flatten (Fig. 7.20c) compared with the ideal, confined aquifer response. The drawdown is less than expected and resembles the response of a leaky aquifer. At later time, the contribution of delayed yield declines and groundwater flow in the aquifer is mainly radial producing a response that can be matched to a Theis type

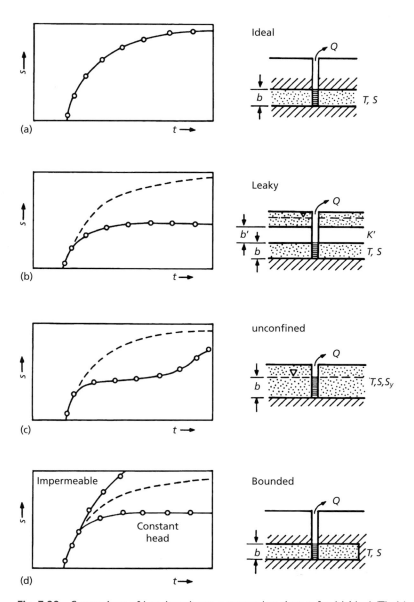

Fig. 7.20 Comparison of log drawdown, s, versus log time, t, for (a) ideal (Theis), (b) leaky, (c) unconfined and (d) bounded aquifer systems (Freeze and Cherry 1979). (*Source:* Freeze, R.A. and Cherry, J.A. (1979) *Groundwater.* Prentice-Hall, Inc., Englewood Cliffs, New Jersey. © 1979, Pearson Education)

curve. Values of storage coefficient calculated for this third segment of the curve provide a value for the specific yield, S_y, of the aquifer (Section 2.10). Graphical methods for interpreting pumping test data in unconfined aquifers which account for the differing aquifer responses are provided by Boulton (1963) and Neuman (1975).

When a well is pumped close to an aquifer boundary, for example an influent river or impermeable geological fault, the assumption that the aquifer is of infinite areal extent is no longer true and the drawdown response is of the type shown in Fig. 7.20d. As shown in Fig. 7.21a, where the boundary is a constant head, for example a surface water body such as the sea, a river, or a lake, the drawdown around the pumping well is less than expected compared with the ideal, confined aquifer of infinite extent, eventually reaching a steady-state condition with the amount of water pumped balanced by the water recharging from the constant head boundary. Where the boundary is an impermeable, no-flow boundary, then a greater drawdown than expected is observed than would be the case if the aquifer were infinite in extent (Fig. 7.21b).

Analysis of pumping test data affected by boundaries requires application of the principle of superposition of drawdown (Section 7.3.2.5). By introducing imaginary, or image wells with the same discharge or recharge rate as the real well, an aquifer of finite extent can be interpreted in terms of an infinite aquifer so that the solution methods described in the previous sections can be applied. For a well close to a constant head boundary, the image well is a recharging well placed at an equal distance from the boundary as the real well, but reflected on the opposite side of the boundary (Fig. 7.21a). For an impermeable boundary, the image well is a discharging well, again placed at an equal distance from the boundary as the real well, but on the opposite side of the boundary (Fig. 7.21b). Further explanation of image well theory is given by Todd (1980) and includes the case of a wedge-shaped

aquifer, such as a valley bounded by two converging impermeable boundaries. For complex regional aquifer situations or for the analysis of multiple well systems, then a numerical modelling approach to the solution of the steady-state or non-equilibrium groundwater flow equations is usually required (Section 7.5).

7.3.3 Tracer tests

In hydrogeology, a tracer is any substance contained in water or a water property that can be measured and used to infer environmental processes. An artificial tracer is deliberately released or injected for this purpose. In contrast, an environmental tracer is either naturally present in groundwater or produced as a result of human activity, but not specifically released for tracing purposes (see Chapter 5) (Cook 2015). The principal applications for tracer tests are in the determination of groundwater flowpaths and residence times, in the measurement of aquifer properties and in the mapping and characterizing of karst conduit networks. Much experience has been gained in karstic aquifers to demonstrate connectivity and measure travel times. In intergranular aquifers, tracer tests are used less frequently because of the slower groundwater velocities and the greater potential for dilution. To be of use, an ideal tracer should be non-toxic and easily measured at very large dilutions. The tracer should either be absent or present in very low concentrations in the groundwater system to be studied. An ideal tracer should also follow the same pathway as the substance to be investigated, whether particulates or solutes and should not react chemically with the groundwater or be absorbed on to the aquifer rock.

There is no tracer available that meets all of these criteria, but there is a wide range of substances and properties of water that can be used as tracers including: temperature; suspended particles (for example, spores, fluorescent microspheres); solutes (sodium chloride, halogen ions); dyes (fluorescent dyes, optical brighteners); gases (noble gases, sulfur hexafluoride);

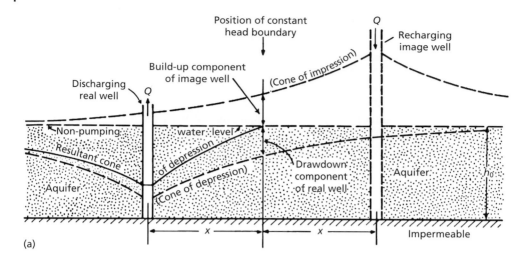

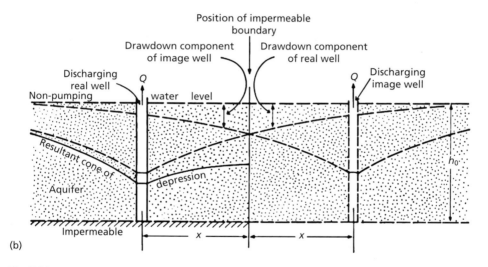

Fig. 7.21 Application of image well theory in the case of pumping wells affected by (a) constant head and (b) impermeable boundaries. The sections show the equivalent hydraulic system required to meet the Theis solution assumption of an aquifer of infinite areal extent (Ferris *et al.* 1962). (*Source:* Ferris, J.G., Knowles, D.B., Browne, R.H. and Stallman, R.W. (1962) Theory of aquifer tests. *United States Geological Survey Water Supply Paper* **1536-E**. United States Department of the Interior, Washington.)

microbes (bacteriophage); and environmental tracers such as chlorofluorocarbons (CFCs), radiocarbon, tritium and the stable isotopes of hydrogen, oxygen, nitrogen, sulphur and carbon (Ward *et al.* 1998). The choice of tracer type will depend on its suitability in terms of detectability, toxicity, relative cost and ease of use (Table 7.4) and the choice of tracer test method will be determined by the hydrogeological properties to be measured (Table 7.5).

The chlorofluorocarbons CFC-11 (trichlorofluoromethane), CFC-12 (dichlorodifluoromethane) and CFC-113 (trichlorotrifluoroethane) have received attention as groundwater tracers with useful reviews of their source, distribution in groundwaters and applications in tracing and age dating modern groundwaters provided by Plummer and Busenberg (2000) and Höhener *et al.* (2003). CFCs are synthetic, halogenated, volatile organic compounds that

Table 7.4 Summary of tracer types and their properties (Hobson 1993).

Tracer Type

	Lycopodium spores	Colloid	Phage	Inorganic salts	Fluorescent tracers	Fluorocarbons	Organic anions	Radio-isotopes	'Natural'
Conservative?	N	N	N	N/C	N/C	N	C	C	N/C
Quantifiable?	N (Yes if dyed)	Y	S	Y	Y	Y	Y	Y	Y
Sampling?	P	S	S	I/S	I/S/P	S	S	I/S	I/S/P
Detectability	?	H	H	L	M	H	M	H	M
Toxicity – actual	A	A	A	A/C	A	A	A	C	A/C
Relative cost (tracer and analysis)	E	E	E	C	C	E	E	E	C
Ease of use	D	D	E	E	E	D	E	D	E
Type of medium for which suitable	K	K/F/G	K/F/G	K/F/G	K/F/G	F/G	K/F/G	K/F/G	K/F/G
Types of test for which suitable*	3	1,6	4	1,2,5,6	1,2,3,4,5,6	6	1,2,4	1,2,5,6	3,4

Conservative?
C – Conservative
N – Non-conservative

Quantifiable?
N – No
Y – Yes
S – Semi-quantifiable

Sampling?
P – Passive detector
S – Sample of water required
I – In situ measurement possible

Detectability
? – Uncertain
L – Low (×10⁵ dilution)
M – Medium (×10⁸ dilution)
H – High (×10¹⁰ dilution)

Toxicity
A – Acceptable
C – Possible concern

Relative cost
E – Expensive
C – Cheap

Ease of use
D – Difficult
E – Easy

Type of medium
K – Karst
F – Fractured
G – Granular

*Types of tracer test for which suitable – see Table 7.5 for key
(*Source:* Adapted from Hobson, G. (1993) Practical use of tracers in hydrogeology. *Geoscientist* **4**, 26–27).

Table 7.5 Hydrogeological properties which may be measured using tracer tests (Ward *et al.* 1998).

Property to be determined	Suitable test methods
Measurement of flow paths	
– Connection between two or more points	3, 4, 6
– Direction of flow	3, 4
Measurement of velocities	
– Average linear water velocity	3, 4, 5, 6
– Specific discharge/darcy velocity	2
– Contaminant migration velocity	3, 4, 6
Measurement of aquifer properties	
– Hydraulic conductivity	2
– Effective porosity	5
– Heterogeneity	4
– Fracture characterization	4
– Matrix diffusion	1, 6, 4
Measurement of solute/contaminant transport properties	
– Dispersion	3, 4, 5, 6
– Sorption	1, 4
– Dilution	3, 4, 6
Measurement of recharge/groundwater catchments	3, 4
Measurement of groundwater age	3, 4

Key
Tracer test methods:
1) Laboratory test
2) Single borehole dilution
3) Natural gradient test (without boreholes)
4) Natural gradient test (multi-well)
5) Drift (injection) and pump back
6) Forced gradient (multi-well)
(*Source:* Ward, R.S., Williams, A.T., Barker, J.A. *et al.* (1998) Groundwater tracer tests: a review and guidelines for their use in British aquifers. *British Geological Survey Report* **WD/98/19**. © 1998, British Geological Survey.)

were manufactured from 1930 for use as aerosol propellants and refrigerants until banned by the Montreal Protocol in 1996. In general, CFC dating is most likely to be successful in rural settings, with shallow water tables, where the groundwater is aerobic and not impacted by local contaminant sources such as septic tanks or industrial applications (Plummer and Busenberg 2000). As discussed in Section 5.5.2, age dating techniques have limitations, with greater confidence in deriving

apparent age obtained when multiple dating techniques are applied to the same sample, for example CFCs and ^{3}H/^{3}He dating (Section 5.5.5).

Some of the most commonly used tracers in groundwater studies are the fluorescent dyes fluorescein (uranine) and rhodamine WT, the optical brightner photine CU and bacteriophage. The detection and assay of fluorescent dyes is made by illuminating the test solution at an appropriate narrow band of wavelengths (the excitation wavelength) and measuring the amount of fluorescent light emitted at a corresponding longer wavelength band (the emission wavelength). Representative excitation and emission spectra are given in Table 7.6 for eight fluorescent dyes. Measurements are made using a fluorometer and standards of known concentration used to derive a calibration curve relating fluorescence in arbitrary units to concentration, typically measured in $\mu g\,L^{-1}$ (Fig. 6.28a).

Smart and Laidlaw (1977) and Atkinson and Smart (1981) evaluated the suitability of different fluorescent dyes for water tracing and note their extensive application in Britain, particularly in the investigation of karstic limestones (Box 7.2). From the topological work of Brown and Ford (1971) and Atkinson *et al.* (1973), tracer tests can be used in mapping and characterizing karst conduit networks. By comparing discharges and masses of tracer at the entrance to and exit from a karst system, it is possible to classify the conduit network into one of five topological types (Fig. 7.22).

The scale of investigation of tracer tests can range from laboratory experiments to the site scale (tens or hundreds of metres) and to the regional scale (kilometres) in karst aquifers (Box 7.2). When compared with the scale of influence of other field methods for determining aquifer properties, Niemann and Rovey (2000) noted, in an area of glacial outwash deposits near Des Moines, Iowa, that the

Table 7.6 Excitation and emission maxima of tracer dyes and filter combinations for their analysis (Smart and Laidlaw 1977).

Dye	Maximum excitation (nm)	Maximum emission (nm)	Primary filter	Mercury line (nm)	Secondary filter
Blue fluorescent dyes					
Amino G acid	355(310)	445	7–37*	365	98[†]
Photine CU	345	435(455)			
Green fluorescent dyes					
Fluorescein	490	520	98[†]	436	55[†]
Lissamine FF	420	515			
Pyranine	455(405)	515			
Orange fluorescent dyes					
Rhodamine B	555	580	$2 \times 1 - 60^{*} + 61^{†}$	546	$4 - 97^{*} + 3 - 66^{*}$
Rhodamine WT	555	580			
Sulfo rhodamine B	565	590			

Notes: Figures in parentheses refer to secondary maxima. For all spectra, pH is 7.0
* Corning filter
[†] Kodak Wratten filter
(*Source:* Smart, P.L. and Laidlaw, I.M.S. (1977) An evaluation of some fluorescent dyes for water tracing. *Water Resources Research* **13**, 15–33. © 1977, John Wiley & Sons.)

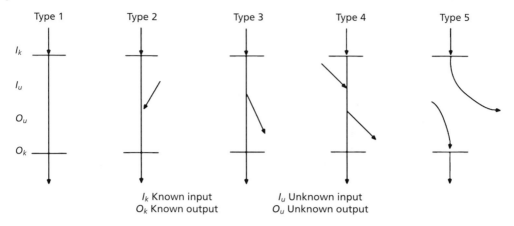

Type 1   Type 2   Type 3   Type 4   Type 5

I_k Known input
O_k Known output

I_u Unknown input
O_u Unknown output

Fig. 7.22 Five topological types of conduit network in karst groundwater systems (Brown and Ford 1971). (*Source:* Brown, M.C. and Ford, D.C. (1971) Quantitative tracer methods for investigation of karst hydrologic systems. *Transactions of the Cave Research Group of Great Britain* **13**, 37–51. © 1971, Cave Research Group)

hydraulic conductivity values of outwash determined from pumping tests by curve matching techniques (Section 7.3.2) can be an order of magnitude larger than values found from a tracer test using the conservative solute chloride. This discrepancy may be caused by the different scale and dimensionality of the two test methods (the cones of depression for the pumping tests were about 30–130 m while the tracer tests related to a zone of influence of less than 30 m) with dispersion of the tracer within the glacial outwash preventing the conservative solute from flowing exclusively within smaller, high permeability paths which have a strong influence on the groundwater flow and hydraulic conductivity measured by pumping tests. Hence, the results from Des Moines suggest that the velocity of a conservative solute plume in an intergranular aquifer may be overestimated if values of hydraulic conductivity derived from pumping tests were used in calculations.

At field and regional scales, tracer tests can be carried out with or without wells or boreholes and can be performed under natural hydraulic gradient conditions or under forced gradient (pumping) conditions (for example, Niemann

and Rovey 2000). The advantages and limitations of this range of tests and detailed protocols for conducting two of the most useful tests, the single borehole dilution method and the convergent radial flow tracer test are discussed by Ward *et al.* (1998). Both these tests can provide measurements of aquifer properties.

Relative to pumping tests, the advantages of the single borehole dilution method conducted under a natural hydraulic gradient are the low cost of materials and equipment and the simplicity of the method in determining values for specific discharge, q and aquifer hydraulic conductivity, K. Given its simplicity, dilution methods have been tested with many different tracers and appropriate detection probes in the past (Pitrak *et al.* 2007). The method requires injection of tracer into the whole water column in the borehole, or a packered interval, so that a well-mixed column of tracer of uniform initial concentration is obtained. The subsequent dilution of tracer is then monitored by means of an in situ detector or by careful depth sampling of the borehole to minimize disturbance of the concentration profile. The rate of change of concentration at any level, z, in the borehole is given by

Box 7.2   Dye-tracer test in the Chepstow Block Carboniferous limestone aquifer, south-west England

As part of an investigation into the risk of landfill leachate contamination of the Chepstow Block Carboniferous limestone aquifer in south-west England, a dye-tracer test was conducted to study the groundwater flow system (Clark 1984). The study area is shown in Fig. 7.23(a) with groundwater discharge from the Chepstow Block almost entirely focused on the Great Spring located in the Severn rail tunnel. The karstic nature of the limestone suggests that groundwater flow is through solutionally widened fissures or joints, with the joints parallel to the main north-westerly oriented faults that trend towards the Great Spring. The spring is an important water resource for the area having a mean flow of about $650 \, L \, s^{-1}$ ($56 \times 10^3 \, m^3 \, day^{-1}$) (Fig. 7.23c).

A dye-tracer test using fluorescein was designed to determine groundwater flow-paths and residence times and confirm the theory of conduit flow. A mass of 30 kg of fluorescein was injected into a sinkhole in the Cas Troggy Brook located to the north-west of the Great Spring in November 1982 and its recovery monitored at the Great Spring (Fig. 7.23b). The recovery curve shows a minimum travel time of 42 days and a peak travel time of 130 days over 7 km giving a groundwater velocity of $54 \, m \, day^{-1}$.

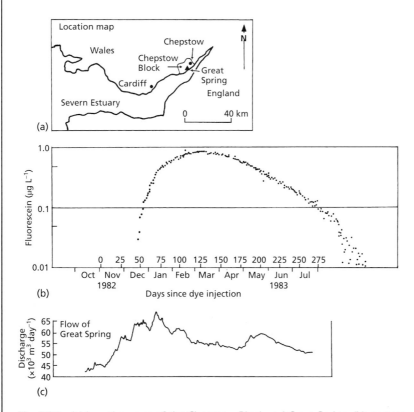

Fig. 7.23   (a) Location map of the Chepstow Block and Great Spring, (b) recovery of fluorescein dye-tracer and (c) variation in discharge at the Great Spring during the tracer test. (*Source:* Courtesy of WRC Plc)

(Continued)

Box 7.2 (Continued)

The curve can be simulated using an advection-dispersion model with a longitudinal dispersivity (Section 8.3.1) of 720 m. This high dispersivity, even allowing for the scale of the test, suggests that groundwater flow is through a complex fissure system, rather than as pure conduit flow (Clark 1984).

The total recovery of the fluorescein tracer was about 23% suggesting losses by

adsorption and degradation of some 77%. The maximum concentration of dye-tracer recovered was 0.92 µg L^{-1} giving a dilution of the dye input of 6×10^7 and leading to the conclusion that the impacts of waste disposal activities on the groundwater quality of the Great Spring will be considerably lessened by dilution in the limestone aquifer.

$$\frac{\partial C(z,t)}{\partial t} = \frac{\partial(Cu(z))}{\partial z} - \frac{CQ_o(z)}{\pi r^2}$$

(eq. 7.40)

where $C(z,t)$ is the concentration at time t, $u(z)$ is the vertical velocity of water in the borehole, $Q_o(z)$ is the volume of water leaving the borehole per unit depth per unit time and r is the borehole radius. The first term on the right-hand of eq. 7.40 is the change in concentration due to vertical flow in the borehole and the second term is the effect of tracer solution leaving the borehole. Solution of eq. 7.40 depends upon the form of the functions $u(z)$ and $Q_o(z)$. If it is assumed that there is no vertical flow in the borehole, then $u(z)$ is equal to zero and eq. 7.40 becomes

$$\frac{\partial C(z,t)}{\partial t} = -\frac{CQ_o(z)}{\pi r^2}$$

(eq. 7.41)

which, for a fixed value of z, yields upon integration:

$$C = C_o \exp\left[-\frac{Q_o t}{\pi r^2}\right]$$

(eq. 7.42)

If it is assumed that the water flowing through the borehole is drawn from a width of aquifer equal to twice the borehole diameter, then, for a one-dimensional flow system with parallel streamlines, $Q_o = 2qd$ and eq. 7.42 becomes (Lewis *et al.* 1966):

$$C = C_o \exp\left[-\frac{8qt}{\pi d}\right]$$

(eq. 7.43)

where C_o is concentration at time $t = 0$, d is the borehole diameter and q is the horizontal specific discharge or darcy velocity (Section 2.6). Hence, the dilution of tracer in the borehole should be exponential, with a time constant related to the specific discharge. By plotting concentration versus time on semi-logarithmic paper, the specific discharge can be calculated by substituting any two values of tracer concentration and the corresponding time interval into eq. 7.43. Calculation of the hydraulic conductivity can then be made if the hydraulic gradient is known using eq. 2.9. An example of the application of the single borehole dilution method is given in Box 7.3.

In a convergent radial flow tracer test, tracer is added to a soakaway, well, or piezometer and the breakthrough monitored at a pumping well. A divergent radial flow tracer test is also possible where tracer is added to an injection well and the forced plume of tracer observed in an array of surrounding observation points. In comparison, a shortcoming of the convergent radial flow test is that the converging flow field counteracts spreading due to dispersion. An example of the convergent radial flow test is given in Box 7.4.

7.3.4 Downhole geophysical techniques

Downhole (borehole) and surface geophysical techniques are now routinely used in hydrogeological investigations and take advantage of

Box 7.3 Single borehole dilution tracer test conducted at a Chalk aquifer site

As part of an investigation into the flow behaviour of a Chalk aquifer at a site in southern England that is subjected to artificial recharge of treated sewage effluent, 8.5 g of the optical brightner amino G acid were injected into a Chalk observation borehole in a single borehole dilution tracer test. Using a depth sampler to obtain groundwater samples, the decrease in concentration of amino G acid was measured daily at intervals of 1 m in the open section of the borehole (Hiscock 1982). Results for the dilution of amino G acid during the six days of the test are shown in Fig. 7.24(a) for the Chalk horizon at 30 m below ground level (m bgl). By choosing C_o = 3600 µg L^{-1} at t = 0 and C = 15 µg L^{-1} at t = 3.6 days, the specific discharge, q, at this level is calculated, using eq. 7.43 for a borehole diameter of 150 mm, as follows:

$$15 = 3600 \exp \left[- \frac{8q3.6}{\pi 0.15} \right] \qquad \text{(eq. 1)}$$

Hence, q = 0.09 m day^{-1}

The calculated specific discharge values for all the sampling levels are shown in Fig. 7.24(b) with the resulting profile indicating the presence of solutionally widened fissures in the Chalk to explain the higher values recorded between 18–21 m bgl and at 26 and 30 m bgl. For a regional hydraulic gradient of 0.005 (Fig. 2.61a), the mean Chalk hydraulic conductivity found using the dye-tracer results is 18 m day^{-1}, ranging from a minimum value of 12 m day^{-1} to a maximum value of 37 m day^{-1}.

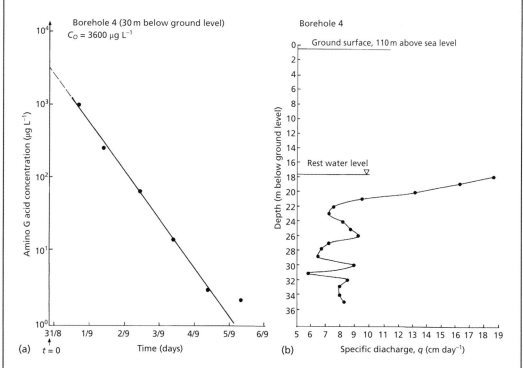

Fig. 7.24 Results of a single borehole dilution tracer test in an open Chalk borehole at Ludgershall, Hampshire (NGR SU 272 497) showing (a) the logarithmic decrease in amino G acid dye-tracer concentration with time at a depth of 30 m below ground level and (b) the profile of specific discharge values, q, obtained from the tracer test.

Box 7.4   Convergent radial flow tracer test conducted at a sand and gravel aquifer site

In an investigation of the risk of road drainage entering a soakaway and contaminating a well pumping from a sand and gravel aquifer, Bateman *et al.* (2001) conducted a convergent radial flow test to establish a connection between the soakaway and well and also to determine a value for the aquifer longitudinal dispersivity (Section 8.3.1). The well has an open base with a single horizontal collecting pipe extending 18.3 m to the north. To test a connection between the road soakaway and the pumping well, a distance of 30 m, 60 g of fluorescein were injected into the soakaway and the recovery monitored at the well. Prior to injection, 500 L of water were trickled into the soakaway to wet the unsaturated zone above the shallow water table. The tracer was flushed into the aquifer following injection by trickling a further 2000 L of water into the soakaway.

The fluorescein tracer recovery (Fig. 7.25) clearly shows that there is a connection between the road soakaway and the well with fluorescein breakthrough beginning at 6–7 days after injection, with the peak arriving at approximately 16 days. A recovered mass of 37.7 g represents 63% of the injected fluorescein, with some of the loss explained by the fact that pumping of the well ceased while recovery of the fluorescein continued.

Analysis of the aquifer dispersivity from the tracer breakthrough data is possible using the numerical method of Sauty (1980) developed for application in either diverging or converging groundwater flow problems. Sauty (1980) produced a set of type curves with tracer results presented in terms of dimensionless concentration, C_R, versus dimensionless time, t_R, for various Peclet numbers, P, with

$$C_R = \frac{C}{C_{max}} \qquad \text{(eq. 1)}$$

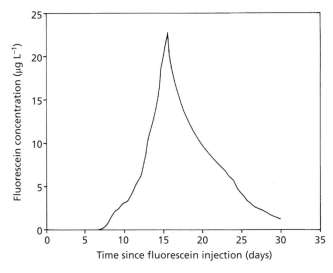

Fig. 7.25   Recovery of fluorescein dye-tracer at Quay Lane well, Southwold, Suffolk (NGR TM 486 774) recorded during a convergent radial flow test in a sand and gravel aquifer (Bateman *et al.* 2001). (Source: Bateman, A.S., Hiscock, K.M. and Atkinson, T.C. (2001) Qualitative risk assessment using tracer tests and groundwater modelling in an unconfined sand and gravel aquifer. In: *New Approaches Characterizing Groundwater Flow* (eds K.-P. Seiler and S. Wohnlich). Swets & Zeitlinger, Lisse, pp. 251–255. © 2001, Taylor & Francis.)

Box 7.4 (Continued)

$$t_R = \frac{t}{t_c} \qquad\qquad \text{(eq. 2)}$$

$$P = \frac{r}{\alpha} \qquad\qquad \text{(eq. 3)}$$

where C_{max} is equal to the peak tracer concentration, t_c is the time to peak concentration, r is radial distance and α is aquifer longitudinal dispersivity. By superimposing the field data on the Sauty type curves

(Fig. 7.26) and notwithstanding the presence of the horizontal collecting pipe that disturbs the assumption of symmetrical flow around the well, the best fit line to the data suggests a Peclet number of between 30 and 100, giving an aquifer dispersivity value of between 1.0 and 3.3 m for a radial distance, r, of 30 m.

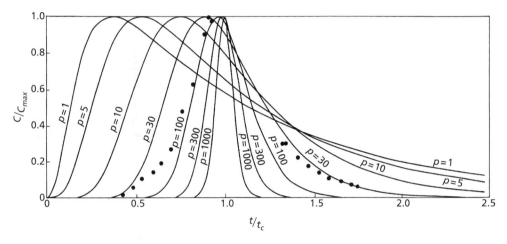

Fig. 7.26 Fluorescein concentration data from the radial flow test at Quay Lane well, Southwold plotted on Sauty (1980) type curves (Bateman *et al.* 2001). (*Source:* Bateman, A.S., Hiscock, K.M. and Atkinson, T.C. (2001) Qualitative risk assessment using tracer tests and groundwater modelling in an unconfined sand and gravel aquifer. In: *New Approaches Characterizing Groundwater Flow* (eds K.-P. Seiler and S. Wohnlich). Swets & Zeitlinger, Lisse, pp. 251–255. © 2001, Taylor & Francis.)

modern techniques and instrumentation that have benefited from advances in electronics and digital technology. Although the two types of geophysics can be considered complementary, surface geophysics is generally employed at an early stage in a hydrogeological investigation, before boreholes are drilled, while logging techniques are employed later to obtain detailed information on aquifer and fluid properties. A downhole geophysical log of physical properties is of value in groundwater investigations in correlating stratigraphic beds bounding aquifer units and in estimating aquifer

properties such as porosity, fluid content and approximate permeability.

All downhole geophysical logging is carried out by lowering a detector or 'sonde' on a multicore cable and recording its output on a recorder at the surface. Several measurements can be made simultaneously using a multichannel recorder. Sondes may be several metres in length with a diameter suited to common borehole diameters in the range 0.1–0.5 m. The sonde may be free to move from side to side of the hole or held centrally. Qualitative and quantitative analysis of the borehole records and computer analysis of digitized logs are used to derive hydrogeological information.

Geophysical borehole logs can provide information on the construction of boreholes and on the geometric and temporal character of the surrounding fluid-saturated rocks. The response of borehole logs is caused by petrophysical factors, the temperature and quality of interstitial fluids and groundwater flow. The benefits of geophysical borehole logging are in collecting continuous, subsurface information for application in the design and location of boreholes, in the definition of geological structure, high porosity and permeable zones and in the identification of groundwater flow patterns, zones of saline water and groundwater sampling levels.

There are three broad areas of interest in the collection of geophysical borehole logs. First, constructional logging provides information on borehole casing location, casing cement bonding and location, borehole dip and orientation, borehole collapses, borehole diameter and the position of fissures/fractures in an open borehole. Sondes for use in constructional logging include the calliper tool for measurement of borehole diameter and location of fractures and fissures and closed-circuit televiewer (CCTV) for qualitative inspection of borehole condition.

Second, fluid logging is carried out using the temperature, conductivity and flow (heat pulse or impeller) sondes and provides information of fluid movement, inflow and outflow horizons, presence of permeable zones, flow rates and water quality in the borehole and surrounding formation. Examples of temperature, fluid electrical conductivity and flow logs are shown in Fig. 8.47 and Plate 7.1.

Third, formation logging reveals geological bed boundaries, thickness and lithology, rock bulk density, permeable zones, porosity and formation water quality. Sondes for use in formation logging include the single-point resistance, multi-electrode electrical resistivity (16-inch 'short' normal and 64-inch 'long' normal),

lateral, spontaneous potential, natural gamma, optical televiewer (OPTV) and acoustic logs. In addition, and under careful supervision, the dual-density (gamma-gamma) and neutron radiation logs are useful for measurement of the bulk density and porosity of formations, respectively.

Electrical logs are of value in aquifer evaluation since they can lead to determination of the formation factor, F, and an estimate of the porosity using the following two relationships:

$$F = \frac{\rho}{\rho_w} \tag{eq. 7.44}$$

where ρ is the resistivity of the water-saturated rock and ρ_w is the resistivity of the water, and

$$F = \frac{a}{\phi^m} \tag{eq. 7.45}$$

where ϕ is porosity and a and m are constants, their value being governed by the nature of the formation. The relationship is often referred to as Archie's law (Archie 1942). Since m has a value close to 2, the formation factor varies approximately inversely to the square of the porosity. Unlike most rock-forming minerals, which are usually non-conductors, conduction of electrical current takes place through clays by the way of weakly bonded surface ions. In this case, eq. 7.45 does not apply to porous rocks containing any appreciable amount of clay minerals (Griffiths and King 1981).

7.3.4.1 Examples of downhole geophysical logging

Examples of the application of downhole geophysical logging in hydrogeology are demonstrated in the following papers:

- geophysical borehole measurements in crystalline rocks using acoustic televiewer and calliper measurements, electrical resistance, thermal techniques and vertical seismic profiling (Wilhelm *et al.* 1994);
- application of formation and fluid logging to characterize salinity distribution in coastal aquifers to illustrate the development of flow

horizons in relation to rock layering, structure and base levels (Buckley *et al.* 2001);

- identification of continental basalt sequences and host aquifers using natural gamma, neutron and gamma–gamma density logs in unsaturated rocks and, additionally, resistivity and velocity logs in saturated rocks (Helm-Clark *et al.* 2004);
- application of geophysical borehole logging and time-domain electromagnetic soundings to identify a shallow, low-resistive and high-gamma radiation layer in the karst aquifer system of the south-eastern Yucatan Peninsula (Gondwe *et al.* 2010); and
- deployment of self-potential measurements in a coastal Chalk borehole that reflect changes in the location of a nearby saline front (MacAllister *et al.* (2018).

For further reading, downhole geophysical techniques are discussed by Griffiths and King (1981), Beesley (1986), BSI (1988), Keys (1988), Chapellier (1992), Sharma (1997), Brassington (2017) and Acworth (2019).

7.3.5 Surface geophysical techniques

Surface geophysical surveys consist of a set of measurements, usually collected in a systematic pattern over the Earth's surface by land, sea or air. The measurements may be of spatial variations of static fields of force, including gradients of electrical, gravitational or magnetic potential, or of characteristics of wave fields, in particular travel times of elastic (seismic) waves and amplitude and phase distortions of electromagnetic waves. These force and wave fields are affected by the physical properties and structure of subsurface rocks. In choosing the geophysical technique to be used to study a problem, the contrasting properties of the formation are important factors to be considered (Griffiths and King 1981).

In hydrogeology, the seismic refraction, electrical resistivity and electromagnetic techniques are useful in groundwater investigations due to the close relationship between electrical

conductivity and some hydrogeological properties of the aquifer (for example, porosity, clay content, mineralization of the groundwater and degree of water saturation). Unlike seismic refraction, the seismic reflection method is generally unsuited to investigations of less than 50 m depth and so has not been widely used in groundwater studies, although the method has been adapted to shallow investigations (Hill 1992; King 1992). Although surface geophysical methods cannot replace test drilling, stratigraphic interpretations based on surface geophysical measurements can be calibrated against test-hole information.

7.3.5.1 Seismic refraction survey method

The seismic refraction method is based on the fact that elastic waves travel through different earth materials at different velocities. The denser the material, the higher the wave velocity. When elastic waves cross a geologic boundary between two formations with different elastic properties, the velocity of wave propagation changes and the wave paths are refracted according to Snell's Law. In seismic exploration, elastic waves are initiated by an energy source, usually a small explosion, at the ground surface. A set of receivers, called geophones, is set up in a line radiating outward from the energy source. Waves initiated at the surface and refracted at the critical angle by a high-velocity layer at depth will reach the more distant geophones more quickly than waves that travel directly through the low-velocity surface layer. The time between the shock and the arrival of the elastic wave at a geophone is recorded on a seismograph. A set of seismograph records can be used to derive a graph of arrival time versus distance from shot point to geophone and this, in turn, with the aid of some simple theory, can be used to calculate layer depths and their seismic velocities (Freeze and Cherry 1979).

The seismic refraction method has been used in groundwater investigations to determine such features as the depth to bedrock, the presence of buried bedrock channels, the thickness of surficial fracture zones in crystalline rock and the areal extent of potential aquifers. The interpretations are most reliable in cases where there is a simple two-layer or three-layer geological configuration in which the layers exhibit a strong contrast in seismic velocity. The velocities of the layers must increase with depth. The method cannot pick up a low-velocity layer (which might well be a porous potential aquifer) that underlies a high-velocity surface layer. The depth of penetration of the seismic method depends on the strength of the energy source. For shallow investigations (say, up to 30 m), geophysicists have often employed hammer seismic methods, in which the energy source is simply a hammer blow on a steel plate set on the ground surface (Freeze and Cherry 1979).

7.3.5.2 Electrical resistivity survey method

The electrical resistivity survey technique depends for its operation on the fact that any subsurface variation in conductivity alters the form of the current flow within the Earth and this affects the distribution of electric potential, the degree to which depends on the size, shape, location and electrical resistivity of the subsurface layers or bodies. It is therefore possible to obtain information about the sub-surface from potential measurements made at the surface. For example, a sink or hollow in limestone filled with clay and concealed by a layer of overburden is a body of relatively good conductivity within a poorly conducting medium. There is a concentration of current flow through the clay fill in preference to the limestone and a corresponding disturbance in the potential distribution in and around the sink, which can be measured at the surface if the sink is large enough or not too deeply buried. Resistivity methods are likely to be employed where the structures are simple and resistivity contrasts well marked. Though investigation to depths of 1 km or more are possible, such measurements require the use of long cable arrays and interpretation can be affected by lateral inhomogeneity. Both above- and below-ground infrastructure (pipes, rails and wires) and complex geology can severely constrain the cable spread lengths and so limit the depth of investigation to less than a hundred metres (Griffiths and King 1981).

With the exception of clays and certain metallic ores, the passage of electrical current through rocks occurs through the groundwater contained in the pores and fissures, with the rock matrix itself being non-conducting. An increase in the salinity of the groundwater leads to a decrease in resistivity. The degree of saturation also affects resistivity, which increases with a decrease in the degree of saturation or pore space. To understand the behaviour of the flow of electrical current in layered media and how this affects the distribution of potential, then the initial equation is Ohm's Law:

$$\frac{V}{I} = R \qquad \text{(eq. 7.46)}$$

where, I is current in a conducting body, V is potential difference between two surfaces of constant potential and R is a constant known as resistance between the surfaces. From this relationship, the electrical resistivity, ρ, of a geological formation is defined as follows:

$$\rho = \frac{RA}{L} \qquad \text{(eq. 7.47)}$$

where R is the resistance to electrical current for a unit block of cross-sectional area, A and length, L. Units of electrical resistivity are given in Ohm.m (symbol, Ωm) and is simply the inverse of electrical conductivity (1 Ωm is equivalent to 1 S m^{-1}). The resistivities of rocks are strongly influenced by fissuring, porosity, groundwater conductivity and saturation so that only approximate ranges can be given for various earth materials. Igneous and metamorphic rocks are typically in the range 10^3–10^5 Ωm, limestones 10^2–10^5 Ωm, sandstones 10^2–10^3 Ωm, shales 10^1–10^3 Ωm and clays 10^1–10^2 Ωm.

In an electrical resistivity survey, an electric current is passed into the ground through a pair of current electrodes and the potential drop is measured across a pair of potential electrodes. The spacing of the electrodes controls the depth of penetration. Conventional electrical resistivity methods (direct current, DC) commonly used for groundwater exploration use various electrode configurations such as Wenner, Schlumberger and dipole–dipole. The type of array depends on the objective of the study, the targets of interest, local geology and the sensitivity of the array to vertical and lateral variations in the subsurface resistivity distribution of the rock matrix. At each setup, an apparent resistivity is calculated on the basis of the measured potential drop, the applied current and the electrode spacing to provide a depth profile.

In vertical resistivity depth profiling or vertical electrical sounding (VES), the centre of the electrode array remains fixed while the inter-electrode spacing is increased between resistance measurements, enabling information relating to greater depths to be obtained (Barker 1986). The electrical sounding data are plotted as a double logarithmic plot with electrode spacing on the *x*-axis (abscissa) and apparent resistivity on the *y*-axis (ordinate). The resulting sounding curve, or apparent resistivity curve, is then interpreted in terms of depths to sub-surface layers and their resistivities by modelling against theoretical curves for simple layered geometries. Constant separation traversing (CST) is used to map lateral variations in resistivity in which the electrode configuration is moved horizontally along a line. CST may be considered as sampling resistivity in a horizontal direction over a fixed depth range that depends on the electrode spacing (Griffiths and King 1981). Horizontal profiling is useful for rapid delineation of horizontal changes in resistivity and can be used to define aquifer limits or to map areal variations in groundwater salinity.

By combining vertical depth profiling with horizontal profiling, software-controlled measurements provide a 'pseudo-depth profile' of resistivity, which can be contoured. A mathematical inversion procedure allows a 2D resistivity-depth section to be produced from these data known as electrical imaging. Electrical imaging, or electrical resistivity tomography (ERT), is widely used for applications in hydrogeology. Typical applications include locating water-bearing fracture zones in bedrock, mapping the limits of municipal landfills and delineating sand and gravel lenses in a clay environment. The method can also be used to locate the saltwater-freshwater interface in coastal aquifers and detect the location of the water table. An example of the use of time-lapse ERT in monitoring groundwater fluctuation associated with quarry dewatering is shown in Plate 7.2.

7.3.5.3 Electromagnetic survey method

Electromagnetic (EM) surveys are similar to resistivity surveys but the methods induce current flows without using electrodes and, therefore, have possible use in aerial surveys. An alternating electrical current set up in the transmitter coil results in a time-varying magnetic field. This magnetic field induces current and thus a secondary magnetic field, that is observed by the receiver coil. The EM method is useful when the surface layer has a high resistivity; otherwise, the method is limited by conductive surface layers. Although less precise than electrical resistivity modelling, EM methods are quick to use, for example in the survey of contaminated groundwater at landfill sites or saline groundwater in coastal locations. The frequency of EM methods range from hundreds to tens of thousands of hertz. EM waves are absorbed by conducting rocks and so penetration depth of the technique increases as frequency decreases. Ground penetrating radar (GPR) with frequencies between 25 and 1000 MHz gives the highest resolution but is only applicable for shallow structures, for example in locating subsurface voids such as abandoned mine workings or solution features in karst aquifers.

7.3.5.4 Gravity survey method

Gravity surveying is less commonly employed in hydrogeology, the main application being in establishing the depth to bedrock for the purpose of finding the aquifer thickness and in assessing groundwater flow boundaries. The aim of gravitational surveying is to detect underground structures by means of the disturbance they produce at the surface in the Earth's gravitational field. The method is straightforward in requiring small field differences in gravity to be measured, but in practice, the measurement techniques employed are highly sophisticated (Griffiths and King 1981). The basis of the method is Newton's Law of Gravitation, which states that every particle of matter exerts a force of attraction on every other particle, the force, F, being proportional to the product of the masses, m_1 and m_2 and inversely proportional to the square of the distance, r, between the masses, as follows:

$$F = G\frac{m_1 m_2}{r^2} \qquad \text{(eq. 7.48)}$$

where G is the universal gravitational constant equal to 6.674×10^{-11} N m^2 kg^{-2}.

The value of gravity has units of gal (symbol, Gal, equal to 1 cm s^{-2}), although units of milligal (mGal) are often used in exploration studies. An alternative unit is the gravity unit, gu, where 1 gu is 1 mm s^{-2}, meaning that 1 mgal = 10 gu. Changes in the gravity field on the Earth's surface are due to the variable density of earth materials. Changes can be as large as 1000 mGal, but also as small as a few mGal (Ackworth 2019).

An application of the gravity survey technique known as microgravity is useful for finding buried shafts and sink holes in limestone terrain and in determining the extent and depth of closed landfill sites (Ackworth 2019). The advent of GRACE satellite data and interpretation of changes in the Earth's gravity field as a result of basin-scale changes in groundwater storage volume has led to new applications for gravity surveying in hydrogeology (see Section 7.4).

7.3.5.5 Examples of surface geophysical surveying

Examples of the application of surface geophysical methods in hydrogeology are included in the following papers:

- Gravity survey and resulting Bouguer anomaly map of the subsurface position of a buried channel in the Chalk aquifer of East Anglia (Barker and Harker 1984);
- Detection of leaks from environmental barriers using electrical current imaging (Binley *et al.* 1997);
- Investigation of saline intrusion using borehole logging, seismic reflection profiling, vertical electrical resistivity soundings (VES) and EM induction surveying in a coastal sand and gravel aquifer (Holman and Hiscock 1998; Holman *et al.* 1999);
- Evaluation of lithological, stratigraphical and structural controls on the distribution of aquifers using VES, transient electromagnetic (TEM), tensorial audio-magnetotelluric (AMT) and nuclear magnetic resonance (NMR) depth sounding and inversion measurements (Meju *et al.* 1999, 2002);
- Application of cross-borehole transmission radar and ERT to characterize groundwater flow and solute transport in the unsaturated (vadose) zone (Binley *et al.* 2002a,b);
- Review of helicopter-borne EM methods for groundwater exploration (Siemon *et al.* 2009);
- Integration of electrical resistivity imaging and GPR to investigate solution features in the Biscayne Aquifer, southeast Florida (Yeboah-Forson *et al.* 2014);
- Targeted application of 2D, high-resolution ERT to delineate potential groundwater zones in complex geological terrain in southern India (Kumar *et al.* 2020); and
- Observations from ERT, seismic refraction tomography and GPR to characterize the subsurface structure and connectivity of an alpine aquifer system in the Canadian Rocky Mountains (Christensen *et al.* 2020).

For further reading, surface geophysical techniques are discussed by Zohdy *et al.*

(1974), Barker (1986), Telford *et al.* (1991), Mussett and Khan (2000), Kearey *et al.* (2002), Burger *et al.* (2006), Kirsch (2006), Reynolds (2011) and Ackworth (2019). Milsom and Eriksen (2011) provide practical guidance on conducting field geophysical surveys.

7.4 Remote sensing methods

Remote sensing and earth observation technologies provide an important means of collecting groundwater-related data on a regional scale and to assess the state of groundwater resources. Satellite remote sensing has drawbacks, but it offers the advantages of global coverage, availability of data, metadata, error statistics and the ability to provide meaningful spatial averages (Green *et al.* 2011). Undoubtedly, satellite remote sensing represents a powerful method for detection and monitoring of environmental and climate change on a global scale. However, the capabilities of remote sensing to 'look below the ground surface' and to detect properties that directly relate to groundwater conditions are limited. A notable exception to this is satellite-based observations of the gravity field which contain key information of changes in groundwater storage. Also, remote sensing can provide essential constraints on surface components of the hydrological cycle, which indirectly influence the subsurface water balance (see Section 6.3.2).

Several satellite and airborne remote sensing technologies can contribute to groundwater monitoring activities. Aerial thermal infrared imaging is being used increasingly for mapping groundwater discharge zones in estuaries, rivers and oceans. Peterson *et al.* (2009) used aerial thermal infrared imaging to reveal that submarine groundwater discharge (SGD, see Section 2.16) along the western coast of the Big Island of Hawaii is often focused as point-source discharges that create buoyant groundwater plumes that mix into the coastal ocean.

Landsat, the Moderate-resolution Imaging Spectroradiometer (MODIS), the Advanced Very High Resolution Radiometer (AVHRR) and certain other instruments can resolve the location and type of vegetation, which can be used to infer a shallow water table. Landsat imagery can also provide geological information where not obscured by vegetation.

Altimetry measurements and Interferometric Synthetic Aperture Radar (InSAR) over time can show where subsidence is occurring (for example, see Box 2.8), which is often an indicator of groundwater depletion. The launch of the NASA-ISRO Synthetic Aperture Radar Mission (NISAR) satellite in 2022 will enable systematic mapping of the Earth's surface using two different radar frequencies (L-band and S-band) to measure changes of less than a centimetre across and provide systematic measurement across entire aquifers. The ability to map surface deformation of a few millimetres monthly over large areas at resolutions of a few tens of metres should open up new possibilities for remote monitoring of groundwater resources (NASA 2016). Furthermore, microwave radar and radiometry measurements can also be used to estimate snow and surface soil water, which further constrain groundwater assessments.

Another valuable remote sensing technology for groundwater investigations is high-precision satellite gravimetry, as enabled by the NASA/GFZ Gravity Recovery and Climate Experiment (GRACE). Unlike radars and radiometers, GRACE is not limited to measurement of atmospheric and near-surface phenomena. Since its launch in 2002, GRACE has been employed to detect small temporal changes in the Earth's gravity field (Ramillien *et al.* 2008). The GRACE satellites comprise two chasing satellites. As the Earth's gravity increases, the leading satellite accelerates such that gravity variations induce distance variations between the satellites. Measured, temporal changes in gravity are primarily caused by changes in total water (mass) storage (*TWS*) in the atmosphere, ocean and at and below the surface of the continents.

Post-launch studies using GRACE data have demonstrated that when combined with ancillary measurements of surface water and soil water content, GRACE is capable of monitoring changes in groundwater storage with reasonable accuracy (temporal resolution 10 days to monthly, spatial resolution 400–500 km and mass change ~9 mm water equivalent) (for example, see Plate 7.3). GRACE is able to measure variations in equivalent height of water over regions of about 200 000 km^2 or larger, with uncertainties of the order of a few centimetres (Wahr *et al.* 2006). Accuracy degrades rapidly as the spatial resolution increases. While this is sufficient for many large-scale hydrological investigations, such as the example of groundwater depletion in North-West India (see Box 7.5), most water resources, meteorological, agricultural and natural hazards applications require higher resolution data (Green *et al.* 2011).

Validation studies have found acceptable agreement between GRACE-derived changes in continental water mass storage and independent inferences from global hydrology models and surface data. Seasonal correlations of 0.8–0.9 were found by comparing GRACE and piezometer-network data for different parts of the USA (Green *et al.* 2011). In other studies, GRACE data have been used to monitor drought impacts in the Murray-Darling Basin, south-eastern Australia (Leblanc *et al.* 2009) and Famiglietti *et al.* (2011) have used GRACE-data to estimate groundwater depletion rates of 20.4 ± 3.9 mm a^{-1} in the Central Valley Aquifer, USA (see Box 2.7).

7.5 Groundwater modelling

Numerical modelling of groundwater flow can be undertaken at the start or end of a hydrogeological investigation: at the start for conceptualizing the main controls on groundwater flow in the model area and in indicating the type and length of field data that will be required to construct a model; and at the end for predicting future aquifer response under different groundwater conditions (Rushton

1986, 2003). With a well-constructed model, the ability to predict groundwater flow patterns, for example the effects of different groundwater abstraction patterns on sensitive aquatic systems (Box 7.6), or the shape of wellhead capture zones for protecting groundwater quality (Fig. 9.15), or future aquifer response to changing recharge amounts under climate change (Section 10.4), makes groundwater modelling an indispensable tool for managing local and regional groundwater resources.

In the process of constructing a groundwater model, the primary aim is to represent adequately the different features of groundwater flow through the aquifer within the model area or domain. In this respect, the important features to consider in governing the response of an aquifer to a change in hydrogeological conditions include: aquifer inflows (recharge, leakage and cross-formational flows); aquifer outflows (abstractions, spring flows and river baseflows); aquifer properties (hydraulic conductivity and storage coefficient); and aquifer boundaries (constant or fixed head, constant flow or variable head, and no-flow boundaries).

Given the complexity of regional groundwater flow problems, the equations of groundwater flow cannot be solved by analytical methods (Section 2.12) and, instead, approximate numerical techniques are used. These techniques require that the space and time co-ordinates are divided into some form of discrete mesh and time interval. Common approaches to defining the space co-ordinates in the model domain are the finite-difference and finite-element approximations. The finite-difference approach is based on a rectilinear mesh, whereas the finite-element approach is more flexible in allowing a spatial discretization that can fit the geometry of the flow problem (see Plate 3.2a). For each cell in the mesh and, for transient simulations, at each time step, the unknown heads are represented by a set of simultaneous equations that can be solved iteratively by specifying initial head conditions. Model runs are performed by a computer program that employs matrix methods for solving

Box 7.5   Groundwater depletion in North-West India

The World Bank has warned that India is on the brink of a severe water crisis (Briscoe 2005) with groundwater abstractions as a percentage of recharge as much as 100% in some states (Fig. 7.27a). Nationally, groundwater accounts for about 50–80% of domestic water use and 45–50% of irrigation water use (Kumar *et al.* 2005; Mall *et al.* 2006). The total irrigated area in India nearly tripled to 33.1 million ha between 1970 and 1999 (Zaisheng *et al.* 2006). The states of Rajasthan, Punjab and Haryana in North-West India are semi-arid to arid, averaging about 50 cm of annual rainfall (Xie and Arkin 1997) and encompass the eastern part of the Thar Desert (Fig. 7.27a). The 114 million residents of the region have benefited from India's 'green revolution', a massive agricultural expansion made possible largely

by increased abstraction of groundwater for irrigation, which began in the 1960s. Wheat, rice and barley are the major crops. The region is underlain by the Indus River Plain Aquifer, a 560 000 km^2 unconfined to semi-confined porous alluvial formation that straddles the border between India and Pakistan (Zaisheng *et al.* 2006).

In a study using terrestrial water storage change observations from the NASA Gravity Recovery and Climate Experiment (GRACE) satellites (Fig. 7.27b), Rodell *et al.* (2009) simulated soil water variations from a data-integrating hydrological modelling system. Groundwater storage variations can be isolated from GRACE data given auxiliary information on the other components of total water (mass) storage (*TWS*), from either in situ observations or land-surface models. Using

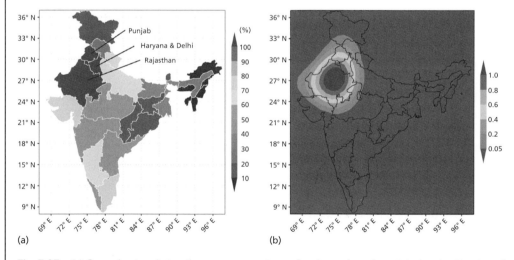

Fig. 7.27   (a) Groundwater abstractions as a percentage of recharge based on state-level estimates of annual abstractions and recharge reported by the Indian Ministry of Water Resources (2006) and (b) map of the unscaled, dimensionless averaging function used to retrieve regional *TWS* time series from GRACE satellite data in order to estimate terrestrial water storage changes in North-West India (Rodell *et al.* 2009). (*Source:* Rodell, M., Velicogna, I. and Famiglietti, J.S. (2009) Satellite-based estimates of groundwater depletion in India. *Nature* **460**, 999–1002. © 2009, Springer Nature)

(*Continued*)

Box 7.5 (Continued)

the second of these approaches, Rodell *et al.* (2009) produced a time series of groundwater storage anomalies (deviations from the mean state) averaged over the area encompassed by Rajasthan (342 239 km^2), Punjab (50 362 km^2) and Haryana (45 695 km^2, including the National Capital Territory of Delhi) between August 2002 and October 2008. This region was chosen because the Indian Ministry of Water Resources estimates that groundwater abstractions in each of the three states exceed recharge. Rodell *et al.* (2009) showed that groundwater is being depleted at a mean rate of 4.0 ± 1.0 cm a^{-1} equivalent depth of water (17.7 ± 4.5 km^3 a^{-1}) over these states, with maximum rates of groundwater depletion centred on Haryana. Assuming a specific yield of 0.12, the regional mean rate of water table decline is estimated to be about 0.33 m a^{-1}. Local rates of water table

decline, which are highly variable, are reported to be as large as 10 m a^{-1} in certain urban areas.

During the period August 2002 to October 2008, a period when annual rainfall was close to normal, groundwater depletion was equivalent to a net loss of 109 km^3 of water, about double the capacity of India's largest surface water reservoir. Although the GRACE observational record is relatively short, the results suggest that unsustainable consumption of groundwater for irrigation and other anthropogenic uses is likely to be the cause. If measures are not taken to ensure sustainable groundwater usage, the consequences for the population of the region may include a reduction of agricultural output and shortages of drinking water, leading to extensive socio-economic stresses (Rodell *et al.* 2009).

---

the large number of unknowns. Successful completion of a model run is obtained when convergence of the head solution is reached, usually determined by a model error criterion set at the beginning of the run.

In all groundwater modelling investigations, for example, the case study presented in Box 7.6, the process of deriving a model for predictive purposes involves the following common steps: conceptualization of the flow mechanisms in the model area based on existing knowledge; acquisition of available field data on groundwater heads, aquifer properties and river flows; discretization of the model domain and construction of the model input file; calibration of the model by comparing simulated steady-state (equilibrium) and transient (time-variant) heads and flows (flow vectors and water balance) against field-measured values; sensitivity analysis of aquifer property values and recharge and boundary conditions; validation of the model against an independent

set of data (for example, groundwater head and river flow data not used during model calibration, or hydrochemical data such as salinity); and, lastly, prediction of aquifer response under changed groundwater conditions.

Traditionally, observations of groundwater levels and surface water discharge are usually used to calibrate groundwater flow models. Model calibration can be a significant challenge given that groundwater flow systems are characterized by spatial and temporal heterogeneity in both physical properties, such as hydraulic conductivity, and the processes driving groundwater flow, such as recharge in response to rainfall and snowmelt events. Generally, the more processes a groundwater flow model simulates, the more complex the numerical solution. In turn, this requirement increases the number of parameters that need to be calibrated. Often, there is a trade-off between the increased model complexity and the benefit of additional observation types,

---

Box 7.6   Groundwater modelling of the Monturaqui-Negrillar-Tilopozo Aquifer, Chile

The Monturaqui-Negrillar-Tilopozo (MNT) Aquifer is located at the foot of the Andes in the extremely arid environment of northern Chile and occupies a north-south oriented graben approximately 60 km long, with surface elevations decreasing from around 3200 m in the southern (Monturaqui) area to 2300 m in the north (Tilopozo wetland) area where the aquifer discharges at the south-eastern margin of the Salar de Atacama (Fig. 7.25a). The MNT Basin is hosted by Palaeozoic rocks and infilled with mainly Tertiary alluvial and volcanic sediments to depths of over 400 m. The Tertiary sediment infill to the basin forms the MNT aquifer with transmissivity values ranging between 500 and 4500 m^2 day^{-1}, confined storage coefficient values of around 1×10^{-3} and an

estimated specific yield of about 0.1. The aquifer is exploited to meet the demands of the important mining industry for potable and ore-processing water. The Tilopozo wetland is the only groundwater-dependent natural feature in the MNT Basin (Anderson *et al.* 2002).

As part of a sustainability assessment of the region, it has been suggested that a decline in water level of approximately 0.25 m, equivalent to a reduction in aquifer outflow of approximately 6%, should not be exceeded in order to protect the flora and fauna of the Tilopozo wetland. The MNT Aquifer and the established wellfields (Fig. 7.28a) have complex flow dynamics in relation to the capture of discharge from the wetland. Given these complexities, a

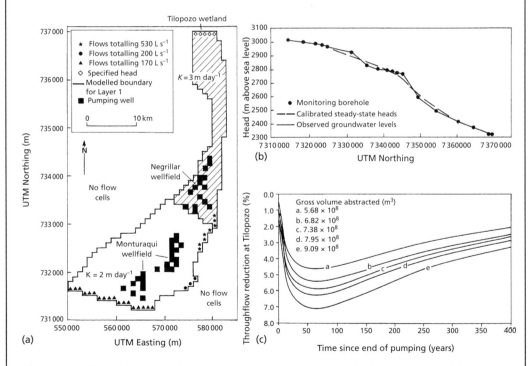

Fig. 7.28 (a) Model domain and boundary conditions for the groundwater flow model of the Monturaqui-Negrillar-Tilopozo Aquifer, Chile; (b) Observed groundwater heads versus modelled pre-abstraction heads of the calibrated steady-state model; (c) Predictive modelling results for different wellfield abstraction volumes leading to simulated groundwater through-flow reductions at the Tilopozo wetland (Anderson *et al.* 2002). (*Source:* Adapted from Anderson, M., Low, R. and Foot, S. (2002) Sustainable groundwater development in arid, high Andean basins. In: *Sustainable Groundwater Development* (eds K.M. Hiscock, M.O. Rivett and R.M. Davison). Geological Society, London, Special Publications **193**, pp. 133–144.)

(Continued)

Box 7.6   (Continued)

three-dimensional, spatially distributed, time-variant numerical flow model using the MODFLOW code (McDonald and Harbaugh 1988) was developed to investigate a groundwater abstraction strategy that would satisfy the sustainability criteria (Anderson *et al.* 2002).

The groundwater flow model extended from the southern limit of the Monturaqui Basin north to the Tilopozo wetland, and used a 1000-m resolution finite-difference grid with two layers. The model domain and boundary conditions are shown in Fig. 7.28(a). Constant inflows on the southern and eastern boundaries were used to represent system recharge, with a range of possible total inflows simulated. The model was refined within the bounds established in the conceptual model so that it reproduced as closely as possible the observed steady-state (pre-abstraction) head distribution in the aquifer (Fig. 7.28b). Following model calibration, a time-variant model for a period of up to 500 years was developed for predictive modelling of impacts. Groundwater abstractions with a range of durations from both the Monturaqui and Negrillar wellfields were represented in the model with the model simulations used to predict outflow changes with time across the northern boundary representing the Tilopozo wetland.

Significant uncertainty in the value of specific yield (varied between 0.05 and 0.2) and recharge rate (between 450 and 1800 L s^{-1}) was addressed at all stages of the modelling through sensitivity analyses.

Analysis of the model output (Fig. 7.28c) showed that the impacts of groundwater abstraction will reach the wetland between 20 and 40 years after the start of pumping from either wellfield, with maximum impacts likely to occur 75–300 years after abstraction ceases. The location of the Monturaqui wellfield upgradient of approximately 60% of the aquifer recharge was found to significantly reduce its relative impacts on groundwater flows at Tilopozo. From an interpretation of the results of sensitivity analyses, a 'worst-case' prediction model was developed which maximized the magnitude of predicted impacts, and model runs executed until a pumping duration of between 10 and 20 years and a gross abstracted volume of 7.38×10^8 m^3 (Fig. 7.28c) were found that limited through-flow reduction at the Tilopozo wetland to 6% (Anderson *et al.* 2002). The resulting model now forms the basis for the sustainable groundwater development of the MNT Aquifer which, fortunately, is supported by the extremely large volume of groundwater storage in the aquifer (approximately 10^{10} m^3).

---

for example temperature and tracer concentration data requiring additional model capability to simulate heat and mass transport. As discussed by Schilling *et al.* (2019), a balance needs to be struck that depends on the specific purpose of the model. To reduce the predictive uncertainty of groundwater flow models, Schilling *et al.* (2019) recommended a more systematic implementation of other observation types together with the application of mathematically robust and automated flow model calibration using parameter estimation software.

A further challenge related to model complexity is coping with the high computational challenge of fully integrated groundwater-surface water flow models, especially when applied at large spatial and temporal scales. Increasingly, running flow simulators on computational cloud infrastructure is suited to practical water management problems. Advantages of cloud-based implementation include the independence from computational infrastructure and the straight-forward integration of cloud-based observation databases with the modelling and data assimilation platform (Kurtz *et al.* 2017).

An example of the application of high-performance computing in groundwater modelling is the presentation by de Graaf *et al.* (2017) of a global-scale, two-layer groundwater model that represents unconfined and confined aquifers based on global datasets of surface geology and hydraulic properties, and topography-based estimates of the vertical structure of aquifer systems (see also Section 6.8.2.3). The model consists of two parts: a hydrological model (PCR-GLOBWB; van Beek *et al.* 2011) that provides recharge values to a groundwater flow model (MODFLOW; McDonald & Harbaugh 1988) with a grid resolution of 5 arc-minute (approximately 10×10 km at the equator). Running MODFLOW in transient mode with a monthly time step for the period 1960–2010 provides a global-scale simulation of lateral groundwater flows and head fluctuations caused by changes in climate or human water use during the model period. The model results estimate a global groundwater depletion during the period 1960–2010 of 7013 km^3, in the range of previous estimates of 5000 km^3 (Wada *et al.* 2012) and 8000 km^3 (Wada *et al.* 2010).

Other groundwater modelling approaches include more specialist applications such as solute transport modelling using the method of characteristics (Konikow and Bredehoeft 1978) or the 'random walk' method (Prickett *et al.* 1981) for solving the advection–dispersion equation (eq. 8.7). In the method of characteristics, advective transport is simulated by particles distributed in a geometrically uniform pattern over the entire model area with each particle assigned an initial concentration associated with the concentration in the cell containing the particle. Dispersive transport is simulated by a finite-difference calculation on the rectangular grid after which the particle concentrations are updated according to the changes in the grid concentrations, and advective transport in the next time step is calculated. In a similar way, the random walk method combines a flow sub-model, usually based on the finite-difference method, with the use of random variables that are Gaussian distributed and applied to particles introduced into the flow field to simulate dispersion. Each of the particles that is advected and dispersed by groundwater flow is assigned a mass which represents a fraction of the total mass of the chemical constituent involved. At the end of the simulation, the total mass of particles within an overlaid grid cell is divided by the product of the cell volume and porosity to give the average cell concentration.

A popular solute transport model developed by Zheng (1990), and available from the United States Environmental Protection Agency, is the code MT3D for simulating reactive mass transport including equilibrium-controlled linear or non-linear sorption (eq. 8.13) and first-order irreversible decay or biodegradation (eq. 5.11). The model uses a mixed Eulerian–Lagrangian approach to solve the three-dimensional advection-dispersion-reaction equation with three basic options based on the method of characteristics.

Techniques for simulating saline intrusion, or situations where density variations are significant, for example the underground disposal of brine wastes, include the US Geological Survey's density-coupled model SUTRA (Voss 1984). Given that hydraulic conductivity and head are functions of density (eqs 2.4 and 2.22), the model provides a numerical solution to the governing groundwater flow equation written in terms of the pressure potential, ψ, and intrinsic permeability. Hence, models simulating density-dependent flow require initial pressure and density distributions in order to find a solution.

For further discussion of finite-difference and finite-element numerical modelling techniques, the reader is referred to the texts by Wang and Anderson (1982), Spitz and Moreno (1996) and Rushton (2003). A popular finite-difference model for application in two- and three-dimensional groundwater flow problems is, as mentioned above, the United States Geological

Survey's code MODFLOW (McDonald and Harbaugh 1988), with demonstrations of this model presented by Chiang and Kinzelbach (2001) and Anderson *et al.* (2015). The proprietary finite-element code, FEFLOW, is widely used and supported and the reader is directed to Diersch (2014) and Anderson *et al.* (2015). A popular parameter estimation software suite, PEST, that includes widely used approaches for parameter estimation with many advanced options, is presented by Doherty (2015).

Further reading

Brandon, T.W. (ed.) (1986) *Groundwater: Occurrence, Development and Protection*. Institution of Water Engineers and Scientists, London.

Cook, P.G. and Herczeg, A.L. (2000) *Environmental Tracers in Subsurface Hydrology*. Kluwer Academic Publishers, Boston.

Dassargues, A. (ed.) (2000) *Tracers and Modelling in Hydrogeology*. IAHS Press, Wallingford, Oxfordshire, International Association of Hydrological Sciences Publication **262**.

Kendall, C. and McDonnell, J.J. (1998) *Isotope Tracers in Catchment Hydrology*. Elsevier, Amsterdam.

MacDonald, A., Davies, J., Calow, R. and Chilton, J. (2005) *Developing Groundwater: A Guide for Rural Water Supply*. ITGD Publishing, Bourton-on-Dunsmore, Warwickshire, UK.

Reynolds, J. (2011) *An Introduction to Applied and Environmental Geophysics* (2nd edn). Wiley-Blackwell, Chichester.

Schwartz, F.W. and Zhang, H. (2003) *Fundamentals of Ground Water*. John Wiley & Sons, Inc., New York.

Shepley, M., Whiteman, M., Hulme, P. and Grout, M. (2012) *Groundwater Resources Modelling: A Case Sstudy from the UK*. Geological Society, London, Special Publications **364**.

References

Ackworth, I. (2019) *Investigating Groundwater*. CRC Press, Boca Raton, FL.

Anderson, M., Low, R. and Foot, S. (2002) Sustainable groundwater development in arid, high Andean basins. In: *Sustainable Groundwater Development* (eds K.M. Hiscock, M.O. Rivett and R.M. Davison). Geological Society, London, Special Publications **193**, pp. 133–144.

Anderson, M.P., Woessner, W.W. and Hunt, R.J. (2015) *Applied Groundwater Modeling: Simulation of Flow and Advective Transport* (2nd edn). Academic Press, Inc., San Diego, California.

Archie, G.E. (1942) The electrical resistivity log as an aid in determining some reservoir characteristics. *Transactions of the AIME* **146**, 389–409.

Atkinson, T.C. and Smart, P.L. (1981) Artificial tracers in hydrogeology. In: *A Survey of British Hydrogeology 1980* (eds C.R. Argent and D.J.H. Griffin). Royal Society, London, pp. 173–190.

Atkinson, T.C., Smith, D.I., Lavis, J.J. and Whitaker, R.J. (1973) Experiments in tracing underground waters in limestones. *Journal of Hydrology* **19**, 323–349.

Barker, R.D. (1986) Surface geophysical techniques. In: *Groundwater: Occurrence, Development and Protection* (ed. T.W. Brandon). Institution of Water Engineers and Scientists, London, pp. 271–314.

Barker, R.D. and Harker, D. (1984) The location of the Stour buried tunnel-valley using geophysical techniques. *Quarterly Journal of Engineering Geology and Hydrogeology* **17**, 103–115.

Bateman, A.S., Hiscock, K.M. and Atkinson, T.C. (2001) Qualitative risk assessment using tracer tests and groundwater modelling in an unconfined sand and gravel aquifer. In: *New Approaches Characterizing Groundwater Flow* (eds K.-P. Seiler and S. Wohnlich). Swets & Zeitlinger, Lisse, pp. 251–255.

Beesley, K. (1986) Downhole geophysics. In: *Groundwater: Occurrence, Development and Protection* (ed. T.W. Brandon). Institution of Water Engineers and Scientists, London, pp. 315–352

Binley, A., Cassiani, G., Middleton, R. and Winship, P. (2002a) Vadose zone flow model parameterisation using cross-borehole radar and resistivity imaging. *Journal of Hydrology* **267**, 147–159.

Binley, A., Daily, W. and Ramirez, A. (1997) Detecting leaks from environmental barriers using electrical current imaging. *Journal of Environmental and Engineering Geophysics* **2**, 11–19.

Binley, A., Winship, P., West, J. *et al.* (2002b) Seasonal variation of moisture content in unsaturated sandstone inferred from borehole radar and resistivity profiles. *Journal of Hydrology* **267**, 160–172.

Boulton, N.S. (1963) Analysis of data from non-equilibrium pumping tests allowing for delayed yield from storage. *Proceedings of the Institution of Civil Engineers* **26**, 469–482.

Bouwer, H. (1989) The Bouwer and Rice slug test – an update. *Ground Water* **27**, 304–309.

Bouwer, H. and Rice, R.C. (1976) A slug test for determining hydraulic conductivity of unconfined aquifers with completely or partially penetrating wells. *Water Resources Research* **12**, 423–428.

Brassington, R. (1998) *Field Hydrogeology* (2nd edn). John Wiley & Sons, Ltd, Chichester.

Brassington, R. (2017) *Field Hydrogeology* (4th edn). John Wiley & Sons, Ltd, Chichester.

Briscoe, J. (2005) *India's Water Economy: Bracing for a Turbulent Future*. Report No. 34750-IN, World Bank, Washington DC, viii–xi.

Brodsky, E.E., Roeloffs, E., Woodcock, D. *et al.* (2003) A mechanism for sustained groundwater pressure changes induced by distant earthquakes. *Journal of Geophysical Research* **108**(B8), 2390.

Brown, M.C. and Ford, D.C. (1971) Quantitative tracer methods for investigation of karst hydrologic systems. *Transactions of the Cave Research Group of Great Britain* **13**, 37–51.

BSI (1988) *British Standard Guide for Geophysical Logging of Boreholes for Hydrogeological Purposes*. British Standards Institution, London.

BSI (1992) *Code of Practice for Test Pumping Water Wells. BS 6316: 1992*. British Standards Institution, London.

Buckley, D.K., Hinsby, K. and Manzano, M. (2001) Application of geophysical borehole logging techniques to examine coastal aquifer palaeohydrogeology. In: *Palaeowaters in Coastal Europe: Evolution of Groundwater since the Late Pleistocene* (eds W.M. Edmunds and C.J. Milne). Geological Society, London, Special Publications **189**, pp. 251–270.

Burger, H., Sheehan, A. and Jones, C. (2006) *Introduction to Applied Geophysics: Exploring the Shallow Subsurface*. W.W. Norton, New York.

Burgess, W.G., Shamsudduha, M., Taylor, R.G. *et al.* (2017) Terrestrial water load and groundwater fluctuation in the Bengal Basin. *Nature Scientific Reports* **7**, 3872, doi: 10.1038/s41598-017-04159-w.

Chambers, J.E., Meldrum, P.I., Wilkinson, P.B. *et al.* (2015) Spatial monitoring of groundwater drawdown and rebound associated with quarry dewatering using automated time-lapse electrical resistivity tomography and distribution guided clustering. *Engineering Geology* **193**, 412–420.

Chapellier, D. (1992) *Well Logging in Hydrogeology*. A.A. Balkema, Rotterdam.

Chiang, W.-H. and Kinzelbach, W. (2001) *3D-Groundwater Modeling with PMWIN: A Simulation System for Modeling Groundwater Flow and Pollution*. Springer-Verlag, Berlin.

Christensen, C.W., Hayashi, M. and Bentley, L.R. (2020) Hydrogeological characterization of an alpine aquifer system in the Canadian Rocky Mountains. *Hydrogeology Journal*, doi: 10.1007/s10040-020-02153-7.

Clark, L. (1977) The analysis and planning of step drawdown tests. *Quarterly Journal of*

Engineering Geology and Hydrogeology **10**, 125–143.

Clark, L. (1984) Groundwater development of the Chepstow Block: a study of the impact of domestic waste disposal on a karstic limestone aquifer in Gwent, South Wales. In: *Proceedings of the IAH International Groundwater Symposium on Groundwater Resources Utilization and Contaminant Hydrogeology*, Vol. **II** (ed. R. Pearson). Atomic Energy of Canada Ltd., Manitoba, Canada, pp. 300–309.

Clark, L. (1988) *The Field Guide to Water Wells and Boreholes*. Open University Press, Milton Keynes, Geological Society of London Professional Handbook Series.

Cook, P.G. (2015) The role of tracers in hydrogeology. *Groundwater* **53**, 1–2.

Cooper, H.H. and Jacob, C.E. (1946) A generalized graphical method for evaluating formation constants and summarizing well field history. *Transactions of the American Geophysical Union* **27**, 526–534.

Cooper, H.H., Bredehoeft, J.D., Papadopulos, I.S. and Bennett, R.R. (1965) The response of well-aquifer systems to seismic waves. *Journal of Geophysical Research* **70**, 3915–3926.

De Graaf, I.E.M., van Beek, R.L.P.H., Gleeson, T. *et al.* (2017) A global-scale two-layer transient groundwater model: development and application to groundwater depletion. *Advances in Water Resources* **102**, 53–67.

Diersch, H.-J.G. (2014) *FEFLOW: Finite Element Modeling of Flow, Mass and Heat Transport in Porous and Fractured Media*. Springer-Verlag, Berlin, Heidelberg.

Doherty, J. (2015) *Calibration and Uncertainty Analysis for Complex Environmental Models*. Watermark Numerical Computing, Brisbane, Australia.

Driscoll, F.G. (1986) *Groundwater and Wells* (2nd edn). Johnson Filtration Systems Inc., St. Paul, Minnesota.

Famiglietti, J.S., Lo, M., Ho, S.L. *et al.* (2011) Satellites measure recent rates of groundwater depletion in California's Central Valley. *Geophysical Research Letters* **38**, L03403, doi:10.1029/2010GL046442.

Farquharson, J.I. and Amelung, F. (2020) Extreme rainfall triggered the 2018 rift eruption at Kīlauea Volcano. *Nature* **580**, 491–495.

Ferris, J.G., Knowles, D.B., Browne, R.H. and Stallman, R.W. (1962) Theory of aquifer tests. *United States Geological Survey Water Supply Paper* **1536**-E. United States Department of the Interior, Washington.

Foulger, G.R., Wilson, M.P., Gluyas, J.G. *et al.* (2018) Global review of human-induced earthquakes. *Earth-Science Reviews* **178**, 438–514.

Freeze, R.A. and Cherry, J.A. (1979) *Groundwater*. Prentice-Hall, Inc., Englewood Cliffs, New Jersey.

Glotfelty, M.F. (2019) *The Art of Water Wells*. NGWA Press, Westerville, Ohio.

Gondwe, B.R.N., Lerer, S., Stisen, S. *et al.* (2010) Hydrogeology of the south-eastern Yucatan Peninsula: new insights from water level measurements, geochemistry, geophysics and remote sensing. *Journal of Hydrology* **389**, 1–17.

Green, T.R., Taniguchi, M., Kooi, H. *et al.* (2011) Beneath the surface of global change. *Journal of Hydrology* **405**, 532–560.

Griffiths, D.H. and King, R.F. (1981) *Applied Geophysics for Geologists and Engineers: The Elements of Geophysical Prospecting* (2nd edn). Pergamon Press, Oxford.

Hantush, M.S. (1956) Analysis of data from pumping tests in leaky aquifers. *Transactions of the American Geophysical Union* **37**, 702–714.

Helm-Clark, C.M., Rodgers, D.W. and Smith, R.P. (2004) Borehole geophysical techniques to define stratigraphy, alteration and aquifers in basalt. *Journal of Applied Geophysics* **55**, 3–38.

Herbert, R. and Kitching, R. (1981) Determination of aquifer parameters from large-diameter dug well pumping tests. *Ground Water* **19**, 593–599.

Hill, I.A. (1992) Field techniques and instrumentation in shallow seismic reflection. *Quarterly Journal of Engineering Geology* **25**, 183–190.

Hiscock, K.M. (1982) *Hydraulic properties of the Chalk at Ludgershall sewage treatment works*. BSc Thesis, University of East Anglia, Norwich.

Hobson, G. (1993) Practical use of tracers in hydrogeology. *Geoscientist* **4**, 26–27.

Höhener, P., Werner, D., Balsiger, C. and Pasteris, G. (2003) Worldwide occurrence and fate of chlorofluorocarbons in groundwater. *Critical Reviews in Environmental Science and Technology* **33**, 1–29.

Holman, I.P. and Hiscock, K.M. (1998) Land drainage and saline intrusion in the coastal marshes of northeast Norfolk. *Quarterly Journal of Engineering Geology and Hydrogeology* **31**, 47–62.

Holman, I.P., Hiscock, K.M. and Chroston, P.N. (1999) Crag aquifer characteristics and water balance for the Thurne catchment, northeast Norfolk. *Quarterly Journal of Engineering Geology and Hydrogeology* **32**, 365–380.

Hvorslev, M.J. (1951) Time lag and soil permeability in ground water observations. *United States Army Corps of Engineers Waterways Experimentation Station, Bulletin* **36**.

Indian Ministry of Water Resources (2006) *Dynamic Ground Water Resources of India (as on March 2004)*. Central Ground Water Board, Government of India, New Delhi.

Ingebritsen, S.E. and Manga, M. (2019) Earthquake hydrogeology. *Water Resources Research* **55**, doi: 10/1029/ 2019WR025341.

Jacob, C.E. (1940) On the flow of water in an elastic artesian aquifer. *Transactions of the American Geophysical Union* **22**, 574–586.

Karami, G.H. and Younger, P.L. (2002) Analysing step-drawdown tests in heterogeneous aquifers. *Quarterly Journal of Engineering Geology and Hydrogeology* **35**, 295–303.

Kearey, P., Brooks, M. and Hill, I. (2002) *An Introduction to Exploration Geophysics* (3nd edn). Blackwell Science Ltd., Oxford.

Keys, W.S. (1988) *Borehole geophysics applied to ground-water investigations*. United States Geological Survey, Denver, Colorado, Open-File Report **87–539**, 305 pp.

King, R.F. (1992) High-resolution shallow seismology: history, principles and problems. *Quarterly Journal of Engineering Geology* **25**, 177–182.

Konikow, L.F. and Bredehoeft, J.D. (1978) *Computer Model of Two-Dimensional Solute Transport and Dispersion in Groundwater. Techniques of Water-Resources Investigations of the United States Geological Survey. Book 7, Chapter C2*. Scientific Software Group, Washington, DC.

Kruseman, G.P. and de Ridder, N.A. (1990) *Analysis and Evaluation of Pumping Test Data* (2nd edn). Pudoc Scientific Publishers, Wageningen, The Netherlands.

Kumar, D., Rajesh, K., Mondal, S. *et al.* (2020) Groundwater exploration in limestone–shale–quartzite terrain through 2D electrical resistivity tomography in Tadipatri, Anantapur district, Andhra Pradesh. *Journal of Earth System Science* **129**, 71, doi: 10.1007/s12040-020-1341-0.

Kumar, R., Singh, R.D. and Sharma, K.D. (2005) Water resources of India. *Current Science* **89**, 794–811.

Kurtz, W., Lapin, A., Schilling, O.S. *et al.* (2017) Integrating hydrological modelling, data assimilation and cloud computing for real-time management of water resources. *Environmental Modelling and Software* **93**, 418–435.

Leblanc, M.J., Tregoning, P., Ramillien, G. *et al.* (2009) Basin-scale, integrated observations of the early 21st century multiyear drought in southeast Australia. *Water Resources Research* **45**, W04408, doi: 10.1029/2008wr007333.

Lewis, D.C., Kriz, G.J. and Burgy, R.H. (1966) Tracer dilution sampling technique to determine hydraulic conductivity of fractured rock. *Water Resources Research* **2**, 533–542.

MacAllister, D.J., Jackson, M.D., Butler, A.P. and Vinogradov, J. (2018) Remote detection of saline intrusion in a coastal aquifer using borehole measurements of self-potential. *Water Resources Research* **54**, 1669–1687, doi: 10.1002/ 2017WR021034.

Mall, R.K., Gupta, A., Singh, R. *et al.* (2006) Water resources and climate change: an Indian perspective. *Current Science* **90**, 1610–1626.

Manga, M. (2020) When it rains, lava pours. *Nature* **580**, 457–458.

McDonald, M.G. and Harbaugh, A.W. (1988) *A Modular Three-Dimensional Finite-Difference Ground-Water Flow Model. Techniques of Water-Resources Investigations of the United States Geological Survey. Book 6, Chapter A1.* Scientific Software Group, Washington, DC.

Meju, M.A., Denton, P. and Fenning, P. (2002) Surface NMR sounding and inversion to detect groundwater in key aquifers in England: comparisons with VES-TEM methods. *Journal of Applied Geophysics* **50**, 95–111.

Meju, M.A., Fontes, S.L., Oliveira, M.F.B. *et al.* (1999) Regional aquifer mapping using combined VES-TEM-AMT/EMAP methods in the semiarid eastern margin of Parnaiba Basin, Brazil. *Geophysics* **64**, 337–356.

Milsom, J. and Eriksen, A. (2011) *Field Geophysics* (4th edn). John Wiley & Sons, Ltd, Chichester.

Misstear, B., Banks, D. and Clark, L. (2017) *Water Wells and Boreholes* (2nd edn). John Wiley & Sons, Ltd, Chichester.

Mussett, A.E. and Khan, K.A. (2000) *Looking into the Earth: An Introduction to Geological Geophysics.* Cambridge University Press, Cambridge.

NASA (2016) *Water: Sustaining Life.* NASA-ISRO SAR Mission. https://nisar.jpl.nasa.gov/system/documents/files/35_NASA_ISRO_SAR_Mission_Water1.pdf (accessed 10 August 2020).

Neuman, S.P. (1975) Analysis of pumping test data from anisotropic confined aquifers considering delayed gravity response. *Water Resources Research* **11**, 329–342.

Niemann, W.L. and Rovey, C.W. (2000) Comparison of hydraulic conductivity values obtained from aquifer pumping tests and conservative tracer tests. *Ground Water Monitoring and Remediation* **20**, 122–128.

Peterson, R.N., Burnett, W.C., Glenn, C. and Johnson, A. (2009) Quantification of point-source groundwater discharges to the ocean from the shoreline of the Big Island,

Hawaii. *Limnology and Oceanography* **54**, 890–904.

Pitrak, M., Mares, S. and Kobr, M. (2007) A simple borehole dilution technique in measuring horizontal ground water flow. *Ground Water* **45**, 89–92.

Plummer, L.N. and Busenberg, E. (2000) Chlorofluorocarbons. In: *Environmental tracers in subsurface hydrology* (eds P. Cook and L. Herczeg). Kluwer Academic Publishers, Norwell, Massachusetts, pp. 441–478.

Prickett, T.A., Naymik, T.G. and Lonnquist, C.G. (1981) A 'Random-Walk' solute transport model for selected groundwater quality evaluations. *Illinois State Water Survey Bulletin* **65**, Champaign, Illinois.

Pugh, J.C. (1975) *Surveying for Field Scientists.* Methuen, London.

Ramillien, G., Famiglietti, J.S. and Wahr, J. (2008) Detection of continental hydrology and glaciology signals from GRACE: a review. *Surveys in Geophysics* **29**, 361–374.

Reynolds, J.M. (2011) *An Introduction to Applied and Environmental Geophysics* (2nd edn). John Wiley & Sons, Ltd, Chichester.

Rodell, M., Velicogna, I. and Famiglietti, J.S. (2009) Satellite-based estimates of groundwater depletion in India. *Nature* **460**, 999–1002.

Rushton, K.R. (1986) Groundwater models. In: *Groundwater: Occurrence, Development and Protection* (ed. T.W. Brandon). Institution of Water Engineers and Scientists, London, pp. 189–228.

Rushton, K.R. (2003) *Groundwater Hydrology: Conceptual and Computational Models.* John Wiley & Sons, Ltd, Chichester.

Sauty, J.-P. (1980) An analysis of hydrodynamic transfer in aquifers. *Water Resources Research* **16**, 145–158.

Schilling, O.S., Cook, P.G. and Brunner, P. (2019) Beyond classical observations in hydrogeology: the advantages of including exchange flux, temperature, tracer concentration, residence time, and soil moisture observations in groundwater model calibration. *Reviews of Geophysics* **57**, 146–182.

Sharma, P.V. (1997) *Environmental and Engineering Geophysics*. Cambridge University Press, Cambridge.

Siemon, B., Christiansen, A.V. and Auken, E. (2009) A review of helicopter-borne electromagnetic methods for groundwater exploration. *Near Surface Geophysics* **7**, 629–646.

Smart, P.L. and Laidlaw, I.M.S. (1977) An evaluation of some fluorescent dyes for water tracing. *Water Resources Research* **13**, 15–33.

Spitz, K. and Moreno, J. (1996) *A Practical Guide to Groundwater and Solute Transport Modeling*. John Wiley & Sons, Inc., New York.

Telford, W.M., Geldart, L.P. and Sheriff, R.E. (1991) *Applied Geophysics* (2nd edn). Cambridge University Press, Cambridge.

Theis, C.V. (1935) The relation between the lowering of the piezometric surface and rate and duration of discharge of a well using groundwater storage. *Transactions of the American Geophysical Union* **2**, 519–524.

Todd, D.K. (1980) *Groundwater Hydrology* (2nd edn). John Wiley & Sons, Inc., New York.

Toynton, R. (1983) The relation between fracture patterns and hydraulic anisotropy in the Norfolk Chalk, England. *Quarterly Journal of Engineering Geology and Hydrogeology* **16**, 169–185.

Voss, C.I. (1984) *A Finite-Element Simulation Model for Saturated-Unsaturated, Fluid-Density – Dependent Groundwater Flow with Energy Transport or Chemically-Reactive Single-Species Solute Transport*. USGS, Washington, DC, Water-Resources Investigations of the United States Geological Survey Report **84-4369**.

Wada, Y., van Beek, L.P.H. and Bierkens, M.F.P. (2012) Nonsustainable groundwater sustaining irrigation: a global assessment. *Water Resources Research* **48**, W00L06, doi:10.1029/2011WR010562.

Wada, Y., van Beek, L.P.H., van Kempen, C.M. et al. (2010) Global depletion of groundwater resources. *Geophysical Research Letters* **37**, L20402, doi: 10.1029/2010GL044571.

Wahr, J., Swenson, S. and Velicogna, I. (2006) The accuracy of GRACE mass estimates. *Geophysical Research Letters* **33**, L06401, doi: 10.1029/2005GL025305.

Walton, W.C. (1960) Leaky artesian aquifer conditions in Illinois. *Illinois State Water Survey Report Investigation* **39**.

Walton, W.C. (1987) *Groundwater Pumping Tests: Design and Analysis*. Lewis Publishers, Chelsea, MI and National Water Well Association, Dublin, Ohio.

Wang, H.F. and Anderson, M.P. (1982) *Introduction to Groundwater Modelling: Finite Difference and Finite Element Methods*. Academic Press, San Diego, California.

Ward, R.S., Williams, A.T., Barker, J.A. et al. (1998) Groundwater tracer tests: a review and guidelines for their use in British aquifers. *British Geological Survey Report* **WD/98/19**.

Wilhelm, H., Rabbel, W., Lüschen, E. et al. (1994) Hydrological aspects of geophysical borehole measurements in crystalline rocks of the Black Forest. *Journal of Hydrology* **157**, 325–347.

Xie, P.P. and Arkin, P.A. (1997) Global precipitation: a 17-year monthly analysis based on gauge observations, satellite estimates, and numerical model outputs. *Bulletin of the American Meteorological Society* **78**, 2539–2558.

Yeboah-Forson, A., Comas, X. and Whitman, D. (2014) Integration of electrical resistivity imaging and ground penetrating radar to investigate solution features in the Biscayne Aquifer. *Journal of Hydrology* **515**, 129–138.

Zaisheng, H., Hao, W. and Rui, C. (2006) *Transboundary Aquifers in Asia with Special Emphasis to China 10–18*. UNESCO, Paris.

Zhang, H., Shi, Z., Wang, G. et al. (2019) Large earthquake reshapes the groundwater flow system: insight from the water-level response to Earth tides and atmospheric pressure in a deep well. *Water Resources Research* **55**, doi: 10.1029/2018WR024608.

Zheng, C. (1990) *MT3D A Modular Three-Dimensional Transport model for Simulation of Advection, Dispersion and Chemical Reactions of Contaminants in Groundwater Systems.* S.S. Papadopulos & Associates, Inc., Rockville, Maryland.

Zohdy, A.A.R., Eaton, G.P. and Mabey, D.R. (1974) *Application of surface geophysics to ground-water investigations. Techniques of Water-Resources Investigations of the United States Geological Survey, Book 2, Chapter D1.* United States Geological Survey, Denver, Colorado, 116 pp.

8

Groundwater quality and contaminant hydrogeology

8.1 Introduction

The occurrence of groundwater contamination is a legacy of past and present land-use practices and poor controls on waste disposal. Many raw materials and chemicals have had a long history of usage before becoming recognized as hazardous. During this time, handling and waste disposal practices have frequently been inadequate. Hence, it must be considered that any industrial site where hazardous materials have been used is now a potential source of contaminated land. In the United States alone, the National Academy of Sciences (1994) reported that there are an estimated 300 000–400 000 hazardous waste sites and that, over the next three decades, US$750 billion could be spent on groundwater remediation at these sites. In the United Kingdom, with its long industrial history, there are estimated to be as many as 100 000 contaminated land sites covering between 50 000 and 200 000 ha, equivalent to an area larger than Greater London. Added to this picture of industrial contamination, the drive towards self-sufficiency in agricultural production and the increasing urbanization of the world's growing population directly threaten the quality of groundwater through the over-application of agrichemicals and the often uncontrolled disposal of human and landfill wastes.

As a definition, contaminated groundwater is groundwater that has been polluted by human activities to the extent that it has higher concentrations of dissolved or suspended constituents than the maximum admissible concentrations formulated by national or international standards for drinking, industrial or agricultural purposes. The main contaminants of groundwater include chemicals such as heavy metals, organic solvents, mineral oils, pesticides and fertilizers, microbiological contaminants such as faecal bacteria and viruses. Table 8.1 is a compilation of the sources and potential characteristics of groundwater contaminants.

This chapter is first concerned with the quality of water intended for drinking and irrigation purposes as determined by international water quality standards. Given the importance of water hardness to consumers and its apparent health benefits, a section provides background information on this water quality parameter. Following this introduction to water quality, the principles of groundwater contaminant transport are then discussed in relation to non-reactive and reactive solutes and their behaviour in homogeneous and heterogeneous aquifer material. The latter part of the chapter provides an overview of major polluting activities, including industrial, mining, agricultural and municipal sources of contaminants, together with a consideration of emerging contamination from microplastic pollution, and concludes with a discussion of the causes and effects of saline intrusion in coastal regions and on small oceanic islands.

Hydrogeology: Principles and Practice, Third Edition. Kevin M. Hiscock and Victor F. Bense.
© 2021 John Wiley & Sons Ltd. Published 2021 by John Wiley & Sons Ltd.
Companion website: www.wiley.com/go/hiscock/hydrogeology3e

Table 8.1 Potential sources of groundwater pollution arising from domestic, industrial and agricultural activities (Jackson 1980).

Contaminant source	Contaminant characteristics
Septic tanks	Suspended solids 100–300 mg/L BOD 50–400 mg/L Ammonia 20–40 mg/L Chloride 100–200 mg/L High faecal coliforms and streptococci. Trace organisms, greases.
Storm water drains	Suspended solids ~1000 mg/L Hydrocarbons from roads, service areas. Chlorides or urea from de-icing. Compounds from accidental spillages. Bacterial contamination.
Industry	
Food and drink manufacturing	High BOD. High suspended solids. Colloidal and dissolved organic substances. Odours.
Textile and clothing	High suspended solids and BOD. Alkaline effluent.
Tanneries	High BOD, total solids, hardness, chlorides, sulphides, chromium.
Chemicals	
○ acids	Low pH.
○ detergents	High BOD.
○ pesticides	High TOC, toxic benzene derivatives, low pH.
○ synthetic resins and fibres	High BOD.
Petroleum and petrochemical	
○ refining	High BOD, chloride, phenols, sulphur compounds.
○ process	High BOD, suspended solids, chloride, variable pH.
Plating and metal finishing	Low pH. High content of toxic metals.
Engineering works	
Power generation	High suspended solids, hydrocarbons, trace heavy metals. Variable BOD, pH. Pulverized fuel ash: sulphate, and may contain germanium and selenium. Fly ash and flue gas scrubber sludges: low pH, disseminated heavy metals.
Deep well injection	Concentrated liquid wastes, often toxic brines. Acid and alkaline wastes. Organic wastes.
Leakage from storage tanks and pipelines	Aqueous solutions, hydrocarbons, petrochemicals, sewage.
Agriculture	
○ arable crops	Nitrate, ammonia, sulphate, chloride and phosphates from fertilizers. Bacterial contamination from organic fertilizers. Organo-chlorine compounds from pesticides.
○ livestock	Suspended solids, BOD, nitrogen. High faecal coliforms and streptococci.
○ silage	High suspended solids, BOD $1–6 \times 10^4$ mg/L. Carbohydrates, phenols.
Mining	
○ coal mine drainage	High TDS (total dissolved solids), suspended solids. Iron. Low pH. Possibly high chloride.
○ metals	High suspended solids. Possibly low pH. High sulphates. Dissolved and particulate metals.
Household wastes	High sulphate, chloride, ammonia, BOD, TOC and suspended solids from fresh wastes. Bacterial contamination. On decomposition: initially TOC of mainly volatile fatty acids (acetic, butyric, propionic acids), subsequently changing to high molecular weight organics (humic substances, carbohydrates).

BOD is biological oxygen demand, TOC is total organic carbon, pH is $-\log_{10}(H^+)$.
(*Source:* Adapted from Jackson, R.E. (ed.) (1980) *Aquifer Contamination and Protection*. UNESCO, Paris.)

8.2 Water quality standards

The chemical composition of natural groundwaters is discussed in Section 4.3. In addition, groundwaters may contain synthetic organic compounds and microbiological organisms, for example organic solvents and pathogenic bacteria, introduced from sources of surface contamination. To limit the possible harmful effects of natural and introduced components of groundwater, various measures have been developed to protect water users. Water quality criteria include the scientific information with which decisions on water quality can be based, for example toxicity data, information relating to the available water treatment technology and environmental degradation rates. With this information, and taking into account political, legal and socio-economic issues, policy makers are able to set water quality objectives for the attainment of good quality.

For groundwater, the setting of chemical quality objectives may not always be the best approach since it gives the impression of an allowed level of pollution. An alternative approach is to state that groundwater should not be polluted at all. In the EU, this precautionary approach to protecting the chemical status of groundwater is adopted and comprises a prohibition on direct discharges to groundwater and, to cover indirect discharges, a requirement to monitor groundwater bodies in order to detect changes in chemical composition and reverse any upward trend in pollution. Under the Directive on the Protection of Groundwater against Pollution Caused by Certain Dangerous Substances (80/68/EEC; Council of European Communities 1980), the most toxic substances are listed under Lists I and II (Appendix 9). List I substances, including organophosphorus compounds, mercury and cadmium, should be prevented from being discharged into groundwater. List II substances, including metals, fluoride and nitrate, should have discharges of these substances into groundwater minimized. Only a few specific directives have been established at European level for particular issues, including the Directive on Diffuse Pollution by Nitrates (91/676/EEC; Council of European Communities 1991). Taken together, it is envisaged that the above measures, which are subsumed under the EU Water Framework Directive (Section 1.7.2.1), should prevent and control groundwater pollution and achieve good groundwater chemical status for the future.

Further to water quality objectives, water quality standards present the detailed rules that govern how the objectives should be met. To be workable, the standards must be relatively simple so that routine monitoring can detect water quality failures. A standard may allow some variability, or derogation, in terms of a given concentration being met for a certain percentage of samples but with no single sample allowed to exceed a maximum allowable concentration. A further consideration is where, in the cycle of water abstraction, treatment and supply, to apply the standard. Standards applied at the tap control the standard of water used for human consumption and are termed drinking water quality standards. A number of large organizations, including the EU, United States EPA and WHO have published drinking water quality standards and these are summarized in Appendix 9.

The hardness of groundwater can become a water quality issue especially where it affects industrial and domestic uses where the water is heated. There has also been a long debate as to the relative health benefits of drinking hard water. In fact, no health-based guideline value is proposed for hardness since it is considered that the available data on the inverse relationship between the hardness of drinking water and cardiovascular disease (CVD) are inadequate to permit the conclusion that the association is causal (World Health Organization 2011). However, a concentration of 500 mg L^{-1} is at the upper limit of aesthetic acceptability. Further discussion follows in the next section and Box 8.1.

Box 8.1   The 'hard-water story'

The history and debate surrounding whether hard water protects against cardiovascular disease (CVD) is often referred to as the 'hard-water story' and started with a Japanese agricultural chemist. Kobayashi (1957) had for many years studied the nature of agricultural irrigation water and found a close relation between the chemical composition of river water and the death rate from 'apoplexy' (cerebrovascular disease). The death rate of apoplexy in Japan was extraordinarily high compared to other countries, and the biggest cause of death in Japan. Kobayashi (1957) found that it was especially the ratio of sulphur to carbonate ($SO_4/CaCO_3$) in drinking water that was related to the death rate from apoplexy and suggested that inorganic acid might induce, or $CaCO_3$ prevent, apoplexy.

Since Kobayashi (1957), and in different parts of the world, many studies have been completed on the relation between Ca and Mg in local drinking water, and CVD mortality. These studies are generally based upon death registers and water quality data at regional or municipality levels. Even with all these studies, the results are not conclusive as to the role of Ca and Mg in drinking water for CVD. However, most of these studies are 'ecological', meaning that the exposure to water constituents is determined at group levels with a high risk of mis-classification. Often, very large groups, for example all inhabitants in large cities or areas, are assigned the same value of water Ca and Mg, despite the presence of several waterworks or private wells. In addition, the disease diagnoses studied are sometimes unspecific, with wide definitions that include both cardiac and cerebrovascular diseases. In some studies, it is also unclear whether the range of Ca and Mg in drinking water is large enough to allow for appropriate analyses.

One of the most comprehensive studies of the geographic variations in cardiovascular mortality was the British Regional Heart Study. The first phase of this study (Pocock

et al. 1980) applied multiple regression analysis to the geographical variations in CVD for men and women aged 35–74 in 253 urban areas in England, Wales and Scotland for the period 1969–1973. The investigation showed that the relationship to water hardness was non-linear, being much greater in the range from very soft to medium-hard water than from medium to very hard water. Geometric means for the standardized mortality ratio (SMR) for CVD were calculated for towns grouped according to water hardness both with and without adjustments (by analysis of covariance) for the effects of four climatic and socio-economic variables (percentage of days with rain, mean daily maximum temperature, percentage of manual workers and car ownership). The adjusted SMR decreased steadily in moving from a hardness of 10–170 mg L^{-1} but changed little between 170 and 290 mg L^{-1} or greater. After adjustment, CVD in areas with very soft water, around 25 mg L^{-1}, was estimated to be 10–15% higher than in areas with medium-hard water, around 170 mg L^{-1}, while any further increase in hardness beyond 170 mg L^{-1} did not additionally lower CVD mortality. Hence, it appeared that the maximum effect on CVD lay principally between the very soft and medium-hard waters. Importantly, adjusting for climatic and socio-economic differences considerably reduced the apparent magnitude of the effect of water hardness (Pocock *et al.* 1980).

A problem with correlation studies such as the British Regional Heart Study, as argued by Jones and Moon (1987), is the failure of much of the research to consider the causal mechanism that links independent variables to the disease outcome. Also, many of the calibrated models presented in the literature are socially blind in including only those variables pertaining to the physical environment, often a large number of water quality elements. Even in those better analyses that have included social variables, as in the case

Box 8.1 (Continued)

of the British Regional Heart Study, the relatively strong correlation found for Ca in England and Wales may be a result of Ca acting as a very good surrogate for social variables. The soft water areas of the north and west of the country equate to the areas of early industrialization, and today, these areas house a disproportionate percentage of the socially disadvantaged (Jones and Moon 1987). Therefore, it is important that further studies undertake the challenge of quantitatively analysing the separate effects of social variables from those of water hardness.

In a later study reported by Rubenowitz-Lundin and Hiscock (2005), three case-control studies were conducted over a decade in a part of southern Sweden, in a relatively

small geographic area where there is a great difference between and also within the municipalities regarding Mg and Ca content in drinking water. The advantage with this limited study area was that the possible risk of such confounding factors as climate, geographical, cultural and socio-economic differences was minimized. In the first study, the relation between death from acute myocardial infarction (AMI) and the level of Mg and Ca in drinking water was examined among men, using mortality registers. A few years later a study with a similar design was made comprising women.

Seventeen municipalities were identified in the counties of Skåne and Blekinge where water quality with respect to water hardness,

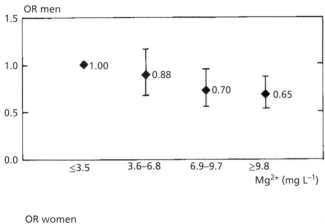

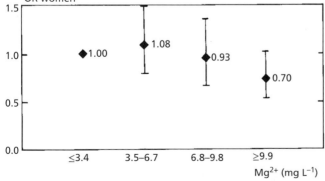

Fig. 8.1 Odds ratios (OR) with 95% confidence intervals for death from acute myocardial infarction in relation to Mg in drinking water (adjusted for age and Ca) in men and women in southern Sweden (Rubenowitz-Lundin and Hiscock 2005). (*Source:* Rubenowitz-Lundin, E. and Hiscock, K. (2005) Water hardness and health effects. In: *Essentials of Medical Geology* (eds O. Selinus *et al.*). Elsevier Academic Press, San Diego, pp. 331–345).

(Continued)

Box 8.1   (Continued)

acidity and treatment procedures had remained basically unchanged (change of hardness <10% and pH <5%) during the most recent 10 years. Cases included were men (*n* = 854) and women (*n* = 378) in 17 municipalities who had died of AMI between ages 50 and 69 years. Controls were men (*n* = 989) and women (*n* = 1368) of the same age group who had died of cancer. Individual water quality data were collected. The subjects were divided into quartiles according to the levels of Mg. Odds ratios were calculated in relation to the group with the lowest exposure. Adjustments for age were made in all analyses. The results show odds ratios of 0.65 for men and 0.70 for women in the quartile with highest Mg levels in the drinking water ($\geq$9.8–9.9 mg L^{-1}) (Fig. 8.1). This means that the risk of dying from AMI was about 30% lower compared with the risk for those

who used drinking water with the lowest levels of Mg.

A few years later a prospective interview study was conducted in the same area, where men and women who suffered from AMI during the years 1994–1996 were compared with population controls. The results showed that Mg in drinking water protected against death from AMI, but the total incidence was not affected. In particular, the number of deaths outside hospitals was lower in the quartile with high Mg levels and supports the hypothesis that Mg prevents sudden death from AMI rather than all coronary heart disease deaths (Rubenowitz *et al.* 2000).

Fewer studies have been carried out in developing countries, but Dissanayake *et al.* (1982), for example, found a negative correlation between water hardness and various forms of CVD and leukaemia in Sri Lanka.

8.2.1   Water hardness

Water hardness is the traditional measure of the capacity of water to react with soap and describes the ability of water to bind soap to form lather, a chemical reaction detrimental to the washing process. Hardness has little significance in terms of hydrochemical studies, but it is an important parameter for water users. Today, the technical significance of water hardness is more concerned with the corrosive effects on water pipes that carry soft water.

Despite the wide usage of the term, the property of hardness is difficult to define exactly. Water hardness is not caused by a single substance but by a variety of dissolved polyvalent metallic ions, predominantly Ca and Mg, although other ions, for example Al, Ba, Fe, Mn, Sr and Zn, also contribute. The source of the metallic ions are typically sedimentary rocks, the most common being limestone ($CaCO_3$) and dolomite ($CaMg(CO_3)_2$). In igneous rock, magnesium is typically a constituent of the dark-coloured ferromagnesian minerals, including olivine, pyroxenes, amphiboles and

dark-coloured micas, and slow weathering of these silicate minerals produces water hardness.

Hardness is normally expressed as the total concentration of Ca^{2+} and Mg^{2+} ions in water in units of mg L^{-1} as equivalent $CaCO_3$. For this purpose, hardness can be determined by substituting the concentration of Ca^{2+} and Mg^{2+}, expressed in mg L^{-1}, in the following equation:

$$\text{Total hardness} = 2.5\left(Ca^{2+}\right) + 4.1\left(Mg^{2+}\right)$$

$$(\text{eq. 8.1})$$

Each concentration is multiplied by the ratio of the formula weight of $CaCO_3$ to the atomic weight of the ion; hence, the factors 2.5 and 4.1 are included in the hardness relation (Freeze and Cherry 1979).

Where reported, carbonate hardness includes that part of the total hardness equivalent to the HCO_3^- and CO_3^{2-} content (or alkalinity). If the total hardness exceeds the alkalinity, the excess is termed the non-carbonate hardness and is a measure of the calcium and magnesium sulphates. In older publications, the terms 'temporary' and 'permanent' are used in place of 'carbonate' and 'non-carbonate'. Temporary hardness reflects the fact that the ions responsible may be precipitated by boiling, such that

$$Ca^{2+} + 2HCO_3^- \rightarrow \underset{\text{'scale'}}{CaCO_3} \downarrow + H_2O + CO_2 \uparrow$$

(eq. 8.2)

In Europe, water hardness is often expressed in terms of degrees of hardness. One French degree is equivalent to $10 \, mg \, L^{-1}$ as $CaCO_3$, one German degree to $17.8 \, mg \, L^{-1}$ as $CaCO_3$ and one English or Clark degree to $14.3 \, mg \, L^{-1}$ as $CaCO_3$. One German degree of hardness (dH) is equal to 1 mg of calcium oxide (CaO) or 0.72 mg of magnesium oxide (MgO) per 100 mL of water.

A number of attempts have been made to classify water hardness. Water with hardness values greater than $150 \, mg \, L^{-1}$ as equivalent $CaCO_3$ is designated as being very hard. Soft water has values of less than $60 \, mg \, L^{-1}$. Groundwaters in contact with limestone or gypsum ($CaSO_4.2H_2O$) rocks can commonly attain levels of $200–300 \, mg \, L^{-1}$. In water from gypsiferous formations, $1000 \, mg \, L^{-1}$ or more of hardness may be present (Hem 1985).

Hardness in water used for domestic purposes does not become particularly troublesome until a level of $100 \, mg \, L^{-1}$ is exceeded. Depending on pH and alkalinity, hardness of about $200 \, mg \, L^{-1}$ can result in scale deposition, particularly on heating, and increased soap consumption. Soft waters with a hardness of less than about $100 \, mg \, L^{-1}$ have a low buffering capacity and may be more corrosive to water pipes resulting in the presence of heavy metals, such as Cd, Cu, Pb and Zn, in drinking water, depending also on the pH and dissolved oxygen content of the water.

In developing countries reliant on groundwater supplies developed in crystalline bedrock aquifers, water hardness is often an important consideration, particularly as silicate weathering not only produces Ca^{2+}, Mg^{2+} and HCO_3^- but can also release elements such as Al, As and F that are hazardous to human health. Studies by Dissanayake (1991) and Rajasooriyar (2003) have highlighted the problem of high fluoride concentrations and associated dental fluorosis in areas of hard water in Sri Lanka.

Rajasooriyar (2003) measured total hardness in dug wells and tube wells in the Uda Walawe Basin of southern Sri Lanka in the wide range $7–3579 \, mg \, L^{-1}$ as $CaCO_3$ with an average of $395 \, mg \, L^{-1}$ (Fig. 8.2). Compared with the

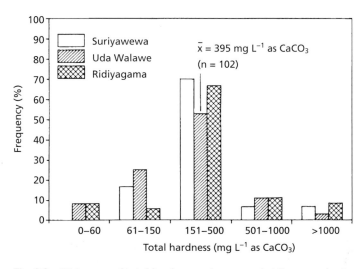

Fig. 8.2 Histogram of total hardness values recorded for groundwaters sampled during the wet season (January and February 2001) from dug wells and tube wells in three sub-catchments of the Uda Walawe Basin of Sri Lanka. (*Source:* Courtesy of L. Rajasooriyar.)

government water quality limit of 600 mg L^{-1}, of the 102 samples collected during the wet season in 2001, 12% of the samples were in excess of the limit and are considered too hard to drink (values above 100–150 mg L^{-1} are locally considered too hard as a water supply). Soft waters are found in areas with a dense irrigation network supplied by rain-fed surface reservoirs. Irrigation canal waters in the Suriyawewa and Uda Walawe sub-catchments were measured in the dry season to have a total hardness in the range 40–90 mg L^{-1} as CaCO$_3$ and it is leakage of this water source that leads to the softening of shallow groundwater. Most groundwater (63% of samples analysed) in the fractured aquifer is very hard with carbonate hardness in the range 151–500 mg L^{-1} contributed by the weathering of ferromagnesian minerals, anorthite, calcite and dolomite. The products of this weathering lead to high concentrations of dissolved Ca^{2+}, Mg^{2+} and HCO$_3^-$ in groundwaters. Exceptionally high values of hardness (>1000 mg L^{-1}) typically occur in non-irrigated areas with additional non-carbonate hardness contributed by pyrite oxidation buffered by weathering of Ca-minerals and, in the case of the coastal Ridiya-gama catchment, by saltwater inputs of Ca^{2+} and SO$_4^{2-}$.

8.2.2 Irrigation water quality

Crop irrigation is the most extensive use of groundwater in the world, and so it is important to consider plant requirements with respect to water quality. The most damaging effects of poor quality irrigation water are excessive accumulation of soluble salts (the salinity hazard) and a high percentage sodium content (the sodium hazard). The salinity hazard increases the osmotic pressure of the soil water and restricts the plant roots from absorbing water, even if the field appears to have sufficient moisture. The result is a physiological drought condition. The salinity hazard is generally determined by measuring the electrical conductivity of the water in µS cm^{-1} and then assessed against the type of criteria given in Table 8.2.

The sodium hazard relates to the accumulation of excessive sodium which causes the physical structure of the soil to breakdown. The replacement by sodium of calcium and magnesium adsorbed on clays results in the dispersion of soil particles. The soil becomes hard and compact when dry and increasingly impervious to water such that the plant roots do not get enough water, even though water may be standing on the surface. The sodium

Table 8.2 Salinity hazard of irrigation water with basic guidelines for water use relative to dissolved salt content.

Salinity hazard	Dissolved salt content (mg L^{-1})	Electrical conductivity (µS cm^{-1})
Water for which no detrimental effects will usually be noticed	500	750
Water that may have detrimental effects on sensitive crops[a]	500–1000	750–1500
Water that may have adverse effects on many crops and requiring careful management practices	1000–2000	1500–3000
Water that can be used for salt-tolerant plants[b] on permeable soils with careful management practices and only occasionally for more sensitive crops	2000–5000	3000–7500

[a] Field beans, string beans, peppers, lettuce, onions, carrots, fruit trees.
[b] Sugarbeet, wheat, barley.

hazard of irrigation water is estimated by the sodium adsorption ratio which relates the proportion of Na^+ to Ca^{2+} and Mg^{2+} in the water as follows:

$$SAR = \frac{Na^+}{\sqrt{\dfrac{Ca^{2+} + Mg^{2+}}{2}}} \qquad \text{(eq. 8.3)}$$

with the ionic concentrations expressed in $meq\,L^{-1}$. Generally, irrigation water with a SAR greater than nine should not be used on crops, even if the total salt content is relatively low. Higher values of SAR may be tolerated if the soil contains an appreciable amount of gypsum ($CaSO_4.2H_2O$) or if gypsum can be added to the soil and so provide a source of soluble calcium to decrease the SAR.

The two aspects of salinity and sodium content can be combined and an irrigation water quality classification obtained. An example classification is shown in Fig. 8.3 based on research by the US Department of Agriculture.

Other water quality considerations for irrigation water include the bicarbonate and carbonate concentrations in water which effectively increase the sodium hazard by precipitating calcium and magnesium carbonates during soil drying, hence increasing the SAR, and the presence of toxic elements, particularly boron and chloride. Excessive levels of boron and chloride are common in groundwater and concentrations of boron greater than $1\,mg\,L^{-1}$ and of chloride above $70\,mg\,L^{-1}$ can lead to injury in sensitive plants.

Additional factors to consider in deciding the usefulness of water for a specific irrigation purpose include soil texture and structure, drainage conditions, gypsum and lime content of the soil and the irrigation method and management. Further information can be found in UNESCO/FAO (1973).

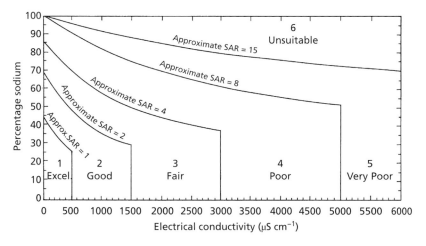

Fig. 8.3 Diagram for classifying irrigation water based on percentage sodium content (sodium hazard) and electrical conductivity (salinity hazard). The sodium adsorption index (SAR) is also shown. The six classes are described as follows: Class 1 (Excellent), suitable for use on all crop types; Class 2 (Good), suitable for use on most crops under most conditions but limiting conditions can develop on poorly draining clayey soils; Class 3 (Fair), suitable for most crops if care is taken to prevent accumulation of soluble salts, including sodium, in the soil; Class 4 (Poor), suitable only in situations having very well-drained soils for production of salt tolerant crops; Class 5 (Very poor), restricted to irrigation of sandy, well-drained soils in areas receiving at least 750 mm of rainfall; Class 6 (Unsuitable), not recommended for crop irrigation. (*Source:* Oklahoma State University Extension Facts F-2401, Classification of Irrigation Water Quality. © Oklahoma State University)

8.3 Transport of contaminants in groundwater

The type of soil, sediment or rock in which a pollution event has occurred and the physico-chemical properties of individual or mixtures of contaminants influence the spread and attenuation of groundwater contaminants. The fundamental physical processes controlling the transport of non-reactive contaminants are advection and hydrodynamic dispersion, which create a spreading pollution plume and cause a dilution in the pollutant concentration. For reactive contaminant species, attenuation of the pollutant transport occurs by various processes including chemical precipitation, sorption, microbially mediated redox reactions and radioactive decay. For the class of contaminants known as non-aqueous phase liquids (NAPLs) both immiscible and dissolved phases of the contaminant need to be considered. To explain these processes, it is convenient to divide the following sections into general contaminant classes, namely non-reactive and reactive dissolved contaminants and non-aqueous phase liquids.

8.3.1 Transport of non-reactive dissolved contaminants

Non-reactive contaminants, such as saline wastes containing chloride, are principally affected by the major processes of advection and hydrodynamic dispersion. Advection is the component of solute movement attributed to transport by the flowing groundwater. The advective velocity of the contaminant is the average linear velocity of the groundwater and can be calculated from a consideration of Darcy's Law (eq. 2.10). Hydrodynamic dispersion of contaminants in porous material occurs as a result of mechanical mixing and molecular diffusion as illustrated in Fig. 8.4. The significance of the dispersive processes is to decrease the contaminant concentration with distance from the source. As shown in Fig. 8.5, a continuous pollution source will produce an elongate plume, whereas a single-point source will

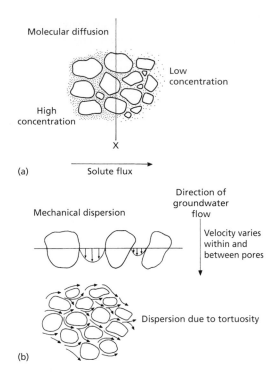

Fig. 8.4 Diagrammatic representation of (a) molecular diffusion and (b) mechanical dispersion which combine to transport solute within a porous material by the process of hydrodynamic dispersion. Notice that mechanical dispersion results from the variation of velocity within and between saturated pore space and from the tortuosity of the flowpaths through the assemblage of solid particles. Molecular diffusion can occur in the absence of groundwater flow, since solute transport is driven by the influence of a concentration gradient, while mechanical dispersion occurs when the contaminant is advected by the groundwater.

produce a slug that grows with time while becoming less concentrated as a result of dispersion as the plume moves in the direction of groundwater flow.

Molecular diffusion of contaminants is not normally of practical consideration where advection and mechanical dispersion are dominant. This is typically the case for shallow groundwater environments, but in situations such as the very long-term deep disposal of waste in stable geological environments of low hydraulic conductivity and low hydraulic gradient, diffusion is significant. The safe disposal of low- to medium-level nuclear wastes

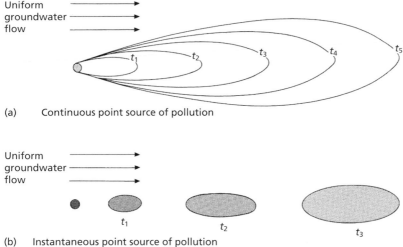

(a)   Continuous point source of pollution

(b)   Instantaneous point source of pollution

Fig. 8.5   Dispersion within an isotropic porous material of (a) a continuous point source of pollution at various times, t, and (b) an instantaneous (single event) point source of pollution. Spreading of the pollution plumes results from hydrodynamic dispersion and acts to dilute the contaminant concentrations, while advection transports the plumes in the field of uniform groundwater flow.

in rock repositories is dependent on the engineered containment of the waste and, should the containment fail over periods of thousands of years, the absence of an advective transport route back to the biosphere.

Diffusion represents the net movement of solute under a concentration gradient (Fig. 8.4) and can be described using Fickian theory. In one-dimension, Fick's second law describes the time-varying change in solute concentration for a change in solute flux as follows:

$$\frac{\partial C}{\partial t} = -D^* \frac{\partial^2 C}{\partial x^2}$$ (eq. 8.4)

where D^* is the molecular diffusion coefficient for the solute in a porous material. The analytical solution for an instantaneous step-change in solute concentration, C, for an infinite aquifer space is given by

$$\frac{C}{C_0} = \operatorname{erfc}\left(\frac{x}{2\sqrt{D^* t}}\right)$$ (eq. 8.5)

where erfc is the complementary error function (see Appendix 8 for tabulated values of erf (error function) and erfc), C_o is the initial concentration at $x = 0$ at time $t = 0$, and C is the concentration measured at position x at

time t. A graphical solution to eq. 8.5 is shown in Fig. 8.6 for values of diffusion coefficient equal to 10^{-10} and 10^{-11} m^2 s^{-1}. Even after 10 000 years, the diffusive breakthrough of contaminant with a relative concentration of 0.01 (or 1% of the initial concentration) has only reached about 25 m from the pollution source.

Of greater importance in terms of contaminant transport in the shallow sub-surface, mechanical dispersion of a dissolved solute in a groundwater flow field is represented by

$$\text{Mechanical dispersion} = \alpha \bar{v}$$

where α is the dispersivity of the porous material and $\bar{v}$ is the advective velocity (average linear velocity) of the groundwater. Dispersivity is a natural physical characteristic of porous material and determines the degree of contaminant spreading. Dispersivity is greatest in the longitudinal direction of groundwater flow but much smaller, typically one-thirtieth to one-fifth of the longitudinal dispersivity, in the direction perpendicular, or transverse, to the flow. Dispersivity is found to be scale dependent. At the micro scale, for example in controlled laboratory experiments using sand-filled columns, longitudinal dispersivity is

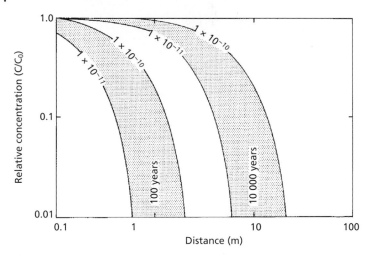

Fig. 8.6 Positions of a contaminant front transported by one-dimensional molecular diffusion at times of 100 years and 1000 years for a source concentration, $C = C_0$, at time, $t > 0$, and a diffusion coefficient, D^*, of 10^{-10} and 10^{-11} m^2 s^{-1}. The curves of relative concentration are calculated using eq. 8.5 (Freeze and Cherry 1979). (*Source:* Freeze, R.A. and Cherry, J.A. (1979) *Groundwater*. Prentice-Hall, Inc., Englewood Cliffs, New Jersey. © 1979, Pearson.)

measured to between 0.1 and 10 mm, and is mainly caused by pore-scale effects. In contrast, tracer experiments at the macro, field scale (Boxes 7.2 and 7.4) give higher values of dispersivity, generally of a few metres and normally less than 100 m, as a result of the physical heterogeneities in the aquifer encountered during transport.

Both laboratory and field experiments show that contaminant mass spreading by dispersion in a porous material conforms to a normal (Gaussian) distribution, with the position of the mean of the concentration distribution representing transport at the advective velocity of the water. The degree of contaminant dispersion about the mean is proportional to the variance (σ^2) of the concentration distribution. The variance tensor can be resolved into three principal components that are approximately aligned with the longitudinal, transverse horizontal and transverse vertical directions. Assuming a constant groundwater velocity, dispersivity can be calculated as one half of the gradient of the linear spatial trend in variance.

To illustrate the dispersive transport of groundwater solutes, Hess *et al.* (2002) conducted a tracer test in the unconfined sand

and gravel aquifer of Cape Cod, Massachusetts, using the conservative tracer bromide (Br$^-$). As shown in Fig. 8.7a, and with increasing time of transport, physical dispersion of the injected mass of tracer caused the solute plume to become elongate. The tracer plume continued to lengthen as it travelled downgradient through the aquifer and should conform to a linear increase in the longitudinal variance with distance travelled. The synoptic results of the tracer test showed that the longitudinal Br$^-$ variance increased at a slow rate early in the test but increased at a larger rate after about 70 m of transport. A linear trend fit to the later results (69–109 m of transport) produced a longitudinal dispersivity estimate for Br$^-$ of 2.2 m (Fig. 8.7b). The number of observations ($n = 4$) on which this estimate is based is small and scatter around the trend is apparent such that the dispersive process may not yet have reached a constant asymptotic value of longitudinal dispersivity (Hess *et al.* 2002). In general, transverse horizontal and vertical dispersivities were much smaller with values of 1.4×10^{-2} m and 5×10^{-4} m, respectively.

Combining the mathematical description of mechanical dispersion with the molecular diffusion coefficient gives an expression for the

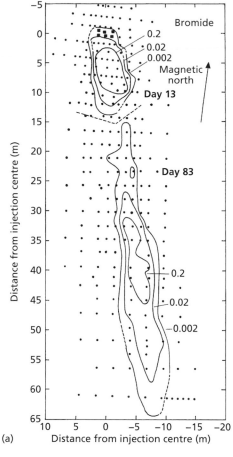

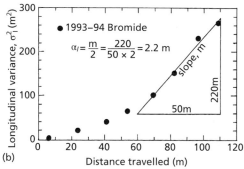

Fig. 8.7 Results of a tracer experiment in the Cape Cod sand and gravel aquifer showing: (a) the distributions of relative concentrations (C/C_o) of Br⁻ observed 13 and 83 days after injection (dashed where inferred); and (b) calculated longitudinal variances of the Br⁻ tracer plume for each synoptic sampling round. Also shown in (a) are locations of multilevel samplers available for groundwater sampling (solid circles) and tracer injection (solid squares) (Hess et al. 2002). (*Source:* Hess, K.M., Davis, J.A., Kent, D.B. and Coston, J.A. (2002) Multispecies

hydrodynamic dispersion coefficient, D, as follows:

$$D_l = \alpha_l \bar{v}_l + D^*$$
$$D_t = \alpha_t \bar{v}_t + D^*$$
(eq. 8.6)

where the subscripts l and t indicate the longitudinal and transverse directions, respectively.

The relative effects of mechanical dispersion and molecular diffusion can be demonstrated from the results of a controlled column experiment. The breakthrough curve for a continuous supply of tracer fed into a column packed with granular material is shown in Fig. 8.8. At low tracer velocity, molecular diffusion is the important contributor to hydrodynamic dispersion, although with little effect in spreading the tracer front. At high velocity, mechanical dispersion dominates and the breakthrough curve adopts a characteristic S-shape with some of the tracer moving ahead of the advancing front and some lagging behind, as controlled by the tortuosity of the flowpaths. The mid-point of the breakthrough curve occurs for a relative concentration, C/C_0, equal to one-half. This point of half-concentration represents the advective behaviour of the solute transport (shown by the vertical dashed line in Fig. 8.8) as if the tracer were moving by a plug-flow type mechanism.

8.3.1.1 One-dimensional solute transport equation

Following from the previous description of solute transport processes, the one-dimensional form of the solute transport equation describing the time-varying change in concentration of non-reactive dissolved contaminants in saturated, homogeneous, isotropic material under steady-state, uniform flow conditions and undergoing advection and hydrodynamic

reactive tracer test in an aquifer with spatially variable chemical conditions, Cape Cod, Massachusetts: dispersive transport of bromide and nickel. *Water Resources Research* **38**, 36-1–36-17. doi:10.1029/2001WR000945. © 2002, John Wiley & Sons.)

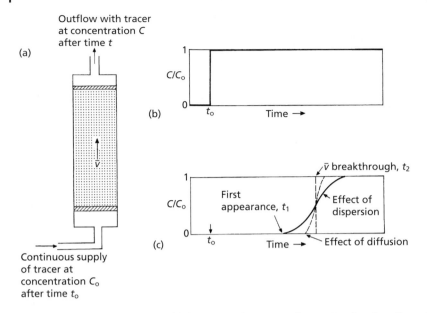

Outflow with tracer
at concentration C
after time t

(a)

(b)

C/C_o

t_o Time →

$\bar{v}$ breakthrough, t_2

First
appearance, t_1

C/C_o

Effect of
dispersion

(c)

t_o Time →   Effect of diffusion

Continuous supply
of tracer at
concentration C_o
after time t_o

Fig. 8.8 Results of a controlled laboratory column experiment showing the effect of longitudinal dispersion of a continuous inflow of tracer in a porous material. (a) experimental set-up; (b) step-function type tracer input relation; (c) relative concentration of tracer in outflow from column. At low tracer velocity, molecular diffusion dominates the hydrodynamic dispersion and the breakthrough curve would appear as the dashed curve. At higher velocity, mechanical dispersion dominates and the solid curve would typically result. The vertical dashed line indicates the time of tracer breakthrough influenced by advective transport without dispersion (Freeze and Cherry 1979). (*Source:* Freeze, R.A. and Cherry, J.A. (1979) *Groundwater*. Prentice-Hall, Inc., Englewood Cliffs, New Jersey. © 1979, Pearson Education.)

dispersion in the longitudinal direction, is given as follows:

$$D_l \frac{\partial^2 C}{\partial l^2} - \bar{v}_l \frac{\partial C}{\partial l} = \frac{\partial C}{\partial t} \qquad \text{(eq. 8.7)}$$

An analytical solution to the advection-dispersion equation (eq. 8.7) was provided by Ogata and Banks (1961) and is written as follows:

$$\frac{C}{C_0} = \frac{1}{2}\left[\operatorname{erfc}\left(\frac{1-\bar{v}_l t}{2\sqrt{D_l t}}\right) + \exp\left(\frac{\bar{v}_l l}{D_l}\right)\operatorname{erfc}\left(\frac{1+\bar{v}_l t}{2\sqrt{D_l t}}\right)\right]$$

(eq. 8.8)

for a step-function concentration input with the following boundary conditions:

$C(l, 0) = 0$	$l \geq 0$
$C(0, t) = C_0$	$t \geq 0$
$C(\infty, t) = 0$	$t \geq 0$

For conditions in which the dispersivity of the porous material is large or when the

longitudinal distance, l, or time, t, is large, the second term on the right-hand side of eq. 8.8 is negligible. Additionally, if molecular diffusion is assumed small compared to mechanical dispersion, then the denominator $\sqrt{D_l t}$ can also be written as $\sqrt{\alpha_l \bar{v}_l t}$. The expression $\sqrt{\alpha_l \bar{v}_l t}$ has the dimensions of length and may be regarded as the longitudinal spreading length or a measure of the spread of contaminant mass around the advective front, represented by the half-concentration, $C/C_0 = 0.5$.

The Ogata–Banks equation can be used to compute the shape of breakthrough curves and concentration profiles. For example, a non-reactive contaminant species is injected into a sand-filled column, 0.4 m in length, in which the water flow velocity is 1×10^{-4} m s^{-1}. If a relative concentration of 0.31 is recorded at a time of 35 minutes, calculate the longitudinal dispersivity, α_l, of the sand. First, taking the simplified version of eq. 8.8 that ignores the second term on the right-hand side, and expressing the denominator as the longitudinal spreading length, then

$$\frac{C}{C_0} = \frac{1}{2}\left[\text{erfc}\left(\frac{1-\bar{v}_l t}{2\sqrt{\alpha_l \bar{v}_l t}}\right)\right] \qquad \text{(eq. 8.9)}$$

and substituting the known values gives

$$0.31 = \frac{1}{2}\text{erfc}\left[\frac{0.4 - 1\times 10^{-4}\times 35\times 60}{2\sqrt{\alpha_l \times 1\times 10^{-4}\times 35\times 60}}\right]$$
$$\text{(eq. 8.10)}$$

which reduces to

$$0.62 = \text{erfc}\left[\frac{0.19}{2\sqrt{\alpha_l 0.21}}\right] \qquad \text{(eq. 8.11)}$$

From Appendix 8, a value of $\beta = 0.35$ produces a $\text{erfc}[\beta] = 0.62$, hence,

$$0.35 = \left[\frac{0.19}{2\sqrt{\alpha_l 0.21}}\right] \qquad \text{(eq. 8.12)}$$

Rearranging and solving for the longitudinal dispersivity results in a calculated value for $\alpha_l = 0.35$ m or 35 cm.

8.3.2 Transport of reactive dissolved contaminants

Reactive substances behave similarly to conservative solute species, but can also undergo a change in concentration resulting from chemical reactions that take place either in the aqueous phase or as a result of adsorption of the solute to the solid matrix of soil, sediment or rock. The chemical and biochemical reactions that can alter contaminant concentrations in groundwater are acid–base reactions, solution–precipitation reactions, oxidation–reduction reactions, ion pairing or complexation, microbiological processes and radioactive decay. Discussion of a number of these processes is included in Chapter 4 in relation to natural groundwater

chemistry and can be equally applied to the fate of dissolved contaminants.

One important type of process affecting the transport of reactive dissolved contaminants is sorption. Sorption processes such as adsorption and the partitioning of contaminants between aqueous and solid phases attenuate, or retard, dissolved solutes in groundwater and are of special relevance to the transport of organic contaminants, particularly hydrophobic compounds.

Attenuation due to sorption can be described by the retardation equation, as follows:

$$R_d = 1 + K_d\frac{(1-n)\rho_s}{\theta} \qquad \text{(eq. 8.13)}$$

where R_d is known as the retardation factor, K_d is the partition or distribution coefficient with units of mL g^{-1} (the reciprocal of density), ρ_s is the solid mass density of the sorbing material and θ is the moisture content of the unsaturated or saturated porous material. Below the water table, θ equates to the porosity and recognizing that $(1-n)\rho_s$ equates to the bulk density, ρ_b, eq. 8.13 can be written as follows:

$$R_d = 1 + K_d\frac{\rho_b}{n} \qquad \text{(eq. 8.14)}$$

Bouwer (1991) presented a simple derivation of the retardation equation and also demonstrated applications describing preferential contaminant movement through macropores (Fig. 6.13) and the shape of the concentration breakthrough curve for macrodispersion caused by layered heterogeneity (Box 8.2). Limitations of the equation are that it assumes ideal, instantaneous sorption and equilibrium between the chemical sorbed and that remaining in solution.

Box 8.2   Macrodispersion caused by layered heterogeneity

The effect of macrodispersion in a layered aquifer system on the transport of a reactive contaminant can be demonstrated by combining Darcy's Law (eq. 2.7) and the retardation equation (eq. 8.13) to derive the shape of the relative concentration breakthrough curve. For the example of layered

heterogeneity shown in Fig. 8.9, and for a well-positioned at 1000 m from the pollution source and pumping at 1.75 m^3 day^{-1}, then assuming saturated flow under a hydraulic gradient of 0.01 in each layer, the relative breakthrough concentration in each layer can be calculated as shown in Table 8.3.

(Continued)

Box 8.2 (Continued)

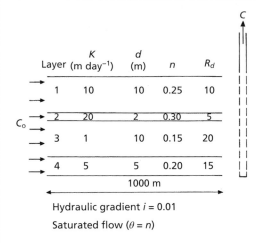

Hydraulic gradient $i = 0.01$

Saturated flow ($\theta = n$)

Fig. 8.9 Set-up of the layered aquifer system and transport parameters used in Table 8.3 to demonstrate macrodispersion of the source concentration, C_o. K is hydraulic conductivity, d is layer thickness, n is porosity and R_d is the retardation factor. C is the concentration of contaminant in the pumped well water.

Notice that the greater sorption, as indicated by the larger R_d values, and longest travel times occur in the lower permeability layers. The first breakthrough at 7.5×10^3 days is in the layer with a hydraulic conductivity of 20 m day^{-1}. At the well, the chemically laden flow in this layer contributes 23% (0.4/1.75) of the actual flow through the aquifer. Thus, the relative concentration of the chemical in the well at this time, which is found from the concentration in the well water, C, as a fraction of the original source concentration, C_o, also equals 0.23. After 25×10^3 days, the flow of contaminated water at the well increases to 1.4 m^3 day^{-1} (0.4 + 1), or 80% of flow from the aquifer. The calculation is continued until 303×10^3 days when the relative concentration C/C_o of the well water equals 1.

The concentration breakthrough curve for this layered aquifer system is shown in Fig. 8.10 and resembles a parabola with

Table 8.3 Calculation of relative breakthrough concentrations, C/C_o, for macrodispersion caused by layered heterogeneity in an aquifer of unit width. The layered aquifer system and associated transport parameters are shown in Fig. 8.9 and the resulting breakthrough curve in Fig. 8.10 (Bouwer 1991).

Layer (in order of breakthrough)	Darcy's Law $v_w = iK/n$ (m day^{-1})	Retardation equation $v_c = v_w/R_t$ (m day^{-1})	$t = 1000/v_c$ ($\times 10^3$ days)	$Q = -diK$ (m^3 day^{-1})	C/C_o
2	0.6667	0.1333	7.5	0.40	(0.40/1.75) = 0.23
1	0.4000	0.0400	25.0	1.00	(1.40/1.75) = 0.80
4	0.2500	0.0167	59.9	0.25	(1.65/1.75) = 0.94
3	0.0667	0.0033	303.0	0.10	(1.75/1.75) = 1.00
Well				1.75	

(*Source:* Bouwer, H. (1991) Simple derivation of the retardation equation and application to preferential flow and macrodispersion. *Ground Water* **29**, 41–46. © 1991, John Wiley & Sons.)

Box 8.2   (Continued)

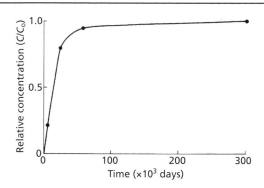

Fig. 8.10   Concentration breakthrough curve demonstrating macrodispersion caused by layered heterogeneity for the set-up shown in Fig. 8.9 and calculations given in Table 8.3.

pronounced tailing, no longer showing the typical symmetrical sigmoidal shape for dispersion in homogeneous porous material (Fig. 8.8c). Also, the point $C/C_o = 0.5$, which is theoretically reached after one pore volume has passed through a homogeneous material with retardation of the chemical, now occurs after about 23×10^3 days. In this time, $1.75 \times 23 \times 10^3 = 40.25 \times 10^3$ m^3 of water are pumped from the well, representing $(40.25 \times 10^3)/(5.6 \times 10^3)$ or 7.2 pore volume flushes of the layered system, where one pore volume is equal to 5.6×10^3 m^3.

The dimensionless retardation factor, R_d, is a measure of the attenuated transport of a reactive contaminant species compared to the advective behaviour of groundwater. As such, the retardation factor can be expressed in three ways as follows:

$$R_d = \frac{\bar{v}_w}{\bar{v}_c} = \frac{l_w}{l_c} = \frac{t_c}{t_w} \qquad \text{(eq. 8.15)}$$

where the subscripts $_w$ and $_c$ indicate the water and dissolved contaminant species, respectively, $\bar{v}$ is the average linear velocity, l is the distance travelled by the water or the central mass of a contaminant plume and t is the arrival time of the water or the mid-point of a contaminant breakthrough curve.

With knowledge of the rate of movement of a non-reactive tracer such as chloride representing the unattenuated flow of water by advection and dispersion, the time axis of a contaminant breakthrough curve can be transformed to a dimensionless time, t/t_{tracer}, where t_{tracer} is the breakthrough time of the tracer. Alternatively, the time axis can be written as the number of pore water flushes (V/V_P, the ratio of feed volume, V, to pore volume, V_P). In doing so, the retardation factor can be read

directly from the dimensionless breakthrough time of contaminant at $C/C_o = 0.5$. Examples of laboratory column breakthrough curves for two aromatic amine compounds are shown in Fig. 8.11. Aromatic amines occur as constituents of industrial waste waters, for example from dye production, and also as degradation products (metabolites) of pesticides.

When inspecting contaminant plumes from field data, it is also useful to choose a non-reactive tracer such as chloride to represent the advective transport of groundwater for comparison with the contaminant behaviour. Examples of retardation factors calculated from both spatial and temporal field data for organic solutes in a sand aquifer are given in Box 8.3. If there is no difference in behaviour, then the retardation factor is equal to one and K_d is zero, in other words, there is no reaction between the contaminant and the soil, sediment or aquifer material. For values of R_d greater than one, attenuation of the contaminant species relative to the groundwater is indicated.

The partition coefficient, K_d, describes the process of contaminant sorption from the aqueous phase to the solid phase. Qualitatively, K_d is

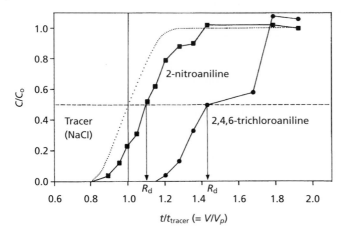

Fig. 8.11   Column breakthrough curves for 2-nitroaniline (C_o = 6 μg L^{-1}) and 2,4,6-trichloroaniline (C_o = 6 μg L^{-1}). t_{tracer} is the breakthrough time of the tracer and V/V_P is the ratio of feed volume, V, to pore volume, V_P. The retardation factor, R_t, can be read directly from the dimensionless breakthrough time (t/t_{tracer}) of contaminant at C/C_o = 0.5 (Worch *et al.* 2002). (*Source:* Worch, E., Grischek, T., Börnick, H. and Eppinger, P. (2002) Laboratory tests for simulating attenuation processes of aromatic amines in riverbank filtration. *Journal of Hydrology* **266**, 259–268. © 2002, Elsevier)

Box 8.3   Controlled field experiments to investigate transport of organic solutes

Hydrophobic sorption of organic solutes was examined in detail as part of two large-scale field experiments to investigate the natural gradient transport of organic solutes in groundwater. The experiments were conducted in an unconfined sand aquifer at the Canadian Forces Base Borden, Ontario, and are described in detail in papers by Mackay *et al.* (1986), Roberts *et al.* (1986), Rivett *et al.* (2001) and Rivett and Allen-King (2003). The aquifer is about 9 m thick and is underlain by a thick, silty clay aquitard. The water table has an average horizontal gradient of about 0.005 that may vary seasonally by as much as a factor of 2 (Rivett *et al.* 2001). The aquifer is composed of clean, well-sorted, fine- to medium-grained sand of glacio-lacustrine origin. Although the aquifer is fairly homogeneous, undisturbed cores reveal distinct bedding features. The bedding is primarily horizontal and parallel, although some cross-bedding and convolute bedding occur. Clay size fractions in the sand are very low and the organic carbon content (0.02%), specific surface area (0.8 m^2 g^{-1}) and cation exchange capacity (0.52 meq (100 g)$^{-1}$) of the aquifer solids are all low. The bulk density of the sand was estimated at 1.81 g cm^{-3} and the average porosity at 0.33. The hydraulic conductivity varies by approximately an order of magnitude with depth as a consequence of layering of the sand material. The geometric means of the hydraulic conductivity for cores samples at two locations were similar at about 10^{-4} m s^{-1} (Mackay *et al.* 1986; Sudicky 1986).

The first experiment, the Stanford–Waterloo natural gradient tracer experiment (Mackay *et al.* 1986; Roberts *et al.* 1986), was designed with nine injection wells, each slotted and screened within the saturated zone (Fig. 8.12). Five halogenated organic

Box 8.3 (Continued)

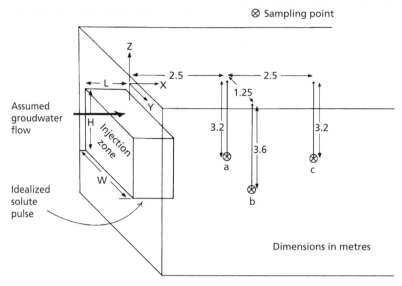

⊗ Sampling point

Z

Assumed
groudwater
flow

Injection
zone

L

X

Y

H

W

2.5

2.5

1.25

3.2

3.6

3.2

a

b

c

Idealized
solute
pulse

Dimensions in metres

Fig. 8.12 Configuration of the injected pulse and the time-series sampling points in the natural gradient tracer experiment. The zone initially permeated by injected water is conceptualized as a rectangular prism with dimensions of length, L, 3.2 m, height, H, 1.6 m, and width, W, 6 m (Roberts *et al.* 1986). (*Source:* Roberts, P.V., Goltz, M.N. and Mackay, D.M. (1986) A natural gradient experiment on solute transport in a sand aquifer 3. Retardation estimates and mass balances for organic solutes. *Water Resources Research* **22**, 2047–2058. © 1986, John Wiley & Sons.)

solutes were chosen representing a range of expected mobilities as measured by their octanol-water partition coefficient (K_{OW}) values (Table 8.4).

At the start of the experiment, approximately 12 m³ of solution were injected over a 15-h period in order not to disturb the natural hydraulic gradient. The injected volume was chosen to be large relative to the scales of heterogeneity of the aquifer, as well as to ensure that dispersion during transport for several years would not too rapidly reduce the solution concentrations to values close to background levels. The monitoring network system consisted of a dense network of 340 multilevel sampling devices. The horizontal spacing of the multilevel wells varied from 1.0 to 4.0 m, while the vertical spacing of

the sampling points varied from 0.2 to 0.3 m, again chosen to be consistent with the estimated scales of hydraulic conductivity.

The closely spaced array of sampling points gave an unparalleled opportunity to study the morphology of the developing solute plumes and their attenuation by sorption and biodegradation by comparison with chloride introduced as a non-reactive tracer. Equal concentration contour plots of vertically averaged solute concentration for the chloride ion at 647 days after injection and carbon tetrachloride (CTET) and tetrachloroethene (PCE) at 633 days after injection are shown in Fig. 8.13. Initially, the plumes were nearly rectangular in plan view. The solute plumes moved at an angle to the field coordinate system and, with time, became

(Continued)

Box 8.3 (Continued)

Table 8.4 Injected organic solutes used in the natural gradient tracer experiment and their associated sorption properties (Mackay *et al.* 1986).

Solute	Injected concentration (mg L^{-1})	Injected mass (g)	Octanol-water partition coefficient K_{OW}
Chloride (tracer)	892	10 700	–
Bromoform (BROM)	0.032	0.38	200
Carbon tetrachloride (CTET)	0.031	0.37	500
Tetrachloroethene (PCE)	0.030	0.36	400
1,2-dichlorobenzene (DCB)	0.332	4.0	2500
Hexachloroethane (HCE)	0.020	0.23	4000

(*Source:* Mackay, D.M., Freyberg, D.L. and Roberts, P.V. (1986) A natural gradient experiment on solute transport in a sand aquifer 1. Approach and overview of plume movement. *Water Resources Research* **22**, 2017–2029. © 1986, John Wiley & Sons.)

progressively more ellipsoidal due to hydrodynamic dispersion. The chloride plume appeared to move at an approximately constant velocity, yet a distinct bimodality developed during the first 85 days. The centre of the chloride plume exhibited a constant advective velocity of 0.09 m day^{-1}, while the organic solutes showed decreasing velocities with time. Significant spreading in the longitudinal direction, and its accompanying dilution, were observed for both the inorganic and organic solute plumes. Relatively little horizontal transverse spreading was evident. As can be seen in Fig. 8.13, the relative mobility of the CTET was significantly less than that of chloride, providing qualitative evidence of retardation due to sorption. The retardation of the other organic solutes was even greater, as observed for PCE in Fig. 8.13, generally in accord with their hydrophobicity.

Retardation factors were estimated by two methods. First, by comparing average travel times estimated from concentration breakthrough responses for the organic solutes with that of chloride, based on time-series sampling at the discrete points shown in Fig. 8.12; and second, by comparing the velocities of the organic solutes with that of the chloride tracer based on analyses taken from the three-dimensional sampling array at a particular time, based on snapshot or synoptic sampling.

A comparison of retardation estimates from temporal and spatial data is given in Table 8.5 and retardation factors estimated from the synoptic sampling data are shown in Fig. 8.14. Retardation factors for the organic solutes relative to chloride ranged from 1.5 to 9.0, being generally greater for the more strongly hydrophobic compounds (Table 8.4). Interestingly, the retardation factors increased over time for the organic solutes, for example CTET increased by 40%, PCE by 120% and 1,2-dichlorobenzene

Box 8.3 (Continued)

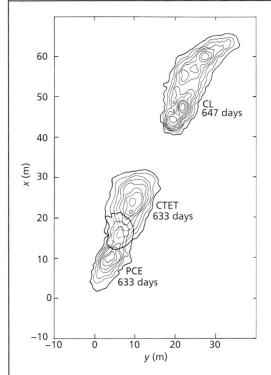

Fig. 8.13 Movement and dispersive spreading of the carbon tetrachloride (CTET, 633 days), tetrachloroethylene (PCE, 633 days) and chloride (CL, 647 days) plumes during the natural gradient tracer experiment. The contour interval for chloride is 5 mg L^{-1} beginning with an outer contour of 10 mg L^{-1}. Contour intervals depicted for CTET and PCE are 0.1 µg L^{-1} beginning with an outer contour of 0.1 µg L^{-1} (Roberts *et al.* 1986). (*Source:* Roberts, P.V., Goltz, M.N. and Mackay, D.M. (1986) A natural gradient experiment on solute transport in a sand aquifer 3. Retardation estimates and mass balances for organic solutes. *Water Resources Research* **22**, 2047–2058. © 1986, John Wiley & Sons.)

(DCB) by 130% in the period 16–650 days. One possibility to explain the increase in retardation behaviour is a gradual, temporal increase in the partition coefficient (K_d) as a result of slow approach to sorption equilibrium, for example owing to a diffusion rate limitation imposed by stratification or

aggregation of the aquifer solids at the particle scale. However, it must be recognized that the temporal behaviour of the retardation factors may also be influenced by the spatial variability of the porous material at the field scale and, as shown in the second experiment described below, non-linear and competitive sorption effects (Rivett and Allen-King 2003).

The second experiment, the emplaced source natural gradient tracer experiment (Rivett *et al.* 2001; Rivett and Allen-King 2003), involved the controlled emplacement below the water table of a block-shaped source of sand with dimensions of length, 0.5 m, height, 1.0 m, and width, 1.5 m, and containing the chlorinated solvents perchloroethene (PCE), trichloroethene (TCE) and trichloromethane (TCM) together with gypsum. Gypsum was added to provide a continuous source of conservative inorganic tracer as dissolved sulphate in the aerobic groundwater of the Borden Aquifer. Unlike the first experiment conducted about 150 m away, and which involved a finite pulse of dissolved organic solutes at low concentrations, the emplaced source experiment was intended to provide a simplified, yet realistic analogue of actual solvent contaminated sites. Such sites commonly contain residual zones of dense non-aqueous phase liquid (DNAPL) that continuously generate dissolved phase organic solute plumes over long time periods.

The gradual dissolution of the residual, multi-component chlorinated solvent source under natural aquifer conditions caused organic solute plumes to develop continuously down-gradient. Source dissolution and three-dimensional plume development were again monitored via a dense array of 173 multilevel sampling wells over a 475-day tracer test period. As shown in Fig. 8.15, organic solute plumes with concentrations spanning 1–700 000 µg L^{-1} were identified.

(Continued)

Box 8.3 (Continued)

Table 8.5 Comparison of retardation estimates for organic solutes from temporal and spatial data from the natural gradient tracer experiment (Roberts *et al.* 1986).

| Organic solute | Temporal data | | | Spatial data | |
| | Retardation factor[a] | | Time range[a] (days) | Instantaneous retardation factor[b] | Ratio of travel distances[c] |
	Mean	Range			
CTET	1.73	1.6–1.8	48–119	2.0–2.1	1.8–1.9
BROM	1.70	1.5–1.8	46–122	2.1–2.3	1.9–2.0
PCE	3.30	2.7–3.9	83–217	3.8–4.7	3.0–3.7
DCB	2.73	1.8–3.7	55–245	5.2–7.2	4.0–5.6

Notes:
[a] Retardation factor and average travel time from time-series sampling data.
[b] Retardation factor from Fig. 8.14 evaluated over the range of times given in column 4 of this table.
[c] Ratio of travel distances (chloride : organic) evaluated to conform to the time interval of column 4 in this table.
(*Source:* Roberts, P.V., Goltz, M.N. and Mackay, D.M. (1986) A natural gradient experiment on solute transport in a sand aquifer 3. Retardation estimates and mass balances for organic solutes. *Water Resources Research* **22**, 2047–2058. © 1986, John Wiley & Sons.)

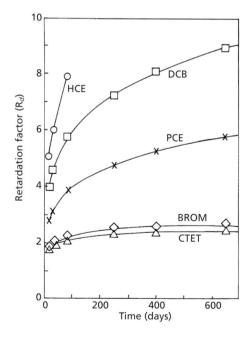

Fig. 8.14 Comparison of retardation factors estimated from synoptic sampling of organic solute movement during the natural gradient tracer experiment (Roberts *et al.* 1986). (*Source:* Roberts, P.V., Goltz, M.N. and Mackay, D.M. (1986) A natural gradient experiment on solute transport in a sand aquifer 3. Retardation estimates and mass balances for organic solutes. *Water Resources Research* **22**, 2047–2058. © 1986, John Wiley & Sons.)

Box 8.3   (Continued)

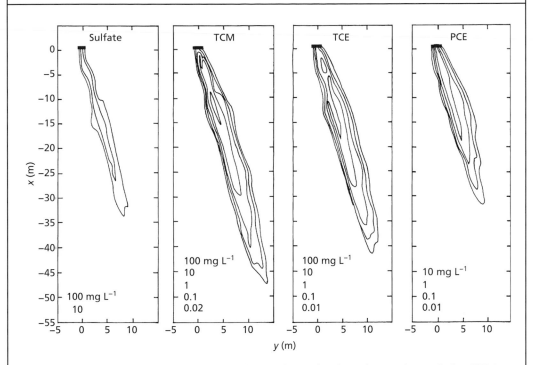

Fig. 8.15 Plan view of the conservative sulphate and organic solute plumes observed after 322 days during the emplaced source tracer experiment and based on maximum concentrations measured in each multilevel sampling well. Contour values are shown at the base of the plots, with the minimum value corresponding to the outer plume contour (Rivett *et al.* 2001). (*Source:* Rivett, M.O., Feenstra, S. and Cherry, J.A. (2001) A controlled field experiment on groundwater contamination by multicomponent DNAPL: creation of the emplaced-source and overview of dissolved plume development. *Journal of Contaminant Hydrology* **49**, 111–149. © 2001, Elsevier.)

The calculated mean groundwater pore velocity until 322 days was 0.085 m day^{-1} inferring a travel distance due to advection alone of 27 m. The dissolved solvent plumes were observed to be narrow (less than 6 m width after 322 days) due to weak transverse dispersion processes and much more elongate (the TCM plume migrated 50 m in 322 days) due to advection and significant longitudinal dispersion. TCM was observed to be the most mobile dissolved organic solute, closely followed by TCE. PCE was comparatively more retarded. Background sulphate concentrations in the aquifer restricted sulphate plume detection to a range of two orders of magnitude, such that the lower detection limit for the organic solute plumes caused these plumes to be detected beyond the conservative sulphate tracer plume.

Hence, in the emplaced source tracer experiment, all the organic solute plumes appeared to have relatively high mobility, including the most retarded solute, PCE. This higher mobility was confirmed by the estimated retardation factors; TCM was essentially conservative (R_d = ~1.0) and TCE almost so (R_d = ~1.1), much lower than values estimated for the organic solutes injected in the Stanford–Waterloo tracer experiment. The sorptive retardation

(Continued)

Box 8.3 (Continued)

of PCE of about 1.4–1.8 (Rivett and Allen-King 2003) was observed to be greater than TCE, consistent with its greater hydrophobicity. Interestingly, PCE, the most retarded solute in the emplaced source experiment, exhibited a retardation factor up to three times lower than observed in the previous Stanford–Waterloo natural gradient tracer experiment. Further laboratory and modelling studies by Rivett and Allen-King (2003) indicated that sorption of PCE was non-linear and competitive with reduced sorption behaviour observed in the presence of co-solvents such as TCE.

Significantly, plume attenuation due to abiotic chemical reactions or biodegradation

processes was not apparent in the emplaced source tracer experiment. The absence of biodegradation was primarily attributed to a lack of additional, easily biodegradable carbon within the emplaced source plumes, and the prevailing aerobic conditions which did not favour sulphate-reducing or methanogenic conditions required for dechlorination. Although attenuation due to biodegradation may be a significant factor at some sites, the emplaced source experiment demonstrated that dissolved DNAPL plumes can be extremely mobile, persistent and very capable of causing extensive aquifer contamination.

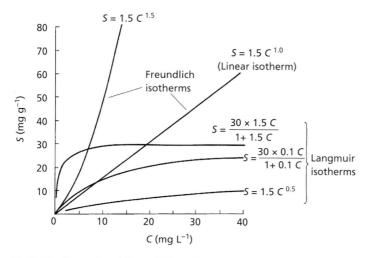

Fig. 8.16 Examples of Freundlich and Langmuir isotherms. S is the mass of chemical sorbed per unit mass of solid and C is the dissolved chemical concentration (Domenico and Schwartz 1998). (*Source:* Domenico, P.A. and Schwartz, F.W. (1998) *Physical and Chemical Hydrogeology* (2nd edn). Wiley, New York. © 1998, John Wiley & Sons.)

---

equal to the mass of chemical sorbed to the solid phase (per unit mass of solid) per concentration of chemical in the aqueous phase. Values of K_d are found empirically from laboratory batch tests conducted at a constant temperature to derive sorption 'isotherms'. As shown in Fig. 8.16, two common relationships describing the sorption of dissolved contaminants are the Freundlich isotherm described by the equation:

$$S = K_d C^n \qquad \text{(eq. 8.16)}$$

and the Langmuir isotherm described by

$$S = \frac{Q^o K_d C}{1 + K_d C} \qquad \text{(eq. 8.17)}$$

where K_d is the partition coefficient reflecting the degree of sorption, S is the mass of chemical sorbed per unit mass of solid, C is the dissolved chemical concentration, Q^o is the maximum sorptive capacity of the solid surface and n is a constant usually between 0.7 and 1.2. A Freundlich isotherm with $n = 1$ is a special case known as the linear isotherm (Fig. 8.16). The gradient of the straight line defining the linear isotherm provides a value of K_d that is the appropriate value for inclusion in the retardation equation (eq. 8.13).

In reality, K_d is not a constant but changes as a function of the mineralogy, grain size and surface area of the solid surface, the experimental conditions of the batch experiments, for example temperature, pressure, pH and Eh conditions, and undetected chemical processes such as mineral precipitation. Not surprisingly, it is difficult to control all the relevant variables in order to give reproducible results, and it is therefore unrealistic to represent all the processes affecting the sorptive behaviour of contaminants in porous material by a simple one-parameter model

defined by K_d. Even so, and as explained in Section 8.3.3.1, for the case of hydrophobic sorption of organic compounds, the attenuation process can be successfully modelled using values of K_d to derive R_d.

The retardation equation can also be used to study cation exchange reactions (Section 4.8), an important consideration in the attenuation of heavy metals (Box 8.4). By defining the partition coefficient as a function of the properties of the exchanger and the solution as found from laboratory experimentation, the partition coefficient can be written as follows:

$$K_d = \frac{K_s \text{CEC}}{\tau} \qquad \text{(eq. 8.18)}$$

where K_s is the selectivity coefficient, CEC is the cation exchange capacity (meq per mass) and τ is the total competing cation concentration in solution (meq per mass). Selectivity coefficients are found from mass-action

---

Box 8.4 Groundwater contamination by heavy metals in Nassau County, New York

The heavy metals of concern in drinking water supplies include Ni, Zn, Pb, Cu, Hg, Cd and Cr. In reducing and acidic waters, heavy metals remain mobile in groundwater; but in soils and aquifers that have a pH buffering capacity, and under oxidizing conditions, heavy metals are readily adsorbed or exchanged by clays, oxides and other minerals. Sources of heavy metals include, in general, the metal processing industries, particularly electro-plating works with their concentrated acidic electrolytes, and other metal surface treatment processes.

Infiltration of metal plating wastes through disposal basins in Nassau County on Long Island, New York, since the early 1940s has formed a plume of contaminated groundwater (Fig. 8.17). The plume contains elevated Cr and Cd concentrations. The area is within an undulating glacial outwash plain, and there are two major hydrogeological units: the Upper glacial aquifer of Late Pleistocene age; and the Magothy aquifer of Late

Cretaceous age, which supplies all local municipal water supplies.

The Upper glacial aquifer is between 24 and 43 m thick, with a water table from 0 to 8 m below ground level. The aquifer comprises medium to coarse sand and lenses of fine sand and gravel. Stratification of the deposits means that the vertical permeability is up to 5–10 times less than the horizontal permeability.

As part of the historical investigations in the South Farmingdale-Massapequa area, a number of test wells were installed in 1962 driven to depths ranging from 2 to 23 m below ground level. Water samples were collected at depth intervals of 1.5 m by hand pump during installation. The results of this investigation are shown in Fig. 8.17, and define a pollution plume that is about 1300 m long, up to 300 m wide, and as much as 21 m thick. The upper surface of the plume is generally less than 3 m below the water table. The plume is thickest along its

(Continued)

Box 8.4 (Continued)

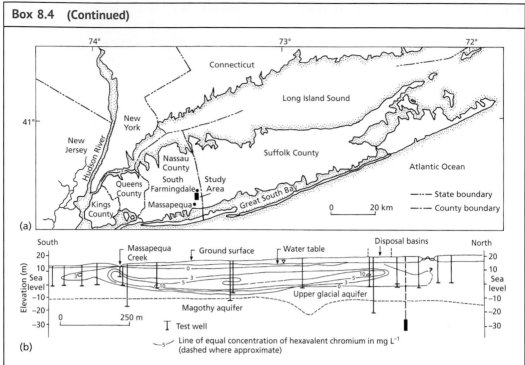

Fig. 8.17 Groundwater contamination by metal plating wastes, Long Island, New York. The location of the investigation area is shown in (a) and a cross-section of the distribution of the hexavalent chromium plume in the Upper glacial aquifer in 1962, South Farmingdale-Massapequa area, Nassau County, is illustrated in (b) (Ku 1980). (*Source:* Adapted from Ku, H.F.H. (1980) Ground-water contamination by metal-plating wastes, Long Island, New York, USA. In: *Aquifer Contamination and Protection* (ed. R.E. Jackson). UNESCO, Paris, pp. 310–317.)

longitudinal axis, the principal path of flow from the basins and is thinnest along its east and west boundaries. The plume appears to be entirely within the Upper glacial aquifer.

Differences in chemical quality of water within the plume may reflect the varying types of contamination introduced in the past. In general, groundwater in the southern part of the plume reflects conditions prior to 1948 when extraction of Cr from the plating wastes, before disposal to the basins, commenced. Since the start of Cr treatment, the maximum observed concentrations in the plume have decreased from about 40 mg L^{-1} in 1949 to about 10 mg L^{-1} in 1962. The WHO guide value for Cr is 0.05 mg L^{-1}. Concentrations of Cd have apparently decreased in some places and increased in others, and peak concentrations do not coincide with those of Cr. These differences are probably due partly to changes in the chemical character of the treated effluent over the years and partly to the influence of hydrogeological factors such as aquifer permeability and sorption characteristics. A test site near the disposal basins recorded as much as 10 mg L^{-1} of Cd in 1964. The WHO guide value for Cd is 0.003 mg L^{-1}.

At the time of operation, the pattern of movement of the plating waste was vertically downwards from the disposal basins, through the unsaturated zone, and into the saturated zone of the Upper glacial aquifer. From here, most of the groundwater contamination moved horizontally southwards, with an average velocity of about 0.5 m day^{-1}, and discharged to the Massapequa Creek.

Box 8.4 (Continued)

Analysis of cores of aquifer material along the axis of the plume showed that the median concentrations of Cr and Cd per kg of aquifer material were, respectively, 7.5 and 1.1 mg, and the maximum concentrations were, respectively, 19 and 2.3 mg. Adsorption occurs on hydrous iron oxide coatings on the aquifer sands. The ability of the aquifer material to adsorb heavy metals complicates the prediction of the movement and concentration of the plume. Furthermore, metals may continue to leach from the aquifer material into the groundwater long after cessation of plating waste discharges, and so necessitating continued monitoring of the site.

---

equations. For example, the cation exchange reaction involving exchangeable ions A and B can be written as follows:

$$aA_{aq} + bB_{ad} = aA_{ad} + bB_{aq} \quad \text{(eq. 8.19)}$$

where a and b are the number of moles and $_{ad}$ and $_{aq}$ indicate the adsorbed and aqueous phases, respectively. The equilibrium coefficient for this reaction is

$$K_{A-B} = K_s = \frac{[A_{ad}]^a [B_{aq}]^b}{[A_{aq}]^a [B_{ad}]^b} \quad \text{(eq. 8.20)}$$

Thus, for a problem involving binary exchange, the retardation equation is now:

$$\frac{\bar{v}_w}{\bar{v}_c} = 1 + \frac{\rho_b K_s \text{CEC}}{\theta \tau} \quad \text{(eq. 8.21)}$$

For some exchange systems involving electrolytes and clays, K_s is found from experimentation to be constant over large ranges of concentrations of the adsorbed cations and ionic strength. For the Mg^{2+}–Ca^{2+} exchange pair, $K_{sMg\text{-}Ca}$ is typically in the range 0.6–0.9, meaning that Ca^{2+} is adsorbed preferentially to Mg^{2+} (Freeze and Cherry 1979).

Clay minerals, metal oxides and organic material exhibit preferential exchange sites for ion occupation and attempts have been made to establish a selectivity sequence, particularly with respect to heavy metals, where equivalent amounts of cations are arranged according to their relative affinity for an exchange site. In general, the greater the charge on a cation, the greater the affinity for an exchange site. Account must also be taken of the pH of contaminant leachate and the competitive effect between the heavy metals present.

Yong and Phadungchewit (1993) demonstrated that a change in soil solution pH results in a corresponding change of the dominant retention mechanism of heavy metals in soils. At high pH values, precipitation mechanisms, for example precipitation of hydroxides and carbonates, dominate. As pH decreases, precipitation becomes less important and cation exchange becomes dominant. It was also shown that the selectivity order governing the retention of heavy metals in soils depends on the soil solution pH. At pH values above 4–5, when precipitation prevails, the selectivity order was found to be Pb>Cu>Zn≥Cd, as demonstrated for illite, montmorillonite and natural clay soils. At lower soil solution pH, the selectivity order was Pb>Cd>Zn>Cu, as shown in the case of kaolinite and montmorillonite.

8.3.3 Transport of non-aqueous phase liquids

The transport of non-aqueous phase liquids (NAPLs) concerns the contamination of groundwater by organic compounds and includes dense non-aqueous phase liquids (DNAPLs) with a density greater than water, for example chlorinated hydrocarbon solvents, and light non-aqueous phase liquids (LNAPLs) with a density less than water, for example hydrocarbons (refined mineral oils). NAPLs have relatively low solubility in water and partition preferentially towards organic material contained in soils, sediments and rocks and in doing so demonstrate hydrophobic sorption behaviour (see Section 8.3.3.1).

The presence of individual phases of water, NAPL and air (in the vadose zone) leads to the condition of multiphase flow where each phase is competing for the available pore space. In the presence of two fluids, wettability is defined as the tendency for a given fluid to be attracted to a surface (solid or liquid) in preference to another fluid. Thus, if fluid A has a higher attraction to a given surface than fluid B, then fluid A is the 'wetting fluid' with respect to fluid B. For the purpose of considering groundwater pollution by organic compounds

within porous material, then water can always be considered the wetting fluid with respect to NAPLs or air.

If a porous material is water-wet and a compound such as oil is introduced, the water will continue to occupy the capillary space in preference to the oil phase (Fig. 8.18). During simultaneous flow of two immiscible fluids, and as shown in Fig. 8.18, part of the available pore space will be filled with water and the remainder with oil such that the cross-sectional area of the pore space available for each fluid is less

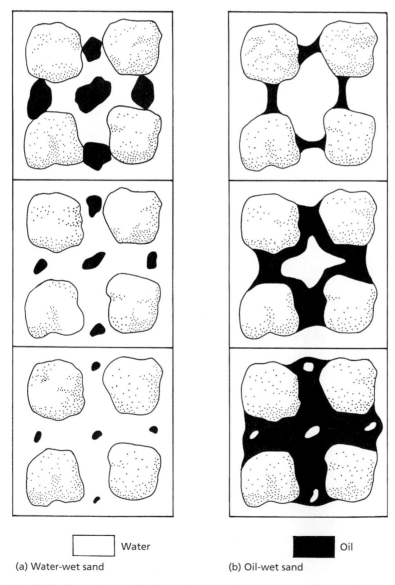

☐ Water   ■ Oil

(a) Water-wet sand   (b) Oil-wet sand

Fig. 8.18 Comparison of fluid wetting states for a porous sand containing water and oil. In (a) water is the wetting fluid and in (b) oil is the wetting fluid (Fetter 1999). (*Source:* Fetter, C.W. (1999) *Contaminant Hydrogeology* (2nd edn). Prentice-Hall, Inc., Upper Saddle River, New Jersey.)

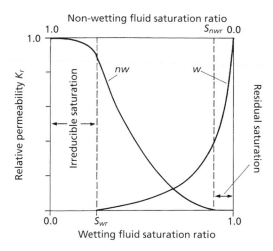

Fig. 8.19 Relative permeability curves for a two-phase system of wetting, w, and non-wetting, nw, liquids (Fetter 1999). (*Source:* Fetter, C.W. (1999) *Contaminant Hydrogeology* (2nd edn). Prentice-Hall, Inc., Upper Saddle River, New Jersey.)

than the total pore space. This situation leads to the concept of relative permeability and is defined as the ratio of the permeability for the fluid at a given saturation to the total permeability of the porous material. A relative permeability exists for both the wetting and non-wetting phases (Fig. 8.19).

Chlorinated solvents such as trichloroethene (TCE), tetrachloroethene (PCE) and 1,1,1-trichloroethane (TCA) are DNAPLs that are volatile and of low viscosity, and consequently are more mobile than water in a porous material (Table 8.6). On infiltrating through the unsaturated zone, solvents leave behind a residual contamination which partitions into a vapour phase that subsequently migrates upwards and laterally by diffusion. The remaining contaminant mass migrates downwards under its own weight and through the water table until halted by the base of the aquifer, or by some other intermediate impermeable barrier (Fig. 8.20a). At the point of reaching an aquitard layer, the pore openings are so small that the weight of DNAPL cannot overcome the pore water pressure. A small, residual amount of solvent, or residual DNAPL

saturation, is left in the pore spaces through which the solvent body has passed (Fig. 8.21). In fractured material, and as long as the weight of DNAPL exceeds the displacement pressure of water contained in a fracture, the DNAPL can potentially migrate to significant depths (Fig. 8.22a).

Refined mineral oils such as petrol, aviation fuel, diesel and heating oils are LNAPLs that behave in a similar manner to chlorinated solvents except, as shown in the Fig. 8.20b, by reason of their density, they float on the water table. The aromatic BTEX compounds, benzene, toluene, ethylbenzene and xylene, are released in significant amounts by petroleum and can be transported by groundwater in the aqueous phase. When spilled at the land surface, oil will migrate vertically in the vadose zone under the influence of gravity and capillary forces, in an analogous manner to water, until it reaches the top of the capillary fringe. Much of the LNAPL will be left trapped in the vadose zone, but on reaching the capillary fringe, the LNAPL will accumulate and an 'oil table' will develop. As the weight of LNAPL increases, the capillary fringe will become thinner until mobile or 'free' product accumulates. Eventually, the capillary fringe may disappear completely and the oil table will rest directly on the water table. In the case of a thick zone of mobile LNAPL, the water table may be depressed by the weight of the LNAPL (Fig. 8.20b). The mobile LNAPL can migrate in the vadose zone, following the slope of the water table, while the dissolved components can disperse with the advecting groundwater. The residual LNAPL phase in the vadose zone can partition into the vapour phase as well as the water phase, with the degree of partitioning dependent on the relative volatility of the hydrocarbon and its solubility in water. In fractured rocks, LNAPL will typically resist migration below the water table, but where there is sufficient weight, LNAPL can penetrate below the water table to a limited extent when the pressure exerted by the LNAPL exceeds the

Table 8.6 Physical and chemical properties of five common chlorinated solvents. Values from Verschueren (1983), Devitt *et al.* (1987) and Schwille (1988).

Chlorinated solvent (abbreviation)	Chemical formula	Molecular weight	Density (g cm^{-3})	Kinematic viscosity (mm^2 s^{-1})	Solubility (mg L^{-1})	Vapour pressure (mm at 20°C) (kPa)	K_{oc}^{a} (cm^3 g^{-1})	Henry's law constant (kPa m^3 mol^{-1} at 25 °C)
Trichloroethene (TCE)	CCl$_2$=CHCl	131.5	1.46	0.4	1100 at 25°C	60 / 8.0	150	1.2
1,1,1-trichloroethane (TCA)	CCl$_3$CH$_3$	133.4	1.35	0.6	4400 at 20°C	100 / 13.3	113	2.8
Tetrachloroethene (Perchloroethene) (PCE)	CCl$_2$CCl$_2$	165.8	1.63	0.5	150 at 25°C	14 / 1.0	364	2.3
Tetrachloromethane (Carbon tetrachloride) (CTET)	CCl$_4$	153.8	1.59	0.6	800 at 20°C	90 / 12.1	110	2.4
Trichloromethane (Chloroform) (TCM)	CHCl$_3$	119.4	1.49	0.4	8000 at 20°C	160 / 32.8	29	0.4

[a] Partition coefficient between organic carbon and water.

(*Sources*: Verschueren, K. (1983) *Handbook of Environmental Data on Organic Chemicals* (2nd edn). Van Nostrand Reinhold Co., New York.; Devitt, D.A., Evans, R.B., Jury, W.A. *et al.* (1987) *Soil Gas Sensing for Detection and Mapping of Volatile Organics.* National Water Well Association, Worthington, Ohio.; Schwille, F. (1988) *Dense Chlorinated Solvents in Porous and Fractured Media: Model Experiments.* Lewis Publishers Inc, Chelsea, Michigan.)

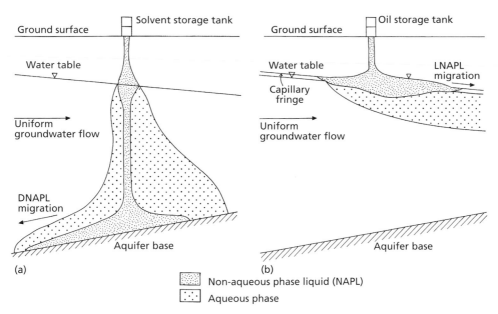

(a)

(b)

Non-aqueous phase liquid (NAPL)

Aqueous phase

Fig. 8.20 Behaviour of organic contaminants in groundwater. The chlorinated solvent contaminant shown in (a) has a density greater than water (a dense non-aqueous phase liquid or DNAPL) and so sinks to the base of the aquifer. Here, transport of the DNAPL is controlled by the slope of the base of the aquifer, while the dissolved aqueous phase moves in the direction of groundwater flow. The hydrocarbon contaminant shown in (b) has a density less than water (a light non-aqueous phase liquid or LNAPL) and so floats on the water-table. In this case, transport of the LNAPL is controlled by the slope of the water table, while the dissolved aqueous phase moves in the direction of groundwater flow.

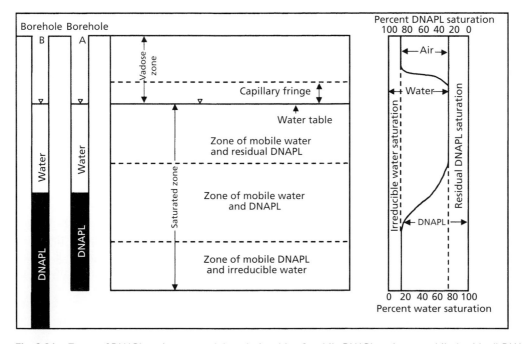

Fig. 8.21 Zones of DNAPL and water, and the relationship of mobile DNAPL and non-mobile (residual) DNAPL to the degree of DNAPL saturation. The relationship of mobile DNAPL thickness to the thickness of DNAPL measured in two monitoring boreholes (A and B) is also shown, with monitoring borehole B giving a false measure of the DNAPL thickness since it contains DNAPL filled below the level of the aquitard top (Fetter 1999). (*Source:* Fetter, C.W. (1999) *Contaminant Hydrogeology* (2nd edn). Prentice-Hall, Inc., Upper Saddle River, New Jersey.)

(a)

(b)

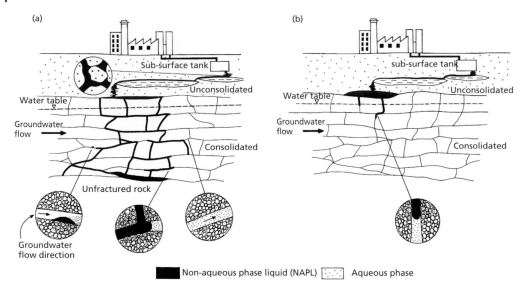

Non-aqueous phase liquid (NAPL)   Aqueous phase

Fig. 8.22 Diagram of (a) DNAPL and (b) LNAPL spills and their transport in fractured rock. (*Source:* Adapted from CL:AIRE 2002. Reproduced with permission.)

displacement pressure of the water in the fractures (Fig. 8.22b).

8.3.3.1 Hydrophobic sorption of non-polar organic compounds

Non-polar organic molecules, for example, low molecular weight volatile organic compounds (VOCs), polycyclic aromatic hydrocarbons (PAHs), polychlorinated benzenes and biphenyls (PCBs) and non-polar pesticides and herbicides, have a low solubility in water, itself a polar molecule (Section 4.2). These immiscible organic compounds tend to partition preferentially into non-polar environments, for example on to small quantities of solid organic carbon such as humic substances and kerogen present as discrete solids or as films on individual grains of soil, sediment and rock. The organic carbon content of sediments varies depending on lithology and can range from a few per cent in the case or organic-rich alluvial deposits to less than 1/10 of a per cent for clean sands and gravels. At low concentrations, the sorption of non-polar compounds on to organic material can often be modelled with a linear isotherm (eq. 8.16 with $n = 1$).

In order to provide a rapid assessment of the sorption behaviour of solid organic carbon and to minimize experimental work, it is useful to find empirical correlations between K_{OC}, the organic carbon-water partition coefficient, and the properties of known substances. In studies of sorption processes, it is useful to correlate K_{OC} with the octanol-water partition coefficient, K_{OW}, a measure of hydrophobicity. Such an approach is possible given that the partitioning of an organic compound between water and organic carbon is not dissimilar to that between water and octanol. An extensive compilation of K_{OC} and K_{OW} values, as well as other physico-chemical properties of organic compounds is provided by Mackay *et al.* (1997). A number of empirical correlations are given in the literature with a selection shown in Table 8.7. The partition coefficient, K_d, for application in the retardation equation (eq. 8.13) can now be normalized to the weight fraction organic carbon content of the sediment, f_{OC}, assuming that adsorption of hydrophobic substances occurs preferentially on to organic matter, as follows:

$$K_{OC} = \frac{K_d}{f_{OC}}$$
(eq. 8.22)

As shown in Table 8.7, a number of studies have proposed non-class specific correlations

Table 8.7 Empirical correlations between $\log_{10}K_{OC}$ and $\log_{10}K_{OW}$ for non-class specific and class specific organic compounds undergoing hydrophobic sorption.

Non class-specific	
(a) $\log_{10}K_{OC} = 0.544 \log_{10}K_{OW} + 1.377$	Kenaga and Goring (1980)
(b) $\log_{10}K_{OC} = 0.679 \log_{10}K_{OW} + 0.663$	Gerstl (1990)
(c) $\log_{10}K_{OC} = 0.909 \log_{10}K_{OW} + 0.088$	Hassett *et al.* (1983)
(d) $\log_{10}K_{OC} = 0.903 \log_{10}K_{OW} + 0.094$	Baker *et al.* (1997)
Class-specific	
(e) Chloro and methyl benzenes	Schwarzenbach and Westall (1981)
$\log_{10}K_{OC} = 0.72 \log_{10}K_{OW} + 0.49$	
(f) Benzene, PAHs	Karickhoff *et al.* (1979)
$\log_{10}K_{OC} = 1.00 \log_{10}K_{OW} - 0.21$	
(g) Polychlorinated biphenyls	Girvin and Scott (1997)
$\log_{10}K_{OC} = 1.07 \log_{10}K_{OW} - 0.98$	
(h) Aromatic amines	Worch *et al.* (2002)
$\log_{10}K_{OC} = 0.42\log_{10}K_{OW} + 1.49$	

between K_{OC} and K_{OW} which should be applicable to many types of solid organic carbon. However, as noted by Worch *et al.* (2002), the parameters of the non-class specific compounds differ significantly and it is not clear which correlation is most reliable. Hence, these correlations which were obtained for different sediments and thus different organic matter composition should only be used as first approximations to the sorption behaviour of specific compounds. For more exact K_{OC} estimations, experimental determinations of the $\log K_{OW}$–$\log K_{OC}$ correlation, or column experiments to determine the retardation coefficient and K_d, for the substance class of interest are required. Currently, however, class-specific correlations exist only for a limited number of substance classes.

As an example calculation of the application of the hydrophobic sorption model, consider a sand aquifer, with a solid density of 2.65 g cm^{-3}, an organic carbon content of 1.5×10^{-4} kg kg^{-1} ($f_{OC} = 0.00015$) and porosity of 0.35, that is contaminated with the aromatic amine 2,4,6-trichloroaniline. The $\log_{10}K_{OW}$ value for this compound found from laboratory experimentation is equal to 3.7 (Worch *et al.*

2002) and this enables an estimate of the retardation factor as follows.

From Table 8.7, the adsorption behaviour of aromatic amines is described by the equation:

$$\log_{10}K_{OC} = 0.42 \log_{10}K_{OW} + 1.49$$
$$\text{(eq. 8.23)}$$

By substituting $\log_{10}K_{OW} = 3.7$ in eq. 8.23, the resulting value for $\log_{10}K_{OC}$ is

$$\log_{10}K_{OC} = (0.42 \times 3.7) + 1.49 = 3.04$$

and

$$K_{OC} = 1107$$

From eq. 8.22, the value of the partition coefficient K_d is

$$K_d = 1107 \times 0.00015 = 0.17 \text{ mL g}^{-1}.$$

and using eq. 8.13, the retardation factor, R_d is

$$R_d = 1 + 0.17\frac{(1-0.35)2.65}{0.35} = 1.84$$

With a retardation factor of less than 2, the retardation of 2,4,6-trichloroaniline by adsorption in the sand aquifer is low and so it is likely that this type of aromatic compound, which is poorly biodegradable, will persist as a groundwater contaminant.

The hydrophobic sorption model is recommended where the solubility of the organic compound is less than about 10^{-3} M. More soluble compounds, such as methanol, show much less affinity for organic carbon and preferentially partition to the aqueous phase. Also, if the fraction of organic carbon, f_{OC}, is small, it is possible that organic compounds will show a tendency to sorb to a small but significant extent on to inorganic surfaces, particularly where a clay fraction presents a large surface area to the contaminant. A further limitation is the possible effect of the presence of mixtures of organic compounds, where co-solvents alter the solubility characteristics of individual compounds and compete in sorption–desorption reactions.

Overall in the above approach, it is assumed that the higher the hydrophobicity of a contaminant, the greater the adsorption behaviour. In advancing models to describe the sorption of hydrophobic organic chemicals to heterogeneous carbonaceous matter in soils, sediments and rocks, Allen-King *et al.* (2002) considered more specifically the relative importance of partitioning and adsorption of hydrophobic contaminants in the environment. Partitioning is likely to control contaminant solid-aqueous behaviour when the solid phase contains a substantial proportion of 'soft' humic substances, such as in modern soils and sediments high in organic carbon content, and/or concentrations of any adsorbing solutes are sufficiently high to effectively saturate all the adsorbent sites present. Under these circumstances, K_d values estimated from K_{OW}, K_{OC} and total f_{OC} values (eq. 8.22) are likely to produce a reasonable estimate of K_d, usually accurate to within a factor of 3. Such estimates can be successfully applied in circumstances such as the attenuation of hydrophobic pesticides in agricultural soils and for chlorinated solvents near a source containing non-aqueous phase liquid. Deeper in the sub-surface environment, other types of organic matter with

high adsorption capacities are present, for example, thermally altered 'hard' carbonaceous material (pieces of coal, soot, char or kerogen), and under these circumstances, it is likely that adsorption will contribute significantly to the total sorption over a broad concentration range of contaminants. Therefore, hydrophobic sorption of organic compounds in soils, sediments and rocks occurs as a combination of phase partitioning and surface adsorption, with the former typically more linearly dependent on aqueous concentration (Allen-King *et al.* 2002).

In the process of remediating a contaminated aquifer, it should be noted that desorption involving the partitioning of an organic compound back into the aqueous phase, yields a different isotherm to the sorption process. This so-called sorption 'hysteresis' is often attributed to slow processes of diffusion within sediment particles and soil aggregates. Usually, desorption exhibits a greater affinity for partitioning to the solid phase than is the case during sorption.

8.3.4 Effects of density and heterogeneity

The processes of advection, dispersion and retardation all influence the pattern of contaminant distribution away from a pollution source. Other considerations include the effects of contaminant density and, moreover, the influence of aquifer heterogeneity, both of which make the overall monitoring and prediction of the extent of groundwater pollution very difficult. A density contrast between the contaminant and groundwater will affect the migration of a pollution plume, with a contaminant denser than water tending to sink steeply downwards into the groundwater flow field as shown in Fig. 8.23.

The description of contaminant transport given in Section 8.3.1 assumes a homogeneous porous material with steady, uniform groundwater flow. This is a simplification of real situations in nature where heterogeneities within

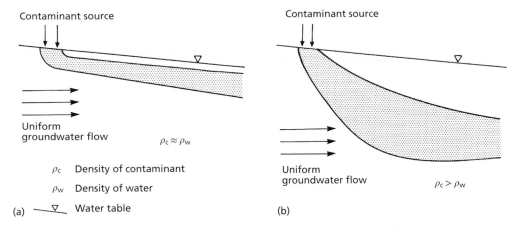

Fig. 8.23 Effect of density on the transport of dissolved contaminants in a uniform groundwater flow field. In (a) the contaminant density (ρ_c) is slightly greater than the density of the groundwater (ρ_w). In (b) the contaminant density is greater than the groundwater density.

the aquifer lithology create a pattern of solute movement considerably different to that predicted by the theory for homogeneous material. If a pollution source contains multiple solutes and occurs within a heterogeneous aquifer containing beds, lenses and fractures of differing hydraulic conductivity, then there will be a number of contaminant fronts and pathways such that the morphology of the resulting plume will be complex (Fig. 8.24).

In fractured material, aquifer properties are spatially variable and are often controlled by the orientation and frequency of fractures. As shown in Fig. 8.25, when contamination occurs in fractures, there is a gradient of contaminant concentration between the mobile groundwater in the fracture and the static water in the adjacent rock matrix. Under this condition, part of the contaminant mass will migrate by molecular diffusion from the fracture into the porewater contained in the rock matrix, so effectively removing it from the flowing groundwater. Such dual-porosity aquifers are notoriously difficult to remediate since the contaminant stored in the matrix can gradually diffuse back into the moving groundwater in the fracture, long after the source of contamination has been removed.

8.4 Sources of groundwater contamination

The sources of groundwater contamination are, as shown in Table 8.1, as varied as the range of polluting activities. The purpose of this section is to introduce the principal sources and classes of groundwater contaminants and their behaviour with respect to urban and industrial contaminants, municipal and septic wastes, agricultural contaminants and saline intrusion in coastal aquifers.

8.4.1 Urban and industrial contaminants

Urban expansion and industrial activity, in some cases in the United Kingdom since the industrial revolution in the 1700s, is accompanied by continual disposal and spillage of potentially polluting wastes. The range of wastes is diverse, ranging from inorganic contaminants associated with mining and foundry wastes to organic compounds produced by the petrochemical and pharmaceutical industries. Accompanying urbanization is the need to dispose of domestic municipal and septic wastes leading to the risk of contamination from toxic materials and

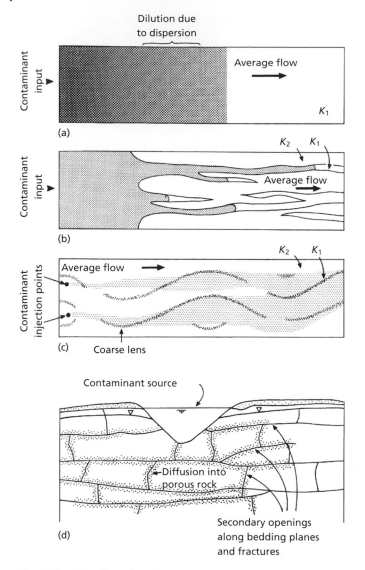

Fig. 8.24 The effect of aquifer heterogeneity on contaminant zones influenced by hydrodynamic dispersion. In (a) dilution occurs in the direction of advancing contaminant in a homogeneous intergranular material. In (b) the presence of higher hydraulic conductivity beds and lenses causes fingering of the contaminant transport ($K_1 > K_2$). In (c) contaminant spreading is created by the presence of irregular lenses of higher hydraulic conductivity ($K_1 > K_2$). In (d) contaminant migration is dispersed throughout the network of secondary openings developed in a fractured limestone with molecular diffusion into the porous rock matrix (Freeze and Cherry 1979). (*Source:* Freeze, R.A. and Cherry, J.A. (1979) *Groundwater*. Prentice-Hall, Inc., Englewood Cliffs, New Jersey. © 1979, Pearson.)

sewage. Other sources of pollution in the urban environment include salt and urea used in de-icing roads, paths and airport runways (Howard and Beck 1993), highway runoff potentially directed to soakaways (Price *et al.* 1989), the application of fertilizers and pesticides in parks and gardens, and the presence of chlorinated compounds such as trihalomethanes caused by leakage of chlorinated mains water that has reacted with organic carbon either in the

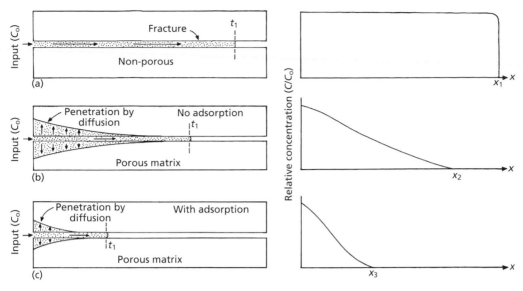

Fig. 8.25 Contaminant transport within porous fractured material. In (a) the solute is advected, without hydrodynamic dispersion, with the groundwater flowing through a fracture where the matrix porosity is insignificant. In (b) the solute transport is retarded by the instantaneous molecular diffusion of the solute into the uncontaminated porous matrix. Further attenuation occurs in (c) where adsorption of a reactive solute occurs, accentuated by the greater surface area of contact resulting from migration of the solute into the porous matrix. The position of the leading edge of the contaminant front within the fracture is shown for time t_1 in each case (Freeze and Cherry 1979). (*Source:* Freeze, R.A. and Cherry, J.A. (1979) *Groundwater*. Prentice-Hall, Inc., Englewood Cliffs, New Jersey. © 1979, Pearson.)

distribution system or in the subsurface. Atmospheric emissions of sulphur dioxide and nitrogen oxides from urban areas contribute to wet and dry deposition of sulphur and nitrogen in adjacent regions that can impact soils, vegetation and freshwaters as a result of acid deposition and eutrophication (NEGTAP 2001).

The regulated control of waste disposal in urban areas is now practised in many developed countries but because of the slow transmission time of contaminants in the unsaturated zone, the legacy of historical, uncontrolled disposal of wastes may present a potential for groundwater pollution. A rise in groundwater levels, caused by a reduction in groundwater abstraction in post-industrial urban centres, may lead to remobilization of this pollution. In the Birmingham Triassic sandstone aquifer in the English Midlands, a region of metal manufacturing and processing and mechanical engineering, samples from shallow piezometers, tunnels

and basements show that groundwater concentrations at shallow depths are often heterogeneous in distribution and much higher than in groundwater pumped from greater depth (Ford and Tellam 1994).

Generalizations as to the likely contaminants to be found in urban areas are not always possible except that the dominant inorganic contaminants are likely to be chloride and nitrate associated with a long history of supply and a wide range of multiple point sources. Other than these two, contamination is normally correlated to land use. In the Birmingham Triassic sandstone aquifer, the highest major ion concentrations (Fig. 8.26) and levels of boron and total heavy metal concentrations are associated with metal working sites (Ford and Tellam 1994). In other areas, the discharge of acidic mine water from disused mine workings and spoil heaps is a potential cause of heavy metal pollution (Box 8.5).

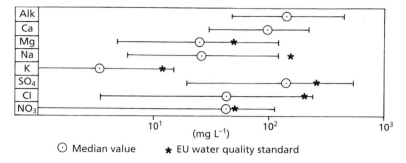

Median value   ★ EU water quality standard

Fig. 8.26 Ranges of major ion concentrations for pumped samples from abstraction boreholes in the urban Birmingham Triassic sandstone aquifer (Ford and Tellam 1994). (*Source:* Ford, M. and Tellam, J.H. (1994) Source, type and extent of inorganic contamination within the Birmingham urban aquifer system, UK. *Journal of Hydrology* **156**, 101–135. © 1994, Elsevier.)

Box 8.5   Mine water pollution

Mine water pollution is a widespread problem in present and former mining districts of the world with numerous cases of severe water pollution having been reported from base metal mines, gold mines and coal mines. In the United Kingdom, concern centres on aquatic pollution from the major coalfields and most of the base metal ore fields, such as those of Cornwall, upland Wales, northern England and Scotland. A general conceptual model for sources of mine water pollution, transport pathways for soluble contaminants and potentially sensitive receiving waters at risk of contamination is shown in Fig. 8.27. Mining activities contribute significantly to the solute load of receiving surface waters and aquifers. To illustrate, the contribution of sulphide mineral weathering associated with mine sites to the sulphate ion load is estimated at 12% of the global fluvial sulphate flux to the world's oceans (Nordstrom and Southam 1997). Associated with this weathering flux are dissolved metal ions (for example, Fe, Zn and Al), sulphate and, in the case of pyrite, acidity. When discharging into the wider environment, acid mine drainage can coat stream beds with orange precipitates of iron hydroxides and

oxyhydroxides ('ochre') as well as white aluminium hydroxide deposits (Gandy and Younger 2003).

The oxidative weathering and dissolution of contaminant source minerals such as pyrite and sphalerite associated with abandoned deep or opencast coal and metal mines and surface spoil heaps is described by the following two equations:

$$FeS_2(s) + \frac{7}{2}O_2(aq) + H_2O$$
$$\rightarrow Fe^{2+} + 2SO_4^{2-} + 2H^+$$

(pyrite weathering)   (eq. 1)

$$ZnS(s) + 2O_2(aq) \rightarrow Zn^{2+} + SO_4^{2-}$$

(sphalerite weathering)   (eq. 2)

The study of long-term changes in the quality of polluted mine water discharges from abandoned underground coal workings in the Midland Valley of Scotland shows that mine water pollution is most severe in the first few decades after a discharge begins (the 'first flush'), and that the largest systems settle down to a lower level of pollution, particularly in terms of iron concentration, within 40 years (Wood *et al.* 1999). As shown in Fig. 8.28, long-term iron concentrations of less than 30 mg L^{-1} are typical, with many

Box 8.5   (Continued)

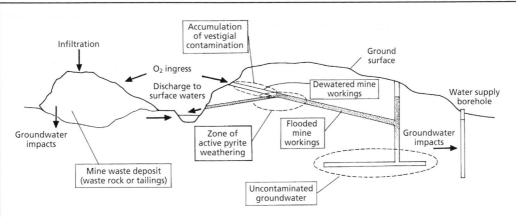

Fig. 8.27 Schematic diagram of contaminant sources, transport pathways and potential receiving waters in mining environments. A distinction is made between juvenile contamination arising from active weathering of sulphide minerals above the water table where oxygen ingress occurs, and vestigial contamination that accumulates as secondary mineral precipitates of metal ions and sulphate that arise from sulphide mineral weathering in dewatered void spaces within mine environments (Banwart *et al.* 2002). (*Source:* Banwart, S.A., Evans, K.A. and Croxford, S. (2002) Predicting mineral weathering rates at field scale for mine water risk assessment. In: *Mine Water Hydrogeology and Geochemistry* (eds P.L. Younger and N.S. Robins). Geological Society, London, Special Publications **198**, pp. 137–157.)

less than $10\,\text{mg L}^{-1}$. In the Scottish coalfield, low pH values do not generally persist due to the rapid buffering of localized acidic waters by carbonates.

The Durham coalfield of north-east England (Fig. 8.29) was one of the first coalfields in the world to be commercially exploited and has left a legacy of acid mine discharge as the mines have closed. The worked Coal Measures comprise Carboniferous strata of fluvio-lacustrine and fluvio-deltaic facies. High-sulphur coals, which might be expected to be prolific generators of acid mine water, are associated with shale bands of marine origin (Younger 1995). In the more easterly districts of the coalfield, the Coal Measures are unconformably overlain by Permian strata including the Magnesian limestone, an important public supply aquifer. Mining ceased in the exposed coalfield to the west

of the Permian scarp in the early 1970s and the deep mines beneath Permian cover on the coast closed in 1993. Following closure, pumping water from the coastal mines ceased and the water table has begun to rebound. In the far west of the coalfield, uncontrolled discharges of acid mine drainage occurs. The water quality of the five most significant discharges is given in Table 8.8.

Remediation technologies for the treatment of acidic mine waters can be divided into active techniques, such as alkali dosing, aeration, flocculation and settlement; and passive techniques, such as constructed wetlands, inorganic media passive systems (Fig. 8.30) and sub-surface flow, bacterial sulphate reduction systems. Unless the metal loadings are particularly high, in which case active treatment can be a cost-effective solution in the long term, passive

(Continued)

Box 8.5 (Continued)

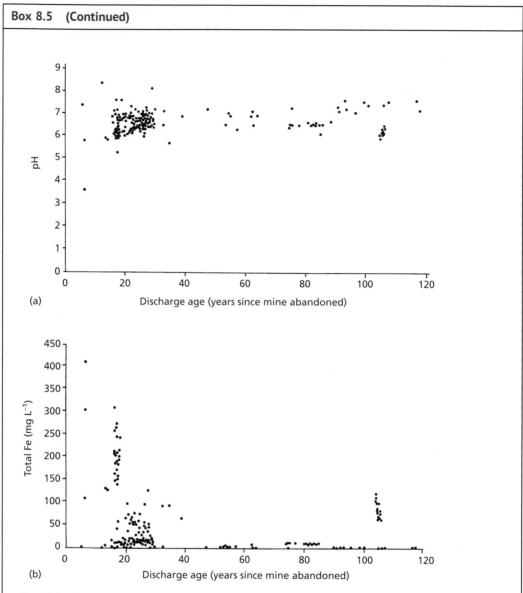

Fig. 8.28 Graphs showing long-term changes in the quality of polluted mine water discharges from abandoned underground coal workings in the Midland Valley of Scotland in terms of the variation of (a) pH and (b) total iron concentration (Wood *et al.* 1999). (*Source:* Wood, S.C., Younger, P.L. and Robins, N. S. (1999) Long-term changes in the quality of polluted minewater discharges from abandoned underground coal workings in Scotland. *Quarterly Journal of Engineering Geology and Hydrogeology* **32**, 69–79.)

treatment is increasingly the preferred option (Gandy and Younger 2003). In practice, mine water remediation should allow for active treatment of discharges for the first decade or two, followed by long-term passive treatment after asymptotic pollutant concentrations are attained (Wood *et al.* 1999).

Box 8.5   (Continued)

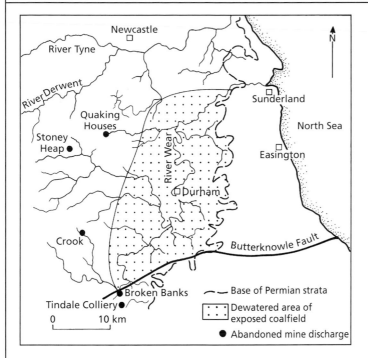

Fig. 8.29   Map showing the location of major uncontrolled discharges from abandoned mine workings in the Durham coalfield, north-east England (Younger 1995). (*Source:* Younger, P.L. (1995) Hydrogeochemistry of minewaters flowing from abandoned coal workings in County Durham. *Quarterly Journal of Engineering Geology and Hydrogeology* **28**, S101–S113.)

Table 8.8   Hydrochemistry of the Durham coalfield mine drainage discharges at first emergence at the locations shown in Fig. 8.29 (Younger 1995).

Site (Grid Reference)	Broken Banks (NZ197295)	Crook (NZ185356)	Quaking Houses (NZ178509)	Stoney Heap (NZ147515)	Tindale Colliery (Brusselton) (NZ197269)
Flow rate $(m^3 s^{-1})$ on 15-4-94	0.14	0.002	0.007	0.0256	0.01
Calcium $(mg L^{-1})$	100.7	185	255	83.6	262
Magnesium $(mg L^{-1})$	61.23	93	103	49.7	107
Sodium $(mg L^{-1})$	26.9	21.5	463.6	27.9	80
Potassium $(mg L^{-1})$	10.7	6.8	57.0	6.7	13

(Continued)

(Continued)

Box 8.5   (Continued)

Table 8.8   (Continued)

Site (Grid Reference)	Broken Banks (NZ197295)	Crook (NZ185356)	Quaking Houses (NZ178509)	Stoney Heap (NZ147515)	Tindale Colliery (Brusselton) (NZ197269)
Iron (total) ($mg\,L^{-1}$)	1.8	79.8	18.0	26.3	1.8
Manganese ($mg\,L^{-1}$)	1.0	6.9	4.8	1.2	1.7
Aluminium ($mg\,L^{-1}$)	0.26	4.2	12.9	0.16	0.04
Zinc ($mg\,L^{-1}$)	0.023	0.045	0.040	0.022	0.0184
Copper ($mg\,L^{-1}$)	0.11	0.23	0.23	0.09	0.01
Alkalinity ($mg\,L^{-1}$ as $CaCO_3$)	364.0	0.0	0.0	188	357
Sulphate ($mg\,L^{-1}$)	137.0	810	1358	325	890
Chloride ($mg\,L^{-1}$)	60	65	1012	102	75
pH	6.5	4.8	4.1	6.3	6.4
Temperature (°C)	10.9	11.8	11.2	10.3	12.0
Eh (mV)	39	264	327	36	−50
Conductivity ($\mu S\,cm^{-1}$)	1177	1563	3560	1134	2360

(*Source:* Younger, P.L. (1995) Hydrogeochemistry of minewaters flowing from abandoned coal workings in County Durham. *Quarterly Journal of Engineering Geology and Hydrogeology* **28**, S101–S113.)

Organic contamination of urban aquifers mainly concerns industrial solvents (DNAPLs) and petroleum hydrocarbons (LNAPLs) and encompasses the chemical classes and sources listed in Table 8.9. Contaminants escape to the sub-surface environment mainly as a result of careless handling, accidental spillages, misuse, poor disposal practices and inadequately designed, poorly maintained or badly operated equipment. These contaminants are of concern in groundwater because of their toxicity and persistence, particularly with respect to solvents.

Chlorinated solvents were developed in the early years of the last century as a safe, non-flammable alternative to petroleum-based degreasing solvents in the metal processing

Fig. 8.30 Limestone filter installation sited upstream of Cwm Rheidol, west Wales, to precipitate metals contained in ochreous acid mine drainage resulting from the flushing of ferrous sulphate and sulphuric acid (the products of oxidation of pyrite and marcasite during dry-working of mines) and issuing from the abandoned mine adits and spoil heaps in the Ystumtuen area. The acidified water dissolved heavy metals such as Pb, Zn, Cd and Al which entered the river system, the effect of which was detected 16 km downstream of the mined area with a pH as low as 2.6 (Fuge *et al.* 1991). (*Source:* Fuge, R., Laidlaw, I.M.S., Perkins, W.T. and Rogers, K.P. (1991) The influence of acid mine and spoil drainage on water quality in the mid-Wales area. *Environmental Geochemistry and Health* **13**, 70–75. doi: 10.1007/BF01734297)

industry. Until about 1970, trichloroethene (TCE) and tetrachloroethene (perchloroethylene, PCE) were predominantly used, the latter also in dry cleaning applications. TCE and PCE degrade extremely slowly, and some of the degradation products may be more toxic, soluble and mobile than the parent compounds. For example, tetrachloroethene can be progressively de-

halogenated, first to trichloroethene, then to dichloroethene, and finally to carcinogenic vinyl chloride. From the mid-1970s, concern was expressed about the potentially carcinogenic effects of TCE, PCE and carbon tetrachloride (CTC) at trace level concentrations in drinking water, and the WHO set guide values of 70 µg L^{-1} for TCE, 40 µg L^{-1} for PCE and 2 µg L^{-1} for CTC (Appendix 9). Since the 1960s, both TCE and PCE have begun to be replaced by the less toxic 1,1,1-trichloroethane (TCA) and 1,1,2-trichlorotrifluoroethane (Freon 113).

Hydrocarbons include petrol, aviation fuel, diesel and heating oils. As a group, their physical characteristics are variable, particularly that of viscosity; but all have a density less than water, and a heterogeneous composition dominated by pure hydrocarbons. In the context of groundwater, regulation is aimed primarily at taste and odour control (guideline values are given in Appendix 9). Sources of contamination include oil storage depots, cross-country oil pipelines, service stations, tanker transport and airfields.

In a further survey of the Birmingham Triassic sandstone aquifer in the English Midlands (Rivett *et al.* 1990), almost half of the 59 supply boreholes sampled that contained chlorinated solvents were located on the sites of metal manufacturing and processing and mechanical engineering industries. The results are shown in Table 8.10 and Fig. 8.31 and indicate that chlorinated solvents are widespread, in particular TCE which is detected in 78% of boreholes. TCE is frequently observed at high levels with 40% of boreholes contaminated above 30 µg L^{-1} to a maximum of 5500 µg L^{-1} (Rivett *et al.* 1990). Occasional high values are also observed for TCA and PCE, the latter associated with dry cleaning laundry sites. Contamination of the Birmingham aquifer by organic chemicals other than chlorinated solvents is low in the supply boreholes and, where present, is often associated with degraded lubricating oils. Several factors were shown by Rivett *et al.* (1990) to explain the distribution of organic contaminants including the historic use of solvents, point source inputs close to sampling

Table 8.9 Sources of organic contaminants found in urban groundwaters (Lloyd *et al.* 1991).

Chemical class	Sources	Examples
Aliphatic and aromatic hydrocarbons (including benzenes, phenols and petroleum hydrocarbons)	Petrochemical industry wastes Heavy/fine chemicals industry wastes Industrial solvent wastes Plastics, resins, synthetic fibres, rubbers and paints production Coke oven and coal gasification plant effluents Urban runoff Disposal of oil and lubricating wastes	Benzene Toluene Iso-octane Hexadecane Phenol
Polynuclear aromatic hydrocarbons	Urban runoff Petrochemical industry wastes Various high-temperature pyrolytic processes Bitumen production Electrolytic aluminium smelting Coal-tar coated distribution pipes	Anthracene Pyrene
Halogenated aliphatic and aromatic hydrocarbons	Disinfection of water and waste water Heavy/fine chemicals industry waste Industrial solvent wastes and dry cleaning wastes Plastics, resins, synthetic fibres, rubbers and paints production Heat-transfer agents Aerosol propellants Fumigants	Trichloroethylene Trichloroethane Para-dichlorobenzene
Polychlorinated biphenyls	Capacitor and transformer manufacture Disposal of hydraulic fluids and lubricants Waste carbonless copy paper recycling Heat transfer fluids Investment casting industries PCB production	Pentachlorobiphenyls
Phthalate esters	Plastics, resins, synthetic fibres, rubbers and paints production Heavy/fine chemicals industry wastes Synthetic polymer distribution pipes	

(*Source:* Lloyd, J.W., Williams, G., Foster, S.S.D. *et al.* (1991) Urban and industrial groundwater pollution. In: *Applied Groundwater Hydrology: A British Perspective* (eds R.A. Downing and W.B. Wilkinson). Clarendon Press, Oxford, pp. 134–148.)

Table 8.10 Summary of chlorinated solvents detected in groundwater in the Birmingham Triassic sandstone aquifer. Solvent abbreviations are given in Table 8.6 (Rivett *et al.* 1990).

Industrial solvents	TCE	TCA	PCE	TCM	CTET
% boreholes with solvent detected	78	46	44	53	37
Proportion of boreholes exceeding (%)					
1 µg L^{-1}	62	22	9	17	2
10 µg L^{-1}	43	13	4	0	0
100 µg L^{-1}	30	5	2	0	0
Maximum concentration detected (µg L^{-1})	5500	780	460	5	1

(*Source:* Rivett, M.O., Lerner, D.N., Lloyd, J.W. and Clark, L. (1990) Organic contamination of the Birmingham aquifer, U.K. *Journal of Hydrology* **113**, 307–323. © 1990, Elsevier)

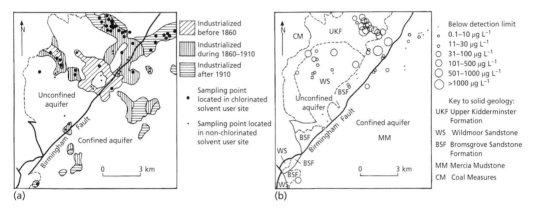

Fig. 8.31 Organic solvent contamination of the urban Birmingham Triassic sandstone aquifer showing: (a) distribution of groundwater sampling points and industrialized areas; and (b) distribution of maximum TCE concentrations (Rivett *et al.* 1990). (*Source:* Rivett, M.O., Lerner, D.N., Lloyd, J.W. and Clark, L. (1990) Organic contamination of the Birmingham aquifer, U.K. *Journal of Hydrology* **113**, 307–323. © 1990, Elsevier.)

boreholes, thickness of the unsaturated zone, the presence or absence of confining deposits above the aquifer, and the depth of groundwater sampling. Higher concentrations of solvents were found at deeper sampling levels, by virtue of their DNAPL behaviour, while the opposite was the case for hydrocarbons (LNAPLs).

In a repeat survey a decade later (1998–2001), Rivett *et al.* (2005) observed that TCE, the principal solvent detected at 78% of abstraction sites in 1987 was reduced to 56% of sites by 1998. The decline in per cent occurrence was attributed to closure of groundwater abstractions associated with former solvent-user industries, rather than signifying an improvement in groundwater quality. Indeed, comparison of abstractions common to both surveys indicated greater contamination in the later survey attributed to the persistence of historical dense nonaqueous-phase liquid (DNAPL) source zones. Persistence of PCE relative to TCE was observed and ascribed to its lower solubility and greater sorption. Biodegradation, although a key attenuation process, appeared to have moderate-to-weak influence regionally on plumes, but was insufficient to prevent parent solvents from being detected at half of the non-solvent user sites sampled in 1998. Nevertheless, solvent-free water can still be abstracted but may become increasingly difficult as industrial decline has caused closure of many solvent-user industry abstractions previously thought to have been undertaking inadvertent pump-and-treat attenuation of plumes. Such plumes are now able to migrate more freely throughout the aquifer. Rivett *et al.* (2012) concluded that challenges for the future development of the Birmingham Triassic sandstone aquifer, and urban aquifers more generally, include the adequacy of groundwater quality monitoring data and uncertainties in contaminant source terms, abstraction well capture zone predictions and plume natural attenuation, in particular degradation rates.

Urban districts can extend over wide areas of aquifer outcrop below which is a large volume of potentially available water; a contrasting situation with adjacent rural areas where groundwater sources are increasingly fully developed (Lerner 2002). However, research in the Nottingham Triassic sandstone aquifer in the English Midlands has shown that although the inorganic quality of the groundwater may be acceptable for water supply, high nitrate concentration and microbial contamination relating to untreated sewage inputs from broken or leaking sewers are causes for concern (Barrett *et al.* 1999; Cronin *et al.* 2003; Powell *et al.* 2003; Fukada *et al.* 2004). Also, research in Berlin has shown that polar pharmaceutical compounds such as clofibric acid, diclofenac, ibuprofen, propyphenazone, primidone and carbamazepine are detectable at individual trace ($\mu g\,L^{-1}$) levels in surface waters and groundwaters with a source in municipal sewage treatment plants (Heberer 2002). Waste discharge to surface water courses infiltrating to groundwater was highlighted as a transport pathway for these soluble polar organic compounds, enabling migration over large distances in impacted aquifers.

8.4.2 Municipal landfill wastes

The transport of potentially toxic contaminants in leachate from both old and modern landfills is a serious environmental problem. In the United States alone, approximately 139.6 million tons of municipal solid waste was disposed

in landfills in 2017 (USEPA 2019). Leachate from municipal landfills can create groundwater contaminant plumes that may last for decades to centuries. The fate of reactive contaminants in aquifers impacted by landfill leachate is dependent on the sustainability of biogeochemical processes affecting contaminant transport (Cozzarelli *et al.* 2011).

The landfilling of municipal wastes is common practice, with older sites up until the late 1980s formerly operated under a 'dilute and disperse' principle where leachate generated in the waste was allowed to migrate away from the site and disperse in groundwater below the water table. This practice is no longer allowed in Europe, with engineered solutions required to contain leachate in the landfill site. A typical solution is to line the site with an artificial liner with a low design permeability of no higher than 10^{-9} m s^{-1} and to install a leachate collection system for subsequent treatment and discharge of the leachate (Department of the Environment 1995). After completion of filling, the landfill site may be restored to agricultural land by sealing the site with a low permeability clay cap that prevents water infiltration into the buried waste.

In the absence of engineered controls, the extent to which leachate from a landfill site contaminates groundwater is dependent on the hydraulic, geochemical and microbiological properties of the hydrogeological system (Nicholson *et al.* 1983). For reviews of the biogeochemistry of landfill leachate plumes, the reader is referred to Christensen *et al.* (2001) and Cozzarelli *et al.* (2011).

Considering the example of a landfill receiving a mixture of municipal, commercial and mixed industrial wastes, landfill leachate may be characterized as a water-based solution containing four groups of pollutants: dissolved organic matter; inorganic macro-components; heavy metals; and xenobiotic organic compounds (Christensen *et al.* 2001). Ranges of values for the first three of these groups are given in Table 8.11. Xenobiotic organic compounds (aromatic hydrocarbons, halogenated

Table 8.11 Table showing the compositional range of landfill leachate (after compilation of Christensen *et al.* 2001). Data for October 1985 and September 1986 describe the leachate composition pumped from containment cell 4 at Compton Bassett landfill site, west of England (Robinson 1989). All results in mg L^{-1}, except pH and electrical conductivity (μS cm^{-1}).

Parameter	Range	October 1985 (Acetogenic or acid phase)	September 1986 (Methanogenic phase)
pH	4.5–9	6.5	7.4
Electrical conductivity	2500–35 000	15 000	10 400
Total solids	2000–60 000	550	230
Organic matter			
Total organic carbon (TOC)	30–29 000		
Biological oxygen demand (BOD)	20–57 000	15 750	580
Chemical oxygen demand (COD)	140–152 000	20 700	2000
BOD/COD ratio	0.02–0.80	0.76	0.29
Organic nitrogen	14–2500		
Inorganic macro-components			
Total phosphorus	0.1–23		
Chloride	150–4500	1710	2020
Sulphate	8–7750		
Bicarbonate	610–7320		
Sodium	70–7700		
Potassium	50–3700		
Ammonium-N	50–2200	930	840
Calcium	10–7200	1410	143
Magnesium	30–15 000		
Iron	3–5500	787	24
Manganese	0.03–1400		
Silica	4–70		
Inorganic trace elements			
Arsenic	0.01–1		
Cadmium	0.0001–0.4		
Chromium	0.02–1.5		
Cobolt	0.005–1.5		
Copper	0.005–10		
Lead	0.001–5		
Mercury	0.00005–0.16		
Nickel	0.015–13		
Zinc	0.03–1000	8.4	2.4

(*Sources:* Christensen, T.H., Kjeldsen, P., Bjerg, P.L. *et al.* (2001) Biogeochemistry of landfill leachate plumes. *Applied Geochemistry* **16**, 659–718.; Robinson, H.D. (1989). Development of methanogenic conditions within landfills. In: *Proceedings of the 2nd International Landfill Symposium*, Cagliari, Sardinia, 9 pp.)

hydrocarbons, phenols and pesticides) originating in household or industrial chemicals are usually present in relatively low concentrations in the leachate (<1 mg L^{-1} for individual compounds).

Landfill leachate contamination of groundwater has a source in the internal, biogeochemical decomposition processes that take place during the breakdown of putrescible materials contained in domestic wastes. Three phases of decomposition are recognized. In the first phase, aerobic decomposition rapidly uses the available oxygen in the wastes. The reaction is common with septic systems (eq. 7 in Table 8.12) with the process usually lasting for up to one month, in which significant quantities of CO_2 and some H_2 are produced. In the second phase, anaerobic and facultative organisms (acetogenic bacteria) hydrolyse and ferment cellulose and other putrescible materials, producing simpler, soluble compounds such as volatile fatty acids. This phase (represented by eq. 3 in Table 8.12) can last for several years producing an acidic leachate (pH of 5 or 6) high in biological oxygen demand (BOD) of greater than 10 000 mg L^{-1}, and NH_4^+ in the range 500–1000 mg L^{-1}. The aggressive leachate assists in the dissolution of other waste components, such that the leachate can contain high levels of Fe, Mn, Zn, Ca and Mg. Gas production consists mainly of CO_2 with lesser quantities of CH_4 and H_2. The third phase experiences slower-growing methanogenic bacteria that gradually consume simple organic compounds, producing a mixture of CO_2 and CH_4 gases which is released as landfill gas that can be recovered and used as an energy source. This phase of methanogenesis (eqs. 6a and 6b in Table 8.12) will continue for many years, if not decades, until the landfill wastes are largely decomposed and atmospheric oxygen can once more diffuse into the landfill. Leachates produced during this last phase are characterized by relatively low BOD values. However, NH_4^+ continues to be released and will be present at high values in the leachate. Inorganic substances such as Fe, Na, K,

sulphate and chloride may continue to dissolve and leach from the landfill for many years.

These various phases of landfill decomposition are represented in the analyses of leachate given in Table 8.11. The landfill cell at Compton Bassett was filled over a two- to three-year period from late 1983 during which waste was filled to a depth of 15–20 m, compacted, capped with clay and restored to grassland in 1987. Table 8.11 demonstrates rapid changes in leachate composition from a strongly acetogenic to methanogenic state during a twelve-month period from October 1985. The change is accompanied by a rise in pH from acid to slightly alkaline conditions, but more obviously by steep declines in BOD and chemical oxygen demand (COD) concentrations as organic compounds such as volatile fatty acids are being degraded at rates faster than they are produced. The less aggressive nature of this leachate is associated with substantial reductions in concentrations of Fe and Ca being solubilized from the wastes. Concentrations of conservative determinands such as chloride remain stable or even increase, demonstrating that dilution is of little importance in the results obtained. High concentrations of NH_4^+ are also maintained, demonstrating the continuing biological activity and rapid acetogenic processes occurring to decompose the solid wastes (Robinson 1989).

In the event of leachate contamination of the unsaturated zone below a landfill site, attenuation of the leachate by physical, chemical and biological processes is possible. However, the extent of penetration of organic and other components of landfill leachate in an aquifer will depend on its buffering capacity leading to the development of favourable conditions for microbial degradation. This buffering effect is shown in Fig. 8.32 and Fig. 8.33 for two contrasting landfill sites located above sandstone and Chalk aquifers, respectively. In the Nottinghamshire Triassic sandstone aquifer at the Burntstump municipal landfill site, sequential profiling of the unsaturated zone shows a downward migration of a conservative

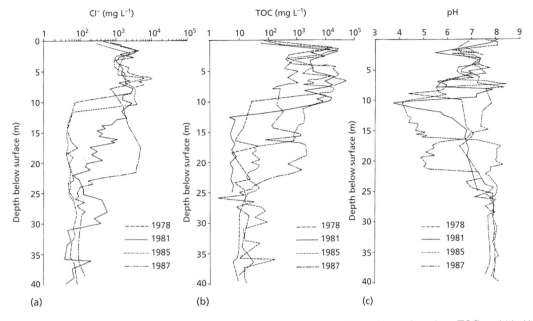

Fig. 8.32 Temporal variation of concentration profiles of (a) chloride, (b) total organic carbon (TOC) and (c) pH in the unsaturated zone of the Triassic sandstone aquifer at the Burntstump landfill site, Nottinghamshire, 1978–1987 (Williams *et al.* 1991). (*Source:* Williams, G.M., Young, C.P. and Robinson, H.D. (1991) Landfill disposal of wastes. In: *Applied Groundwater Hydrology* (eds R.A. Downing and W.B. Wilkinson). Clarendon Press, Oxford, pp. 114–133. © 1991, Oxford University Press.)

Fig. 8.33 Temporal variation in TOC : Cl ratios in the unsaturated zone of the Chalk aquifer at the Ingham landfill site, Suffolk, 1975–1986. TOC is total organic carbon content (Williams *et al.* 1991). (*Source:* Williams, G.M., Young, C.P. and Robinson, H.D. (1991) Landfill disposal of wastes. In: *Applied Groundwater Hydrology* (eds R.A. Downing and W.B. Wilkinson). Clarendon Press, Oxford, pp. 114–133. © 1991, Oxford University Press.)

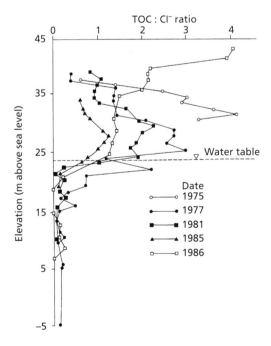

chloride front over a nine-year period (Fig. 8.32). A similar movement is observed for the organic compounds as indicated by the TOC (total organic carbon) values, mainly attributable to total volatile acids. A zone of reduced pH values migrated downwards consistent with the TOC, with the lower edge of the low pH zone corresponding closely with the leading edge of the TOC front. The profiles for TOC and pH are interpreted to result from the combination of the low buffering capacity of the sandstone due to its limited carbonate content, the high concentrations of organic acids and the dissolution of CO_2 generated within the wastes, all of which depress the pH of the interstitial water at the leachate front. Hence, at this site, there is significant penetration of the organic and other components of leachate due to the low buffering capacity and the persistence of conditions unfavourable to microbial degradation at low pH (Williams et al. 1991).

In contrast, the landfill site at Ingham, Suffolk, in eastern England and situated on the Cretaceous Chalk aquifer gave the sequential profiles shown in Fig. 8.33. Reinstatement of the site to agricultural soil was completed in 1977 after which the chloride front effectively stagnated as a result of the effectiveness of the completed cap in limiting infiltration to very low values. During an eight-year period of stagnation until 1986, a persistent decrease in the TOC : Cl ratio occurred in the unsaturated zone indicating the continuous removal of organic carbon. Early surveys of organic compounds at the site showed the presence of phenols (absent post-1984), readily degradable volatile fatty acids (absent post-1977), mineral oils and halogenated solvents; the latter remaining at levels of up to 50 $\mu g\,L^{-1}$ beneath the landfill. The results showed that the high buffering capacity of the Chalk is conducive to microbial metabolism to explain the decrease in TOC : Cl ratios with time and the disappearance of readily degradable organic compounds. However, the organic solvents remained as persistent contaminants (Williams et al. 1991).

Below the water table, the general shape of a landfill leachate plume is determined by the advective–dispersive nature of groundwater flow in the aquifer, the amount of recharge from the leachate mound developed below the landfill, and the increased density of the leachate (Fig. 8.23). In most cases, the leachate plumes are relatively small: a few hundred metres wide, corresponding to the width of the landfill; and restricted to less than 1000 m in length as a result of attenuation processes within the leachate plume, although potentially longer if contamination occurs in fissured or fractured material. According to Christensen et al. (2001), the infiltrating leachate creates a sequence of redox zones in the groundwater, with methanogenic conditions close to the landfill and oxidized conditions at the outer boundary of the plume. The anaerobic zones are driven by microbial utilization of dissolved organic matter in the leachate in combination with reduction of oxidized species in the aquifer, particularly iron oxides, which provides substantial redox buffering by reducing iron oxides and precipitating reduced iron species. Other important attenuating mechanisms include dilution, ion exchange, complexation and precipitation, such that heavy metals are not normally considered a major groundwater pollution problem in leachate plumes. The attenuation mechanism for NH_4^+ is not well understood but probably involves anaerobic oxidation of this persistent pollutant in landfill leachate. Xenobiotic organic compounds in leachate are not extensively attenuated by sorption processes, but there is increasing evidence that many organic compounds are degradable in the strongly anaerobic iron-reducing zone of leachate plumes (Christensen et al. 2001).

To illustrate the sequence of redox reactions in a landfill leachate plume, Fig. 8.34 shows the Villa Farm landfill site near Coventry in the English Midlands. The site became disused in 1980 having received a wide variety of industrial wastes over 30 years, including oil/water mixtures and effluent treatment sludges containing

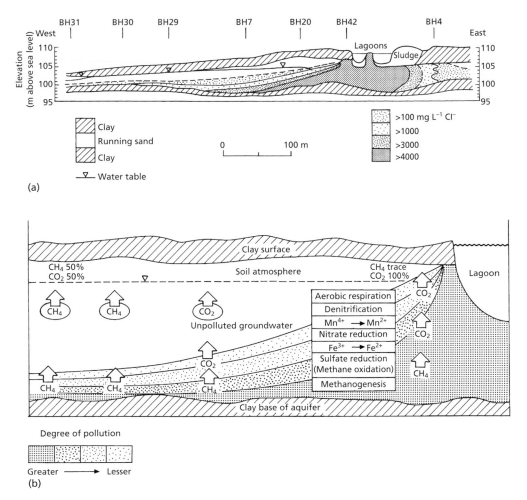

Fig. 8.34 Landfill leachate contamination of groundwater in a lacustrine sand aquifer at Villa Farm, West Midlands, showing (a) the distribution of chloride concentration along a vertical cross-section of the pollution plume as determined by borehole (BH) sampling and (b) a schematic diagram showing the transition from oxidizing conditions in the background, uncontaminated groundwater to the heavily polluted and highly reducing zone near the lagoons at the base of the aquifer (Williams *et al.* 1991). (*Source:* Williams, G.M., Young, C.P. and Robinson, H.D. (1991) Landfill disposal of wastes. In: *Applied Groundwater Hydrology* (eds R.A. Downing and W.B. Wilkinson). Clarendon Press, Oxford, pp. 114–133. © 1991, Oxford University Press.)

heavy metals, acids, alkalis, organic solvents and paint wastes. These liquid wastes were disposed of directly into lagoons in hydraulic continuity with a shallow lacustrine sand aquifer. The inorganic reactions and organic compounds observed in the groundwater are identical to those reported for domestic waste and co-disposal sites. The results of extensive monitoring at the site are shown schematically in Fig. 8.34 with chloride acting as a conservative tracer to delimit the extent of the groundwater leachate plume (Fig. 8.34a). A geochemical zonation, based on redox reactions is observed in the transition from oxidizing conditions in the background, uncontaminated groundwater to the heavily polluted and highly reducing zone near the lagoons at the base of the aquifer (Fig. 8.34b). These zones follow the theoretical sequence of redox reactions predicted from thermodynamic considerations in a closed, organically polluted system (Table 4.15). Heavy metals are attenuated near the lagoon as carbonates and as sulphides in the zone where sulphate is reduced to sulphide.

Measurements of TOC suggested little gross change in the organic carbon content of the pollution plume at Villa Farm, although biotransformations did occur. Aromatic hydrocarbons were broken down with increasing distance from the lagoons as suggested by the presence of benzoic acid derivatives which were not present in the lagoons, but synthesized from the primary disposal of phenol. The highly reducing conditions throughout the plume below the zone of sulphate reduction led to the production of CH_4. However, at the same time, CH_4 appeared to be consumed during oxidation to CO_2 under anaerobic conditions in the overlying zone of sulphate reduction. Where sulphate reduction was limited, such as at the leading edge of the plume, CH_4 was probably able to diffuse upwards without oxidation, and so explaining the CH_4 found in the soil atmosphere at concentrations of up to 55% by volume (Williams *et al.* 1991). Other studies also reveal the temporal variability in constituent concentrations near plume boundaries related in part to hydrologic changes at various time scales. The upper boundary of a plume is a particularly active location where redox reactions respond to recharge events and seasonal water table fluctuations (Cozzarelli *et al.* 2011).

8.4.3 Faecal, domestic and cemetery wastes

Since the pioneering work of John Snow (1813–1858) into the transmission of cholera in London (Price 2004; Plate 8.1), water has been recognized as an effective vehicle for the spread of many gastrointestinal human pathogens (Barrett *et al.* 1999). The microbiological contamination of groundwater from cess pits and leaking sewage and the transmission of disease has profound and severe implications for public health, particularly in small communities and developing countries where

groundwater is often the preferred source of drinking water. Contaminated groundwater can contribute to high morbidity and mortality rates from diarrhoeal diseases and sometimes lead to epidemics (Pedley and Howard 1997). Globally, there are nearly 1.7 billion cases of childhood diarrhoeal disease every year, responsible annually for the deaths of around 525 000 children under the age of five years (World Health Organization 2020).

The most frequently used method for the determination of faecal pollution of a water sample, and hence sewage recharge to groundwater, is to test for the presence of faecal indicator bacteria. These are groups of bacteria that are present in high numbers in faeces, but are absent from other sources. Historically, thermotolerant (formerly faecal) coliforms, *Escherichia coli*, total coliforms and faecal streptococci have been used as indicators. However, many bacteria with the characteristics of total coliforms can be isolated from environmental sources that are free from faecal contamination. Yet, despite the many limitations associated with the use of faecal indicator bacteria, they continue to be the most widely used groups of indicator organisms by the water industry. Viruses, such as enterovirus, may also enter the groundwater system as a result of faecal contamination. These are not generally used as indicator species due to their problematic isolation and quantification. An alternative to standard virus analysis is the use of bacteriophage. These are viruses which infect a host bacteria, enabling easier isolation and quantification. An example is coliphage, a virus that uses *E. coli* as a host organism (Barrett *et al.* 1999).

The disposal of excreta using land-based systems is a key issue for groundwater quality and public health protection. The use of inappropriate water supply and sanitation technologies in peri-urban areas leads to severe and long-term public health risks. The use of poorly constructed sewage treatment works and land application of sewage can lead to groundwater contamination close to water supply sources. Microbiological, in particular virus survival in

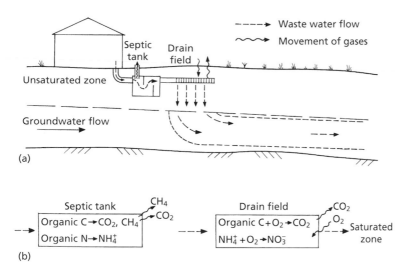

Fig. 8.35 Diagram showing (a) a schematic cross-section of a conventional septic system, including septic tank, distribution pipe and groundwater plume; and (b) the sequence of simplified redox reactions in the two principal zones of a conventional septic system, the septic tank and drain field (Wilhelm *et al.* 1994). (*Source:* Wilhelm, S.R., Schiff, S.L. and Cherry, J.A. (1994) Biogeochemical evolution of domestic waste water in septic systems: 1. Conceptual model. *Ground Water* **32**, 905–916. © 1994, John Wiley & Sons.)

these circumstances is not well understood, but there are indications of extended pathogen survival and therefore increased public health risk (Pedley and Howard 1997).

On-site septic systems for the disposal of domestic waste are common in rural areas without a connected sewerage system. In the United States, it is estimated that on-site systems dispose of approximately one-third of the population's domestic waste water. Since the domestic waste water in septic systems contains many environmental contaminants, septic systems in North America constitute approximately 20 million potential point sources for groundwater contamination (Wilhelm *et al.* 1994). A conceptual model of the biogeochemical evolution of domestic waste water in conventional on-site septic systems is given in Fig. 8.35. As described by Wilhelm *et al.* (1994), the evolution of waste water is driven by microbially catalysed redox reactions involving organic carbon and nitrogen and occurs in as many as three different redox zones (Table 8.12).

Anaerobic digestion of organic matter and production of CO_2, CH_4 and NH_4^+

predominate in the first zone, which consists mainly of the septic tank. In the second zone, gaseous diffusion through the unsaturated sediments of the drain field supplies oxygen for aerobic oxidation of organic carbon and NH_4^+ with a consequent decrease in waste water alkalinity. The nitrate formed by NH_4^+ oxidation in this zone is the primary and generally unavoidable adverse impact of septic systems at most sites (Fig. 8.36). In the third zone, nitrate is reduced to N_2 by the anaerobic process of denitrification, although rarely found below septic systems due to a lack of labile organic carbon in natural settings. Without natural attenuation by denitrification, it is quite likely that in unconfined sand aquifers common in North America, the typical minimum permissible distance between a well and septic tank (25–35 m) will not be sufficient to provide protection against nitrate contamination by dispersive dilution alone (Robertson *et al.* 1991).

In a later study, Robertson *et al.* (2012) identified anaerobic ammonium oxidation (anammox) using nitrite and nitrate as electron acceptors (eqs. 12a and 12b in Table 8.12) as

Table 8.12 Hydrochemical and biogeochemical reactions in septic systems (Wilhelm *et al.* 1994).

Anaerobic zone (septic tank and biological mat):

Organic molecule hydrolysis: (1)
Proteins + H_2O → Amino acids
Carbohydrates + H_2O → Simple sugars
Fats + H_2O → Fatty acids and glycerol

Ammonium release:
Urea $[CO(NH_3^+)_2]$ + H_2O → $2NH_4^+$ + CO_2 (2a)
Amino acids + H_2O → NH_4^+ + organic compounds (2b)

Fermentation:
Amino acids, simple sugars → H_2, acetate (CH_3OO^-), other organic acids (3)

Anaerobic oxidation:
Fatty acids + H_2O → H_2, CH_3OO^- (4)

Sulphate reduction:
SO_4^{2-} + $2CH_2O^a$ + $2H^+$ → H_2S + $2CO_2$ + $2H_2O$ (5)

Methanogenesis:
CH_3OO^- (acetate) + H^+ → CH_4 + CO_2 (6a)
CO_2 + $4H_2$ → CH_4 + $2H_2O$ (6b)

Aerobic zone (unsaturated zone and saturated zone to lesser extent):

Organic matter oxidation:
CH_2O + O_2 → CO_2 + H_2O (7)

Nitrification:
NH_4^+ + $2O_2$ → NO_3^- + $2H^+$ + H_2O (8)

Sulphide oxidation:
H_2S (or organic sulfide) + $2O_2$ → SO_4^{2-} + $2H^+$ (9)

Carbonate buffering:
H^+ + HCO_3^- → H_2CO_3 (10a)
$CaCO_3$ + H^+ → Ca^{2+} + HCO_3^- (10b)
$CaCO_3$ + CO_2 + H_2O → Ca^{2+} + $2HCO_3^-$ (10c)

Second anaerobic zone (saturated or near-saturated conditions):

Denitrification[b]:
$4NO_3^-$ + $5CH_2O$ + $4H^+$ → $2N_2$ + $5CO_2$ + $7H_2O$ (11)

Anammox[c]
NH_4^+ + NO_2^- → N_2 + $2H_2O$ (12a)
$3NO_3^-$ + $5NH_4^+$ → $4N_2$ + $9H_2O$ + $2H^+$ (12b)

---

[a] Organic matter is simplified as CH_2O throughout. Actual organic matter contains C of various oxidation states and other elements such as N, P and S, and therefore actual reaction products vary.
[b] Nitrate reduction can also be accomplished via oxidation of reduced sulphur compounds.
[c] After Robertson *et al.* (2012)
(*Source:* Wilhelm, S.R., Schiff, S.L. and Cherry, J.A. (1994) Biogeochemical evolution of domestic waste water in septic systems: 1. Conceptual model. *Ground Water* **32**, 905–916. © 1994, John Wiley & Sons.)

an important process in nitrogen attenuation at the site shown in Fig. 8.36. In a suboxic zone of the contaminant plume where anammox is active, total inorganic nitrogen of approximately 100 mg L^{-1} is attenuated by 80%, including removal of NH_4^+-N. This zone has both NO_3^--N and NH_4^+-N at >5 mg L^{-1}. Bacterial community composition was assessed with molecular, DNA-based analysis and demonstrated that groundwater bacterial populations were predominantly composed of bacteria from two *Candidatu*s genera of the Planctomycetales (Brocadia and Jettenia) that are found in a variety of aquatic environments

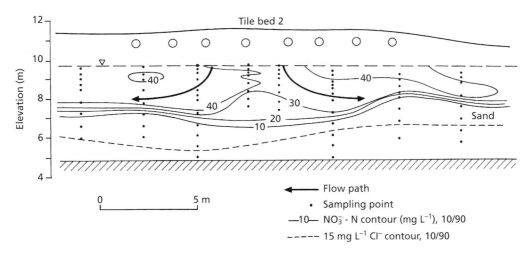

Fig. 8.36 Distribution of nitrate concentration in an unconfined medium sand aquifer below a large septic system located at Long Point on the north shore of Lake Erie, Ontario. The water table depth below the infiltration pipes of Tile bed 2 is about 1.2 m and is sufficient to allow for almost complete oxidation of the sewage constituents during migration through the sandy vadose zone (Aravena and Robertson 1998). (*Source:* Aravena, R. and Robertson, W.D. (1998) Use of multiple isotope tracers to evaluate denitrification in ground water: study of nitrate from a large-flux septic system plume. *Ground Water* **36**, 975–982. © 1998, John Wiley & Sons.)

(Francis *et al.* 2007). Their dominance in groundwater communities provided further evidence for autotrophic anammox growth and activity. Robertson *et al.* (2012) concluded that anammox has the potential to play an important role in the natural attenuation of nitrogen from septic system plumes, particularly in low carbon environments where denitrification may be less active, a finding also supported by Sbarbati *et al.* (2018) for a polluted industrial site above an unconfined coastal aquifer.

Other waste water constituents are influenced by the major changes in redox and pH conditions that occur in the reaction zones of septic systems. Calcite ($CaCO_3$) is often dissolved in drain fields in order to buffer the acidity released during NH_4^+ oxidation and this results in increased Ca^{2+} concentrations in the effluent. Other cations may also be released from the solid phase during buffering reactions such as mineral dissolution or cation exchange. Wilhelm *et al.* (1994) also identified trace metal cations such as Cu, Cr, Pb and Zn in concentrations in the range of 2–300 $\mu g\,L^{-1}$ in many domestic waste waters as a result of the

changes in redox and pH. Although their specific behaviour in septic systems is less well understood, a large fraction of trace metals is likely to be retained in particulate matter in the septic tank where they form insoluble sulphides.

As described above, domestic waste water can contain pathogenic bacteria and viruses and overflow and seepage of waste water is a major cause of disease outbreaks (Craun 1985; Pedley and Howard 1997). Bacteria are retained in septic systems primarily by straining in the biological mat; the layer of accumulated organic matter found directly beneath the distribution pipes. In general, the mobility of bacteria and viruses is much greater in saturated than unsaturated flow, making unsaturated conditions below septic systems desirable for both oxygen supply and pathogen retention.

In terms of hydrophobic organic contaminants such as halogenated aliphatics and aromatics, partitioning on to the accumulated organic matter in the septic tank and the drain field will act to retain these contaminants.

Cemetery operations are a further potential source of groundwater contamination. Detailed studies within cemeteries in Australia (Knight and Dent 1998; Dent 2002) have principally identified forms of not only nitrogen but also sodium, magnesium, strontium, chloride, sulphate and forms of phosphorus as characterizing cemetery groundwaters. Cemetery functions are best understood conceptually as a special kind of landfill operation that is strongly influenced by temporal and spatial variability of cemetery practices. Dent (2002) found that the amounts of decomposition products leaving cemeteries are very small and that well-sited and managed cemeteries have a low environmental impact and are a sustainable activity. The most serious pollution situation is for the escape of pathogenic bacteria or viruses into the environment. The potential for such contamination can only be assessed by a comprehensive hydrogeological investigation.

8.4.4 Microplastic contamination

Microplastic contamination is ubiquitous in ecosystems worldwide. The world's oceans are contaminated with microplastics. Eriksen *et al.* (2014) estimated that there are 4.85 trillion microplastic particles in the global ocean weighing 35,540 tons. The sorption of toxicants to plastic while transported through the environment has led to the claim that synthetic polymers in the ocean should be regarded as hazardous waste (Rochman *et al.* 2013). Given their pervasive and persistent nature, microplastics have become a global environmental concern and a potential risk to human populations (Hurley *et al.* 2018).

Plastic polymers are used widely due to their ease of manufacture, low cost and stable chemical properties. Microplastics are generally characterized as water-insoluble, solid polymer particles that are less than 5 mm in size. Particles below 1 μm are usually referred to as nanoplastics, rather than microplastic. Microplastics are generally divided into two major categories: primary and secondary. Primary microplastics are intentionally manufactured for use in cosmetics, personal care products, industrial processing (e.g. sandblasting), textile applications, production of synthetic clothes and in domestic and industrial washing processes of fabrics. Being too small to be filtered by waste water treatment plants, primary microplastics can be introduced directly into oceans through direct runoff. Secondary microplastics are typically generated by degradation and fragmentation of larger pieces of plastics due to exposure to ultraviolet light from the sun and/or by mechanical means such as tidal waves. Secondary microplastics are often associated with marine litter (Re 2019).

Although microplastics are often detected in the environment, the risks they pose are debated and largely unknown. Waste water such as septic effluent can contain many thousands of micro-fibres made up of fine polymers (polyester and polyethylene) and fibres from the washing of synthetic fleece garments. Ecological concerns related to microplastics include their ability to adsorb persistent organic pollutants (POPs), which can be transferred to animal tissues, affecting bioaccumulation of POPs, and irritation of digestive tissues following ingestion (Panno *et al.* 2019).

One key challenge in assessing the risks of microplastics to humans and the environment relates to the variability of the physical and chemical properties, composition and concentration of the particles. Further, microplastics in the environment are difficult to identify and standardized methods do not exist. The dominant source of microplastics often is the fragmentation of larger plastics or product wear, yet the rate of fragmentation under natural conditions is unknown. These challenges and unknowns hamper the prospective assessment of exposure and risk (Koelmans *et al.* 2019).

Very few studies have examined the presence, abundance or environmental drivers of microplastics in groundwater systems. In a study of two karst aquifers in Illinois in the United States, Panno *et al.* (2019) sampled eight spring and three shallow (<65 m) wells in the Driftless Area region and six springs in the Salem Plateau region and found that groundwater samples contained microplastics and other anthropogenic contaminants. Sixteen of the 17 groundwater samples contained microplastics with a median concentration of 6.4 particles L^{-1} and a maximum of 15.2 particles L^{-1}. Four out of 20 microplastic samples were identified as polyethylene. The presence of microplastic was consistent with other parameters, including phosphate, chloride and triclosan, a widely used antibacterial and antifungal agent found in consumer products, suggesting septic effluent as a source, although surface runoff may also be a source given the susceptibility of karst aquifers to contamination by surface-borne pollutants (Panno et al. 2019).

In Lower Saxony in north-west Germany, Mintenig et al. (2019) analysed microplastics in large volumes of drinking water derived from the purification of groundwater pumped from a sand and gravel aquifer. The samples were taken at different points in the supply chain, ranging from groundwater wells at five

drinking water treatment plants to drinking water from conventional household taps to assess if and where contamination with microplastics would occur. In addition, groundwater from three wells with an approximate depth of 30 m was sampled in the area of Holdorf. To identify potential microplastics, Fourier transform infrared (FTIR) microscopy coupled to a focal plane array (FPA) detector was applied, which enabled the identification of microplastic particles down to a size of 20 µm.

Mintenig et al. (2019) observed that the concentrations of microplastics ranged from 0 to 7 particles m^{-3} in raw groundwater and drinking water, with an overall mean of 0.7 particles m^{-3} (Fig. 8.37). In 14 of the 24 water samples, no microplastic particles were detected. Less than 1 particle m^{-3} was determined in five of the samples, another four samples contained between 0 and 3 particles m^{-3}, and one sample 7 particles m^{-3}. In total, five different polymer types were identified in raw groundwater and drinking water (see Fig. 8.37). The majority of particles (62%) was made of polyester and detected mainly in two water samples. All

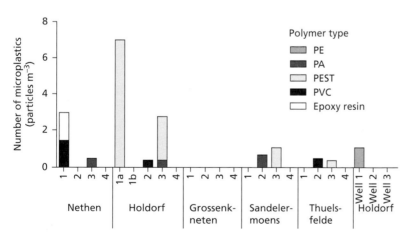

Fig. 8.37 Microplastic particles identified in (1) raw groundwater at the drinking water treatment plant inlet, (2) drinking water at the plant outlet, and (3) drinking water at the water meter and (4) a conventional water tap in a selected household in five drinking water supply areas in Lower Saxony, north-west Germany. In addition, groundwater samples extracted from three wells (30 m depth) were tested in the area around Holdorf. Note polymer types: PE – polyethylene, PA – polyamide, PEST – polyester, PVC – polyvinylchloride (Mintenig et al. 2019). (Source: Mintenig, S.M., Löder, M.G.J., Primpke, S. and Gerdts, G. (2019) Low numbers of microplastics detected in drinking water from ground water sources. Science of the Total Environment **648**, 632–635.)

microplastic particles were in a size range of 50 and 150 μm. Mintenig *et al.* (2019) concluded that their results indicated only minor contamination by microplastics of drinking water derived from the purification of groundwater. Even though plastic is a resistant and durable material, the abrasion of plastic equipment used during water purification or transport is a likely explanation for the plastic particles detected in the water samples (Mintenig *et al.* 2019).

Currently, there is little evidence of the behaviour of microplastics in porous, intergranular media. Chu *et al.* (2019) conducted column experiments using carboxyl-modified polystyrene latex colloids (density $1.055 \, \text{g cm}^{-3}$, average diameter 1 μm) as model microplastic colloids. The latex colloids were added to NaCl electrolyte solution at different ionic strengths (0.001, 0.01, 0.1 and 0.2 M), resulting in microplastic suspensions for column transport experiments with a concentration of 20 mg L^{-1}. Transport experiments of microplastics were carried out in acrylic columns (inner diameter 3 cm, length 10 cm) that were wet-packed with clean glass beads ranging in size from 250 to 300 μm. All column transport experiments adopted a flow velocity of $3 \times 10^{-5} \, \text{m s}^{-1}$. Different pore volumes of microplastic influent suspensions were injected into the glass bead-packed columns and dissections performed to measure microplastic retention profiles. The results showed that the variation in the concentrations of retained microplastics with depth changed from monotonic to non-monotonic with increased pore volume of the injected influent suspension and solution ionic strength. The non-monotonic retention was attributed to blocking of microplastics and transfer of these colloids among glass bead surfaces in the down-gradient direction. The microplastic breakthrough curves were modelled successfully by the convection–diffusion–equation (eq. 8.7) including two types of first-order kinetic deposition (i.e., reversible and irreversible attachment). However, the model was unable to simulate the non-monotonic retention profiles due to the fact that the transfer of colloids among glass beads was not considered (Chu *et al.* 2019).

The above studies provide a basis for further research. Urgent assessment of the occurrence and fate of microplastic fibres in porous media and the study of the role of microplastics as carriers of contaminants within aquifers are required. Only by addressing these key challenges will it be possible to tackle this contaminant of emerging concern in groundwater (Re 2019).

8.4.5 Agricultural contaminants

Agricultural contaminants include nitrate and pesticides used in intensive farming practices that often affect wide areas of aquifer outcrop. As such, nitrate and pesticides are diffuse pollutants in the environment and can lead to serious consequences for the quality of groundwater resources and surface waters receiving contaminated groundwater. Other sources of contamination arise from livestock and poultry farming through the intensive management of grazing pasture and the operation of concentrated animal feeding operations (CAFOs) (Mallin and Cahoon 2003).

Nitrate contamination of groundwater has resulted from the desire for greater self-sufficiency in food supply and has resulted from the ploughing-up of grassland and the application of nitrogen-based fertilizers and organic manure. Ploughing-up grassland stimulates the natural process of mineralization and the release of nitrate from the organic-bound nitrogen in the soil zone (Fig. 8.38). The use of fertilizers during the period of crop growth often coincides with the onset of the rainfall season and the use of irrigation water. If fertilizer and manure applications are not applied following good agricultural practice (Section 9.3.5), soil nitrate leaching losses can occur causing high nitrate concentrations in the unsaturated (Fig. 5.14b) and saturated zones (Box 8.6) of aquifers.

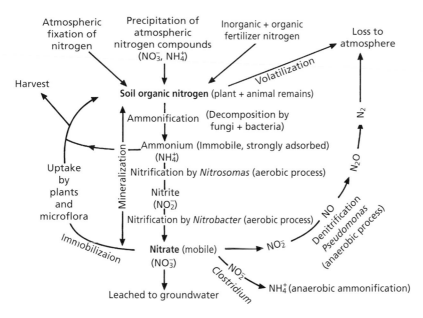

Fig. 8.38 The nitrogen cycle showing sources of nitrogen incorporated in the soil zone and the principal pathways of nitrate production (mineralization) and consumption (denitrification). Any excess nitrate not taken by plants and microflora or removed by denitrification can potentially leach to groundwater below the soil zone.

Box 8.6   Nitrate contamination of the Jersey bedrock aquifer

The island of Jersey is the largest of the British Channel Islands. The island setting comprises a plateau, formed largely of Precambrian crystalline rocks, with a steep topographic rise along the coastline. The temperate maritime climate encourages early flowers and vegetables, with intensive agricultural production sustained by large fertilizer applications. During winter and early spring, applications of nitrogen fertilizer to early cropping potatoes and horticultural crops may exacerbate the problem of nutrient leaching, with estimates of leaching losses of up to $100 \, kg \, N \, ha^{-1}$ expected from Jersey potato crops.

Mains water supply is principally from surface water storage, but there are large areas, particularly in the rural north of the island, that are reliant on well and borehole supplies, typically yielding less than $0.5 \, L \, s^{-1}$, to meet domestic, agricultural and light industrial demands. The main aquifer and isolated perched aquifers occur within a shallow zone of weathering in the bedrock, up to 25 m in depth below the water table surface, with groundwater flow almost entirely dependent on secondary permeability, imparted by dilated fractures.

The chemical composition of groundwater is controlled by maritime recharge inputs and water–rock interaction, although the effects of anthropogenic pollution, particularly from nitrate, are in places severe. A 1995 survey of groundwater quality at 46 locations across the island produced the regional distribution of nitrate shown in Fig. 8.39. Elevated nitrate concentrations occurred across much of the island and ranged from undetected to $215 \, mg \, L^{-1}$, with a mean value of $71 \, mg \, L^{-1}$ (Fig. 8.40). Of the total, 67% of samples exceeded the European Union water quality

(Continued)

Box 8.6   (Continued)

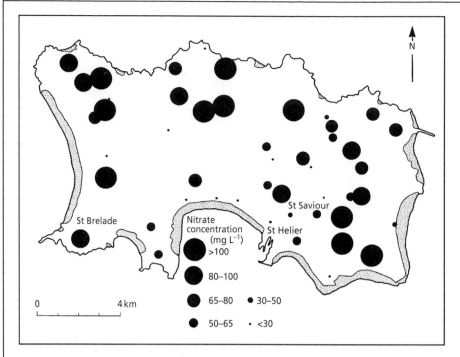

Fig. 8.39   Regional distribution of groundwater nitrate concentrations in the Jersey bedrock aquifer sampled in June 1995. The absence of a regional pattern is due to differences in land use and inorganic nitrogen fertilizer and cattle manure inputs, variable recharge rates across the island and the physical heterogeneity of the weathered bedrock aquifer that leads to unpredictable borehole yields and groundwater flowpaths (Green *et al.* 1998). (*Source:* Green, A.R., Feast, N.A., Hiscock, K.M. and Dennis, P.F. (1998) Identification of the source and fate of nitrate contamination of the Jersey bedrock aquifer using stable nitrogen isotopes. In: *Groundwater Pollution, Aquifer Recharge and Vulnerability* (ed. N.S. Robins). Geological Society, London, Special Publications **130**, pp. 23–35. © 1998, Geological Society of London.)

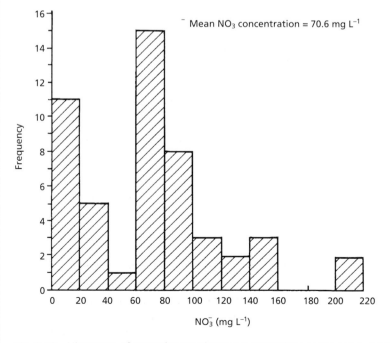

Fig. 8.40   Histogram of groundwater nitrate concentrations in the Jersey bedrock aquifer sampled in June 1995 (Green *et al.* 1998). (*Source:* Green, A.R., Feast, N.A., Hiscock, K.M. and Dennis, P.F. (1998) Identification of the source and fate of nitrate contamination of the Jersey bedrock aquifer using stable nitrogen isotopes. In: *Groundwater Pollution, Aquifer Recharge and Vulnerability* (ed. N.S. Robins). Geological Society, London, Special Publications **130**, pp. 23–35. © 1998, Geological Society of London.)

Box 8.6 (Continued)

standard for nitrate in drinking water of 50 mg L^{-1}. The high nitrate concentrations appear to decrease in a coastal direction, especially in those samples from the central southern area and in the valley areas around St. Saviour, probably as a result of localized denitrification (Green *et al.* 1998). Otherwise, local variations in well and borehole depths and local land-use and agricultural practices, combined with the physical heterogeneity of

the aquifer, produced no obvious pattern in the distribution of nitrate. It is concluded that the source of dissolved nitrate in the Jersey bedrock aquifer is primarily a result of the intensive agricultural and horticultural practices and high livestock densities on the island. In some areas, domestic pollution from septic tank discharges is a further potential hazard.

It is estimated that inputs of reactive nitrogen into the terrestrial biosphere are currently more than double pre-industrial levels due to modern agricultural practices and application of nitrogen fertilizers (Galloway *et al.* 2004). Globally, it is estimated that about 57% of anthropogenic nitrogen fixation results from the manufacture of nitrogen-containing fertilizers, 29% from cultivation of nitrogen-fixing crops and 14% from burning fossil fuels (Erisman *et al.* 2005).

Natural, background concentrations of nitrate in groundwater are very low, with most of the nitrate found in groundwater being of anthropogenic origin and mostly related to agricultural activities. Van Drecht *et al.* (2003) estimated that the total leaching of nitrogen to groundwater at the global scale is 55 Tg a^{-1}, of which 40% is expected to reach river outlets. The contribution of deep aquifers mostly affected by historical use of fertilizers was estimated at 10% of the total nitrogen load.

In aquifers, the slow travel time for solutes in the vadose zone can lead to the storage of significant amounts of dissolved reactive nitrogen, resulting in a lag between any changes in agricultural practices at the land surface to reduce nitrogen loadings and subsequent improvements in groundwater and surface water quality (Wang *et al.* 2012). The global spatiotemporal distribution of nitrate in the vadose zone has been mapped by Ascott *et al.*

(2017) through the application of numerical models and published datasets of nitrate leaching, depth to groundwater, recharge rate and porosity. The modelling showed a continuous increase in the amount of nitrogen stored in the vadose zone between 1900 and 2000. For the year 2000, and based on a sensitivity analysis of model parameters, Ascott *et al.* (2017) estimated that the total storage of nitrate in the vadose zone is 605–1814 TgN. This total vadose zone storage of nitrogen is small (<3%) in comparison to estimates of total soil nitrogen (68 000–280 000 TgN; Lin *et al.* 2000, Xu-Ri and Prentice 2008), but potentially more significant in comparison to estimates of more labile soil inorganic nitrogen (NO_3^- and NH_4^+) of 940–25 000 Tg (Lin *et al.* 2000, Xu-Ri and Prentice 2008). The modelled spatiotemporal distribution of nitrate stored in the vadose zone shows substantial increases between 1950 and 2000 associated with increased global use of nitrogen fertilizers and subsequent leaching. Basins in North America, China and Central and Eastern Europe have developed large amounts of nitrate stored in the vadose zone due to thick vadose zones, slow travel times and high nitrate loadings (Ascott *et al.* 2017).

The health effects of high nitrate concentrations in excess of European and WHO water quality standards (50 mg L^{-1} as nitrate) are concerned with methaemoglobinaemia ('blue baby' syndrome) in infants (Walton 1951;

Craun *et al.* 1981) and gastric cancer (National Academy of Sciences 1981; Nomura 1996). Environmental impacts associated with excessive nitrate in the aquatic environment are eutrophication of inland and coastal waters and the consequent loss of biodiversity (Hecky and Kilham 1988; European Environment Agency 2003; Sutton and van Grinsven (2011). Approaches to controlling diffuse contamination of groundwater by nitrate include 'end of pipe' technological solutions such as blending a contaminated source with a low-nitrate water, biological reactor beds and anion exchange resins (Hiscock *et al.* 1991). To prevent further contamination and achieve environmental standards, as for example set by the EU Nitrates Directive, approaches include reductions in fertilizer and manure applications following good agricultural practice and, more radically, changing land use from arable cropping to low intensity grassland or forestry (Section 9.3.5).

Pesticides refer to the group of synthetic organic chemicals used mainly as fungicides, herbicides and insecticides. Herbicides are used in the largest quantities and generally have much greater water solubility compared with insecticides such that critical concentrations, in excess of water quality guidelines and standards, may be exceeded. The European Union has adopted a maximum admissible concentration of $0.1\ \mu g\ L^{-1}$ for any individual pesticide and $0.5\ \mu g\ L^{-1}$ for the sum of all individual pesticides, and in North America guidelines are applied to individual pesticides (Appendix 9). The factors affecting the leaching of pesticides from soils include the timing of application, the quantity reaching the target area and the physical and chemical properties of the soil. Newer formulations of pesticides are tailored to have short half-lives of less than one month in the soil through retention and elimination of compounds by hydrophobic sorption (Section 8.3.3.1) and degradation by chemical hydrolysis and bacterial oxidation. Caution is required, however, in that quoted half-lives appropriate to a fertile clay-loam soil may not be representative of permeable sandy soils developed on aquifer outcrops. Below the soil zone, pesticide mobility will again be affected by the availability of sorption sites for attenuation and the viability of micro-organisms for bacterial degradation.

Sorption is promoted by organic carbon, iron oxides and clay minerals and is a significant mechanism in the attenuation of pesticides with depth such that the amount of pesticide leached to groundwater is generally less than the amount lost to surface runoff (Rodvang and Simpkins 2001). Total herbicide losses in subsurface drainage on fine-textured soils are usually less than 0.3%, but occasionally 1.5% of the amount applied.

Contamination of groundwater by pesticides is common in agricultural and urban areas. In a survey of groundwater in 20 of the major hydrological basins in the United States in which 90 pesticide compounds (pesticides and degradates) were analysed, one or more pesticide compounds were detected at 48% of the 2485 sites sampled. The pesticide concentrations encountered were generally low, with the median total concentration being $0.05\ \mu g\ L^{-1}$. Pesticides were commonly detected in shallow groundwater beneath both agricultural (60%) and urban (49%) areas and so highlighting urban areas as a potential source of pesticides (Kolpin *et al.* 2000).

In Iowa, which has some of the most intensive applications of herbicides in the United States, herbicide compounds were detected in 70% of 106 municipal wells sampled; with degradation products comprising three of the four most frequently detected compounds (Kolpin *et al.* 1997). The highest herbicide concentrations in groundwater were found in areas of greatest intensity of herbicide use (Table 8.13). Factors explaining the distribution of herbicides included an inverse relation to well depth and a positive correlation with dissolved oxygen concentration that appear to relate to groundwater age, with younger groundwater likely to contain herbicide compounds. The occurrence of herbicide

Table 8.13 Pesticides and their degradation products in samples collected during the summer of 1995 from 106 municipal wells in Iowa (Kolpin *et al.* 1997).

Compound	Per cent detection	Reporting limit ($\mu g\,L^{-1}$)	Maximum concentration ($\mu g\,L^{-1}$)	Maximum contaminant level[a] ($\mu g\,L^{-1}$)	Health advisory level[a] ($\mu g\,L^{-1}$)	Use or origin
Alachlor–ESA[b]	65.1	0.10	14.8	–	–	Alachlor degradation product
Atrazine	40.6	0.05	2.13	3.0	3.0	Herbicide
Deethylatrazine	34.9	0.05	0.59	–	–	Triazine degradation product (atrazine, propazine)
Cyanazine amide	19.8	0.05	0.58	–	–	Cyanazine degradation product
Metolachlor	17.0	0.05	11.3	–	70	Herbicide
Prometon	15.1	0.05	1.0	–	100	Herbicide
Deisopropylatrazine	15.1	0.05	0.44	–	–	Triazine degradation product (atrazine, cyanazine, simazine)
Alachlor	7.5	0.05	0.63	2.0	–	Herbicide
Cyanazine	5.7	0.05	0.30	–	1.0	Herbicide
Acetochlor	0.9	0.05	0.77	–	–	Herbicide
Metribuzin	0.9	0.05	0.27	–	100	Herbicide
Ametryn	0.0	0.05	–	–	60	Herbicide
Prometryn	0.0	0.05	–	–	–	Herbicide
Propachlor	0.0	0.05	–	–	90	Herbicide
Propazine	0.0	0.05	–	–	10	Herbicide
Simazine	0.0	0.05	–	4	4	Herbicide
Terbutryn	0.0	0.05	–	–	–	Herbicide

[a] US Environmental Protection Agency.
[b] Alachlor ethanesulfonic acid.
(*Source:* Kolpin, D.W., Kalkhoff, S.J., Goolsby, D.A. *et al.* (1997) Occurrence of selected herbicides and herbicide degradation products in Iowa's ground water, 1995. *Ground Water* **35**, 679–688. © 1997, John Wiley & Sons.)

compounds was substantially different among the major aquifer types across Iowa, being detected in 83% of the alluvial, 82% of the bedrock/karst region, 40% of the glacial till and 25% of the bedrock/non-karst region aquifers. Again, the observed distribution was partially attributed to variations in groundwater age among these aquifer types. A significant, inverse relationship was identified between total herbicide compound concentrations in groundwater and the average soil slope within a 2-km radius of the sampled wells. Steeper soil slopes may increase the likelihood of surface runoff occurring rather than transport to groundwater by infiltration (Kolpin *et al.* 1997).

In the United Kingdom, isoproturon has been used extensively as a herbicide with over 3×10^6 ha treated in 1996 (Thomas *et al.* 1997). Concentrations of isoproturon greater than the European Union limit have been found in groundwater abstracted from the major Chalk Aquifer (Table 8.14). Although concentrations are generally low, there is concern that significant quantities of isoproturon may be moving through the unsaturated zone only to contaminate groundwater in the future. Clark and Gomme (1992) recovered unsaturated Chalk cores for pore water analysis and showed that the uron herbicides (isoproturon, chlortoluron and linuron) left the base of the profile at very low concentrations (Table 8.15) and had not penetrated beyond 2 m into the unsaturated zone. If correct, these results would suggest that pollution of Chalk groundwater by the uron herbicides through intergranular flow in the Chalk matrix is unlikely except in areas where the water table is close to the surface. Where uron herbicides are detected in Chalk

Table 8.14 Summarized analytical results of pesticides in Chalk boreholes in the Granta catchment, Cambridgeshire (Gomme *et al.* 1992).

Site	Number of samples	Pesticides detected	Number of detections	Concentration range ($\mu g\,L^{-1}$)
Babraham	3	Atrazine	3	<l.q.[a] – 0.07
		Simazine	2	<l.q.
Sawston	3	Atrazine	3	<l.q. – 0.13
		Simazine	3	<l.q. – 0.07
Linton	3	Atrazine	1	<l.q.
Fleam Dyke	3	n.d.[b]	–	–
45/12	1	Atrazine	1	0.31
		Simazine	1	0.40
45/17	1	n.d.	–	–
54/28	1	n.d.	–	–
54/99	1	n.d.	–	–
54/101	1	Simazine	1	<l.q.
54/112	1	Atrazine	1	<l.q.
54/116	3	Atrazine	2	0.05–0.06
		Chlortoluron	2	0.17–0.35
		Isoproturon	2	0.49–0.61
		Simazine	3	0.09–0.12
54/119	1	Chlortoluron	1	<l.q.
		Isoproturon	1	<l.q.
		Simazine	1	0.05
55/84	1	Simazine	1	<l.q.
64/40	1	n.d.	–	–

[a] <l.q. = below limit of quantification.
[b] n.d. = none detected.
(*Source:* Gomme, J., Shurvell, S., Hennings, S.M. and Clark, L. (1992) Hydrology of pesticides in a Chalk catchment: groundwaters. *Journal of the Institution of Water and Environmental Management* **6**, 172–178.)

Table 8.15 Analytical results of Chalk core profiling, Granta catchment, Cambridgeshire (Clark and Gomme 1992).

Depth of sample (m below ground)	Isoproturon concentration (μg kg^{-1})	Chlortoluron concentration (μg kg^{-1})	Linuron concentration (μg kg^{-1})
0.0–0.5	0.11	0.53	–
0.5–1.0	–	–	–
1.0–1.5	<0.03	0.21	–
1.5–2.0	–	3.04	–
2.5–3.0	–	–	n.r.[a]
4.5–5.0	–	–	n.r.
6.5–7.0	–	–	n.r.
9.5–10.0	–	–	n.r.

[a] n.r. not recorded
– below detection limit (0.03 μg L^{-1})
(*Source:* Clark, L. and Gomme, J. (1992) Pesticides in a Chalk catchment in Eastern England. *Hydrogéologie* **4**, 169–174.)

groundwater, it is possible that pesticide transport has occurred by flow through the fissure system. Support for these results is provided by Besien *et al.* (2000) who measured recovery rates of isoproturon of 48–61% in laboratory column experiments using Chalk cores eluted with non-sterile groundwater containing an initial mass of 1.5 mg of isoproturon. The column results also illustrated the importance of microbial degradation in removing isoproturon during the 162-day experiment.

Livestock farming produces waste containing many pathogenic micro-organisms associated with serious gastrointestinal disease, that include bacteria such as *E. coli* and *Streptococcus*, viruses such as enterovirus, and protozoa such as *Cryptosporidium* and *Giardia*. The presence of faecal coliform bacteria indicates that other disease-causing organisms may be present. In Ontario, 17 and 20% of farm wells in coarse- and fine-textured sediments were contaminated with faecal coliform and *E. coli*, respectively (Goss *et al.* 1998). Coliform bacteria were also present in 17% of domestic water wells in loess and till deposits in eastern Nebraska (Gosselin *et al.* 1997).

Not all strains of *E. coli* are harmful but some strains, such as O157:H7, are serious

pathogens. A stark example illustrating that pathogen occurrence is not only restricted to developing countries, is the case of Walkerton, Ontario, when in May 2000, *E. coli* O157:H7 and *Campylobacter jejuni* contaminated the drinking water supply leading to the death of seven individuals and illness in over 2,000 others. *E. coli* bacteria were found to have entered the Walkerton drinking water supply through a well which had been contaminated by cattle manure spread on a nearby farm. Normally, water can be treated using chlorine which acts to kill *E. coli* bacteria, but in the case of the Walkerton outbreak chlorine levels had not been sufficiently maintained. Also, exceptional environmental factors contributed to the outbreak with heavy rainfall in early May that assisted transport of the contaminants to Well 5, located in a shallow fractured aquifer vulnerable to surface-derived contamination. A full judicial inquiry was set-up by the Ontario Provincial Government into the circumstances surrounding the outbreak and also acted to introduce a new drinking water regulation to protect water supplies (Holme 2003).

Cryptosporidiosis is a significant cause of gastroenteritis in the United Kingdom with an estimated 42 000 cases in England and

Wales in 1995 (Adak *et al.* 2002). Cryptosporidiosis is caused by the protozoan pathogen *Cryptosporidium parvum* which is widespread in the environment and is found in the intestinal regions of most humans and animals. It is excreted from infected individuals as an oocyst which can survive for long periods in the environment and is resistant to disinfection by conventional water treatment. Outbreaks of cryptosporidiosis have occurred due to oocyst-contaminated groundwater supplied by wells, mainly in hydrogeological settings characterized by fractured material. These outbreaks have occurred in karst limestone aquifers, for example the Edwards Aquifer in Texas (Bergmire-Sweat *et al.* 1999) and the Chalk Aquifer in the north London Basin (Willocks *et al.* 1999). *Cryptosporidium* contamination hazard assessment and risk management for British groundwater sources are discussed by Morris and Foster (2000).

8.4.6 Saline water intrusion in coastal aquifers

Intrusion of saltwater into an aquifer occurs where seawater displaces or mixes with fresh groundwater. The intrusion of saltwater is one of the most common pollutants of fresh groundwater (Todd 1980; Custodio 1987) and often results from human activities which reduce groundwater flow towards the sea. In an aquifer where freshwater is flowing towards the sea, the Ghyben–Herzberg relation predicts, for freshwater and seawater densities (ρ_f and ρ_s) of 1000 and 1025 kg m^{-3}, respectively, that the depth below sea level to the saline water interface, z_s, is approximately 40 times the height of the freshwater table above sea level, z_f. This can be shown with reference to Fig. 8.41a and assuming simple hydrostatic conditions in a homogeneous, unconfined coastal aquifer in which:

$$\rho_s g z_s = \rho_f g (z_f + z_s) \qquad \text{(eq. 8.24)}$$

or

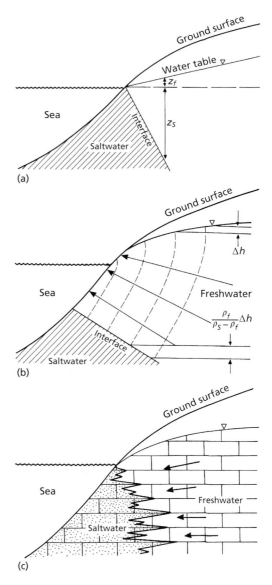

Fig. 8.41 Development of a saline interface in an unconfined coastal aquifer under (a) a hydrostatic condition and (b) a condition of steady-state seaward freshwater flow. In (c) the absence of a simple saline interface is caused by complex flow conditions in a fissured aquifer.

$$z_s = \frac{\rho_f}{\rho_s - \rho_f} z_f \qquad \text{(eq. 8.25)}$$

which for $\rho_f = 1000$ kg m^{-3} and $\rho_s = 1025$ kg m^{-3} gives the Ghyben–Herzberg relation:

$$z_s = 40z_f \qquad \text{(eq. 8.26)}$$

The Ghyben–Herzberg relation can also be applied to confined aquifers by substituting the water table by the potentiometric surface.

It can be seen from eq. 8.26 that small variations in the freshwater head will have a large effect on the position of the saltwater interface. If the water table in an unconfined aquifer is lowered by 1 m, the saltwater interface will rise 40 m. The freshwater–saltwater equilibrium established requires that the water table (or potentiometric surface) lies above sea level and that it slopes downwards towards the sea. Without these conditions, for example when groundwater abstraction reduces the freshwater table in coastal boreholes below sea level, seawater will advance directly inland causing saline intrusion to occur.

It can be shown that where the groundwater flow is nearly horizontal, the Ghyben–Herzberg relation gives satisfactory results, except near the coastline where vertical flow components are more pronounced leading to errors in the position of the predicted saltwater interface. In most real situations, the Ghyben–Herzberg relation underestimates the depth to the saltwater interface. Where freshwater flow to the sea occurs, a more realistic picture is shown in Fig. 8.41b for steady-state outflow

to the sea. The exact position of the interface can be determined for any given water table configuration by graphical flow net construction (Box 2.3), noting the relationships shown in Fig. 8.41b for the intersection of equipotential lines on the freshwater table and at the interface (Freeze and Cherry 1979).

The saltwater interface shown in Fig 8.41a and b is assumed to be a sharp boundary, but in reality a brackish transition zone of finite thickness separates the freshwater and saltwater. This zone develops from dispersion caused by the flow of freshwater and unsteady movement of the interface by external influences such as tides, groundwater recharge and pumping wells. In general, the thickest transition zones are found in highly permeable coastal aquifers subject to large abstractions. An important consequence of the development of a transition zone and its seaward flow is the cyclic transport of saline water back to the sea (Fig. 8.42). This saline water component originates in the underlying saline water and so, from continuity considerations, there exists a small landward flow in the saltwater wedge (Todd 1980).

Groundwater systems respond hydraulically to sea level rise over a continuum depending on two principal conceptual models: (1) flux-controlled or recharge-limited systems, in

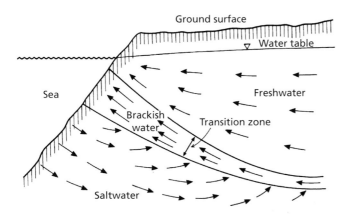

Fig. 8.42 Vertical cross-section showing flow patterns of freshwater and saltwater in an unconfined coastal aquifer illustrating the development of a brackish transition zone and the cyclic flow of saline water to the sea (Todd 1980). (*Source:* Todd, D.K. (1980) *Groundwater Hydrology* (2nd edn). Wiley, New York. © 1980, John Wiley & Sons.)

which groundwater discharge to the sea is persistent despite changes in sea level; and (2) head-controlled or topography-limited systems, whereby groundwater abstractions or surface features maintain the head condition in an aquifer despite sea-level changes (Werner and Simmons, 2009). The hydrogeologic setting, which combines the influence of geology and climate, controls the hydraulic type and the vulnerability of the aquifer to seawater intrusion, the amount of fresh groundwater flowing through the aquifer, and the rate of submarine groundwater discharge and its role in transporting terrestrial chemicals to coastal waters (Michael *et al.* 2013; Befus *et al.* 2020). Unconfined aquifers in hydraulic connection with a rising sea level can experience shoaling of water tables as the intrusion of denser marine water forces an increase in the water table. As the water table rises, groundwater discharge to receiving drainage networks may be initiated or intensified (Rotzoll and Fletcher 2013).

Saline intrusion problems are known from around the world with well-documented examples including the Biscayne aquifer in Florida (Klein and Hull 1978), the Quaternary sand aquifers of Belgium and The Netherlands (De Breuck 1991), the Chalk Aquifer of South Humberside (Howard and Lloyd 1983), the Llobregat Delta confined aquifer of Spain (Iribar *et al.* 1997) (Box 8.7) and the unconfined coastal aquifers of California (Befus *et al.* 2020).

Box 8.7 Saltwater intrusion in the Llobregat Delta aquifer, Spain

The Lower Llobregat aquifer system is formed by the Lower Valley and deep delta aquifers located a few kilometres south-west of Barcelona (Fig. 8.43). The Lower Valley aquifer is formed from Quaternary sands and coarse gravels and extends over an area of 100 km². The aquifer formation continues below the present morphological delta towards the coast (Fig. 8.44). At the sides of the delta, the aquifer materials change to sediments from local creek alluvial fans and beach deposits. The deep delta aquifer is formed by these deposits and the deep formation shown in Fig. 8.44. This formation extends seawards with decreasing thickness and outcrops on the sea floor at around 100 m depth and 4–5 km offshore. The deep delta aquifer is confined by wedge-shaped clay, silt and fine sand sediments that act as an aquitard. Above this aquitard, the shallow delta aquifer is formed by sands, gravels and silt.

Since 1950, the Lower Valley and the deep aquifers have been intensively exploited for groundwater as an important supply source and an emergency reserve. However, over-exploitation has caused a depression of the potentiometric surface in the central area (Fig. 8.44) and the salinization of 30% of the confined aquifer below the delta. Wells near the greatest part of the potentiometric depression draw water both supplied as recharge from the Lower Valley and groundwater from the seaward margin. Monitoring of water levels and chloride concentrations is well documented at the position where the Lower Valley aquifer meets the deep delta aquifer (Fig. 8.45).

The displacement of the 1000 mg L^{-1} isochlor (often chosen to delineate a saltwater front) is indicated for the Lower Valley and delta by comparing the time evolution of chloride content in numerous wells and piezometers (Fig. 8.46). Saline water has penetrated inland from the sea following three preferential paths with the plumes pointing towards the main extraction wells of the delta. These plumes relate to the

Box 8.7 (Continued)

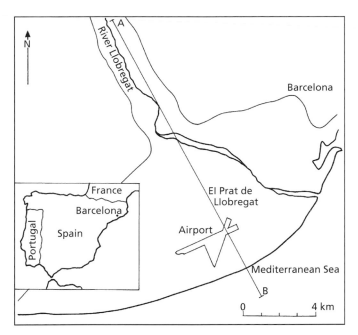

Fig. 8.43 Location map of the lower valley and delta of the River Llobregat, Spain (Iribar *et al.* 1997). (*Source:* Iribar, V., Carrera, J., Custodio, E. and Medina, A. (1997) Inverse modelling of seawater intrusion in the Llobregat delta deep aquifer. *Journal of Hydrology* **198**, 226–244. © 1997, Elsevier.)

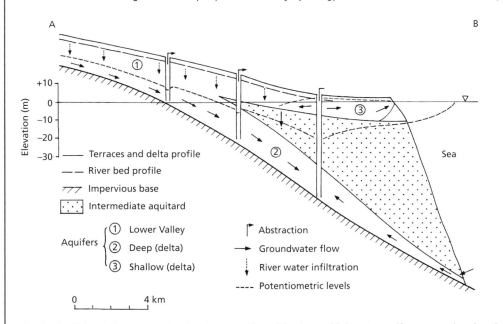

Fig. 8.44 Schematic hydrogeological cross-section of the Lower Llobregat aquifer system showing the flow pattern in the Lower Valley and delta aquifers. The line of section A–B is shown in Fig. 8.43 (Iribar *et al.* 1997). (*Source:* Iribar, V., Carrera, J., Custodio, E. and Medina, A. (1997) Inverse modelling of seawater intrusion in the Llobregat delta deep aquifer. *Journal of Hydrology* **198**, 226–244. © 1997, Elsevier.)

(Continued)

Box 8.7 (Continued)

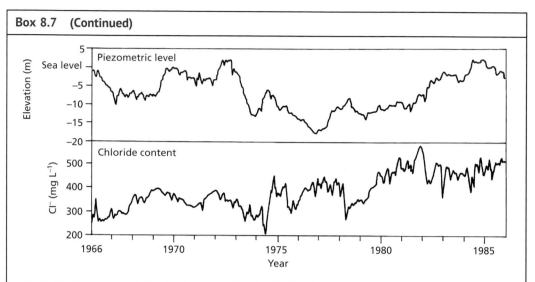

Fig. 8.45 Temporal variation in piezometric level and chloride content at the position where the Lower Valley aquifer meets the deep delta aquifer (Iribar *et al.* 1997). (*Source:* Iribar, V., Carrera, J., Custodio, E. and Medina, A. (1997) Inverse modelling of seawater intrusion in the Llobregat delta deep aquifer. *Journal of Hydrology* **198**, 226–244. © 1997, Elsevier.)

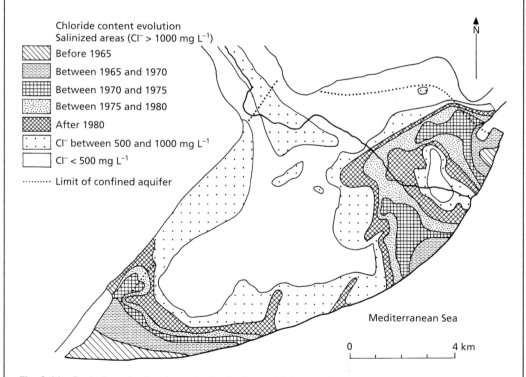

Fig. 8.46 Evolution of saline intrusion in the Lower Llobregat deep delta aquifer as indicated by the progressive encroachment of water with a chloride content above 1000 mg L^{-1} (Iribar *et al.* 1997). (*Source:* Iribar, V., Carrera, J., Custodio, E. and Medina, A. (1997) Inverse modelling of seawater intrusion in the Llobregat delta deep aquifer. *Journal of Hydrology* **198**, 226–244. © 1997, Elsevier.)

Box 8.7	(Continued)

sedimentological features of the delta. The plume in the central part of the delta intrudes through a high permeability zone, coinciding with the pre-glaciation palaeovalley of the Llobregat River. At the eastern boundary of the delta, the deep aquifer is covered by a sandy formation, and seawater penetration is hindered by thin muddy deposits present on the sea bed. At the south-west delta boundary, a former saline water body existed as the remnant of incomplete flushing of marine water by freshwater. Since 1965, abstractions along this south-west boundary have reversed the process of slow flushing and saline water has now penetrated towards the main wells in the area, causing the

unconfined aquifer in this coastal zone to become brackish. Two of the plumes merge at the delta centre, leaving a freshwater pocket surrounded by saline water, the surface area of which is decreasing owing to groundwater abstractions within it.

A very wide saltwater-freshwater transition zone with little or no vertical salinity stratification is evident as a result of the high aquifer permeability and dispersivity in the heterogeneous aquifer sediments, the small aquifer thickness of about 5 m compared to the flowpath lengths, and the long displacement of saline water inside a confined area without flushing (Iribar *et al.* 1997).

Methods for controlling saline intrusion are described by Todd and Mays (2005) and include (1) changing the locations of pumping wells, typically by moving them inland; (2) artificial recharge from a supplemental water source to raise groundwater levels; (3) an extraction barrier created by a continuous pumping trough with a line of wells adjacent to the sea; (4) an injection barrier to maintain a pressure ridge along the coast by a line of recharge wells injected with high quality imported water; and (5) an impermeable subsurface barrier constructed parallel to the coast and through the vertical extent of the aquifer.

An example of the first of the above methods is practised in the Cretaceous Chalk aquifer on the Sussex coast of southern England. For fissured aquifers such as the Chalk, there is no simple saline interface and over-abstraction can induce seawater to invade the aquifer along discrete fissure zones, often for discernible distances (Fig. 8.41c). This complex form of intrusion is illustrated by the downhole fluid electrical conductivity logs for a coastal borehole at Brighton (Fig. 8.47). The geophysical borehole logs reveal freshwater moving

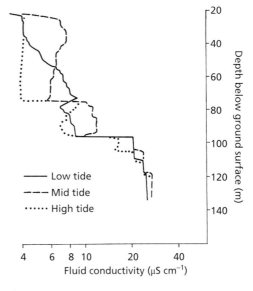

Fig. 8.47 Fluid electrical conductivity logs recorded by downhole geophysical borehole logging at a site on the Brighton sea front, Sussex, illustrating the inland penetration of saline water along discrete horizontal fissures in the Chalk aquifer (Headworth and Fox 1986). (*Source:* Headworth, H.G. and Fox, G.B. (1986) The South Downs Chalk aquifer: its development and management. *Journal of the Institution of Water Engineers* **40**, 345–361.)

seawards and saltwater moving inland along discrete horizontal fissures extending to 100 m below sea level. Below a depth of about 130 m the fluid logs indicate the existence of a saline water zone. Analysis of porewater in the Chalk between fissured zones revealed it to be composed predominantly of freshwater. Analysis of seasonal changes in the fluid log profiles showed that the salinity increases in response to the natural depletion of groundwater storage in the Chalk during the summer and can respond rapidly to changes in abstraction rates from wells located as much as 6 km inland (Headworth and Fox 1986).

Groundwater management to limit saline water intrusion in the Brighton Chalk block on the south coast is based on abstracting from pumping stations located around the margins of the aquifer in order to intercept outflows from the aquifer, while at the same time reducing abstractions from inland pumping stations in order to conserve aquifer storage. In addition, aquifer losses to the sea in winter are reduced as far as possible to assist inland storage levels to recover to be able to support increased output in summer. In drought years, following winters with below-average recharge, coastal outflows from the aquifer decline, and inland storage levels, increased as a result of the operating policy, allow greater use to be made of inland pumping stations so as to meet high summer demand for water in the coastal resorts (Headworth and Fox 1986; Miles 1993).

Carretero *et al.* (2019) compared groundwater abstraction methods for a freshwater aquifer occurring in coastal dunes in the eastern region of the Province of Buenos Aires, Argentina, and developed a decision–tree tool, constructed on the basis of hydrogeological characteristics (water table depth, freshwater thickness, land surface area required and distance between wells), for the optimal management of unconfined coastal aquifers of limited saturated thickness. The decision tree is intended to assist water resources managers choose the most suitable groundwater

extraction system, whether vertical wells or well points, as well as identifying areas that are unsuitable for sustainable groundwater extraction. The methodology may be of particular use for decision-makers in developing countries who are not specialists in hydrogeology and yet must make urgent decisions concerning the supply of water to small localities.

The protection of coastal groundwater resources under the threat of changes in freshwater recharge and sea level rise under climate change, in combination with increased anthropogenic activities, will affect coastal groundwater systems throughout the world. In a study of the low-lying Dutch Delta, Oude Essink *et al.* (2010) showed from numerical modelling that along the southwest coast of the Netherlands, salt loads will double in some parts of the deep and large polders by the year 2100 due to sea level rise. Similarly, modelling of the karstic Apulian aquifer of southern Italy by Romanazzi *et al.* (2015) showed that the high-quality groundwater of this large, deep coastal aquifer will deteriorate as a result of combined overexploitation, climate change and seawater intrusion. The model results showed a dramatic decrease in piezometric head of more than 2.5 m by 2060 compared to steady-state or natural conditions. Simulated salinity is predicted to increase by greater than 5000 mg L^{-1} by 2060 compared to steady-state conditions, primarily located along the coastal zone, especially in the western Ionian Sea area as a result of the enlargement and worsening of the upconing effect observed around areas of groundwater abstraction (Romanazzi *et al.* 2015).

8.4.7 Saline water intrusion on small oceanic islands

The majority of naturally occurring freshwater on small oceanic islands is groundwater, primarily recharged by precipitation (Falkland 1991) within lenses developed in relatively permeable aquifers, typically consisting of sand, lava, coral or limestone. A freshwater lens is formed by the radial flow of freshwater towards the coast. The lens floats on the underlying saltwater with its thickness decreasing from the centre coastwards (Fig. 8.48). Starting from the Dupuit assumptions and the Ghyben–Herzberg relation, an approximate freshwater boundary can be determined. With reference to Fig. 8.48, and assuming a circular island of radius, R, receiving an effective rainfall recharge rate, W, the outward flow, Q, at radius, r, is found from:

$$Q = 2\pi r K(z + h)\frac{dh}{dr} \qquad \text{(eq. 8.27)}$$

where K is the hydraulic conductivity and h and z are as defined in Fig. 8.48.

Noting that $h = (\Delta\rho/\rho)z$ and that from a consideration of continuity, $Q = \pi r^2 W$, then

$$z\,dz = \frac{Wr\,dr}{2K\left[1 + \frac{\Delta\rho}{\rho}\right]\left[\frac{\Delta\rho}{\rho}\right]} \qquad \text{(eq. 8.28)}$$

Integrating and applying the boundary condition that $h = 0$ when $r = R$ gives

$$z^2 = \frac{W(R^2 - r^2)}{2K\left[1 + \frac{\Delta\rho}{\rho}\right]\left[\frac{\Delta\rho}{\rho}\right]} \qquad \text{(eq. 8.29)}$$

Thus, the depth to the saline interface at any location is a function of the rainfall recharge,

the size of the island and the aquifer hydraulic conductivity.

To avoid the risk of entrainment of saline water, small island wells should be designed for minimal disturbance of the fresh–saltwater equilibrium due to the close proximity of the transition zone to the water table. To avoid the upconing of saline water into a well, wells should just skim freshwater from the top of the lens. In areas where the water table is shallow, an infiltration gallery, consisting of a horizontal collecting tunnel at the water table, is desirable. Examples of the occurrence and development of freshwater on oceanic island settings in the Pacific are discussed by White and Falkland (2010), in Barbados and Honolulu by Todd and Mays (2005) and in Malta by Stuart *et al.* (2010) and Heaton *et al.* (2012).

The Republic of Malta mostly comprises two islands, Malta (area, 246 km²) and Gozo (67 km²), and is densely populated with high demand for water for domestic, tourist, industrial and farming activities. The demand is met predominantly by groundwater abstraction (56%) and seawater desalinization (34%), together with limited use of harvested rainfall runoff and treated sewage effluent (Sapiano *et al.* 2006). A schematic hydrogeological cross-section is shown in Fig. 8.49. The Tertiary

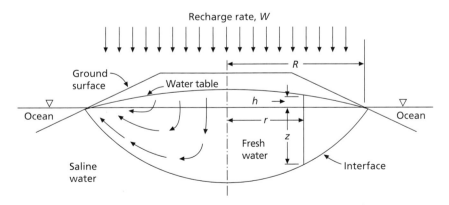

Fig. 8.48 Freshwater lens developed in a small oceanic island under natural conditions (Todd and Mays 2005). (*Source:* Adapted from Todd, D.K. and Mays, L.W. (2005) *Groundwater Hydrology* (3rd edn.) Wiley, Chichester.)

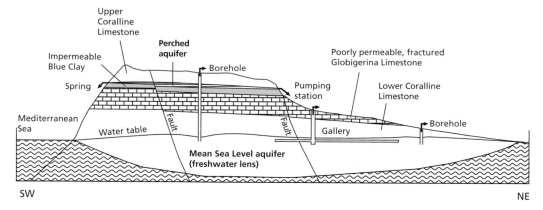

Fig. 8.49 Schematic hydrogeological cross-section of the perched and Mean Sea Level (MSL) aquifers of Malta. Abstraction from pumping station galleries and boreholes controls the water level of the MSL aquifer and also leads to saline water upconing (Stuart *et al.* 2010). (*Source:* Stuart, M.E., Maurice, L., Heaton, T.H.E. *et al.* (2010) Groundwater residence time and movement in the Maltese islands – a geochemical approach. *Applied Geochemistry* **25**, 609–620.)

geology essentially comprises two limestone sequences separated by an impermeable clay-marl known as the Blue Clay. Above the Blue Clay, the top of the succession is formed by Upper Coralline Limestone in which perched groundwater conditions are developed. Faulting and erosion of the limestone means that groundwater is present as several hydrologically separate perched aquifers that provide water for agricultural use from shallow boreholes or natural springs (Heaton *et al.* 2012). Stratigraphically below the Blue Clay, Globigerina and Lower Coralline limestones comprise the major rock formations and support a freshwater lens floating on seawater as determined by the Ghyben–Herzberg relation. The freshwater is abstracted from boreholes and long galleries that take water from the top of the lens and maintain its water table at just a few metres above mean sea level (Heaton *et al.* 2012). The 'Mean Sea Level' aquifers of Malta and Gozo supply over 80% of Malta's groundwater for domestic and agricultural use. Concentrations of Na^+ and Cl^- are substantially higher than in the perched aquifers at approximately 420 and 900 mg Cl L^{-1} in Malta and Gozo, respectively, reflecting intrusion of seawater

into the fresh water lens, largely in response to intensive abstraction (Sapiano *et al.* 2006).

In small oceanic islands, the nature of fresh groundwater lenses and their host aquifers coupled with anthropogenic pressures from surface contamination and changes in rainfall recharge and sea level rise under climate change make these precious groundwater resources some of the most vulnerable aquifer systems in the world. Holding *et al.* (2016) applied climate change projections and found that of 43 small island developing states distributed worldwide, 44% of islands are in a state of water stress. While recharge is projected to increase by as much as 117% on 12 islands situated in the western Pacific and Indian Oceans, recharge is projected to decrease by up to 58% on the remaining 31 islands. Of great concern is the lack of enacted groundwater protection legislation for many small island states identified as highly vulnerable to current and future conditions (Holding *et al.* 2016). Recognizing this challenge, White and Falkland (2010) argued that the vulnerability of small island freshwater lenses requires careful assessment, vigilant monitoring, appropriate development and expert management to protect the future groundwater resources of small islands.

Further reading

Appelo, C.A.J. and Postma, D. (2005) *Geochemistry, Groundwater and Pollution* (2nd edn.). A.A. Balkema, Leiden, The Netherlands.

Bitton, G. and Gerba, C.P. (eds) (1984) *Groundwater Pollution Microbiology*. Wiley, New York.

Fetter, C.W., Boving, T. and Kreamer, D. (2018) *Contaminant Hydrogeology* (3rd edn.). Waveland Press, Long Grove, Illinois.

Hemond, H.F. and Fechner, E.J. (1994) *Chemical Fate and Transport in the Environment*. Academic Press, Inc., San Diego, California.

Jiao, J. and Post, V. (2019) *Coastal Hydrogeology*. Cambridge University Press, Cambridge.

Pankow, J.F. and Cherry, J.A. (eds) (1996) *Dense Chlorinated Solvents and Other DNAPLs in Groundwater: History, Behaviour, and Remediation*. Waterloo Press, Portland, Oregon.

Williams, P.T. (1998) *Waste Treatment and Disposal*. Wiley, Chichester.

Younger, P.L., Banwart, S.A. and Hedin, R.S. (2002) *Mine Water: Hydrology, Pollution, Remediation*. Kluwer Academic Publishers, Dordrecht.

References

Adak, G.K., Long, S.M. and O'Brien, S.J. (2002) Trends in indigenous foodborne disease and deaths, England and Wales: 1992–2000. *Gut* **51**, 832–841.

Allen-King, R.M., Grathwohl, P. and Ball, W.P. (2002) New modeling paradigms for the sorption of hydrophobic organic chemicals to heterogeneous carbonaceous matter in soils, sediments, and rocks. *Advances in Water Resources* **25**, 985–1016.

Aravena, R. and Robertson, W.D. (1998) Use of multiple isotope tracers to evaluate denitrification in ground water: study of nitrate from a large-flux septic system plume. *Ground Water* **36**, 975–982.

Ascott, M.J., Gooddy, D.C., Wang, L. *et al.* (2017) Global patterns of nitrate storage in the vadose zone. *Nature Communications* **8**:1416. DOI: 10.1038/s41467-017-01321-w.

Baker, J.R., Mihelcic, J.R., Luehrs, D.C. and Hickey, J.P. (1997) Evaluation of estimation methods for organic carbon normalized sorption coefficients. *Water Environment Research* **69**, 136–145.

Banwart, S.A., Evans, K.A. and Croxford, S. (2002) Predicting mineral weathering rates at field scale for mine water risk assessment. In: *Mine Water Hydrogeology and Geochemistry* (eds P.L. Younger and N.S. Robins). Geological Society, London, Special Publications **198**, pp. 137–157.

Barrett, M.H., Hiscock, K.M., Pedley, S. *et al.* (1999) Marker species for identifying urban groundwater recharge sources: a review and case study in Nottingham UK. *Water Research* **33**, 3083–3097.

Befus, K.M., Barnard, P.L., Hoover, D.J. *et al.* (2020) Increasing threat of coastal groundwater hazards from sea-level rise in California. *Nature Climate Change* **10**, 946–952. DOI: 10.1038/s41558-020-0874-1.

Bergmire-Sweat, D., Wilson, K., Marengo, L. *et al.* (1999). Cryptosporidiosis in Brush Creek: describing the epidemiology and causes of a large outbreak in Texas, 1998. In: *Proceedings of the International Conference on Emerging Infectious Diseases*, Milwaukee, Wisconsin. American Water Works Association, Denver, Colorado.

Besien, T.J., Williams, R.J. and Johnson, A.C. (2000) The transport and behaviour of isoproturon in unsaturated chalk cores. *Journal of Contaminant Hydrology* **43**, 91–110.

Bouwer, H. (1991) Simple derivation of the retardation equation and application to preferential flow and macrodispersion. *Ground Water* **29**, 41–46.

Carretero, S.C., Rodrigues Capítulo, L. and Kruse, E.E. (2019) Decision tree as a tool for the management of coastal aquifers of limited saturated thickness. *Quarterly Journal of Engineering Geology and Hydrogeology* **53**, 189–200.

Christensen, T.H., Kjeldsen, P., Bjerg, P.L. *et al.* (2001) Biogeochemistry of landfill leachate plumes. *Applied Geochemistry* **16**, 659–718.

Chu, X., Li, T., Li, Z. *et al.* (2019) Transport of microplastic particles in saturated porous media. *Water* **11**, 2474. DOI: 10.3390/w11122474.

CL:AIRE (2002). *Introduction to an Integrated Approach to the Investigation of Fractured Rock Aquifers Contaminated with Non-Aqueous Phase Liquids. Technical Bulletin* **TB1**, Contaminated Land: Applications in Real Environments, London.

Clark, L. and Gomme, J. (1992) Pesticides in a Chalk catchment in Eastern England. *Hydrogéologie* **4**, 169–174.

Council of the European Communities (1980) *Directive on the Protection of Groundwater Against Pollution Caused by Certain Dangerous Substances (80/68/EEC).* Official Journal of the European Communities, Brussels, L20.

Council of the European Communities (1991) *Directive Concerning the Protection of Waters Against Pollution Caused by Nitrates from Agricultural Sources (91/676/EEC).* Official Journal of the European Communities, Brussels, L375.

Cozzarelli, I.M., Böhlke, J.K., Masoner, J. *et al.* (2011) Biogeochemical evolution of a landfill leachate plume, Norman, Oklahoma. *Ground Water* **49**, 663–687.

Craun, G.F. (1985) A summary of waterborne illness transmitted through contaminated groundwater. *Journal of Environmental Health* **48**, 122–127.

Craun, G.F., Greathouse, D.G. and Gunderson, D. H. (1981) Methemoglobin levels in young children consuming high nitrate well water in the United States. *International Journal of Epidemiology* **10**, 309–317.

Cronin, A.A., Taylor, R.G., Powell, K.L. *et al.* (2003) Temporal trends in the depth-specific hydrochemistry and sewage-related microbiology of an urban sandstone aquifer, Nottingham, United Kingdom. *Hydrogeology Journal* **11**, 205–216.

Custodio, E. (ed.) (1987) *Groundwater Problems in Coastal Areas: A Contribution to the International Hydrological Programme.* Studies and Reports in Hydrology **45**. UNESCO, Paris.

De Breuck, W. (ed.) (1991) *Hydrogeology of Salt Water Intrusion.* International Contributions to Hydrogeology **11**. Verlag Heinz Heise, Hannover.

Dent, B.B. (2002). The hydrogeological context of cemetery operations and planning in Australia. PhD Thesis, University of Technology, Sydney.

Department of the Environment (1995) *Landfill Design, Construction and Operational Practice. Waste Management Paper 26B.* HMSO, London.

Devitt, D.A., Evans, R.B., Jury, W.A. *et al.* (1987) *Soil Gas Sensing for Detection and Mapping of Volatile Organics.* National Water Well Association, Worthington, Ohio.

Dissanayake, C.B. (1991) The fluoride problem in the groundwater of Sri Lanka – environmental management and health. *The International Journal of Environmental Studies* **38**, 137–156.

Dissanayake, C.B., Senaratne, A. and Weerasooriya, V.R. (1982) Geochemistry of well water and cardiovascular diseases in Sri Lanka. *The International Journal of Environmental Studies* **19**, 195–203.

Domenico, P.A. and Schwartz, F.W. (1998) *Physical and Chemical Hydrogeology* (2nd edn). Wiley, New York.

Eriksen, M., Lebreton, L.C.M., Carson, H.S. *et al.* (2014) Plastic pollution in the world's oceans: more than 5 trillion plastic pieces weighing over 250,000 tons afloat at sea. *PLoS One* **9**, e111913. DOI: 10.1371/journal.pone.0111913.

Erisman, J.W., Domburg, N., de Vries, W. et al. (2005) The Dutch N-cascade in the European perspective. *Science in China. Series C, Life Sciences* **48**, 827–842.

European Environment Agency (2003). *Europe's Water: An Indicator-Based Assessment. Summary.* European Environment Agency, Copenhagen.

Falkland, A. ed. (1991) *Hydrology and water resources of small islands: a practical guide.* UNESCO, Paris, 435 pp.

Fetter, C.W. (1999) *Contaminant Hydrogeology* (2nd edn). Prentice-Hall, Inc., Upper Saddle River, New Jersey.

Ford, M. and Tellam, J.H. (1994) Source, type and extent of inorganic contamination within the Birmingham urban aquifer system, UK. *Journal of Hydrology* **156**, 101–135.

Francis, C.A., Beman, J.M., and Kuypers, M.M.M. (2007) New processes and players in the nitrogen cycle: the microbial ecology of anaerobic and archaeal ammonia oxidation. *The ISME Journal* **1**, 19–27.

Freeze, R.A. and Cherry, J.A. (1979) *Groundwater.* Prentice-Hall, Inc., Englewood Cliffs, New Jersey.

Fuge, R., Laidlaw, I.M.S., Perkins, W.T. and Rogers, K.P. (1991) The influence of acid mine and spoil drainage on water quality in the mid-Wales area. *Environmental Geochemistry and Health* **13**, 70–75.

Fukada, T., Hiscock, K.M. and Dennis, P.F. (2004) A dual-isotope approach to the nitrogen hydrochemistry of an urban aquifer. *Applied Geochemistry* **19**, 709–719.

Galloway, J.N., Dentener, F.J., Capone, D.G. *et al.* (2004) Nitrogen cycles: past, present and future. *Biogeochemistry* **70**, 153–226.

Gandy, C.J. and Younger, P.L. (2003) Effect of a clay cap on oxidation of pyrite within mine spoil. *Quarterly Journal of Engineering Geology and Hydrogeology* **36**, 207–215.

Gerstl, Z. (1990) Estimation of organic chemical sorption by soils. *Journal of Contaminant Hydrology* **6**, 357–375.

Girvin, D.C. and Scott, A.J. (1997) Polychlorinated biphenyl sorption by soils: measurement of soil-water partition coefficients at equilibrium. *Chemosphere* **35**, 2007–2025.

Gomme, J., Shurvell, S., Hennings, S.M. and Clark, L. (1992) Hydrology of pesticides in a Chalk catchment: groundwaters. *Journal of the Institution of Water and Environmental Management* **6**, 172–178.

Goss, M.J., Barry, D.A.J. and Rudolph, D.L. (1998) Contamination in Ontario farmstead wells and its association with agriculture. I. Results from drinking water wells. *Journal of Contaminant Hydrology* **32**, 63–90.

Gosselin, D.C., Headrick, J., Tremblay, R. *et al.* (1997) Domestic well water quality in rural Nebraska: focus on nitrate-nitrogen, pesticides, and coliform bacteria. *Ground Water Monitoring and Remediation* **17**, 77–87.

Green, A.R., Feast, N.A., Hiscock, K.M. and Dennis, P.F. (1998) Identification of the source and fate of nitrate contamination of the Jersey bedrock aquifer using stable nitrogen isotopes. In: *Groundwater Pollution, Aquifer Recharge and Vulnerability* (ed. N.S. Robins). Geological Society, London, Special Publications **130**, pp. 23–35.

Hassett, J.J., Banwart, W.L. and Griffen, R.A. (1983) Correlation of compound properties with sorption characteristics of nonpolar compounds by soil and sediments: concepts and limitations. In: *Environmental and Solid Wastes Characterization, Treatment and Disposal* (eds C.W. Francis and S.I. Auerbach). Butterworth Publishers, Newton, Massachusetts.

Headworth, H.G. and Fox, G.B. (1986) The South Downs Chalk aquifer: its development and management. *Journal of the Institution of Water Engineers* **40**, 345–361.

Heaton, T.H.E., Stuart, M.E., Sapiano, M. and Sultana, M.M. (2012) An isotope study of the sources of nitrate in Malta's groundwater. *Journal of Hydrology* **414–415**, 244–254.

Heberer, T. (2002) Tracking persistent pharmaceutical residues from municipal sewage to drinking water. *Journal of Hydrology* **266**, 175–189.

Hecky, R.E. and Kilham, P. (1988) Nutrient limitation of phytoplankton in freshwater and marine environments: a review of recent evidence on the effects of enrichment. *Limnology and Oceanography* **33**, 796–822.

Hem, J.D. (1985). Study and interpretation of the chemical characteristics of natural water

(3rd edn). United States Geological Survey Water Supply Paper 2254, 263 pp.

Hess, K.M., Davis, J.A., Kent, D.B. and Coston, J. A. (2002) Multispecies reactive tracer test in an aquifer with spatially variable chemical conditions, Cape Cod, Massachusetts: dispersive transport of bromide and nickel. *Water Resources Research* **38**, 36-1–36-17. doi:10.1029/2001WR000945.

Hiscock, K.M., Lloyd, J.W. and Lerner, D.N. (1991) Review of natural and artificial denitrification of groundwater. *Water Research* **25**, 1099–1111.

Holding, S., Allen, D.M., Foster, S. *et al.* (2016) Groundwater vulnerability on small islands. *Nature Climate Change* **6**, 1100–1103. DOI: 10.1038/NCLIMATE3128.

Holme, R. (2003) Drinking water contamination in Walkerton, Ontario: positive resolutions from a tragic event. *Water Science and Technology* **47**, 1–6.

Howard, K.W.F. and Beck, P.J. (1993) Hydrogeochemical implications of groundwater contamination by road deicing chemicals. *Journal of Contaminant Hydrology* **12**, 245–268.

Howard, K.W.F. and Lloyd, J.W. (1983) Major ion characterization of coastal saline ground waters. *Ground Water* **21**, 429–437.

Hurley, R., Woodward, J. and Rothwell, J.J. (2018) Microplastic contamination of river beds significantly reduced by catchment-wide flooding. *Nature Geoscience* **11**, 251–257. DOI: 10.1038/s41561-018-0080-1.

Iribar, V., Carrera, J., Custodio, E. and Medina, A. (1997) Inverse modelling of seawater intrusion in the Llobregat delta deep aquifer. *Journal of Hydrology* **198**, 226–244.

Jackson, R.E. (ed.) (1980) *Aquifer Contamination and Protection*. UNESCO, Paris.

Jones, K. and Moon, G. (1987) *Health, Disease and Society: A Critical Medical Geography.* Routledge and Kegan Paul, London, pp. 134–140.

Karickhoff, S.W., Brown, D.S. and Scott, T.A. (1979) Sorption of hydrophobic pollutants on natural sediments. *Water Research* **13**, 241–248.

Kenaga, E.E. and Goring, C.A.I. (1980) Relationship between water solubility, soil sorption, octanol-water partitioning, and concentration of chemicals in biota. In: *Aquatic Toxicology ASTM Special Technical Publication* **707** (eds J.G. Eaton, P.R. Parrish and A.C. Hendricks). American Society for Testing and Materials, Philadelphia, Pennsylvania, pp. 78–115.

Klein, H. and Hull, J.E. (1978). Biscayne Aquifer, Southeast Florida. United States Geological Survey Water-Resources Investigations 78–107, 52 pp.

Knight, M.J. and Dent, B.B. (1998). Sustainability of waste and groundwater management systems. In: *Proceedings of the Congress of the International Association of Hydrogeologists on Groundwater: Sustainable Solutions,* Melbourne, Australia, 8–13 February. International Association of Hydrogeologists, Melbourne, 359–374.

Kobayashi, J. (1957) On geographical relations between the chemical nature of river water and death rate from apoplexy. *Berichte des Ohara Instituts für Landwirtschaftliche Biologie, Okayama University* **11**, 12–21.

Koelmans, A.A., Nor, N.H.M., Hermsen, E. *et al.* (2019) Microplastics in freshwaters and drinking water: Critical review and assessment of data quality. *Water Research* **155**, 410–422.

Kolpin, D.W., Barbash, J.E. and Gilliom, R.J. (2000) Pesticides in ground water of the United States, 1992–1996. *Ground Water* **38**, 858–863.

Kolpin, D.W., Kalkhoff, S.J., Goolsby, D.A. *et al.* (1997) Occurrence of selected herbicides and herbicide degradation products in Iowa's ground water, 1995. *Ground Water* **35**, 679–688.

Ku, H.F.H. (1980) Ground-water contamination by metal-plating wastes, Long Island, New York, USA. In: *Aquifer Contamination and Protection* (ed. R.E. Jackson). UNESCO, Paris, pp. 310–317.

Lerner, D.N. (2002) Identifying and quantifying urban recharge: a review. *Hydrogeology Journal* **10**, 143–152.

Lin, B.-L., Sakoda, A., Shibasaki, R. *et al.* (2000) Modelling a biogeochemical nitrogen cycle in terrestrial ecosystems. *Ecological Modelling* **135**, 89–110.

Lloyd, J.W., Williams, G., Foster, S.S.D. *et al.* (1991) Urban and industrial groundwater pollution. In: *Applied Groundwater Hydrology: A British Perspective* (eds R.A. Downing and W. B. Wilkinson). Clarendon Press, Oxford, pp. 134–148.

Mackay, D., Shiu, W.-Y. and Ma, K.-C. (1997). *Illustrated Handbook of Physical-Chemical Properties and Environmental Fate for Organic Chemicals. Volume I Monoaromatic Hydrocarbons, Chlorobenzenes, and PCBs; Volume II Polynuclear Aromatic Hydrocarbons, Polychlorinated Dioxins, Dibenzofurans; Volume III Volatile Organic Chemicals; Volume IV Oxygen, Nitrogen, and Sulfur Containing Compounds; Volume V Pesticide Chemicals.* Lewis Publishers, Boca Raton, Florida.

Mackay, D.M., Freyberg, D.L. and Roberts, P.V. (1986) A natural gradient experiment on solute transport in a sand aquifer 1. Approach and overview of plume movement. *Water Resources Research* **22**, 2017–2029.

Mallin, M.A. and Cahoon, L.B. (2003) Industrialized animal production – a major source of nutrient and microbial pollution to aquatic ecosystems. *Population and Environment* **24**, 369–385.

Michael, H.A., Russoniello, C.J. and Byron, L.A. (2013) Global assessment of vulnerability to sea-level rise in topography-limited and recharge-limited coastal groundwater systems. *Water Resources Research* **49**, 2228–2240.

Miles, R. (1993) Maintaining groundwater supplies during drought conditions in the Brighton area. *Journal of the Institution of Water and Environmental Management* **7**, 382–386.

Mintenig, S.M., Löder, M.G.J., Primpke, S. and Gerdts, G. (2019) Low numbers of microplastics detected in drinking water from ground water sources. *Science of the Total Environment* **648**, 632–635.

Morris, B.L. and Foster, S.S.D. (2000) *Cryptosporidium* contamination hazard assessment and risk management for British groundwater sources. *Water Science and Technology* **41**, 67–77.

National Academy of Sciences (1981) *The Health Effects of Nitrate, Nitrite and N-Nitroso Compounds.* Part 1 of a two-part study by the Committee on Nitrite and Alternative Curing Agents in Food. National Academy Press, Washington, DC.

National Academy of Sciences (1994) *Alternatives for Ground Water Cleanup.* Report of the National Academy of Sciences Committee on Ground Water Cleanup Alternatives. National Academy Press, Washington, DC.

NEGTAP (2001) *Transboundary Air Pollution: Acidification, Eutrophication and Ground-Level Ozone in the UK.* Report prepared by the National Expert Group on Transboundary Air Pollution. Centre for Ecology and Hydrology, Edinburgh.

Nicholson, R.V., Cherry, J.A. and Reardon, E.J. (1983) Migration of contaminants in groundwater at a landfill: a case study. 6. Hydrogeochemistry. *Journal of Hydrology* **63**, 131–176.

Nomura, A. (1996) Stomach Cancer. In: *Cancer Epidemiology and Prevention* (2nd edn, eds D. Schottenfeld and J.F. Fraumeni). Oxford University Press, New York, pp. 707–724.

Nordstrom, D.K. and Southam, G. (1997) Geomicrobiology of sulfide mineral oxidation. Geomicrobiology: interactions between microbes and minerals. *Reviews in Mineralogy* **35**, 361–390.

Ogata, A. and Banks, R.B. (1961). A solution of the differential equation of longitudinal dispersion in porous media. United States Geological Survey Professional Paper 411-A, 7 pp.

Oude Essink, G.H.P., van Baaren, E.S. and de Louw, P.G.B. (2010) Effects of climate change on coastal groundwater systems: a modelling study in the Netherlands. *Water Resources Research* **46**, W00F04. DOI:10.1029/2009WR008719.

Panno, S.V., Kelly, W.R., Scott, J. *et al.* (2019) Microplastic contamination in karst groundwater systems. *Groundwater* **57**, 189–196.

Pedley, S. and Howard, G. (1997) The public health implications of microbiological contamination of groundwater. *Quarterly Journal of Engineering Geology and Hydrogeology* **30**, 179–188.

Pocock, S.J., Shaper, A.G., Cook, D.G. *et al.* (1980) British Regional Heart Study: geographic variations in cardiovascular mortality, and role of water quality. *British Medical Journal* **280**, 1243–1249.

Powell, K.L., Taylor, R.G., Cronin, A.A. *et al.* (2003) Microbial contamination of urban sandstone aquifers in the UK. *Water Research* **37**, 339–352.

Price, M. (2004). Dr John Snow and an early investigation of groundwater contamination. In: *200 Years of British Hydrogeology* (ed. J.D. Mather). Geological Society, London, Special Publications **225**, pp. 31–49.

Price, M., Atkinson, T.C., Wheeler, D. *et al.* (1989). Highway drainage to the Chalk aquifer: the movement of groundwater in the Chalk near Bricket Wood, Hertfordshire, and its possible pollution by drainage from the M25. British Geological Survey Technical Report WD/89/3. British Geological Survey, Keyworth, Nottingham, 64 pp.

Rajasooriyar, L.D. (2003). A study of the hydrochemistry of the Uda Walawe Basin, Sri Lanka, and the factors that influence groundwater quality. PhD Thesis, University of East Anglia, Norwich, 175 pp.

Re, V. (2019) Shedding light on the invisible: addressing the potential for groundwater contamination by plastic microfibers. *Hydrogeology Journal* **27**, 2719–2727.

Rivett, M.O. and Allen-King, R.M. (2003) A controlled field experiment on groundwater contamination by a multicomponent DNAPL: dissolved-plume retardation. *Journal of Contaminant Hydrology* **66**, 117–146.

Rivett, M.O., Feenstra, S. and Cherry, J.A. (2001) A controlled field experiment on groundwater contamination by multicomponent DNAPL: creation of the emplaced-source and overview of dissolved plume development. *Journal of Contaminant Hydrology* **49**, 111–149.

Rivett, M.O., Lerner, D.N., Lloyd, J.W. and Clark, L. (1990) Organic contamination of the Birmingham aquifer, U.K. *Journal of Hydrology* **113**, 307–323.

Rivett, M.O., Shepherd, K.A., Keeys, L.L. and Brennan, A.E. (2005) Chlorinated solvents in the Birmingham aquifer, UK: 1986–2001. *Quarterly Journal of Engineering Geology and Hydrogeology* **38**, 337–350.

Rivett, M.O., Turner, R.J., Glibbery, P. and Cuthbert, M.O. (2012) The legacy of chlorinated solvents in the Birmingham aquifer, UK: Observations spanning three decades and the challenge of future urban groundwater development. *Journal of Contaminant Hydrology* **140–141**, 107–123.

Roberts, P.V., Goltz, M.N. and Mackay, D.M. (1986) A natural gradient experiment on solute transport in a sand aquifer 3. Retardation estimates and mass balances for organic solutes. *Water Resources Research* **22**, 2047–2058.

Robertson, W.D., Cherry, J.A. and Sudicky, E.A. (1991) Ground-water contamination from two small septic systems on sand aquifers. *Ground Water* **29**, 82–92.

Robertson, W.D., Moore, T.A., Spoelstra, J. *et al.* (2012) Natural attenuation of septic system nitrogen by anammox. *Ground Water* **50**, 541–553.

Robinson, H.D. (1989). Development of methanogenic conditions within landfills. In: *Proceedings of the 2nd International Landfill Symposium*, Cagliari, Sardinia, 9 pp.

Rochman, C.M., Browne, M.A., Halpern, B.S. *et al.* (2013) Classify plastic waste as hazardous. *Nature* **494**, 169–171.

Rodvang, S.J. and Simpkins, W.W. (2001) Agricultural contaminants in Quaternary aquitards: a review of occurrence and fate in North America. *Hydrogeology Journal* **9**, 44–59.

Romanazzi, A., Gentile, F. and Polemio, M. (2015) Modelling and management of a Mediterranean karstic coastal aquifer under the effects of seawater intrusion and climate change. *Environmental Earth Sciences* **74**, 115–128.

Rotzoll, K. and Fletcher, C.H. (2013) Assessment of groundwater inundation as a consequence of sea-level rise. *Nature Climate Change* **3**, 477–481. DOI: 10.1038/NCLIMATE1725.

Rubenowitz, E., Molin, I., Axelsson, G. and Rylander, R. (2000) Magnesium in drinking water in relation to morbidity and mortality from acute myocardial infarction. *Epidemiology* **11**, 416–421.

Rubenowitz-Lundin, E. and Hiscock, K. (2005) Water hardness and health effects. In: *Essentials of Medical Geology* (eds O. Selinus *et al.*). Elsevier Academic Press, San Diego, pp. 331–345.

Sapiano, M., Mangion, J. and Batchelor, C. (2006) *Malta Water Resources Review*. Food and Agriculture Organization of the United Nations, Rome.

Sbarbati, C., Colombani, N., Mastrocicco, M. *et al.* (2018) Reactive and mixing processes governing ammonium and nitrate coexistence in a polluted coastal aquifer. *Geosciences* **8**, 210. DOI: 10.3390/geosciences8060210.

Schwarzenbach, R.P. and Westall, J. (1981) Transport of nonpolar organic compounds from surface water to groundwater. Laboratory sorption studies. *Environmental Science & Technology* **15**, 1360–1367.

Schwille, F. (1988) *Dense Chlorinated Solvents in Porous and Fractured Media: Model Experiments*. Lewis Publishers Inc, Chelsea, Michigan.

Stuart, M.E., Maurice, L., Heaton, T.H.E. *et al.* (2010) Groundwater residence time and movement in the Maltese islands – a geochemical approach. *Applied Geochemistry* **25**, 609–620.

Sudicky, E.A. (1986) A natural gradient experiment on solute transport in a sand aquifer: spatial variability of hydraulic conductivity and its role in the dispersion process. *Water Resources Research* **22**, 2069–2082.

Sutton, M.A. and van Grinsven, H. (2011) Summary for policy makers. In: *The European Nitrogen Assessment: Sources, Effects and Policy Perspectives* (eds M.A. Sutton, C.M. Howard, J. W. Erisman *et al.*). Cambridge University Press, Cambridge, pp. xxiv–xxxiv.

Thomas, M.R., Garthwaite, D.G. and Banham, A.R. (1997) *Pesticide Usage Survey Report 141: Arable Farm Crops in Great Britain 1996*. Ministry of Agriculture, Fisheries & Food and Scottish Office Agriculture, Environment & Fisheries Department, MAFF Publications, London.

Todd, D.K. (1980) *Groundwater Hydrology* (2nd edn). Wiley, New York.

Todd, D.K. and Mays, L.W. (2005) *Groundwater Hydrology* (3rd edn.) Wiley, Chichester.

UNESCO/FAO (1973) *Irrigation, Drainage and Salinity: An International Source Book*. Hutchinson & Co. (Publishers) Ltd., London.

USEPA (2019). *Advancing Sustainable Materials Management: 2017 Fact Sheet*. United States Environmental Protection Agency. https://www.epa.gov/sites/production/files/2019-11/documents/2017_facts_and_figures_fact_sheet_final.pdf (accessed 23 May, 2020).

Van Drecht, G., Bouwman, A.F., Knoop, J.M., Beusen, A.H.W. and Meinardi, C.R. (2003) Global modeling of the fate of nitrogen from point and nonpoint sources in soils, groundwater, and surface water. *Global Biogeochemical Cycles* **17**, 26–31.

Verschueren, K. (1983) *Handbook of Environmental Data on Organic Chemicals* (2nd edn). Van Nostrand Reinhold Co., New York.

Walton, G. (1951) Survey of literature relating to infant methemoglobinemia due to nitrate-contaminated water. *American Journal of Public Health* **41**, 986–996.

Wang, L., Stuart, M.E., Bloomfield, J.P. *et al.* (2012) Prediction of the arrival of peak nitrate concentrations at the water table at the regional scale in Great Britain. *Hydrological Processes* **26**, 226–239.

Werner, A.D. and Simmons, C.T. (2009) Impact of sea-level rise on sea water intrusion in coastal aquifers. *Ground Water* **47**, 197–204.

White, I. and Falkland, T. (2010) Management of freshwater lenses on small Pacific islands. *Hydrogeology Journal* **18**, 227–246.

Wilhelm, S.R., Schiff, S.L. and Cherry, J.A. (1994) Biogeochemical evolution of domestic waste water in septic systems: 1. Conceptual model. *Ground Water* **32**, 905–916.

Williams, G.M., Young, C.P. and Robinson, H.D. (1991) Landfill disposal of wastes. In: *Applied Groundwater Hydrology* (eds R.A. Downing and W.B. Wilkinson). Clarendon Press, Oxford, pp. 114–133.

Willocks, L., Crampin, A., Milne, L. *et al.* (1999) A large outbreak of cryptosporidiosis associated with a public water supply from a deep chalk borehole. *Communicable Disease and Public Health* **1**, 239–243.

Wood, S.C., Younger, P.L. and Robins, N.S. (1999) Long-term changes in the quality of polluted minewater discharges from abandoned underground coal workings in Scotland. *Quarterly Journal of Engineering Geology and Hydrogeology* **32**, 69–79.

Worch, E., Grischek, T., Börnick, H. and Eppinger, P. (2002) Laboratory tests for simulating attenuation processes of aromatic amines in riverbank filtration. *Journal of Hydrology* **266**, 259–268.

World Health Organization (2011). *World Health Organization Guidelines for Drinking-water Quality (4th edn)*. World Health Organization, Geneva, Switzerland, 541 pp.

World Health Organization (2020). Diarrhoea in the Western Pacific. https://www.who.int/westernpacific/health-topics/diarrhoea (accessed 28 June, 2020).

Xu-Ri and Prentice, I.C. (2008) Terrestrial nitrogen cycle simulation with a dynamic global vegetation model. *Global Change Biology* **14**, 1745–1764.

Yong, R.N. and Phadungchewit, Y. (1993) pH influence on selectivity and retention of heavy-metals in some clay soils. *Canadian Geotechnical Journal* **30**, 821–833.

Younger, P.L. (1995) Hydrogeochemistry of minewaters flowing from abandoned coal workings in County Durham. *Quarterly Journal of Engineering Geology and Hydrogeology* **28**, S101–S113.

9

Groundwater pollution remediation and protection

9.1 Introduction

Compared with the consequences of groundwater contamination described in Chapter 8, it is without doubt better to prevent pollution from occurring in the first instance in order to avoid expensive aquifer remediation costs, damage to the environment and the necessity of finding alternative water supplies. In other words, prevention is better than cure. Strategies for preventing groundwater pollution are typically divided between protecting individual groundwater sources and the wider aquifer resource. Differences are also apparent in dealing with point sources of contamination, for example from waste disposal sites and diffuse sources such as agrochemicals. Groundwater contamination is a global problem and so it should be recognized that strategies developed in more technologically advanced countries will need adapting for application in developing countries.

The following sections of this chapter address the above issues and discuss groundwater remediation techniques and risk assessment methods for deciding the location of potentially polluting activities and the selection of remedial measures. Methods for protecting groundwater sources and aquifer resources (groundwater bodies), including a section on groundwater vulnerability assessment and mapping for the protection of karstic aquifers, are also presented. The last section presents examples of how spatial planning, including fundamental changes in land use, can be used to protect groundwater from long-term contamination from diffuse agricultural contaminants.

9.2 Groundwater pollution remediation techniques

The remediation of groundwater can involve an attempt at the total clean-up of a contaminated aquifer or the containment of a groundwater pollution source. In certain circumstances, for example in areas of low risk of human or environmental exposure to contaminants, a further option may be to leave the aquifer to recover through natural attenuation processes.

The successful remediation of contaminated groundwater must address both the source of pollution and remediation of the contaminant plume. Conventional remediation techniques employ pump-and-treat methods, but these have been shown to be less successful, particularly with respect to the clean-up of pools of trapped organic pollutants such as crude oil and chlorinated solvents that act as long-term sources of groundwater contamination. Newer technologies include soil vapour extraction, air sparging and bioremediation for the enhanced removal of organic pollutants. A useful overview is provided by Fetter (1999). Of increasing interest are passive techniques such as permeable reactive barriers that provide an innovative, cost-effective and low-maintenance solution to the clean-up of contaminated land and groundwater.

Hydrogeology: Principles and Practice, Third Edition. Kevin M. Hiscock and Victor F. Bense.
© 2021 John Wiley & Sons Ltd. Published 2021 by John Wiley & Sons Ltd.
Companion website: www.wiley.com/go/hiscock/hydrogeology3e

The following sections provide an introduction to the pump-and-treat and passive techniques of groundwater remediation and also review the case for monitored natural attenuation. The final choice of remediation technique for a given pollution incident will be decided on the basis of a thorough site investigation giving consideration to the type of pollution source, the hydrogeological characteristics and natural attenuation capacity of the affected aquifer, and a cost–benefit analysis to achieve an acceptable reduction in the environmental risks.

9.2.1 Pump-and-treat

The conventional pump-and-treat method of aquifer clean-up is to extract the contaminated groundwater and, following treatment to remove and possibly recover the contaminant source, for the treated water to be either injected into the aquifer or, if a discharge consent is obtained, released to a surface water course. Once the cause of the groundwater pollution has been eliminated, and depending on the shape, extent and concentration distribution of a contaminant plume, the following design criteria must be considered in order to choose the least expensive pumping arrangement for capturing the plume (Javandel and Tsang 1986):

1) What is the optimum number of pumping wells or boreholes required?
2) Where should the wells or boreholes be sited so that no contaminated water can escape between the pumping wells?
3) What is the optimum pumping rate for each well or borehole?
4) What is the optimum water treatment method?
5) How should the treated water be disposed of?

Depending on available resources, a detailed site investigation, including the installation and testing of monitoring wells to provide information on aquifer properties and contaminant distribution, may result in a numerical groundwater model for the site. The model can then be used with particle tracking methods to simulate capture zones (Section 7.5) for one or more pumping wells that encompass the zone of contamination. At an earlier stage in the investigation, a desk study using the following straightforward method may assist in the initial selection of design criteria. The method is presented by Javandel and Tsang (1986) and is based on the application of complex potential theory to provide an analytical solution to the problem of flow to a fully penetrating well in a homogeneous and isotropic aquifer of uniform thickness. Uniform and steady regional groundwater flow is also assumed. The simplest case is to assume a single pumping well, although the method can be applied to any number of wells, with the solution to a problem obtained using type curves for either single-, double-, three- or four-well capture zones.

In terms of design criteria, the objective of the analytical method is to select the type curve that encompasses the specified concentration contour that delimits the contaminant plume within the capture zone of the well or array of wells. The following procedure, demonstrated for the problem given in Box 9.1, and using the type curves for a single well given in Appendix 10 (Fig. A10.15), explains the method:

1) Prepare a site map using the same scale as the type curves and showing the direction of regional groundwater flow and the contour of the maximum allowable concentration in the aquifer of a given contaminant that defines the contour line of the plume.
2) Superimpose the map on the set of type curves making sure that the direction of regional groundwater flow is aligned with the direction of regional flow shown in the type curves. Now move the contour line of the plume towards the head of the capture zone type curves and read the value of Q/bq from the particular curve which

completely encompasses the contour line of the plume.

3) Calculate the value of Q, the well discharge rate, by multiplying Q/bq obtained in the previous step by bq, the product of the aquifer thickness, b, and the magnitude of regional groundwater flow, or specific discharge, q (eq. 2.9), to provide a value for the pumping rate for a single well.

4) If the well is able to produce the required discharge rate Q, then a solution has been reached and a single well, with the location copied directly from the position of the well on the type curves to the site map at the matching position, is the optimum design.

5) If the single well is unable to produce at the calculated rate, then the above procedure has to be followed using the type curves for two or more wells (see type curves in Javandel and Tsang 1986) with the optimum distance between two wells given by $Q/\pi bq$.

Box 9.1 Pump-and-treat system design using capture zone type curves

In this hypothetical example, a pump-and-treat system is under consideration for the restoration of an aquifer at the site of a former sand quarry used for industrial waste disposal (Fig. 9.1). The site, which is now closed, is unlined and is known to have received drums of the organic solvent TCE for disposal at the site. Some of these drums have ruptured and monitoring wells around the site show a plume of dissolved TCE contaminating the sand aquifer. The aquifer is confined by a clay aquitard, except where this

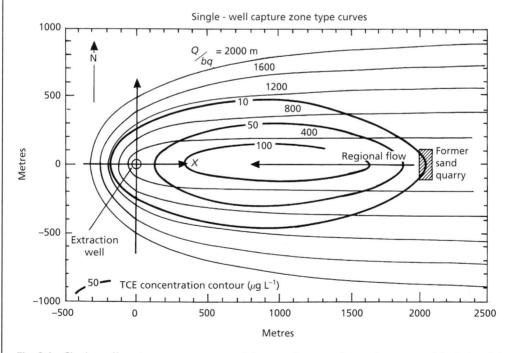

Fig. 9.1 Single-well capture zone curves overlain on a site map of a sand quarry used for industrial waste disposal. The contour line of the 10 μg L^{-1} concentration of TCE defines the limit of a pollution plume emerging from the waste site and is encompassed by the matching position of the capture zone curve with a value of Q/bq = 1200 m.

(Continued)

Box 9.1 (Continued)

has been removed by quarrying, and preliminary site investigation has revealed that the natural hydraulic gradient in the aquifer is 0.001 from east to west. A short constant-rate pumping test using a pair of monitoring wells has yielded a value of aquifer transmissivity, T, of 10^{-3} m^2 s^{-1} which gives a hydraulic conductivity value of 10^{-4} m s^{-1} for a general saturated aquifer thickness, b, of 10 m in the vicinity of the site. The storage coefficient, S, calculated from the pumping test results is 2×10^{-4}.

Given the above information, find the position and required pumping rate, Q, of a single extraction well to capture the TCE plume. The target clean-up standard for TCE dissolved in water is 10 µg L^{-1}.

Following the procedure outlined in Section 9.2.1, by superimposing the single-well type curves given in Appendix 10, Fig. A10.15, and moving the contour line of the plume represented by the 10 µg L^{-1} TCE concentration towards the head of the curves, the chosen type curve that encompasses the contaminant plume has a value of Q/bq = 1200 m. The result of the overlay of the type curves on to the site map is represented in Fig. 9.1. In performing this overlay operation, the x-axis of the type curves should be parallel with the direction of regional groundwater flow shown on the site map.

The specific discharge, q, for the regional flow is calculated from Darcy's Law (eq. 2.9) as follows:

$$q = -K\frac{dh}{dl} = 0.001 \times 10^{-4} = 1 \times 10^{-7}\, \text{m s}^{-1}$$

(eq. 1)

Now, given that Q/bq = 1200 m, then with the calculated value of q and for a given aquifer thickness of 10 m, the required discharge rate for the extraction well is found from

$$Q = 1200 \times bq = 1200 \times 10 \times 1 \times 10^{-7}\, \text{m s}^{-1}$$
$$= 1.2 \times 10^{-3}\, \text{m}^3\, \text{s}^{-1}\, \text{or}\, 1.2\, \text{L s}^{-1}$$

(eq. 2)

The location of the extraction well is identified by copying directly from the position of the well on the type curves to the site map at the matching point, as indicated in Fig. 9.1. To check that this pumping rate produces an acceptable water level drawdown at the well, then with a chosen well radius, r_w, of 0.075 m, the following non-equilibrium radial flow equation (see Section 7.3.2.3) can be applied for large values of time, t, of say 1 year:

$$s = \frac{2.3Q}{4\pi T} \log_{10} \frac{2.25\ Tt}{r_w^2 s}$$

(eq. 3)

which, on substitution of these values gives

$$s = \frac{2.3 \times 1.2 \times 10^{-3}}{4\pi \times 1 \times 10^{-3}}$$

$$\log_{10} \frac{2.25 \times 1 \times 10^{-3} \times 3.1536 \times 10^7}{0.075^2 \times 2 \times 10^{-4}} = 2.37\text{m}$$

(eq. 4)

This is an acceptable drawdown, but for the required discharge rate, the time required to remove the estimated volume of 2.92×10^6 m^3 of contaminated water from within the 10 µg L^{-1} contour line of the plume (for an aquifer porosity of 0.20) is 77 years, assuming that no water with a concentration below 10 µg L^{-1} is extracted by the well. Therefore, the pump-and-treat system will require a long investment of time, on-going maintenance and water treatment costs, and substantial energy inputs to maintain this remediation approach.

These procedures can also be used to explore the position of an injection well for the treated water at the upper end of the plume in order to force the contaminated water towards the extraction well, and so shorten the total clean-up time of the aquifer.

Limitations of this analytical method are the basic assumptions of a homogeneous and isotropic aquifer and a fully penetrating well open over the entire thickness of the aquifer. The method can be applied to unconfined aquifers where the amount of drawdown relative to the total saturated thickness of the aquifer is small, but in heterogeneous aquifers such as fluvial deposits with low permeability clay lenses and high permeability gravel beds, the technique may give erroneous results.

An example of the successful application of the pump-and-treat method for groundwater remediation is described in Box 9.2 for an airport site that experienced a leak of kerosene. This example of organic contamination is common worldwide, but it is now recognized that non-aqueous phase liquids (NAPLs), such as oil products and organic solvents, are not treated satisfactorily by the pump-and-treat approach. In a review of the technology, Mackay and Cherry (1989) considered that the rate of contaminant mass removal by extraction wells is exceedingly slow compared with the often large mass of the contaminant source. In such cases, the pump-and-treat option is best considered as a method of hydraulic manipulation of the aquifer to prevent continuation of contaminant migration. The pump-and-treat method often effectively shrinks the plume towards its source, but for the shrinkage to persist, it is necessary for pumping to continue. However, the long-term cost of such pumping becomes expensive and without sufficient detail concerning the mass of NAPL and its distribution at or below the water table in heterogeneous aquifer material, then it becomes difficult to predict reliably the time necessary for permanent clean-up. Hence, groundwater remediation by pump-and-treat may last for a very long time. It is these shortcomings that have provided a strong incentive for the development of alternative remediation technologies, such as passive treatment using permeable reactive barriers.

Box 9.2 Jet fuel clean-up at Heathrow International Airport

This case study concerns hydrocarbon contamination of groundwater adjacent to Technical Block L at Heathrow Airport. Heathrow Airport is built on the Taplow Terrace adjacent to the River Thames floodplain. The geology is formed by 4.5 m of coarse clean gravels overlying low permeability London Clay. The water table is shallow, about 2.5 m from the surface, with groundwater flow southwards beneath the airport towards the River Thames at Shepperton. A leak of jet fuel (kerosene, a light non-aqueous phase liquid, or LNAPL) occurred from a cracked fuel pipe leading to an engine maintenance facility, the leak having occurred over a number of years. The leak was discovered when fuel was observed floating on drainage water in a manhole north of Technical Block M (Fig. 9.2). In response, a large concrete-lined well (Well 1), about 1.5 m in diameter, was installed close to the manhole and revealed about 10 cm of kerosene floating on the water table. As a first step in remediating the contaminated site, the leak was traced to the cracked pipe and the fracture repaired.

(Continued)

Box 9.2   (Continued)

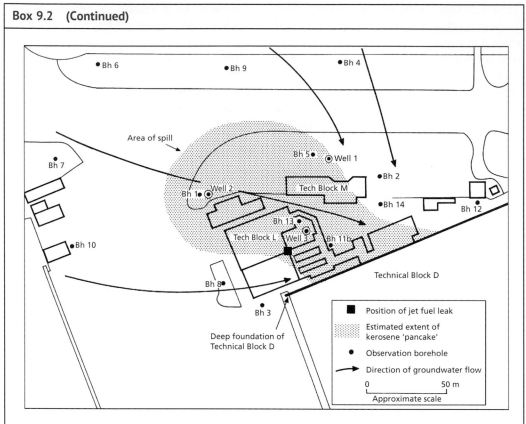

Fig. 9.2   Site of a jet fuel leak adjacent to Technical Block L at Heathrow Airport showing the estimated extent of the kerosene 'pancake' resting on the gravel aquifer water table (Clark and Sims 1998). (*Source:* Clark, L. and Sims, P.A. (1998) Investigation and clean-up of jet-fuel contaminated groundwater at Heathrow International Airport, UK. In *Groundwater Contaminants and Their Migration* (eds J. Mather, D. Banks, S. Dumpleton and M. Fermor). Geological Society, London, Special Publications **128**, pp. 147–157. © 1998, Geological Society of London.)

A detailed site investigation, including the installation of 14 monitoring boreholes, showed that the 'pancake' of floating kerosene was about 100 m in diameter and at its thickest point measured a depth of 0.95 m in borehole 5, with further 'free product' measured in boreholes 1, 11 and 13 and in Well 1 (Fig. 9.2). Odour was reported during the drilling of boreholes 3, 9 and 10, indicative of kerosene. No kerosene was detected in the outlying observation boreholes, including borehole 12.

The basic remediation structures used included large diameter wells lined with perforated concrete rings about 1.5 m in diameter (Fig. 9.3). Wells 1 and 2, installed close to borehole 1 where a considerable thickness of fuel was shown to be floating on the water-table, were used to begin recovery of the floating kerosene. The kerosene was removed by floating oil-skimmer pumps. Surface-mounted centrifugal pumps, installed with their intakes in the two wells, were also used to lower the water-table and encourage the kerosene to move towards the recovery wells. The waste water pumped from the wells was discharged to the drainage system of the airport, which leads to a balancing reservoir before flowing into the River Thames. The balancing reservoir

Box 9.2 **(Continued)**

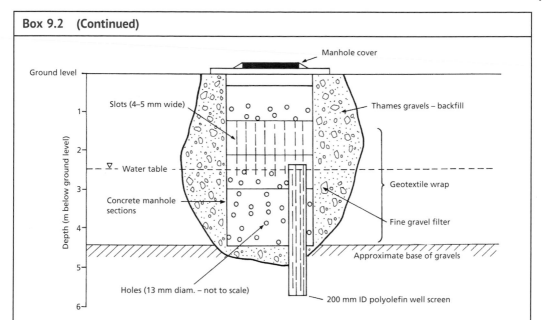

Fig. 9.3 Construction details of recovery Well 2. The well was installed by excavating a pit by back-hoe as deeply as possible, about 2 m below the water table, lowering the perforated concrete rings into position and then backfilling around the rings using the gravel excavated from the pit (Clark and Sims 1998). (*Source:* Clark, L. and Sims, P.A. (1998) Investigation and clean-up of jet-fuel contaminated groundwater at Heathrow International Airport, UK. In *Groundwater Contaminants and Their Migration* (eds J. Mather, D. Banks, S. Dumpleton and M. Fermor). Geological Society, London, Special Publications **128**, pp. 147–157. © 1998, Geological Society of London.)

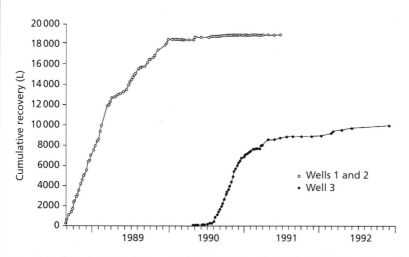

Fig. 9.4 Cumulative recovery of hydrocarbon from Wells 1, 2 and 3 at Heathrow Airport during clean-up of the gravel aquifer (Clark and Sims 1998). (*Source:* Clark, L. and Sims, P.A. (1998) Investigation and clean-up of jet-fuel contaminated groundwater at Heathrow International Airport, UK. In *Groundwater Contaminants and Their Migration* (eds J. Mather, D. Banks, S. Dumpleton and M. Fermor). Geological Society, London, Special Publications **128**, pp. 147–157. © 1998, Geological Society of London.)

(*Continued*)

Box 9.2 (Continued)

provided settlement and dilution of the remediation waste water.

Groundwater levels were monitored regularly, and within two months, the cone of depression in the water table produced by the pumping of Wells 1 and 2 encompassed the estimated area of the kerosene 'pancake'. Initially, the recovery rate was such that 19 200 L of kerosene were removed (Fig. 9.4) and sold to be blended into commercial heating oil. The recovery rate then dropped substantially, yet the kerosene layer in borehole 11 still remained unaltered, suggesting that Wells 1 and 2 were not affecting the southern part of the kerosene 'pancake'.

Later, Well 3 was installed. It is believed that Well 3 tapped a 'pool' of kerosene isolated from the effects of Wells 1 and 2 by the foundations of the Technical Blocks. A further 10 100 L of kerosene were removed from Well 3 (Fig. 9.4), making a total recovery of 29 300 L of kerosene in four years at which point the removal of the original kerosene 'pancake' was considered to be complete. Although active remediation by pump-and-treat ended at this time, kerosene recovery using passive collectors, for example absorbent mops, in the three wells continued for about another year. The clean-up project officially ended in 1994 (Clark and Sims 1998).

9.2.2 Permeable reactive barriers

Following recognition that the pump-and-treat approach can prove expensive and in many cases ineffective, research since the late 1980s has focused on alternative, in situ approaches such as permeable reactive barriers (PRBs). In outline, PRBs are constructed by excavating a portion of the aquifer and then replacing the material excavated with a permeable mixture designed to react with the contaminant. Typically, PRBs are installed in trenches, but barriers have also been constructed by jetting reactive materials into the ground, or by generating fractures within an aquifer and filling the fractures with reactive materials (Hocking *et al.* 2000; Richardson and Nicklow 2002).

The reactive material contained in the barrier is selected to retain the contaminant within the barrier. PRBs containing zero-valent iron (iron filings) have been used to treat hexavalent chromium, uranium and technetium (Blowes *et al.* 2000) and chlorinated ethenes (PCE and TCE) (O'Hannesin and Gillham 1998). Solid-phase organic carbon in the form of municipal compost has been used to remove dissolved constituents associated with acid mine drainage, including sulphate, iron, nickel, cobalt and zinc. Dissolved nutrients, including nitrate and phosphate have also been removed from domestic septic-system effluent and agricultural drainage in this way (Blowes *et al.* 2000).

In treating inorganic and organic contaminants, a range of processes have been used such as manipulation of the redox potential to enhance biological reductive dechlorination and to change the chemical speciation of metals; chemical (abiotic) degradation; precipitation; sorption to promote organic matter partitioning and ion exchange; and biodegradation. Further information and guidance on the use of PRBs for the remediation of contaminated groundwater is given by Carey *et al.* (2002).

The most common design of PRBs are the 'funnel and gate' and 'continuous wall' reactive barriers illustrated in Fig. 9.5. Funnel and gate PRBs are described by Starr and Cherry (1994) and consist of low hydraulic conductivity cut-off walls such as sheet piles and slurry walls with gaps that contain in situ reactors for removal of contaminants. Funnel and gate systems can be installed in front of plumes to prevent further plume growth, or immediately downgradient of contaminant source areas to prevent contaminants from creating plumes. Cut-off walls (the funnel) modify the groundwater flow pattern so that groundwater flows primarily through the high conductivity gaps of the gates. Continuous PRBs transect the

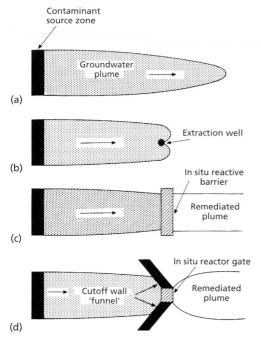

Fig. 9.5 Three options for remediation of contaminated groundwater. (a) Unremediated contaminant plume. (b) Pump-and-treat system. (c) In situ 'continuous wall' reactive barrier. (d) In situ 'funnel and gate' reactive barrier. In the case of the funnel and gate system, a balance must be achieved between maximizing the size of the capture zone for a gate and maximizing the retention time of contaminated groundwater in the gate. In general, capture zone size and retention time are inversely related (Starr and Cherry 1994). (*Source:* Starr, R.C. and Cherry, J.A. (1994) In situ remediation of contaminated ground water: the funnel-and-gate system. *Ground Water* **32**, 465–476. © 1994, Ground Water Publishing Company.)

contaminant plume with an unbroken wall of permeable material which is combined with the reactive material, for example a pea-gravel and reagent-filled trench. The majority of PRBs have been placed at relatively shallow depths, around 10–20 m deep, although a few have been placed to depths of 40 m.

To be successful, it is likely that PRBs will need to be operated over extended periods, possibly decades, but unlike the pump-and-treat method of remediation, the low operating and maintenance requirements of PRBs make such long-term clean-up a possibility. The treatment process may result in a change in

the in situ biological and geochemical environment within and downgradient of the PRB, for example a change from oxidizing to reducing conditions, that may cause secondary reactions including precipitation of mineral phases such as hydroxides and carbonates. The long-term hydraulic and chemical performance of PRBs can be affected by biofouling, chemical precipitation and the production of gases, and long-term studies are required for the full evaluation of PRBs (Box 9.3).

9.2.3 Monitored natural attenuation

The US Environmental Protection Agency (1997) defined natural attenuation as a variety of physical, chemical or biological processes that, under favourable conditions, act without human intervention to reduce the mass, toxicity, mobility, volume or concentration of contaminants in soil or groundwater. These in situ processes include biodegradation, dispersion, dilution, sorption, volatilization, radioactive decay, and chemical or biological stabilization, transformation or destruction of contaminants (Fig. 9.8).

As an alternative to more expensive pump-and-treat and engineered solutions to groundwater contamination, reliance on monitored natural attenuation (MNA) appears attractive but opponents claim that natural attenuation conveniently avoids the high costs of installing clean-up systems. The feasibility of MNA as a strategy depends on whether the regulatory aim is to clean-up a contaminant plume to drinking water standards or whether a less stringent, risk-based goal applies, such as preventing a plume from spreading. Since the mid-1990s, the use of MNA as a remedial solution for benzene, toluene, ethylbenzene and xylene (BTEX compounds) has increased dramatically (National Academy of Sciences 2000). Natural attenuation has been proposed for chlorinated solvents, nitroaromatics, heavy metals, radionuclides and other contaminants for which further research and scientific understanding is required before the technique can be considered robust (Bekins *et al.* 2001a).

Box 9.3   *In situ* permeable reactive barrier for remediation of chlorinated solvents

A field demonstration of a 'continuous wall' in situ permeable reactive barrier (PRB) was conducted in the Borden sand aquifer, Ontario, and downgradient of the emplaced source of mixed chlorinated solvents (perchloroethene (PCE), trichloroethene (TCE) and trichloromethane (TCM)) described in Box 8.3. At the time of the long-term test of the PRB, the plume from the emplaced source was approximately 1 m thick and 1 m wide near the source, with peak PCE and TCE concentrations of about 50 and 270 mg L^{-1}, respectively. However, most of the TCM had been dissolved from the source, resulting in very low concentrations. As shown in Fig. 9.6, the PRB was installed 5.5 m downgradient from the emplaced source and positioned below the water table. The reactive material used to construct the wall consisted of 22% by weight of granular iron mix with 78% by weight of coarse sand, and had dimensions of 5.5 × 1.6 × 2.2 m, giving a volume of 19.4 m^3. Permeameter measurements on samples of the iron-sand mixture gave hydraulic conductivity values of 4.37 × 10^{-4} m s^{-1} (O'Hannesin and Gillham 1998).

Metal-enhanced degradation of chlorinated organic compounds is an abiotic redox reaction involving reduction of the organic compound and oxidation of the metal (Johnson *et al.* 1996). The reaction appears to be pseudo-first order with respect to the organic concentration and the products of the reaction are chloride, iron (Fe^{2+}) and non-chlorinated, or less chlorinated hydrocarbons (dichloroethene and vinyl chloride). In the case of chlorinated ethenes such as PCE and TCE, dechlorination is complete with ethene and ethane as the final carbon-containing compounds, while for CTET, a fraction of the parent compound persists as dichloromethane.

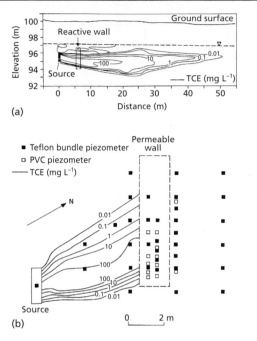

(a)

(b)

Fig. 9.6   In (a) a cross-section of the emplaced source of chlorinated solvents, reactive wall and TCE plume are shown for the Borden aquifer test site, Ontario. In (b) a plan view is shown of the test site, monitoring network and TCE plume (O'Hannesin and Gillham 1998). (*Source:* O'Hannesin, S.F. and Gillham, R.W. (1998) Long-term performance of an in situ 'iron wall' for remediation of VOCs. *Ground Water* **36**, 164–170. © 1998, Ground Water Publishing Company.)

The results of the field experiment are shown in Fig. 9.7 and show that for both TCE and PCE, there is a substantial decline in concentration of the core of the contaminant plume at the position of the first sampling fence (50 cm into the wall), followed by a gradual decrease with further distance into the wall. As a result, TCE declined from an influent concentration of 268 000 µg L^{-1} to an effluent value measured at the 7.5 m fence (50 cm downgradient of the wall) of 23 350 µg L^{-1}. Similarly, PCE declined from

Box 9.3   (Continued)

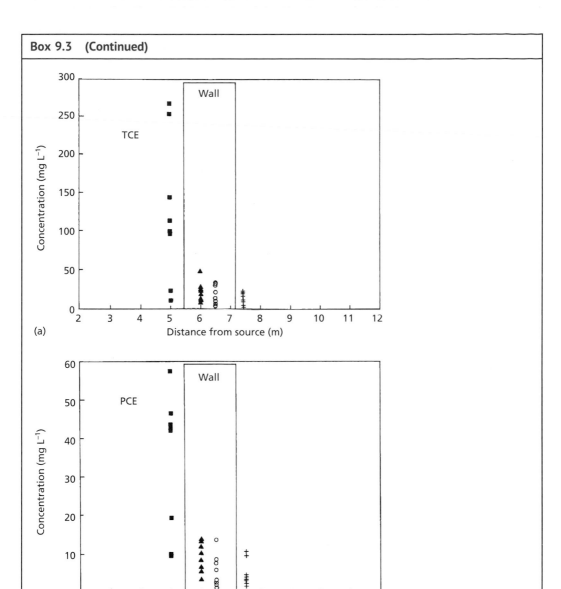

Fig. 9.7 Longitudinal section through the Borden aquifer test site showing the maximum chlorinated solvent concentrations obtained for 10 sampling sessions over a five-year period along the flowpath of (a) the TCE and (b) the PCE plumes. Note that the variation in concentration at a particular distance does not reflect variation over time but is a consequence of plume position (O'Hannesin and Gillham 1998). (*Source:* O'Hannesin, S.F. and Gillham, R.W. (1998) Long-term performance of an in situ 'iron wall' for remediation of VOCs. *Ground Water* **36**, 164–170. © 1998, Ground Water Publishing Company.)

58 000 to 10 970 $\mu g\,L^{-1}$. Thus, based on the maximum observed concentrations at each sampling fence, 91% of the TCE and 81% of the PCE were removed from solution with passage through the reactive material. It is reasonable to expect that, had a higher percentage of iron been used in the iron-sand mixture, or had a more reactive material been used, then the quality of the effluent leaving the wall could have been further improved.

Potential chlorinated degradation products were also analysed, including chloride, 1,1-dichloroethene (1,1-DCE), *trans*-1,2-dichloroethene (tDCE), *cis*-1,2-dichloroethene (cDCE) and vinyl chloride (VC). Of these, the major product within the iron-sand mixture was cDCE

(*Continued*)

Box 9.3   (Continued)

(2110 µg L^{-1}), with substantially lesser amounts of 1,1-DCE (453 µg L^{-1}) and tDCE (146 µg L^{-1}); the sum of which is equivalent to about 1% of the influent TCE. However, measurements of the effluent leaving the wall showed that the DCE isomers were also degraded within the PRB. Concentrations of VC above the limit of the analytical method were not detected.

Changes in water chemistry as a result of abiotic reduction of organic compounds involve the oxidation of zero valent iron (Fe^0) by water producing Fe^{2+}, an increase in H$^+$ and OH$^-$ (eqs. 1 and 2), and a decrease in redox potential and dissolved oxygen. The H$^+$ forms hydrogen gas and the OH$^-$ remaining in solution causes an increase in pH that can cause precipitation of iron hydroxides and carbonate minerals (eqs. 4–6).

$$2Fe^0 + O_2 + 2H_2O \rightarrow 2Fe^{2+} + 4OH^-$$

$$\text{(aerobic conditions)} \qquad \text{(eq. 1)}$$

$$Fe^0 + 2H_2O \rightarrow Fe^{2+} + H_2 + 2OH^-$$

$$\text{(anaerobic conditions)} \qquad \text{(eq. 2)}$$

$$Fe^{2+} + 2OH^- \rightarrow Fe(OH)_2 \quad \text{(iron hydroxide)}$$
$$\text{(eq. 3)}$$

$$HCO_3^- + OH^- \rightarrow CO_3^{2-} + H_2O \quad \text{(carbonate)}$$
$$\text{(eq. 4)}$$

$$Fe^{2+} + CO_3^{2-} \rightarrow FeCO_3 \quad \text{(siderite)}$$
$$\text{(eq. 5)}$$

$$Ca^{2+} + CO_3^{2-} \rightarrow CaCO_3 \quad \text{(calcite)}$$
$$\text{(eq. 6)}$$

The field results showed that iron concentrations entering the wall were <0.5 mg L^{-1}, while within the PRB concentrations were generally within the range of 5–10 mg L^{-1} before decreasing to <0.5 mg L^{-1} downgradient of the wall. Dissolved oxygen and Eh values within the treatment zone were nearly always recorded as zero and within the range −200 to −350 mV, respectively. The pH increased from a background value of 8.0–8.7 in the PRB as a result of the reduction of water. After four years of operation, only trace amounts of iron oxides and calcium and iron carbonates were found within the first few millimetres of the wall at the upgradient interface between the aquifer sand and the wall, such that there was no evidence of a decline in performance of the PRB over the duration of the study.

Overall, O'Hannesin and Gillham (1998) concluded that the results of this long-term field study provide good evidence that in situ use of granular iron can provide a long-term, low-maintenance cost solution for groundwater contamination problems.

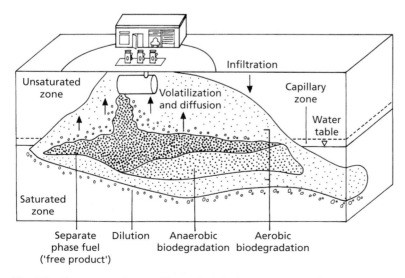

Fig. 9.8   Conceptual diagram illustrating the important natural attenuation processes that affect the fate of petroleum hydrocarbons in aquifers (Bekins *et al.* 2001a). (*Source:* Bekins, B., Rittmann, B. and MacDonald, J. (2001a) Natural attenuation strategy for groundwater cleanup focuses on demonstrating cause and effect. *Eos* **82**(53), 57–58. © 2001, John Wiley & Sons.)

In considering the case for application of MNA at a contaminated site, a substantial degree of understanding of the sub-surface processes must be developed. Thus, the major expense is likely to shift from the design and operation of an active pump-and-treat or passive PRB to detailed investigation and modelling of the site in order to understand the natural groundwater flow and biogeochemical reactions responsible for attenuating the contamination (Box 9.4).

Key to the future success of MNA is further research into the practical issues regarding the performance of natural attenuation over long time periods and should include the effects of active remediation efforts on the natural attenuation process; the design of long-term monitoring networks to verify that natural attenuation is working and proving

the natural attenuation capacity of the aquifer over the lifetime of the contaminant source.

9.3 Groundwater pollution protection strategies in developed countries

9.3.1 Groundwater vulnerability mapping and aquifer resource protection

As illustrated in Fig. 9.10, the vulnerability of groundwater to surface-derived pollution is a function of the nature of the overlying soil cover, the presence and nature of overlying superficial deposits, the nature of the geological strata forming the aquifer and the depth of the unsaturated zone or thickness of

---

Box 9.4   Monitored natural attenuation of a crude oil spill, Bemidji, Minnesota

A demonstration site for monitored natural attenuation (MNA) within the US Geological Survey Toxic Substances Hydrology Program (http://toxics.usgs.gov) is located near Bemidji, Minnesota, where a buried pipeline located in a glacial outwash plain ruptured in 1979 spilling crude oil into the subsurface. The oil is entrapped as a residual non-aqueous phase in the vadose zone and also forms two bodies of oil floating on the water table. The largest oil body was estimated to contain 147 000 L of oil in 1998. As shown in Fig. 9.9a, the oil forms a long-term, continuous source of hydrocarbon contaminants that dissolve in and are transported with the groundwater. Microbial degradation of the petroleum hydrocarbons in the plume has resulted in the growth of aquifer microbial populations dominated by aerobes, iron-reducers, fermentors and methanogens (Fig. 9.9b). The biodegradation reactions cause a number of geochemical changes near

the dissolved aqueous plume which include decreases in concentrations of oxygen and hydrocarbons and increases in concentrations of dissolved iron, manganese and methane (Fig. 9.9c).

Modelling of the natural attenuation processes simulates initial aerobic degradation followed by the development of an anoxic zone in which manganese and iron reducers and methanogens begin to grow, consuming solid phase $Mn(IV)$ and $Fe(III)$ and releasing dissolved $Mn(II)$, $Fe(II)$ and methane (Fig. 9.9c). The modelling predicts that 40% of the hydrocarbon degradation occurs aerobically and 60% anaerobically. Combined with field data and the measurement of microbial populations, the results suggest that the natural attenuation capacity of the glacial outwash sands is being slowly consumed by depletion of the intrinsic, electron-accepting capacity of the aquifer (Bekins *et al.* 2001b; Cozzarelli *et al.* 2001).

(Continued)

Box 9.4 (Continued)

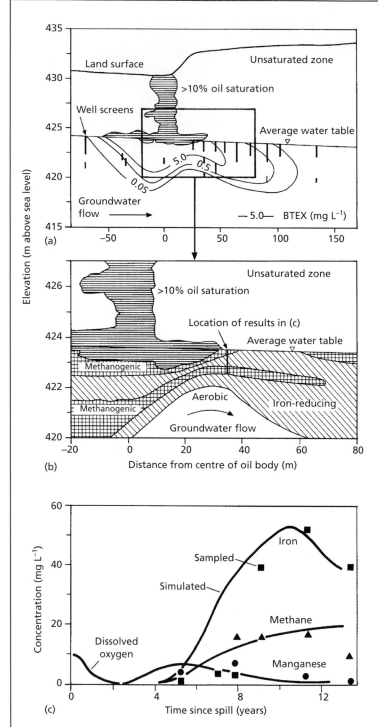

Fig. 9.9 Illustration of natural attenuation of crude oil contamination by aerobic and anaerobic biodegradation in a glacial outwash aquifer located near Bemidji, Minnesota. In (a) the 1995 concentration of BTEX compounds define the extent of contamination in the aquifer for a vertical cross-section along the plume axis. In (b) the cross-section shows the distribution of microbial populations inferred from most probable number data. In (c) modelled and observed concentrations versus time plots for a well positioned at the water table, 36 m downgradient from the contaminant source and illustrating the loss of oxygen and production of reduced electron acceptors (Fe(II), Mn(II) and methane) during the temporal evolution of redox conditions in the aquifer (Bekins *et al.* 2001a). (*Source:* Bekins, B., Rittmann, B. and MacDonald, J. (2001a) Natural attenuation strategy for groundwater cleanup focuses on demonstrating cause and effect. *Eos* **82**(53), 57–58. © 2001, John Wiley & Sons.)

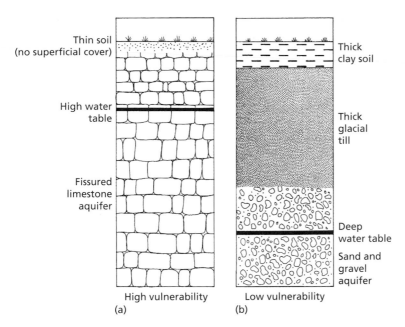

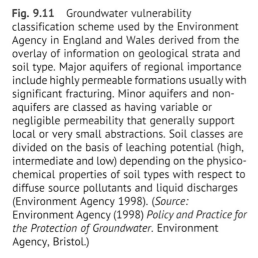

Fig. 9.10 Illustration of two situations of contrasting groundwater vulnerability to surface-derived pollution. In (a) the unconfined, fissured limestone aquifer with a permeable soil cover and high water table (thin unsaturated zone) has a high apparent vulnerability. In (b) the sand and gravel aquifer, overlain by a low permeability soil and glacial till cover has a low apparent vulnerability (Environment Agency 1998). (*Source:* Environment Agency (1998) *Policy and Practice for the Protection of Groundwater.* Environment Agency, Bristol.)

Fig. 9.11 Groundwater vulnerability classification scheme used by the Environment Agency in England and Wales derived from the overlay of information on geological strata and soil type. Major aquifers of regional importance include highly permeable formations usually with significant fracturing. Minor aquifers and non-aquifers are classed as having variable or negligible permeability that generally support local or very small abstractions. Soil classes are divided on the basis of leaching potential (high, intermediate and low) depending on the physico-chemical properties of soil types with respect to diffuse source pollutants and liquid discharges (Environment Agency 1998). (*Source:* Environment Agency (1998) *Policy and Practice for the Protection of Groundwater.* Environment Agency, Bristol.)

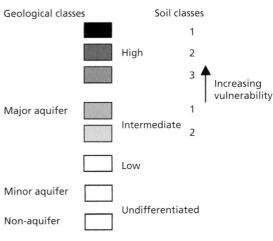

confining deposits. This approach has been used by the Environment Agency in England and Wales to produce a series of 53 regional groundwater vulnerability maps showing vulnerability classes determined from the overlay of soils and hydrogeological information at a scale of 1 : 100 000 (Fig. 9.11). The maps form part of the Environment Agency's strategy for protecting groundwater resources (Robins *et al.* 1994; Environment Agency 1998, 2012) with the intention of encouraging the development of potentially polluting activities in those areas where it will present least concern. As regional maps, the control of diffuse pollution can be readily related to zones of aquifer vulnerability. The overlay operation of soils and hydrogeological information can be conveniently manipulated within a geographical information system (GIS) to provide specific groundwater vulnerability maps, such as the nitrate vulnerability map shown in Fig. 9.12.

In the United States, the Environmental Protection Agency has developed a similar methodology to evaluate groundwater vulnerability

Specific vulnerability variant 3

■ Class 1	■ Class 5	▨ Class 9
■ Class 2	■ Class 6	▨ Non aquifer
■ Class 3	■ Class 7	
■ Class 4	▨ Class 8	

▨ Environment agency regional boundaries

0 25 50 75 100 125 km

Fig. 9.12 Specific groundwater nitrate vulnerability map for England and Wales. The vulnerability classes are derived from a GIS overlay operation of (a) simulated mean nitrate concentrations in land drainage assuming a uniform nitrogen loading of 100 kg N ha^{-1}, (b) soil types, (c) presence or absence of low permeability superficial deposits and (d) aquifer types. Regions of high groundwater vulnerability to nitrate pollution (classes 1–4) are in areas of major aquifers (compare with Fig. 2.68) (Lake *et al.* 2003). (*Source:* Lake, I.R., Lovett, A.A., Hiscock, K.M. *et al.* (2003) Evaluating factors influencing groundwater vulnerability to nitrate pollution: developing the potential of GIS. *Journal of Environmental Management* **68**, 315–328. © 2003, Elsevier.)

designed to permit the systematic evaluation of the groundwater pollution potential at any given location (Aller *et al.* 1987). The system has two major components: first, the designation of mappable units, termed hydrogeologic settings; and second, the superposition of a relative rating system having the acronym DRASTIC. Inherent in each hydrogeologic setting are the physical characteristics that affect groundwater pollution potential. The most important mappable factors considered to control the groundwater pollution potential are depth to water (*D*); net recharge (*R*); aquifer media (*A*); soil media (*S*); topography (slope) (*T*); impact of the vadose zone (*I*) and hydraulic conductivity of the aquifer (*C*). The numerical ranking system which is applied to the DRASTIC factors contains three significant parts: weights, ranges and ratings. Weights relate to the relative importance of each of the seven factors on a scale of 1–5, where five is the most important. Each factor is divided into ranges (or significant media types) depending on the impact each has on pollution potential. Each range is then assigned a rating (1–10) to differentiate the significance of each range with respect to pollution potential. The factor for each range receives a single value, except the ranges for factors *A* and *I*, for which a typical rating and a variable rating have been provided. The following equation is then used to provide each hydrogeologic setting with a relative numerical value:

$$\text{Pollution potential} = D_R D_W + R_R R_W + A_R A_W + S_R S_W + T_R T_W + I_R I_W + C_R C_W$$

(eq. 9.1)

where *R* is rating and *W* is weight. The greater the DRASTIC score the greater the pollution potential. The scores are applied to their respective hydrogeologic settings and are mapped.

A limitation of the above approaches to mapping apparent and specific groundwater vulnerability is that they provide a regional picture that is insufficiently detailed to demonstrate the actual threat to the groundwater resource at a local scale. The true vulnerability can only be established with confidence through supporting, site-specific field investigations. Even so, groundwater vulnerability maps are instrumental in conveying groundwater pollution potential to planners and can help achieve water quality objectives by influencing land-use management.

9.3.2 Source protection zones

In the definition of groundwater source protection zones, the proximity of a hazardous activity to a point of groundwater abstraction (including springs, wells and boreholes) is one of the most important factors in assessing the pollution threat to an existing groundwater source. In principle, the entire recharge area in the vicinity of a groundwater source should be protected, but this is unrealistic on socio-economic grounds. In this situation, a system of zoning of the recharge area, or protection area, is desirable and this approach has been adopted in Europe and the United States. For example, in the Netherlands, abstraction of drinking water supplies is concentrated in wellfields tapping mainly uniform, horizontally layered aquifers of unconsolidated sands and clays. As illustrated in Fig. 9.13 and Table 9.1, the zoning system includes a first zone based on a delay time of 60 days from any point below the water table in order to protect against pathogenic bacteria and viruses and rapidly degrading chemicals. This zone typically extends some 30–150 m from an individual borehole. For the continuity of water supplies in the event of a severe pollution incident requiring remedial action, and in order to exclude public health risks, a delay time of at least 10 years is needed in the next zone. In many cases, even 10 years is not sufficient to guarantee the continuity of safe water supplies, and a protection zone of 25 years is necessary. The 10- and 25-year protection zones extend to about 800 and 1200 m from the borehole, respectively, and constitute the source protection area.

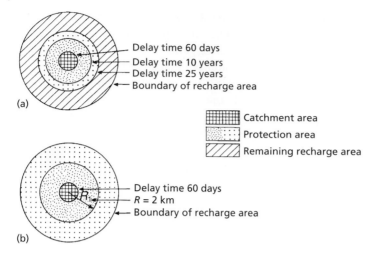

Fig. 9.13 Examples of protection zones for groundwater sources in (a) a porous, permeable aquifer and (b) a fissured, karstic aquifer. See Table 9.1 for land-use restrictions applied in each area (van Waegeningh 1985). (*Source:* van Waegeningh, H.G. (1985) Protection of groundwater quality in porous permeable rocks. In: *Theoretical Background, Hydrogeology and Practice of Groundwater Protection Zones* (eds G. Matthess, S.S.D. Foster and A.C. Skinner). Verlag Heinz Heise, Hanover, pp. 111–121.)

Table 9.1   Land-use restrictions for the source protection zones shown in Fig. 9.13 (van Waegeningh 1985).

Catchment area	Protection area	Remaining recharge area
60 days and ≥30 m	10- and 25-year delay-time or 2 km	
Protection against pathogenic bacteria and viruses and against chemical pollution sources	Protection against hardly degradable chemicals	Soil and groundwater protection rules
Only activities in relation to water supply are admissible	As a rule, the following are not admissible: • Transport and storage of dangerous goods • Industrial sites • Waste disposal sites • Building • Military activities • Intensive agriculture and cattle breeding • Quarrying • Waste water disposal	

(*Source:* van Waegeningh, H.G. (1985) Protection of groundwater quality in porous permeable rocks. In: *Theoretical Background, Hydrogeology and Practice of Groundwater Protection Zones* (eds G. Matthess, S.S.D. Foster and A.C. Skinner). Verlag Heinz Heise, Hanover, pp. 111–121.)

In the United States, the Wellhead Protection Program (US Environmental Protection Agency 1993) aims to delineate the area from which an abstraction well obtains its water and then limit potentially hazardous activities from taking place in this area. The first area, the zone of influence (ZOI), is almost synonymous with the cone of depression while the second area, the well capture zone or zone of contribution (ZOC), is defined as the region surrounding a pumping well that encompasses all areas or features that supply groundwater to the well (Fig. 9.14). The size and shape of the ZOI and ZOC are dependent on well design, aquifer properties and boundaries, and the position and hydraulic loading of the contaminant source. The ZOC can be further delineated by the zone of contaminant transport (ZOT), generally presented as isochrones (contours of equal travel time) that indicate the time required for a contaminant to reach a pumping well from a source within the

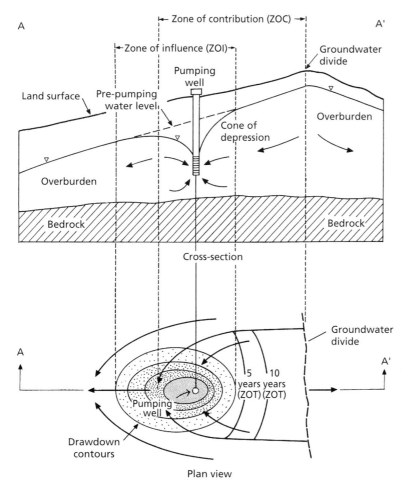

(ZOI)   Zone of influence
(ZOC)   Zone of contribution
(ZOT)   Zone of transport (contaminant)
←——   Groundwater flow direction

Fig. 9.14 Conceptual model of a wellhead protection area and associated terminology. As shown, the zone of contribution (ZOC) and the zone of influence (ZOI) do not coincide. For the ZOC, groundwater is removed from the pumping well from only a relatively small portion of the downstream area of the well, but it may extend as far as the groundwater divide on the upgradient side of the well. In contrast, the downgradient portion of the groundwater within the ZOI is not drawn towards the pumping well but continues downgradient, while the ZOI does not extend to the upgradient limit of the ZOC. However, experience shows that if the ZOC is small, then the ZOC and ZOI will generally overlap (Livingstone *et al.* 1995). (*Source:* Livingstone, S., Franz, T. and Guiguer, N. (1995) Managing ground-water resources using wellhead protection programs. *Geoscience Canada* **22**, 121–128. © 1995, Geoscience Canada.)

ZOC (Fig. 9.14). The time of travel depends on the groundwater flow velocity, the contaminant characteristics and the properties and composition of the aquifer material (Livingstone *et al.* 1995).

Mapping of wellhead protection area (WHPA) criteria can be performed at different costs and levels of complexity, ranging from arbitrary radii to numerical flow and transport models (Fig. 9.15), including the capacity of the aquifer to assimilate contaminants (Livingstone *et al.* 1995). The overall objectives of wellhead protection are to produce a remedial action zone, an attenuation zone and a management zone. Specific guidance is available in more complex hydrogeologic settings such as in confined aquifers (US Environmental Protection Agency 1991a) and fractured rocks (US Environmental Protection Agency 1991b).

In England and Wales, the Environment Agency (1998, 2012) has established source protection zones (SPZs) that are applied to public water supplies (there are nearly 2000 major sources) and private water supplies, including bottled water, and commercial food and drink production. As illustrated schematically in Fig. 9.16, the orientation, shape and size of the SPZs are determined by the hydrogeological characteristics and the direction of groundwater flow around each source. Steady-state groundwater flow modelling is used to define three zones (Zones I, II and III) in each SPZ and a set of groundwater protection policy statements set out the acceptability of various polluting activities in each zone, for example landfill operations and the application of liquid effluents to land.

Zone I, or the inner source protection zone, is located immediately adjacent to the groundwater source and is designed to protect against the impacts of human activity which might have an immediate effect upon the source. The area is defined by a 50-day travel time from any point below the water table to the source and as a minimum 50-m radius from the source. This rule of thumb is used in other countries and is based on the presumed time taken for

biological contaminants to decay in groundwater. The land immediately adjacent to the source and controlled by the operator of the source is included within this zone.

Zone II, or the outer protection zone, is the area around the source defined by a 400-day travel time and is based on the requirement to provide delay and attenuation of slowly degrading pollutants. In high groundwater storage aquifers, such as sandstones, it is necessary, in order to provide adequate attenuation, to define further the outer protection zone to be the larger of either the 400-day travel time area or the recharge catchment area calculated using 25% of the long-term abstraction rate for the source.

Zone III, or the source catchment, is the remaining catchment area of a groundwater source and is defined as the area needed to support an abstraction from long-term annual groundwater recharge (effective rainfall). For wells and boreholes, the source catchment area is defined by the authorized abstraction rate while, for springs, it is defined by the best known value of average annual total discharge. In practice, the size of Zone III will vary from tens to a few thousands of hectares depending on the volume of groundwater abstraction and the amount of recharge. In areas where the aquifer is confined beneath impermeable cover, the source catchment may be some distance from the actual abstraction.

9.3.3 Risk assessment methods

Of increasing relevance to managing aquifers, risk assessment methods are applied in the decision-making process, both with reference to the choice of aquifer remediation technology in cases where pollution has already occurred, for example in areas of contaminated land, and also in the siting of new containment facilities, such as municipal landfills. Influential publications concerning the definition of risk assessment include the US National Research Council (1983) and the Royal Society (1992). Petts *et al.* (1997) stated that risk assessment

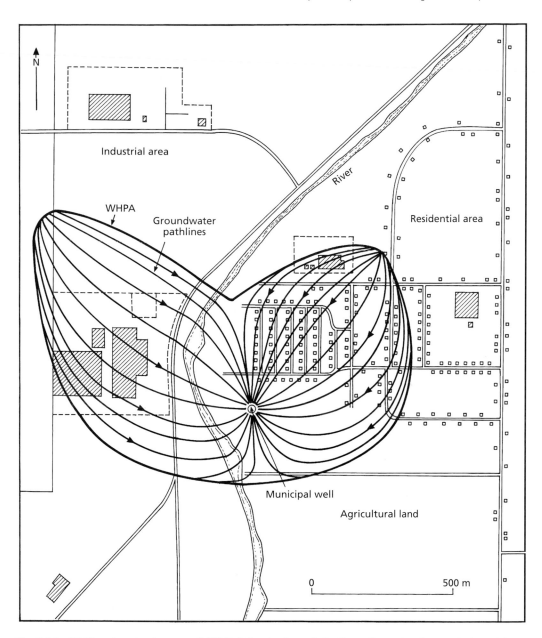

Fig. 9.15 Wellhead protection area (WHPA) defined using a fully three-dimensional numerical model (MODFLOW and MODPATH) to simulate groundwater flow and contaminant migration using particle tracking. In this hypothetical example, the municipal well pumps at a rate of 250 m^3 day^{-1}. The aquifer comprises sand and gravel with a transmissivity of 26 m^2 day^{-1} and is 30 m thick. The well partially penetrates the aquifer with the well screen extending 20–30 m below ground surface. The well is located close to a river that is situated within an area of alluvial deposits with a transmissivity of 0.1 m^2 day^{-1} and a thickness of 10 m. The groundwater model explicitly represents the alluvial unit overlying the sand and gravel aquifer near the river. The results of the simulation define a capture zone configuration that suggests that groundwater protection efforts should be concentrated in the areas to the west and east of the river (Livingstone *et al.* 1995). (*Source:* Livingstone, S., Franz, T. and Guiguer, N. (1995) Managing ground-water resources using wellhead protection programs. *Geoscience Canada* **22**, 121–128. © 1995, Geoscience Canada.)

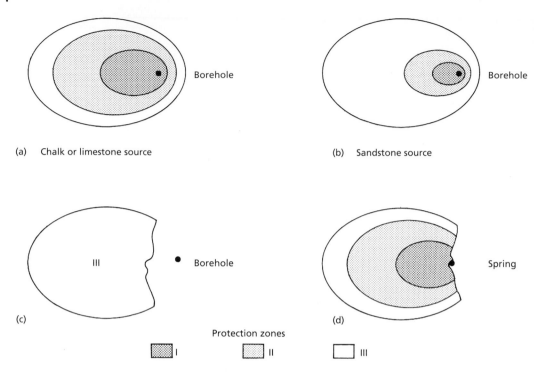

(a)   Chalk or limestone source

(b)   Sandstone source

(c)

(d)

Protection zones

I   II   III

Fig. 9.16 Schematic illustration of source protection zones showing the relationship between Zones I, II and III and the groundwater source in four idealized hydrogeological situations representing: (a) a low effective porosity limestone aquifer; (b) a high effective porosity sandstone aquifer; (c) a confined aquifer; and (d) a spring. In reality, the size, shape and relationship of the zones will vary significantly depending on the soil, geology, amount of recharge and volume of water abstracted. See text for an explanation of the definitions of Zones I, II and III (Environment Agency 1998). (*Source:* Environment Agency (1998) *Policy and Practice for the Protection of Groundwater.* Environment Agency, Bristol.)

is a process comprising hazard identification, hazard assessment, risk estimation and risk evaluation and, in general, is the study of decisions subject to uncertain consequences.

A basic risk assessment calculation can be performed by the determination of a toxicological index, I_{tox}, for a given site using the following equation:

$$I_{tox} = \sum_{i=1}^{n} c_i / LAC_i \qquad \text{(eq. 9.2)}$$

where $i = 1 \dots n$ represents the contaminant constituent, c_i is the measured concentration of constituent i, and LAC is the limit of the admissible concentration of constituent i. An example calculation of I_{tox} values for a large pulp and paper mill complex in north-west Russia is given in Table 9.2.

In calculating I_{tox}, constituents are chosen arbitrarily, mainly as a function of laboratory and financial capabilities, such that the importance of different compounds in terms of their hazard potential is not evaluated. The results of the risk assessment allow a comparative, quantitative assessment of analytical results for different measurement points but are neither source- nor target-related. For the calculations shown in Table 9.2, it is clear that all three samples are predicted to be at a high potential risk, given the values of I_{tox} in excess of 1 and would therefore suggest that remedial action is necessary. However, shortcomings of the data presented in Table 9.2 are that substances with no toxicological potential, for example chemical oxygen demand (COD) for sample 1, can determine the outcome of the toxicological index

Table 9.2 Calculation of the toxicological index, I_{tox}, for three sites located at the pulp and paper mill complex at Sjasstroj, north-west Russia. Analytical results are given in mg L^{-1} (Schoenheinz et al. 2002).

Sample number	Sample date	Constituent, i	BOD	COD	SO$_4^{2-}$	Cl$^-$	NH$_4^+$	NO$_2^-$	NO$_3^-$	Fe	Al	Phenols	Surfactants	I_{tox}
		LAC$_i$	3	30	500	350	2	3	45	0.3	0.5	0.001	0.1	1
I	09/99		3	3044	160	15	5.6	0.04	1.0	5	90	0.001	0.1	301
II	11/99		5.4	61	13	42.7	3.5	0.005	0.45	53	0.2	0.001	0.1	185
	02/00		0.9	44	6.9	83	5	0.002	0.15	2.3	4.6	0.003	0.13	25
III	02/00		0.6	7.7	5.8	4	1.45	0.005	0.0	0.6	4.5	0.002	0	14

Notes: I, excess sludge; II, groundwater close to active sludge basin in upper sand aquifer; III, groundwater in lower aquifer; LAC$_i$, Russian limit of the admissible concentration of constituent, i; BOD, biochemical oxygen demand; COD, chemical oxygen demand.

(*Source:* Schoenheinz, D., Grischek, T., Worch, E. et al. (2002) Groundwater pollution at a pulp and paper mill at Sjasstroj near Lake Ladoga, Russia. In: *Sustainable Groundwater Development* (eds K.M. Hiscock, M.O. Rivett and R.M. Davison). Geological Society, London, Special Publications **193**, pp. 277–291. © 2002, Geological Society of London.)

calculation, and that concentrations of phenols, surfactants and the biological oxygen demand (BOD) were not always available (Schoenheinz *et al.* 2002).

More sophisticated approaches to groundwater pollution risk assessment recognize a source-mobilisation-pathway-target paradigm and adopt a cost-effective, tiered approach to risk assessment. In contaminated land studies, risk assessment identifies the pathway term as the route that the contaminant takes from the pollution source to a receiving well or borehole receptor. A refinement is to divide the pathway into an environmental pathway between the source and groundwater receptor and an exposure pathway between the receptor and an ecological or human target. The objective of assessing the effects of the environmental pathway is to determine the concentration of the contaminant at an abstraction site while the exposure pathway assesses the effects of exposure to contaminated water.

To gain the most efficient use of time and financial constraints, tiered approaches to risk assessment are widely applied. Increasing tiers equate to increasing levels of sophistication with respect to site-specific risk assessment. In general, four levels of assessment can be identified:

Tier 1 *Preliminary investigations*: qualitative desk study; determination of source-mobilisation-pathway-target chains; limited intrusive investigation and sampling.
Tier 2 *Site characterization*: semi-quantitative; some intrusive site investigation and sampling; prioritization and screening methods.
Tier 3 *Generic risk assessment*: qualitative/semi-quantitative; comparison of estimated contaminant concentrations with generic guidelines and standards; assessment of environmental pathway; intrusive site investigations; computer simulations/modelling; stochastic approaches.
Tier 4 *Quantitative risk assessment*: use of derived contaminant concentrations for exposure pathway assessment; exposure assessment models; integrated approaches to risk assessment.

This type of tiered approach is the basis of the ASTM (2001) standard guide to risk-based corrective action (RBCA) for site investigation and remediation at petroleum contaminated sites, although the process can be applied to any contaminant and release scenario. The framework has a site characterization stage followed by a three-tier approach where the human health and environmental risks are equally accounted for. As each tier is completed, there is an evaluation to determine if more information is required. If so, then the next level of assessment is conducted; if not, then an assessment of the corrective action is required. A tiered approach is also used by the Environment Agency in England and Wales in the hydrogeological risk assessment of landfills (Leeson *et al.* 2003; Environment Agency 2012) as outlined in Fig. 9.17.

A practical example of a Tier 3 risk assessment is provided by Davison *et al.* (2002) who presented a management tool to identify the best use for urban groundwater pumped from a user-defined location. A probabilistic catchment zone model and land-use model are combined to provide contaminant source data that are then used in a pollution risk model to calculate a probability distribution for the concentration of a contaminant at a selected pumped borehole for comparison with water quality standards.

9.3.4 Groundwater vulnerability assessment and mapping for the protection of carbonate (karstic) aquifers

Carbonate (karstic) aquifers are highly vulnerable to surface contamination because water can move rapidly through fissures widened by dissolution, sinking streams can provide direct entry points to groundwater with little or no attenuation of contaminants and the soil cover is often thin or absent. In karst areas, rapid infiltration in sinkholes and swallets allows focused flow through the epikarst and vadose zone that often results in reduced travel times. Therefore, special strategies are required

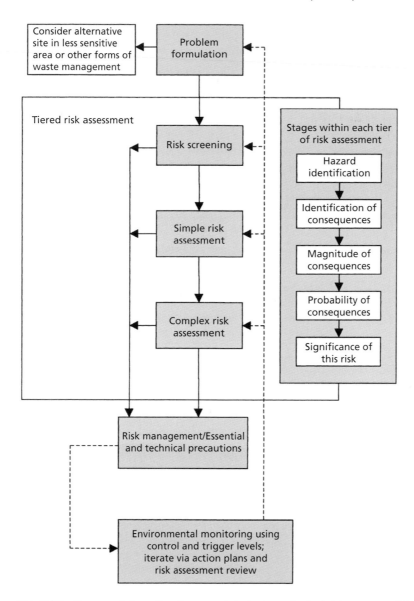

Fig. 9.17 Framework for a tiered approach to hydrogeological risk assessment for application in waste management. Two levels of risk assessment are recognized in the scheme. Simple risk assessment consists of quantitative calculations, typically deterministic analytical solutions using conservative (worst-case) input parameters, assumptions and methods. Complex risk assessment consists of quantitative, stochastic (probabilistic) techniques applied to analytical solutions using site-specific characterization data (Environment Agency 2003). (*Source:* European Environment Agency (2003) *Europe's Water: An Indicator-Based Assessment. Summary.* European Environment Agency, Copenhagen.)

in order to preserve the optimum quantity and quality of karst waters (Daly *et al.* 2002). An example of an approach to karst groundwater vulnerability assessment is the so-called 'European approach', developed under the framework of European Commission COST Action

620, and known as the concentration-overburden-precipitation (COP) method (Daly *et al.* 2002; Zwahlen 2004; Vías *et al.* 2006; Iván and Mádl-Szonyi 2017) (Fig. 9.18). The COP method models intrinsic vulnerability of groundwater using a semi-quantitative

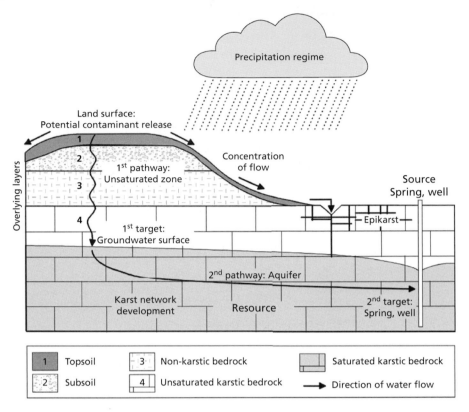

Fig. 9.18 The European approach to groundwater vulnerability assessment in carbonate (karstic) aquifers based on an origin-pathway-target conceptual model. Possible contamination events are assumed to originate at the land surface. For resource protection, the groundwater surface in the aquifer is the target, for source protection, the spring or well is the target. The pathway consequently consists of the passage through the overlying layers for resource protection and includes the passage through the aquifer for source protection. The main factors for vulnerability assessment are the **P**recipitation regime, **O**verlying layers, lateral **C**oncentration of flow and **K**arst network development (Zwahlen 2004). (*Source:* Adapted from Zwahlen, F. ed. (2004) *Vulnerability and Risk Mapping for the Protection of Carbonate (karst) Aquifers.* European Commission COST Action 620. European Communities, Luxembourg.)

approach where the properties and location of an individual contaminant are not considered (Daly *et al.* 2002). As presented by Jones *et al.* (2019), the model contains three basic components: the concentration of flow factor (C_{score}), the overlying layers factor (O_{score}) and the precipitation factor (P_{score}).

The *C* factor accounts for the location of sinking streams and swallow holes (sinkholes), which are assumed to concentrate surface water into groundwater recharge points. The *C* factor also incorporates vegetation cover and slope into its calculations, which influence overland flow amounts and patterns. The final

C factor for karst regions is calculated using the following equation:

$$C_{score} = dh \times ds \times sv \qquad \text{(eq. 9.3)}$$

where *dh* is the distance from the recharge area to a swallow hole (sinkhole) (value between 0 and 1), *ds* is the distance to a sinking stream (value between 0 and 1), and *sv* is the slope-vegetation value (between 0.75 and 1) determined by the amount of vegetation cover and degree of slope.

The *O* factor estimates the protectiveness of layers of rock and soil overlying a given aquifer. Daly *et al.* (2002) proposed a subdivision of four layers: topsoil, subsoil, non-karstic rocks and unsaturated karstic rocks (numbered 1–4 in Fig. 9.18). In the final COP method, only two layers with important hydrogeological roles are used in order to evaluate the *O* factor: soils and the lithological layers of the unsaturated zone. In general, thicker and less permeable lithic layers are considered more protective than thin, highly permeable and/or karst-forming layers. Soil content and thickness are also considered, with thick, clay-rich soil being most protective. The equation for calculating the *O* factor is

$$O_{score} = [O_s] + [O_L] \qquad \text{(eq. 9.4)}$$

where O_s is the soil sub-factor value (between 0 and 5) corresponding to attenuation of infiltration due to the texture, grain size distribution and thickness of soil cover. O_L is the lithology sub-factor, which is the sum of values assigned to each rock layer overlying the aquifer describing the protectiveness of the unsaturated rock units above the aquifer based on rock type, thickness, existence of secondary permeability (such as fracturing), and whether the units are confined or unconfined.

The *P* factor accounts for precipitation quantity and temporal distribution with the equation:

$$P_{score} = P_Q + P_I \qquad \text{(eq. 9.5)}$$

where P_Q is the precipitation quantity value, a value between 0.2 and 0.4 describing the quantity of rainfall per year, and P_I is the temporal distribution value, with a value between 0.2 and 0.6 representing the temporal distribution determined by the precipitation per year divided by the number of days with precipitation per year.

Once all *C*, *O* and *P* scores are calculated, the three scores are multiplied together to evaluate the intrinsic vulnerability of a groundwater resource ($COP_{index} = C \times O \times P$). The COP_{index} is then categorized into a final COP map showing the vulnerability classes. A COP_{index} score between 0 and 0.5 has very high vulnerability,

a score between 0.5 and 1 has high vulnerability, a score between 1 and 2 has moderate vulnerability, a score between 2 and 4 has low vulnerability, and a score between 4 and 15 has very low vulnerability. The COP method assigns any region within 500 m of a sinkhole a value of zero, resulting in a categorization of 'very high vulnerability' for that region, regardless of other factors. All other regions are assumed to be controlled by diffuse recharge. An example application of the COP method of groundwater vulnerability assessment to two carbonate aquifers in the South of Spain, the Sierra de Líbar (a conduit flow system) and the Torremolinos (a diffuse flow system), is presented by Vías *et al.* (2006).

For source intrinsic vulnerability mapping, where the target is a karstic source (spring or borehole), a *K* factor should be taken into account for the mostly horizontal flow path in the saturated zone. The *K* factor represents the degree of karst network development and is based on a general description of the bedrock, giving a range of possibilities from non-karstified carbonate rocks with only inter-granular porosity to karst aquifers with fast active conduit systems (Daly *et al.* 2002; Zwahlen 2004). The means of classifying the karst network factor include information relating to geology and geomorphology, cave and karst maps, groundwater tracing results, pumping tests results, spring hydrograph and chemograph analyses, remote sensing and geophysical prospecting, borehole data and geophysical logging results, bedrock sampling and laboratory experiments and calibrated modelling results. Other indicators that may provide information on the underground characteristics include drainage density, and soil and vegetation type. Once the *K* factor is determined, the source vulnerability map is consequently obtained by a combination of the factors *C*, *O*, *P* and *K*. Together, the resulting source and resource vulnerability maps can be used as a basis for the delineation of source and resource protection zones, respectively. Together with a hazard map, the approach can also be used for risk assessment (Zwahlen 2004).

The COP method is regarded as a detailed, easy-to-use, accurate method for modelling karst vulnerability in humid regions. However, it is limited by the assumption that surface karst features, such as sinkholes, have a direct path to the underlying aquifer (Jones *et al.* 2019). Also, the COP method does not consider the presence of structural features or the variations of recharge capacity of individual sinkholes that may affect aquifer vulnerability. These factors are potentially critical in evaluating the vulnerability of aquifers located in dry environments that often have negligible diffuse recharge and naturally thick overburden. In semi-arid and arid environments, aquifers are often deep below the surface and recharge is primarily focused along ephemeral stream channels, topographic depressions, and zones of faulted and fractured rock. Therefore, as explained by Jones *et al.* (2019), if surface karst exists, direct connections from surface karst to aquifers cannot be assumed, especially when non-carbonate strata are present.

To address the above shortcomings in the application of the COP method in arid and semi-arid environments to assess the overall vulnerability of karst aquifers, Jones *et al.* (2019) compared the original COP methodology with a modified method that uses sinkhole density as well as the location of faulted and fractured rock to model intrinsic vulnerability in deep, multi-layered aquifer systems that have abundant surface karst. Two models were developed for the Kaibab Plateau, Arizona, the primary catchment area supplying springs along the north side of the Grand Canyon (Tobin *et al.* 2018). Together, as shown in Plate 9.1, the modified model created greater spatial variation in vulnerability class predictions that better reflect recharge patterns for this region containing a layered perched (Coconino aquifer) and deep (Redwall-Muav aquifer) aquifer system (Jones *et al.* 2019).

9.3.5 Spatial planning and groundwater protection

The problem of nitrate and pesticides leaching from regions of intensively managed arable farming and cattle grazing affects wide areas of Europe and other regions of the world, and it has become clear that more stringent controls on land-use activities, integrated into local and regional spatial planning, are required if groundwater quality is to improve. Such an approach has been applied in the United Kingdom and Denmark (Box 9.5) and provides experience for the future formulation of strategies for reducing the impacts of diffuse groundwater contamination.

Box 9.5   The Drastrup Project, north Jutland, Denmark

A unique example of spatial planning to achieve sustainable land use in a groundwater catchment area with the aim of improving and protecting groundwater quality is provided by the Drastrup Project in north Jutland, Denmark. The motivation for the project was decreasing water quality from the diffuse input of nitrate and pesticides, with concentrations of nitrate up to 125 mg L^{-1} in groundwater. The Drastrup Area, covering about 870 ha, is one of two large groundwater catchment areas for the city of Aalborg, contributing 20%, or approximately 1.7×10^6 m^3, of the annual municipal water supply demand (Fig. 9.19).

To protect the underlying Chalk aquifer and demonstrate the effects of land-use change on nitrate leaching, it was decided to apply a municipal planning act to execute voluntary land distribution among farmers in

Box 9.5   (Continued)

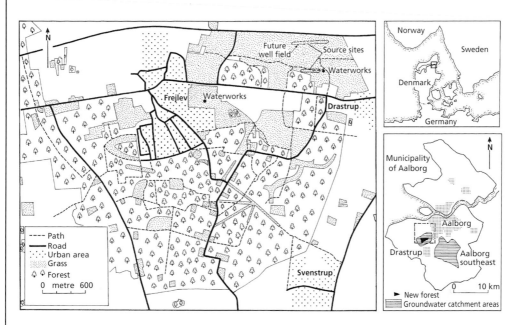

Fig. 9.19   The Drastrup Area, Denmark, showing the area of new forest planted to protect groundwater from the leaching of nitrate (Municipality of Aalborg 2001). (*Source:* Municipality of Aalborg (2001). Sustainable land-use in ground water catchment areas. Technical Final Report and Layman's Report. EU LIFE Project Number LIFE97 ENV/DK/000347. The Municipality of Aalborg, Aalborg, 56 pp. © European Union, 1995–2012.)

the groundwater catchment area near Frejlev and Drastrup. The farm owners received compensation in the form of either payments or as land outside of the groundwater catchment area. The project also had the objective of establishing a recreational area close to Aalborg for the benefit of its citizens through the creation of a new, 230-ha recreational forest comprising a non-rotation, mixed age forest with some permanent grass cover with low intensity grazing (Fig. 9.19). Planting began in the late 1990s and the new forest was inaugurated in September 2001 (Municipality of Aalborg 2001).

The initial preparation of soil in the new forest area by deep ploughing caused an initial flush of nitrate (up to 40 mg L^{-1} as N), but it is considered that as the trees grow, soil nitrate concentrations will begin to decline

after 4–6 years. Trees demand a high nitrogen uptake during the first 15 years of growth, and so effectively reducing soil nitrate leaching, but later soil nitrate concentrations may increase as a result of atmospheric inputs of nitrogen to the ageing forest canopy. In areas of grassland, soil monitoring has shown that conversion to grassland quickly reduces nitrate concentrations to effectively zero within the first 2–3 years of conversion (Fig. 9.20a).

Since the implementation of land-use change in 1994, depth sampling of groundwater may now be showing signs of improvement in water quality, although further years of monitoring are required to identify unambiguously an improvement in water quality from natural background variation in nitrate content (Fig. 9.20b).

(Continued)

Box 9.5 (Continued)

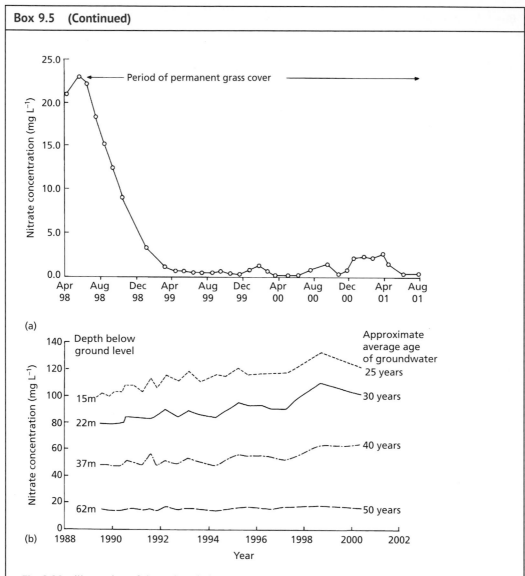

Fig. 9.20 Illustration of the reduced nitrate concentrations achieved in (a) soil water beneath an area converted from arable to permanent grass and (b) in depth samples of groundwater samples in an area of new forest in the Drastrup area, Denmark (Municipality of Aalborg 2001). (*Source:* Municipality of Aalborg (2001). Sustainable land-use in ground water catchment areas. Technical Final Report and Layman's Report. EU LIFE Project Number LIFE97 ENV/DK/000347. The Municipality of Aalborg, Aalborg, 56 pp. © European Union, 1995–2012.)

In England, the Pilot Nitrate Sensitive Area (NSA) scheme was started in 1990, and by the time of the Main NSA scheme in 1998, 80% of the land area comprising 35 000 ha in 32 catchments was included in this voluntary, compensated agri-environment scheme. Two levels of payment were offered to farmers entering the scheme with higher compensation given under the Premium scheme for arable land conversion to grass with total nitrogen inputs of less

Table 9.3 Summary of measured soil nitrate losses prior to and during the Pilot Nitrate Sensitive Area (NSA) scheme. Fluxes have been adjusted to mean rainfall conditions (ADAS 2003).

Crop type	Winter 1990/91			Mean N of winters 1992/3, 1993/4, 1994/5		
N loss:	kg N ha^{-1} (adjusted)	mg L^{-1} as NO$_3^-$	n	kg N ha^{-1} (adjusted)	mg L^{-1} as NO$_3^-$	n
Potatoes, sugar beet	92 ± 23	228 ± 52	10	38 ± 4	87 ± 9	46
Cereals	40 ± 4	102 ± 11	61	53 ± 3	85 ± 5	142
Grass	44 ± 7	102 ± 16	27	42 ± 8	69 ± 14	79
Premium scheme	–	–	0	6 ± 4	9 ± 6	70
All sites	65 ± 12	163 ± 29	108	47 ± 3	79 ± 5	380

(*Source:* ADAS (2003). Assessment of the effectiveness of the nitrate sensitive areas scheme in reducing nitrate concentrations. Technical report for R&D project P2-267/U/2. Environment Agency, Bristol, 79 pp.)

than 150 kg ha^{-1}. Under the alternative, Basic scheme, farmers were paid less for sowing winter cover crops, restricting organic manure inputs and their timing, and limiting nitrogen fertilizer inputs to below the recommended optimum. Table 9.3 summarizes the results of long-term monitoring of the effectiveness of the scheme in the first 10 catchments included in the Pilot NSA scheme with respect to measured soil nitrate losses prior to and during the scheme. The results indicate an overall reduction of about 50% in nitrate concentrations and 28% in nitrogen fluxes leaving the root zone compared to baseline values. Corresponding values for the further catchments included in the Main NSA scheme are 34% and 16%, respectively, between 1994–1996 and 1998–2000 (ADAS 2003). Take-up and conversion of arable land under the Premium scheme option was limited but was the most effect in reducing nitrate leaching losses by at least 80% and made an important contribution to the total reduction in nitrate losses in a catchment.

In response to the EU Directive on Diffuse Pollution by Nitrates (91/676/EEC; Council of the European Communities 1991), a Nitrate Vulnerable Zone (NVZ) scheme has been implemented in England (Fig. 9.21) with mandatory, uncompensated measures based on 'good agricultural practice' now applied to about 59% of the land area of England (DEFRA 2012). The NVZ areas

have been identified as lands draining into polluted waters, including surface waters and groundwaters which contain or could contain, if preventative action is not taken, nitrate concentrations in excess of 50 mg L^{-1}, and natural freshwater lakes, or other freshwater bodies, estuaries, coastal waters and marine waters which are eutrophic or may become so in the near future if action is not taken (DEFRA 2002).

It is likely that the impact of NVZs on reducing nitrate leaching will be modest and probably smaller than the 16–28% reduction achieved in nitrate leaching fluxes reported for the earlier NSA scheme. In critical areas, land-use change similar to the Premium NSA scheme will be needed if nitrate concentrations in surface waters and groundwater are to meet the EU drinking water quality standard of 50 mg L^{-1}.

9.4 Groundwater protection strategies in developing countries

Groundwater is extensively used for drinking water supplies in developing countries, especially in smaller towns and rural areas, where it is often the cheapest and safest source. Waste water disposal is often by means of unsewered,

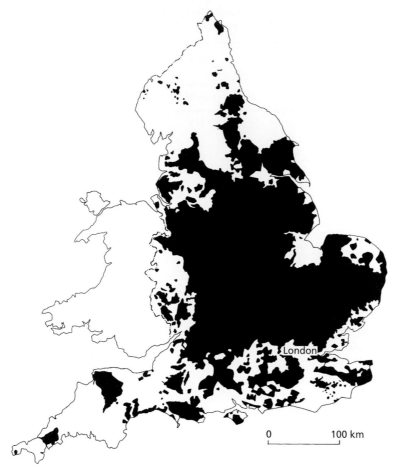

Fig. 9.21 Areas of NVZs in England. The following rules apply under this mandatory, uncompensated scheme based on 'good agricultural practice': (a) closed periods for inorganic nitrogen (fertilizer) applications during the autumn and winter and for organic nitrogen (manure) applications during the autumn for both arable and grass lands; (b) nitrogen limits applied for inorganic and organic nitrogen applied to arable and grass lands that do not exceed crop requirements; (c) spreading controls to restrict fertilizer and manure applications on steep slopes or close to water courses; (d) slurry storage for manure during the autumn closed period and (e) record-keeping of agricultural practices for at least five years. Further, specific details are provided by DEFRA (2002). (*Source:* DEFRA (2002). *Guidelines for Farmers in NVZs – England.* Department for Environment, Food and Rural Affairs, London. http://archive.defra.gov.uk/environment/quality/water/waterquality/diffuse/nitrate/help-for-farmers.htm (accessed 10 June, 2013).

pour-flush pit latrines that provide adequate waste disposal at a much lower cost than main sewerage systems. In cases where thin soils are developed on aquifer outcrops, there is the risk of direct migration of pathogenic microbes, especially viruses, to adjacent groundwater sources. The inevitable result will be the transmission of water-borne diseases. A further problem with human wastes is the organic nitrogen content which can cause widespread and persistent problems of nitrate in water, even where dilution and biological reduction processes occur.

Groundwater pollution problems are exacerbated in less-developed areas without significant regional groundwater flow to provide dilution and by the use of inorganic fertilizers and pesticides in an effort to secure self-sufficiency in food production. Also, the use of irrigation to provide crop moisture requirements poses the risk of leaching of nutrients, especially from thin, coarse-textured soils. Increases in chloride, nitrate and trace elements will result from excessive land application of waste water, sewage effluent and sludge, and animal slurry.

Other pollution sources occur in urban areas where increasing numbers of small-scale industries, such as textiles, metal processing, vehicle maintenance and paper manufacture are located. The quantities of liquid effluent generated by these industries will generally be discharged to the soil, especially in the absence of specific control measures and the prohibitive cost of waste treatment. Larger industrial plants generating large volumes of process water will commonly have unlined surface impoundments for the handling of liquid effluents.

Unless shallow dug wells have adequate protection from surface water runoff and are sufficiently distant from pit latrines, this type of groundwater source is vulnerable to both water table decline in drought periods and to contamination. Although simple measures such as boiling can help combat water-borne diseases, it is understandable that the large aid programmes in the last few decades have focused on drilling deeper boreholes and installing simple pumping apparatus. As a result, hand-pumped tube wells are very common across much of Africa and Asia, but even these sources are now associated with problems, as illustrated graphically by the natural occurrence of arsenic in groundwater in Bangladesh and West Bengal in India (see Box 9.6).

Even so, the natural soil profile can be effective in purifying human wastes, including the elimination of faecal microbes, and also in the adsorption, breakdown and removal of many chemicals. Given the potential for groundwater pollution in developing countries, protection of water supplies requires a broad-based approach that should include a strategy of minimum separations, depending on the hydrogeological situation, between a groundwater supply source and pit latrines for microbiological protection. The water laws and codes of practice of many countries require a minimum spacing between groundwater supply source and waste disposal unit of 15 m. There is, however, considerable pressure to reduce this permitted spacing to as little as 5 m in some developing countries such as Bangladesh and parts of India and Sri Lanka (Table 9.6), often resulting from the lack of space in very densely populated settlements. This example of law governing the location of waste disposal units demonstrates that criteria for groundwater pollution protection is rather arbitrary, based on limited or no technical data.

Other practical recommendations include the delineation of dilution zones of modified land use to alleviate the impact of polluting activities (Foster 1985) and the replacement of unsanitary municipal dumpsites, or tips, by controlled landfills using simple technology at a sustainable and realistic cost appropriate to gross domestic product (GDP). For example, in Tanzania and the Gambia, controlled but unlined landfills at existing quarry sites have been proposed that will operate on a dilute and disperse basis. Risk assessments demonstrate that local aquifers are not at risk such that some local groundwater contamination is acceptable in return for major improvements in health and hygiene resulting from the removal of the current dumpsites (Griffin and Mather 1998).

The application of groundwater vulnerability mapping using available information and scientific knowledge is a valuable aid in educating people about the potential risks of groundwater pollution and explaining the need for appropriate land-use management. An example groundwater vulnerability map for the Uda Walawe Basin in Sri Lanka is shown in Fig. 9.23. The map was developed with regard to the locally important risk parameters of fluoride, arsenic and nitrate in groundwater from dug wells and tube wells (Rajasooriyar 2003). Microbiological factors, although important, were not included in the derivation of the map due to

Box 9.6 Arsenic pollution of groundwater in southern Bangladesh

The Quaternary alluvial aquifers of Bangladesh provide drinking water for 95% of the population and also most of the water used for irrigation (Rahman and Ravenscroft 2003). Relative to surface water, the groundwater is bacteriologically safe, and its increased exploitation since the late 1970s has probably saved many millions of lives that would otherwise have been lost to water-borne diseases resulting from the use of contaminated surface water sources. Prolonged exposure to inorganic arsenic in water causes a variety of ailments including melanosis (a darkening of the skin), keratosis (a thickening of the skin, mostly on hands and feet), damage to internal organs and, ultimately, cancer of the skin or lungs.

Arsenic was first detected in groundwater in Bangladesh in 1993, when analysis was prompted by increasing reports of contamination and sickness in the adjoining state of West Bengal in India. Groundwater studies have demonstrated the wide extent of arsenic occurrence in Bangladesh (Dhar *et al.*

1997; BGS & DPHE 2001) at concentrations greater than the Bangladesh regulatory limit for arsenic in drinking water of 50 μg L^{-1} and the World Health Organization (1994) recommended limit of 10 μg L^{-1}. These regional surveys have shown that aquifers of the Ganges, Meghna and Brahmaputra floodplains are all affected in parts, making this the most extensive occurrence of groundwater pollution in the world (Table 9.4, Fig. 9.22). It is estimated that at least 21 million people are presently drinking water containing more than 50 μg L^{-1} of arsenic, while probably more than double this number are drinking water containing more than 10 μg L^{-1} of arsenic (DPHE 2000; Burgess *et al.* 2002; Nickson *et al.* 2000). The number of persons who must be considered 'at risk' of arsenic poisoning is even higher because testing of the 5–10 million tubewells in Bangladesh will take years to complete.

The cause of the elevated arsenic concentrations in the Ganges–Meghna–Brahmaputra deltaic plain is thought to relate to the

Table 9.4 Frequency distribution of total arsenic concentrations in groundwater in three sub-districts (upazilas) of Bangladesh: CN, Chapai Nawabganj (all aquifers, n = 94); F, Faridpur (all depths, n = 64) and Lakshmipur (all depths, n = 77). Aquifer depths are not divided and locations of the three sub-districts are shown in Fig. 9.22b (DPHE 2001).

Total arsenic concentration class[a] (μg L^{-1})	Percentage in concentration class			Cumulative percentage in or below concentration class		
	CN	F	L	CN	F	L
<2	52	9	24	52	9	24
2–10	16	22	7	68	31	31
10–50	15	29	13	83	60	44
50–100	4	9	23	87	69	67
100–1000	12	30	33	99	99	100
>1000	1	1		100	100	

[a] WHO recommended limit = 10 μg L^{-1}; Bangladesh regulatory limit = 50 μg L^{-1}.
(*Source:* DPHE (2001) *Groundwater Studies for Arsenic Contamination in Bangladesh. Final Report, Rapid Investigation Phase.* Department of Public Health Engineering, Government of Bangladesh. British Geological Survey & MottMacDonald.)

Box 9.6   (Continued)

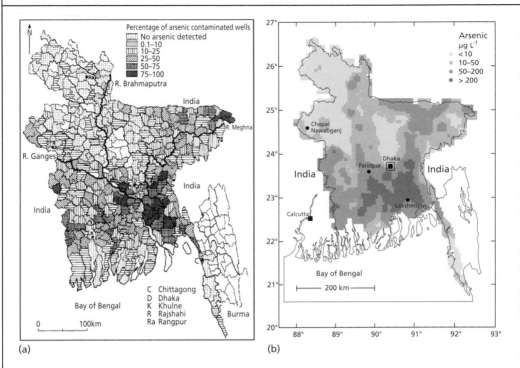

Fig. 9.22   Distribution of arsenic pollution in the main aquifer system in Bangladesh. In (a) the map represents the results of the DPHE (1999) survey showing the percentage of contaminated wells in the sub-districts (upazilas) of Bangladesh that exceed the Bangladesh regulatory limit for arsenic in drinking water of 50 µg L^{-1}. The map may misrepresent the true percentage of wells that are contaminated by arsenic because of sampling bias (many samples were collected at wells where arsenic poisoning was suspected) and measurement inaccuracies (field test kits where used do not reliably indicate values exceeding 50 µg L^{-1} when arsenic concentrations are in the range 50–200 µg L^{-1}). Although the affected areas are unlikely to change, the percentages shown may be revised in the future. (*Source:* DPHE (1999) *Groundwater Studies for Arsenic Contamination in Bangladesh. Final Report, Rapid Investigation Phase.* Department of Public Health Engineering, Government of Bangladesh. British Geological Survey and MottMacDonald, London). In (b) the interpolated surface generated by disjunctive kriging using laboratory data for 3208 groundwater samples from the shallow Holocene aquifer (<150 m depth) shows the spatial distribution of groundwater arsenic concentrations in Bangladesh. Also shown are the locations of the three sub-districts included in Table 9.4. (*Source:* BGS & DPHE (2001) Arsenic contamination of groundwater in Bangladesh (eds D.G. Kinniburgh & P.L. Smedley). Volume 2, *Final Report. British Geological Survey Report WC/00/19.* British Geological Survey, Keyworth. https://www2. bgs.ac.uk/groundwater/health/arsenic/Bangladesh/reports.html (accessed 11 August, 2021).

microbial reduction of iron oxy-hydroxides contained in the fine grained Holocene sediments and the release of the adsorbed load of arsenic to groundwater. It has been proposed that the reduction is driven by microbial metabolism of buried peat deposits (McArthur *et al.* 2001; Ravenscroft *et al.* 2001). The presence of abundant organic matter is expected in deltaic or fluvial areas that supported peat formation during climatic optimums.

Severely polluted aquifers are all of Holocene age, although at a local scale the

(Continued)

Box 9.6 (Continued)

distribution of arsenic pollution is very patchy. There are cases of grossly polluted boreholes, pumping groundwater with arsenic at concentrations greater than 1000 μg L^{-1} being separated spatially by only a few tens of metres from boreholes pumping groundwater with arsenic at concentrations less than 10 $\mu g\,L^{-1}$ (Burgess *et al.* 2002). However, there are many cases where almost all wells in a village contain more than 50 $\mu g\,L^{-1}$.

Data presented by BGS & DPHE (2001) showed that the highest percentage of wells that contain arsenic concentrations above the regulatory limits of 10 and 50 $\mu g\,L^{-1}$ occur at depths above 60 m (Table 9.5). Hand-dug wells are mostly <5 m deep and are usually unpolluted by arsenic, but the risk of bacteriological contamination is high. Below 60 m there is a decrease in the percentage of wells that are polluted, but the risk remains significant until well depths exceed 150 m, the maximum depth of river channel incision during the Last Glacial Maximum at 18 ka. Even so, across much of southern Bangladesh, more than 50% of boreholes in

the shallow aquifer have arsenic levels that comply with the 50 $\mu g\,L^{-1}$ limit and so continued development of the alluvial aquifers may still be possible, at least in the medium term (Burgess *et al.* 2002; Ravenscroft *et al.* 2005).

The problem of arsenic in groundwater is not only confined to Bangladesh and West Bengal. According to Smedley and Kinniburgh (2002), the areas with large-scale problems of arsenic in groundwater tend to be found in two types of environment: inland or closed basins in arid and semi-arid areas; and strongly chemically reducing alluvial aquifers. The hydrogeological situation in these areas is such that the aquifers are poorly flushed and any arsenic released from the sediments following burial tends to accumulate in the groundwater.

Areas containing high-arsenic groundwaters are well-known in Argentina, Chile, Mexico, China and Hungary, but the problems in Bangladesh, West Bengal and, additionally, Vietnam are more recent (Smedley and Kinniburgh 2002). In Vietnam, the capital Hanoi is situated at the upper end of the Red River

Table 9.5 Distribution of mean arsenic concentrations in groundwater as a function of well depth (DPHE 2001).

Depth interval (m)	Number of wells	Percentage of wells	Mean arsenic concentration ($\mu g\,L^{-1}$)	Percentage of wells with >50 $\mu g\,L^{-1}$
<15	287	8	58	25
15–30	1180	33	76	31
30–60	1258	36	56	26
60–90	317	9	33	21
90–150	165	5	45	35
150–200	32	1	7	1
>200	295	8	3	1
All	3534	100	278	140

(*Source:* DPHE (2001) *Groundwater Studies for Arsenic Contamination in Bangladesh. Final Report, Rapid Investigation Phase.* Department of Public Health Engineering, Government of Bangladesh. British Geological Survey & MottMacDonald.)

Box 9.6   (Continued)

Delta and analysis of raw groundwater pumped from the lower Quaternary alluvial aquifer gave arsenic concentrations of 240–320 µg L^{-1} in three of the city's eight treatment plants and 37–82 µg L^{-1} in another five plants (Berg *et al.* 2001). In surrounding rural districts, high arsenic concentrations found in tubewells in the upper aquifers (48% above 50 µg L^{-1} and 20% above 150 µg L^{-1}) indicate that several million people consuming untreated groundwater might be at a high risk of chronic arsenic poisoning (Berg *et al.* 2001).

As in Bangladesh, the source of arsenic in the Red River Delta sediments is believed to be associated with iron oxy-hydroxides that release arsenic to groundwater under chemically reducing conditions. A characteristic feature of arsenic contamination of wells in both Bangladesh and Vietnam is the large degree of spatial variability in arsenic concentrations at a local scale. As a result, it is difficult to know when to take action to provide arsenic-free water sources. For now, it appears safer to analyse each well until further research has been completed into the sources, controls and distribution of arsenic in susceptible areas.

insufficient information. The following factors were combined, using expert judgement, to classify qualitatively areas as low, medium and high groundwater vulnerability:

1) Geological factors: mineralogy (fluoride- and arsenic-bearing minerals); geological structure (divided into the fractured Highland Series and less-fractured Eastern Vijayan Complex).
2) Hydrogeological factors: shallow regolith (weathered) aquifers; deep, hard rock (fractured) aquifers with low/moderate and high transmissivity zones.
3) Recharge conditions: areas subject to low rainfall recharge only; areas subject to high rainfall and irrigation recharge.
4) Salt water mixing: areas close to the coast subject to saline intrusion and sea-salt spray; unaffected areas away from the coast.

Interestingly, it was found that nitrate and phosphate do not pose an immediate groundwater pollution risk in areas of banana and paddy

Table 9.6 Percentage distributions of distances between pit latrines and dug wells and length of lining in dug wells in the Jaffna Peninsula, Sri Lanka (Rajasooriyar *et al.* 2002).

Jaffna municipal area		Valigamam region		Valigamam region	
Distance (m)	% dug wells	Distance (m)	% dug wells	Lining (m)	% dug wells
<1.5	5.7	<10	13.6	<1.0	7.4
1.6–3.0	8.0	10.1–20.0	48.2	1.1–3.0	41.5
3.1–4.5	5.7	>20.1	38.2	>3.1	38.5
4.6–6.0	6.8			Damaged	12.6
>6.1	73.8				

(*Source:* Rajasooriyar, L., Mathavan, V., Dharmagunawardhane, H.A. and Nandakumar, V. (2002) Groundwater quality in the Valigamam region of the Jaffna Peninsula, Sri Lanka. In: *Sustainable Groundwater Development* (eds K. M. Hiscock, M.O. Rivett and R.M. Davison). Geological Society, London, Special Publications **193**, pp. 181–197.)

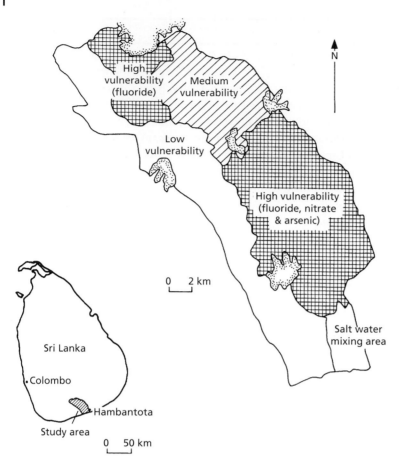

Fig. 9.23   Groundwater vulnerability map for the lower Walawe River Basin, Sri Lanka, derived qualitatively with regard to the locally important human health risk parameters of fluoride, arsenic and nitrate in groundwater from dug wells and tube wells (Rajasooriyar *et al.* 2003). (*Source:* Rajasooriyar, L.D., Boelee, E., Prado, M.C.C.M., Hiscock, K.M. (2003). Mapping the potential human health implications of groundwater pollution in southern Sri Lanka. *Water Resources and Rural Development,* **1–2**, 27–42. Figure 6.)

cultivation in the Highland Series and Eastern Vijayan Complex regions, even though these crops are subject to irrigation recharge and the application of large amounts of fertilizer input. It appears that denitrification occurs where waterlogged paddy field soils are present to explain the absence of widespread nitrate contamination (Rajasooriyar 2003).

Further reading

Addiscott, T.M., Whitmore, A.P. and Powlson, D. S. (1991) *Farming, Fertilizers and the Nitrate Problem.* C.A.B International, Wallingford, Oxon.

Ellis B. (ed.) (1999) Impacts of urban growth on surface water and groundwater quality. *Proceedings of IUGG 99, Symposium HS5, University of Birmingham,* July 1999. IAHS Publ. No. **259**.

Fetter, C.W., Boving, T. and Kreamer, D. (2018) *Contaminant Hydrogeology* (3rd edn.). Waveland Press, Long Grove, Illinois.

Hemond, H.F. and Fechner, E.J. (1994) *Chemical Fate and Transport in the Environment.* Academic Press, Inc., San Diego, California.

Morris, B.L., Lawrence, A.R.L., Chilton, P.J.C. et al. (2003). *Groundwater and Its Susceptibility to Degradation: A Global Assessment of the Problem and Options for Management.* Early

Warning and Assessment Report Series, RS. 03-3. United Nations Environment Programme, Nairobi, Kenya.

United States Environmental Protection Agency (1994) *Handbook–Ground Water and Wellhead Protection*. EPA Report /625/R-94/001, Office of Research and Development, Cincinnati, OH, and Office of Water, Washington, DC, 269 pp.

Ward, C.H., Cherry, J.A. and Scalf, M.R. (1997) *Subsurface Restoration*. CRC Press, Boca Raton, FL.

Wickramanayake, G.B., Gavaskar, A.R. and Chen, A.S.C. (eds) (2000) *Chemical Oxidation and Reactive Barriers: Remediation of Chlorinated and Recalcitrant Compounds*. Battelle Press, Columbus, Ohio.

References

ADAS (2003). Assessment of the effectiveness of the nitrate sensitive areas scheme in reducing nitrate concentrations. Technical report for R&D project P2-267/U/2. Environment Agency, Bristol, 79 pp.

Aller, L., Bennett, T., Lehr, J.H. *et al.* (1987). *DRASTIC: A Standardized System for Evaluating Ground Water Pollution Potential using Hydrogeologic Settings*. NWWA/EPA Series. EPA/600/2-87/035. United States Environmental Protection Agency, Washington, DC.

ASTM (2001) *Risk-Based Corrective Action (RBCA) (E2081-00) Standard Guide for Risk-Based corrective Action*. American Society for Testing and Materials, Philadelphia, PA.

Bekins, B., Rittmann, B. and MacDonald, J. (2001a) Natural attenuation strategy for groundwater cleanup focuses on demonstrating cause and effect. *Eos* **82**(53), 57–58.

Bekins, B.A., Cozzarelli, I.M., Godsy, E.M. *et al.* (2001b) Progression of natural attenuation processes at a crude oil spill site: II. Controls on spatial distribution of microbial populations. *Journal of Contaminant Hydrology* **53**, 387–406.

Berg, M., Tran, H.C., Nguyen, T.C. *et al.* (2001) Arsenic contamination of groundwater and drinking water in Vietnam: a human health threat. *Environmental Science & Technology* **35**, 2621–2626.

BGS & DPHE (2001) Arsenic contamination of groundwater in Bangladesh (eds D.G. Kinniburgh & P.L. Smedley). Volume 2, *Final Report. British Geological Survey Report WC/00/19*. British Geological Survey, Keyworth.

Blowes, D.W., Ptacek, C.J., Benner, S.G. *et al.* (2000) Treatment of inorganic contaminants using permeable reactive barriers. *Journal of Contaminant Hydrology* **45**, 123–137.

Burgess, W.G., Burren, M., Perrin, J. *et al.* (2002) Constraints on sustainable development of arsenic-bearing aquifers in southern Bangladesh. Part 1: a conceptual model of arsenic in the aquifer. In: *Sustainable Groundwater Development* (eds K.M. Hiscock, M.O. Rivett and R.M. Davison). Geological Society, London, Special Publications **193**, pp. 145–163.

Carey, M.A., Fretwell, B.A., Mosley, N.G. and Smith, J.W.N. (2002) *Guidance on the Use of Permeable Reactive Barriers for Remediating Contaminated Groundwater*. National Groundwater and Contaminated Land Centre report NC/01/51. Environment Agency, Bristol.

Clark, L. and Sims, P.A. (1998) Investigation and clean-up of jet-fuel contaminated groundwater at Heathrow International Airport, UK. In *Groundwater Contaminants and Their Migration* (eds J. Mather, D. Banks, S. Dumpleton and M. Fermor). Geological Society, London, Special Publications **128**, pp. 147–157.

Council of the European Communities (1991) Directive concerning the protection of waters against pollution caused by nitrates from agricultural sources (91/676/EEC). *Official Journal of the European Communities* L375, 1–8. Brussels.

Cozzarelli, I.M., Bekins, B.A., Baedecker, M.J. *et al.* (2001) Progression of natural attenuation processes at a crude oil spill site: I Geochemical evolution of the plume. *Journal of Contaminant Hydrology* **53**, 369–385.

Daly, D., Dassargues, A., Drew, D. *et al.* (2002) Main concepts of the "European approach" to karst-groundwater-vulnerability assessment and mapping. *Hydrogeology Journal* **10**, 340–345.

Davison, R.M., Prabnarong, P., Whittaker, J.J. and Lerner, D.N. (2002) A probabilistic management system to optimize the use of urban groundwater. In: *Sustainable Groundwater Development* (eds K.M. Hiscock, M.O. Rivett and R.M. Davison). Geological Society, London, Special Publications **193**, pp. 265–276.

DEFRA (2002). *Guidelines for Farmers in NVZs – England*. Department for Environment, Food and Rural Affairs, London. http://archive. defra.gov.uk/environment/quality/water/ waterquality/diffuse/nitrate/help-for-farmers. htm (accessed 10 June, 2013).

DEFRA (2012). *Nitrate Vulnerable Zones*. Department for Environment, Food and Rural Affairs, London. http://archive.defra.gov.uk/ environment/quality/water/waterquality/ diffuse/nitrate/nvz2008.htm (accessed 10 June, 2013).

Dhar, R.K., Biswas, B.K., Samanta, G. *et al.* (1997) Groundwater arsenic calamity in Bangladesh. *Current Science* **73**, 48–59.

DPHE (1999) *Groundwater Studies for Arsenic Contamination in Bangladesh. Final Report, Rapid Investigation Phase*. Department of Public Health Engineering, Government of Bangladesh. British Geological Survey and MottMacDonald, London.

DPHE (2000). *Groundwater Studies for Arsenic Contamination in Bangladesh. Supplemental Data to Final Report, Rapid Investigation Phase*. Department of Public Health Engineering, Government of Bangladesh. British Geological Survey. http://www.bgs.ac.uk/arsenic/ Bangladesh/home.htm (accessed 10 June, 2013).

DPHE (2001) *Groundwater Studies for Arsenic Contamination in Bangladesh. Final Report, Rapid Investigation Phase*. Department of

Public Health Engineering, Government of Bangladesh. British Geological Survey & MottMacDonald.

Environment Agency (1998) *Policy and Practice for the Protection of Groundwater*. Environment Agency, Bristol.

Environment Agency (2012) *Groundwater Protection: Principles and Practice*. Environment Agency, Bristol.

European Environment Agency (2003) *Europe's Water: An Indicator-Based Assessment. Summary*. European Environment Agency, Copenhagen.

Fetter, C.W. (1999) *Contaminant Hydrogeology* (2nd edn). Prentice-Hall, Inc., Upper Saddle River, New Jersey.

Foster, S.S.D. (1985) Groundwater protection in developing countries. In: *Theoretical Background, Hydrogeology and Practice of Groundwater Protection Zones* (eds G. Matthess, S.S.D. Foster and A.C. Skinner). Verlag Heinz Heise, Hanover, pp. 167–200.

Griffin, A.R. and Mather, J.D. (1998) Landfill disposal of urban wastes in developing countries: balancing environmental protection and cost. In: *Geohazards in Engineering Geology* (eds J.G. Maund and M. Eddleston). Geological Society, London, Engineering Geology Special Publications **15**, pp. 339–348.

Hocking, G., Wells, S.L. and Ospina, R.I. (2000) Deep reactive barriers for remediation of VOCs and heavy metals. In: *Chemical Oxidation and Reactive Barriers: Remediation of Chlorinated and Recalcitrant Compounds* (eds G.B. Wickramanayake, A.R. Gavaskar and A.S.C. Chen). Battelle Press, Columbus, Ohio, pp. 307–314.

Iván, V. and Mádl-Szonyi, J. (2017) State of the art of karst vulnerability assessment: overview, evaluation and outlook. *Environmental Earth Sciences* **76**, 112. DOI: 10.1007/s12665-017-6422-2.

Javandel, I. and Tsang, C.-F. (1986) Capture-zone type curves: a tool for aquifer cleanup. *Ground Water* **24**, 616–625.

Johnson, T.L., Scherer, M.M. and Tratnyek, P.G. (1996) Kinetics of halogenated organic compound degradation by iron metal.

Environmental Science & Technology **30**, 2634–2640.

Jones, N.A., Hansen, J., Springer, A.E. *et al.* (2019) Modeling intrinsic vulnerability of complex karst aquifers: modifying the COP method to account for sinkhole density and fault location. *Hydrogeology Journal* **27**, 2857–2868.

Lake, I.R., Lovett, A.A., Hiscock, K.M. *et al.* (2003) Evaluating factors influencing groundwater vulnerability to nitrate pollution: developing the potential of GIS. *Journal of Environmental Management* **68**, 315–328.

Leeson, J., Edwards, A., Smith, J.W.N. and Potter, H.A.B. (2003) *Hydrogeological Risk Assessment for Landfills and the Derivation of Groundwater Control and Trigger Levels.* Environment Agency, Bristol.

Livingstone, S., Franz, T. and Guiguer, N. (1995) Managing ground-water resources using wellhead protection programs. *Geoscience Canada* **22**, 121–128.

Mackay, D.M. and Cherry, J.A. (1989) Groundwater contamination: pump-and-treat remediation. *Environmental Science & Technology* **23**, 630–636.

McArthur, J.M., Ravenscroft, P., Safiulla, S. and Thirwall, M.F. (2001) Arsenic in groundwater: testing pollution mechanisms for sedimentary aquifers in Bangladesh. *Water Resources Research* **37**, 109–117.

Municipality of Aalborg (2001). Sustainable land-use in ground water catchment areas. Technical Final Report and Layman's Report. EU LIFE Project Number LIFE97 ENV/DK/000347. The Municipality of Aalborg, Aalborg, 56 pp.

National Academy of Sciences (2000) *Natural Attenuation for Groundwater Remediation.* National Academy Press, Washington, DC.

Nickson, R.T., McArthur, J.M., Ravenscroft, P. *et al.* (2000) Mechanism of arsenic release to groundwater, Bangladesh and West Bengal. *Applied Geochemistry* **15**, 403–413.

O'Hannesin, S.F. and Gillham, R.W. (1998) Long-term performance of an in situ 'iron wall' for remediation of VOCs. *Ground Water* **36**, 164–170.

Petts, J., Cairney, T. and Smith, M. (1997) *Risk-Based Contaminated land Investigation and Assessment.* Wiley, Chichester.

Rahman, A.A. and Ravenscroft, P. (eds) (2003) *Groundwater Resources and Development in Bangladesh: Background to the Arsenic Crisis, Agricultural Potential and the Environment.* University Press Ltd., Dhaka

Rajasooriyar, L., Mathavan, V., Dharmagunawardhane, H.A. and Nandakumar, V. (2002) Groundwater quality in the Valigamam region of the Jaffna Peninsula, Sri Lanka. In: *Sustainable Groundwater Development* (eds K.M. Hiscock, M.O. Rivett and R.M. Davison). Geological Society, London, Special Publications **193**, pp. 181–197.

Rajasooriyar, L.D. (2003). A study of the hydrochemistry of the Uda Walawe Basin, Sri Lanka, and the factors that influence groundwater quality. PhD Thesis, University of East Anglia, Norwich.

Rajasooriyar, L.D., Boelee, E., Prado, M.C.C.M., Hiscock, K.M. (2003). Mapping the potential human health implications of groundwater pollution in southern Sri Lanka. *Water Resources and Rural Development*, **1–2**, 27–42.

Ravenscroft, P., Burgess, W.G., Ahmed, K.M. *et al.* (2005) Arsenic in groundwater of the Bengal Basin, Bangladesh: distribution, field relations, and hydrogeological setting. *Hydrogeology Journal* **13**, 727, doi:10.1007/s10040-003-0314-0.

Ravenscroft, P., McArthur, J.M. and Hoque, B.A. (2001) Geochemical and palaeohydrological controls on pollution of groundwater by arsenic. In: *Arsenic Exposure and Health Effects* (eds W.R. Chappell, C.O. Abernathy and R.L. Calderon). Elsevier Science, Amsterdam, pp. 53–77.

Richardson, J.P. and Nicklow, J.W. (2002) *In situ* permeable reactive barriers for groundwater contamination. *Soil and Sediment Contamination* **11**, 241–268.

Robins, N.S., Adams, B., Foster, S.S.D. and Palmer, R.C. (1994) Groundwater vulnerability

mapping: the British perspective. *Hydrogéologie*
3, 35–42.

Royal Society (1992) *Risk: Analysis, Perception
and Management.* The Royal Society, London.

Schoenheinz, D., Grischek, T., Worch, E. *et al.*
(2002) Groundwater pollution at a pulp and
paper mill at Sjasstroj near Lake Ladoga,
Russia. In: *Sustainable Groundwater
Development* (eds K.M. Hiscock, M.O. Rivett
and R.M. Davison). Geological Society,
London, Special Publications **193**, pp. 277–291.

Smedley, P.L. and Kinniburgh, D.G. (2002) A
review of the source, behaviour and
distribution of arsenic in natural waters.
Applied Geochemistry **17**, 517–568.

Starr, R.C. and Cherry, J.A. (1994) In situ
remediation of contaminated ground water: the
funnel-and-gate system. *Ground Water* **32**,
465–476.

Tobin, B.W., Springer, A.E., Kreamer, D.K. and
Schenk, E. (2018) Review: the distribution, flow,
and quality of Grand Canyon Springs, Arizona
(USA). *Hydrogeology Journal* **26**, 721–732.

United States Environmental Protection Agency
(1991a) *Wellhead Protection Strategies for
Confined-Aquifer Settings.* EPA Report /570-9-
91/008. Office of Ground Water and Drinking
Water, Washington, DC.

United States Environmental Protection Agency
(1991b) *Delineation of Wellhead Protection
Areas in Fractured Rocks.* EPA Report /570-9-
91/009. Office of Ground Water and Drinking
Water, Washington, DC.

United States Environmental Protection Agency
(1993) *Guidelines for Delineation of Wellhead
Protection Areas.* EPA Report /440-5-93/001,
Office of Ground Water, Office of Ground
Water Protection. Washington, DC.

United States Environmental Protection
Agency (1997) *Use of Monitored Natural
Attenuation at Superfund, RCRA Corrective
Action, and Underground Storage Tank Sites.*
Directive 9200.4-17P. Office of Solid
Waste and Emergency Response,
Washington, DC.

United States National Research Council (1983)
*Risk Assessment in the Federal Government:
Managing the Process.* National Academy Press,
Washington, DC.

van Waegeningh, H.G. (1985) Protection of
groundwater quality in porous permeable
rocks. In: *Theoretical Background,
Hydrogeology and Practice of Groundwater
Protection Zones* (eds G. Matthess, S.S.D. Foster
and A.C. Skinner). Verlag Heinz Heise,
Hanover, pp. 111–121.

Vías, J.M., Andreo, B., Perles, M.J. *et al.* (2006)
Proposed method for groundwater
vulnerability mapping in carbonate (karstic)
aquifers: the COP method. *Hydrogeology
Journal* **14**, 912–925.

World Health Organisation (1994) *Guidelines for
Drinking Water Quality. Volume 1:
Recommendations* (2nd edn). World Health
Organisation, Geneva.

Zwahlen, F. ed. (2004) *Vulnerability and Risk
Mapping for the Protection of Carbonate
(karst) Aquifers.* European Commission COST
Action 620. European Communities,
Luxembourg.

10

Groundwater resources, governance and management

10.1 Introduction

The development of groundwater resources for public, agricultural and industrial uses can create environmental conflicts. Groundwater abstractions capture recharge water that might otherwise flow to springs and rivers and so diminishing the freshwater habitats dependent on groundwater discharge (Section 6.8). In the current era of integrated river basin management (Section 1.7.2.1), sufficient volume of water is required to maintain freshwater (or saline) ecosystems. In this way, the fraction of available recharge needed for environmental benefits is accounted for, together with the fraction required for human and economic benefits in order to achieve sustainable groundwater development (Fig. 1.1).

In this chapter, examples of sustainable and non-sustainable groundwater resources development schemes at large and regional scales are discussed together with examples of modern groundwater management techniques including artificial storage and recovery and riverbank filtration schemes. The next section illustrates the adverse environmental impacts of groundwater exploitation on ecosystems with reference to the sensitivity of wetlands to changes in groundwater inputs. The following section discusses possible changes in the quantity and quality of groundwater resources as a result of climate change and includes consideration of the contribution of groundwater pumping to greenhouse gas emissions and the impacts of climate change of cold-region hydrogeology. Given the drive to net-zero carbon emissions by 2050, the role and interaction of groundwater in the exploitation of energy resources, including renewable resources and shale gas, is reviewed. Finally, this chapter concludes with a description of approaches to groundwater governance and management to ensure the future, long-term sustainability of groundwater resources.

10.2 Groundwater resources schemes

The assessment and development of groundwater resources is central to hydrogeology. Groundwater resources have a number of positive advantages compared to surface reservoir developments that include: (1) a large storage volume that can be developed in stages as demand for water arises; (2) resilience to drought conditions because of the large storage volume; (3) relatively low environmental impact of wellfield developments; and (4) no loss of storage volume to evaporation. With current awareness that surface water and groundwater resources should be managed together, it is useful to conceive this approach in the context of a water balance equation (Section 6.2) that equates demand for water against abstraction requirements and environmental needs.

Hydrogeology: Principles and Practice, Third Edition. Kevin M. Hiscock and Victor F. Bense.
© 2021 John Wiley & Sons Ltd. Published 2021 by John Wiley & Sons Ltd.
Companion website: www.wiley.com/go/hiscock/hydrogeology3e

10.2.1 Large-scale groundwater development schemes

The understanding of aquifer conditions and the compilation of a water balance are central to water resources management. An example of a groundwater budget for a large-scale aquifer system, in which the primary inflows and outflows to the aquifer are tabulated, is shown in Table 10.1 for the United States High Plains Aquifer. The High Plains Aquifer consists mainly of near-surface deposits of late Tertiary or Quaternary age forming one unconfined aquifer and underlies 450 660 km^2 in parts of eight States within the Great Plains physiographic province. The Ogallala Formation of Miocene age, which underlies 347 060 km^2 is the principal hydrogeological unit and consists of a heterogeneous sequence of clay, silt, sand and gravel. Use of the High Plains Aquifer as a source of irrigation water has transformed the mid-section of the United States into one of the major agricultural regions of the world. Principal crops are cotton, alfalfa and grains, especially wheat, sorghum and maize. Grains provide feed for the 15 million cattle and the 4.25 million swine (1997) that are raised over the aquifer. In addition, the aquifer provides drinking water to 82% of the people who live within the aquifer boundaries (Dennehy *et al.* 2002).

Table 10.1 Groundwater budget information for the High Plains Aquifer based on the United States Geological Survey's Regional Aquifer System Analysis groundwater model[a] (Luckey *et al.* 1986).

Budget parameter	Northern High Plains	Central High Plains	Southern High Plains
Primary inflows (in cubic metres per year)			
Recharge from precipitation on rangeland and streams	[b]5.98 × 10^8	[b] 4.66 × 10^8	1.97 × 10^8
Recharge from precipitation on agricultural land	[c] 2.89 × 10^9	–	[c] 1.43 × 10^9
Groundwater irrigation return (pumpage minus crop demand)	2.31 × 10^9	2.07 × 10^9	3.61 × 10^9
Recharge from other human activities (e.g. seepage from reservoirs and canals)	2.31 × 10^9	–	–
Recharge from other aquifers across subunit boundary	–	[d]1.88 × 10^7	–
Totals	8.11 × 10^9	2.55 × 10^9	5.24 × 10^9
Primary outflows (in cubic metres per year)			
Total pumpage	6.48 × 10^9	[e]6.89 × 10^9	8.59 × 10^9
Discharge to streams and shallow water-table areas	2.87 × 10^9	4.15 × 10^8	–
Discharge along eastern boundary	–	6.97 × 10^7	1.05 × 10^8
Totals	9.35 × 10^9	7.37 × 10^9	8.70 × 10^9
Net residual	−1.24 × 10^9	−4.82 × 10^9	−3.46 × 10^9

[a] Assumptions: Inflow/Outflow values determined using 1960–1980 estimates; base of aquifer modelled as no-flow boundary; vertical flow in aquifer considered negligible on regional scale.
[b] Recharge distributed unevenly based on soil type.
[c] Additional recharge from precipitation on agricultural land because of changes in soil character due to tillage.
[d] Flow only from northern and southern subunits to central subunit.
[e] Municipal and industrial pumpage is 3.2% of this amount.
(*Source:* Luckey, R.R., Gutentag, E.D., Heimes, F.J. and Weeks, J.B. (1986). Digital simulation of ground-water flow in the High Plains aquifer in parts of Colorado, Kansas, Nebraska, New Mexico, Oklahoma, South Dakota, Texas, and Wyoming. United States Geological Survey Professional Paper 1400-D, 57 pp.)

According to Dennehy *et al.* (2002), groundwater flow in the High Plains Aquifer is generally from west to east, discharging naturally to springs and streams and is subject to evapotranspiration (ET) in areas where the water table is close to the land surface. Pumping from numerous irrigation wells is, however, the principal mechanism of groundwater discharge. Abstractions greatly exceed recharge in many areas, causing large declines in water levels; for example, declines of 30–43 m since the 1940s to 1980 in parts of Kansas, New Mexico, Oklahoma and Texas.

From the groundwater budget shown in Table 10.1, it is apparent that each region of the High Plains Aquifer is in a deficit situation with outflows greater than inflows. In the southern and central regions, outflows are about twice the inflows. Prior to irrigation development, precipitation recharged the aquifer at an average rate of 15 mm a^{-1} and small quantities of water discharged to the springs and rivers. Irrigation represents a largely consumptive use of water and since development, groundwater abstraction has removed about 7% of the original total water volume from the aquifer. The water budget indicates that the decrease in storage would have been worse if 30–40% of the water pumped for irrigation

had not infiltrated to the aquifer as irrigation return flows each year. Declining groundwater levels are a direct threat to the current way of life of the area, and Dennehy *et al.* (2002) detail attempts that are being made to introduce more water-efficient irrigation and best-management farming practices including, ultimately, a shift from irrigated agriculture to dryland farming with its attendant far-reaching implications for the local economy.

Although the High Plains Aquifer presents a case of non-sustainable development of groundwater, in practice, there is no fundamental reason why the temporary over-exploitation of aquifer storage for a given benefit should not be allowed as part of a logical water resources management strategy as long as the groundwater system is sufficiently well-understood to evaluate impacts. An example of the deliberate mining of groundwater resources for the benefit of a national economy is the Great Man-made River project (GMRP) in Libya (Box 10.1). The GMRP involves the abstraction of fossil groundwater recharged during a pluvial period of the last ice age from beneath the Libyan Desert and its transfer to coastal cities and towns. Clearly the groundwater supply is not a permanent, renewable resource but for a period of at least decades this

---

Box 10.1 Great Man-made River Project

Libya covers an area of some 1.8×10^6 km^2 bounded in the north by the Mediterranean coastline of approximately 1600 km in length (Fig. 10.1). The climate varies from Mediterranean along the coast with winter rainfall totals of about 270 mm in Tripoli and Benghazi to a desert climate in the south where rain seldom falls. Rainfall is higher at about 400 mm a^{-1} in the mountainous areas to the south of Tripoli and east of Benghazi. A total of 80% of Libya's population of about 4.7 million is concentrated in the coastal area with approximately 2.5 million residents in the major towns and cities of Tripoli and Benghazi. The majority (80%) of Libya's agricultural production is centred on the coastal plains and adjacent uplands, but arable crops are only viable with irrigation. Importation of cereals, sheep meat and other food stuffs are costly in terms of foreign exchange and has given impetus to improving the productivity of the land through irrigation and creating employment opportunities in agriculture. However, the relatively shallow aquifers in the coastal areas and the pressure of competing municipal and industrial demands for water pose a limitation on further agricultural development.

(Continued)

Box 10.1 (Continued)

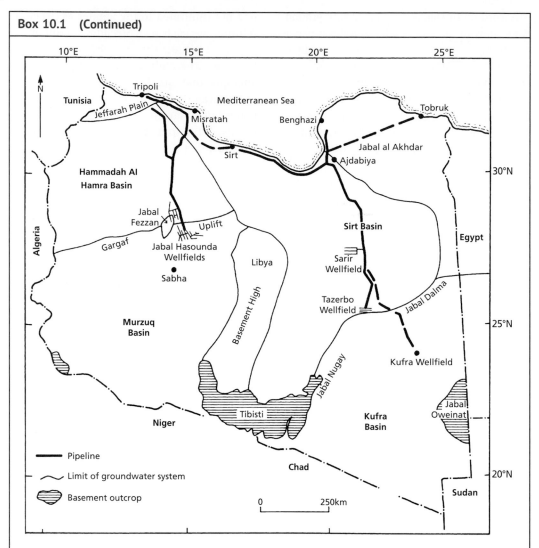

Fig. 10.1 Major aquifer basins of Libya and location of wellfields and conveyance systems of the Great Man-made River Project (McKenzie and Elsaleh 1994; Pim and Binsariti 1994). (*Sources:* McKenzie, H.S. and Elsaleh, B.O. (1994) The Libyan Great man-made river project paper 1. Project overview. *Proceedings of the Institution of Civil Engineers – Water Maritime and Environment* **106**, 103–122.; Pim, R.H. and Binsariti, A. (1994) The Libyan great man-made river project paper 2. The water resource. *Proceedings of the Institution of Civil Engineers – Water Maritime and Environment* **106**, 123–145.)

During the 1960s, exploratory drilling for oil in the Libyan desert established the presence of extensive groundwater reserves in three to four major basins, but an unwillingness of people to move to the Libyan Desert to utilize this resource for irrigation gave rise to the state decision to initiate the Great Man-made River Project (GMRP), one of the world's major groundwater developments. The total capacity of the major basins, which is estimated to be 35 000 km^3, is immense

and the only technical constraints on development are considered to be local ones of water quality and aquifer hydraulics.

The principal features of the GMRP are described by McKenzie and Elsaleh (1994) and include wellfields and conveyance systems comprising long lengths of very large diameter (4 m) pipelines required for the transfer of water to coastal districts. Phase I wellfields at Sarir and Tazerbo (Fig. 10.1) are situated 381 and 667 km south of

---

Box 10.1 (Continued)

Ajdabiya in the Sirt and Kufra Basins, respectively. The Sirt Basin contains a post-Eocene thickness of 1600 m of continental sands with increasing amounts of limestone found northwards. The important aquifer sediments of the Kufra Basin are Lower Devonian well-sorted, uncemented sand and sandstone. Regional groundwater contours indicate a small natural groundwater flow from south to north which is believed to be fed mainly from degradation of a mound of fossil groundwater located below the northern flank of the Tibisti Mountains. This palaeo-groundwater was recharged during a major pluvial period at the time of the last ice age between 14 000 and 38 000 years ago. Present recharge from rainfall is effectively zero in the central part of the Sirt Basin and Kufra Basin. Therefore, the eastern region wellfield developments are dependent on groundwater mining with only a minor contribution from interception of throughflow (Pim and Binsariti 1994).

The Sarir wellfield contains 136 production wells drilled at 450 m depth, each able to deliver 92 L s^{-1}. Pumping tests indicated that transmissivity ranges from 400 to 6000 m^2 day^{-1}. The pumped water quality ranges from 530 to 1367 mg L^{-1} of total dissolved solids with an average of 815 mg L^{-1}. At Tazerbo, 118 production wells are drilled at depths varying between 380 and 600 m depth, with each borehole able to deliver 102 L s^{-1}. Modelled drawdowns for the Tazerbo

wellfield are 95 m after 50 years with the greatest contribution (86%) to abstraction from vertical leakage (Pim and Binsariti 1994). Assuming an average annual demand of 10 000 m^3 ha^{-1} for irrigation water, the limit for irrigated agriculture for the Phase I supply is about 70 000 ha.

Phases II and III wellfield developments in the western region are planned on a similar scale in the Hammadah Al Hamra and Murzuq Basins which lie north and south of the Gargaf uplift, respectively (Fig. 10.1). Natural groundwater flows in the thick and complex sequence of layered and interconnected Palaeozoic and Mesozoic limestones and sandstones are generally from south to north, indicating that the southern mountains were a major recharge area during the last pluvial period. As in the eastern region, the source of present throughflow is considered to be the slowly depleting groundwater mound below the area of ancient recharge (Pim and Binsariti 1994).

The fully developed GMRP is expected to achieve a total groundwater abstraction of 6.18 × 10^6 m^3 day^{-1} and support approximately 200 000 ha of irrigated agricultural development with a minimum design life of 50 years. The total abstraction in 50 years is calculated to be 113 km^3 or 0.3% of the total groundwater resource. The estimated total cost of the GMRP is large at US$20 billion with an implementation period from the mid-1980s to 2005.

---

fossil water can supply most of the population of Libya with an essential resource. As argued by Price (2002), the GMRP appears to meet the ethical requirements of groundwater mining, namely that: (a) evidence is available that pumping can be maintained for a long period; (b) the negative impacts of development are smaller than the benefits; and (c) the users and decision-makers are aware that the resource will be eventually depleted. However,

Price (2002) noted that using non-renewable groundwater for inefficient agriculture in a region where there is effectively no recharge is not a sensible long-term plan.

10.2.2 Regional-scale groundwater development schemes

The previous two examples of large-scale groundwater developments in support of irrigated agriculture tip the sustainability equation towards economic gain. At the regional scale, effective river basin management can develop the large storage volume in aquifers in conjunction with surface resources. This concept is not new and several schemes, principally the Thames, Great Ouse, Severn, Itchen and Waveney Schemes, were developed in England and Wales following the Water Resources Act of 1963 (Section 1.7.1.1). Pilot studies were undertaken to assess the feasibility of regulating rivers by pumping groundwater into them for abstraction for water supply in the lower reaches, while maintaining an acceptable flow in the river to meet all other environmental requirements (Downing 1993).

As an example of a river regulation scheme, the Great Ouse Groundwater Scheme in eastern England included two well groups in a pilot area of 71.5 km^2 in the River Thet catchment (Fig. 10.2): riverside wells and a second group more remote from the river on higher ground. A control area was established in the adjacent River Wissey catchment. Eighteen boreholes were drilled into the Chalk aquifer and yielded 70×10^3 m^3 day^{-1}. When all the wells were pumped for 250 days, the pumping rate was three times the average infiltration across the pilot area. In six months, all the baseflow to the River Thet was intercepted and groundwater levels fell below the river bed over 80% of its length. Groundwater levels fell by an average of 2.5 m with stable conditions achieved after 200 days due to induced groundwater flow across the eastern boundary of the pilot area. In that the riparian zone is underlain by relatively low permeability peats, silts and clays, leakage through the river bed was 10% of the groundwater pumped. Concerns were expressed that pumping the Chalk aquifer could affect the Breckland Meres situated within the pilot area. These meres are groundwater-fed lakes of notable scientific and ecological importance (Herrera-Pantoja *et al.* 2012). However, drawdown of the groundwater level by more than 1 m due to extensive pumping in 1970 was relatively limited within the pilot area (Backshall *et al.* 1972).

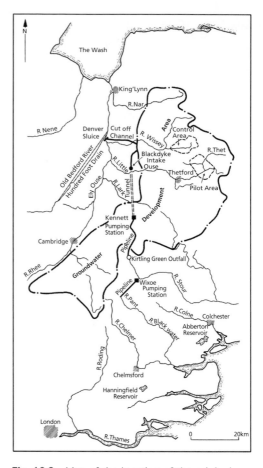

Fig. 10.2 Map of the location of the original development area of the Great Ouse Groundwater Scheme showing the various linked components of the current Ely Ouse Essex Water Transfer Scheme.

The success of a river regulation scheme can be expressed by the net gain to the river. The actual net gain at any given time depends on the aquifer properties and the distance of boreholes from natural spring discharges. During 1970 and 1971, and as shown in Fig. 10.3, the net gain in the Great Ouse Groundwater Scheme varied, but during summer periods, values were about 70% of the quantity pumped. The scheme was therefore successful in maintaining river flows at a high proportion of the mean flow. After pumping stopped, river flows and groundwater levels steadily recovered towards natural levels during the following

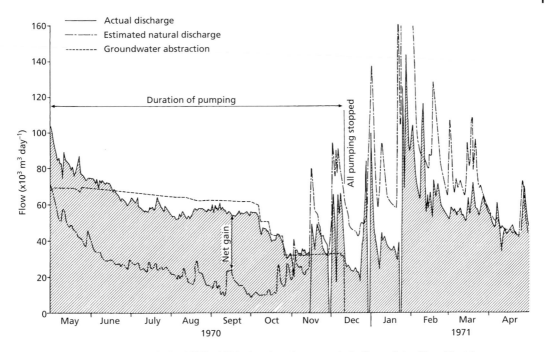

Fig. 10.3 Hydrograph records for 1970–1971 showing the net gain in flow of the River Thet from groundwater abstracted from Chalk boreholes in the pilot area of the Great Ouse Groundwater Scheme (see Fig. 10.2 for location). As shown, the net gain is the difference between the actual discharge with groundwater abstraction and the estimated natural discharge without abstraction. As a result of measurements taken for identical 24-h periods without allowance for travel times (or temporary storage effects), occasional zero flows (or even apparent losses) can arise. (Backshall *et al.* 1972.)

winter. Natural river flows were re-established in time for the beginning on the following base-flow recession period (Fig. 10.3). The delay in recovery is a feature of river regulation schemes and is due to winter groundwater recharge first replenishing the groundwater storage depleted by pumping during the summer regulation period. By definition, the net gain to the river over a long period of years must be zero if environmental impacts are not to be permanent.

Full development of the Great Ouse Groundwater Scheme was envisaged to include almost 350 new boreholes distributed over an area of 2500 km². The possibility of developing the Chalk groundwater resource in conjunction with surface runoff was estimated to yield 360×10^3 m³ day^{-1} of water for export to demand areas outside of the basin. The drawdown of the Chalk aquifer for the fully

developed scheme was predicted to be 4 m in a drought period with a 1 in 50-year return period. Such a large engineering programme of works was not carried forward partly because of conflicts with other users for Chalk groundwater, for example irrigation water for valuable vegetable crops, and also the change in philosophy towards protection of the environment, especially the Breckland Meres, from the effects of groundwater drawdown. However, the boreholes drilled for the pilot study are today incorporated in a much larger scheme, the Ely Ouse Essex Water Transfer Scheme, to provide groundwater to the large public water supply demand in the south-east of England.

The Ely Ouse Essex Water Transfer Scheme enables the transfer of surplus water from the Ely Ouse (the River Great Ouse) to the heads of Essex Rivers in the south-east of the Anglian

Region (Fig. 10.2) thereby making extra water available to the Essex Rivers. The county of Essex experiences conditions of low-effective precipitation of <125 mm a^{-1}, yet has a large and expanding population to the north of London. One of the great merits of the scheme is that it augments existing reservoir capacity, thus avoiding the loss of agricultural land to create new reservoirs. Under the scheme, surplus water from the eastern part of the catchment, including Chalk groundwater resources available for regulation of the westward flowing Chalk rivers (the Rivers Lark, Little Ouse and Thet), is transferred to the flood protection scheme Cut-Off Channel at Denver (Fig. 10.4). At this point, all upstream river needs have been met and so any surplus water that would otherwise be lost to tidal waters and eventually to the Wash is potentially available for transfer. The Impounding Sluice gate at Denver is designed to enable the water level in the river channel to be raised approximately 0.6 m, thereby producing a reversal of flow. The water is sent in a reverse direction approximately 25 km south-east to the Blackdyke Intake. Here it is drawn off into a 20-km long tunnel which terminates at Kennett, whereupon the water is pumped through a 14.3-km long pipeline to the River Stour at Kirtling Green. Part of this discharge is drawn off at Wixoe, 13.7-km

Fig. 10.4 Transfer of water from the Ely Ouse River to the Cut-Off Channel at Denver. Water transferred from here travels to Abberton and Hanningfield reservoirs in support of water supply and river flows in the drier south-east of England (Fig. 10.2).

downstream and pumped 10.3 km to the River Pant. The water transferred from Denver travels 141 km to Abberton reservoir and 148 km to Hanningfield reservoir. For two-thirds of these distances, use is made of exiting watercourses.

10.2.3 Managed aquifer recharge

Managed aquifer recharge (MAR) describes intentional banking and treatment of water in aquifers (Dillon 2005) and refers to a suite of methods that is increasingly used to maintain, enhance and secure groundwater systems under stress. River-bank filtration for drinking water supplies was firmly established in Europe by the 1870s and the first infiltration basins in Europe appeared in 1897 in Sweden and in 1899 in France (Dillon *et al.* 2019). Since the 1960s, implementation of MAR has accelerated at a rate of 5% a^{-1}. Currently, MAR has reached an estimated 10 km^3 a^{-1}, about 2.4% of groundwater abstraction in countries reporting MAR (or about 1% of global groundwater abstraction) (Dillon *et al.* 2019). In the future, and based on experience where MAR is more advanced, MAR is likely to exceed 10% of global groundwater abstraction (Dillon *et al.* 2019).

MAR can be applied at the local and basin level, to make more efficient use of water resources, assist conjunctive management of surface water and groundwater resources, to buffer against increasing intensity of climate extremes, particularly drought, and to protect and improve water quality in aquifers (Dillon *et al.* 2019). MAR is widely used in many countries internationally to enhance water supplies, particularly those not only in semi-arid and arid areas, but also in humid areas, primarily for water quality improvement. MAR has been successfully implemented to increase groundwater storage, improve water quality, restore groundwater levels, prevent saltwater intrusion, manage water distribution systems and enhance ecological benefits (Stefan and Ansems 2018).

To better understand the role of MAR in sustainable water management and adaptation to climate and land-use change, about 1200 case studies from 62 countries were collected and analysed with respect to historical development, site characterization, operational scheme, objectives and methods used, as well as quantitative and qualitative characterisation of inflow and outflow of water. The data were compiled into a global inventory of MAR schemes (Stefan and Ansems 2018), now available through the IGRAC Managed Aquifer Recharge Portal (IGRAC 2018).

The basic concept of MAR is the artificial recharge of an aquifer, either via an injection well or a recharge channel or pond, with water from another source, in order to be abstracted at a later date. A wide variety of methods are used for managing aquifer recharge under four broad categories of streambed channel modifications, bank filtration, water spreading and recharge wells (Dillon *et al.* 2019); while a fifth category, runoff harvesting, is used by IGRAC (2018). In describing the growing variety of methods used for MAR, Dillon *et al.* (2009) list the following schematic types, which are also illustrated in Fig. 10.5:

a) Aquifer storage and recovery (ASR): injection of water into a well for storage and recovery from the same well. This is useful in brackish aquifers, where storage is the primary goal and water treatment is a smaller consideration (see Section 10.2.3.1).

b) Aquifer storage, transfer and recovery (ASTR): injection of water into a well for storage, then recovery from a different well. This method achieves additional water treatment by extending the groundwater residence time in the aquifer beyond that of a single well.

c) Dry wells: typically shallow wells where water tables are very deep, allowing infiltration of very high quality water to the unconfined aquifer at depth.

d) Percolation tanks or recharge weirs: dams built in ephemeral streams detain water which infiltrates through the bed to enhance storage in unconfined aquifers for extraction down-valley.

e) Rainwater harvesting for aquifer storage: roof runoff is diverted into a well, sump or caisson filled with sand or gravel and allowed to percolate to the water table where it is collected by pumping from a well.

f) Bank filtration: extraction of groundwater from a well or caisson near or under a river or lake to induce infiltration from the surface water body, thereby improving and making more consistent the quality of water recovered (see Section 10.2.3.2).

g) Infiltration galleries: buried trenches (containing polythene cells or slotted pipes) in permeable soils that allow infiltration through the unsaturated zone to an unconfined aquifer.

h) Dune filtration: infiltration of water from ponds constructed in dunes and extraction from wells or ponds at lower elevation for water quality improvement and to balance supply and demand.

i) Infiltration ponds: diversion of surface water into off-stream basins and channels that allow water to soak through an unsaturated zone to the underlying unconfined aquifer.

j) Soil aquifer treatment (SAT): treated sewage effluent is intermittently infiltrated through infiltration ponds to facilitate nutrient and pathogen removal during passage through the unsaturated zone for recovery by wells after residence in the unconfined aquifer.

k) Underground dams: in ephemeral streams where basement highs constrict flows, a trench is constructed across the streambed, keyed to the basement and backfilled with low permeability material to help retain flood flows in saturated alluvium for stock and domestic use.

l) Sand dams: built in ephemeral stream beds in arid areas on low permeability lithology to trap sediment when flow occurs. Following successive floods, the sand dam is raised

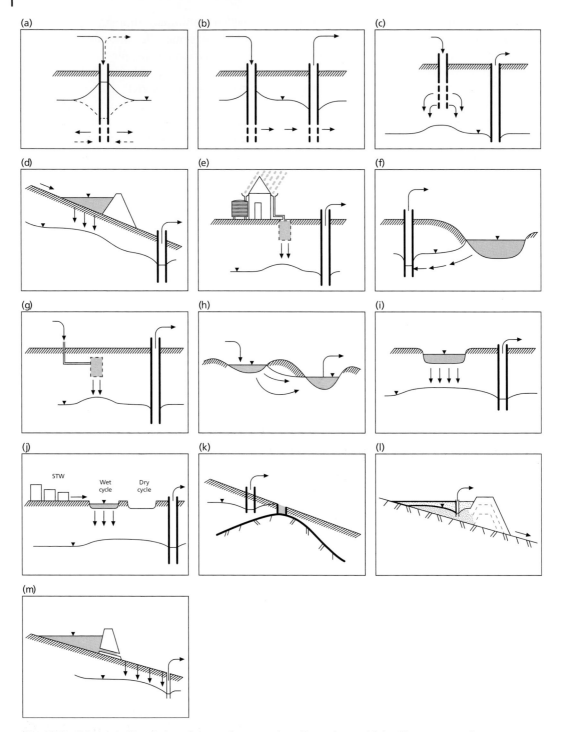

Fig. 10.5 Schematic illustration of types of managed aquifer recharge. (a) Aquifer storage and recovery, ASR; (b) Aquifer storage, transfer and recovery, ASTR; (c) Dry well; (d) Percolation tank; (e) Rainwater harvesting; (f) Bank filtration; (g) Infiltration gallery; (h) Dune filtration; (i) Infiltration pond; (j) Soil aquifer treatment (SAT); (k) Underground dam; (l) Sand dam; (m) Recharge releases. See text for further description of scheme types (Dillon *et al.* 2009). (*Source:* Adapted from Dillon, P., Pavelic, P., Page, D. *et al.* (2009). *Managed Aquifer Recharge: An Introduction.* Waterlines Report Series **13**, Australian Government, National Water Commission, Canberra, 64 pp.)

to create an 'aquifer' which can be tapped by wells in dry seasons.

m) Recharge releases: dams on ephemeral streams are used to detain flood water. Applications include slow release of water into the streambed downstream to match the capacity for infiltration into underlying aquifers, thereby significantly enhancing recharge.

Selection of suitable sites for MAR and choice of method will depend on the topography, hydrology, hydrogeology and land use of the area. The water used for recharge usually comes from not only a lake, river or other aquifer but also reclaimed water (Song *et al.* 2019). The use of reclaimed water offers increased sustainability of water resources use. Therefore, MAR schemes are beneficial in allowing the storage of water in an aquifer in times of surplus, and then abstraction from aquifers in the time of need, for example in the summer when water supplies are in demand for agricultural use (Rawluk *et al.* 2013).

In order for MAR to be feasible in an area, a number of criteria have to be met for it to be viable (Fuentes and Vervoort 2020). The hydraulic conductivity is an important factor for MAR in terms of infiltration and later abstraction. Generally, high hydraulic conductivity and low specific yield of the aquifer, natural boundaries to stop groundwater escaping horizontally and vertically, and low salinity of existing groundwater are favourable characteristics (Knapton *et al.* 2019). Soil properties are critical for recharge, including soil percolation rate, root zone residence time, chemical limitations and soil surface condition, especially in agricultural applications (O'Geen *et al.* 2015; Kourakos *et al.* 2019). Naturally sandy soils are better than clay-rich soils, and those with a higher organic matter content have a beneficial effect on hydraulic conductivity. Further criteria in the selection of the recharge zone are areas with both proximity to a raw water source and conditions for water storage (Dahlqvist *et al.* 2019). In tropical areas, the capacity of an aquifer to receive recharge

during the wet season is considered a primary constraint (Knapton *et al.* 2019).

The means by which MAR has been employed vary greatly, depending on the availability of water sources, underlying geology and user need. In China, stormwater MAR has been used, although reliant on quantity and distribution of rainfall (Song *et al.* 2019). In Darwin, Australia, MAR has been considered to alleviate the impact of the dry season decline in groundwater levels in the early to mid-dry season (Knapton *et al.* 2019). Kourakos *et al.* (2019) considered MAR over a very large agricultural area in the Sacramento River basin, California, to improve drought resilience and baseflow with stream diversion to recharge zones. In Bangladesh, MAR was introduced in the form of collecting rooftop rainwater and pond water, redirecting into a sand filter, then infiltrated into the aquifer, where it would later be abstracted via standard tube wells for drinking water (Hasan *et al.* 2019). The scheme was successful, providing water in sufficient quantities throughout the year, resilient against cyclone and contamination hazards, and installed at lower cost compared with other methods, and relatively easy to operate (Hasan *et al.* 2019). In a reclaimed tidal flat area in the south-western Netherlands, Pauw *et al.* (2015) developed a controlled artificial recharge and drainage system to increase the volume of the freshwater lenses below creek ridges to increase freshwater supply and meet the irrigation demand.

MAR projects, particularly in urban areas, can meet additional objectives, other than for water supply. MAR schemes can provide multiple economic, social and environmental benefits and often it is the combination of these benefits which provides the basis for investing in MAR. In both rural and urban areas, MAR has been used successfully to reduce salinity of groundwater and protect crops where irrigation water was causing salinization, and it can be used to protect coastal aquifers from saline intrusion (Dillon *et al.* 2009).

Potential disadvantages of MAR include clogging of the recharge site and decline of permeability attributed to chemical, biological and physical processes (Song *et al.* 2019). Xinqiang *et al.* (2019) found that colloids play an important role, with even a small concentration having a significant impact. The cost of a MAR scheme also requires consideration, although compared to alternative resource options can be favourably low (Hasan *et al.* 2019). Furthermore, the risk of aquifer contamination by pollutants is a concern, especially when reclaimed water is used, with pre-treatment potentially needed to ensure safe drinking water supplies. Even though natural filtration removes many contaminants, pathogens and trace chemicals can still remain (Yuan *et al.* 2019). It is important that aquifer contamination is avoided if the MAR scheme is not to be rendered inoperable for many years.

Even though MAR can be the most economic, most benign, most resilient and most socially acceptable solution to water supply shortages, it has frequently not been implemented due to lack of awareness, inadequate knowledge of aquifers, immature perception of risk and incomplete policies for integrated water management, including linking MAR with demand management (Dillon *et al.* 2020). The growth in research has enlarged the scope of MAR, especially using wells, widened the types of source waters for recharge, reduced the costs of water treatment for sustainable operations, improved the quality and quantity of recovered water and given greater certainty for safe and efficient operation of MAR systems (Dillon *et al.* 2019). In spite of these advances, there remain a number of basic steps that would improve efficiency of investment in MAR and underpin the uptake of MAR where this is currently low. These can be expressed in the categories of extending case study information to include economic evaluations, extending research on fundamental processes to better locate, design, operate and monitor MAR schemes, and to translate scientific evidence into governance arrangements for water allocations and water quality protection (Dillon *et al.* 2019).

10.2.3.1 Artificial storage and recovery schemes

As surface water and groundwater schemes reach full development, the final stage is artificial recharge where water, often treated wastewater, is recharged through basins and returned to the aquifer. Although practised in other countries, often on an uncontrolled basis and potentially threatening longer-term groundwater quality (for example in China and Mexico; Foster *et al.* 1999), artificial recharge is not typically practised in the United Kingdom where treated water is returned to rivers from sewage treatment works. A relatively recent development of artificial recharge is aquifer storage and recovery (ASR) to meet peak demands for water. The ASR technique, shown schematically in Fig. 10.6, works on the principle of using boreholes to recharge drinking water-quality water into aquifers and to subsequently recover the stored water from the same boreholes during times of peak demand or drought periods. Such schemes operate by displacing the native groundwater, effectively creating an underground reservoir of near drinking water-quality water. The volume of water recovered from the aquifer for supply purposes is generally close to but not more than the volume injected (Eastwood and Stanfield 2001). ASR schemes are therefore considered to offer a sustainable means of groundwater development.

The criteria for deciding how much water can be recovered from an ASR scheme are normally based on water quality parameters. The recovery efficiency is calculated as the volume of water recovered when the water quality parameter is reached compared to the volume of water injected. It is not uncommon for recovery efficiencies to approach 100%. Full details of ASR programme development, system design and technical and non-technical issues are presented by Pyne (1995).

Wet season - aquifer storage replenishment

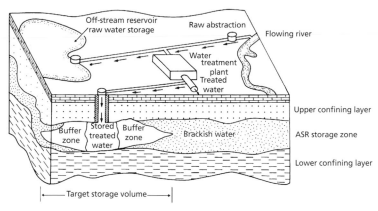

Dry season - aquifer storage depletion

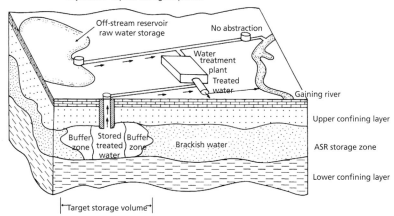

Fig. 10.6 Schematic diagram showing the operation of an aquifer storage and recovery (ASR) scheme. ASR involves storage of available, principally surface or drinking water-quality water during the wet season injected through boreholes completed in a brackish-water aquifer, with subsequent pumping from these same boreholes during dry periods. The freshwater forms a 'bubble' within the aquifer around the ASR well and can be retrieved when needed. The hydrogeological characteristics of a successful ASR storage zone include moderate aquifer hydraulic conductivity; confinement above and below by low permeability strata; and water quality as fresh as possible to limit mixing with the native brackish water (United States Geological Survey 2002). (*Source:* United States Geological Survey (2002). *SOFIA-SFRSF-Hydrology-What is aquifer storage and recovery (ASR)?* http://sofia.usgs.gov/sfrsf/rooms/hydrology/ASR/index.html (accessed 10 June 2013). (http://sofia.usgs.gov/).)

By the mid-1990s more than 20 ASR schemes were operational in the United States with many more planned. ASR is popular in that improvements in water quality in the native aquifer result from the injection of high-quality water, which then allows these aquifers to be used for supply purposes at a lower cost than other resource options. One ambitious scheme is the application of ASR technology as part of the Everglades Restoration project in South Florida. Between 300 and 330 ASR boreholes were originally planned with a combined capacity exceeding 6×10^6 m^3 day^{-1}, storing freshwater in the deep brackish Upper Floridan Aquifer (the ASR zone) using surface water that is currently discharged to the tide during the wet season. The stored water will be recovered during drought periods to sustain delivery of adequate high-quality water to the Everglades (Box 10.5). ASR has not previously been implemented on this scale and key uncertainties include the compatibility of the injected water with the aquifer water; effects of large volumes of injected water on the confining unit; efficiency in terms of how much water will be recovered; and the effects of the recovered water on the environment (National Research Council 2015). To answer these concerns, a phased approach to ASR implementation is being adopted to evaluate its feasibility and effectiveness as a regional water storage option.

Attempts at ASR in the United Kingdom have met with mixed success. A full-scale trial of ASR on a confined Chalk aquifer in Dorset in the south of England proved that there were no detrimental environmental effects; for any injection and recovery scenario, the impacts are absorbed by aquifer storage. However, a fluoride concentration of $2\,\mathrm{mg\,L^{-1}}$, equal to half of the background groundwater fluoride concentration, led to disappointingly low recovery efficiencies (<15%) (Eastwood and Stanfield 2001). In contrast to the Dorset scheme, ASR has been successfully operated in the confined Chalk and Tertiary Basal Sands aquifer in North London (Box 10.2).

Box 10.2 The North London Artificial Recharge Scheme

The North London Artificial Recharge Scheme and its relationship to existing surface water resources in the Lea Valley are shown in Fig. 10.7. In average rainfall years, flows in the Rivers Lea and Thames, with the associated pumped-storage reservoirs, are sufficient to meet current demands. Normally, there is surplus water which can be used to increase aquifer storage. During a drought, when river flows and associated storage levels in the reservoirs become critical, stored groundwater can be abstracted for supply. Abstracted water from the Enfield-Haringey boreholes is discharged to the New River, an aqueduct built in 1613, where it is transferred to the Coppermills water treatment works (Fig. 10.7). All groundwater, including water abstracted from the Lea Valley wells and boreholes and discharged directly into the surface reservoirs, is blended with raw surface water and treated at Coppermills, thus minimizing capital and operational costs. This is an important consideration in that the scheme has been designed for infrequent use with long periods of relatively small-scale recharge, followed by shorter periods of large-scale abstraction (O'Shea and Sage 1999).

The North London scheme utilizes fully treated drinking water as the source of the gravity-fed artificial recharge water, via the normal distribution system. The recharge water quality is similar to the background groundwater in the aquifer. The Enfield-Haringey boreholes vary in depth from 80 to 130 m and can provide a total yield of $90 \times 10^3\,\mathrm{m^3\,day^{-1}}$. The Lea Valley wells and boreholes can supply $60 \times 10^3\,\mathrm{m^3\,day^{-1}}$. Hence, the design yield is $150\,\mathrm{m^3\,day^{-1}}$ for a drought period of 200 days and is expected to give only small declines in regional groundwater levels.

During the dry years of the 1990s, the North London Artificial Recharge Scheme was used on several occasions in 1995, 1996 and 1997 to support low river and reservoir levels in the Lea Valley. Cycles of abstraction and recharge from June to November 1997 recorded individual daily rates of abstraction averaging about $100 \times 10^3\,\mathrm{m^3\,day^{-1}}$. This abstraction rate allowed a decrease in support for the Lea Valley system from the River Thames which, in turn, decreased the rate of decline in the Thames stored-water system, while conserving aquifer storage. In total during this period, $10.7 \times 10^6\,\mathrm{m^3}$ was withdrawn

Box 10.2   (Continued)

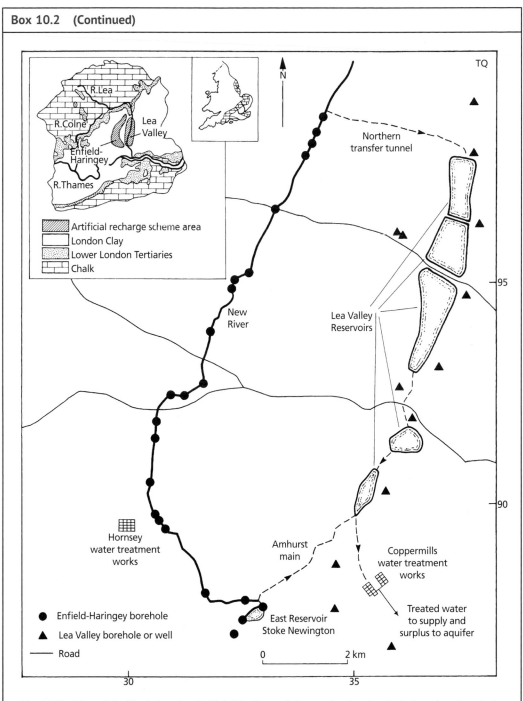

Fig. 10.7   Map of the North London Artificial Recharge Scheme showing borehole locations in relation to the New River, strategic transfer water mains and Lea Valley reservoirs (O'Shea and Sage 1999). (*Source:* O'Shea, M.J. and Sage, R. (1999) Aquifer recharge: an operational drought-management strategy in North London. *Journal of the Institution of Water and Environmental Management* **13**, 400–405.)

(*Continued*)

Box 10.2   (Continued)

from groundwater storage in North London, equivalent to 25% of the useable capacity of the Lea Valley reservoirs (O'Shea and Sage 1999).

This innovative artificial storage and recovery scheme is therefore considered successful in providing good quality water with the

environmental benefit of balancing groundwater abstraction with natural and artificial recharge with no net effect on long-term groundwater levels. In addition, the confined nature of the combined Chalk and Basal Sands aquifer ensures that abstraction has no impact upon the overlying river system.

10.2.3.2 Riverbank filtration schemes

In many countries of the world, alluvial aquifers hydraulically connected to a water course are preferred sites for drinking water production, given the relative ease of shallow groundwater exploitation, the generally high production capacity and the proximity to demand areas (Doussan *et al.* 1997). Although proximity to a river can ensure significantly higher recharge and pumping rates, water quality problems may be encountered during

exploitation of riverbank well-fields (Bertin and Bourg 1994). Even with these problems, groundwater derived from infiltrating river water provides 50% of potable supplies in the Slovak Republic, 45% in Hungary, 16% in Germany and 5% in The Netherlands. In Germany, riverbank filtration supplies 75% of the water supply to the City of Berlin and is the principal source of drinking water in Düsseldorf, situated on the Rhine (Box 10.3).

Box 10.3   Riverbank filtration at the Düsseldorf waterworks, River Rhine, Germany

The Düsseldorf waterworks has been using riverbank filtration since 1870 and is the most important source for public water supply in this densely populated and industrialized region (Fig. 10.8). There have been several threats to this supply in the last few decades that have included poor river water quality, heavy clogging of the river bed and accidental pollution, all of which have been overcome (Schubert 2002). Until about 1950, the Düsseldorf riverbank filtration scheme, like others in the Lower Rhine region, experienced good quality that permitted drinking water production without further treatment, other than disinfection. After 1950, the quality of the river water began to deteriorate gradually as the impacts of increasing quantities and insufficient treatment of industrial and municipal effluents became apparent. The consequences of these discharges were a noticeable drop in the oxygen concentration, an increase in the load of organic pollutants and the development of

anoxic conditions in the adjacent aquifer. Moreover, clogging of the river bed with particulate organic matter threatened the well yield of the riverbank filtration scheme. At this time, the pumped well water had to be treated to remove iron, manganese and ammonium and, furthermore, micro-pollutants. These changes prompted field studies to understand the riverbank filtration process and to better manage the water supply through the development of calibrated numerical models for the simulation of flow and transport.

An investigation of the river bed conditions and flow and transport processes was carried out in 1987 at the Flehe waterworks, Düsseldorf. As shown in Fig. 10.8, the production wells are situated on an outer bend of the River Rhine between 730.7 and 732.1 km down-river of the source. There is one horizontal collector well at 730.7 km situated approximately 80 m from the waterline (at mean discharge) and 70 vertical filter wells

Box 10.3 (Continued)

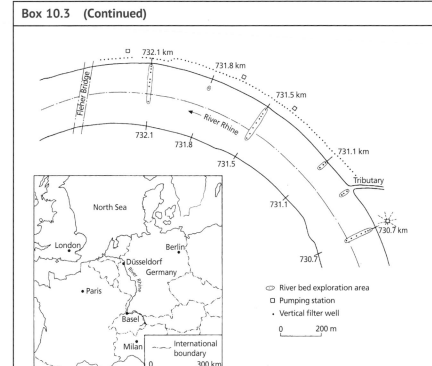

Fig. 10.8 Location map of the River Rhine at the Flehe waterworks, Düsseldorf (Schubert 2002). (*Source:* Schubert, J. (2002) Hydraulic aspects of riverbank filtration – field studies. *Journal of Hydrology* **266**, 145–161. © 2002, Elsevier.)

connected by siphon pipes that form a well gallery parallel to the riverbank at about 50 m from the waterline. The total length of the well gallery is 1400 m with the pumping rate during the field studies roughly constant between 3.0 and 3.4×10^4 m^3 day^{-1}. The aquifer consists of Pleistocene sandy gravel sediments with a hydraulic conductivity of between 4×10^{-3} and 2×10^{-2} m s^{-1} and a thickness of approximately 20 m. The aquifer is overlain by a 0.5–2 m thick meadow loam and is underlain by nearly impermeable Tertiary fine sands. The riverbank slope is coated by a 0.5 m thick clay layer above the mean water level of the river and is protected by basalt blocks below the mean water level.

With the aid of a diving cabin, a 1987 field study identified three different zones on the river bed (Fig. 10.9). Zone 1 nearest the production wells is 80 m wide and is comprised of fixed ground that is fully clogged by suspended solids that form a silt layer. This silt layer is formed by mechanical clogging and is almost impermeable with a hydraulic conductivity of 10^{-8} m s^{-1}. No chemical clogging due to mineral precipitation was observed under the aerobic conditions present in the aquifer. Zone 2 is also fixed ground but is only partially clogged allowing good permeability for infiltrating river water. This zone is 50–80 m wide and has a hydraulic conductivity of 3×10^{-3} m s^{-1}. Zone 3, the region between the middle of the river and the opposite bank, is formed by moveable ground which is shaped by normal river flow but mainly by flood events. The hydraulic

(Continued)

Box 10.3 (Continued)

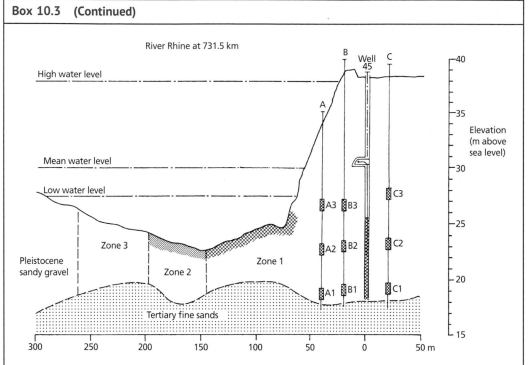

Fig. 10.9 Vertical section through the Pleistocene sandy gravel aquifer below the River Rhine in front of the Flehe waterworks, Düsseldorf, showing the location of monitoring well installations A, B and C at position 731.5 km (see Fig. 10.8) and the three zones of classification of the river bed (Schubert 2002). (*Source:* Schubert, J. (2002) Hydraulic aspects of riverbank filtration – field studies. *Journal of Hydrology* **266**, 145–161. © 2002, Elsevier.)

conductivity in Zone 3 is higher than the other zones with a value in the range 4×10^{-3} to 2×10^{-2} m s^{-1}.

Insight into the flow processes is provided by data for chloride that acts as a tracer for the bank filtrate. Figure 10.10 shows the results from a later investigation between March and May 1990 when nearly steady-state conditions characterized the river level (line WL in Fig. 10.10). Observed weekly variations in chloride concentrations result due to inputs from salt mining and its effluents. Chloride concentration data for the River Rhine (RH) and sampling points A1, A2 and A3 in monitoring well A (see Fig. 10.9 for position) show a clear succession; flow first appears in A3 (the upper layer), second, in A2 (the middle layer) and third, in A1

(the lower layer). Under the prevailing steady-state conditions, the travel times can be calculated; infiltrated river water takes approximately 10 days to flow to A3, 30 days to reach A2 and 60 days to pass A1 (Schubert 2002). The important finding from these field studies is the marked age stratification of the bank filtrate between the river and wells.

The effect of age stratification is such that water pumped from a bank filtrate well (for example, Well 45 in Fig. 10.9) is of mixed composition and this explains the almost total equalization of the fluctuating solute concentrations between the river and production wells. This well-known balancing effect provides safe conditions for further water treatment and is also significant in mitigating the peaks of accidental river water

Box 10.3   (Continued)

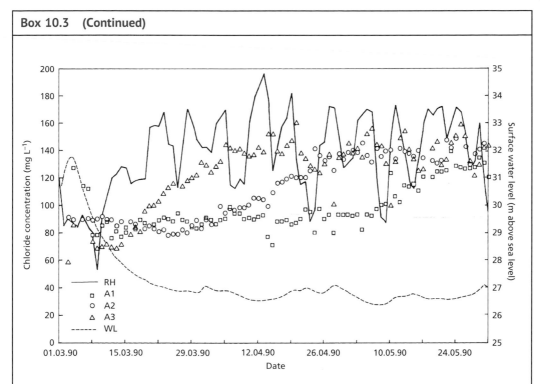

Fig. 10.10   Graph showing the fluctuation in chloride concentrations in the River Rhine (RH) and in monitoring wells A1 (lower layer), A2 (middle layer) and A3 (upper layer) at Flehe waterworks, Düsseldorf at position 731.5 km (see Fig. 10.8 and Fig. 10.9). WL is the water level (Schubert 2002). (*Source:* Schubert, J. (2002) Hydraulic aspects of riverbank filtration – field studies. *Journal of Hydrology* **266**, 145–161. © 2002, Elsevier.)

pollution. It is for this reason that the Düsseldorf waterworks has been able to overcome extreme conditions with respect to poor river water quality between 1950 and 1975 and also to withstand the Sandoz accident involving a pesticides manufacturing plant in Basel in November 1986 (Schubert 2002).

In the United States, the water supply industry has adopted the broadly defined regulatory concept of 'groundwater under the direct influence' (GWUDI) of surface water (variably defined and implemented in response to local conditions by each state, tribe or other regulatory agent). Groundwater sources in this category are considered at risk of being contaminated with surface water-borne pathogens (specifically disinfection-resistant pathogenic protozoa such as *Cryptosporidium*). Efforts to address the regulation of riverbank filtration focus on the removal of microbial pathogens and are contained in the Long Term 2 Enhanced Surface Water Treatment Rule (LT2ESWTR) of the United States Environmental Protection Agency (2006).

Riverbank filtration can occur under natural conditions or be induced by lowering the groundwater table below the surface water level by abstraction from adjacent boreholes. Typical flow conditions associated with different types of riverbank filtration schemes are shown in Fig. 10.11. For the quantitative and

the dynamic hydrology and cannot be regarded as constant, particularly following periods of flooding. The hydraulic conductivity of the riverbed is therefore a principle factor determining the volume of bank filtrate.

Compared with conventional surface water abstraction, the natural attenuation processes of riverbank filtration can provide the following advantages: elimination of suspended solids, particles, biodegradable compounds, bacteria, viruses and parasites; part elimination of adsorbable compounds and the equilibration of temperature changes and concentrations of dissolved constituents in the bank filtrate. Undesirable effects of riverbank filtration on water quality can include increases in hardness, ammonium and dissolved iron and manganese concentrations and the formation of hydrogen sulphide and other malodorous sulphur compounds as a result of changing redox conditions (Fig. 10.12).

The beneficial attenuation processes result mainly from mixing, biodegradation and sorption within two main zones: the biologically active colmation layer, where intensive degradation and adsorption processes occur within a short residence time; and along the main flowpath between the river and abstraction borehole where degradation rates and sorption capacities are lower and mixing processes greater. In general, the distance between production wells and the river or lake bank is more than 50 m with typical travel times of between 20 and 300 days (Grischek et al. 2002). Travel times based on measurements of specific conductance ranged from approximately 20 hours to 3 months at the site of a riverbank filtration scheme in south-west Ohio (Sheets et al. 2002) with shorter and more consistent travel times obtained under conditions of continuous pumping.

Diffuse sources of contamination in catchment runoff, especially from agricultural activities, can adversely affect river water quality and therefore bank filtrate. The study by Grischek et al. (1998) demonstrated the potential for denitrification at a sand and gravel aquifer site on the River Elbe in eastern Germany where both dissolved and solid organic carbon within the aquifer act as electron donors. Verstraeten et al. (2002) reported changes in concentrations of triazine and acetamide herbicides at a well-field on the River Platte in Nebraska and showed that parent compounds were reduced by 76% of the river water value (with a third of this due to riverbank filtration), but that increases in concentrations of specific metabolite compounds were identified after riverbank filtration and ozonation treatment.

Polar organic molecules are an increasingly problematic class of contaminants for riverbank filtration schemes and include several pharmaceutically active compounds (PhACs) that are discharged almost unchanged from municipal sewage treatment plants. Heberer (2002) reported monitoring studies carried out in Berlin where PhACs such as clofibric acid (a blood lipid regulator used in human medical care), diclofenac and ibuprofen were detected at individual concentrations of up to several $\mu g\,L^{-1}$ in groundwater samples from aquifers near to contaminated water courses. It is therefore apparent that several drug residues are not eliminated during recharge through the subsoil (Heberer 2002).

For the future optimization and protection of riverbank filtration schemes, further research is required to quantify chemical reaction rates and microbial degradation, especially in the river bed, and to include the effects of pH and redox controls, the behaviour and importance of biofilms, the fate of micro-pollutants and persistent compounds such as PhACs, and the mobility, adsorption and inactivation of viruses, pathogens and protozoa. The development of risk quantification methods, for example identification of relevant compounds and metabolites, and appropriate alarm systems are also required to ensure the long-term sustainability of riverbank filtration schemes (Hiscock and Grischek 2002).

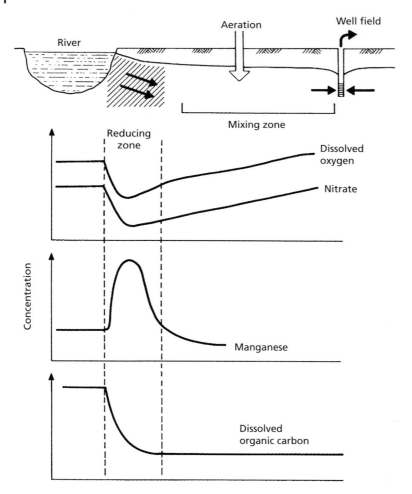

Fig. 10.12 Schematic diagram depicting the evolution of dissolved oxygen, nitrate, dissolved manganese and dissolved organic carbon along a flowpath during riverbank filtration. Dissolved oxygen becomes significantly depleted in the river bed sediments after a few metres of infiltration. Under these anoxic conditions, the microbial activity of denitrifying bacteria further decreases the groundwater redox potential leading to mobilization of the surface coatings of manganese and iron oxy-hydroxides causing a significant reduction in water quality. With further distance from the river bed, microbial activity decreases as a result of a decline in available electron donors. If the groundwater abstraction is non-continuous, or there are strong fluctuations in the river water level, a zone could potentially develop near the well where riverbank filtrate or groundwater flow temporarily results in fluctuations in microbial activity and redox conditions that can affect manganese and iron reduction and precipitation (Tufenkji *et al.* 2002). (*Source:* Tufenkji, N., Ryan, J.N. and Elimelech, M. (2002) The promise of bank filtration. *Environmental Science & Technology* **36**, 423A–428A. © 2002, American Chemical Society.)

10.2.4 Horizontal well schemes

It has long been recognized that well yields can be increased by driving horizontal tunnels (adits) below the water table which radiate away from a well or borehole shaft. Systems of adits, typically 1.2 m wide and 1.8 m high, are associated with many large groundwater sources in the Chalk of south-east England, Belgium and the Netherlands. The water supply to Brighton on the south coast of England includes 13.6 km of adits and in east London, there are 18 km. Generally, the adits were driven to intersect the principal fissure or fracture directions in the Chalk aquifer (Downing *et al.* 1993). Groundwater flow in an adit may be pipe or open channel flow. Adits in the Chalk of south-east England are normally full of water contained under pressure. In this situation, Darcy's Law (eq. 2.7) is not applicable and alternative methods are required for modelling flow in aquifer-adit systems (Zhang and Lerner 2000).

More recently, and with advances in drilling technology, horizontal and slanted wells have been investigated for various hydrogeological situations (Chen *et al.* 2003; Park and Zhan 2003) and also for environmental applications such as vapour extraction in contaminated aquifers (Plummer *et al.* 1997; Zhan and Park 2002). Horizontal wells have screened sections that can be positioned parallel to the horizontal flow direction. These wells have several advantages including interception of vertical components of groundwater flow; greater control over the dynamics of the water table; better contact between well screens and horizontal aquifer units; easier drilling operations close to ground surfaces that are obstructed by infrastructure (airport runways, roads, buildings, etc.) and the possibility of installing long screen sections in aquifers of limited thickness.

10.3 Wetland hydrogeology

The global extent of wetlands is estimated to be from 7 to 8×10^6 km² and, compared to other ecosystems, are an extremely productive part of the landscape with an estimated average annual production of 1.125 kg C m⁻² a⁻¹ (Mitsch *et al.* 1994). The relatively high productivity and biological diversity of wetlands support an important landscape role in nutrient recycling, species conservation and plant and animal harvest. Although very much smaller in extent compared to marine habitats, inland water habitats exhibit greater variety in their physical and chemical characteristics.

Wetlands, with their often abundant and highly conspicuous bird species, are protected by national and international agreements and legislation. Notable wetland protected areas include the Moremi Game Reserve in the Okavango Delta, Botswana, the Camargue National Reserve in France, the Keoladeo (Bharatpur) National Park in India, Doñana National Park in Spain and the Everglades National Park in the United States (Groombridge and Jenkins 2000). Inland water ecosystems are unusual in that an international convention, the 1975 Convention of Wetlands of International Importance especially as Waterfowl Habitat (the Ramsar Convention; Navid 1989), is dedicated specifically to them.

Inland water habitats can be divided into running or *lotic* systems (rivers) and standing or *lentic* systems (lakes and ponds). Wetlands are typically heterogeneous habitats of permanent or seasonal shallow water dominated by large aquatic plants and broken into diverse microhabitats occupying transitional areas between terrestrial and aquatic habitats (Groombridge and Jenkins 2000). The four major wetland habitat types are bogs, fens, marshes and swamps. Bogs are peat producing wetlands in moist climates where organic matter has accumulated over long periods. Water and nutrient input is entirely through precipitation. Bogs are typically acid and deficient in nutrients and are often dominated by *Sphagnum* moss. Fens are peat producing wetlands that are influenced by soil nutrients flowing through the system and are typically supplied by mineral-rich groundwater. Grasses and sedges, with mosses, are the dominant vegetation. Marshes are inundated areas with emergent herbaceous vegetation, commonly dominated by grasses, sedges and reeds, which are either permanent or seasonal and are fed by groundwater or river water, or both. Swamps are forested freshwater wetlands on waterlogged or inundated soils where little or no peat accumulation occurs. Like marshes, swamps may be either permanent or seasonal.

Various attempts have been made to classify wetlands and a variety of subdivisions have been recognized based on broad features such as substratum type, base status, nutrient status and water source, water level and successional stage. The development of the main wetland habitat categories and terms, in relation to the main ecological gradients, has been reviewed by Wheeler and Proctor (2000). Other approaches include hydrological and hydrogeological classifications based on the main inflows and outflows of water, flowpaths and water level fluctuations (Lloyd *et al.* 1993; Gilvear and McInnes 1994) and a hydromorphological (or hydrotopographical) classification based on the shape of the wetland and its situation with respect to apparent sources of water (Goode 1977). A simplification of the hydrogeological classification is shown in Fig. 10.13 to illustrate the influence of topography, geology and water source in maintaining wetlands.

10.3.1 Impacts of groundwater exploitation on wetlands

A change in the factors controlling the source of water to a wetland can have potentially devastating consequences for the fen community, particularly a change in flow direction and volume. Whiteman *et al.* (2010) presented a procedure based on experience in England and Wales to determine whether a wetland is groundwater-dependent and damaged or at risk of damage as a result of groundwater quality or quantity pressures. The procedure first determines which sites are thought to be critically dependent on groundwater and to quantify the level of risk of damage, with the results verified by local ecologists and hydrogeologists who are able to advise on the cause or causes of damage to each site, and the level of risk and confidence in that judgement. Second, the procedure classifies each groundwater body in terms of good or poor status, using condition assessments of sites protected for nature conservation to identify actual ecological damage, and a seven-step process to assess whether the environmental supporting conditions for

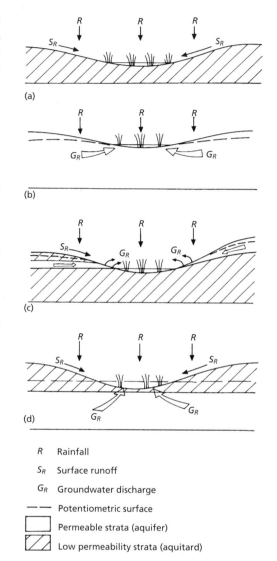

R — Rainfall
S_R — Surface runoff
G_R — Groundwater discharge
− − — Potentiometric surface
▭ — Permeable strata (aquifer)
▨ — Low permeability strata (aquitard)

Fig. 10.13 Simple hydrogeological classification of wetland types. In (a) surface runoff is fed by rainfall and collects in a topographic hollow (for example, valley bottom, pingo or kettle hole) underlain by a low permeability layer. In (b) rainfall recharge to an unconfined aquifer supports a wetland in a region of low topography and groundwater discharge. In (c) superficial deposits, both unconfined and semi-confined, and underlain by a low permeability layer, contribute groundwater seepage in addition to surface water runoff. In (d) surface water runoff is in addition to artesian groundwater discharge from a semi-confined aquifer.

each site are met and, if not, whether the departure is due to poor groundwater quality or inadequate quantity.

If managed well, and as part of an ecosystem services approach to catchment management, wetlands can provide cost-effective mitigation for a number of catchment pressures, particularly the reduction of sedimentation and eutrophication (Whiteman *et al.* 2010). In a further approach, integrated constructed wetlands have been demonstrated to provide an effective alternative to conventional tertiary treatment of sewage effluent and mitigate the risk of eutrophication of downstream receiving waters (Cooper *et al.* 2020).

An example of the impact of groundwater abstraction on the freshwater habitat of a valley

fen and the measures taken to restore the fen is given in Box 10.4. When land drainage and competing demands for water for wetlands, agriculture and public supply conflict, wide-scale destruction of wetland habitat can occur, as illustrated graphically by the Florida Everglades (Box 10.5). A further pressure on wetlands is the effect of drier conditions under climate change as demonstrated by the example of the Chalk groundwater-fed meres in Breckland, East Anglia (Section 10.2.2) (Herrera-Pantoja *et al.* 2012).

Box 10.4   Impact of groundwater abstraction on Redgrave and Lopham Fen, East Anglia, England

Redgrave and Lopham Fen is an internationally important British calcareous valley fen situated on the Norfolk and Suffolk border in the peat-filled headwaters of the River Waveney (see Figs. 10.14 and 10.15 for location and general aspect). The fen, covering 123 ha, is the largest fen of its type in lowland Britain and was declared a Ramsar site in 1991. The largest part of the fen is covered by shallow peat supporting a complex mosaic of reed and sedge beds, mixed species fen and spring flushes. The fen is noted for its rare and precarious community of fen raft spiders. For nearly 40 years, the fen experienced substantial ecological change, principally due to a change in the groundwater flow regime relating to an adjacent water company borehole.

The general geology of the fen consists of Cretaceous Chalk covered by glacial till, sands and gravels. The Chalk surface is incized by a deep buried channel which is thought to be about 1 km wide in the vicinity of the fen. With reference to Fig. 10.13, the fen is a combination of wetland types (c) and (d). Before the late 1950s, calcareous and nutrient-poor water rose under artesian

pressure from the semi-confined Chalk aquifer and seeped into the fen both around the fen margins and within the peats (Fig. 10.16a). The extreme heterogeneity of the superficial Quaternary deposits resulted in great spatial variation in the quantity of rising Chalk water. The interaction of base-poor water from marginal sands with the calcareous and acid peats produced local variation in soil chemistry that supported a diverse mosaic of fen plant communities of high botanical interest.

In 1957, two Chalk abstraction boreholes were installed adjacent to the fen (see Fig. 10.14 for location) for public water supplies and licensed in 1965 to abstract 3600 m^3 day^{-1}. Warby's Drain and the River Waveney were deep-dredged at this time, substantially increasing channel capacity, with a sluice at the downstream end of Redgrave Fen installed to control outflows. As shown schematically in Fig. 10.16b, the operation of the water company source led to the elimination of vertical groundwater seepage and the frequent drying out of Warby's Drain. The normal condition of perennial, high

(Continued)

Box 10.4 (Continued)

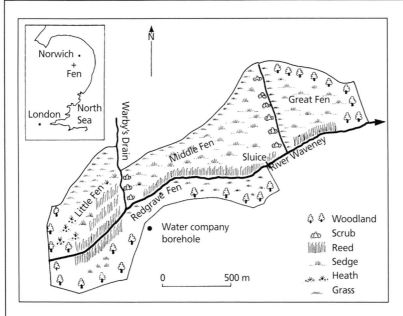

Fig. 10.14 Location and site map of Redgrave and Lopham fen in East Anglia showing the position of the former operating water company borehole.

Fig. 10.15 General aspect of Redgrave and Lopham Fen looking north-east across Great Fen from the position of the Sluice (see Fig. 10.14 for location).

Box 10.4 **(Continued)**

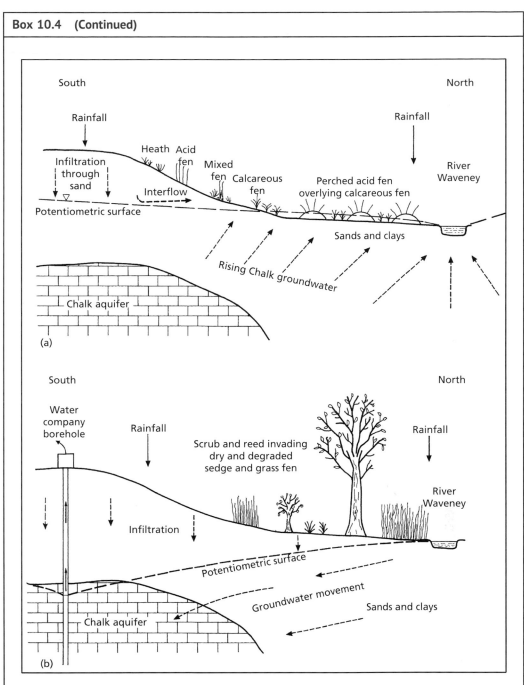

Fig. 10.16 Schematic cross-section through Redgrave and Lopham Fen illustrating groundwater and ecological conditions: (a) before groundwater pumping; and (b) after several years of groundwater pumping from the water company borehole (see Fig. 10.14 for location) (Burgess 2002). (*Source:* Adapted from Burgess, D.B. (2002) Groundwater resource management in eastern England: a quest for environmentally sustainable development. In: *Sustainable Groundwater Development* (eds K.M. Hiscock, M.O. Rivett and R.M. Davison). Geological Society, London, Special Publications **193**, pp. 53–62.)

(*Continued*)

X

Box 10.4 (Continued)

water levels with Chalk groundwater discharging through the fen, thus maintaining a *soligenous* hydrology (where wetness of the site is maintained by water flow through soil), was replaced by a seasonal downward movement of surface water. The hydrology of the fen had now become controlled by rainfall patterns and river levels thus producing a *topogenous* hydrology (where wetness is maintained by the valley topography). During the summer, the fen dried out more frequently with groundwater heads reduced to a metre below the fen surface. Test pumping and radial flow modelling suggested that about a quarter of the pumped groundwater was at the expense of spring flow into the fen (Burgess 2002).

These hydrogeological changes caused by groundwater abstraction were matched to a deterioration of the flora and fauna at the site (Harding 1993). From a comparative study of botanical records, Harding (1993) showed that great changes had occurred to the ecological character of the fen as a result of

the drying out, namely the invasion of scrub. The reduction in the water table altered the balance of competition towards dry fen species and the expansion of *Phragmites* and *Molinia*, which are tolerant of low water levels, while previously dominant species such as *Cladium* and *Schoenus* contracted. The loss of calcareous and base-poor seepage water and the increased fertility from the sudden release of large amounts of stored nitrates through peat wastage under a lower water level also benefited *Phragmites*.

To reverse the environmental damage, the groundwater pumping was relocated to a borehole 3.5 km east and downstream of the fen that became operational in 1999. The total cost of the replacement supply was of the order of £3.3 million, which included the cost of the investigation, source works, pipeline and restoration work on the fen, principally the removal of scrub and the regeneration of peat areas.

Box 10.5 The Florida Everglades: a region under environmental stress

The Florida Everglades comprises part of the south Florida ecosystem that formed during the last several thousand years during the Holocene epoch. The ecosystem consists primarily of wetlands and shallow-water habitats set in a subtropical environment (McPherson and Halley 1997). The south Florida region is underlain by a thick sequence of shallow marine carbonate sediments deposited from the Cretaceous through to the early Tertiary as a carbonate platform. Younger Tertiary deposits consist of shallow marine sandy limestone, marls and sands. As shown in Fig. 10.17, the marine carbonate sediments contain three major aquifer systems: the Floridan; the intermediate and the surficial. The surficial aquifer system includes the highly permeable Biscayne aquifer. The Biscayne aquifer is more than 60 m thick under parts of the Atlantic Coastal

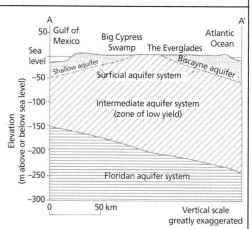

Fig. 10.17 Generalized hydrogeological cross-section of south Florida (the line of the section is shown in Fig. 10.18) showing the three major aquifer systems (McPherson and Halley 1997). (*Source:* McPherson, B.F. and Halley, R. (1997). The South Florida Environment – A Region Under Stress. United States Geological Survey Circular 1134, 61 pp.)

Box 10.5   (Continued)

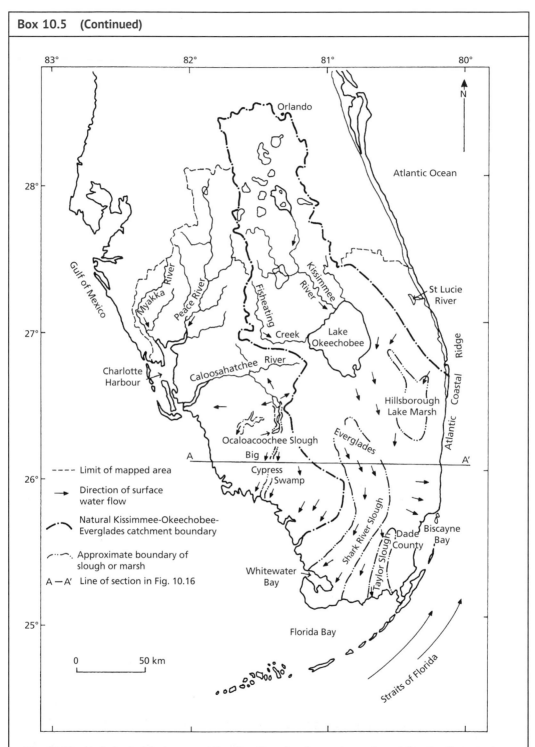

Fig. 10.18 Hydrological features and the direction of surface water and coastal water flows under natural conditions in south Florida (McPherson and Halley 1997). (*Source:* McPherson, B.F. and Halley, R. (1997). The South Florida Environment – A Region Under Stress. United States Geological Survey Circular 1134, 61 pp.)

(Continued)

Box 10.5 (Continued)

Ridge and wedges out about 65 km to the west in the Everglades. The shallow aquifer of southwest Florida is about 40 m thick along the Gulf Coast and wedges out in the eastern Big Cyprus Swamp. The surficial aquifer system is recharged by abundant rainfall that under natural conditions favoured the expansion of coastal and freshwater wetlands during the Holocene and the deposition of thick layers of peat.

Wetlands are the predominant landscape feature of south Florida. Before development of the area, the natural functioning of the wetlands depended on several weeks of flooded land following the wet season. For example, the Kissimmee–Okeechobee–Everglades catchment, an area of about 23 000 km^2, once extended as a single hydrological unit from present-day Orlando to Florida Bay, about 400 km to the south (Fig. 10.18). In the northern half of the catchment, the Kissimmee River and other tributaries drained slowly through large areas of wetlands into Lake Okeechobee, a shallow lake of about 1900 km^2. The lake periodically spilled water south into the Everglades, a vast wetland of about 12 000 km^2. Under high water level conditions, water in the Everglades moved slowly to the south by sheet flow, thus forming the area known as the River of Grass before discharging into Florida Bay and the Gulf of Mexico.

The Everglades was formerly a complex mosaic of wetland plant communities and

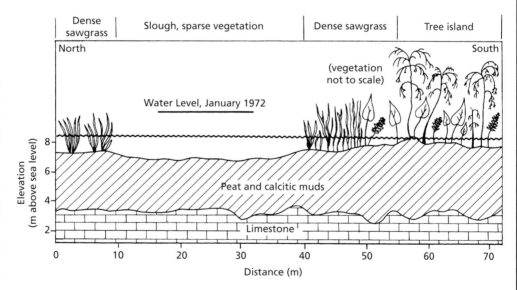

Fig. 10.19 Generalized section of the Everglades wetlands in the Shark River Slough (see Fig. 10.18 for location). Peat develops in wetlands that are flooded for extensive periods during the year and calcitic muds develop in wetlands where the periods of flooded land are shorter and limestone is near the surface. The Everglades has been a dynamic environment with numerous shifts between marl- and peat-forming marshes and between sawgrass marshes and water-lily sloughs (McPherson and Halley 1997). (*Source:* McPherson, B.F. and Halley, R. (1997). The South Florida Environment – A Region Under Stress. United States Geological Survey Circular 1134, 61 pp.)

Box 10.5 **(Continued)**

landscapes with a central core of peatland that extended from Lake Okeechobee to mangrove forest that borders Florida Bay. The peatland was covered by a swamp forest along the southern shore of Lake Okeechobee and by a vast plain of monotypic sawgrass to the south and east of the swamp forest. Further south-east, the sawgrass was broken by sloughs and small tree islands in Shark River Slough and Hillsborough Lake Slough (Fig. 10.19).

Prior to development, water levels in the Everglades fluctuated over a wider range, but water management has tended to reduce peak and minimum water levels and to lessen flooding and drought (Fig. 10.20). Water management, principally drainage for agricultural development, has altered most of south Florida and caused severe environmental changes including large losses of soil through oxidation and subsidence, degradation of water quality, nutrient enrichment, contamination by pesticides and mercury, fragmentation of the landscape, large losses of wetlands and wetland functions, and widespread invasion by exotic species. Additionally, the large and growing human population and the agricultural development in the region are in intense competition with the natural system for freshwater resources (McPherson and Halley 1997).

Public water supplies for the 5.8 million people in south Florida are abstracted from shallow aquifers, the most productive and widespread of which are the Biscayne aquifer in the south-east and the shallow aquifer in the south-west (Fig. 10.17). Freshwater abstractions within south Florida were about 15.6×10^6 m^3 day^{-1} in 1990 with most of this water used for public supply (22%) and agriculture (67%). Groundwater supplied 94% (3.3×10^6 m^3 day^{-1}) of the water used for public supply in 1990. Water abstracted for agricultural purposes is divided between groundwater

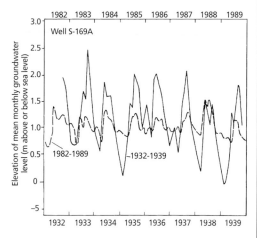

Fig. 10.20 Long-term hydrograph showing water level fluctuations at a well in southern Dade County (see Fig. 10.18 for approximate location), 1932–1939 and 1982–1989. Drainage of the Everglades began in the early 1880s and continued into the 1960s with the purpose of reducing the risk of flooding and drought and so opening land for agricultural development south of Lake Okeechobee. The effect of water management in the Everglades has been to reduce peak and minimum water levels in the Everglades, as illustrated in the well hydrograph (McPherson and Halley 1997). (*Source:* McPherson, B.F. and Halley, R. (1997). The South Florida Environment – A Region Under Stress. United States Geological Survey Circular 1134, 61 pp.)

and surface water. In 1990, groundwater accounted for 4.7×10^6 m^3 day^{-1} and surface water accounted for 5.7×10^6 m^3 day^{-1} of the agricultural requirement.

To contribute to the restoration of the Everglades, a major effort is required to understand the hydrology, geology and ecology of the region and to monitor modifications to the land drainage and flood control structures. Better land management to improve water quality and the development of more sustainable water supplies are also an integral part of the solution. The Comprehensive Everglades Restoration Plan (CERP) is a multibillion dollar project that aims to increase freshwater

(Continued)

Box 10.5 (Continued)
storage, improve water quality and re-establish the natural water flow through the greater Everglades ecosystem. If successful, these efforts will help protect underlying	aquifers from saltwater intrusion, delay the impacts of sea level rise along the coast and allow time for wildlife to adapt to the changing environment (NPS 2021).

10.3.2 Hydrogeology of dune slacks

Dune slack (or pond) habitats are a type of wetland that appear as damp or wet hollows left between sand dunes, where the groundwater reaches or approaches the surface of the sand (Tansley 1949). Dune slacks are a unique type of wetland ecosystem, highly ranked on the international conservation agenda due to the occurrence of many rare and endangered plant species and their associated fauna (Grootjans *et al.* 1998). One of the most distinctive features of dune slacks is a seasonally fluctuating water table, which usually reaches a maximum in winter and spring and declines in summer (Lammerts *et al.* 2001). Two types of dune slacks can be distinguished on the basis of their geomorphological history: primary and secondary slacks (Boorman *et al.* 1997). Primary slacks originate in sandy beaches, which have been partially or fully cut off from the influence of the sea by new fore-dunes, particularly in prograding systems. Exceptionally, slacks may also form from salt marshes, as dune sand encroaches upon them. Secondary slacks result from blow outs or the landward movement of dune ridges in eroding systems.

Hydrological and hydrochemical controls on humid dune slacks are dependent on several factors including climatic setting, coastal geomorphology, hydrogeological conditions and substrate mineralogy. The hydrological regime of a dune slack is critical to the functioning of the dune slack ecosystem (Plate 10.1). To demonstrate the interconnection of these controls, Fig. 10.21 provides a general conceptual model showing the position of dune slacks in a coastal zone connected to an inland area with an underlying regional aquifer or aquitard (Davy *et al.* 2006). Five types of dune slacks are shown (A–E). Not all types will be present at a location, for example Type D is not expected in island situations. Two groundwater flow systems are shown: a local circulation of fresh groundwater in the dune system and recharged directly by precipitation; and a regional groundwater flow system originating in the inland area and discharging in the coastal zone. The degree of influence of the regional flow system will depend on the extent and nature of the underlying geological unit as to whether it forms an aquifer of good hydraulic conductivity or an aquitard of poor hydraulic conductivity. Local groundwater flow circulation in the dune system will occur at depth if the sand is freely draining with a shingle base but flow may be restricted vertically by the presence of clay lenses or peat layers.

The topographic elevation of the dune system and inland areas determines the shape of the water table. As show in Fig. 10.21, the inland area and dune system are unconfined with groundwater discharge directed towards topographic hollows. Hollows in the dunes may be formed by blow outs or as a result of successive dune formation landwards. Dune slacks of Types B and C are fed solely by precipitation infiltrating the dune sands. In Type B, groundwater flow is directed towards the slack and water is lost by evapotranspiration. An example is shown in Fig. 10.22a. In Type C, groundwater flows into the upgradient edge of the slack, flows through the slack and then exits the slack at the downgradient edge before continuing to flow in the direction of the hydraulic gradient (Stuyfzand 1993). In a

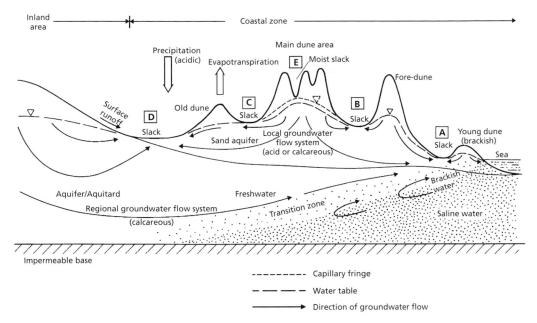

Fig. 10.21 Conceptual model of hydrological and hydrogeological controls on humid dune slack formation in coastal areas. See text for an explanation of dune slack Types A, B, C, D and E (Davy *et al.* 2006). (*Source:* Adapted from Davy, A.J., Grootjans, A.P., Hiscock, K. and Petersen, J. (2006) Development of eco-hydrological guidelines for dune habitats – Phase 1. *English Nature Research Reports* **696**, 78 pp.)

dune area with several dune slacks in close proximity, slight differences in water level between the slacks may initiate groundwater flow from one slack to another (Kenoyer and Anderson 1989; Grootjans *et al.* 1998).

Type D is at the boundary between the dune system and inland area and is fed by both the regional and local groundwater flow systems and, as shown in Fig. 10.21, may receive some surface water runoff. An example is shown in Fig. 10.22b. Type E represents a moist dune slack situated at a high elevation in the main dune area. Moisture in the capillary fringe above the water table keeps the base of the dune slack moist with only occasional flooding when the water table is high in wet years. An example is shown in Fig. 10.22c.

The dune slacks inland of the large fore-dune (Types B, C, D and E) are above the brackish water body in the subsurface and so are fresh. In contrast, the seaward Type A dune slack is in reach of the transition zone between the circulation of fresh and saline groundwaters and so may be subject to brackish conditions. Dune slack A is in the most dynamic part of the coastal environment and is considered only temporary as the developing dune system moves inland.

Commonly, the sand below the dune slack is initially calcareous (from the input of shelly material) and the groundwater is typically base-rich, although a range of more acidic systems can also develop. As in dunes generally, the nutrient availability (particularly nitrogen and phosphorus) in pristine systems is low and this is undoubtedly one of the keys to their species diversity (Willis 1985). Slacks are often part of the dynamic mosaic of dune habitats both spatially and temporally, being subject both to internal successional processes and the movements of the surrounding dunes (Ranwell 1960). In the dynamic, successional setting of most sand dune systems, the characteristic slack communities are maintained at least partly by disturbances, including fluctuations in the water table, blown sand, the effects of nutrient limitation and grazing.

(a)

(b)

(c)

Fig. 10.22 Examples of coastal dune slacks: (a) Type B dune slack and (b) Type E moist dune slack at Winterton Dunes, north-east Norfolk, eastern England; and (c) Type D dune slack at Holme Dunes, north-west Norfolk, eastern England, as observed in September 2020.

Humid dune slacks are refuges for certain 'flagship' endangered species: in Great Britain, these include the Natterjack toad *Epidalea calamita*, the fen orchid *Liparis loeselii* and the thalloid liverwort *Petallophyllum ralfsii*. In Europe, these species are protected by the EU Habitats Directive (Council of the European Communities 1992) and as such are the subject of conservation measures set out in the Directive. Historically, the integrity and biodiversity of British dune systems have been better maintained than many other habitats, often because they are relatively remote and usually unsuitable for intensive agriculture or industrial development. They have, however, been popular areas for recreation, especially holiday tourism and golf courses. In North West Europe, dune systems generally face larger-scale threats: they have been used for abstraction of drinking water for large cities, for example on the Dutch mainland coast, have experienced increasing atmospheric nitrogen deposition from industrial and agricultural activities (Whiteman *et al.* 2017); and have been subject to afforestation (van Dijk and Grootjans 1993; Clarke and Sanitwong Na Ayutthaya 2010). In the Netherlands, where sand dunes are regarded as some of the last remnants of natural ecosystems, active restoration projects have been initiated to restore dune ecosystems to high biodiversity value (Grootjans *et al.* 2002; Geelen *et al.* 2017).

10.4 Climate change and groundwater resources

The Intergovernmental Panel on Climate Change has highlighted the implications of accelerated climate change for groundwater (IPCC 2007). Changes in rainfall, evaporation and soil moisture conditions (Plate 10.2) leading to changing patterns of recharge and runoff are expected to add to the resource management burden for both groundwater depletion and rising water tables, depending on the region. However, the magnitude of these impacts should be compared with the stresses placed on groundwater systems by current socio-economic drivers. This interplay between climate change and its effects on the natural environment and socio-economic systems has been evident throughout history as exemplified by the fortunes of human populations in the Middle East, as recorded extensively in ancient written records (see Box 10.6).

Box 10.6 History of climate change and groundwater impacts on human civilization in the Fertile Crescent

The impact of climate change on the natural environment and socio-economic systems is most apparent in the Middle East as recorded in extensively studied ancient written records. In general, the Middle East is influenced by a Mediterranean climate, positioned in the transition zone between the Westerlies low-pressure system to the north and the Azores high-pressure zone of the subtropics to the south. The Westerlies are associated with rain-bearing low-pressure depressions in the winter, causing humid conditions along the south and south-eastern regions bordering the Mediterranean. In summer, the high-pressure belt moves north, bringing higher temperatures and aridity to these regions. In years when the high-pressure belt remains over the area, the winter rainstorms are less frequent, and there is drought. The pattern of annual rainfall reflects a general increase in aridity from north towards the south and east, and becomes even more pronounced in the Jordan – Dead Sea – Arava Rift Valleys which are located in the rain shadow of the

(Continued)

Box 10.6 (Continued)

mountain range extending from north to south along the central part of Israel and Palestinian territory (Issar 2008).

Two ancient cities, Jericho and Arad, are located approximately on the $200 \, mm \, a^{-1}$ rainfall isohyet, bordering the more humid and arid parts of Israel, with this line having shifted during historical periods in response to global climate change during the Pleistocene and Holocene. As explained by Issar (2003), climate change throughout history has had an impact on the welfare of people living in the Fertile Crescent, the comparatively fertile regions of Mesopotamia and the Levant in western Asia. Correlating the major climate change periods with historical and archaeological data show that during cold, humid periods the Fertile Crescent flourished, while warm, dry periods caused socio-economic crises and the desertification of urban centres, especially along the desert margins, often determined by the fate of the

water supply. Urban and rural centres that obtained their water supply from non-perennial streams or perched local aquifers, such as Arad, did not survive. On the other hand, sites such as Jericho, believed to be the oldest continuously inhabited city in the world with archaeological evidence for human settlements dating back 11 000 years, survived because of a reliable supply of water from rivers and springs fed by regional aquifers, with only short periods of total desertification. As shown in Fig. 10.23, for the example of Jericho, the perennial springs, including Ain es-Sultan (*Ain* being Arabic for 'spring'), which mitigated the impact of periods of dryness and also enabled rapid recovery, emerge from limestone and dolomite aquifers of Mesozoic and Eocene age (Por 2004; Issar 2008). These aquifer-bearing Mesozoic rocks form the backbone of the anticlinal structures of the Taurides, Zagroids and Syrian arch which comprise mountainous

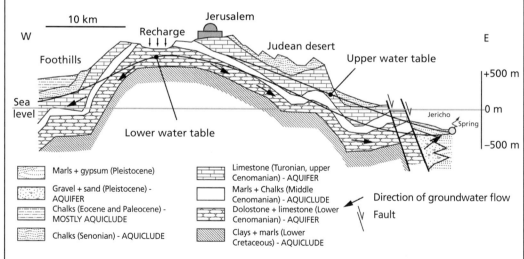

Fig. 10.23 Hydrogeological cross-section explaining the emergence of perennial springs feeding the oasis of Jericho. The ancient settlement of Jericho has been able to withstand the effects of climate change throughout history because its supply of water for irrigation derives from springs which are fed by limestone aquifers drawing from a large area of the eastern Judean Mountains (Issar 2008). (*Source:* Issar, A.S. (2008) A tale of two cities in ancient Canaan: how the groundwater storage capacity of Arad and Jericho decided their history. In: *Climate Change and Groundwater* (eds W. Dragoni and B.S. Sukhija). Geological Society, London, Special Publications **288**, pp. 137–143.)

Box 10.6   (Continued)

regions receiving high amounts of precipitation and recharge to the regional aquifers. The recharge of these aquifers emerges as perennial springs feeding the main rivers of the Fertile Crescent.

Hence, it can be concluded that favourable hydrogeological conditions helped the civilizations of the Fertile Crescent to survive during periods of warm and dry climate. As well as the beneficial natural conditions, archaeological evidence shows that human resourcefulness also delayed or averted a crisis resulting from a decline in spring flow. Adaptation was achieved by various methods for augmenting and lengthening the period of flow, by tunnelling, or by inventing methods of pumping from wells. In general, and as shown for the case of human settlements in the Middle East, the ability of a society to withstand the impact of climate change and its consequences depends on the total resilience of its natural and social systems, yet ultimately it is the magnitude and duration of the impact that has the decisive effect (Issar 2008).

The global climate is undoubtedly changing. The Earth's global surface temperature in 2019 was the second warmest since 1880 (second only to 2016) and was 0.98 °C warmer than the 1951–1980 mean (Plate 10.3). The five years 2015–2019 have been the warmest of the past 140 years (NASA 2020). It is likely that the last 100 years was the warmest century in the last millennium (Hulme *et al.* 2002) with the Arctic having warmed three times faster than the rest of the planet since 1970 (NASA 2020). Evidence for changes in global climate includes continued mass loss of ice from Greenland and Antarctica, a diminishing extent in winter Arctic Ocean sea ice, a near world-wide decrease in mountain glacier extent and ice mass, and more intense rainfall events over many Northern Hemisphere mid-to-high latitude land areas.

In central England, the thermal growing season for plants has lengthened by one month since 1900 and winters over the last 200 years have become wetter relative to summers throughout the United Kingdom. Also, a larger proportion of winter precipitation in all regions now falls on heavy rainfall days than was the case 50 years ago. Around the United Kingdom, and adjusting for natural land movements, average sea level is now about 10 cm higher than the level in 1900 (Hulme *et al.* 2002).

Climate change is influenced by both natural and human causes. The Earth's climate varies naturally as a result of interactions between the ocean and atmosphere, changes in the Earth's orbit, fluctuations in incoming solar radiation and volcanic activity. The main human cause is the increasing emissions of greenhouse gases such as carbon dioxide, methane, nitrous oxide and chlorofluorocarbons. Currently, about $6.5 \times 10^9 \, t \, a^{-1}$ of carbon are emitted globally into the atmosphere, mostly through the burning of fossil fuels. Changes in land use, including the clearance of tropical rainforest, contribute a further net emission of $1-2 \times 10^9 \, t \, a^{-1}$. The carbon dioxide level was about 285 parts per million in 1880 when global temperature records began. By 1960, the average concentration had risen to 315 parts per million, and since 2015 has exceeded 400 ppm (NOAA 2019). Increasing concentrations of greenhouse gases in the atmosphere in the last 200 years (Table 10.2) have trapped outgoing long-wave radiation in the lower atmosphere, leading to an increase in global temperature (see Fig. 10.24). Opposite effects that act to cool the climate include other atmospheric pollutants such as sulphate aerosols that absorb and scatter incoming solar radiation back to space. Under the Kyoto Protocol, countries of the European Union

Table 10.2 Major sources, concentrations and residence times of important greenhouse gases in the atmosphere (IPCC 2001).

Greenhouse gas	Pre-industrial concentration (1750)	Present concentration (1998)	Residence time	Annual rate of increase[a]	Major sources
Water vapour	3000 ppm	3000 ppm	10–15 days	n.a.	Oceans
CO_2 (Carbon dioxide)	~280 ppm	~365 ppm	5–200[b] years	1.5[c] ppm a^{-1}	Combustion of fossil fuels, deforestation
CH_4 (Methane)	~700 ppb	1745 ppb	12[d] years	7.0[c] ppb a^{-1}	Rice production, cattle rearing, industry
N_2O (Nitrous oxide)	~270 ppb	314 ppb	114[d] years	0.8 ppb a^{-1}	Agriculture, industry, biomass burning
CFC-11 (Chlorofluorocarbon-11)	0	268 ppt	45 years	−1.4 ppt a^{-1}	Aerosols, refrigeration
HFC-23 (Hydrofluorocarbon-23)	0	14 ppt	260 years	0.55 ppt a^{-1}	Industrial by-product
CF_4 (Perfluoromethane)	40 ppt	80 ppt	>50 000 years	1 ppt a^{-1}	Aluminium industry

n.a. not applicable.

[a] Rate is calculated over the period 1990–1999.

[b] No single lifetime can be defined for CO_2 because of the different rates of uptake by different removal processes.

[c] Rate has fluctuated between 0.9 and 2.8 ppm a^{-1} for CO_2 and between 0 and 13 ppb a^{-1} for CH_4 over the period 1990–1999.

[d] This lifetime has been defined as an 'adjustment' time that takes into account the indirect effect of the gas on its own residence time.

(*Source*: IPCC (2001). *Climate Change 2001. The Scientific Basis.* Cambridge University Press, Cambridge. A Report of Working Group I of the Intergovernmental Panel on Climate Change.)

committed to an 8% reduction in emissions from 1990 levels of a 'basket' of six gases, including carbon dioxide, methane and nitrous oxide by 2008–2012, and since the 2016 Paris Agreement, to limit global warming to below 2°C, and preferably to 1.5°C, compared to pre-industrial levels.

The hydrological cycle is an integral part of the climate system and is therefore involved in many of the interactions and feedback loops that give rise to the complexities of the system (Askew 1987). Climate change during the next 100 years is expected to lead to an intensification of the global hydrological cycle and have major impacts on regional water resources (IPCC 1998). A summary of the likely impacts of climate change on natural hydrological systems is shown in Fig. 10.25. The potential water resources impacts of climate change are generally negative, such as a shorter precipitation season and an increase in hydrological extremes such as floods and droughts (see Box 10.7 for a classification and assessment of drought severity). A shorter precipitation season, possibly coupled with heavier precipitation events and a shift from snow precipitation to rainfall, would generate larger volumes of runoff over shorter time intervals. More runoff would occur in winter and less runoff would result from spring snowmelt. This complicates the storage and

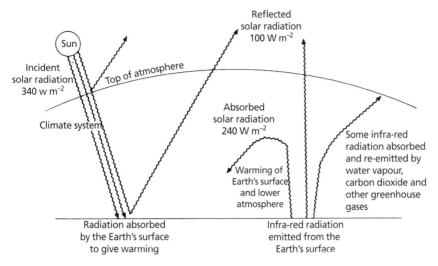

Fig. 10.24 Schematic representation of the global radiation budget. Averaged over the globe, there is 340 W m^{-2} of incident solar radiation at the top of the atmosphere. The climate system absorbs 240 W m^{-2} of solar radiation so that under equilibrium conditions, it must emit 240 W m^{-2} of infra-red radiation. The carbon dioxide radiative forcing for a doubling of carbon dioxide concentrations constitutes a reduction in the emitted infra-red radiation of 4 W m^{-2} producing a heating of the climate system known as global warming. This heating effect acts to increase the emitted radiation in order to re-establish the Earth's radiation balance (Climate Change: The IPCC Scientific Assessment 1990). (*Source:* IPCC (1990) *Climate Change. The Intergovernmental Panel on Climate Change Scientific Assessment* (eds J.T. Houghton, G.J. Jenkins and J.J. Ephraums). Cambridge University Press, Cambridge. © 1990, IPCC.)

routing of floodwater, both for the purpose of protecting the human environment as well as for meeting water supply targets. It may also complicate the conjunctive use of surface water and groundwater, as the opportunity for groundwater recharge is reduced under these conditions. In tropical latitudes, water resources are not likely to suffer changes under the predicted climate change impacts. Water quality is also affected by changes in temperature, rainfall and sea level rise that affect the volume of river flow and the degree of saline intrusion (Loáiciga *et al.* 1996).

Hu *et al.* (2019) presented a review based on published modelling studies of projected hydrologic changes under the Representative Concentration Pathway (RCP) 8.5 greenhouse gas emissions scenario, the most extreme scenario with radiative forcing of 8.5 W m^{-2} by 2100 leading to a rise in global average temperatures of between 4 and 6 °C above pre-industrial levels. Hu *et al.* (2019) concluded

that the general pattern of climate change impact on groundwater resources follows the precipitation pattern with a depletion in tropical and/or sub-tropical regions and an increase in the high-latitude regions of the Northern and Southern Hemispheres. However, regional variability is evident and corresponds to the heterogeneous impact of climate change on the regional distribution of precipitation and evapotranspiration, localized interaction between surface water and groundwater and distances from the oceans with rising sea level. Hu *et al.* (2019) considered that the decline in the water table in many areas may seriously reduce irrigated crop production and adversely impact groundwater-dependent ecosystems.

Hydrological models of varying degrees of complexity in representing current and future climatic conditions provide an objective approach to estimating hydrological responses to climate change. The linking of physically based hydrological models to output from

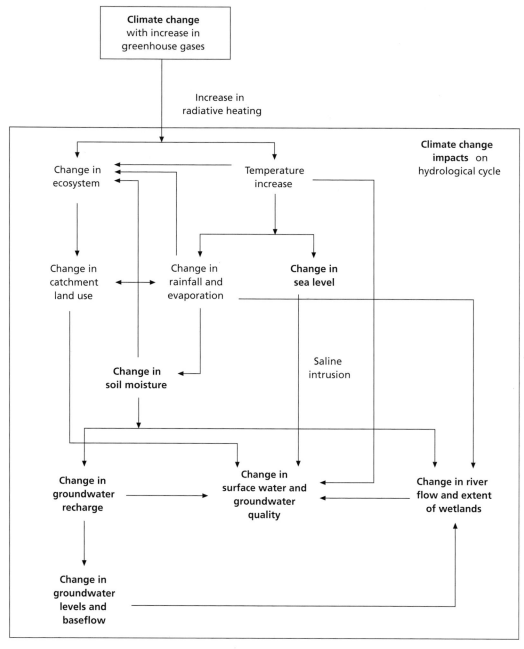

Fig. 10.25 Impacts of increasing greenhouse gas concentrations on the natural hydrological cycle emphasizing changes in hydrogeological conditions (Arnell 1996). (*Source:* Adapted from Arnell, N. (1996) *Global Warming, River Flows and Water Resources.* Wiley, Chichester. © 1996, John Wiley & Sons.)

global circulation models (GCMs) enables the study of a variety of climate change effects (Conway 1998). A note of caution is required in dealing with the scale effects which complicate the prediction of regional hydrological changes. Runoff and precipitation at regional scales are highly variable, with 10- to 20-year averages commonly fluctuating in the range ±25% of their long-term means. At shorter timescales, the problem is likely to be worse,

---

Box 10.7   Assessment of drought severity

The impact of droughts on groundwater is a major threat to water security affecting socio-economic activity and the sustainability of natural ecosystems, which may be exacerbated by future environmental change (Ascott *et al.* 2020). Droughts and floods are characterized by the extremes of the frequency, intensity and amounts of precipitation (Trenberth *et al.* 2003). Interestingly, there is no single definition of drought. Beran and Rodier (1985) considered, without specifying when a period of dry weather becomes a drought, that the chief characteristic of a drought is a decrease of water availability in a particular period over a particular area. Meteorological drought is defined in terms of a deficit of precipitation. Agricultural drought relates mostly to deficiency of soil moisture, while hydrological drought relates to deficiencies in, for example, lake levels, river flows and groundwater levels. The greatest severity and extent of drought in the United States occurred during the Dust Bowl years of the 1930s, particularly during 1934 and 1936. The decades of the 1950s and 1960s were also characterized by episodes of widespread, severe drought, while the 1970 and 1980s as a whole were unusually wet. Drought conditions that began in the west of the United States in 1998 persisted in many areas through to the summer of 2003 (Trenberth *et al.* 2004). In the United Kingdom, long time-series data of daily precipitation and monthly groundwater level provide an indicator of hydrological drought periods, as illustrated by comparing the Central England annual precipitation record and the mean annual Chalk groundwater level for Chilgrove House in southern England (Fig. 10.26).

Globally, drought areas increased more than 50% throughout the twentieth century, largely due to the drought conditions over the Sahel and Southern Africa during the latter part of the century, while changes in wet areas were relatively small (Trenberth *et al.* 2004). The most significant cause of drought worldwide is the El Niño Southern Oscillation (ENSO) which also emphasises the concurrent nature of floods and droughts, with droughts favoured in some areas during an ENSO event, while wet areas are favoured in others. These areas tend to switch during La Niña in the tropics and subtropics. Palaeoclimatological studies show evidence of dramatic changes in drought and the hydrological cycle over many parts of the world, with droughts lasting several decades not uncommon (Box 6.1). The full range of drought variability is probably much larger than has been experienced in the last 100 years (Trenberth *et al.* 2004).

The shortage of water experienced by water users and the aquatic environment during a drought is a complex issue. Potentially, the full characterization of events that vary regionally and temporally can involve the assessment of a wide range of hydrological indices (for example precipitation amount, river flows, surface reservoir storage and groundwater levels). According to Mawdsley *et al.* (1994), there are essentially two types of users impacted by the shortage of water. Firstly, there are those who are affected directly by a deficit of precipitation, possibly compounded by high evaporation, and second, those consumers who are affected indirectly due to the way in which water storage facilities are managed during a drought. Following from this, and including environmental water requirements, it is recommended that droughts are classified either as environmental or water supply droughts. An environmental drought measures the significance of the drought for those water uses directly affected by a shortage of precipitation, for

(Continued)

Box 10.7 (Continued)

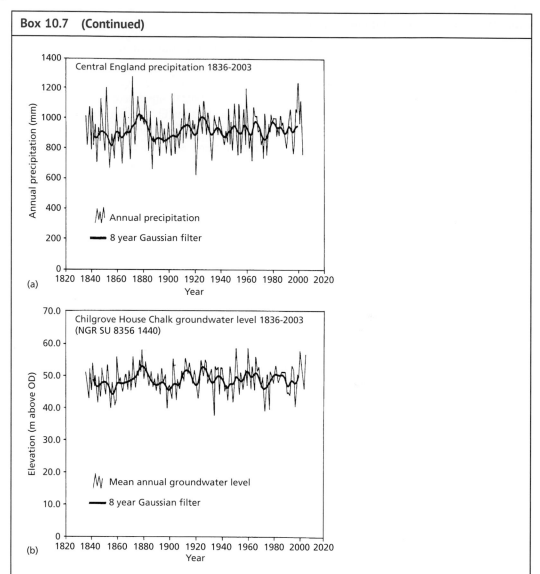

Fig. 10.26 Comparison of (a) the Central England annual precipitation record (*Source:* Hadley Centre Central England Temperature (HadCET) dataset, www.metoffice.gov.uk/hadobs/hadcet/, contains public sector information licensed under the Open Government Licence v1.0.) with (b) the mean annual Chalk groundwater level record at Chilgrove House (*Source:* Data from Chilgrove House, UKRI 2020. Available at www.bgs.ac.uk/research/groundwater/datainfo/levels/sites/ChilgroveHouse.html.) for the period 1836–2003. The data are presented with a smoothed line calculated using an eight-year Gaussian filter (Fritts 1976) applied as a moving average to the annual values. Statistical analysis of the record for Chilgrove House gives mean and median groundwater levels of 48.81 and 48.96 m above Ordnance Datum, respectively, with a standard deviation of 4.03 m. The record indicates that the first and second highest groundwater levels (58.49 and 58.31 m above Ordnance Datum) occurred in 1960 and 1951, respectively. The two lowest groundwater levels (38.48 and 39.51 m above Ordnance Datum) occurred in 1934 and 1973, respectively, giving a maximum range of groundwater level fluctuation over the length of the record of 20.01 m. Applying the formula of Gringorten (1963) for recurrence interval or return period, T, where $T = (n + 0.12)/(m - 0.44)$ with n equal to the number of events and m equal to the event ranking, events ranked first and second in the mean annual groundwater level series have return periods of 300 and 108 years, respectively. Comparison of the records shown in (a) and (b) highlights the hydrological droughts that occurred in the 1840s, 1850s, 1900s, 1940s, 1970s and 1990s.

Box 10.7   (Continued)

example aquatic ecology, fisheries, low river flows, reduced spring flows and groundwater levels, agriculture and horticulture. A water supply drought measures the significance for those indirectly affected, for example the risk of water demand reductions imposed on domestic and industrial water consumers. In many cases, a drought will be classified as having both environmental and water supply impacts (Mawdsley *et al.* 1994).

The communication of the severity of a drought can be simply presented according to its intensity and duration and classified as moderate, serious or severe (Table 10.3). The duration of an environmental drought is of particular relevance to groundwater resources. If a region is largely dependent on groundwater, and if the available aquifer storage is relatively insensitive to droughts only lasting one recharge period, then this situation is likely to result in a short environ

mental drought without significant implications for water supplies and the aquatic environment. However, if the drought is of long duration during which there is a reduction in seasonal recharge intensity, then a moderate drought may result with significant implications for water supplies, water levels and river flows dependent on groundwater.

A quantitative measure of environmental droughts is to calculate a drought severity index. In the United States, the Palmer Drought Severity Index (PDSI), as devised by Palmer (1965), is used to represent the severity of dry and wet periods based on weekly or monthly temperature and precipitation data as well as the soil water holding capacity at a location (Dai *et al.* 1998). Areas experiencing a severe drought have a PDSI score of −4.0, while areas with severe moisture surplus score +4.0. Between these two extremes, 11 categories of wet and dry conditions are defined.

Table 10.3   Classification scheme for environmental droughts and consequences for groundwater resources (Mawdsley *et al.* 1994).

Class of drought	Duration	Return period	Groundwater impacts
Moderate	Short	5–20 years	Reduced spring and river flows; drying out of floodplain areas
	Long[a]	5–20 years	Reduced spring and river flows; drying out of floodplain areas and wetlands; well yields may decrease
Serious	Short	20–50 years	Reduced spring and river flows; wetlands and ponds dry up
	Long[a]	20–50 years	Reduced spring and river flows; rivers become influent; wetlands and ponds dry up; saline intrusion in coastal aquifers
Severe	Short	>50 years	Springs and rivers dry up; wetlands and ponds dry up; well yields decrease as groundwater levels fall
	Long[a]	>50 years	Springs and rivers dry up; wetlands and ponds dry up; well yields fail as groundwater levels fall substantially; saline intrusion in coastal aquifers

[a] Longer than one groundwater recharge season.

(*Source:* Mawdsley, J., Petts, G. and Walker, S. (1994) *Assessment of Drought Severity*. Institute of Hydrology, Wallingford. British Hydrological Society Occasional Paper No. 3.)

(Continued)

Box 10.7   (Continued)

A more generally applicable drought severity index (DSI), which may be considered for different hydrological data, is to calculate an accumulated monthly deficit relative to the mean for a standard period (Bryant *et al.* 1994). In this approach, based on available long-term hydrological records, a drought is considered to end when, for example the three-monthly precipitation total exceeds the three-monthly mean for these months. It must be noted that the choice of termination criterion must be considered carefully since different rules can produce different impressions of drought severity that are potentially inappropriate for some long-duration events. A disadvantage of this approach is that a DSI based only on precipitation data does not directly indicate the impact on the environment since this will depend on the antecedent soil moisture conditions in a catchment. Instead, drought severity may be better determined using effective precipitation, or selected local river flow and groundwater level data, even though such records may only be available for a short time relative to precipitation records. Potentially, effective precipitation (the balance between precipitation and evapotranspiration (ET), eq. 6.2) could be a useful indicator of environmental drought, particularly in regions with significant groundwater resources.

For the prediction of climate change impacts on future minimum groundwater levels across the southern half of England, Bloomfield *et al.* (2003) developed a statistical method based on a multiple linear regression model of monthly rainfall totals for a given period against values of minimum annual groundwater levels for the same period. Application of the model to synthetic rainfall data from climate change scenarios to simulate changes in future annual minimum groundwater levels showed that there is a small reduction in annual minimum groundwater levels for a specific return period and that changes in the seasonality and frequency of extreme events could lead to an increase in the frequency and intensity of groundwater droughts in some areas of the United Kingdom. Bloomfield *et al.* (2003) concluded that the Chalk aquifer in southern and eastern England might be most susceptible to these effects.

In further work, Bloomfield and Marchant (2013) developed the Standardized Groundwater level Index (SGI) as a method for standardizing groundwater level time series and characterizing groundwater droughts. The SGI accounts for differences in the form and characteristics of groundwater level and precipitation time series and is estimated using a non-parametric normal scores transform of groundwater level data for each calendar month. These monthly estimates are then merged to form a continuous index. Analysing standardized indices of monthly groundwater levels, precipitation and temperature using two unique groundwater level datasets from the Chalk aquifer in England (Chilgrove House and Dalton Holme), for the period 1891–2015, Bloomfield *et al.* (2019) showed that precipitation deficits are the main control on groundwater drought formation and propagation. Interestingly, long-term changes in groundwater drought frequency, magnitude and intensity were shown to be associated with anthropogenic warming over the study period. In the absence of long-term changes in precipitation deficits, Bloomfield *et al.* (2019) inferred that the changing nature of groundwater droughts is due to changes in ET associated with anthropogenic warming, facilitated by the thick capillary fringe found in the Chalk aquifer, which may enable ET to be supported by groundwater through major episodes of groundwater drought. Bloomfield *et al.* (2019) concluded that this may be a globally important process affecting groundwater drought formation and propagation. Wherever droughts in shallow groundwater systems and/or aquifers with relatively thick capillary fringes are influenced by ET, it is inferred that they may be susceptible to changes due to anthropogenic warming.

as inherent hydrological variability increases over a shorter time average (Loáiciga *et al.* 1996).

An early example of hydrological simulation under climate change is presented by Vaccaro (1992) who studied the sensitivity of groundwater recharge estimates for a semi-arid basin, located on the Columbia Plateau, Washington, USA, to historic and projected climatic regimes. Recharge was estimated for pre-development and current (1980s) land-use conditions using a daily energy-soil-water balance model. A synthetic daily weather generator was used to simulate lengthy sequences with parameters estimated from subsets of the historical record that were unusually wet and unusually dry. Estimated recharge in the basin was found to be sensitive to climatic variability in the historical record, especially the precipitation variability. For GCM-projected climate changes to carbon dioxide doubling, the variability in the estimated annual recharge was less than that estimated from the historic data. In addition, the median annual recharge, for the case of a climate scenario averaged from three different GCM simulation runs, was less than 75% of the median annual recharge for the historical simulation. This large change reflects the potential importance of climate on groundwater recharge.

In a more recent study, Loáiciga *et al.* (2000) and Loáiciga (2003) generated climate change scenarios from scaling factors derived from several GCMs to assess the likely impacts of aquifer pumping on the water resources of the Edwards Balcones Fault Zone aquifer system in south-central Texas. This karst aquifer, formed in the Edwards limestone formation, is one of the most productive regional aquifers in the United States and is a primary source of agricultural and municipal water supplies. The Edwards Aquifer has been identified as one of the regional catchments most vulnerable to climate change impacts in the United States. Groundwater recharge to the aquifer takes place almost exclusively as stream seepage, with groundwater discharge directed to major springs that support an important and threatened groundwater ecosystem. To explore the vulnerability of these springs to combined pumping and climate change impacts, several pumping scenarios were combined with $2 \times CO_2$ climate scenarios. The climate change scenarios were linked to the surface hydrology to provide future recharge values which were then used as input to a numerical groundwater flow model. The scenarios and model results are shown in Tables 10.4 and 10.5 and summarize the effects on spring flows in the Edwards Aquifer. In the Comal Springs and, to a lesser

Table 10.4 Scenarios of climate and groundwater use change effects considered for the Edwards Aquifer, south-central Texas (Loáiciga 2003).

		Scenario		
	I Base	II Climate change effect	III Groundwater use change effect	IV Total effects
Climate (recharge)	$R_{1978-1989}$[a]	$R_{1978-1989}(2 \times CO_2/ 1 \times CO_2)$[b]	$R_{1978-1989}$	$R_{1978-1989}(2 \times CO_2/1 \times CO_2)$
Groundwater use	1978–1989 use[c]	1978–1989 use	2050 use[d]	2050 use

Notes:
[a] Historical recharge (R) during 1978–1989 (mean = 0.949×10^9 m^3 a^{-1}).
[b] Historical recharge scaled to $2 \times CO_2$ climate conditions.
[c] Average groundwater use between 1978 and 1989 = 0.567×10^9 m^3 a^{-1}.
[d] Groundwater use forecast for 2050 = 0.784×10^9 m^3 a^{-1}.
(*Source:* Loáiciga, H.A. (2003) Climate change and ground water. *Annals of the Association of American Geographers* **93**, 30–41. © 2003, Taylor & Francis.)

Table 10.5 Simulated minimum spring flows in the Edwards Aquifer, south-central Texas, for climate and groundwater use change effects listed in Table 10.4 (Loáiciga 2003).

Climate change and groundwater use scenario	Edwards Aquifer springs	
	Comal	San Marcos
I	4.84	4.84
II	12.7 (+162%)	5.67 (+17%)
III	0 (−100%)	3.79 (−22%)
IV	1.31 (−73%)	4.79 (−1%)

Notes:
Spring flows are given in 10^6 m^3 month^{-1}.
The numbers in parentheses represent the percentage increase (+) or decrease (−) caused by a scenario relative to the base condition (I).
(*Source:* Loáiciga, H.A. (2003) Climate change and ground water. *Annals of the Association of American Geographers* **93**, 30–41.)

extent, the San Marcos Springs, there is a simulated increase in spring flow under a doubling of CO_2 for an unchanged groundwater use condition (Scenario II). For the case of Scenario III (base climate and increased groundwater use), there are negative impacts on spring flow. Combining the effects of both climate change and increased groundwater use (Scenario IV) simulates not only a serious depletion of 73% in spring flow for the Comal Springs but also a marginal decrease (1%) for the San Marcos Springs. This study highlights that the primary threat to groundwater use in the Edwards Aquifer is from the potential rise in groundwater use caused by population growth and not from climate change, although protracted droughts under climate change will accentuate the competition between human and ecological water uses. Hence, and as concluded by Loáiciga (2003), aquifer strategies in the Edwards Aquifer must be adapted to climate variability and climate change.

Uncertainty is inherent in the development of climate change scenarios due to unknown future emissions of greenhouse gases and aerosols, uncertain global climate sensitivity and to the difficulty of simulating the regional characteristics of climate change. It is because of these uncertainties that the regional changes in climate derived from GCM experiments are termed *scenarios* or projections and cannot be considered predictions. Comprehensive reviews of climate change scenarios, including standards and construction, can be found in Carter and La Rovere (2001). The United Kingdom Climate Impacts Programme (UKCIP, available at www.ukcip.org.uk) provides projections of future changes to climate up to the end of this century based on simulations from climate models using a methodology that allows a measure of uncertainty in future climate projections to be included in the information.

In a study of the impacts of climate change on groundwater resources in the Chalk aquifer of eastern England, Yusoff *et al.* (2002) developed scaling factors for changes in precipitation (P) and potential ET (PE). The factors, as percentage changes in P and PE, were defined by comparing the monthly average P and PE values for a control run of the UK Hadley Centre HadCM2 model with the monthly average values for future Medium-low (ML) and Medium-high (MH) greenhouse gas emissions scenarios defined for the 2020s and 2050s. In this approach, it is assumed that the monthly factors can be applied equally to each year of the observed historical record to obtain calculated future groundwater recharge values. The derived recharge values were then used as input to a numerical groundwater model calibrated against an historical record of groundwater levels and river baseflow. A limitation of this linked GCM-hydrological modelling approach is that the climate variability represented within the historic record is preserved in the future scenarios. However, the approach gives a general indication of the possible range of changes in hydrological regimes.

The most noticeable and consistent result of the climate change impact simulations carried out by Yusoff *et al.* (2002) was a decrease in groundwater recharge expected in autumn for all scenarios as a consequence of the smaller amount of summer precipitation and increased autumn potential ET. For the 2050MH scenario, these conditions lead to a 42% increase in autumn soil moisture deficit and a 26% reduction in recharge. Hence, eastern England can expect longer and drier summers and a delay in the start of groundwater recharge in the autumn and winter period. The drier conditions will have relatively little effect on summer groundwater levels (generally a 1–2% decrease), but a modelled decrease of up to 14% in autumn baseflow volume for the 2050MH scenario indicates that Chalk groundwater-fed rivers may show environmental impacts with potential conflicts with other water demands.

Other scenarios for climate change impacts on groundwater relate to water quality and the balance of ocean salinity. Younger *et al.* (2002) examined the possibility that carbonate aquifers may act as a possible sink (or source) for atmospheric carbon dioxide and therefore have important consequences for the calcium carbonate content, or hardness, of groundwater. Younger *et al.* (2002) modelled increases in calcium concentrations of $\leq 10\,\mathrm{mg\,L^{-1}}$ for two European carbonate aquifers over a 50-year simulation period to 2045. These increases are negligible in water resources terms but draw attention to the possibility that the world's carbonate aquifers may represent a sink for atmospheric carbon dioxide and a

slowing of global warming over long time scales.

Direct groundwater discharges to the world's oceans and seas from inland catchments are estimated to be $2220\,\mathrm{km^3\,a^{-1}}$ (Section 1.5.3). Zektser and Loaiciga (1993) argued that a hypothetical 10% increase in global precipitation of $2\,\mathrm{mm\,a^{-1}}$ for greenhouse warming would provide an additional direct groundwater discharge of approximately $222\,\mathrm{km^3\,a^{-1}}$. Although this additional flux is small, the salt load, with a dissolved solids content of $585\,\mathrm{mg\,L^{-1}}$ (Section 1.5.4), would add a further $1.3 \times 10^8\,\mathrm{t\,a^{-1}}$ of salts to the world's oceans and seas. Whether this additional salt load would increase the salinity of the oceans is uncertain and depends on the balance of water inputs to and from the oceans under an intensified hydrological cycle and any volume change in the oceans as a result of thermal expansion and ice melting. It should also be considered that these effects of changes in direct groundwater flow might only stabilize over timescales of hundreds to thousands of years.

10.4.1 Groundwater response time to climate change

Deciphering the direct and indirect influence of climate on groundwater is more complex than with surface water, as residence times range from days to tens of thousands of years or more, and so a detectable response to climate change can be delayed, especially in deep aquifers (Gurdak 2017). Water levels in shallow aquifers are expected to respond readily to sub-annual and annual climate variability due to the physical constraints on recharge. In contrast, water levels in deep aquifers are more sensitive to inter-annual and longer-term climate variability. In an analysis using analytical groundwater model results in combination with hydrologic datasets, Cuthbert *et al.* (2019) found that nearly half of global groundwater fluxes could equilibrate with recharge variations due to climate change on human (~100 year) timescales, and that areas where water tables are most sensitive to changes are also those that have the longest groundwater response times. In particular, groundwater fluxes in arid regions are shown to be less responsive to climate variability than in humid regions. Given the variation in response time to climate change, adaptation strategies that rely on groundwater should take account of lag times and include appropriately long time-scales for water resources decision-making (Cuthbert *et al.* 2019).

Correlations to multi-annual or decadal climate patterns, such as the Pacific Decadal Oscillation (PDO) and El Niño Southern Oscillation (ENSO) have been observed in several aquifers (Gurdak *et al.* 2009; Holman *et al.* 2011; Kuss and Gurdak 2014). Although climate-induced variations of deep aquifer natural recharge are expected to have multi-annual time lags, the study by Russo and Lall (2017) found that deep groundwater levels can respond indirectly to climate over time-scales of less than one year. Such a quick response is considered to be the result of climate-induced pumping variability, as observed in response to persistent drought in the irrigated agricultural areas of the western United States, including parts of the High Plains Aquifer and Mississippi Embayment (Russo and Lall 2017).

10.4.2 Groundwater pumping and greenhouse gas emissions

Groundwater pumping for irrigation can itself contribute to the problem of global warming from greenhouse gas emissions given the energy required to lift groundwater to the surface. In a study of the water-energy nexus in China, where approximately 70% of the irrigated area in northern regions is supplied by groundwater, Wang *et al.* (2012) estimated that energy consumption associated with groundwater pumping for irrigation produced an equivalent of 33.1 Mt CO_2 annually, representing around 3% of agricultural emissions and 0.58% of total emissions nationally. As noted

by Wang *et al.* (2012), water scarcity in China is already driving policies to improve water conservation with the potential for considerable co-benefits from water and energy savings.

In considering other greenhouse gases that are more potent than carbon dioxide, for example methane (CH_4) with a global warming potential of 25 over 100 years ($GWP_{100} = 25$, compared with CO_2 with a $GWP_{100} = 1$), Kulongoski and McMahon (2019) quantified CH_4 emissions from groundwater pumping in the Principal Aquifers of the United States as 0.044 Tg a^{-1} in the year 2000 using the average of CH_4 concentrations in groundwater and annual groundwater pumping volumes. This flux represents a small percentage (~0.2%) of total annual CH_4 emissions in the United States, but a previously unquantified flux in the global CH_4 budget. Globally, CH_4 emissions from groundwater pumping were estimated by Kulongoski and McMahon (2019) to be 0.53 Tg a^{-1}, representing 0.2% of global CH_4 emissions based on a global estimate for groundwater extraction and an average CH_4 concentration in older groundwater of 0.44 mg L^{-1}. Using a similar approach, Hiscock *et al.* (2003) estimated the total emissions of nitrous oxide, N_2O ($GMP_{100} = 298$), from the major aquifers in the United Kingdom of 3.45×10^5 kg N_2O-N a^{-1}, or about 0.6% of the modelled fertilizer-induced emissions of N_2O from arable and grassland soils. The magnitude of this flux suggests that indirect losses of N_2O from regional aquifers are much less significant (<1%) than direct emissions of N_2O from agricultural soils.

10.4.3 Impact of climate change on cold-region hydrogeology

The hydrogeology of cold-regions has long been considered to be in a non-active state (Walvoord and Kurylyk 2016). This is because in these areas, the near-surface occurrence of permafrost, ground which is on average below freezing for two or more consecutive years, would prohibit both the deep infiltration (recharge) or seepage (discharge) of groundwater. Under these conditions, hydrogeological activity is limited to the unfrozen 'active' layer that develops to a depth of at most a few metres during the summer (Fig. 10.27). In recent years, however, it has been increasingly recognized that climate warming, which is occurring at a rate which is amplified in polar regions (Plate 10.3), will trigger this presumed dormant state to come to an end. It is expected that permafrost will start to degrade and eventually disappear in many of these regions, which will consequently start to transition into hydrogeological regimes which are more similar to those that can be observed in more temperate climates.

The degradation and eventual loss of permafrost from currently cold-regions can be expected to have profound impacts on both the water cycle and associated biogeochemical cycles of these regions. In particular, the potential for the release of carbon into the atmosphere through the emission of the greenhouse gases CO_2 and CH_4 from these areas is cause for grave concern (Schädel *et al.* 2016). These areas are often peaty due to the low permeability of the shallow subsurface favouring the ponding of surface water in summer, but often also contain vast volumes of organic matter in the upper tens of metres of the subsurface. As the surface warms, the deepening of groundwater flow paths and the increasing intensity of pore water movement

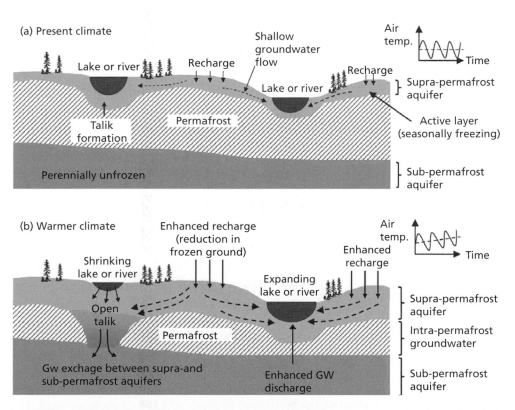

Fig. 10.27 (a) Profile view of a permafrost landscape with shallow (supra-permafrost) and deep (sub-permafrost) groundwater (GW) flow. (b) Warming atmospheric conditions lead to permafrost thaw and increased hydrologic and hydrogeologic connections between lakes, rivers and shallow and deep aquifers that can result in some landscapes becoming dryer while others become wetter (Walvoord and Kurylyk 2016). (*Source:* Adapted from Walvoord, M.A. and Kurylyk, B.L. (2016) Hydrologic impacts of thawing permafrost—a review. *Vadose Zone Journal* **15**. DOI: 10.2136/vzj2016.01.0010.)

that will result from the retreat of the perma-frost table (resulting in the development of a 'supra-permafrost' aquifer) can possibly medi-ate the release of C, but also that of toxic metals such as Hg that are currently sequestered in the subsurface (Walvoord and Striegl 2007; Colombo *et al.* 2018).

Classic observational data of groundwater level and water quality are usually scarce in these permafrost areas and therefore cannot be relied upon for groundwater systems analy-sis. Hence, hydrogeologists have, for example, turned to the interpretation of long-term series of river flow records to infer baseflow compo-nents as an indication of the vigour of regional groundwater flow (Section 6.7.1) and decadal time-scale changes therein. For example, an analysis of annual peak- and low-flow records of the Lena River Basin in Siberia that reach back to the 1920–1930s (Fig. 10.28) has sug-gested that regional groundwater flow has been intensifying over the latter part of the twentieth century (Ye *et al.* 2009). Whether or not perma-frost degradation is entirely responsible for this trend in combination with other climatological factors such as shifts in precipitation and/or evaporation and transpiration has become

the subject of ongoing debate (Rawlins *et al.* 2019). While river flow observations at a drain-age basin outlet such as in the Lena Basin can potentially provide an indication of changing groundwater flow conditions, geophysical techniques using ground-based or airborne instruments (Minsley *et al.* 2012) to image per-mafrost thickness distributions and the pres-ence of any permafrost-free zones (so-called taliks; Fig. 10.27) can also help establish the hydrogeological state of a permafrost-dominated groundwater flow system.

The numerical modelling of permafrost hydrogeological systems requires the consider-ation of coupled heat and groundwater flow including the heat flow impacts of the latent heat effects to thaw and freeze water stored in the pore space (Bense *et al.* 2009). Also, hydraulic conductivity needs to be parameter-ized as a function of temperature and be reduced by at least several orders of magnitude at temperatures below freezing. Thus, the kind of modelling scheme that is required goes beyond the capability of most standard suites of groundwater flow modelling (e.g. MOD-FLOW; Section 7.5). Therefore, in the past dec-ade, a substantial research effort has been put

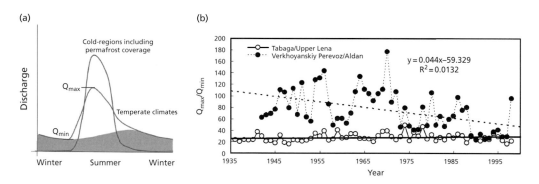

Fig. 10.28 (a) Conceptual river discharge hydrographs dominated by spring/summer snowmelt peaks but showing shifting ratios between annual peak-flow (Q_{max}) and low-flow (Q_{min}) for a river in permafrost terrain *versus* areas with a temperate climate where permafrost is absent and baseflow contributions are more prominent. (b) Analysis of annual peak- and low-flow records (as the ratio Q_{max}/Q_{min}) at the hydrological stations Tabaga on the Upper Lena and Verkhoyanskiy Perevoz on the Aldan within the Lena River Basin, Siberia. Over the Aldan sub-basin, the maximum/minimum discharge ratios significantly decrease during 1942–1998 due to an increase in baseflow, with this change consistent in general with permafrost degradation over eastern Siberia (Ye *et al.* 2009). (*Source:* Adapted from Ye, B., Yang, D., Zhang, Z. and Kane, D.L. (2009) Variation of hydrological regime with permafrost coverage over Lena Basin in Siberia. *Journal of Geophysical Research* **114**, D07102. DOI:10.1029/2008JD010537.)

into the development of new codes, or via an extension of the capabilities of existing codes, to include the coupling of processes required to simulate and forecast the hydrogeological response of permafrost areas undergoing climate warming at an unprecedented pace (Grenier *et al.* 2018).

10.4.4 Adaptation to climate change

Even without considering climate change, sustainable development of groundwater is a major challenge given that groundwater is a widely distributed resource responding at the basin scale, but is affected by local users and polluters (municipalities, industrial enterprises and farmers) whose behaviours are greatly influenced by national policies determining land and water use. Hence, in general, governance systems, resource policies, innovation incentives, data collection and information provision need to relate to a wide range of scales, with different adaptive management approaches in rural and urban environments (IAH 2006).

Climate change challenges the traditional assumption that past hydrological experience provides a good guide to future conditions (IPCC 2007). In times of surface water shortages during droughts, a typical response is for groundwater resources to be abstracted as an emergency supply (Table 10.6). Under conditions of climate change, this response is

Table 10.6 Types of adaptation options for surface water and groundwater supply and demand (IPCC 2008).

Supply-side	Demand-side
Increase storage capacity by building reservoirs and dams	Improve water-use efficiency by recycling water
Desalinate seawater	Reduce water demand for irrigation by changing the cropping calendar, crop mix, irrigation method and area planted
Expand rain-water storage	
Remove invasive non-native vegetation from riparian areas	Promote traditional practices for sustainable water use
Prospect and extract groundwater	Expand use of water markets to reallocate water to highly valued uses
Develop new wells and deepen existing wells	
Maintain well condition and performance	Expand use of economic incentives including metering and pricing to encourage water conservation
Develop aquifer storage and recovery systems	
Develop conjunctive use of surface water and groundwater resources	
Develop surface water storage reservoirs filled by wet season pumping from surface water and groundwater	Introduce drip-feed irrigation technology
	License groundwater abstractions
	Meter and price groundwater abstractions
Develop artificial recharge schemes using treated wastewater discharges	
Develop riverbank filtration schemes with vertical and inclined bank-side wells	
Develop groundwater management plans that manipulate groundwater storage, e.g. resting coastal wells during times of low groundwater levels	
Develop groundwater protection strategies to avoid loss of groundwater resources from surface contamination	
Manage soils to avoid land degradation to maintain and enhance groundwater recharge	

(*Source:* IPCC (2008) *Technical Paper on Climate Change and Water (Finalized at the 37th Session of the IPCC Bureau).* Intergovernmental Panel on Climate Change, Geneva. © 2008, IPCC.)

likely to be unsustainable, especially in those areas expected to experience an increase in drought frequency and duration. Also, rising sea levels under climate change will further threaten coastal freshwater aquifers, especially those already experiencing salinization due to over-exploitation. In this section, and to address possible future adaptive responses to climate change, reference is made to the literature on the mitigation and adaptation responses that apply to water resources in general, together with specific consideration of groundwater.

Adaptation approaches can be preventative or reactive and apply to natural and social systems. Ensuring the sustainability of investments in, for example groundwater resources planning and development, over the entire lifetime of a scheme and taking explicit account of changing climate, is referred to as *climate proofing* (CEC 2007). At minimum, and in the absence of reliable projections of future changes in hydrological variables, adaptation processes and methods can be implemented, such as improved water use efficiency and water demand management, offering no-regrets options to cope with climate change. For example the Netherlands, like the rest of the world's delta regions, is vulnerable to climate change and sea-level rise and associated groundwater quality (and quantity) related challenges. In response, the Dutch are investing in climate proofing that uses hard infrastructure and softer measures, such as insurance schemes or evacuation planning, to reduce the risks of climate change and hydrologic variability to a quantifiable level that is acceptable to society (Kabat *et al.* 2005).

According to the IPCC (2007), the array of potential adaptive responses available to human societies is very large, ranging from purely technological (e.g. deepening of existing boreholes), through behavioural (e.g. altered groundwater use), to managerial (e.g. altered farm irrigation practices), and to policy (e.g. groundwater abstraction licensing regulations). The IPCC (2007) argued that while most technologies and strategies are known and developed in some countries (e.g. demand-management through the conjunctive use of surface water and groundwater resources; see Section 10.2.2), the effectiveness of various options to fully reduce risks for vulnerable water-stressed areas, particularly at higher levels of warming and related impacts, is not yet known.

For water resources management, there are generally two types of decisions to be considered: those dealing with new investments and those dealing with the operation and maintenance of existing systems. In order to inform these decisions, information is needed about future water availability and demand, both of which are affected by climate change at the river-basin scale (Ballentine and Stakhiv 1993). Table 10.6 summarizes supply-side and demand-side adaptation options designed to ensure supplies of water and groundwater during average and drought conditions. As explained by the IPCC (2008), supply-side options generally involve increases in storage capacity or water abstraction. Demand-side adaptation options rely on the combined actions of individuals (industry users, farmers (especially irrigators) and individual consumers) and may be less reliable. Indeed, some options, for example those incurring increased pumping and treatment costs, may be inconsistent with climate change mitigation measures because they involve high energy consumption.

One of the major challenges facing water resources managers is coping with climate change uncertainties in the face of real-world decision-making, particularly where expensive investment in infrastructure such as well-field design, construction and testing and laying of pipelines is required. As discussed by Dessai and Hulme (2007), this challenge presents a number of new questions, for example how much climate change uncertainty should we adapt to? Are robust adaptation options socially, environmentally and economically acceptable and how do climate change uncertainties compare with other uncertainties such as changes in demand? The answers to these questions leading to robust adaptation decisions will require the development of probability distributions of specified outcomes (Wilby and Harris 2006) and negotiation between decision-makers and stakeholders involved in the adaptation process (Dessai and Hulme 2007). For lower-income countries, availability of resources and building adaptive capacity are particularly important in order to meet water shortages and salinization of fresh waters (IPCC 2007).

Examples of current adaptation to observed and anticipated climate change in the management of groundwater resources are few, with groundwater typically considered as part of an integrated water-supply system. Here, three examples serve to highlight the difference in approach in technically advanced and developing country contexts. The ability of California's water supply system to adapt to long-term climate and demographic changes is examined by Tanaka *et al.* (2006) using a state-wide economic-engineering optimization model of water supply management and considering two climate warming scenarios for the year 2100. The results suggested that California's water supply system appears physically capable of adapting to significant changes in climate and population, albeit at significant cost. Such

adaptations would entail large changes in the operation of California's large groundwater storage capacity, significant transfers of water among water users and some adoption of new technologies. In a further study, in the Sacramento Valley, California, Purkey *et al.* (2007) used four climate time series to simulate agricultural water management with adaptation in terms of improvements in irrigation efficiency and shifts in cropping patterns during dry periods leading to lower overall water demands in the agricultural sector, together with associated reductions in groundwater pumping and increases in surface water allocations to other water use sectors. Land-use adaptation to projected climate change may include management changes within land-use classes (e.g. alternative crop rotations) or changes in land classification (e.g. converting annual cropping systems to perennial grasslands or forests). Soil and water conservation programmes already encourage some of these types of land-use changes.

A similar technological approach to that demonstrated for California is presented for the Mediterranean region of Europe. This region is experiencing rapid social and environmental changes with increasing water scarcity problems that will worsen with climate change (see Box 10.8). Iglesias *et al.* (2007) found that these pressures are

---

Box 10.8   Climate change impacts on European groundwater resources
Groundwater is a significant economic resource in Europe. As shown in Table 1.6, large quantities of groundwater are abstracted in Europe, with about 75% of the population dependent on groundwater for their water supply. Furthermore, in Mediterranean regions, groundwater is a valuable resource in meeting the high agricultural irrigation demand (Krinner *et al.* 1999; UNEP 2003). In common with other areas of the world, European groundwater resources are threatened by over-abstraction and contamination from surface-derived pollutants, with these pressures potentially exacerbated by the additional anthropogenically driven threat of climate change. The interpretation of climate change on these important groundwater resources is difficult to predict but will be dependent on regional hydrogeological characteristics, as well as socioeconomic conditions that will determine future water supply demand (Holman *et al.*

(Continued)

Box 10.8 (Continued)

2005; Holman 2006). In Europe, records show that over the last century (1901–2001), the average temperature has risen by 0.95 °C, and that climate change has caused a steepening of precipitation and temperature gradients resulting in wetter conditions in northern regions and drier conditions in southern areas (IPCC AR4 WGII, Alcamo *et al.* 2007).

In a study of the impacts of future climate change on European groundwater resources, Hiscock *et al.* (2011) used climatological, geological and land-use data to characterise five study areas in northern and southern Europe, centred on the Å (northern Denmark), Medway (southern England), Seine (northern France), Guadalquivir (southern Spain) and Po (northern Italy) river basins. To analyse the impacts of climate change on groundwater resources in these areas, four Global Circulation Models (GCMs) were used to predict future precipitation and temperature trends based on a 'high' (SRES A1FI) gas emissions scenario for the 2020s, 2050s and 2080s. Using a methodology similar to that presented by Herrera-Pantoja and Hiscock (2007), the future precipitation and temperature values were used in a soil moisture balance model to calculate future potential groundwater recharge (Table 10.7).

Most of the GCMs predicted that by the end of this century, northern Europe will receive more winter rainfall, leading to increased groundwater recharge but during a shorter time period, potentially leading to an increased risk of groundwater flooding (Ireson and Butler 2011). Summers are predicted to be drier with a longer period of limited or no groundwater recharge. Hence, the key to climate change adaptation in northern Europe in order to cope with longer, drier summer periods is to capture the winter recharge and use it efficiently through good demand management in which surface water and groundwater resources are used conjunctively, especially in the densely populated areas in southern England and northern France.

In southern Europe, groundwater supplies are already heavily stressed with climate change expected to further reduce potential groundwater recharge and available groundwater resources throughout the year, with groundwater recharge virtually disappearing in some areas such as southern Spain by the 2080s. Any increase in winter recharge is unlikely to compensate for reduced autumn groundwater recharge. In combination with water requirements that are projected to increase under a drier climate, severe water shortages are a distinct possibility in southern Europe by the end of this century. Under these conditions, adaptation responses to climate change are to conserve groundwater resources by controlling the irrigation of crops in the most water-stressed areas in order that domestic water supplies can be met. Furthermore, in Mediterranean coastal areas where water scarcity is aggravated by high population densities and intense economic activities leading to acute seasonal water demands, Salgot and Torrens (2008) recommended the increasing use of non-conventional water resources of lower quality, such as waste water reclamation and reuse that can augment groundwater resources and so meet additional demands (other than for drinking water) for urban, industrial and agricultural applications. Ultimately, the achievement of long-term sustainability of groundwater resources in water-stressed regions can only be met by effective integration of regional water and agricultural policies that, for example, control illegal groundwater abstractions, establish water banking and promote cropping diversification and modern irrigation methods, to ensure ecological protection and human development at an acceptable social cost (Varela-Ortega *et al.* 2011).

Box 10.8   (Continued)

Table 10.7   Percentage change in mean annual potential groundwater recharge values calculated using four GCMs (HadCM3, CGCM2, CSIRO2 and PCM) for the 2020s, 2050s and 2080s 'high' gas emissions scenarios compared with the baseline period, 1961–1990, for five study areas (Denmark, England, France, Spain and Italy). Negative percentage changes indicating a decrease in annual groundwater recharge are shown in bold (Hiscock *et al.* 2011).

	Baseline	2020s	% change	2050s	% change	2080s	% change
HadCM3							
Denmark	279.8	312.0	11.5	333.6	19.2	302.3	8.0
England	286.8	301.5	5.1	284.8	**−0.7**	347.2	21.0
France	140.7	159.0	13.1	175.3	24.6	235.5	67.4
Spain	30.6	24.5	**−20.1**	30.2	**−1.4**	5.1	**−83.4**
Italy	494.0	370.6	**−25.0**	330.3	**−33.1**	346.2	**−29.9**
CGCM2							
Denmark	279.8	333.5	19.2	384.3	37.3	376.3	34.5
England	286.8	303.6	5.8	289.0	0.8	340.4	18.7
France	140.7	148.2	5.4	147.5	4.9	184.4	31.1
Spain	30.6	33.4	9.0	46.9	53.0	12.1	**−60.6**
Italy	494.0	392.1	**−20.6**	362.0	**−26.7**	383.4	**−22.4**
CSIRO2							
Denmark	279.8	330.9	18.3	376.3	34.5	369.3	32.0
England	286.8	308.6	7.6	294.4	2.7	349.2	21.7
France	140.7	162.0	15.1	173.1	23.0	232.9	65.6
Spain	30.6	43.2	41.0	70.6	130.5	35.8	16.9
Italy	494.0	420.6	**−14.9**	424.6	**−14.1**	493.9	0.0
PCM							
Denmark	279.8	318.4	13.8	348.3	24.5	318.5	13.8
England	286.8	297.2	3.6	268.1	**−6.5**	302.7	5.5
France	140.7	138.9	**−1.2**	117.7	**−16.3**	138.1	**−1.8**
Spain	30.6	37.5	22.4	57.0	86.1	22.0	**−28.1**
Italy	494.0	406.6	**−17.7**	390.3	**−21.0**	438.4	**−11.3**

(*Source:* Hiscock, K.M., Sparkes, R. and Hodgson, A. (2011) Evaluation of future climate change impacts on European groundwater resources. In: *Climate Change Effects on Groundwater Resources: A Global Synthesis of Findings and Recommendations* (eds T. Holger, J.-L. Martin-Bordes and J. Gurdak). CRC Press, Baton Rouge, FL, pp. 27. IAH International Contributions to Hydrogeology. © 2011, Taylor & Francis.)

heterogeneous across the region or water use sectors and adaptation strategies to cope with water scarcity include technology, use of strategic groundwater and better management based on preparedness rather than a crisis approach. Iglesias *et al.* (2007) also advocated the importance of local management at the basin level but with the potential benefits dependent on the appropriate multi-institutional and multi-stakeholder coordination.

In contrast to the examples from North America and Europe, Ojo *et al.* (2003) discussed the downward trends in rainfall and groundwater levels and increases in water deficits and drought events affecting water resources availability in West Africa. Here, the response strategies needed to adapt to climate change emphasize the need for water supply–demand adaptations. Moreover, the mechanisms needed to implement adaptation measures include: building the capacity and manpower of water institutions in the region for hydro-climatological data collection and monitoring; the public participation and involvement of stakeholders; and the establishment of both national and regional co-operation.

Further to the challenges presented by climate change, water resources management has a clear association with many other policy areas such as energy, land use and nature conservation. In this context, groundwater is part of an emerging integrated water resources management approach that recognizes society's views, reshapes planning processes, co-ordinates land and water resources management, recognizes water quantity and quality linkages, manages surface water and groundwater resources conjunctively and protects and restores natural systems while including a consideration of climate change. This integrated approach presents new challenges for groundwater. For example, better understanding is needed of leakage processes associated with carbon capture and storage if the potential degradation of groundwater quality is to be avoided. Also, insight is needed into the effects of large-scale plantations of commercial energy crops on groundwater recharge quantity and quality (IPCC 2008).

In summary, groundwater resources stored in aquifers can be managed, given reasonable scientific knowledge, adequate monitoring and sustained political commitment and provision of institutional arrangements. Although

there is no single approach to relieving pressures on groundwater resources, given the intrinsic variability of both groundwater systems and socio-economic situations, incremental improvements in resource management and protection can be achieved now and in the future under climate change. Future sustainable development of groundwater will only be possible by approaching adaptation through the effective engagement of individuals and stakeholders at community, local government and national policy levels.

10.5 Groundwater and energy resources

The imperative to combat climate change requires adaptation to cleaner energy resources to reduce global emissions of greenhouse gases and limit the effects of global warming. Decarbonization of space and water heating is essential if greenhouse gas emissions are to be reduced to net zero by 2050 in a cost-effective and environmentally acceptable way (Lucon *et al.* 2014; Boon *et al.* 2019). In this section, the role of groundwater in the development of geothermal energy resources, ground source heat pump systems and shale gas exploration are discussed.

10.5.1 Geothermal energy

Geothermal energy provides an extensively used, renewable energy worldwide and has the advantage over other renewable energies of being available year-round. This, and its large resource, make geothermal energy an attractive option as a sustainable future energy supply. Geothermal energy can be exploited with various technologies and generally involves drilling and pumping water from depth (Hurter and Schellschmidt 2003).

The growth of geothermal energy, mainly for electrical power and direct use, is restricted by its geographic availability (as dictated by geologic controls), high upfront costs of drilling and risk of not finding a viable resource after incurring possibly significant expenditure of time and money (Boden 2016). The cost-effectiveness of a geothermal scheme depends upon the amount of heat that can be recovered, which is a function of the flow rate, the heat capacity of the fluid and the temperature drop as the heat is extracted by heat exchangers and heat pumps (Downing and Gray 1986).

Geothermal energy can be used over a range of temperatures: electrical power generation can occur at high temperatures; direct use, such as for space and water heating, can take advantage of moderate temperatures; and geothermal heat pumps can operate at ambient Earth temperatures (Boden 2016). Geothermal energy plays a significant role for both power generation and direct use in countries with limited or no conventional fossil fuel resources such as in Iceland, Japan and New Zealand. In Europe, the geothermal resources of most countries have been estimated and compiled in an atlas, with maps that permit a first-order evaluation of the geothermal potential in terms of technical and economic viability (Hurter and Schellschmidt 2003).

Geothermal systems can be classified by a variety of criteria, such as the nature of heat transfer (conductive or convective), the presence or absence of recent magmatism or volcanic activity, the particular geologic setting (for example, type of volcanic environment) or tectonic setting (for example, type of plate boundary or intraplate geologic hot spot), and temperature (low-, moderate- and high-enthalpy systems) (Boden 2016). Hot spot tectonic settings, such as Yellowstone or Hawaii, are typically magmatic systems, whereas those in extended crust, such as in much of the Basin and Range Province of Nevada, are mainly amagmatic. A liquid-dominated system can be used for flash or binary power generation, combined power and heat, or direct use, whereas geologically rare vapour-dominated systems are used for power generation (Boden 2016).

Outside of volcanic areas, deep carbonate rock aquifers, most of which are to some degree karstified, are probably the most important thermal water resources (Goldscheider *et al.* 2010). Deep and hot sedimentary aquifers occurring at depths of 3–4 km can achieve temperatures of the contained fluids ranging from about 150 °C to as much as 200 °C. Regional fault and fracture zones are often the most productive zones, but can be difficult to locate, resulting in relatively high exploration uncertainty. These hypogenic karst systems (as opposed to relatively shallow, cold-water karst aquifers, or epigenic systems) are related to deep regional groundwater circulation systems and are probably much more widespread than previously suspected. Groundwater is the main driver for the creation of porosity in deep carbonate rock aquifers, summarized under the term 'hypogenic speleogenesis'. Many different hydrogeochemical reactions are involved such as mixing corrosion, retrograde calcite solubility, dissolution due to geogenic acids from deep sources, and other processes (Goldscheider *et al.* 2010).

The best examples of geothermal power plants in deep carbonate rocks are found in Central Europe, particularly in the Upper Malm (Upper Jurassic) of the Northern Alpine Molasse Basin, the Upper Malm and Dogger of the Paris Basin and the Upper Muschelkalk (Middle Triassic) in the Upper Rhine Graben. Operational geothermal systems using hot water from these reservoirs are summarized in Table 10.8. For example, the deep sedimentary aquifer of the Paris Basin has been tapped for geothermal heating of buildings since the early 1970s. At present, approximately 150 000 buildings are heated from 34 operating geothermal facilities consisting mainly of doublet boreholes (one well for production and one well for injection) with a total thermal energy production of 12 500 MWh. The geothermal reservoir is the mid-Jurassic carbonate Dogger aquifer that lies at a depth of between 1500 and 1900 m and has temperatures ranging between 65 and 85 °C. Flow rates average about 200 m^3 $hour^{-1}$ but can be as much as 600 m^3 $hour^{-1}$ (Boissier *et al.* 2009). A single doublet well

Table 10.8 Summary of geothermal installations producing geothermal energy for district heating and electric power generation from selected karst aquifers in Central Europe (compiled from various sources). NA denotes not applicable (for example, injection data for a single-well system) and ND denotes no data available (Goldscheider *et al.* 2010).

Location	Temperature (°C) Production/ Injection	Flow rate (L s^{-1}) Pump/ No pump	Well depth (m) Production/ Injection	Aquifer	Thermal power (MW)
Altheim (Austria)	106/65	ca. 50/46	2300/2165	Upper Malm	11.5
Bad Blumau (Austria)	110/50	ca. 80/30	2843/2583	Upper Malm	7.6
Bad Waltersdorf (Austria)	63/55	NA/17	1400/1061	Upper Malm	2.3
Geinberg (Austria)	105/35	ND/25	2225/NA	Upper Malm	7.8
Simbach Braunau (Austria)	81/ND	74/30	2200/1848	Upper Malm	9.3
Paris Basin (France) (31 operating doublets)	50–85/ca. 45	40–170/ND	1400–2000/ 900–2000	Dogger	NA
Riehen (Switzerland)	66.4/52.2	18/ND	1547/1247	Upper Muschelkalk	3.6
Erding (Germany)	65/ND	55/ND	2350/2060	Upper Malm	8
Pullach (Germany)	107/ND	50/ND	3443/3370	Upper Malm	6
Riem (Germany)	93/ca. 50	75/ND	2746/3020	Upper Malm	9
Unterföhring (Germany)	86/ND	50/ND	2512/2120	Upper Malm	ND
Unterhaching (Germany)	123/ND	150/ND	3346/3590	Upper Malm	40
Unterschleissheim (Germany)	81/ND	90/ND	1960/2000	Upper Malm	13
Kirchstockach (Germany)	139/ND	145/ND	3750/3730	Upper Malm	4.5

(*Source:* Adapted from Goldscheider, N., Mádl-Szőnyi, J., Erőss, A. and Schill, E. (2010) Review: thermal water resources in carbonate rock aquifers. *Hydrogeology Journal* **18**, 1303–1318.)

system with a 250 m^3 hour^{-1} flow rate, a production temperature of 70 °C and an injected fluid temperature of 45 °C can serve about 4000 dwellings. Despite the long production of some of the doublet systems (35–40 years), production well temperatures have remained steady. Nonetheless, the net heat flux is considered insufficient to maintain the current production temperatures indefinitely. Most numerical models indicate cooling of doublet systems by 1.5–3.5 °C after 40 years due to the continued reinjection of cool brine at 40–45 °C (Lopez *et al.* 2010).

In Germany, geothermal regions consist of deep (3–4 km), hot (>100–150 °C) sedimentary aquifers, such as in the North German Basin, the Upper Rhine Valley in south-west Germany and the combined heat and geothermal power facilities in the Northern Alpine Molasse Basin in southern Germany (Herzberger *et al.* 2009). The largest geothermal power station in Germany is located at Unterhaching, near

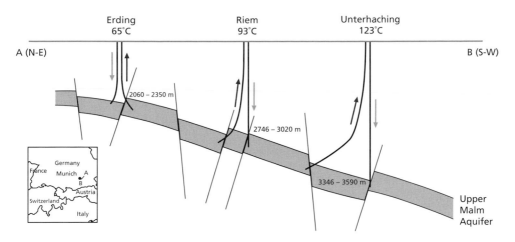

Fig. 10.29 Schematic NE–SW profile illustrating geothermal resources in the Upper Malm Limestone aquifer below the northern foreland of the Alps (Bavaria, Germany), with increasing depth and temperature towards the Alps (SW). Three examples of geothermal installations are shown (see Table 10.8 for technical details), which exploit the fault-bound thermal water reservoirs in this aquifer (Goldscheider *et al.* 2010). (*Source:* Goldscheider, N., Mádl-Szőnyi, J., Erőss, A. and Schill, E. (2010) Review: thermal water resources in carbonate rock aquifers. *Hydrogeology Journal* **18**, 1303–1318.)

Munich (Fig. 10.29) and exploits thermal water from Upper Jurassic (Malm) limestone below the Molasse Basin, the northern foreland basin of the Alps (Berge and Veal 2005). This deep karst aquifer is considered to be the largest thermal water resource in Central Europe, accessible mainly via drilled wells. The power plant is located at a major NNW–SSE striking fault zone. The injection well was drilled through a fault with a vertical displacement of 238 m and illustrates the high heterogeneity of carbonate rock aquifers and the important role of fault zones (Wolfgramm *et al.* 2007). The production well at Unterhaching is 3346 m deep and produces about 540 m^3 hour^{-1} of 123 °C hot water, used for heating and electric power generation (Table 10.8; Goldscheider *et al.* 2010). At the more recently completed Kirchstockach geothermal facility near Munich, production and injection well depths are in the region of 3750 m. The production well produces fluid at 139 °C at a flow rate of 522 m^3 hour^{-1} and the installed power capacity is 5.5 MW (Table 10.8; Boden 2016).

In the United Kingdom, low-temperature, hot-water resources in the range 40–100 °C occur in Permo-Triassic sandstones in several deep sedimentary basins, including the East Yorkshire and Lincolnshire Basin, Wessex Basin, Worcester Basin, Cheshire Basin, and Ballycastle and Larne Basins in Northern Ireland (Downing and Gray 1986; Barker *et al.* 2000). From a consideration of measured temperature data, a geothermal gradient of 28 °C km^{-1} has been calculated for the upper 1 km of the sedimentary crust (Busby *et al.* 2011). Temperatures of more than 60 °C at a depth of 2 km and possibly 100 °C at 3 km are suitable for direct heating applications for space heating, industrial processes and horticulture (Downing and Gray 1986). Elevated temperatures have been observed mainly in eastern and southern England, attributed to convection within some of the thicker Permo-Triassic sandstones and the thermal blanketing effect of Triassic and Jurassic argillaceous rock. Productive sandstones vary from a few tens of metres to hundreds of metres thick resulting in productive transmissivities and an estimated Inferred

Geothermal Resource of 201–328 × 10^{15} kJ (Busby 2014). However, exploitation of this resource requires the coincidence of heat loads with the resource and so far has only been achieved in Southampton in the Wessex Basin where geothermal energy contributes some 2 MW to a 12 MW district-heating scheme (Barker *et al.* 2000).

Low-enthalpy resources exist in Upper Palaeozoic aquifers but development is hindered by the difficulty of forecasting the position of extensive fracture systems at the depth necessary to give high temperatures. Natural fracture systems in the Carboniferous limestone at Bath and Bristol (see Box 2.12), Buxton and Matlock in the Peak District (Brassington 2007) and in the Taff Valley, South Wales (Farr and Bottrell 2013) support warm springs that have been developed for various purposes including bathing, hydrotherapy and bottling.

The potential of radiothermal granites for Hot Dry Rock (HDR) development has been investigated in the Carnmenellis granite (see Box 4.12) at Rosemanowes in Cornwall. Three boreholes were drilled in the granite to depths of over 2 km and connected by developing natural fractures. Water circulation between the boreholes and through the fractured rock was demonstrated successfully. However, the ability to develop natural joints in granite to produce a HDR reservoir with satisfactory hydraulic and thermal characteristics is a major challenge to overcome before the technique can be exploited commercially (Downing and Gray 1986).

As noted by Downing and Gray (1986), very large amounts of thermal energy stored in groundwater at less than 40 $^{\circ}$C should not be overlooked, with an estimated 245 × 10^{15} kJ of thermal energy available in the UK in the temperature range 20–40 $^{\circ}$C. In many areas, shallow groundwater at depths of less than 30–40 m, and at low but very advantageous temperatures (for example, 11 $^{\circ}$C in the UK), is available as an energy source. These low temperature waters can only be developed with the assistance of heat pumps as discussed in the next section.

10.5.2 Ground source heat pumps

Ground source heat pumps (GSHP) provide an example of an efficient way to reduce reliance on fossil fuels. GSHP systems exchange heat with the sub-surface and can usefully provide space heating or cooling in the residential and non-residential sectors.

GSHP systems can either utilize a closed-loop heat exchanger, or an open configuration. Groundwater-based open-loop systems exchange heat directly with groundwater and can be more efficient than closed-loop systems owing to the water generally maintaining a constant temperature, whereas in closed-loop systems the ground is affected by heat extraction or injection (Abesser *et al.* 2014; Garcia-Cespedes *et al.* 2020). As shown in Fig. 10.30, a groundwater-based open-loop system usually involves two wells, one to abstract and one to reinject groundwater, often called a 'well doublet' (Banks 2015). Open-loop GSHP systems have traditionally targeted sedimentary bedrock aquifer sources at depths in excess of 100 m below ground surface (Birks *et al.* 2015). However, high yields from deeper aquifers incur increased drilling costs that can make open-loop schemes too expensive for smaller projects. Shallower open-loop options are possible, for example in unconsolidated Quaternary sedimentary deposits or in flooded mine workings with relatively stable water temperature where ground conditions are suitable (Hall *et al.* 2011). Yet, uncertainty remains regarding the technical feasibility, operational efficiency and long-term environmental impacts of shallow open-loop GSHP systems.

The main hydrogeological uncertainties associated with designing open-loop GSHP systems are aquifer yield and recharge capacity (Birks *et al.* 2015; Banks and Birks 2020), although these technical risks can be reduced by conducting geological and hydrogeological investigations prior to design and installation. Other identified risks are impacts on groundwater quality and ground stability in karstic and evaporitic environments. Thermal interference and thermal feedback between

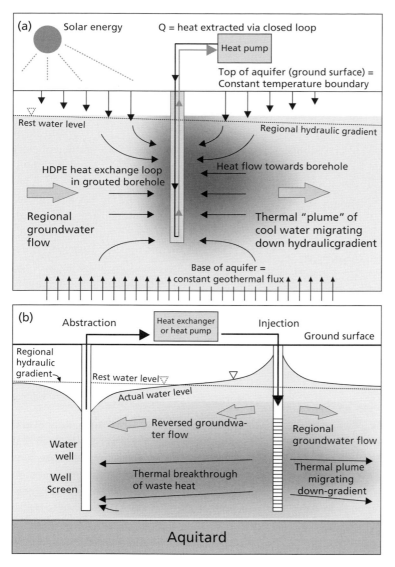

Fig. 10.30 Schematic diagrams of (a) a closed-loop borehole heat exchanger installed in an aquifer, and (b) an open-loop well-doublet heat exchange system, or Groundwater Heat Pump (GWHP), rejecting waste heat to an aquifer. In both cases the regional groundwater gradient is from left to right. Thin arrows depict heat flows and thick arrows show groundwater flows (Banks 2009). (*Source:* Adapted from Banks, D. (2009) An introduction to 'thermogeology' and the exploitation of ground source heat. *Quarterly Journal of Engineering Geology and Hydrogeology* **42**, 283–293.)

adjacent GSHP systems are also potential problems for heat pumps used in heating mode, in that a reduction in source temperature of 1.5 °C reduces system efficiency by around 5–10% (Banks 2009). For future application, adequate resource planning of systems and regulation will be essential as the density of systems, especially in urban areas, increases as a response to the drive towards net-zero carbon emissions by 2050 (Boon *et al.* 2019).

An experimental, open-loop GSHP scheme, retrofitted to a nursery school building in the

City of Cardiff, Wales, is presented by Boon *et al.* (2019). The city lies on moderately flat, low-lying, riverine and coastal floodplain and glacio-fluvial terraces. Locally, the glaciofluvial deposits are up to 30 m thick in buried valleys and typically comprise highly permeable sands and gravels suitable for open-loop GSHP systems. Local groundwater temperature mapping suggests average temperatures are 12.6 °C, resulting from the subsurface urban heat island effect.

The theoretical geothermal potential (or potential heat content) of a subsurface urban heat island can be estimated using the following equation (Zhu *et al.* 2010):

$$Q = Q_w + Q_s = VnC_w\Delta T + V(1-n)C_s\Delta T$$
$$\text{(eq. 10.1)}$$

where Q (kJ) is the total theoretical potential heat content of the aquifer, V (m^3) is the aquifer volume, n is porosity, C_w and C_s (kJ m^{-3} K^{-1}) are the volumetric heat capacity of water and solid, respectively, Q_w and Q_s (kJ) are the heat content stored in groundwater and solid, respectively, and ΔT (K) is the temperature reduction of the whole aquifer. C_w, for water is 4150 kJ m^{-3} K^{-1}, and C_s has a range depending on sediment types (for example, sand and gravel, carbonate rock) between 2100 and 2400 kJ m^{-3} K^{-1}.

In respect of the Cardiff open-loop GSHP scheme, and following initial site investigations, a 22 kW peak output well-doublet type system was installed. Production and reinjection wells were drilled to depths of 22 and 18 m, respectively, with screened response zones in a confined sand and gravel aquifer (Boon *et al.* 2019). Average aquifer thermal degradation in the first three years of the scheme was kept below 2 °C, with a maximum change of 4 °C measured during the heating season. During testing of the pilot installation, the Seasonal Performance Factor, SPF$_{H4}$ (in heating mode, SPF$_{H4}$ is the ratio of thermal energy delivered to electrical energy used, calculated over a year for the total heat pump system) was found to be 4.5 in the three-year monitoring period.

Assessment of the aquifer volume suggests that the pore water contains a thermal energy heat source of between 793 and 856 GWh$_{th}$ (2.86 × 10^{11} kJ and 3.09 × 10^{11} kJ), assuming the aquifer temperature is kept above 5 °C. Lowering the temperature of the aquifer by 8 °C, which would be at the limit of the recommended UK guideline value (Environment Agency 2011), could generate sufficient heat content equivalent to 26% of Cardiff's predicated 2020 heat demand. In reality, the achievable heat extraction would be less and further groundwater investigations are needed to understand the physical limits of abstraction and reinjection across the city. However, Boon *et al.* (2019) concluded that large parts of the sand and gravel aquifer can sustain shallow open-loop GSHP systems, as long as the local ground conditions support the required groundwater abstraction and reinjection rates.

10.5.3 Groundwater and shale gas exploration

Natural gas (primarily methane, ethane and propane) has emerged as an energy source that offers the opportunity for a number of regions around the world to reduce their reliance on energy imports or strive towards energy independence (Vidic *et al.* 2013). Natural gas is also presented as a potential transition fuel to bridge the shift from coal to renewable energy resources, while helping to reduce emissions of CO$_2$. The driving force behind this shift is that it has become economically feasible to extract unconventional sources of gas, such as shale gas, that were previously considered inaccessible through the advent of horizontal drilling and high-volume hydraulic fracturing (HVHF). Deep and long horizontal wells, up to 3 km in length, combined with multi-stage hydraulic fracturing can now effectively exploit geographically extensive, often relatively thin (thicknesses of tens of metres) formations that contain unconventional hydrocarbon resources (Jackson *et al.* 2013).

Conventional gas is typically extracted from porous sandstone and carbonate formations, where it has generally been trapped under impermeable caprocks after migration from its original source rock. In contrast, unconventional gas is usually recovered from low-permeability reservoirs or the source rocks themselves, including coal seams, tight sand formations and fine-grained, organic-rich shales. While conventional gas extraction activities are typically focused in well-defined, spatially restricted areas above oil and gas reservoirs, unconventional gas fields are developed on the premise that the gas-bearing formations are widely distributed in the subsurface. Thus, unconventional gas wells are typically drilled in densely spaced and equidistant surface patterns that cover far larger subsurface areas than conventional plays. Although recent advances in directional drilling technology permit over 20 horizontal wells to be drilled from a single well pad, large numbers of well pads can still create extensive regional footprints (Jackson *et al.* 2013).

Unconventional gas formations are characterized by low permeabilities that limit the recovery of the gas and require additional techniques to achieve economical flow rates (Vidic *et al.* 2013). Hydraulic fracturing (fracking) is a process used to recover commercial quantities of oil and gas from low permeability unconventional reservoirs such as shale. Pressurized fluid, usually consisting of water, chemical additives and a proppant (for example, sand) is injected via a borehole into a low permeability rock to induce hydraulic fractures. These fractures, kept open by the proppant, increase the permeability of the rock and allow oil and gas to flow back to the surface via the borehole. While the technology of fracking has been used for decades, its widespread use to exploit shale reservoirs with multiple stages in horizontal boreholes is a more recent development (Wilson *et al.* 2017).

A leading example of rapidly increasing shale gas development is the Marcellus Shale in the Appalachian region of the eastern United States. Intensive gas extraction began in 2005, and it is one of the top five unconventional gas reservoirs in the United States and one of the world's largest known shale-gas deposits. The shale play extends from upstate New York, as far south as Virginia, and as far west as Ohio, underlying 70% of the state of Pennsylvania and much of West Virginia. The formation consists of black and dark grey shales, siltstones and limestones. On the basis of a geological study of natural fractures in the formation, it is estimated with a 50% probability that the Marcellus will ultimately yield $1.39 \times 10^{13}\,\text{m}^3$ of natural gas (Vidic *et al.* 2013).

Interest in shale gas development is increasing across the world (for example in the United Kingdom, Poland, Ukraine, Australia and Brazil). The current global estimate of natural gas reserves in unconventional shale is approximately $7.16 \times 10^{14}\,\text{m}^3$ (Vengosh *et al.* 2014). In the United Kingdom, the onshore shale gas reserve was first estimated to be $1.50 \times 10^{11}\,\text{m}^3$, with the best shale gas potential mainly associated with the Carboniferous Bowland Shale of the Pennine Basin in northern England and the Jurassic Kimmeridge Clay of the Weald Basin in southern England (Harvey and Gray 2010). A later study by Andrews (2013) using 3D modelling revised the previous estimates, with a best estimate (50% probability) of a total in-place gas resource of $3.76 \times 10^{13}\,\text{m}^3$ in the Bowland Shale across central Britain. Although the UK government has granted several licence areas for shale gas exploration and development, only small-scale exploratory activities (three test wells) have been reported in the Bowland Shale of the Lancashire licence areas (Cai and Ofterdinger 2014).

Alongside hydraulic fracturing and the development of unconventional gas resources, concerns have been raised about accompanying environmental, health and socio-economic issues. Primary concerns include air pollution, greenhouse gas emissions, radiation and groundwater and surface water contamination. These concerns have been heightened in the

United States because the 2005 Energy Policy Act exempts hydraulic fracturing operations from the Safe Drinking Water Act (SDWA). The only exception to the exemption is the injection of diesel fuel. Additionally, because environmental oversight for most oil and gas operations is conducted by state rather than federal agencies, the regulation, monitoring and enforcement of various environmental contamination issues related to unconventional shale gas development are highly variable throughout the United States (Vengosh *et al.* 2014).

Vengosh *et al.* (2014) identified four potential modes of water resources degradation that are illustrated schematically in Fig. 10.31, and include: (1) shallow aquifers contaminated by fugitive natural gas (that is, stray gas

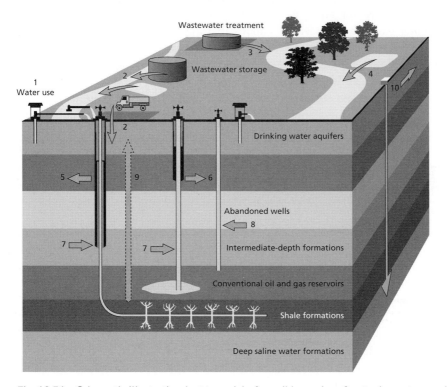

Fig. 10.31 Schematic illustration (not to scale) of possible modes of water impacts associated with shale gas development: (1) overuse of water that could lead to depletion and water quality degradation, particularly in water-scarce areas; (2) surface water and shallow groundwater contamination from spills and leaks of wastewater storage and open pits near drilling sites; (3) disposal of inadequately treated wastewater to local streams and accumulation of contaminant residues in disposal sites; (4) leaks of storage ponds that are used for deep-well injection; (5) shallow aquifer contamination by stray gas that originated in the target shale gas formation through leaking well casing, potentially followed by salt and chemical contamination from hydraulic fracturing fluids and/or formational waters; (6) shallow aquifer contamination by stray gas through leaking of conventional oil and gas well casing; (7) shallow aquifer contamination by stray gas that originated in intermediate geological formations through annulus leaking of either shale gas or conventional oil and gas wells; (8) shallow aquifer contamination through abandoned oil and gas wells; (9) flow of gas and saline water directly from deep formation waters to shallow aquifers; and (10) shallow aquifer contamination through leaking of injection wells (Vengosh *et al.* 2014). (*Source:* Adapted from Vengosh, A., Jackson, R.B., Warner, N. *et al.* (2014) A critical review of the risks to water resources from unconventional shale gas development and hydraulic fracturing in the United States. *Environmental Science and Technology* **48**, 8334–8348. DOI: 10.1021/es405118y.)

contamination) from leaking shale gas and conventional oil and gas wells, potentially followed by water contamination from hydraulic fracturing fluids and/or formation waters from the deep formations; (2) surface water contamination from spills, leaks and the disposal of inadequately treated wastewater or hydraulic fracturing fluids; (3) accumulation of toxic and radioactive elements in soil and the sediments of rivers and lakes exposed to wastewater or fluids used in hydraulic fracturing; and (4) the overuse of water resources, which can compete with other water uses such as agriculture in water limited environments.

Water consumption for shale gas development in the United States varies from 8–100×10^3 m^3 per conventional well. In geographic areas with drier climates and/or higher aquifer consumption, groundwater exploitation for hydraulic fracturing can lead to local water shortages (Nicot and Scanlon 2012; Murray 2013) and subsequent degradation of water quality. While the hydraulic fracturing revolution has increased water use and wastewater production in the United States (for example, from 2012 to 2014, the annual water use rates were 708×10^6 and 232×10^6 m^3 a^{-1} for the major unconventional shale gas and oil formations, respectively), its water use and produced water intensity is lower than other energy extraction methods and represents only a fraction of total industrial water use nationwide (Kondash and Vengosh 2015).

One particular environmental concern is whether fracking fluids injected into the subsurface at depths of several kilometres can migrate via natural geological pathways to shallow aquifers at depths of tens to hundreds of metres. In the Marcellus Shale play in the United States, the deepest reported drinking water level is about 600 m below the surface (Fisher and Warpinski, 2012). In England, groundwater abstractions typically do not descend more than 200 m below ground level and it has been suggested that a reasonable maximum depth of about 400 m may be considered for conventional freshwater aquifers

(Wilson *et al.* 2017). Loveless *et al.* (2018) modelled the vertical separation between different pairs of aquifers and shales that are present across England and Wales and concluded that the risk of aquifer contamination from shale exploration will vary greatly between shale–aquifer pairs and between regions, and that vertical separations will need to be considered carefully as part of the risk assessment and management of any shale gas development. English regulations ensure that fracking only occurs 1000 m below the surface, and 1200 m below the surface in specified groundwater areas, National Parks, Areas of Outstanding National Beauty and World Heritage Sites (Wilson *et al.* 2017).

In reality, observed sub-surface groundwater contamination from fracking sites is rare and disputed. In aquifers overlying the Marcellus and Utica shale formations of northeastern Pennsylvania and upstate New York, Osborn *et al.* (2011) documented systematic evidence for methane contamination of drinking water associated with shale gas extraction. In active gas extraction areas, with one or more gas wells within 1 km, average and maximum methane concentrations in drinking water wells increased with proximity to the nearest gas well and were 19.2 and 64 mg L^{-1} ($n = 26$), respectively, concentrations that present a potential explosion hazard. In contrast, dissolved methane samples in neighbouring non-extraction sites (no gas wells within 1 km) within similar geologic formations and hydrogeologic regimes averaged only 1.1 mg L^{-1} ($n = 34$). Average δ^{13}C–CH$_4$ values of dissolved methane in shallow groundwater were significantly less negative for active extraction than for non-extraction sites ($-37 \pm 7‰$ and $-54 \pm 11‰$, respectively). These δ^{13}C–CH$_4$ data are consistent with deeper thermogenic methane sources such as the Marcellus and Utica Shales at the active sites and matched the gas geochemistry from gas wells nearby. In contrast, samples with lower concentrations from shallow groundwater at non-extraction sites had isotopic signatures reflecting a more biogenic or mixed biogenic/thermogenic methane source.

Surface water and groundwater contamination is usually considered as being associated with treated shale gas waste, well casing integrity issues or surface leaks (for example, Olmstead *et al.* 2013; Llewellyn *et al.* 2015) and not from the migration of fracking fluids from the deep to shallow sub-surface along natural geological pathways. It is difficult to determine whether hydraulic fracturing of shale formations has affected groundwater quality, because it requires baseline conditions for comparison and detailed information of well drilling and casing. Jackson *et al.* (2013) identified the need for field-focused research to: (a) conduct baseline geochemical mapping; and (b) field testing of potential mechanisms and pathways by which hydrocarbon gases, reservoir fluids and fracturing chemicals might potentially invade and contaminate potable groundwater. McIntosh *et al.* (2019) proposed the development of new, naturally occurring isotope tracers (for example, clumped isotopes of hydrocarbons), high-resolution data sets of natural gases and associated fluids, and incorporation of noble gas geochemistry and microbiology as promising analytical tools for identifying sources of subsurface fluids to provide critical baseline information. In the absence of a long-term and extensive monitoring programme, a practical, short-term solution to understanding complex flow and solute transport through induced and natural fractures is to apply numerical modelling techniques (Cai and Ofterdinger 2014; Wilson *et al.* 2017).

10.6   Future challenges for groundwater governance and management

The Sustainable Development Goals (UN 2015) include consideration of the sustainability of water resources to underpin secure and safe access for human populations. Groundwater, as a key resource to achieve Sustainable Development Goal 6 (SDG6) to 'Ensure availability and sustainable management of water and sanitation for all,' though implicitly included, does not receive specific attention in terms of an assessment of its sustainable use. As argued by Villholth and Conti (2018), good groundwater governance is required as a prerequisite to providing a comprehensive overarching framework to accommodate and support the management of groundwater resources globally.

Groundwater governance has emerged as a relatively new concept and can be difficult to distinguish from groundwater management. From their analysis, Villholth and Conti (2018) defined groundwater governance as a framework encompassing the processes, interactions and institutions in which stakeholders (i.e. governmental, public sector, non-governmental, private sector, civil society, etc.) participate and decide on the management of groundwater within and across multiple geographic (i.e. sub-national, national, transboundary and global) and institutional/sectoral levels, as applicable. In comparison, groundwater management is the specific day-to-day actions taken to ensure the strategic use and/or protection of groundwater resources. Groundwater management includes the activities various stakeholders undertake within the governance framework related to the development and protection of groundwater. In practice, the range of stakeholders that participate in groundwater management and the scope of activities involved are often far narrower than those involved in governance (Villholth and Conti 2018).

Groundwater governance is an evolving concept. It remains somewhat intangible and not consciously and actively pursued by many stakeholders and managers, yet it is paramount to proper groundwater development and management. Ideally, as awareness increases, groundwater governance will become a deliberate, conscious, explicit and targeted concept, typically associated with aspirational goals (such as sustainability and equity) and core guiding governance tenets (transparency, accountability, integrity, fairness, etc.) (Villholth and Conti 2018).

Three aquifer characteristics determine whether groundwater resources will ultimately prove sustainable: vulnerability to pollution under contaminant pressure from the land surface; susceptibility to irreversible degradation from excessive exploitation; and renewability of storage reserves under current and future climate regimes. These characteristics vary widely by aquifer type and hydrogeologic setting. Vulnerability to pollution is generally linked to the accessibility of an aquifer. Aquifers that are shallow and readily recharged are more likely to suffer pollution from agrichemicals and urbanization (in particular, from low-cost wastewater disposal and careless disposal of industrial chemicals). Groundwater development and effluent disposal for urban water supply have far-reaching implications for public health, municipal planning and resource sustainability. In Europe, land-use zoning is now used to protect vulnerable key aquifers that provide municipal water supply (see Box 9.5), as well as developing deeper, confined groundwater sources that are naturally protected from urban pollution (WWAP 2009).

The tension between private and public services derived from aquifers remains. More convergent and sustainable resource use will be achieved only through substantial investment in management operations on the ground, working primarily through community consultation and cross-sectoral policy dialogue (WWAP 2009). Such dialogue is supported by shared knowledge and common understanding of the current situation and future options. Good and reliable groundwater information is crucial to facilitate co-operation among stakeholders. All stakeholders should have easy access to reliable data on abstractions, water quality and groundwater levels. Adopting an

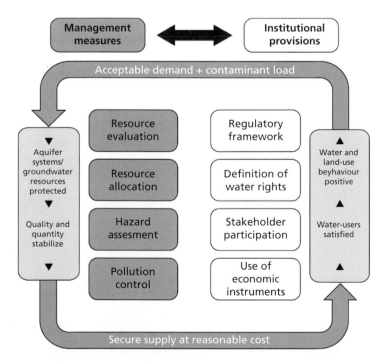

Fig. 10.32 Integrated, adaptive management scheme for the protection of groundwater resources (IAH 2006). (*Source:* IAH (2006). Groundwater for life and livelihoods – the framework for sustainable use. *4th World Water Forum Invitation and Briefing*. International Association of Hydrogeologists, Kenilworth. © 2006, Taylor & Francis.)

adaptive management approach (Fig. 10.32) it should be possible to establish mutually acceptable regulations, agreed by all parties, based on a holistic definition of the aquifer system and understanding of the impacts of abstraction and contamination.

A significant challenge for the future development of groundwater sources is to raise political awareness of the issues involved. Unfortunately, increased scientific understanding of groundwater has not yet had a significant influence on resource policy-making or featured prominently in global or national water policy dialogues, with discussion too often on groundwater development rather than groundwater management. Also, governance and practical management are not well funded and, as a consequence, opportunities for utilizing groundwater resources sustainably and conjunctively are being lost and insufficient attention is being paid to the interrelationship between groundwater and land-use planning (IAH 2006). Often, decisions on groundwater development and management objectives and the allocation of human, financial and environmental resources to meet these objectives, are made by leaders in government, the private sector and civil society, and not by groundwater professionals alone. Therefore, hydrogeologists must help inform the decisions of these leaders outside the water domain on such issues as spatial and development planning, demographic planning, health, education, agriculture, industry, energy, economic development and the environment (WWAP 2009).

Water resources management has a clear and rapidly developing association with many other policy areas such as energy, land use and nature conservation. In this context, groundwater is part of an emerging integrated water resources management approach that recognizes society's views, reshapes planning processes, coordinates land and water resources management, recognizes water quantity and quality linkages, manages surface water and groundwater resources conjunctively and protects and restores natural systems while considering climate change.

To achieve a sustainable groundwater system, in which abstraction can continue safely, water resources managers should adopt the definition of safe yield as the maximum prolonged pumping that meets all logistic, environmental, legal, socio-economic and physical constraints (Gorelick and Zheng 2015). It is essential that there is a complete understanding of the future hydrogeologic system so that policies are set now that determine abstraction rates that can achieve the goal of groundwater sustainability through the process of adaptive management (Gleeson *et al.* 2012; Rhode *et al.* 2017). Further, Gleeson *et al.* (2020) considered that purely physically based definitions of groundwater sustainability founded on the concept of safe yield or physical sustainability are too narrow in that they do not include diverse social and environmental aspects. Gleeson *et al.* (2020) suggested a new definition of groundwater sustainability as maintaining long-term, dynamically stable storage of high-quality groundwater using inclusive, equitable and long-term governance and management. This definition implies that part of groundwater natural capital stocks is non-substitutable, but also allows for significant regional control through equitable governance and management and defining goals, targets and objectives (Gleeson *et al.* 2020). Finally, for groundwater sustainability, it is important to emphasize that a long-term perspective is required given the long response times of groundwater.

Further reading

Alley, W.M. and Alley, R. (2017) *High and Dry: Meeting the Challenges of the World's Growing Dependence on Groundwater*. Yale University Press, Newhaven.

Banks, D. (2008) *An Introduction to Thermogeology: Ground Source Heating and Cooling*. Blackwell Publishing Ltd., Oxford.

Burke, J.J. and Moench, M.H. (2000) *Groundwater and society: resources, tensions and opportunities*. United Nations Department of Economic and Social Affairs and Institute for

Social and Environmental Transition, United Nations, New York.

Cook, H.F. (1998) *The Protection and Conservation of Water Resources: A British Perspective*. Wiley, Chichester.

Dragoni, W. and Sukhija, B.S. (eds) Climate Change and Groundwater. Geological Society, London, Special Publications **288**.

Hiscock, K.M., Rivett, M.O. and Davison, R.M. (eds) (2002). Sustainable Groundwater Development. Geological Society, London, Special Publications, **193**.

Jones, J.A.A. (ed.) (2011) *Sustaining Groundwater Resources: A Critical Element in the Global Water Crisis*. Springer, Dordrecht.

Keddy, P.A. (2000) *Wetland Ecology: Principles and Conservation*. Cambridge University Press, Cambridge.

Mitsch, W.J. and Gosselink, J.G. (2000) *Wetlands* (3rd edn). Wiley, New York.

Ofterdinger, U., MacDonald, A.M., Comte, J.-C. and Young, M.E. eds (2019). *Groundwater in Fractured Bedrock Environments: Managing Catchment and Subsurface Resources*. Geological Society, London, Special Publications **479**.

Sharp, J.M., Green, R.T. and Schindel, G.M. (eds) (2019). *The Edwards Aquifer: The Past, Present, and Future of a Vital Water Resource. Geological Society of America Memoirs* **215**. DOI: 10.1130/MEM215.

Treidel, H., Martin-Bordes, J.-L. and Gurdak, J. (eds) (2012). *Climate Change Effects on Groundwater Resources: A Global Synthesis of Findings and Recommendations*. IAH International Contributions to Hydrogeology, 27. CRC Press, Baton Rouge, Florida.

Villholth, K.G., López-Gunn, E., Conti, K.I., Garrido, A. and van der Gun, J. (eds) (2018) *Advances in Groundwater Governance*. CRC Press, Boca Raton, Florida, 594 pp.

References

Abesser, C., Lewis, M.A., Marchant, A.P. and Hulbert, A.G. (2014) Mapping suitability for open-loop ground source heat pump systems: a screening tool for England and Wales, UK.

Quarterly Journal of Engineering Geology and Hydrogeology **47**, 373–380.

Alcamo, J., Moreno, J.M., Nováky, B. *et al.* (2007) Europe. In: *Climate Change 2007: Impacts, Adaptation and Vulnerability. Contribution of Working Group II to the Fourth Assessment Report of the Intergovernmental Panel on Climate Change* (eds M.L. Parry, O.F. Canziani, J.P. Palutikof *et al.*). Cambridge University Press, Cambridge, pp. 541–580.

Andrews, I.J. (2013) *The Carboniferous Bowland Shale gas study: geology and resource estimation*. British Geological Survey for the Department of Energy and Climate Change, London, UK, 56 pp.

Arnell, N. (1996) *Global Warming, River Flows and Water Resources*. Wiley, Chichester.

Ascott, M.J., Bloomfield, J.P., Karapanos, I. *et al.* (2020) Managing groundwater supplies subject to drought: perspectives on current status and future priorities from England (UK). *Hydrogeology Journal* DOI: 10.1007/s10040-020-02249-0.

Askew, A.J. (1987) Climate change and water resources. *IAHS Publication* **168**, 421–430.

Backshall, W.F., Downing, R.A. and Law, F.M. (1972) Great Ouse groundwater study. *Water and Water Engineering* **06**, 3–11.

Ballentine, T.M. and Stakhiv, E.Z. (eds) (1993) Climate change and water resources management. *Proceedings of the First National Conference*, US Army Corps of Engineers Institute of Water Resources. IWR Report 93-R-17. USACE, Institute for Water Resources, Alexandria, VA.

Banks, D. (2009) An introduction to 'thermogeology' and the exploitation of ground source heat. *Quarterly Journal of Engineering Geology and Hydrogeology* **42**, 283–293.

Banks, D. (2015) A review of the importance of regional groundwater advection for ground heat exchange. *Environmental Earth Sciences* **73**, 2555–2565.

Banks, D. and Birks, D. (2020) Heat from the ground. *Geoscientist* **30**, 12–17.

Barker, J.A., Downing, R.A., Gray, D.A. *et al.* (2000) Hydrogeothermal studies in the United Kingdom. *Quarterly Journal of Engineering Geology and Hydrogeology* **33**, 41–58.

Bense, V. F., Ferguson, G. and Kooi, H. (2009) Evolution of shallow groundwater flow systems in areas of degrading permafrost. *Geophysical Research Letters* **36**, L22401. DOI: 10.1029/2009GL039225.

Beran, M. and Rodier, J.A. (1985). Hydrological aspect of drought. Studies and Report in Hydrology 39. UNESCO-WMO, Paris.

Berge, T.B. and Veal, S.L. (2005) Structure of the Alpine foreland. *Tectonics* **24**, TC5011. DOI: 10.1029/2003TC001588.

Bertin, C. and Bourg, A.C.M. (1994) Radon-222 and chloride as natural tracers of the infiltration of river water into an alluvial aquifer in which there is significant river/groundwater mixing. *Environmental Science & Technology* **28**, 794–798.

Birks, D., Coutts, C., Younger, P. and Parkin, G. (2015) Development of a groundwater heating and cooling scheme in a Permo-Triassic sandstone aquifer in South-West England and approach to managing risks. *Proceedings of the Ussher Society; Geoscience in South-West England* **13**, 428–436.

Bloomfield, J.P. and Marchant, B.P. (2013) Analysis of groundwater drought building on the standardised precipitation index approach. *Hydrology and Earth System Sciences* **17**, 4769–4787. DOI: 10.5194/hess-17-4769-2013.

Bloomfield, J.P., Gaus, I. and Wade, S.D. (2003) A method for investigating the potential impacts of climate-change scenarios on annual minimum groundwater levels. *Journal of the Chartered Institution of Water and Environmental Management* **17**, 86–91.

Bloomfield, J.P., Marchant, B.P. and McKenzie, A.A. (2019) Changes in groundwater drought associated with anthropogenic warming. *Hydrology and Earth System Sciences* **23**, 1393—1408. DOI: 10.5194/hess-23-1393-2019.

Boden, D.R. (2016) *Geologic Fundamentals of Geothermal Energy*. CRC Press, Taylor & Francis Group, Boca Raton, FL.

Boissier, F., Lopez, S., Desplan, A. and Lesueur, H. (2009). 30 years of exploitation of the geothermal resource in Paris Basin for district heating. *Transactions - Geothermal Resources Council* **33**, 355–359.

Boon, D.P., Farr, G.J., Abesser, C. *et al.* (2019) Groundwater heat pump feasibility in shallow urban aquifers: experience from Cardiff, UK. *Science of the Total Environment* **697**, 133847. DOI: 10.1016/j.scitotenv.2019.133847.

Boorman, L.A., Londo, G. and van der Maarel, E. (1997) Communities of dune slacks. In: *Ecosystems of the World: Part 2C, Dry Coastal Ecosystems* (ed. E. van der Maarel). Elsevier, Amsterdam, 275–293.

Brassington, F.C. (2007) A proposed conceptual model for the genesis of the Derbyshire thermal springs. *Quarterly Journal of Engineering Geology and Hydrogeology* **40**, 35–46.

Bryant, S.J., Arnell, N.W. and Law, F.M. (1994). The 1988–92 drought in its historical perspective. *Journal of the Institution of Water Environment and Management* **8**, 39–51.

Burgess, D.B. (2002) Groundwater resource management in eastern England: a quest for environmentally sustainable development. In: *Sustainable Groundwater Development* (eds K.M. Hiscock, M.O. Rivett and R.M. Davison). Geological Society, London, Special Publications **193**, pp. 53–62.

Busby, J. (2014) Geothermal energy in sedimentary basins in the UK. *Hydrogeology Journal* **22**, 129–141.

Busby, J., Kingdon, A. and Williams, J. (2011) The measured shallow temperature field in Britain. *Quarterly Journal of Engineering Geology and Hydrogeology* **44**, 373–387.

Cai, Z. and Ofterdinger, U. (2014) Numeric assessment of potential impacts of hydraulically fractured Bowland Shale on overlying aquifers. *Water Resources Research* **50**, 6236–6259. DOI:10.1002/2013WR014943.

Carter, T.R. and La Rovere, E.L. (2001) Developing and applying scenarios. In: *Climate Change 2001: Impacts, Adaptation and Vulnerability* (eds J.J. McCarthy, O.F. Canziani, N.A. Leary *et al.*). Cambridge University Press, Cambridge, pp. 145–190.

CEC (2007). Adapting to climate change in Europe – Options for EU Action. Green Paper from the Commission to the Council, The European Parliament, The European Economic and Social Committee and the Committee of the Regions. Commission of the European Communities, Brussels, 27 pp.

Chen, C.X., Wan, J.W. and Zhan, H.B. (2003) Theoretical and experimental studies of coupled seepage-pipe flow to a horizontal well. *Journal of Hydrology* **281**, 159–171.

Clarke, D. and Sanitwong Na Ayutthaya, S. (2010) Predicted effects of climate change, vegetation and tree cover on dune slack habitats at Ainsdale on the Sefton Coast, UK. *Journal of Coastal Conservation* **14**, 115–125.

Collins, W.D., Friedlingstein, P., Gaye, A.T. *et al.* (2007) Global climate projections, Chapter 10. In: *Climate Change 2007: The Physical Science Basis. Contribution of Working Group I to the Fourth Assessment Report of the Intergovernmental Panel on Climate Change* (eds S. Solomon, D. Qin, M. Manning *et al.*). Cambridge University Press, Cambridge, pp. 747–846.

Colombo, N., Salerno, F., Gruber, S. *et al.* (2018) Review: impacts of permafrost degradation on inorganic chemistry of surface fresh water. *Global and Planetary Change* **162**, 6983. DOI: 10.1016/j.gloplacha.2017.11.017.

Conway, D. (1998) Recent climate variability and future climate change scenarios for Great Britain. *Progress in Physical Geography* **22**, 350–374.

Cooper, R.J., Hawkins, E., Locke, J. *et al.* (2020) Assessing the environmental and economic efficacy of two integrated constructed wetlands at mitigating eutrophication risk from sewage effluent. *Water Environment Journal* DOI:10.1111/wej.12605.

Council of the European Communities (1992) Directive on the conservation of natural habitats and of wild fauna and flora (92/43/EEC). *Official Journal of the European Communities*, **L206**, 7–50. Brussels.

Cuthbert, M.O., Gleeson, T., Moosdorf, N. *et al.* (2019) Global patterns and dynamics of climate-groundwater interactions. *Nature Climate Change* **9**, 137–141. DOI: 10.1038/s41558-018-0386-4.

Dai, A., Trenberth, K.E. and Karl, T. (1998) Global variations in droughts and wet spells: 1900–1995. *Geophysical Research Letters* **25**, 3367–3370.

Dahlqvist, P., Sjöstrand, K., Lindhe, A. *et al.* (2019). Potential benefits of managed aquifer recharge MAR on the Island of Gotland, Sweden. *Water* **11**, 2164. DOI: 10.3390/w11102164.

Davy, A.J., Grootjans, A.P., Hiscock, K. and Petersen, J. (2006) Development of eco-hydrological guidelines for dune habitats – Phase 1. *English Nature Research Reports* **696**, 78 pp.

Dennehy, K.F., Litke, D.W. and McMahon, P.B. (2002) The High Plains Aquifer, USA: groundwater development and sustainability. In: *Sustainable Groundwater Development* (eds K.M. Hiscock, M.O. Rivett and R.M. Davison). Geological Society, London, Special Publications **193**, pp. 99–119.

Dessai, S. and Hulme, M. (2007) Assessing the robustness of adaptation decisions to climate change uncertainties: a case study on water resources management in the East of England. *Global Environmental Change* **17**, 59–72.

Dillon, P. (2005) Future management of aquifer recharge. *Hydrogeology Journal* **13**, 313–316.

Dillon, P., Fernández Escalante, E., Megdal, S.B. and Massmann, G. (2020) Managed aquifer recharge for water resilience. *Water* **12**, 1846. DOI:10.3390/w12071846.

Dillon, P., Pavelic, P., Page, D. *et al.* (2009). *Managed Aquifer Recharge: An Introduction*. Waterlines Report Series **13**, Australian Government, National Water Commission, Canberra, 64 pp.

Dillon, P., Stuyfzand, P., Grischek, T. *et al.* (2019) Sixty years of global progress in managed aquifer recharge. *Hydrogeology Journal* **27**, 1–30.

Doussan, C., Poitevin, G., Ledoux, E. and Detay, M. (1997) River bank filtration: modelling of the changes in water chemistry with emphasis on nitrogen species. *Journal of Contaminant Hydrology* **25**, 129–156.

Downing, R.A. (1993) Groundwater resources, their development and management in the UK: an historical perspective. *Quarterly Journal of Engineering Geology* **26**, 335–358.

Downing, R.A. and Gray, D.A. (1986) Geothermal resources of the United Kingdom. *Journal of the Geological Society, London* **143**, 499–507.

Downing, R.A., Price, M. and Jones, G.P. (1993) The making of an aquifer. In: *The Hydrogeology of the Chalk of North-West Europe* (eds R.A. Downing, M. Price and G.P. Jones). Clarendon Press, Oxford, pp. 1–13.

Eastwood, J.C. and Stanfield, P.J. (2001) Key success factors in an ASR scheme. *Quarterly Journal of Engineering Geology and Hydrogeology* **34**, 399–409.

Environment Agency (2011) *Environmental Good Practice Guide for Ground Source Heating and Cooling.* Report GEHO0311BTPA-E-E, Environment Agency, Bristol, 34 pp.

Farr, G. and Bottrell, S.H. (2013) The hydrogeology and hydrochemistry of the thermal waters at Taffs Well, South Wales, UK. *Cave and Karst Science* **40**, 5–12.

Fisher, M.K. and Warpinski, N.R. (2012) Hydraulic-fracture-height growth: real data. *SPE Production & Operations* **27**, 8–19. DOI: 10.2118/145949-PA.

Foster, S., Morris, B., Lawrence, A. and Chilton, J. (1999) Groundwater impacts and issues in developing cities – an introduction. In: *Groundwater in the Urban Environment: Selected City Profiles* (ed. J. Chilton). A.A. Balkema, Rotterdam, pp. 3–16.

Fritts, H.C. (1976) *Tree Rings and Climate.* Academic Press, Inc., London.

Fuentes, I. and Vervoort, R.W. (2020). Site suitability and water availability for a managed aquifer recharge project in the Namoi basin, Australia. *Journal of Hydrology: Regional Studies* **27**, 100657. DOI:10.1016/j.ejrh.2019.100657.

García-Céspedes, J., Arnó, G., Herms, I. and de Felipe, J.J. (2020) Characterisation of efficiency losses in ground source heat pump systems

equipped with a double parallel stage: a case study. *Renewable Energy* **147**, 2761–2773.

Geelen, L.H.W.T., Kamps, P.T.W.J. and Olsthoorn, T.N. (2017) From overexploitation to sustainable use, an overview of 160 years of water extraction in the Amsterdam dunes, the Netherlands. *Journal of Coastal Conservation* **21**, 657–668.

Gilvear, D.J. and McInnes, R.J. (1994) Wetland hydrological vulnerability and the use of classification procedures: a Scottish case study. *Journal of Environmental Management* **42**, 403–414.

Gleeson, T., Cuthbert, M., Ferguson, G. and Perrone, D. (2020) Global groundwater sustainability, resources, and systems in the Anthropocene. *Annual Review of Earth and Planetary Sciences* **48**, 17.1–17.33. DOI: 10.1146/annurev-earth-071719-055251.

Gleeson, T., Alley, W.M., Allen, D.M. *et al.* (2012) Towards sustainable groundwater use: setting long-term goals, backcasting, and managing adaptively. *Ground Water* **50**, 19–26.

Goldscheider, N., Mádl-Szőnyi, J., Erőss, A. and Schill, E. (2010) Review: thermal water resources in carbonate rock aquifers. *Hydrogeology Journal* **18**, 1303–1318.

Goode, D. (1977) Peatlands. In: *A Nature Conservation Review* (Vol. **1**, ed. D.A. Ratcliffe). Cambridge University Press, Cambridge, pp. 249–287.

Gorelick, S.M. and Zheng, C. (2015) Global change and the groundwater management challenge. *Water Resources Research* **51**, 3031–3051. DOI: 10.1002/2014WR016825.

Grenier, C., Anbergen, A., Bense, V. *et al.* (2018) Groundwater flow and heat transport for systems undergoing freeze-thaw: intercomparison of numerical simulators for 2D test cases. *Advances in Water Resources* **114**, 196–218.

Gringorten, I.I. (1963) A plotting rule for extreme probability paper. *Journal of Geophysical Research* **68**, 813–814.

Grischek, T., Hiscock, K.M., Metschies, T. *et al.* (1998) Factors affecting denitrification during infiltration of river water into a sand aquifer in

Saxony, eastern Germany. *Water Research* **32**, 450–460.

Grischek, T., Schoenheinz, D., Worch, E. and Hiscock, K. (2002) Bank filtration in Europe – an overview of aquifer conditions and hydraulic controls. In: *Management of Aquifer Recharge for Sustainability* (ed. P.J. Dillon). A. A. Balkema Publishers, Lisse, pp. 485–488.

Groombridge, B. and Jenkins, M.D. (2000) *Global Biodiversity: Earth's Living Resources in the 21st Century*. UNEP – World Conservation Monitoring Centre, World Conservation Press, Cambridge.

Grootjans, A.P., Ernst, W.H.O. and Stuyfzand, P.J. (1998) European dune slacks: strong interactions of biology, pedogenesis and hydrology. *Trends in Ecology & Evolution* **13**, 96–100.

Grootjans, A.P., Geelen, H.W.T., Jansen, A.J.M. and Lammerts, E.J. (2002) Restoration of coastal dune slacks in the Netherlands. *Hydrobiologia* **478**, 181–203.

Gurdak, J.J. (2017) Climate-induced pumping. *Nature Geoscience* **10**, 71–72. www.nature.com/naturegeoscience.

Gurdak, J.J., Hanson, R.T. and Green, T.R. (2009). *Effects of climate variability on groundwater resources of the United States*. United States Geological Survey Fact Sheet, 2009–3074, 4 pp.

Hall, A., Ashley Scott, J. and Shang, H. (2011) Geothermal energy recovery from underground mines. *Renewable and Sustainable Energy Reviews* **15**, 916–924.

Harding, M. (1993) Redgrave and Lopham Fens, East Anglia, England: a case study of change in flora and fauna due to groundwater abstraction. *Biological Conservation* **66**, 35–45.

Harvey, T. and Gray, J. (2010) *The unconventional hydrocarbon resources of Britain's onshore basins – shale gas*. Department of Energy and Climate Change, London, 35 pp.

Hasan, M.B., Driessen, P.P.J., Majumder, S. *et al.* (2019). Factors affecting consumption of water from a newly introduced safe drinking water system: the case of managed aquifer recharge (MAR) systems in Bangladesh. *Water* **11**, 2459. DOI: 10.3390/w11122459.

Heberer, T. (2002) Tracking persistent pharmaceutical residues from municipal sewage to drinking water. *Journal of Hydrology* **266**, 175–189.

Herrera-Pantoja, M. and Hiscock, K.M. (2007) The effects of climate change on potential groundwater recharge in Great Britain. *Hydrological Processes* **22**, 73–86.

Herrera-Pantoja, M., Hiscock, K.M. and Boar, R. R. (2012) The potential impact of climate change on groundwater-fed wetlands in Eastern England. *Ecohydrology* **5**, 401–413.

Herzberger, P., Kölbel, T. and Münch, W. (2009) Geothermal resources in the German basins. *Transactions – Geothermal Resources Council* **33**, 352–354.

Hiscock, K.M. and Grischek, T. (2002) Attenuation of groundwater pollution by bank filtration. *Journal of Hydrology* **266**, 139–144.

Hiscock, K.M., Sparkes, R. and Hodgson, A. (2011) Evaluation of future climate change impacts on European groundwater resources. In: *Climate Change Effects on Groundwater Resources: A Global Synthesis of Findings and Recommendations* (eds T. Holger, J.-L. Martin-Bordes and J. Gurdak). CRC Press, Baton Rouge, FL, pp. 27. IAH International Contributions to Hydrogeology.

Hiscock, K.M., Bateman, A.S., Mühlherr, I.H., Fukada, T. and Dennis, P.F. (2003) Indirect emissions of nitrous oxide from regional aquifers in the United Kingdom. *Environmental Science and Technology* **37**, 3507–3512.

Holman, I.P. (2006) Climate change impacts on groundwater recharge-uncertainty, shortcomings, and the way forward? *Hydrogeology Journal* **14**, 637–647.

Holman, I.P., Rivas-Casado, M., Bloomfield, J.P. and Gurdak, J.J. (2011) Identifying non-stationary groundwater level response to North Atlantic ocean-atmosphere teleconnection patterns using wavelet coherence. *Hydrogeology Journal* **19**, 1269–1278.

Holman, I.P., Nicholls, R.J., Berry, P.M. *et al.* (2005) A regional, multi-sectoral and integrated assessment of the impacts of climate and socio-economic change in the UK: part 2 results. *Climatic Change* **71**, 43–73.

Hu, B., Teng, Y., Zhang, Y. *et al.* (2019) Review: the projected hydrologic cycle under the scenario of 936 ppm CO_2 in 2100. *Hydrogeology Journal* **27**, 31–53.

Hulme, M., Turnpenny, J. and Jenkins, G. (2002). *Climate Change Scenarios for the United Kingdom. The UKCIP02 Briefing Report.* Tyndall Centre for Climate Change, University of East Anglia, Norwich, and Hadley Centre for Climate Prediction and Research, UK Meteorological Office.

Hurter, S. and Schellschmidt, R. (2003) Atlas of geothermal resources in Europe. *Geothermics* **32**, 779–787.

IAH (2006). Groundwater for life and livelihoods – the framework for sustainable use. *4th World Water Forum Invitation and Briefing.* International Association of Hydrogeologists, Kenilworth.

Iglesias, A., Garotte, L., Flores, F. and Moneo, M. (2007) Challenges to manage the risk of water scarcity and climate change in the Mediterranean. *Water Resources Management* **21**, 775–788.

IGRAC (2018). *Managed aquifer recharge portal.* International Groundwater Resources Assessment Centre. https://apps.geodan.nl/igrac/ggis-viewer/globalmar/public/default (accessed 5 July 2020).

IPCC (1990) *Climate Change. The Intergovernmental Panel on Climate Change Scientific Assessment* (eds J.T. Houghton, G.J. Jenkins and J.J. Ephraums). Cambridge University Press, Cambridge.

IPCC (1998) *The Regional Impacts of Climate Change: An Assessment of Vulnerability. A Special Report of the IPCC Working Group II* (eds R.T. Watson, M.C. Zinyowera and R.H. Moss). Cambridge University Press, Cambridge.

IPCC (2001). *Climate Change 2001. The Scientific Basis.* Cambridge University Press, Cambridge. A Report of Working Group I of the Intergovernmental Panel on Climate Change.

IPCC (2007) Summary for policymakers. In: *Climate Change 2007: Impacts, Adaptation and Vulnerability. Contribution of Working Group II to the Fourth Assessment Report of the Intergovernmental Panel on Climate Change* (eds M.L. Parry, O.F. Canziani, J.P. Palutikof *et al.*). Cambridge University Press, Cambridge, pp. 7–22.

IPCC (2008) *Technical Paper on Climate Change and Water (Finalized at the 37th Session of the IPCC Bureau).* Intergovernmental Panel on Climate Change, Geneva.

Ireson, A.M. and Butler, A.P. (2011) Controls on preferential recharge to Chalk aquifers. *Journal of Hydrology* **398**, 109–123.

Issar, A.S. (2003) *Climate Change during the Holocene and Their Impact on Hydrological Systems. UNESCO International Hydrological Series.* Cambridge University Press, Cambridge.

Issar, A.S. (2008) A tale of two cities in ancient Canaan: how the groundwater storage capacity of Arad and Jericho decided their history. In: *Climate Change and Groundwater* (eds W. Dragoni and B.S. Sukhija). Geological Society, London, Special Publications **288**, pp. 137–143.

Jackson, R.E., Gorody, A.W., Mayer, B. *et al.* (2013) Groundwater protection and unconventional gas extraction: the critical need for field-based hydrogeological research. *Groundwater* **51**, 488–510.

Kabat, P., van Vierssen, W., Veraart, J. *et al.* (2005) Climate proofing the Netherlands. *Nature* **438**, 283–284.

Kenoyer, G.J. and Anderson, M.P. (1989) Groundwater's dynamic role in regulating acidity and chemistry in a precipitation-dominated lake. *Journal of Hydrology* **109**, 287–306.

Knapton, A., Page, D., Vanderzalm, J. *et al.* (2019). Managed aquifer recharge as a strategic storage and urban water management tool in Darwin, Northern Territory, Australia. *Water* **11**, 1869. DOI: 10.3390/w11091869.

Kondash, A. and Vengosh, A. (2015) Water footprint of hydraulic fracturing. *Environmental Science & Technology Letters* **2**, 276–280. DOI: 10.1021/acs.estlett.5b00211.

Kourakos, G., Dahlke, H. and Harter, T. (2019). Increasing groundwater availability and seasonal base flow through agricultural managed aquifer recharge in an irrigated basin. *Water Resources Research* **55**, 7464–7492.

Krinner, W., Lallana, C., Estrela, T. *et al.* (1999) *Sustainable Water Use in Europe – Part 1: Sectoral Use of Water*. European Environment Agency, Copenhagen.

Kulongoski, J.T. and McMahon, P.B. (2019) Methane emissions from groundwater pumping in the USA. *npj Climate and Atmospheric Science* **2**. DOI: 10.1038/s41612-019-0068-6.

Kuss, A.J.M. and Gurdak, J.J. (2014) Groundwater level response in U.S. principal aquifers to ENSO, NAO, PDO, and AMO. *Journal of Hydrology* **519**, 1939–1952.

Lammerts, E.J., Maas, C. and Grootjans, A.P. (2001) Groundwater variables and vegetation in dune slacks. *Ecological Engineering* **17**, 33–47.

Llewellyn, G.T., Dorman, F., Westland, J.L. *et al.* (2015). Evaluating a groundwater supply contamination incident attributed to Marcellus Shale gas development. *Proceedings of the National Academy of Sciences* **112**, 6325–6330.

Lloyd, J.W., Tellam, J.H., Rukin, N. and Lerner, D.N. (1993) Wetland vulnerability in East Anglia; a possible conceptual framework and generalised approach. *Journal of Environmental Management* **37**, 87–102.

Loáiciga, H.A. (2003) Climate change and ground water. *Annals of the Association of American Geographers* **93**, 30–41.

Loáiciga, H.A., Valdes, J.B., Vogel, R. *et al.* (1996) Global warming and the hydrologic cycle. *Journal of Hydrology* **174**, 83–127.

Loáiciga, H.A., Maidment, D.R. and Valdes, J.B. (2000) Climate-change impacts in a regional karst aquifer, Texas, USA. *Journal of Hydrology* **227**, 173–194.

Lopez, S., Hamm, V., Le Brun, M. *et al.* (2010) 40 years of Dogger aquifer management in Ile-de-France, Paris Basin, France. *Geothermics* **39**, 339–356.

Loveless, S.E., Bloomfield, J.P., Ward, R.S. *et al.* (2018) Characterising the vertical separation of shale-gas source rocks and aquifers across England and Wales (UK). *Hydrogeology Journal* **26**, 1975–1987.

Luckey, R.R., Gutentag, E.D., Heimes, F.J. and Weeks, J.B. (1986). Digital simulation of ground-water flow in the High Plains aquifer in parts of Colorado, Kansas, Nebraska, New Mexico, Oklahoma, South Dakota, Texas, and Wyoming. United States Geological Survey Professional Paper 1400-D, 57 pp.

Lucon, O., Ürge-Vorsatz, D., Zain Ahmed, A. *et al.* (2014) Buildings. In: *Climate Change 2014: Mitigation of Climate Change. Contribution of Working Group III to the Fifth Assessment Report of the Intergovernmental Panel on Climate Change* (eds O. Edenhofer, R. Pichs-Madruga, Y. Sokona *et al.*). Cambridge University Press, Cambridge, United Kingdom and New York, NY, USA.

Mawdsley, J., Petts, G. and Walker, S. (1994) *Assessment of Drought Severity*. Institute of Hydrology, Wallingford. British Hydrological Society Occasional Paper No. 3.

McIntosh, J.C., Hendry, M.J., Ballentine, C. *et al.* (2019) A critical review of state-of-the-art and emerging approaches to identify fracking-derived gases and associated contaminants in aquifers. *Environmental Science and Technology* **53**, 1063–1077. DOI: 10.1021/acs.est.8b05807.

McKenzie, H.S. and Elsaleh, B.O. (1994) The Libyan Great man-made river project paper 1. Project overview. *Proceedings of the Institution of Civil Engineers – Water Maritime and Environment* **106**, 103–122.

McPherson, B.F. and Halley, R. (1997). The South Florida Environment – A Region Under Stress. United States Geological Survey Circular 1134, 61 pp.

Minsley, B. J., Abraham, J.D., Smith, B.D. *et al.* (2012). Airborne electromagnetic imaging of discontinuous permafrost. *Geophysical Research Letters* **39**, L02503. DOI: 10.1029/2011GL050079.

Mitsch, W.J., Mitsch, R.H. and Turner, R.E. (1994) Wetlands of the Old and New Worlds: ecology and management. In: *Global Wetlands: Old World and New* (ed. W.J. Mitsch). Elsevier Science B.V., Amsterdam, pp. 3–56.

Murray, K.E. (2013) State-scale perspective on water use and production associated with oil and gas operations, Oklahoma, U.S. *Environmental Science and Technology* **47**, 4918–4925. DOI: 10.1021/es4000593.

National Research Council (2015). *Review of the Everglades Aquifer Storage and Recovery Regional Study*. The National Academies Press, Washington, DC, 57 pp. DOI: 10.17226/21724.

Navid, D. (1989) The international law of migratory species: the Ramsar Convention. *Natural Resources Journal* **29**, 1001–1016.

NASA (2020). *2019 was the second warmest year on record*. https://earthobservatory.nasa.gov/images/146154/2019-was-the-second-warmest-year-on-record (accessed 13 September 2020).

Nicot, J.-P. and Scanlon, B.R. (2012) Water use for shale-gas production in Texas, U.S. *Environmental Science and Technology* **46**, 3580–3586. DOI: 10.1021/es204602t.

NOAA (2019). *Trends in Atmospheric Carbon Dioxide*. National Oceanic and Atmospheric Administration https://www.esrl.noaa.gov/gmd/ccgg/trends/ (accessed 12 August 2019).

NPS (2021) *Restoring the Everglades*. National Park Service. https://www.nps.gov/features/ever/climatechange/ever705/ (accessed 22 January 2021).

O'Geen, A.T., Saal, M.B.B., Dahlke, H. *et al.* (2015) Soil suitability index identifies potential areas for groundwater banking on agricultural lands. *California Agriculture* **69**, 75–84.

Ojo, O., Oni, F. and Ogunkunle, O. (2003) Implications of climate variability and climate change on water resources availability and water resources management in West Africa. In: *Water Resources Systems – Water Availability and Global Change* (eds S. Franks, S. Bloschl, M. Kumagai *et al.*). International Association of Hydrological Sciences, Wallingford, pp. 37–47.

Olmstead, S.M., Muehlenbachs, L.A., Shih, J.-S. *et al.* (2013) Shale gas development impacts on surface water quality in Pennsylvania. *Proceedings of the National Academy of Sciences* **110**, 4962–4967.

Osborn, S. G., Vengosh, A., Warner, N.R. and Jackson, R.B. (2011) Methane contamination of drinking water accompanying gas-well drilling and hydraulic fracturing. *Proceedings of the National Academy of Sciences* **108**, 8172–8176.

O'Shea, M.J. and Sage, R. (1999) Aquifer recharge: an operational drought-management strategy in North London. *Journal of the Institution of Water and Environmental Management* **13**, 400–405.

Palmer, W.C. (1965) *Meteorological Drought*. Research Paper No. 45. Department of Commerce, Washington, DC.

Park, E. and Zhan, H.B. (2003) Hydraulics of horizontal wells in fractured shallow aquifer systems. *Journal of Hydrology* **281**, 147–158.

Pauw, P.S., van Baaren, E.S., Visser, M. *et al.* (2015) Increasing a freshwater lens below a creek ridge using a controlled artificial recharge and drainage system: a case study in the Netherlands. *Hydrogeology Journal* **23**, 1415–1430.

Pim, R.H. and Binsariti, A. (1994) The Libyan great man-made river project paper 2. The water resource. *Proceedings of the Institution of Civil Engineers – Water Maritime and Environment* **106**, 123–145.

Plummer, C.R., Nelson, J.D. and Zumwalt, G.S. (1997) Horizontal and vertical well comparison for in situ air sparging. *Ground Water Monitoring and Remediation* **17**, 91–96.

Por, D.F. (2004). The Levantine waterway, riparian archaeology, paleolimnology, and conservation. In: *Human Paleoecology in the Levantine Corridor*, *Oxbow Books* (eds I. Goren-Inbar and J. Speth). Oxford, pp. 5–20.

Price, M. (2002) Who needs sustainability? In: *Sustainable Groundwater Development* (eds K. M. Hiscock, M.O. Rivett and R.M. Davison). Geological Society, London, Special Publications **193**, pp. 75–81.

Purkey, D.R., Joyce, B., Vicuna, S. *et al.* (2007) Robust analysis of future climate change impacts on water for agriculture and other sectors: a case study in the Sacramento Valley. *Climatic Change* **87** (Suppl), S109–S122.

Pyne, R.D.G. (1995) *Groundwater Recharge and Wells: A Guide to Aquifer Storage and Recovery*. CRC Press, Inc., Boca Raton, Florida.

Ranwell, D.S. (1960) Newborough Warren, Anglesey. II. Plant associes and succession cycles of the sand dune and dune slack vegetation. *Journal of Ecology* **48**, 117–141.

Rawlins, M.A., Cai, L., Stuefer, S.L. and Nicolsky, D. (2019) Changing characteristics of runoff and freshwater export from watersheds draining northern Alaska. *The Cryosphere* **13**, 3337–3352.

Rawluk, A., Curtis, A., Sharp, E. *et al.* (2013). Managed aquifer recharge in farming landscapes using large floods: an opportunity to improve outcomes for the Murray-Darling Basin? *Australasian Journal of Environmental Management* **20**, 34–48.

Rhode, M.M., Froend, R. and Howard, J. (2017). A global synthesis of managing groundwater dependent ecosystems under sustainable groundwater policy. *Groundwater* **55**, 293–301.

Russo, T.A. and Lall, U. (2017) Depletion and response of deep groundwater to climate-induced pumping variability. *Nature Geoscience* **10**, 105–108. DOI: 10.1038/NGEO2883.

Salgot, M. and Torrens, A. (2008) Impacts of climatic change on water resources: the future of groundwater recharge with reclaimed water in the south of Europe. In: *Climate Change and Groundwater* (eds W. Dragoni and B.S. Sukhija). Geological Society, London, Special Publications **288**, pp. 145–168.

Schädel, C., Bader, M.K.-F., Schuur, E.A.G. *et al.* (2016) Potential carbon emissions dominated by carbon dioxide from thawed permafrost soils. *Nature Climate Change* **6**, 950–953. DOI: 10.1038/nclimate3054.

Schubert, J. (2002) Hydraulic aspects of riverbank filtration – field studies. *Journal of Hydrology* **266**, 145–161.

Sheets, R.A., Darner, R.A. and Whitteberry, B.L. (2002) Lag times of bank filtration at a well field, Cincinnati, Ohio, USA. *Journal of Hydrology* **266**, 162–174.

Song, Y., Du, X. and Ye, X. (2019). Analysis of potential risks associated with urban stormwater quality for managed aquifer recharge. *International Journal of Environmental Research and Public Health* **16**, 3121. DOI: 10.3390/ijerph16173121.

Stefan, C. and Ansems, N. (2018) Web-based global inventory of managed aquifer recharge applications. *Sustainable Water Resources Management* **4**, 153–162.

Stuyfzand, P.J. (1993). *Hydrochemistry and hydrology of the coastal dune areas of the Western Netherlands*. PhD thesis, Free University of Amsterdam.

Tanaka, S.K., Zhu, T.J., Lund, J.R. *et al.* (2006) Climate warming and water management adaptation for California. *Climatic Change* **76**, 361–387.

Tansley, A.G. (1949) *The British Islands and their vegetation*. Cambridge University Press, Cambridge.

Trenberth, K., Overpeck, J. and Soloman, S. (2004) Exploring drought and its implications for the future. *Eos* **85**, 27.

Trenberth, K.E., Dai, A., Rasmussen, R.M. and Parsons, D.B. (2003) The changing character of precipitation. *Bulletin of the American Meteorological Society* **84**, 1205–1217.

Tufenkji, N., Ryan, J.N. and Elimelech, M. (2002) The promise of bank filtration. *Environmental Science & Technology* **36**, 423A–428A.

UN (2015) *Transforming our World: The 2030 Agenda for Sustainable Development*. Resolution A/RES/70/1. United Nations, New York, 35 pp.

United States Environmental Protection Agency (2006). *National Primary Drinking Water Regulations: Long Term 2 Enhanced Surface Water Treatment Rule*. Environmental

Protection Agency, 40 CFR Parts 9, 141 and 142: 654-786. Washington, DC.

UNEP (2003) *Groundwater and its Susceptibility to Degradation: a Global Assessment of the Problem and Options for Management.* United Nations Environment Programme, Nairobi, Kenya. Early Warning and Assessment Report Series, RS 03-3.

United States Geological Survey (2002). *SOFIA-SFRSF-Hydrology-What is aquifer storage and recovery (ASR)?* http://sofia.usgs.gov/sfrsf/rooms/hydrology/ASR/index.html (accessed 10 June 2013).

Vaccaro, J.J. (1992) Sensitivity of groundwater recharge estimates to climate variability and change, Columbia Plateau, Washington. *Journal of Geophysical Research* **97**, 2821–2833.

Van Dijk, H.W.J. and Grootjans, A.P. (1993) Wet dune slacks: decline and new opportunities. *Hydrobiologia* **265**, 281–304.

Varela-Ortega, C., Blanco-Gutiérrez, I., Swartz, C. H. and Downing, T.E. (2011) Balancing groundwater conservation and rural livelihoods under water and climate uncertainties: an integrated hydro-economic modelling framework. *Global Environmental Change* **21**, 604–619.

Vengosh, A., Jackson, R.B., Warner, N. *et al.* (2014) A critical review of the risks to water resources from unconventional shale gas development and hydraulic fracturing in the United States. *Environmental Science and Technology* **48**, 8334–8348. DOI: 10.1021/es405118y.

Verstraeten, I.M., Thurman, E.M., Lindsey, M.E. *et al.* (2002) Changes in concentrations of triazine and acetamide herbicides by bank filtration, ozonation, and chlorination in a public water supply. *Journal of Hydrology* **266**, 190–208.

Vidic, R.D., Brantley, S.L., Vandenbossche, J.M. *et al.* (2013) Impact of shale gas development on regional water quality. *Science* **340**, 1235009. DOI: 10.1126/science.1235009.

Villholth, K.G. and Conti, K.I. (2018) Groundwater governance: rationale, definition, current state and heuristic framework. In *Advances in Groundwater Governance* (eds K.G. Villholth, E. López-Gunn, K.I. Conti, A. Garrido and J. van der Gun). CRC Press, Boca Raton, Florida, pp. 3–31.

Walvoord, M.A. and Kurylyk, B.L. (2016) Hydrologic impacts of thawing permafrost—a review. *Vadose Zone Journal* **15**. DOI: 10.2136/vzj2016.01.0010.

Walvoord, M.A. and Striegl, R.G. (2007) Increased groundwater to stream discharge from permafrost thawing in the Yukon River basin: potential impacts on lateral export of carbon and nitrogen. *Geophysical Research Letters* **34**, L12402. DOI: 10.1029/2007GL030216.

Wang, J., Rothausen, S.G.S.A., Conway, D., Zhang, L., Xiong, W., Holman, I.P. *et al.* (2012) China's water-energy nexus: greenhouse-gas emissions from groundwater use for agriculture. *Environmental Research Letters* **7**. DOI: 10.1088/1748-9326/7/1/014035.

Wheeler, B.D. and Proctor, M.C.F. (2000) Ecological gradients, subdivisions and terminology of north-west European mires. *Journal of Ecology* **88**, 187–203.

Whiteman, M., Brooks, A., Skinner, A. and Hulme, P. (2010) Determining significant damage to groundwater-dependent terrestrial ecosystems in England and Wales for use in implementation of the water framework directive. *Ecological Engineering* **36**, 1118–1125.

Whiteman, M.I., Farr, G., Jones, P.S. *et al.* (2017) Merthyr Mawr: a case study for the assessment of nitrate at humid dunes in England and Wales. *Journal of Coastal Conservation* **21**, 669–684.

Wilby, R.L. and Harris, I. (2006) A framework for assessing uncertainties in climate change: low-flow scenarios for the River Thames, UK. *Water Resources Research* **42**, W02419, doi: 10.1029/2005WR004065.

Willis, A.J. (1985) Dune water and nutrient regimes - their ecological relevance. In: *Sand Dunes and their Management* (ed. P. Doody). Focus on Nature Conservation **13**, 159–174. Nature Conservancy Council, Peterborough.

Wilson, M.P., Worrall, F., Davies, R.J. and Hart, A. (2017). Shallow aquifer vulnerability from subsurface fluid injection at a proposed shale gas hydraulic fracturing site. *Water Resources Research* **53**, 9922–9940. DOI: 10.1002/2017WR021234.

Wolfgramm, M., Bartels, J., Hoffmann, F. *et al.* (2007). Unterhaching geothermal well doublet: structural and hydrodynamic reservoir characteristic; Bavaria (Germany). *Proceedings European Geothermal Congress 2007*, Unterhaching, Germany, (30 May–1 June 2007), 6 pp.

WWAP (2009) *The United Nations World Water Development Report 3: Water in a Changing World*. World Water Assessment Programme, UNESCO, Paris and Earthscan, London.

Xinqiang, D., Yalin, S., Xueyan, Y. and Ran, L. (2019). Colloid clogging of saturated porous media under varying ionic strength and roughness during managed aquifer recharge. *Journal of Water Reuse and Desalination* **9**, 225–231.

Ye, B., Yang, D., Zhang, Z. and Kane, D.L. (2009) Variation of hydrological regime with permafrost coverage over Lena Basin in Siberia. *Journal of Geophysical Research* **114**, D07102. DOI:10.1029/2008JD010537.

Younger, P.L., Teutsch, G., Custodio, E. *et al.* (2002) Assessments of the sensitivity to climate change of flow and natural water quality in four major carbonate aquifers of Europe. In: *Sustainable Groundwater Development* (eds K.M. Hiscock, M.O. Rivett and R.M. Davison). Geological Society, London, Special Publications **193**, pp. 303–323.

Yuan, J., Van Dyke, M.I. and Huck, P.M. (2019). Selection and evaluation of water pretreatment technologies for managed aquifer recharge (MAR) with reclaimed water. *Chemosphere* **236**, 124886. DOI: 10.1016/j.chemosphere.2019.124886.

Yusoff, I., Hiscock, K.M. and Conway, D. (2002) Simulation of the impacts of climate change on groundwater resources in eastern England. In: *Sustainable Groundwater Development* (eds K.M. Hiscock, M.O. Rivett and R.M. Davison). Geological Society, London, Special Publications **193**, pp. 325–344.

Zektser, I.S. and Loaiciga, H.A. (1993) Groundwater fluxes in the global hydrologic cycle: past, present and future. *Journal of Hydrology* **144**, 405–427.

Zhan, H.B. and Park, E. (2002) Vapor flow to horizontal wells in unsaturated zones. *Soil Science Society of America Journal* **66**, 710–721.

Zhang, B.Y. and Lerner, D.N. (2000) Modeling of ground water flow to adits. *Ground Water* **38**, 99–105.

Zhu, K., Blum, P., Ferguson, G. *et al.* (2010) The geothermal potential of urban heat islands. *Environmental Research Letters* **5**, 044002. DOI: 10.1088/1748-9326/6/1/019501.

Appendix 1

Conversion factors

Table A1.1 Table of conversion factors.

Quantity	Unit	SI Equivalent
Length	1 yard (yd)	0.914 m
	1 foot (ft)	0.305 m
	1 inch (in)	25.4 mm
	1 mile	1.609 km
Area	1 yd^2	0.836 m^2
	1 ft^2	9.290×10^{-2} m^2
	1 in^2	6.452×10^2 mm^2
	1 mile^2	2.590 km^2
	1 acre	4.407×10^3 m^2
	1 hectare (ha)	1×10^4 m^2
Volume	1 ft^3	2.832×10^{-2} m^3
	1 UK gallon	4.546×10^{-3} m^3
	1 US gallon	3.784×10^{-3} m^3
Mass	1 pound (lb)	4.536×10^{-1} kg
	1 UK ton	1.0165×10^3 kg
	1 US ton	9.072×10^2 kg
	1 tonne (t)	1000 kg
Density	1 lb ft^{-3}	16.0185 kg m^{-3}
Force and weight	1 poundal (pdl)	0.138 N
	1 poundal-force (lb$_f$)	4.448 N
Pressure	1 lb$_f$ in^{-2} (psi)	6.895 kPa
	1 lb$_f$ ft^{-2}	4.788×10^{-2} kPa
	1 mm Hg	133.322 Pa
	1 ft H$_2$O	2.989 kPa
	1 m H$_2$O	9.807 kPa

Hydrogeology: Principles and Practice, Third Edition. Kevin M. Hiscock and Victor F. Bense.
© 2021 John Wiley & Sons Ltd. Published 2021 by John Wiley & Sons Ltd.
Companion website: www.wiley.com/go/hiscock/hydrogeology3e

Table A1.1 (Continued)

Quantity	Unit	SI Equivalent
	$1\,kg\,cm^{-2}$	98.067 kPa
	1 bar	100 kPa
	1 atmosphere	101.325 kPa
Velocity	$1\,ft\,s^{-1}$	$3.048 \times 10^{-1}\,m\,s^{-1}$
	$1\,km\,hr^{-1}$	$2.778 \times 10^{-1}\,m\,s^{-1}$
	$1\,mile\,hr^{-1}$	$4.470 \times 10^{-1}\,m\,s^{-1}$
Viscosity	$1\,lb.ft^{-1}s^{-1}$	$1.488\,N\,s\,m^{-2}$
	1 Poise	$0.1\,N\,s\,m^{-2}$
Viscosity (kinematic)	1 stoke	$10^{-4}\,m^2s^{-1}$
Energy	1 kWh	3.6 MJ
	1 Btu	1.055 kJ
	1 Therm	105.506 MJ
	1 kcal	4.187 kJ
	1 foot poundal	$4.214 \times 10^{-2}\,J$
Power	1 horse power	$7.457 \times 10^{-1}\,kW$
	$1\,Btu\,ft^{-2}$	$11.357\,kJ\,m^{-2}$
	$1\,cal\,cm^{-2}$	$41.868\,kJ\,m^{-2}$
Discharge	$ft^3\,s^{-1}$	$2.832 \times 10^{-2}\,m^3\,s^{-1}$
	$US\,gallon\,min^{-1}$	$6.309 \times 10^{-5}\,m^3\,s^{-1}$
Intrinsic permeability	ft^2	$9.290 \times 10^{-2}\,m^2$
	darcy	$9.870 \times 10^{-13}\,m^2$
Hydraulic conductivity	$ft\,s^{-1}$	$3.048 \times 10^{-1}\,m\,s^{-1}$
	$US\,gallon\,day^{-1}\,ft^{-2}$	$4.720 \times 10^{-7}\,m\,s^{-1}$
Transmissivity	$ft^2\,s^{-1}$	$9.290 \times 10^{-2}\,m^2\,s^{-1}$
	$US\,gallon\,day^{-1}\,ft^{-1}$	$1.438 \times 10^{-7}\,m^2\,s^{-1}$

Appendix 2

Properties of water in the range 0–100°C

Table A2.1   Values of the density, viscosity, vapour pressure and surface tension for liquid water in the range 0–100°C. All values (except vapour pressure) refer to a pressure of 100 kPa (1 bar). (Source: Lide, D.R. (ed.) (1991) *Handbook of Chemistry and Physics* (72nd ed). CRC Press Inc., Boca Baton. Taylor & Francis. © 1991, Taylor & Francis.)

Temperature (°C)	Density (kg m^{-3})	Viscosity ($\times 10^{-6}$ N s m^{-2})	Vapour pressure (kPa)	Surface tension (mN m^{-1})
0	999.84	1793	0.6113	75.64
10	999.70	1307	1.2281	74.23
20	998.21	1002	2.3388	72.75
30	995.65	797.7	4.2455	71.20
40	992.22	653.2	7.3814	69.60
50	988.03	547.0	12.344	67.94
60	983.20	466.5	19.932	66.24
70	977.78	404.0	31.176	64.47
80	971.82	354.4	47.373	62.67
90	965.35	314.5	70.117	60.82
100	958.40	281.8	101.325	58.91

Hydrogeology: Principles and Practice, Third Edition. Kevin M. Hiscock and Victor F. Bense.
© 2021 John Wiley & Sons Ltd. Published 2021 by John Wiley & Sons Ltd.
Companion website: www.wiley.com/go/hiscock/hydrogeology3e

Appendix 3

The geological timescale

Table A3.1 Table A3.1 The geological timescale. (Source: Adapted from the International Chronostratigraphic Chart published by the International Commission on Stratigraphy, version 2019/05 (see www.stratigraphy.org).)

Phanerozoic Eonothem or Eon		
(~541 Ma–Present)		
The period of time between the end of the Precambrian Eon and today. The Phanerozoic Eon begins with the Cambrian Period, 541.0 Ma. The Phanerozoic Eon encompasses the period of abundant, complex life on Earth.		

Erathem or Era	System or Period		Series or Epoch
Cenozoic *(66.0 Ma–Present)*	**Quaternary** *(2.6 Ma–Present)*		**Holocene** *(11 700 years–Present)*
			Pleistocene *(2.6 million–11 700 years)*
	Neogene *(23.0–2.6 Ma)*		**Pliocene** *(5.3–2.6 Ma)*
			Miocene *(23.0–5.3 Ma)*
	Palaeogene *(66.0–23.0 Ma)*		**Oligocene** *(33.9–23.0 Ma)*
			Eocene *(56.0–33.9 Ma)*
			Palaeocene *(66.0–56.0 Ma)*
Mesozoic *(251.9–66.0 Ma)*	**Cretaceous** *(~145.0–66.0 Ma)*		
	Jurassic *(201.3–145.0 Ma)*		
	Triassic *(251.9–201.3 Ma)*		
Palaeozoic *(541.0–251.9 Ma)*	**Permian** *(298.9–251.9 Ma)*		
	Carboniferous *(358.9–298.9 Ma)*	**Pennsylvanian**[a] *(323.2–298.9 Ma)*	
		Mississippian[a] *(358.9–323.2 Ma)*	

(Continued)

Hydrogeology: Principles and Practice, Third Edition. Kevin M. Hiscock and Victor F. Bense.
© 2021 John Wiley & Sons Ltd. Published 2021 by John Wiley & Sons Ltd.
Companion website: www.wiley.com/go/hiscock/hydrogeology3e

Table A3.1 (Continued)

Devonian *(419.2–358.9 Ma)*	
Silurian *(443.8–419.2 Ma)*	
Ordovician *(485.4–443.8 Ma)*	
Cambrian *(541.0–485.4 Ma)*	

Precambrian Eonothem or Eon

(Beginning of the Earth, ∼4600–∼541 Ma)

All geological time before the beginning of the Palaeozoic Era. This includes about 90% of all geological time and spans the time from the beginning of the Earth, ∼4600 Ma, to ∼541 Ma. The Precambrian Eon is usually divided into the Hadean (∼4600–4000 Ma), Archaen (4000–2500 Ma) and Proterozoic (2500–∼541 Ma), all three being commonly called eons.

Notes:

Ma = million years before present.

[a] North American usage.

Appendix 4

Symbols, atomic numbers and atomic weights

Table A4.1 Table of symbols, atomic numbers and atomic weights of the elements.

Atomic number	Element	Symbol	Atomic weight
1	Hydrogen	H	1.008
2	Helium	He	4.003
3	Lithium	Li	6.939
4	Beryllium	Be	9.012
5	Boron	B	10.81
6	Carbon	C	12.01
7	Nitrogen	N	14.01
8	Oxygen	O	16.00
9	Fluorine	F	19.00
10	Neon	Ne	20.18
11	Sodium	Na	22.99
12	Magnesium	Mg	24.31
13	Aluminium	Al	26.98
14	Silicon	Si	28.09
15	Phosphorus	P	30.97
16	Sulphur	S	32.06
17	Chlorine	Cl	35.45
18	Argon	Ar	39.95
19	Potassium	K	39.10
20	Calcium	Ca	40.08
21	Scandium	Sc	44.96
22	Titanium	Ti	47.90
23	Vanadium	V	50.94

(*Continued*)

Hydrogeology: Principles and Practice, Third Edition. Kevin M. Hiscock and Victor F. Bense.
© 2021 John Wiley & Sons Ltd. Published 2021 by John Wiley & Sons Ltd.
Companion website: www.wiley.com/go/hiscock/hydrogeology3e

Table A4.1 (Continued)

Atomic number	Element	Symbol	Atomic weight
24	Chromium	Cr	52.00
25	Manganese	Mn	54.94
26	Iron	Fe	55.85
27	Cobalt	Co	58.93
28	Nickel	Ni	58.71
29	Copper	Cu	63.54
30	Zinc	Zn	65.37
31	Gallium	Ga	69.72
32	Germanium	Ge	72.59
33	Arsenic	As	74.92
34	Selenium	Se	78.96
35	Bromine	Br	79.91
36	Krypton	Kr	83.80
37	Rubidium	Rb	85.47
38	Strontium	Sr	87.62
39	Yttrium	Y	88.91
40	Zirconium	Zr	91.22
41	Niobium	Nb	92.91
42	Molybdenum	Mo	95.94
43	Technetium[a]	Tc	97
44	Ruthenium	Ru	101.07
45	Rhodium	Rh	102.91
46	Palladium	Pd	106.4
47	Silver	Ag	107.87
48	Cadmium	Cd	112.41
49	Indium	In	114.82
50	Tin	Sn	118.69
51	Antimony	Sb	121.75
52	Tellurium	Te	127.60
53	Iodine	I	126.90
54	Xenon	Xe	131.30
55	Caesium	Cs	132.91
56	Barium	Ba	137.33
57	Lanthanium	La	138.91
58	Cerium	Ce	140.12
59	Praseodymium	Pr	140.91
60	Neodymium	Nd	144.24
61	Promethium[a]	Pm	145

Table A4.1 (Continued)

Atomic number	Element	Symbol	Atomic weight
62	Samarium	Sm	150.35
63	Europium	Eu	151.96
64	Gadolinium	Gd	157.25
65	Terbium	Tb	158.93
66	Dysprosium	Dy	162.50
67	Holmium	Ho	164.93
68	Erbium	Er	167.26
69	Thulium	Tm	168.93
70	Yytterbium	Yb	173.04
71	Lutetium	Lu	174.97
72	Hafnium	Hf	178.49
73	Tantalum	Ta	180.95
74	Tungsten	W	183.85
75	Rhenium	Re	186.21
76	Osmium	Os	190.2
77	Iridium	Ir	192.22
78	Platinum	Pt	195.09
79	Gold	Au	196.97
80	Mercury	Hg	200.59
81	Thallium	Tl	204.37
82	Lead	Pb	207.19
83	Bismuth	Bi	208.98
84	Polonium[a]	Po	209
85	Astatine[a]	At	210
86	Radon[a]	Rn	222
87	Francium[a]	Fr	223
88	Radium[a]	Ra	226.03
89	Actinium[a]	Ac	227.03
90	Thorium[a]	Th	232.04
91	Protactinium[a]	Pa	231.04
92	Uranium[a]	U	238.03
93	Neptunium[a]	Np	237
94	Plutonium[a]	Pu	244
95	Americium[a]	Am	243
96	Curium[a]	Cm	247
97	Berkelium[a]	Bk	247
98	Californium[a]	Cf	251

(*Continued*)

Table A4.1 (Continued)

Atomic number	Element	Symbol	Atomic weight
99	Einsteinium[a]	Es	254
100	Fermium[a]	Fm	253
101	Mendelevium[a]	Md	256
102	Nobelium[a]	No	253
103	Lawrentium[a]	Lw	257

[a] Radioactive element.

Appendix 5

Composition of seawater and rainwater

A5.1 Seawater composition

Ocean water consists of a complex solution with a total salt content, or salinity, of 3.5% (usually stated as 35‰ or 35 parts per thousand). The composition of seawater is mainly controlled by a balance between the addition of dissolved material from river water and groundwater, and various processes of removal. Important removal processes include losses to accumulating sediments by precipitation, sorption and organic activity, and reactions with basalt at mid-ocean ridges. Of the dissolved materials listed in Table A5.1, the most abundant and most constant in concentration in all parts of the ocean are the conservative species Na^+, K^+, Ca^{2+}, Mg^{2+}, Cl^- and SO_4^{2-}. Some of the less abundant, non-conservative species, notably HCO_3^-, SiO_2 and the ions of N and P, participate in biological processes and therefore show widely varying concentrations depending on the local abundance of marine organisms and supply of organic carbon. Seawater pH is remarkably constant, normally in the range 7.8–8.4 and is buffered principally by inorganic reactions involving carbonate species (Krauskopf and Bird 1995).

A5.2 Rainwater composition

Rainwater can be described as a weakly acidic, dilute solution, with a pH in the range 4–6 and a total salt content of just a few milligrams per litre. Evaporation into the atmosphere results in separation of water molecules from dissolved salts in surface waters. The resulting water vapour ultimately condenses to form rain, and the overall process can be viewed as purification by natural distillation. However, solid particles and gases in the atmosphere are dissolved in rainwater resulting in a wide range in chemical composition, as well as variation in pH. Broadly, and as shown in Tables A5.2 and A5.3, rainwater species derived from terrestrial sources are mainly dominated by Ca^{2+}, K^+, NH_4^+ and NO_3^-, and from marine sources the main species are Cl^-, Na^+, Mg^{2+} and SO_4^{2-}. Elements in rain that result from rainout (determined by the composition of nucleating aerosols) show little change or a slight rise in concentration with time. In contrast, elements contributed by washout (determined by the composition of soluble trace gases) exhibit a sharp decrease in concentration with time as the air is essentially cleaned during the rainfall event (Berner and Berner 1987).

Hydrogeology: Principles and Practice, Third Edition. Kevin M. Hiscock and Victor F. Bense.
© 2021 John Wiley & Sons Ltd. Published 2021 by John Wiley & Sons Ltd.
Companion website: www.wiley.com/go/hiscock/hydrogeology3e

Table A5.1 Table of average abundance of elements in seawater (Goldberg 1963 and Li 1991). (Sources: Adapted from Goldberg, E.D. (1963) The oceans as a chemical system. In: *The Sea: Ideas and Observations on Progress in the Study of the Sea. Vol. 2 The Composition of Seawater; Comparative and Descriptive Oceanography*, (ed. M.N. Hill). John Wiley & Sons, Inc., New York, pp. 3–25.; Li, Y.-H. (1991) Distribution patterns of the elements in the ocean: a synthesis. *Geochimica et Cosmochimica Actas* **55**, 3223–3240.)

Element	Abundance (mg L^{-1})	Principal species
O	880 000	H_2O; $O_2(g)$; SO_4^{2-} and other anions
H	110 000	H_2O
Cl	18 800	Cl^-
Na	10 800	Na^+
Mg	1290	Mg^{2+}; $MgSO_4$
S	900	SO_4^{2-}
N	670	NO_3^-; NO_2^-; NH_4^+; $N_2(g)$; organic compounds
Ca	450	Ca^{2+}; $CaSO_4$
K	390	K^+
Br	67	Br^-
C	28	HCO_3^-; H_2CO_3; CO_3^{2-}; organic compounds
Sr	7.8	Sr^{2+}; $SrSO_4$
B	4.5	$B(OH)_3$; $B(OH)_2O^-$
Si	2.5	$Si(OH)_4$; $Si(OH)_3O^-$
F	1.3	F^-
Ar	0.6	$Ar(g)$
Li	0.18	Li^+
Rb	0.12	Rb^+
P	0.09	HPO_4^{2-}; $H_2PO_4^-$; PO_4^{3-}; H_3PO_4
I	0.058	IO_3^-; I^-
Ba	0.015	Ba^{2+}; $BaSO_4$
Mo	0.01	MoO_4^{2-}
U	0.0032	$UO_2(CO_3)_3^{4-}$
Al	0.003	
Fe	0.003	$Fe(OH)_3(s)$
V	0.0022	$VO_2(OH)_3^{2-}$
As	0.0017	$HAsO_4^{2-}$; $H_2AsO_4^-$; H_3AsO_4; H_3AsO_3
Ni	0.0005	Ni^{2+}; $NiSO_4$
Cr	0.0003	
Cs	0.0003	Cs^+
Kr	0.0003	$Kr(g)$
Zn	0.0003	Zn^{2+}; $ZnSO_4$
Cu	0.0002	Cu^{2+}; $CuSO_4$
Mn	0.0002	Mn^{2+}; $MnSO_4$

Table A5.1 (Continued)

Element	Abundance (mg L^{-1})	Principal species
Sb	0.0002	
Se	0.0002	SeO_4^{2-}
Ne	0.0001	$Ne(g)$
Ti	0.0001	
W	0.0001	WO_4^{2+}
Xe	0.0001	$Xe(g)$
Cd	8×10^{-5}	Cd^{2+}; $CdSO_4$
Zr	2×10^{-5}	
Nb	1×10^{-5}	
Tl	1×10^{-5}	Tl^+
Y	1×10^{-5}	
La	6×10^{-6}	
He	5×10^{-6}	$He(g)$
Au	4×10^{-6}	$AuCl_4^-$
Ge	4×10^{-6}	$Ge(OH)_4$; $Ge(OH)_3O^-$
Nd	4×10^{-6}	
Ag	3×10^{-6}	$AgCl_2^-$; $AgCl_3^{2-}$
Hf	3×10^{-6}	
Pb	3×10^{-6}	Pb^{2+}; $PbSO_4$
Ce	2×10^{-6}	
Dy	2×10^{-6}	
Ga	2×10^{-6}	
Ta	2×10^{-6}	
Yb	2×10^{-6}	
Co	1×10^{-6}	Co^{2+}; $CoSO_4$
Er	1×10^{-6}	
Gd	1×10^{-6}	
Pr	9×10^{-7}	
Sc	9×10^{-7}	
Sm	8×10^{-7}	
Sn	6×10^{-7}	
Ho	5×10^{-7}	
Hg	4×10^{-7}	$HgCl_3^-$; $HgCl_4^{2-}$
Lu	3×10^{-7}	
Pt	3×10^{-7}	
Tm	3×10^{-7}	
Be	2×10^{-7}	

(Continued)

Table A5.1 (Continued)

Element	Abundance (mg L^{-1})	Principal species
Eu	2×10^{-7}	
Tb	2×10^{-7}	
In	1×10^{-7}	
Pd	7×10^{-8}	
Th	5×10^{-8}	
Bi	4×10^{-9}	
Ra	1×10^{-10}	Ra^{2+}; $RaSO_4$
Rn	6×10^{-17}	$Rn(g)$

Table A5.2 Chemical composition of rainwater samples from land, marine, island and coastal sites. All concentrations and pH are given as mean values (Cornell 1996). (Source: Cornell, S.E. (1996) *Dissolved organic nitrogen in rainwater*. PhD Thesis, University of East Anglia, Norwich, 134–138.)

Location (no. of samples)	pH (μmol L^{-1})	NH_4^+	Na^+	K^+	Ca^{2+}	Mg^{2+}	NO_3^-	Cl^-	SO_4^{2-}	$NssSO_4^{2-}$
Land sites										
Norwich, UK (n = 25) 52°38′N 1°17′E	4.7	31.6	55.0	5.4	11.3	15.3	37.0	48.0	46.5	43.2
Norfolk, UK (n = 12) 52°50′N 1°0′E	5.0	35.8	76.4	3.0	5.9	n.d.	39.7	27.9	71.4	79.2
Fichtelberg, Czech Republic (n = 5) 50°10′N 12°0′E	4.0	9.4	3.0	2.0	4.9	2.1	10.1	8.2	39.7	39.4
Cullowhee-NC, USA (n = 8) 34°54′N 82°24′W	4.9	2.1	8.4	9.2	8.9	2.0	6.3	11.4	18.1	19.4
Maraba-PA, Brazil (n = 5) 5°20′S 49°5′W	4.7	54.7	59.5	2.9	2.1	6.2	21.5	17.0	6.6	10.6
Marine, island and coastal sites										
North Atlantic (n = 8) 50°60′N ~30°W and 38°52′N ~30°W	4.9	0.6	856.1	19.8	12.7	74.1	3.1	754.5	144.7	11.0
BBSR, Bermuda (n = 18) 32°35′N 8°25′W	5.0	8.0	58.2	2.6	6.9	3.1	2.9	88.2	35.8	25.8
Recife, Brazil (n = 9) 8°0′S 35°0′W	5.3	4.9	41.5	2.7	1.9	5.3	3.6	35.6	15.5	13.0
Tahiti (n = 16) 17°37′S 149°27′W	5.2	2.1	138.9	1.8	2.5	4.9	0.6	48.6	3.8	3.2

n.d. – not determined

$NssSO_4^{2-}$ – concentration of SO_4^{2-} from sources other than sea-salt.

Table A5.3 Volume weighted mean concentrations of trace elements in rainwater from various sites in and around the central southern North Sea. (Kane *et al.* 1994). (Source: Kane, M.M., Rendell, A.R. and Jickells, T.D. (1994) Atmospheric scavenging processes over the North Sea. *Atmospheric Environment* **28**, 2523–2530. © 1994, Elsevier.)

	Pellworm Island, Germany	Mannington, Norfolk, UK	North Sea
Fe	88	17	31
Mn	3.8	2.7	3.6
Cu	2.3	4.0	1.0
Zn	13	2.9	7.6
Pb	4.0	4.1	3.5
Cd	0.7	0.25	0.08
Na	–	1200	82 100
Ca	388	126	3396
NO_3^-	–	1240	3236
SO_4^{2-}	–	2790^a	3273^b
NH_4^+	–	4070	668

Concentrations in $\mu g\ L^{-1}$

[a] Predominantly non-sea-salt SO_4^{2-}

[b] Non-sea-salt SO_4^{2-}

References

Berner, E.K. and Berner, R.A. (1987) *The Global Water Cycle: Geochemistry and Environment.* Prentice Hall, Englewood Cliffs, New Jersey, pp. 62–70.

Cornell, S.E. (1996) *Dissolved organic nitrogen in rainwater.* PhD Thesis, University of East Anglia, Norwich, 134–138.

Goldberg, E.D. (1963) The oceans as a chemical system. In: *The Sea: Ideas and Observations on Progress in the Study of the Sea. Vol. 2 The Composition of Sea-water; Comparative and Descriptive Oceanography*, (ed. M.N. Hill). John Wiley & Sons, Inc., New York, pp. 3–25.

Kane, M.M., Rendell, A.R. and Jickells, T.D. (1994) Atmospheric scavenging processes over the North Sea. *Atmospheric Environment* **28**, 2523–2530.

Krauskopf, K.B. and Bird, D.K. (1995) *Introduction to Geochemistry (3rd edn).* McGraw-Hill, Inc., New York, pp. 309–317.

Li, Y.-H. (1991) Distribution patterns of the elements in the ocean: a synthesis. *Geochimica et Cosmochimica Acta* **55**, 3223–3240.

Appendix 6

Values of W(u) for various values of u

Table A6.1 Values of $W(u)$ for various values of u.

u		1.0	2.0	3.0	4.0	5.0	6.0	7.0	8.0	9.0
x	1	0.219	0.049	0.013	0.0038	0.0011	0.00036	0.00012	0.000038	0.000012
x	10^{-1}	1.82	1.22	0.91	0.70	0.56	0.45	0.37	0.31	0.26
x	10^{-2}	4.04	3.35	2.96	2.68	2.47	2.30	2.15	2.03	1.92
x	10^{-3}	6.33	5.64	5.23	4.95	4.73	4.54	4.39	4.26	4.14
x	10^{-4}	8.63	7.94	7.53	7.25	7.02	6.84	6.69	6.55	6.44
x	10^{-5}	10.94	10.24	9.84	9.55	9.33	9.14	8.99	8.86	8.74
x	10^{-6}	13.24	12.55	12.14	11.85	11.63	11.45	11.29	11.16	11.04
x	10^{-7}	15.54	14.85	14.44	14.15	13.93	13.75	13.60	13.46	13.34
x	10^{-8}	17.84	17.15	16.74	16.46	16.23	16.05	15.90	15.76	15.65
x	10^{-9}	20.15	19.45	19.05	18.76	18.54	18.35	18.20	18.07	17.95
x	10^{-10}	22.45	21.76	21.35	21.06	20.84	20.66	20.50	20.37	20.25
x	10^{-11}	24.75	24.06	23.65	23.36	23.14	22.96	22.81	22.67	22.55
x	10^{-12}	27.05	26.36	25.96	25.67	25.44	25.26	25.11	24.97	24.86
x	10^{-13}	29.36	28.66	28.26	27.97	27.75	27.56	27.41	27.28	27.16
x	10^{-14}	31.66	30.97	30.56	30.27	30.05	29.87	29.71	29.58	29.46
x	10^{-15}	33.96	33.27	32.86	32.58	32.35	32.17	32.02	31.88	31.76

$W(u)$, the well function,

$$= \int_u^\infty \frac{e^{-u}du}{u} = \left[-0.5772 - \log_e u + u - u^2/2.2! + u^3/3.3! - u^4/4.4! + \cdots \right] \qquad \text{(eq. A6.1)}$$

where

$$u = \frac{r^2 S}{4Tt} \qquad \text{(eq. A6.2)}$$

Hydrogeology: Principles and Practice, Third Edition. Kevin M. Hiscock and Victor F. Bense.
© 2021 John Wiley & Sons Ltd. Published 2021 by John Wiley & Sons Ltd.
Companion website: www.wiley.com/go/hiscock/hydrogeology3e

Appendix 7

Values of *q/Q* and *v/Qt* corresponding to selected values of *t/F* for use in computing the rate and volume of stream depletion by wells and boreholes

Table A7.1 Values of *q/Q* and *v/Qt* corresponding to selected values of *t/F*. (Source: Based on Jenkins, C.T. (1968) Computation of rate and volume of stream depletion by wells. *Techniques of Water-Resources Investigations of the United States Geological Survey, Chapter D1. Book 4, Hydrologic Analysis and Interpretation*. United States Department of the Interior, Washington, 17 pp.)

t/F	*q/Q*	*v/Qt*
0	0	0
0.07	0.008	0.001
0.10	0.025	0.006
0.15	0.068	0.019
0.20	0.114	0.037
0.25	0.157	0.057
0.30	0.197	0.077
0.35	0.232	0.097
0.40	0.264	0.115
0.45	0.292	0.134
0.50	0.317	0.151
0.55	0.340	0.167
0.60	0.361	0.182
0.65	0.380	0.197
0.70	0.398	0.211
0.75	0.414	0.224
0.80	0.429	0.236
0.85	0.443	0.248
0.90	0.456	0.259
0.95	0.468	0.270
1.0	0.480	0.280

(*Continued*)

Hydrogeology: Principles and Practice, Third Edition. Kevin M. Hiscock and Victor F. Bense.
© 2021 John Wiley & Sons Ltd. Published 2021 by John Wiley & Sons Ltd.
Companion website: www.wiley.com/go/hiscock/hydrogeology3e

Table A7.1 (Continued)

t/F	q/Q	v/Qt
1.1	0.500	0.299
1.2	0.519	0.316
1.3	0.535	0.333
1.4	0.550	0.348
1.5	0.564	0.362
1.6	0.576	0.375
1.7	0.588	0.387
1.8	0.598	0.398
1.9	0.608	0.409
2.0	0.617	0.419
2.2	0.634	0.438
2.4	0.648	0.455
2.6	0.661	0.470
2.8	0.673	0.484
3.0	0.683	0.497
3.5	0.705	0.525
4.0	0.724	0.549
4.5	0.739	0.569
5.0	0.752	0.587
5.5	0.763	0.603
6.0	0.773	0.616
7	0.789	0.640
8	0.803	0.659
9	0.814	0.676
10	0.823	0.690
15	0.855	0.740
20	0.874	0.772
30	0.897	0.810
50	0.920	0.850
100	0.944	0.892
600	0.977	0.955

Appendix 8

Complementary error function

Table A8.1 Values of error function, erf (β), and complementary error function, erfc (β), for positive values of β.

β	erf (β)	erfc (β)
0	0	1.0
0.05	0.056372	0.943628
0.1	0.112463	0.887537
0.15	0.167996	0.832004
0.2	0.222703	0.777297
0.25	0.276326	0.723674
0.3	0.328627	0.671373
0.35	0.379382	0.620618
0.4	0.428392	0.571608
0.45	0.475482	0.524518
0.5	0.520500	0.479500
0.55	0.563323	0.436677
0.6	0.603856	0.396144
0.65	0.642029	0.357971
0.7	0.677801	0.322199
0.75	0.711156	0.288844
0.8	0.742101	0.257899
0.85	0.770668	0.229332
0.9	0.796908	0.203092
0.95	0.820891	0.179109
1.0	0.842701	0.157299
1.1	0.880205	0.119795
1.2	0.910314	0.089686

(Continued)

Hydrogeology: Principles and Practice, Third Edition. Kevin M. Hiscock and Victor F. Bense.
© 2021 John Wiley & Sons Ltd. Published 2021 by John Wiley & Sons Ltd.
Companion website: www.wiley.com/go/hiscock/hydrogeology3e

Table A8.1 (Continued)

β	erf (β)	erfc (β)
1.3	0.934008	0.065992
1.4	0.952285	0.047715
1.5	0.966105	0.033895
1.6	0.976348	0.023652
1.7	0.983790	0.016210
1.8	0.989091	0.010909
1.9	0.992790	0.007210
2.0	0.995322	0.004678
2.1	0.997021	0.002979
2.2	0.998137	0.001863
2.3	0.998857	0.001143
2.4	0.999311	0.000689
2.5	0.999593	0.000407
2.6	0.999764	0.000236
2.7	0.999866	0.000134
2.8	0.999925	0.000075
2.9	0.999959	0.000041
3.0	0.999978	0.000022

$$\mathrm{erf}(\beta) = \frac{2}{\pi} \int_0^\beta \mathrm{e}^{-\varepsilon^2} \mathrm{d}\varepsilon$$

$$\mathrm{erf}(-\beta) = -\mathrm{erf}(\beta)$$

$$\mathrm{erfc}(\beta) = 1 - \mathrm{erf}(\beta)$$

$$\mathrm{erfc}(-\beta) = 1 + \mathrm{erf}(\beta)$$

Appendix 9

Drinking water quality standards and Lists I and II substances

Table A9.1 Drinking water quality guidelines and standards. (Sources: Council of European Communities (1998) Directive on the quality of water intended for human consumption (98/83/EC). *Official Journal of the European Communities*, L330. Brussels; United States Environmental Protection Agency (2002) *National Primary and Secondary Drinking Water Regulations*. Report EPA-816-F-02-013. www.epa.gov/safewater/mcl.html; World Health Organization (1996) *Guidelines for Drinking-Water Quality* (2nd edn), volume 2 Health criteria and other supporting information and Addendum to volume 2 (1998). World Health Organization, Geneva. www.who.int/water_sanitation_health/dwq/gdwq3rev/en/.)

Parameter	World Health Organisation guideline value	European Union parametric value	United States Environmental Protection Agency primary and secondary standards	
			NPDWR	
			MCLG	*MCL or TT*
Microbiological parameters				
E. Coli or thermotolerant coliform bacteria	0 in 100 mL sample	0 in 100 mL sample		
Total coliforms[a] (including faecal coliform and *E. coli*)		0 in 100 mL sample	0	No more than 5% of samples total coliform – positive in a month
Enterococci		0 in 100 mL sample		
Cryptosporidium			0	TT
Giardia lamblia			0	TT
Heterotrophic plate count			n/a	TT
Legionella			0	TT
Viruses (enteric)			0	TT
Turbidity[b]	5NTU	n/a	n/a	TT
Disinfection by-products				
Bromate	$0.025 \, \text{mg L}^{-1}$	$0.01 \, \text{mg L}^{-1}$	$0 \, \text{mg L}^{-1}$	$0.010 \, \text{mg L}^{-1}$
Chlorite	0.2		0.8	1.0
Haloacetic acids	<0.1		n/a	0.060

(Continued)

Hydrogeology: Principles and Practice, Third Edition. Kevin M. Hiscock and Victor F. Bense.
© 2021 John Wiley & Sons Ltd. Published 2021 by John Wiley & Sons Ltd.
Companion website: www.wiley.com/go/hiscock/hydrogeology3e

Table A9.1 (Continued)

Parameter	World Health Organisation guideline value	European Union parametric value	United States Environmental Protection Agency primary and secondary standards	
Total trihalomethanes	^{c}	0.1	n/a	0.080
Disinfectants			***MRDLG***	***MRDL***
Chloramines (as Cl_2)	$3\,mg\,L^{-1}$		$4\,mg\,L^{-1}$	$4.0\,mg\,L^{-1}$
Chlorine (as Cl_2)	5		4	4.0
Chlorine dioxide (as ClO_2)	n/a		0.8	0.8
Radionuclides			***MCLG***	***MCL or TT***
Alpha particles	$0.1\,Bq\,L^{-1}$		0	$15\,pCi\,L^{-1}$
Beta particles	$1\,Bq\,L^{-1}$		0	$4\,millirems\,a^{-1}$
Radium-226, -228 (combined)			0	$5\,pCi\,L^{-1}$
Uranium	$2\,(P)\,\mu g\,L^{-1}$		0	$30\,\mu g\,L^{-1}$
Tritium		$100\,Bq\,L^{-1}$		
Total indicative dose		$0.10\,mSv\,a^{-1}$		
Pesticides			***MCLG***	***MCL or TT***
Alachlor	$20\,\mu g\,L^{-1}$	$0.1\,\mu g\,L^{-1}$ for an individual pesticide (except where indicated); $0.5\,\mu g\,L^{-1}$ for the sum of all individual pesticides	$0\,\mu g\,L^{-1}$	$2\,\mu g\,L^{-1}$
Aldicarb	10			
Aldrin/dieldrin	0.03			
Atrazine	2		3	3
Bentazone	300			
Carbofluran	7		40	40
Chlordane	0.2		0	2
Chlorotoluron	30			
Cyanazine	0.6			
DDT	2			
1,2-dibromo-3-chloropropane	1		0	0.2
1,2-dibromoethane	0.4–15 (P)			
2,4-dichlorophenoxyacetic acid (2,4-D)	30		70	70
1,2-dichloropropane (1,2-DCP)	40 (P)		0	5
1,3-dichloropropane	NAD			
1,3-dichlorpropene	20			
Diquat	10 (P)		20	20
Heptachlor and heptachlorepoxide	0.03	0.03	0	0.4 and 0.2
Hexachlorobenzene	1		0	1
Isoproturon	9			

Table A9.1 (Continued)

Parameter	World Health Organisation guideline value	European Union parametric value	United States Environmental Protection Agency primary and secondary standards	
Lindane	2		0.2	0.2
MCPA	2			
Methoxychlor	20		40	40
Metolachlor	10			
Molinate	6			
Pendimethalin	20			
Pentachlorophenol	9 (P)		0	1
Permethrin	20			
Propanil	20			
Pryidate	100			
Simazine	2		4	4
Terbuthylazine (TBA)	7			
Trifluralin	20			
2,4-DB	90			
Dichlorprop	100			
Fenoprop	9			
MCPB	NAD			
Mecoprop	10			
2,4,5-T	9			
Organic constituents			*MCLG*	*MCL or TT*
Acrylamide	0.5 µg L^{-1}	0.10 µg L^{-1}	0 µg L^{-1}	TT
Benzene	10	1.0	0	5 µg L^{-1}
Benzo (a) pyrene (PAHs)	0.7	0.10 (sum of specified PAHs)	0	0.2
Carbon tetrachloride	2		0	5
Chlorobenzene			100	100
Dalapon			200	200
o-Dichlorobenzene	1000		600	600
p-Dichlorobenzene	300		75	75
1,2-Dichloroethane	30	3.0	0	5
1,1-Dichloroethylene	30		7	7
cis-1,2-Dichloroethylene	50		70	70
trans-1,2-Dichloroethylene			100	100
Dichloromethane	20		0	5
Di(2-ethylhexyl) adipate	80		400	400
Di(2-ethylhexyl) phthalate	8		0	6

(*Continued*)

Table A9.1 (Continued)

Parameter	World Health Organisation guideline value	European Union parametric value	United States Environmental Protection Agency primary and secondary standards	
Dinoseb			7	7
Dioxin (2,3,7,8-TCDD)			0	0.00003
Endothall			100	100
Endrin			2	2
Epichlorohydrin		0.10	0	TT
Ethylbenzene	300		700	700
Ethylene dibromide			0	0.05
Fluoranthene	U			
Glyphosate	U		700	700
Hexachlorocyclopentadiene			50	50
Microcystin-LR	1(P)			
Oxamyl (Vydate)			200	200
Polychlorinated biphenyls (PCBs)			0	0.5
Picloram			500	500
Styrene	20		100	100
Tetrachloroethene	40	10	0	5
Toluene	700		1000	1000
Toxaphene			0	3
2,4,5-TP (Silvex)			50	50
1,2,4-Trichlorobenzene	20 (total)		70	70
1,1,1-Trichloroethane	2000 (P)		200	200
1,1,2-Trichloroethane			3	5
Trichloroethene	70 (P)	10	0	5
Vinyl chloride	5	0.5	0	2
Xylenes (total)	500		10000	10000
Inorganic constituents			*MCLG*	*MCL or TT*
Antimony	0.005 (P) mg L^{-1}	0.005 mg L^{-1}	0.006 mg L^{-1}	0.006 mg L^{-1}
Arsenic	0.01 (P)	0.01	0	0.01
Asbestos (fibre > 10 μm)	U		7 MFL	7 MFL
Barium	0.7		2	2
Beryllium	NAD		0.004	0.004
Boron	0.05 (P)	1.0		
Cadmium	0.003	0.005	0.005	0.005
Chromium (total)	0.05 (P)	0.05	0.1	0.1
Copper	2 (P)	2	1.3	TT (action level = 1.3)

Table A9.1 (Continued)

Parameter	World Health Organisation guideline value	European Union parametric value	United States Environmental Protection Agency primary and secondary standards	
Cyanide (as free cyanide)	0.07	0.05	0.2	0.2
Fluoride	1.5	1.5	4.0	4.0
Lead	0.01	0.01	0	TT (action level = 0.015)
Molybdenum	0.07			
Mercury	0.001	0.001	0.002	0.002
Nickel	0.02 (P)	0.02		
Nitrate (as N)	11.3 (acute)	11.3	10	10
Nitrite (as N)	0.91 (acute), 0.06 (P) (chronic)	0.15	1	1
Selenium	0.01	0.01	0.05	0.05
Thallium			0.0005	0.002
Indicator parameters			***NSDWR***	
Aluminium	$0.2 \, mg \, L^{-1}$	$0.2 \, mg \, L^{-1}$	$0.05–0.2 \, mg \, L^{-1}$	
Ammonium	1.5	0.5		
Chloride	250	250	250	
Clostridium perfringens		0 in 100 mL		
Colour	15 TCU	n/a/c	15 colour units	
Conductivity		$2500 \, \mu S \, cm^{-1}$ at 20°C		
Dissolved oxygen	n/a			
Hardness	n/a			
Hydrogen sulphide	$0.05 \, mg \, L^{-1}$			
Hydrogen ion concentration (pH)	n/a	≥ 6.5 and ≤ 9.5	6.5–8.5	
Iron	$0.3 \, mg \, L^{-1}$	$0.2 \, mg \, L^{-1}$	$0.3 \, mg \, L^{-1}$	
Manganese	0.5 (P)	0.05	0.05	
Odour	n/a	n/a/c	3 threshold odour number	
Oxidizability		$5.0 \, mg \, L^{-1} \, O_2$		
Silver	U		$0.1 \, mg \, L^{-1}$	
Sodium	$200 \, mg \, L^{-1}$	$200 \, mg \, L^{-1}$		
Sulphate	250	250	250	

(Continued)

Table A9.1 (Continued)

Parameter	World Health Organisation guideline value	European Union parametric value	United States Environmental Protection Agency primary and secondary standards
Taste	n/a	n/a/c	
Temperature	n/a		
Tin	U		
Colony count 22°		n/a/c	
Total organic carbon		n/a/c	
Total dissolved solids	1000 mg L^{-1}		500 mg L^{-1}
Zinc	3		5

Notes and abbreviations: [a] Coliforms are naturally present in the environment and are not a health threat in itself; the parameter is used to indicate whether other potentially harmful bacteria are present.
[b] Turbidity is a measure of cloudiness of water that is used to indicate water quality and filtration effectiveness.
[c] The sum of the ratio of the concentration of each to its respective guideline value should not exceed 1.
n/a   Not applicable (should be acceptable).
n/a/c   No abnormal change.
MCL   Maximum contaminant level: the highest level of a contaminant that is allowed in drinking water. MCLs are set close to MCLGs as feasible using the best available treatment technology and taking cost into consideration. MCLs are enforceable standards.
MCLG   Maximum contaminant level goal: the level of a contaminant in drinking water below which there is no known or expected risk to health. MCLGs allow for a margin of safety and are non-enforceable public health goals.
MFL   ×10^6 fibres L^{-1}.
MRDL   Maximum residual disinfectant level: the highest level of a disinfectant allowed in drinking water. There is evidence that addition of a disinfectant is necessary for control of microbial contaminants.
MRDLG   Maximum residual disinfectant level goal: the level of a drinking water disinfectant below which there is no known or expected risk to health. MRDLGs do not reflect the benefits of the use of disinfectants to control microbial contaminants.
NAD   No adequate data to permit recommendation of a health-based guideline value.
NPDWR   National Primary Drinking Water Regulations (or primary standards): legally enforceable standards applied to public water systems. Primary standards protect public health by limiting the levels of contaminants in drinking water.
NSDWR   National Secondary Drinking Water Regulations: non-enforceable guidelines regulating contaminants that may cause cosmetic effects (such as skin or tooth discolouration) or aesthetic effects (such as taste, odour or colour) in drinking water.
NTU   Nephelometric turbidity unit.
(P)   Provisional guideline value.
TCU   Trust colour unit.
TT   Treatment technique: a required process intended to reduce the level of a contaminant in drinking water.
U   Not hazardous to human health at concentrations normally found in drinking water.

Table A9.2 List I and II substances as defined by EC Groundwater Directive (80/68/EEC). (Sources: Council of European Communities (1980) Directive on the protection of groundwater against pollution caused by certain dangerous substances (80/68/EEC). *Official Journal of the European Communities*, L20. Brussels; Environment Agency (1998) *Policy and Practice for the Protection of Groundwater*. Environment Agency, Bristol.)

List I of families and groups of substances

These substances should be prevented from being discharged into groundwater.

List I contains the individual substances which belong to the families and groups of substances specified below, with the exception of those which are considered inappropriate to List I on the basis of a low risk toxicity, persistence and bioaccumulation.

Such substances which with regard to toxicity, persistence and bioaccumulation are appropriate to List II are to be classed in List II.

1) Organohalogen compounds and substances which may form such compounds in the aquatic environment.
2) Organophosphorus compounds.
3) Organotin compounds.
4) Substances which possess carcinogenic, mutagenic or teratogenic properties in or via the aquatic environment[a].
5) Mercury and its compounds.
6) Cadmium and its compounds.
7) Mineral oils and hydrocarbons.
8) Cyanides.

List II of families and groups of substances

Discharges of these substances into groundwater should be minimized.

List II contains the individual substances and the categories of substances belonging to the families and groups of substances listed below which could have a harmful effect on groundwater.

1) The following metalloids and metals and their compounds:

Zinc	Tin
Copper	Barium
Nickel	Beryllium
Chrome	Boron
Lead	Uranium
Selenium	Vanadium
Arsenic	Cobalt
Antimony	Thallium
Molybdenum	Tellurium
Titanium	Silver

2) Biocides and their derivatives not appearing in List I.
3) Substances which have a deleterious effect on the taste and/or odour of groundwater and compounds liable to cause the formation of such substances in such water and to render it unfit for human consumption.
4) Toxic or persistent organic compounds of silicon and substances which may cause the formation of such compounds in water, excluding those which are biologically harmless or are rapidly converted in water into harmless substances.
5) Inorganic compounds of phosphorus and elemental phosphorus.
6) Fluorides.
7) Ammonia and nitrates.

[a] Where certain substances in List II are carcinogenic, mutagenic or teratogenic they are included in Category 4 of List I

Appendix 10

Review questions and exercises

Questions

A10.1   Hydrological cycle

1) Explain the relative size of the store of groundwater in the global hydrological cycle and its significance in terms of water and material fluxes to the world's oceans.

2) Given that the average annual fluxes of precipitation and evaporation over the global land surface are 0.110×10^6 and 0.073×10^6 km^3, respectively, and that the storage of water in rivers and shallow groundwater (above a depth of $\sim$750 m) is 0.0017×10^6 and 4.2×10^6 km^3, respectively, answer the following:

   a) What is the balance of average annual precipitation and evaporation (the effective precipitation or total runoff) from the global land surface?

   b) If the total runoff calculated in (a) is partitioned between surface runoff (94%) and groundwater runoff (6%), estimate the residence (flushing) times of water held in storage in rivers and shallow groundwater.

   c) Why is the figure obtained in (b) for groundwater residence time only likely to give a very rough estimate?

   d) An estimate of the direct flux of groundwater to the oceans is 2×10^3 km^3 a^{-1}. Assuming an average total dissolved solids content for groundwater of 585 mg L^{-1}, estimate and comment on the annual material flux to the oceans from groundwater.

A10.2   Physical hydrogeology, groundwater potential and Darcy's law

1) a) Define intrinsic permeability and explain how it differs from hydraulic conductivity.

   b) Explain why typical hydraulic conductivity (K) values of the four following aquifer rock types vary by nine orders of magnitude: igneous rocks, 10^{-12} m s^{-1}; sandstone, 10^{-8} m s^{-1}; fractured limestone, 10^{-5} m s^{-1}; and karst limestone, 10^{-3} m s^{-1}.

   c) A 50-m thick unconfined aquifer has a transmissivity (T) of 8640 m^2 day^{-1}. Calculate the hydraulic conductivity and state the likely lithology using the typical K values listed in part (b) as a guide.

   d) Describe, using appropriate terminology, the likely spatial properties of hydraulic conductivity in a 30-m thick aquifer with the following lithological description: bedded, poorly sorted sandstones and conglomerates that dip at 20°; beds typically 1 m thick; and bedding separated by 5-cm thick, iron-cemented layers.

Hydrogeology: Principles and Practice, Third Edition. Kevin M. Hiscock and Victor F. Bense.
© 2021 John Wiley & Sons Ltd. Published 2021 by John Wiley & Sons Ltd.
Companion website: www.wiley.com/go/hiscock/hydrogeology3e

2) Write concise definitions of the following, briefly explaining the significance of each in hydrogeology:

   a) Porosity;
   b) Effective porosity;
   c) Voids ratio;
   d) Specific yield;
   e) Storativity;
   f) Hydraulic conductivity;
   g) Transmissivity;
   h) Reynolds number.

3) Discuss how topographical and geological setting, both in terms of the variety of rock types and geological structure, influence aquifer conditions and groundwater flow regimes.

4) For an unconfined aquifer condition in arid, frigid and humid climatic settings, discuss how groundwater flow and surface drainage networks are connected.

5) From first principles, derive the following expression for the groundwater potential at any point in a porous material:

   $$\Phi = gh$$

   where Φ is groundwater potential, g is acceleration due to gravity and h is hydraulic head measured above an arbitrary datum. Also show that the hydraulic head at a point is equal to the sum of the pressure head, ψ, and the elevation head, z.

6) A cross-section through a layered aquifer-aquitard system is shown in Fig. A10.1.

The hydraulic gradient of the potentiometric surface across the sandy gravel unit is 0.002 and across the sand unit is 0.004. The sandy gravel and sand have effective porosity values of 25 and 20%, respectively. Both the sandy gravel and sand are 100 m thick. Using the information presented, and for a given aquifer width of 3500 m and uniform groundwater flow rate of 6500 m^3 day^{-1}, calculate the following terms for both the sandy gravel and sand deposits:

   a) Hydraulic conductivity in units of m s^{-1};
   b) Transmissivity in units of m^2 s^{-1};
   c) Average linear velocity in units of m s^{-1}.

7) A technique for estimating surface-water interaction through streambed sediments is by installation of piezometers and applying Darcian principles. Another method is by installation of direct seepage meters on the streambed. Both methods were used in a field campaign at one location that included measurement of hydraulic head in a piezometer nest consisting of two tubes (1 and 2), as given in the following table, with all elevations relative to a reference datum:

Piezometer nest	Elevation of base of tube (cm)	Elevation of water level in tube (cm)
Tube 1	30.6	35.8
Tube 2	33.7	35.1

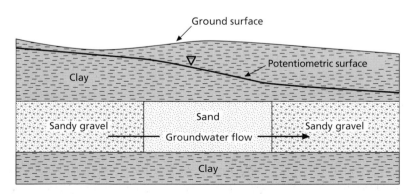

Ground surface

Potentiometric surface

Clay

Sand

Sandy gravel

Groundwater flow

Sandy gravel

Clay

Fig. A10.1 Cross-section through a layered aquifer-aquitard system comprising sand, sandy gravel and clay.

Additionally, a seepage metre installed on the stream bed at the same field location covers an area of 30×30 cm and has a volume of 10 L. The field data showed that the seepage metre filled completely from empty in 4 hours.

a) Draw a sketch to demonstrate that the direction of groundwater seepage in the streambed is upwards.

b) From the set of field data, estimate the hydraulic conductivity of the streambed sediments at this location in units of $m\,s^{-1}$.

c) If the porosity of the stream bed sediments is 30%, what is the groundwater seepage rate in the streambed in units of $m\,s^{-1}$?

8) The following field notes were taken at a nest of piezometers installed in close proximity to each other at a site:

	Piezometer		
	a	*b*	*c*
Elevation at surface (m above sea level)	350	350	350
Depth of piezometer (m)	115	80	40
Depth to water (m)	37	28	21

If *A*, *B*, *C* refer to the points of measurement at the tips of piezometers *a*, *b*, *c*, calculate the following:

a) The hydraulic head at *A*, *B* and *C* (m);

b) The pressure head at *A*, *B* and *C* (m);

c) The elevation head at *A*, *B* and *C* (m);

d) The hydraulic gradients between *A* and *B*, and *B* and *C*.

Is the piezometer nest in a recharge area or a discharge area?

9) Define Darcy's law and discuss its applicability to groundwater flow in porous materials.

10) Give a brief classification of voids in rocks and sediments and describe the relationships between the voids, the hydraulics of groundwater flow and the bulk flow patterns in the following:

a) unconsolidated alluvial sediments;

b) jointed porous sandstone;

c) fractured granitic rock;

d) karstic limestone.

11) A sand filter is enclosed within a cylinder 2 m long and with a 50 cm internal diameter. The cylinder is inclined such that the axis of the upper end lies 75 cm above the axis of the lower end. Water enters the apparatus at one end and flows out of the other end. Sensors monitor the pressure as the water enters and leaves the apparatus. The sensors are also separated by a 75 cm elevation difference.

If the lower pressure reading is 120.5 kPa and the upper reading is 117.6 kPa, is the water flowing up or down through the cylinder? Assume a density of water of 1 g cm^{-3}, so that one metre of water exerts a pressure of 9.81 kPa.

If the apparatus operates at a temperature of 20°C and the specific discharge was measured as $4.0 \times 10^{-5}\,m\,s^{-1}$, answer the following:

a) What is the hydraulic conductivity of the sand?

b) What is the intrinsic permeability of the sand?

c) What volume of water could the apparatus transmit in a day?

d) If the operating temperature were to be increased to 40°C, what pressure differential would be required to keep the flow rate equal to that at 20°C?

Values of the density and viscosity of water at different temperatures are given in Appendix 2, Table A2.1.

12) As part of a site investigation of the area shown in Fig. A10.2, 13 boreholes have been drilled. The area is underlain by a bed of sand above which is a low permeability clay. Figure A10.2 shows structure contours of the contact between the sand and this clay. Three of the boreholes were deep enough to prove the existence of a second low permeability clay beneath the sand. A single-borehole tracer test

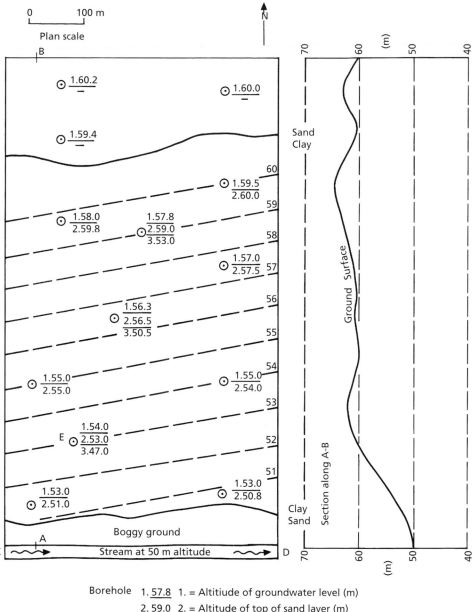

0   100 m

Plan scale

B

1.60.2

1.60.0

1.59.4

Sand
Clay

60

1.59.5
2.60.0

59

1.58.0
2.59.8

1.57.8
2.59.0
3.53.0

58

1.57.0
2.57.5 57

56

1.56.3
2.56.5
3.50.5

55

54

1.55.0
2.55.0

1.55.0
2.54.0

53

1.54.0
E 2.53.0
3.47.0

52

51

1.53.0
2.51.0

1.53.0
2.50.8

Clay
Sand

Boggy ground

A

C   Stream at 50 m altitude   D

70   60   (m)   50   40

Ground Surface

Section along A-B

70   60   50   40
(m)

Borehole 1. <u>57.8</u> 1. = Altitiude of groundwater level (m)
2. <u>59.0</u> 2. = Altitude of top of sand layer (m)
3. <u>53.0</u> 3. = Altitude of base of sand layer (m)

— <u>53</u> — Stucture contour on clay - sand contact with altitude in m

Fig. A10.2   Site investigation area of a sand aquifer.

conducted at borehole E gave a value of hydraulic conductivity for the sand of 1 m day^{-1}.

Prepare a short explanation of the hydrogeology of the area using the above information. Illustrate your account by first completing a copy of the map showing contours of the groundwater level and then the cross-section showing geological structure and the potentiometric surface.

Indicate the parts of the area in which the groundwater is recharged to and discharged from the ground surface, and the unconfined and confined parts of the sand aquifer. Estimate the transmissivity of the sand and the discharge of groundwater from the mapped area to the length of stream C–D.

It has been proposed to drain the boggy ground adjacent to the stream by means of ditches. Comment on the likely success or failure of this approach, giving your reasons.

13) Figure A10.3 shows (a) longitudinal and (b) transverse sections of an alluvial valley formed in a low-permeability bedrock. The average temperature of groundwater in the valley aquifer is 20°C. Continuous recharge to the upstream section of the

valley creates uniform, steady groundwater flow in a downgradient direction. The alluvium has an intrinsic permeability of 30 darcies and an effective porosity of 0.16. If the static water level elevation in Well 1 is 83.4 m relative to datum, and in Well 2, 82.9 m relative to datum, calculate the following:

a) The groundwater discharge (in m^3 day^{-1}) across the width of the aquifer.

b) Relative to your answer in part (a), what is the percentage change in groundwater discharge if the average groundwater temperature were decreased by 50%?

c) For an average groundwater temperature of 20°C, and stating your assumptions, how long would it take for a

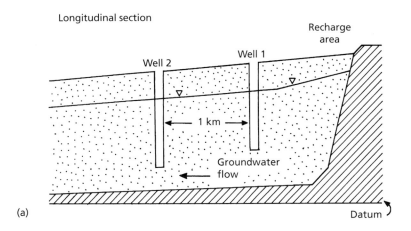

Longitudinal section

(a)

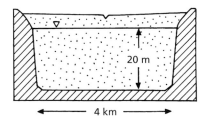

Transverse section

(b)

Fig. A10.3 Longitudinal and transverse sections of an alluvial valley. (a) Longitudinal section. (b) Transverse section.

contaminant introduced in Well 1 to reach Well 2?

Note: the gravitational acceleration, $g = 9.81 \text{ m s}^{-2}$. Values of the density and viscosity of water at different temperatures are given in Table A2.1.

14) Figure A10.4 shows a generalized section along the direction of flow in a confined sandstone aquifer with shale confining beds. The topographic scarp between A and B lies parallel to the strike of the aquifer. Springs and seepages occur frequently towards the base of the scarp in a zone which is parallel with the strike of the aquifer. The marshy pools lie on the surface trace of a fault which is also parallel to the strike.

Consider a strip of aquifer in the plane of the diagram. If the hydraulic conductivity of the sandstone is 200 m day^{-1}, find the following:

a) the discharge per unit width of the spring zone, S;

b) the discharge per unit width of the marshy pools zone;

c) the transmissivity of the fault plane beneath the spring zone;

d) the transmissivity of the fault plane beneath the marshy pools.

On a copy of Fig. A10.4, sketch the flow pattern of the aquifer. You can assume that the springs and marshy pools are the only

outlet for groundwater. Write short notes with your answers stating the assumptions you are making and justifying your calculations.

15) a) A vertical fissure in a limestone aquifer is 2 m high and 0.1 m wide. A tracer moves along the fissure at a velocity of 100 m hour^{-1}. Is the flow laminar or turbulent?

b) A well 0.5 m in diameter in a confined sand aquifer 5 m thick is pumped at 50 L s^{-1}. If the sand has a median grain size of 1.25 mm, estimate the Reynolds number at the following distances from the centre of the well: 0.25 m (i.e. the outer boundary of the well screen), 0.5, 1.0, 2.0, 4.0 and 10.0 m.

c) Using your answers obtained in part (b), plot a graph of Reynolds number against radius. Within what radius of the well would you expect non-Darcian flow to occur and briefly explain why?

16) Describe how and why changes in storage occur in confined and unconfined aquifers in response to both natural and artificial causes.

17) The Salt River Valley in central Arizona is an alluvial valley forming an unconfined aquifer covering an area of about 100 000 ha. The alluvial aquifer is heavily pumped but receives little replenishment

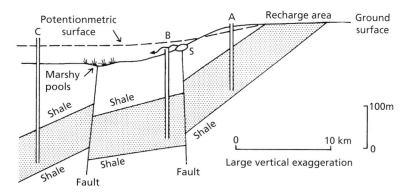

Fig. A10.4 Section through a sandstone aquifer with shale confining beds.

from groundwater recharge. Groundwater abstractions total about $500 \times 10^6 \ m^3 \ a^{-1}$ with an associated water table decline of about $3 \ m \ a^{-1}$. Assuming no groundwater recharge, estimate the specific yield of the alluvial aquifer?

18) Explain the basic principles of steady groundwater flow in porous materials starting from a consideration of fluid potential. Demonstrate how variations in hydraulic conductivity between different geological formations lead to groundwater flow lines being refracted at their boundaries. Discuss how such refraction patterns can be used to classify layered systems of geological formations into aquifers and aquitards, and describe the basis for classifying aquifers into confined, semi-confined (leaky) and unconfined.

19 Figure A10.5 is a flow net of seepage under a dam through an isotropic sandstone with a porosity of 20%. A conservative tracer injected at P takes 5 days to reach Q.

a) What are the assumptions used to construct the flow net?

b) On a copy of Fig. A10.5, sketch and label piezometric levels in the nested piezometers at A, B and C?

c) What is the hydraulic conductivity of the sandstone?

d) Leaky drums of hazardous aqueous waste were dumped in the reservoir at sites D and E. How long will it take for pollution to seep out on the downgradient side of the dam from each of the two sites?

e) Why may the actual breakthrough times of the pollutant differ from those predicted?

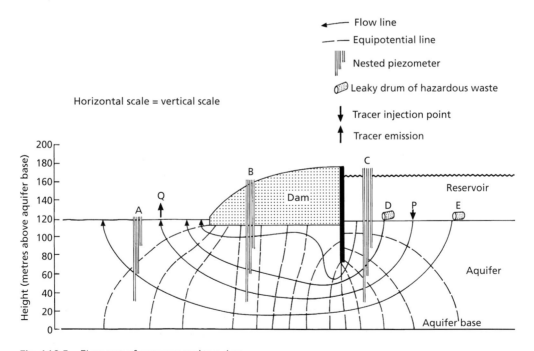

Fig. A10.5 Flow net of seepage under a dam.

A10.3 Chemical hydrogeology

1) Describe three graphical methods of hydrochemical data presentation and define the term *hydrochemical facies*, explaining the value of this concept in determining groundwater flow patterns.

2) Describe the classic sequence of Chebotarev (1955) to explain the chemical evolution of natural waters along a groundwater flowpath. Illustrate your answer with diagrams relating to at least one specific example.

3) Compare and contrast the carbonate chemistry of groundwaters evolving under open-system and closed-system conditions with respect to dissolved CO_2.

4) Figure A10.6 shows the geographical location and topography of Kuwait. Fresh and brackish groundwater fields are shown together with a new area known as the Al-Wafra well-field. The stratigraphic sequence of Kuwait is divided into clastic sediments known as the Kuwait Group and the underlying carbonate sediments of the Hasa Group. The Dammam aquifer occurs in the uppermost part of the Hasa

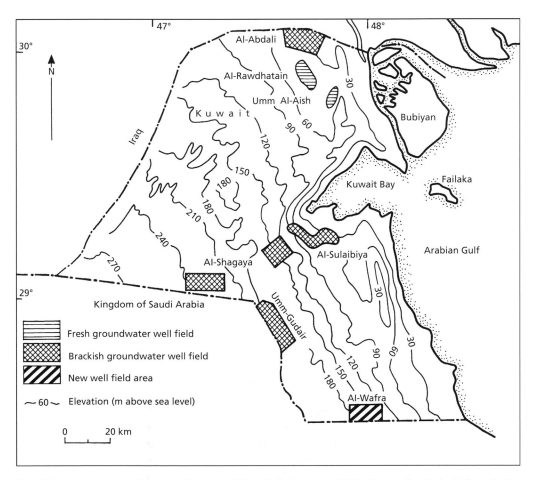

Fig. A10.6 Location and topographic map of Kuwait (Saleh *et al.* 1999). (*Source:* Saleh, A., Al-Ruwaih, F. and Shehata, M. (1999) Hydrogeochemical processes operating within the main aquifers of Kuwait. *Journal of Arid Environments* **42**, 195–209.)

Group and consists of 200 m of porous lime-stones. The top of the aquifer is marked by the presence of a hard, siliceous layer and shale horizon. Table A10.1 is a summary of the mean and standard deviation of hydrochemical data for the Dammam lime-stone aquifer.

a) From inspection of the concentration data given in Table A10.1, assign a hydrochemical water type to the groundwater in the Dammam limestone aquifer?

b) What reasons might explain the high Ca^{2+}, Na^+, HCO_3^- and SO_4^{2-} concentrations in the aquifer?

c) Calculate logarithmic values of saturation indices for calcite ($CaCO_3$) and halite (NaCl) using the mean concentration data, and predict whether these two minerals have reached chemical equilibrium in the Dammam limestone groundwater.

Note: Thermodynamic equilibrium constants in pure water at 25°C and 1 bar total pressure: $K_{HCO3-} = 10^{-10.33}$; $K_{calcite} = 10^{-8.48}$; and $K_{halite} = 10^{0.773}$.

5) Consider the following two redox half-reactions involving the reduction of nitrate (NO_3^-) to gaseous nitrogen (N_2) and sulfate (SO_4^{2-}) to bisulfide (HS^-):

$$\frac{1}{5}NO_3^- + \frac{6}{5}H^+ + e^- = \frac{1}{10}N_{2(g)} + \frac{3}{5}H_2O$$

$$\frac{1}{8}SO_4^{2-} + \frac{9}{8}H^+ + e^- = \frac{1}{8}HS^- + \frac{1}{2}H_2O$$

A groundwater with nitrate and sulphate concentrations of 2×10^{-3} and 1×10^{-3} mol L^{-1}, respectively, and a pH of 8 experiences denitrification followed by sulphate reduction. The reactions produce a partial pressure of dissolved nitrogen of 1×10^{-3} bar and a bisulphide concentration of 1×10^{-3} mol L^{-1}. For an ideal system, which can be described by reversible thermodynamics, calculate the *pe* and *Eh* values produced by each of these redox half-reactions. Assume a temperature of 25°C and that the activities of the dissolved species are equal to their concentrations.

If, in the process of denitrification, ferrous (Fe^{2+}) iron is oxidized to ferric (Fe^{3+}) iron (eq. 4.29 for $pe^o = +13.0$), calculate the

Table A10.1 Kuwait Dammam limestone aquifer hydrochemical data.

Parameter	Concentration	
	Mean	Standard deviation
TDS	5980	1123
pH	7.4	0.2
K^+	56	9.7
Na^+	1180	305.2
Ca^{2+}	540	80.7
Mg^{2+}	164	22.4
HCO_3^-	173	7.1
Cl^-	2365	614.2
SO_4^{2-}	1032	65.6

Note: Concentration units in mg L^{-1} except pH.
Source: Saleh, A., Al-Ruwaih, F. and Shehata, M. (1999) Hydrogeochemical processes operating within the main aquifers of Kuwait. *Journal of Arid Environments* **42**, 195–209. © 1999, Elsevier.

amount of ferric iron produced by this redox reaction in an aquifer containing 1×10^{-5} mol L^{-1} of ferrous iron.

Note:

$$pe = pe^o - \frac{1}{n} \log_{10} \frac{[\text{reductants}]}{[\text{oxidants}]}$$

where pe is equal to $-\log_{10}[e^-]$ and describes the electron activity; pe^o is equal to $\log_{10} K$, where K is an equilibrium constant; and n is the number of electrons involved in the redox half-reaction. Values of pe^o are given in Table 4.14. $Eh = \frac{0.059}{n} pe$ at 25°C.

6) Fresh groundwater in coastal areas is typically dominated by Ca^{2+} and HCO_3^- ions, as a result of calcite dissolution. Consequently, cation exchangers in the aquifer mostly have Ca^{2+} adsorbed on their surfaces. In seawater Na^+ and Cl^- are the dominant ions, and sediment in contact with seawater will adsorb Na^+. Given this information, write two equations that illustrate the expected concentrations of Na^+ and Ca^{2+} resulting from simple mixing and cation exchange between freshwater and seawater when: (a) seawater intrudes a coastal freshwater aquifer; and (b) freshwater flushes a saline aquifer.

The hydrochemical data in Table A10.2 are for samples of freshwater, seawater and a mixed CaCl$_2$ groundwater obtained for a coastal aquifer experiencing seawater intrusion. Using these data, calculate the quantities of Na^+ and Ca^{2+} involved in cation exchange during seawater intrusion.

Briefly discuss your results, stating any assumptions you make in your calculations.

7) Figure A10.7 shows a location map of the Fife and Kinross area of Scotland. Northern Fife is underlain by Devonian Old Red Sandstone and southern Fife by Carboniferous rocks. To the north of the Old Red Sandstone outcrop are the Ochil Hills which consist of low permeability volcanic rocks. To the south, a roughly east-west trending belt through Central Fife is formed of sandstone with conglomerate, including the Knox Pulpit Formation, the most productive aquifer in the Old Red Sandstone (capable of borehole yields in excess of 30 L s^{-1}). The Knox Pulpit Formation is a fine to medium-grained and weakly cemented sandstone. There is also a narrow coastal strip of sandstone faulted against the volcanic rocks to the north of the Ochil Hills and west of easting 30.

The Carboniferous rocks of southern Fife are a complex sequence of lithologies interspersed with intrusive and extrusive igneous rocks. In east Fife there are sandstones and subordinate shales, thin limestones, seatearths and coal. In central Fife there are sandstones, mudstones, subordinate limestones and coals. In south and west Fife, sandstones, fireclays and some coals are present with a single basal limestone. All these strata contain some groundwater, and there is often hydraulic continuity across poorly permeable horizons via cracks, joints and old workings. Borehole yields up to 5 L s^{-1} are not uncommon.

Table A10.2 Hydrochemical data for mixing of freshwater and seawater.

Ion	Seawater	Mixed CaCl$_2$ groundwater	Freshwater
Na^+	485.0	374.8	0
Ca^{2+}	10.7	10.0	3.0
Cl^-	566.0	440.0	0

All concentrations in mmol L^{-1}.

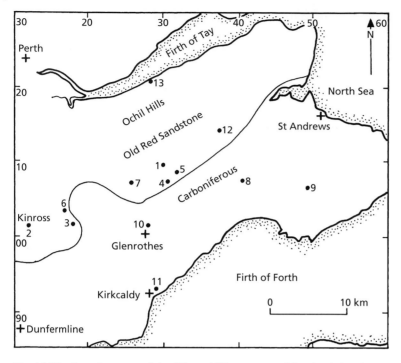

Fig. A10.7 Location map of the Fife and Kinross area of Scotland showing sampling sites (Robins 1986). (Robins, N.S. (1986) Groundwater chemistry of the main aquifers in Scotland. *British Geological Survey Report* **18**, 2. © 1986, British Geological Survey.)

Table A10.3 details the analytical hydrochemical data for 13 samples collected from the sites shown in Fig. A10.7. Using these data, complete a copy of the tri-linear (Piper) diagram given in Fig. A10.8 for the 13 samples given in Table A10.3.

Based on your Piper diagram, and using all the available data, provide both a description and explanation of the hydrochemical characteristics of groundwaters found in Fife and Kinross.

A10.4 Environmental isotope hydrogeology

1) Explain the meaning of the age of groundwater and indicate the range of ages encountered in porous geological materials.

2) Discuss the application of ^{14}C and ^{3}H radiometric dating methods in hydrogeological investigations. Pay particular attention to basic principles, the appropriate methods required to correct for 'dead' carbon (^{14}C method) and the limitations of each technique.

3) Discuss how the stable isotope ratios $^{18}O/^{16}O$ and $^{2}H/^{1}H$ and noble gases (Ne, Ar, Kr and Xe) can be used to characterize the recharge history and subsequent evolution of groundwaters. Illustrate your answer with appropriate examples.

4) Llandrindod Wells in mid-Wales (see Box 4.2) was an important spa town in Victorian times. The mineralized spring waters, of varying composition, all lie in close proximity to each other in the River Irthon valley in an area known as the Rock Park and centred on the Rock Park Fault. The geology

Table A10.3 Hydrochemical data for Fife and Kinross. (Robins 1986). (*Source:* Robins, N.S. (1986) Groundwater chemistry of the main aquifers in Scotland. *British Geological Survey Report* **18**, 2.)

Sample number	Geological formation	Borehole depth (m)/spring	E.C. ($\mu S\,cm^{-1}$)	Temp (°C)	pH	Eh (mV)	Ca^{2+}	Mg^{2+}	Na^+	K^+	HCO_3^-	SO_4^{2-}	Cl^-	$NO_3^- - N$	Diss. O_2	Total Fe	Tritium (TU)
							($mg\,L^{-1}$)										
1	ORS	13	510	8.5	7.7	+390	29	30	16	0.4	113	39	35	18	8.0	<0.3	nd
2	ORS	100	508	9.0	7.4	+350	46	24	16	2.3	205	23	35	4	6.0	<0.3	nd
3	ORS	Spring	329	9.0	7.6	+345	37	16	8	1.1	157	16	14	3	10.0	<0.3	nd
4	ORS	80	828	9.5	7.7	+400	53	39	24	2.0	187	53	41	12	3.0	<0.3	27.6
5	ORS	70	495	9.0	7.7	+400	49	25	18	1.7	173	46	42	9	6.1	<0.3	nd
6	ORS	120	343	8.0	7.2	+370	48	12	7	2.4	124	22	16	8	11.8	<0.3	nd
7	ORS	75	283	8.5	7.6	+390	42	11	7	1.6	137	19	12	2	sat	<0.3	19.6
8	Carb	40	865	8.5	6.8	−90	100	60	13	4.1	323	203	26	3	0	1.12	nd
9	Carb	69	635	8.0	6.6	+270	76	30	17	1.5	234	113	20	1	0	<0.3	nd
10	Carb	50	325	10.0	7.7	+110	45	19	12	2.4	194	59	14	<0.1	1.0	<0.3	nd
11	Carb	130	1560	13.0	6.6	+210	105	107	47	12.5	591	374	42	<0.1	0	4.90	nd
12	ORS	60	965	9.0	7.3	+320	61	43	48	3.7	274	71	83	0.2	6.2	<0.1	6.0
13	ORS	86	1040	10.0	7.4	+130	115	15	72	6.9	210	308	76	0.1	3.9	0.10	nd

Note: ORS = Old Red Sandstone, Carb = Carboniferous; nd = not determined; sat = saturated

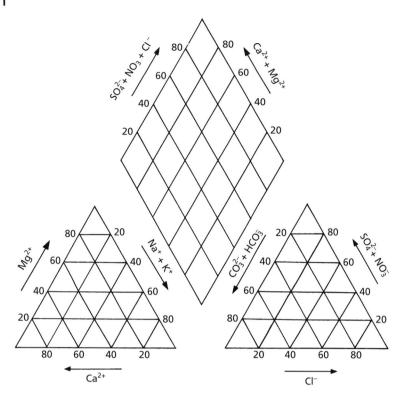

Fig. A10.8 Tri-linear (Piper) diagram.

of both the immediately local and recharge areas is dominated by the Builth Volcanics, a low porosity sequence that is extensively fractured and faulted. The relief of the area is moderate with the springs cropping out at 190 m above sea level, and the surrounding hills rising to 300 m in the immediate vicinity and 400 m at distances of 10 km in all directions. A schematic cross-section is given in Fig. A10.9.

The hydrogeochemistry ($\delta^{18}O$, $\delta^{2}H$, Na^{+} and Cl^{-}) of selected springs, local precipitation and surface waters are given in Table A10.4. Using these data, present a model that accounts for the origin and subsequent modification of the spring water compositions and the relationships between them. Use a copy of Fig. A10.9 to provide a conceptual model for the evolution of the mineral spring waters.

5) As part of a hydrogeological investigation of a porous, fissured limestone aquifer both stable isotope and radioisotope data have been collected at three sites (Table A10.5). Site 1 is located on the aquifer outcrop and Sites 2 and 3 are located downgradient in the confined section of the aquifer. Assuming a $\delta^{13}C_{lithic}$ value of +2‰ and a $\delta^{13}C_{biogenic}$ value of −26‰, and using the information in Table A10.5, calculate:
   a) the corrected groundwater ages at the three sites;
   b) an indicative groundwater velocity in the aquifer.

Comment on the groundwater age dates and the palaeohydrogeology of the limestone aquifer as indicated by the stable isotope ($\delta^{18}O$ and $\delta^{2}H$) data.

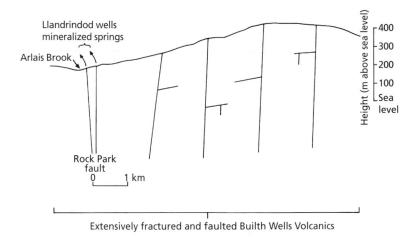

Fig. A10.9 Schematic cross-section of the Builth Volcanics, Llandrindod Wells.

Table A10.4 Stable isotope and hydrochemical data for Llandrindod Wells (Merrin 1996). (*Source:* Merrin, P. D. (1996) *A geochemical investigation into the nature and occurrence of groundwaters from the Llandrindod Wells Ordovician Inlier, Wales.* MSc Thesis, University of East Anglia, Norwich.)

Spring	$\delta^{18}O$ (‰VSMOW)	δ^2H (‰VSMOW)	Na^+ (mg L^{-1})	Cl^- (mg L^{-1})
Chalybeate spring	−8.15	−54	1565	3890
Saline spring	−7.85	−51.5	1105	2950
Sulphur spring	−7.20	−47	490	1270
Magnesium spring	−8.10	−53.5	183	310
Eye Well 1	−6.50	−47.2	1325	3670
Eye Well 2	−5.45	−43	1530	4220
Eye Well 3	−6.65	−44	62	130
Arlais Brook (surface stream)	−6.80	−43	80	140
Local precipitation	−6.70	−43	2	4
Seawater	0	0	10 500	19 000

Table A10.5 Stable isotope and radioisotope data for a limestone aquifer.

Site	Distance from outcrop (km)	Tritium (TU)	^{14}C (pmc)	$\delta^{13}C$ (‰PDB)	$\delta^{18}O$ (‰VSMOW)	δ^2H (‰VSMOW)
1	0	57	60.8	−13.2	−7.1	−46.0
2	15.1	4	4.1	−4.5	−7.4	−48.0
3	16.3	0	2.2	−0.9	−7.9	−51.0

A10.5 Stream gauging, infiltration measurements and groundwater recharge estimation

1) Discuss the range of applications of, and operational constraints on, various methods that are available for measuring stream discharge.
2) Provide a description of the physical storage and movement of water in the soil zone and explain why an understanding of infiltration is important in hydrology.
3) Why is knowledge of groundwater recharge important in the estimation of catchment water resources? Describe three methods for calculating average annual groundwater recharge, highlighting the advantages and disadvantages of each.
4) Write concise definitions for the following hydrological parameters and describe methods for the determination of the parameters listed as (a), (b) and (c):
   a) Volumetric moisture content;
   b) Tension head;
   c) Infiltration capacity;
   d) Field capacity.

5) As part of a groundwater investigation, the discharge of a karst spring was measured in a channel a short distance downstream of its point of emergence. The following data were collected at a well-maintained, horizontal, sharp-crested, thin-plate weir:
Width of crest, $L = 1.67$ m;
Number of side constrictions, $n = 2$;
Ventilation of falling water $=$ free;
Head above crest of weir, measured on a stage board calibrated in centimetres and positioned in the approach pool, $H = 0.12$ m;
Velocity of approach $=$ low.
   Determine the spring discharge, Q, in the channel from these data, using the following stage–discharge relationship:

$$Q = 1.84(L - 0.1\,nH)H^{3/2}$$

Downstream of the weir, the data given in Table A10.6 were collected using a well-calibrated current meter in a straight, uniform part of the channel with a roughly rectangular cross-section. Calculate the discharge in the channel from these measurements, showing your working.

Table A10.6 Current metering data.

Object measured	Distance of object from left bank (m)	Current metre depth below surface (m)	Water velocity (m s^{-1})
Current meter	0.12	0.30	0.015
"	0.38	0.31	0.035
"	0.62	0.32	0.08
"	0.88	0.31	0.12
"	1.12	0.30	0.23
"	1.38	0.30	0.25
"	1.62	0.32	0.15
"	1.88	0.29	0.10
"	2.12	0.28	0.07
"	2.38	0.26	0.04
Right bank	2.55	–	–

Briefly discuss the magnitudes of error involved in these two sets of measurements, and determine whether the discrepancy between them is likely to be due to random or systematic error.

6) The accuracy of a stage-discharge calibration at a current metered gauging station is being checked by dilution gauging.

A slug of 15 kg of rhodamine WT tracer was injected into the river and samples were taken 1 km downstream. No tributaries entered the river over the reach. Given the results in Table A10.7, calculate the discharge for comparison with the value obtained from the rating curve ($110\,\mathrm{m^3\,s^{-1}}$).

What could account for the discrepancy between the two discharge values?

7) A tracer is injected continuously into a stream from a Mariotte vessel at the top of a long, straight reach with a constant rectangular cross section with average water depth of 0.3 m, average width of 2.4 m and

slope of 0.001. The tracer solution has a concentration of $1000\,\mathrm{mg\,L^{-1}}$ and the injection rate is $1.00\,\mathrm{mL\,s^{-1}}$. Three hundred metres downstream, the results shown in Table A10.8 are obtained from water samples taken in the centre of the stream and at its banks.

a) Find the discharge in $\mathrm{m^3\,s^{-1}}$ using the constant rate injection method.

b) Find the discharge in $\mathrm{m^3\,s^{-1}}$ when the depth of water is 0.5 m using the Manning formula, assuming the Manning's n value remains constant for the stream.

8) The initial rate of infiltration in a catchment is estimated as 53 mm hour^{-1}, the final rate is 5.1 mm hour^{-1}, and the time constant, k, is 0.4 hour^{-1}. Use Horton's equation to find:

a) The infiltration rate at $t = 2$ hours and $t = 6$ hours.

b) The total volume of infiltration over the 6-hour period.

Table A10.7 Dilution gauging data.

Time (hours)	0	1	2	3	4	5	6	7	8
Concentration (ppb)	0	1	5	10	7	5	3	1	0

Table A10.8 Results from a constant rate injection dilution-gauging test to evaluate stream discharge.

Time after commencement of injection (minutes)	Concentration in centre ($\mu g\,L^{-1}$)	Concentration at banks ($\mu g\,L^{-1}$) Left	Right
0	0	–	–
5	0	–	–
10	0	–	–
15	2.0	–	–
20	4.5	–	–
25	5.0	4.95	5.05
30	5.0	5.00	5.00

9) Two small catchments of similar area drain an upland region. One catchment is covered by grassland developed on a calcareous till deposit, and the other is covered by forest developed on acidic soil. A field experiment, conducted during the wet season, has been completed to determine the component of total (peak) stream discharge (Q_T) contributed by storm runoff (event water, Q_E), and baseflow (pre-event water, Q_P) in each catchment. With reference to the chemical and stable isotope data given in Table A10.9, and applying a two-component mixing model, rearrange the following equation to express the ratio Q_P/Q_T in terms of the total discharge, event and pre-event stream water tracer concentrations (C_T, C_E and C_P, respectively):

$$C_T Q_T = C_P Q_P + C_E Q_E$$

and answer the following:

a) If the total stream discharge measured at the outlets from the grassland and forested catchments during a rainfall event are 2.5 and 1.8 $m^3 s^{-1}$, respectively, calculate values for the baseflow discharge components using, in turn, the chloride and $\delta^{18}O$ measurements given in Table A10.9. What proportion of each total flow is contributed by storm runoff?

b) From the results obtained in part (a), comment on any observed differences in the hydrograph separations for the two catchments.

10) Table A10.10 contains information on volumetric moisture content and soil water potential recorded beneath a plot of ground of area 1 m^2 by means of a neutron probe and tensiometers. The soil is a homogeneous, structureless clay soil with a cover of short-rooting vegetation and is 90 cm in depth.

a) Plot the data provided on arithmetic graph paper and, having identified the position of the zero flux plane (ZFP), calculate the evaporation and drainage losses from the profile during the 7-day period separating the readings, assuming no rain fell during the period. Express your results in millimetres of water.

b) For the depth range 30–40 cm, estimate the unsaturated hydraulic conductivity value for the clay soil on 8 July 2003.

c) In the event of rainfall, what would be the effect of prior intense drying of the soil upon the accuracy of your method for calculating the drainage loss?

11) Table A10.11 gives the available rainfall and potential evapotranspiration data for a catchment with an area of 200 km^2. The catchment is in a remote area with

Table A10.9 Two-component mixing model data.

Water type	Cl^- (mg L^{-1})	$\delta^{18}O$ (‰VSMOW)
Rain event	2.0	−6.00
Grassland stream		
Total discharge	6.0	−7.00
Pre-event water	10.0	−8.00
Forest stream		
Total discharge	6.0	−7.00
Pre-event water	7.0	−7.25

Table A10.10   Clay soil volumetric moisture content and soil water potential measurements during July 2003.

Depth below ground (cm)	Volumetric moisture content, θ		Soil water potential, Φ^* ($\times 10^3$ cm of water)	
	1.7.03	8.7.03	1.7.03	8.7.03
-5	0.356	0.349	-4.00	-4.20
-15	0.375	0.363	-3.50	-3.75
-25	0.394	0.378	-3.00	-3.40
-35	0.418	0.390	-2.30	-2.75
-45	0.428	0.390	-1.50	-2.00
-55	0.413	0.390	–	–
-70	0.399	0.387	–	–
-85	0.385	0.378	–	–

* Soil water potential data not recorded below 45 cm depth.

Table A10.11   Rainfall (P) and potential evapotranspiration (PE) data.

	J	F	M	A	M	J	J	A	S	O	N	D
2002												
P (mm)	65	69	34	14	29	66	38	66	128	73	142	39
PE (mm)	17	21	37	90	110	105	112	88	54	31	16	20
2003												
P (mm)	90	32	78	43	73	23	32	18	128	28	71	34
PE (mm)	16	17	33	60	84	124	122	102	54	24	8	7

currently no abstraction from the underlying unconfined sand and gravel aquifer. Land use in the catchment is predominantly grassland. A single observation borehole in the catchment recorded an increase in the level of the water table of 2.75 m during the winter recharge period 2002–2003.

a) Assuming a root constant for short-rooted vegetation of 75 mm, use the Penman–Grindley method to calculate the groundwater recharge (hydrological excess water, HXS) expected for the period 2002–2003. Start your calculation on 1 April 2002 when the soils in the catchment were at field capacity (i.e. soil moisture deficit, $SMD = 0$).

b) If the recharge calculated in (a) entirely replenishes the sand and gravel aquifer, calculate a storage coefficient value for the aquifer and estimate the change in groundwater storage volume that occurred over the catchment.

c) A river draining the catchment has an average annual flow of 1.843 m^3 s^{-1}. As a measure of the baseflow index of this river, calculate the fraction of the 2002–2003 groundwater recharge relative to the average annual flow.

d) In a fuller assessment of the catchment water balance, what other considerations would you include in estimating the safe yield of future groundwater abstractions?

12) Agricultural production in the coastal plain of North Africa is supported by drip-feed irrigation. As part of a new development, an area of 26 000 ha is to be irrigated using groundwater pumped from an inland desert area from an unconfined aquifer with a recharge area of 200 km². From the information below, and using the chloride budget method, evaluate the sustainable supply of water available for agricultural use in units of litres per hectare of agricultural land per day (L ha⁻¹ day⁻¹).

Notes:

Mean annual precipitation: 120.1 mm a⁻¹;
Mean concentration of chloride in rainwater: 4.33 mg L⁻¹;
Mean concentration of chloride in interstitial pore water in the unsaturated zone: 8.00 mg L⁻¹.

If the unsaturated zone in the recharge area is 78 m thick, estimate the residence time of water in the unsaturated zone, if the average moisture content is 0.65%.

A10.6   Groundwater resources, pumping tests and stream depletion analysis

1) How can an analysis of river and well hydrographs assist in the assessment of catchment water resources?

2) Describe the repercussions of aquifer overexploitation and discuss how different methods of aquifer management can address the impacts.

3) You are instructed to prepare a report on the groundwater resources of an area comprising (a) folded limestones overlain in part by (b) horizontal thick basaltic lava flows. The area is cut by a mature drainage system with (c) sandy point-bar and silty mud overbank deposits. What type of aquifers might you expect to occur in the area? What type of information would you expect to find in a hydrogeological report on the area which would aid you in this task?

4) Explain the principles by which the transmissivity and storativity of an aquifer can be determined by the Theis method of pumping test analysis. In your answer, describe how changes in storage occur in confined and unconfined aquifers.

5) A borehole penetrates a 25 m thick, confined aquifer and is pumped at a steady rate of 0.2 m³ s⁻¹ until a steady-state cone of depression is formed. At this time, two observation boreholes, 350 and 1000 m away, have water levels of 29.50 and 30.84 m above the top of the aquifer, respectively. Assuming that the initial piezometric surface was horizontal, estimate the transmissivity and hydraulic conductivity of the aquifer.

6) In developing a new unconfined borehole source in a Permo-Triassic sandstone aquifer, a water company has undertaken a long-term equilibrium pumping test. The new source was pumped at 1×10^4 m³ day⁻¹. Groundwater elevations measured in two observation boreholes within the cone of depression were: 321.96 m at a distance of 50 m from the source; and 325.00 m at a distance of 1100 m from the source. Both elevations were measured relative to the horizontal impermeable base of the aquifer as shown in Fig. 7.13b.

a) Calculate a value for the hydraulic conductivity, K, of the sandstone aquifer.

b) Calculate the intrinsic permeability, k_i, of the sandstone assuming that the temperature of groundwater in the aquifer is 15°C. During the development of the borehole, a core was cut completely through the aquifer. Intrinsic permeability measurements on representative samples of core material gave an average intrinsic permeability value of 60 millidarcies.

c) Comment on the comparison between the average intrinsic permeability determined for the core material and the intrinsic permeability of the sandstone calculated from the pumping test data found in part (b).

Note: at 15°C the density, ρ, of water is 999.099 kg m^{-3} and its viscosity, μ, is 1.1404×10^{-3} kg m^{-1} s^{-1}. Acceleration due to gravity, g, is 9.81 m s^{-2}. 1 darcy = 9.87×10^{-13} m^2.

7) Describe the essential procedures for carrying out a two-well, constant rate, time-variant pumping test in the field. Show how the measurements that are necessary relate to the requirements and assumptions of using the Theis formula for interpreting the results.

The data in Table A10.12 were obtained during a short pumping test of a confined Chalk aquifer at a site in Norfolk, England. Figure A10.10 shows geological logs of the two boreholes used. The distance from the abstraction borehole to the observation borehole, r, is 218 m and the abstraction borehole discharge, Q, was 295 m^3 hour^{-1}. Use the Theis method to determine the transmissivity and storativity of the Chalk aquifer at this site. To plot the field curve you will need a sheet of log 3 cycles by 4 cycles graph paper. A type curve for copying is provided as Fig. 7.14.

8) A river support borehole abstracts groundwater from a lower, confined limestone aquifer which is successively overlain by a low permeability clay aquitard and an upper, unconfined limestone aquifer. A site plan (a) and generalized cross-section (b) are shown in Fig. A10.11. The support

Table A10.12 Constant discharge pumping test data.

Time since start of pumping (minutes)	Drawdown (m)
7	0.15
7.5	0.16
8	0.17
8.5	0.19
9	0.21
9.5	0.23
10	0.25
11	0.30
12	0.36
13	0.42
14	0.48
15	0.55
16	0.62
17	0.68
18	0.75
19	0.82
20	0.89
22	1.03
24	1.16
26	1.29
28	1.42
30	1.54
35	1.84
40	2.10
44.5	2.32

borehole was test pumped at a constant discharge of 2000 m^3 day^{-1}, with the groundwater fed into the adjacent river. Two observation boreholes, situated at a distance of 165 m from the support borehole, and positioned in the lower limestone aquifer (borehole A) and upper limestone aquifer (borehole B), produced the drawdown data given in Table A10.13.

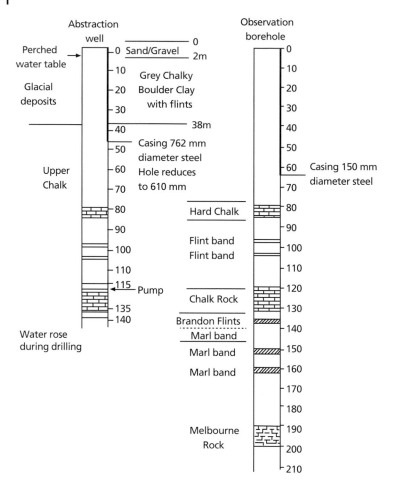

Fig. A10.10 Geological logs of two chalk boreholes at Hargham, Norfolk, England.

a) Using the Cooper–Jacob method of pumping test analysis, and the drawdown versus time data for observation borehole A (Table A10.13), calculate the transmissivity and storativity values for the confined limestone aquifer.

b) The record of drawdown in observation borehole B shows that the unconfined limestone aquifer experienced a rise in water level during the period of the pumping test. Provide an explanation of the possible cause of this response at borehole B.

9) The water supply to a large town is normally supplied by a surface storage reservoir. However, during dry summer months, the reservoir is unreliable, and it is planned to switch the water supply to a new wellfield capable of pumping 5×10^4 m^3 day^{-1} from a highly productive sandstone aquifer. The sandstone is unconfined and has a transmissivity of 1500 m^2 day^{-1} and a specific yield of 0.10. The wellfield is to be situated in the floodplain of a major river that is in direct hydraulic contact with the aquifer. To avoid damage to the river ecosystem, it

Site plan

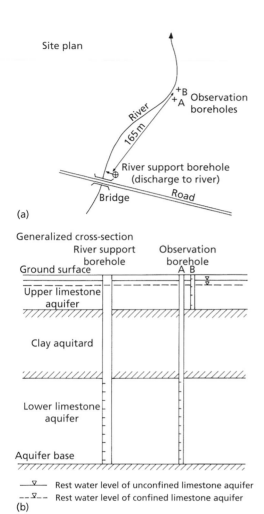

(a)

Generalized cross-section

(b)

——▽—— Rest water level of unconfined limestone aquifer
--▽-- Rest water level of confined limestone aquifer

Fig. A10.11 Site plan and generalized cross-section for a layered limestone aquifer system.

is essential that the river flow is not reduced by more than 2×10^4 m^3 day^{-1} as a result of leakage caused by groundwater abstraction. The wellfield is to be operated for 90 days each summer.

Using this information and the table of values provided in Appendix 7, calculate a safe distance for the position of the centre of the wellfield from the river in order to avoid the effects of stream depletion, explaining the assumptions and limitations of your answer.

Table A10.13 Pumping test data for Hullavington river support borehole, Wiltshire, England.

Time since start of test (seconds)	Drawdown* (m)	
	Borehole A	Borehole B
0	0	0
120	0.315	0
240	0.555	0
360	0.774	−0.005
480	0.953	−0.007
600	1.105	−0.009
720	1.245	−0.010
840	1.370	−0.011
960	1.485	−0.015
1080	1.595	−0.015
1200	1.693	−0.017
1500	1.915	−0.020
1800	2.115	−0.020
2100	2.295	−0.025
2400	2.495	−0.029
2700	2.615	−0.030
3000	2.755	−0.030
3300	2.895	−0.031
3600	3.015	−0.032
3900	3.145	−0.033
4200	3.255	−0.035
4500	3.365	−0.035
4800	3.485	−0.036
5100	3.585	−0.040
5400	3.685	−0.040

* A negative drawdown indicates a rise in water level.

A10.7 Contaminant hydrogeology

1) In the context of contaminant hydrogeology, provide brief definitions of the following terms:
   a) Advection;
   b) Hydrodynamic dispersion;
   c) Retardation;
   d) Sorption isotherm.

2) Provide a definition of the hydrodynamic dispersion coefficient, D_l, where l denotes the longitudinal direction, and explain the meaning of the terms given in the following equation:

$$D_l = \alpha_l \bar{v} + D^*$$

3) Describe how the lithological characteristics of different rock types can affect the transport and attenuation of dissolved, reactive contaminant species.

4) Describe the main features and patterns of groundwater contamination around point sources of pollutants, such as landfill sites, wastewater lagoons, or spillage sites. What general recommendations for landfill site selection and management would you advise to limit such pollution?

5) Choosing either septic tanks or municipal landfill sites, discuss the importance of redox processes in the development of pollution plumes that may occur down-gradient of the point of groundwater contamination.

6) Provide, with illustrations, a general description of the physico-chemical processes that influence the migration of hydrocarbons and organic solvents in porous materials.

7) Explain the hydrophobic sorption model in predicting the attenuation of organic compounds in saturated, porous materials, outlining the applications and limitations of the model.

8) Outline the principal industrial and agricultural activities that present a threat to groundwater quality. In your answer, describe the nature of each contaminant source (whether point or diffuse pollution) and the potential for natural attenuation once contamination has occurred.

9) Devise a strategy for the protection of groundwater resources from diffuse sources of agricultural pollution explaining the benefits and disadvantages of your approach.

10) Define what is meant by the *risk* of groundwater pollution. What action would you take to protect groundwater resources from the risk of groundwater pollution in (a) an industrialized country, and (b) an agrarian, developing country?

11) A former industrial site is to be developed as a retail and entertainment complex known as Waterside. The site is at a distance of 3 km from a river, and there is concern that 'hotspots' of contaminated groundwater in the underlying sand aquifer will migrate towards the river if left undisturbed. However, commercial pressures are such that groundwater remediation at the site will only be considered if a contaminant concentration of greater than or equal to 1% of the original source concentration is likely to reach the river within the 30-year lifespan of the Waterside development. Stating your assumptions, use the following site information to determine whether you would recommend remediation.

Hydraulic gradient between Waterside and river $= 1 \times 10^{-4}$

Hydraulic conductivity of sand aquifer $= 100 \, \text{m day}^{-1}$

Effective porosity of sand aquifer $= 0.115$

Longitudinal dispersivity, α_l, of sand aquifer $= 100 \, \text{m}$

Note: Values of the complementary error function, erfc(β), for positive values of β are given in Appendix 8.

12) Figure A10.12 shows a cross-section through a former sand and gravel quarry that is now used as a landfill for the co-disposal of municipal and industrial wastes. The site is covered daily with low-permeability clay and leachate develops within the waste. Although the site has an impermeable base, horizontal movement of leachate through the sand and gravel aquifer has been induced by groundwater

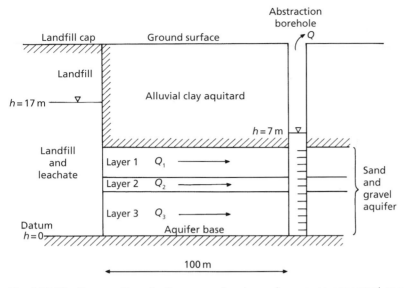

Fig. A10.12 Cross-section of a former sand and gravel quarry now operated as a landfill site.

abstraction from a borehole located at a distance of 100 m from the site. The borehole completely penetrates the aquifer. The aquifer comprises three layers, each with homogeneous and isotropic hydraulic properties. Table A10.14 details the aquifer characteristics of the three layers.

Analysis of the leachate reveals that the chlorinated solvent trichloroethene (TCE) is present as a dissolved phase. Assuming steady-state groundwater flow conditions and a partition coefficient between organic carbon and water, K_{oc}, for TCE of 150 cm^3 g^{-1}, find the following:

a) a distribution coefficient, K_d, and retardation factor, R_d, for each of the three aquifer layers;
b) the breakthrough time for TCE in each layer at the abstraction borehole;
c) the resulting concentration breakthrough curve for TCE arriving at the borehole;
d) the breakthrough time and number of pore volume flushes of the layered aquifer that have occurred once the concentration of TCE in the borehole is half of the original concentration in the leachate (i.e. $C/C_o = 0.5$).

Table A10.14 Layered sand and gravel aquifer characteristics.

	Hydraulic conductivity, K (m day^{-1})	Layer thickness, d (m)	Porosity, n	Bulk density, ρ_b (g cm^{-3})	Fraction of organic carbon, f_{oc}
Layer 1 (sand)	10	2	0.20	1.60	0.015
Layer 2 (gravel)	50	1	0.40	1.50	0.015
Layer 3 (fine sand)	2	3	0.15	1.65	0.015

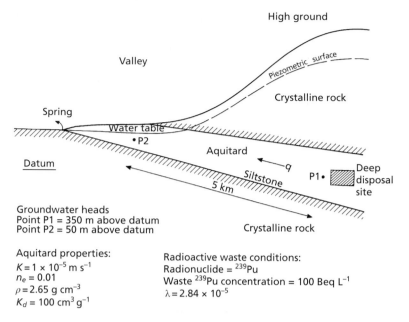

High ground

Valley

Piezometric surface

Crystalline rock

Spring

Water table

• P2

Datum

Aquitard

Siltstone

5 km

P1 •

Deep disposal site

Crystalline rock

Groundwater heads
Point P1 = 350 m above datum
Point P2 = 50 m above datum

Aquitard properties:
$K = 1 \times 10^{-5}$ m s^{-1}
$n_e = 0.01$
$\rho = 2.65$ g cm^{-3}
$K_d = 100$ cm^3 g^{-1}

Radioactive waste conditions:
Radionuclide = ^{239}Pu
Waste ^{239}Pu concentration = 100 Beq L^{-1}
$\lambda = 2.84 \times 10^{-5}$

Fig. A10.13 Cross-section through a proposed deep disposal, radioactive waste site.

13) A site has been located in an aquitard that is suitable for the deep disposal of radioactive waste. Figure A10.13 shows the position of the engineered disposal site within a low permeability siltstone layer confined between two very low permeability layers composed of crystalline rock. The siltstone has been subjected to minor fracturing that imparts a small secondary porosity and hydraulic conductivity of 0.01 and 1×10^{-5} m s^{-1}, respectively. A large groundwater head is developed in the high topography above the position of the disposal site such that there is the potential for groundwater flow in the siltstone layer to migrate to the ground surface where the aquitard reaches outcrop.

   Residents in the adjacent valley fear that radioactive plutonium, ^{239}Pu, will contaminate local groundwater supplies for future generations. Using the information given in Fig. A10.13, assess the likelihood that a measurable concentration of ^{239}Pu will reach the position of P2 close to the shallow water table in the valley. Comment on any assumptions you make in your calculation and whether you regard your answer as a conservative estimate or not.

14) A truck containing saline wastewater has spilled its load above an unconfined gravel aquifer pumped for public water supply from a borehole. The wastewater contains high concentrations of sodium chloride (NaCl). The distance from the spill to the borehole is 240 m. Previous testing of the borehole has provided a value for the hydraulic conductivity of the gravel of 120 m day^{-1}. The hydraulic gradient between the spill and borehole under pumping conditions is 0.005. The effective porosity of the gravel, which also contains some sand and clay, is 0.25.

   a) Calculate the time for the spill to reach the pumped borehole once the wastewater has entered the water table. State the assumptions used in your calculations.

b) Comment on whether the dissolved sodium (Na^+) and chloride (Cl^-) contained in the wastewater will arrive at the borehole at the same time.

15) A leak of the dry cleaning fluid perchlororethene (PCE) has occurred below a dry cleaning shop and has entered the water table in the underlying unconfined Chalk aquifer. A pumped water supply borehole is situated 15 km down-gradient of the dry cleaning shop and is at risk of contamination from the leak.

a) Using the information below, determine how long it will take for the plume of PCE solvent to reach the borehole.

b) What might cause the PCE to arrive at the borehole sooner than calculated in part (a)?

The hydraulic properties of the Chalk aquifer are hydraulic conductivity, $K = 50 \, m \, day^{-1}$; effective porosity, $n_e = 0.05$; and water table hydraulic gradient, $dh/dl = 0.015$. The retardation factor, R_d, for perchloroethene $= 3.3$, where $R_d = t_c/t_w$ or the ratio of the contaminant to groundwater travel times.

16) A railroad tanker with a cargo of trichloroethene (TCE) has derailed, spilling 35 000 L of the solvent directly on an unconfined alluvial sand aquifer. The solvent has entered the water table and a contaminant plume of dissolved TCE has developed in the direction of groundwater flow. The groundwater flow is uniform and horizontal under a hydraulic gradient of 0.005 and discharges to a river 400 m from the site of the derailment. The aquifer has a hydraulic conductivity of $2 \times 10^{-3} \, m \, s^{-1}$, an effective porosity of 0.25, an organic carbon content of 0.5% and a bulk density of $2.10 \, g \, cm^{-3}$. Ignoring dispersion of the dissolved contaminant in the alluvial sand aquifer, answer the following, explaining any assumptions you make:

a) What is the maximum concentration of dissolved TCE expected in the aquifer?

b) What is the average linear groundwater velocity in the sand aquifer?

c) Estimate the average velocity of the contaminated groundwater and the time taken for dissolved TCE to arrive at the river.

Notes:
a) Solubility of TCE at $20°C = 1 \times 10^3 \, mg \, L^{-1}$
b) $\log_{10}K_{OW}$ for TCE at $25°C = 2.42$
c) $K_d = K_{OC}.f_{OC}$
d) $\log_{10}K_{OC} = 0.49 + 0.72 \, \log_{10}K_{OW}$ (Schwarzenbach and Westall 1981 equation)

17) Figure A10.14 shows (a) the plan view and (b) a generalized cross-section centred on a Bottling Plant taking water from a spring that issues from a limestone aquifer at the junction with an underlying clay aquitard. The limestone has a transmissivity of $5000 \, m^2 \, day^{-1}$ and an average thickness of 20 m. On average, the Bottling Plant abstracts $2.5 \times 10^3 \, m^3$ of water each day from the spring for process use. In addition, and under the abstraction licence agreement, the Bottling Plant must leave $0.5 \times 10^3 \, m^3 \, day^{-1}$ of spring water to compensate the downstream water course. Recently, the spring water has shown increasing concentrations of ammonium and biological oxygen demand, causing fear that the source is vulnerable to contamination from one or more of the three dairy farms located upgradient from the spring. The regional groundwater hydraulic gradient is 6×10^{-4}, oriented in a north to south direction.

Using the capture zone type curves given in Fig. A10.15, and stating your assumptions, determine which, if any, of the three dairy farms may be posing a threat to the security of the Bottling Plant.

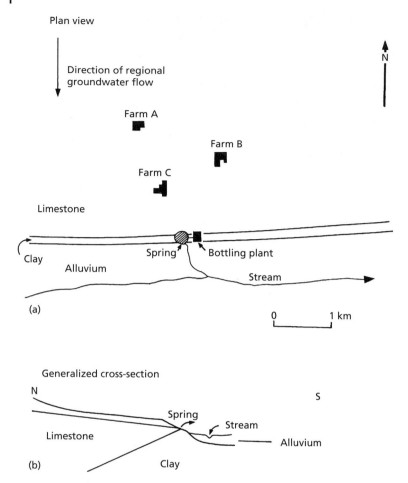

Fig. A10.14 Plan view and generalized cross-section showing the location of a bottling plant and associated spring.

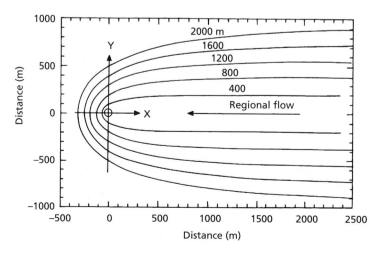

Fig. A10.15 Type curves showing the capture zones of a single pumping well located at point (0, 0) for various values of Q/bq_a.

Worked answers to exercises

The following worked answers apply to the problems provided in this Appendix 10. Answers to the review questions are not given since these can be understood by reading material contained in the textbook and extended by further reading using the list of references supplied. Where appropriate, references to sections of the book where further information can be found in answering some parts of the numerical problems are indicated.

A10.1   Hydrological cycle

2) a) The balance of average annual precipitation and evaporation, or total runoff, from the global land surface is:

$$(0.110 - 0.073) \times 10^6 = 0.037 \times 10^6 \ km^3 \ a^{-1}$$

b) Of the total runoff, 94% or $0.0348 \times 10^6 \ km^3 \ a^{-1}$ is surface runoff and 6% or $0.0022 \times 10^6 \ km^3 \ a^{-1}$ is groundwater discharge.

 The storage of water in either rivers or the ground divided by the runoff/discharge rate provides an estimate of flushing time as follows:

 For rivers: $(0.0017 \times 10^6)/(0.0348 \times 10^6) = 0.0489$ years or 17.8 days

 For groundwater: $(4.2 \times 10^6)/(0.0022 \times 10^6) = 1909$ or about 2000 years

c) The answer in (b) is only very approximate given the expected range of groundwater residence times from a few days in karst aquifer systems to in excess of 10 000 s years in low-permeability strata and large, continental-scale aquifer systems, for example the Great Artesian Basin of Australia.

d) The annual dissolved material flux to the oceans from groundwater is estimated to be equal to $(2 \times 10^3 \times 10^9 \times 10^3)$ L $\times$ (585×10^{-9}) t L^{-1} $= 1.17 \times 10^9$ t a^{-1} and is roughly equivalent to half of the dissolved mineral flux contributed by rivers to the world's oceans.

A10.2   Physical hydrogeology, groundwater potential and Darcy's law

1) a) The intrinsic permeability, k_i, is related to the size of the openings through which the fluid moves and is an intrinsic property of the rock or sediment 'skeleton'. Generally:

$$k_i = Cd^2$$

 where d is the mean pore diameter. C represents a dimensionless 'shape factor' representing the shape of the pore openings, as influenced by the relationship between the pore and grain sizes, d, and their effect on the tortuosity of fluid flow.

b) In igneous rock, values relate only to sparse fractures. In sandstones, there are some fractures and potentially primary porosity. In limestones, there is fracture porosity plus primary porosity and in karst limestones, fracture conduits are enlarged by solution.

c) Converting for the number of seconds in a day (8640/86 400) gives a transmissivity, T, of 0.1 m^2 s^{-1}. Dividing by 50 m (aquifer thickness) gives a hydraulic conductivity, K, of 2×10^{-3} m s^{-1}, in the size range for karst limestone.

d) Aspects of aquifer properties to consider are the following: poorly sorted sandstones are likely to be heterogeneous; the bedding defined by iron-cemented layers and the dip will impart strong anisotropy; and compartmentalisation can be caused by iron-rich layers being aquitards.

6) a) *Sandy gravel unit*

dh/dl (hydraulic gradient) = 0.002, b (thickness) = 100 m, Q (groundwater flow rate) = 6500 m³ day⁻¹, aquifer width = 3500 m, n (effective porosity) = 0.25. Cross-sectional area of flow = (3500 × 100) m²

From Darcy's law, $Q = -AiK$

Rearranging, $K = Q/Ai$

Therefore, the hydraulic conductivity,

$K = 6500/[(3500 \times 100) \times 0.002]$

Hence, $K = 9.29$ m day⁻¹ (or 1.07×10^{-4} m s⁻¹)

Sand unit

dh/dl (hydraulic gradient) = 0.004, b (thickness) = 100 m, Q (groundwater flow rate) = 6500 m³ day⁻¹, aquifer width = 3500 m, n (effective porosity) = 0.25. Cross-sectional area of flow = (3500 × 100) m²

From Darcy's law, $Q = -AiK$

Rearranging, $K = Q/Ai$

Therefore, the hydraulic conductivity,

$K = 6500/[(3500 \times 100) \times 0.004]$

Hence, $K = 4.64$ m day⁻¹ (or 5.37×10^{-5} m s⁻¹)

b) *Sandy gravel unit*

Transmissivity, $T = Kd$

Therefore, using the value of hydraulic calculated above, $T = 1.07 \times 10^{-4} \times 100$ m² s⁻¹

Hence, $T = 1.07 \times 10^{-2}$ m² s⁻¹ (or 924.48 m² day⁻¹)

Sand unit

Transmissivity, $T = Kd$

Therefore, using the value of hydraulic calculated above, $T = 5.37 \times 10^{-5} \times 100$ m² s⁻¹

Hence, $T = 5.37 \times 10^{-3}$ m² s⁻¹ (or 463.97 m² day⁻¹)

c) *Sandy gravel unit*

From Darcy's law, the average linear groundwater velocity, $\bar{v} = q/n$, where the specific discharge, $q = -iK$

Therefore, after substitution, $\bar{v} = iK/n$

Then, using the value of hydraulic conductivity calculated above,

$\bar{v} = (0.002 \times 1.07 \times 10^{-4})/0.25$

Hence, $\bar{v} = 8.56 \times 10^{-7}$ m s⁻¹ (or 7.40×10^{-2} m day⁻¹)

Sand unit

From Darcy's law, the average linear groundwater velocity, $\bar{v} = q/n$, where the specific discharge, $q = -iK$

Therefore, after substitution, $\bar{v} = iK/n$

Then, using the value of hydraulic conductivity calculated above,

$\bar{v} = (0.004 \times 5.37 \times 10^{-5})/0.20$

Hence, $\bar{v} = 1.07 \times 10^{-6}$ m s⁻¹ (or 9.28×10^{-2} m day⁻¹)

7) a) From inspection of the piezometric head data, the hydraulic head gradient is *upwards* since groundwater flows from the region of higher head (35.8 cm) at the position of the base of Tube 1 (elevation, 30.6 cm) towards the region of lower head (35.1 cm) at the position of the base of Tube 2 (elevation, 33.7 cm), as shown in the following sketch (Fig. AS.10.1):

b) From inspection of the sketch, with dh = 0.7 cm and dz = 3.1 cm, then the vertical hydraulic gradient, i, or $dh/dz = 0.7/3.1 = 0.2258$

From the dimensions of the seepage meter, the seepage area, $A = 0.3 \times 0.3 = 0.09$ m²

The time taken for the seepage meter to fill is 4 hours. Hence, $Q = 10$ L/4 hours = 0.0025 m³ hour⁻¹

From Darcy's law, $Q = -iAK$ and, on rearranging, $K = Q/iA$

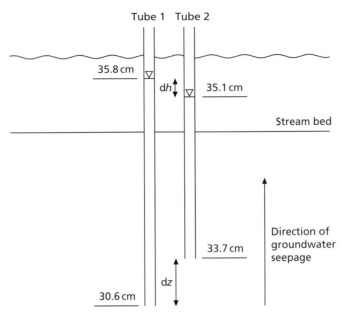

Fig. AS.10.1 Sketch of relative water levels in a piezometer nest installed within streambed sediments.

On substitution of values, $K = 0.0025/$
$(0.2258 \times 0.09) = 0.123$ m hour^{-1}
Hence, $K = 3.417 \times 10^{-5}$ m s^{-1}

c) The effective porosity, $n = 0.30$

From Darcy's law, seepage rate (or average linear velocity), $\bar{v} = q/n$, where $q = -iK$. Hence, $\bar{v} = iK/n$

On substitution of values,
$\bar{v} = (0.2258 \times 3.417 \times 10^{-5})/0.30$

Therefore, seepage rate, $\bar{v} = 2.572$ m s^{-1} (or 2.222 m day^{-1})

8) With reference to Fig. AS.10.2, for piezometer a, the depth to water, m, equals 37 m, the depth of the piezometer, $(m + n)$, equals 115 m and the elevation of the ground surface above the datum level, q, equals = 350 m. Hence, for piezometer a, the following are calculated at the measurement point A:

a) Hydraulic head $= (q - m) = 350 - 37 = 313$ m

b) Pressure head $= n = ((m + n) - m) = (115 - 37) = 78$ m

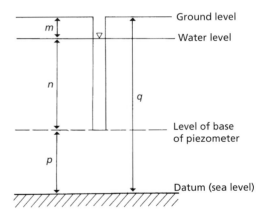

Fig. AS.10.2 Set-up used to calculate the hydraulic head, pressure head and elevation head of groundwater measured in a piezometer.

c) Elevation head $= p = (q - (m + n)) = (350 - 115) = 235$ m

Similarly, calculations for the two other piezometers, b and c, are as shown in Table AS10.1.

Table AS10.1 Head values measured in three piezometers.

Measurement	Piezometer tip A (m)	Piezometer tip B (m)	Piezometer tip C (m)
Hydraulic head	313	322	329
Pressure head	78	52	19
Elevation head	235	270	310

d) The hydraulic gradient between measurement points A and B is equal to the difference in hydraulic head divided by the distance between the piezometer tips (which is equal to the difference in elevation head) $= (322 - 313)/(270 - 235) = 9/35 = 0.257$. Similarly, the hydraulic gradient between measurement points B and C $= (329 - 322)/(310 - 270) = 7/40 = 0.175$. These calculations show that the hydraulic head increases with increasing elevation of the measurement points. The piezometers are, therefore, in a recharge area.

11) First, the method is to recalculate the pressure reading for the upper sensor to give the pressure value that would be obtained at the elevation of the lower sensor. Sensors are 0.75 m apart in the vertical direction, thus a recalculated upper reading should be 117.6 kPa plus the pressure exerted by 0.75 m of water $= 117.6 + (9.81 \times 0.75) = 125.0$ kPa. Hence, the upper value exceeds the lower value and so water flows downwards through the cylinder. The difference in pressure readings $(125.0 - 120.5 = 4.5$ kPa) divided by 9.81 gives the hydraulic head difference of 0.46 m. A head loss of 0.46 m over a distance of 2 m is a hydraulic gradient of 0.23.

a) The hydraulic conductivity of the sand, K, is related to specific discharge, q, thus, from Darcy's law, $q = 4 \times 10^{-5}$ m s^{-1} $= iK$, where i is the hydraulic gradient $(= 0.23)$.

Therefore, at 20°C, $K = 4 \times 10^{-5}/0.23 = 1.739 \times 10^{-4}$ m s^{-1}

b) The intrinsic permeability of the sand is obtained from the relationship $K = k_i \frac{\gamma}{\mu}$ where k_i is the intrinsic permeability, γ is the specific weight $(= \rho g$, where ρ is fluid density and g is gravitational acceleration), and μ is the fluid viscosity. From values of density and viscosity provided in Appendix 2, at 20°C, and taking a value for the acceleration due to gravity of 9.81 m s^{-2}, the value of intrinsic permeability for the sand with a hydraulic conductivity of 1.739 $\times$ 10^{-4} m s^{-1} is found from:

$$k_i = 1.739 \times 10^{-4}$$
$$\times \left(1002 \times 10^{-6}/(998.21 \times 9.81)\right)$$
$$= 1.78 \times 10^{-11} \text{ m}^2 \text{ or 18 darcies}$$

c) The volume of water transmitted by the apparatus in one day is found from $Q = AKi$, where the cross-sectional area of flow $= \pi r^2$ with a radius, r, of the cylinder of 0.25 m.

Hence, $Q = \pi \times 0.25^2 \times 1.739 \times 10^{-4} \times 0.23 \times 86\,400 = 0.68$ m^3 day^{-1}

d) The hydraulic conductivity at 40°C is found from $K = k_i \frac{\gamma}{\mu}$ and, using new values for density and viscosity from Appendix 2, is equal to $(1.78 \times 10^{-11}) \times (992.22 \times 9.81/653.2 \times 10^{-6}) = 2.653 \times 10^{-4}$ m s^{-1}. Now, at 20°C, $Q_{20} = AK_{20}i_{20}$ and at 40°C, $Q_{40} = AK_{40}i_{40}$. Hence, if the flow rate is to remain the same at the higher temperature, then $K_{20}i_{20} = K_{40}i_{40}$. Since, the only unknown is i_{40}, it is possible to calculate the new hydraulic gradient from $i_{40} = K_{20}i_{20}/K_{40} = (1.739 \times 10^{-4} \times 0.23)/2.653 \times 10^{-4}) = 0.15$ corresponding to $\Delta h/\Delta l$, where Δl is the length of the column (2 m). Thus, the new head difference, Δh, is equal to $0.15 \times 2 = 0.30$.

Thus, the pressure head difference between the two sensors to meet the flow conditions

at 40°C needs to be $(0.30 \times 9.81 = 2.94 \text{ kPa})$ for the hydraulic head difference minus $(0.75 \times 9.81 = 7.36 \text{ kPa})$ for the elevation head difference = 4.42 kPa.

12) Refer to the completed Fig. AS.10.3. Contours on top of the sand aquifer are given. Construction of the 55, 54, 53 and 52 m contours suggests that the top of the sand is a planar, isoclinal surface.

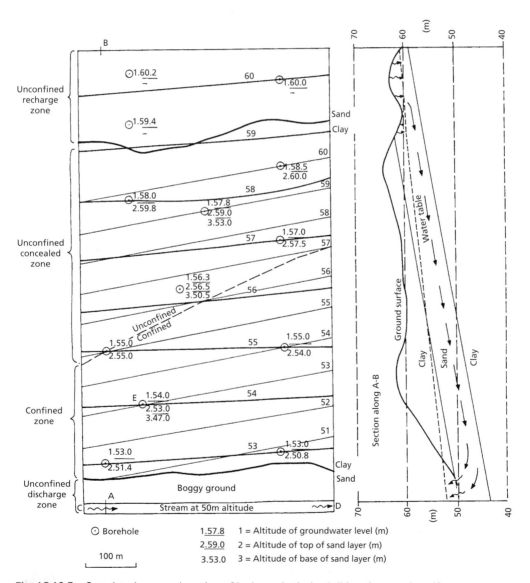

Fig. AS.10.3 Completed map and section of hydrogeological conditions in a sand aquifer.

The rest of the contours can be drawn in readily on the basis of this assumption. The clay-sand contact dips towards 169° at a gradient of 9 m in 550 m or 1°.

The three boreholes that fully penetrate the sand layer all show it to be 6 m thick (i.e. $b = 6$ m).

The confined/unconfined boundary is established where the structure contours and groundwater level contours cross at 55 and 56 m and sketched-in to the edges of the map.

In completing the cross-section, the aquifer is divisible into four zones:

a) Recharge or outcrop zone, which is unconfined.

b) An unconfined zone without recharge where the sand is concealed by clay.

c) A confined zone in which groundwater levels are within the clay.

d) A discharge zone corresponding to the boggy ground in which overflowing artesian groundwater conditions occur despite the unconfined nature of the aquifer.

The stream is 500 m long from C to D. Within the confined zone, around borehole E, the hydraulic gradient is 2 m in 220 m. Since $T = Kb$ and $K = 1.0$ m day^{-1}, then $T = 6$ m^2 day^{-1}.

Now, using Darcy's law, the total discharge to the stream is calculated from

$$Q = AK \frac{dh}{dl} = 6 \times 500 \times 1.0 \times \frac{2}{220}$$
$$= 27.3 \text{ m}^3 \text{ day}^{-1} \text{ or } 0.32 \text{ L s}^{-1}$$

The area is underlain by three strata dipping uniformly to 169° at 1°. The lowest stratum is clay, which is overlain by a 6 m thick sand, which is in turn overlain by clay. The clays are effectively impermeable but the sand, with a hydraulic conductivity of 1.0 m day^{-1}, forms an aquifer. This aquifer receives recharge in the north of the area, where the sand outcrops at the ground surface. Groundwater flow is generally southwards with a discharge zone into the stream CD and into the boggy ground on its north bank. Groundwater conditions in this boggy ground are overflowing artesian and attempts at drainage by means of ditches are unlikely to be successful because the hydraulic head of the sand is above the ground surface. In addition, groundwater may present problems in construction and excavation in parts of the recharge zone where the water table is shallow.

13) a) The groundwater discharge across the width of the aquifer is found using Darcy's law. The hydraulic gradient is found from the head difference measured between Wells 1 and 2 and is equal to $(83.4 - 82.9)/1000 = 5 \times 10^{-4}$. The value of hydraulic conductivity for groundwater at a temperature of 20°C can be found from the relationship:

$$K = k_i \frac{\gamma}{\mu}$$

where k_i is the intrinsic permeability, γ is the specific weight ($=\rho g$, where ρ is fluid density and g is gravitational acceleration), and μ is the fluid viscosity. From values for density and viscosity provided in Appendix 2, at 20°C, and taking a value for the acceleration due to gravity of 9.81 m s^{-2}, the value of hydraulic conductivity, K, for an aquifer with an intrinsic permeability of 30 darcies ($=30 \times 9.87 \times 10^{-13}$ m^2) is found from

$$K = 30 \times 9.87 \times 10^{-13}.$$
$$\frac{(998.21 \times 9.81/1002 \times 10^{-6})}{}$$
$$= 2.89 \times 10^{-4} \text{ m s}^{-1} \text{ or } 25 \text{ m day}^{-1}$$

Now, using Darcy's law, the discharge, Q, across the area of the aquifer (20×4000 m^2) is equal to $80\,000 \times 25 \times 5 \times 10^{-4} = 1000$ m^3 day^{-1}

b) If the average groundwater temperature is decreased by 50% (i.e. to 10°C), the new groundwater discharge is recalculated, using values for changed density and viscosity from Appendix 2, as follows:

$$Q = 80\ 000 \times 5 \times 10^{-4}$$
$$\times \left(30 \times 9.87 \times 10^{-13}\right.$$
$$\cdot \left(999.70 \times 9.81/1307 \times 10^{-6}\right))$$
$$= 8.89 \times 10^{-3} = 768 \ \text{m}^3 \ \text{day}^{-1}$$

Therefore, a temperature decrease of 50% has caused a decrease in groundwater discharge of 23%. Such dramatic changes in temperature, and therefore, hydraulic conductivity, are unlikely in an alluvial valley but may be possible in geothermal systems.

c) The length of time taken for a contaminant introduced in Well 1 to reach Well 2 can be found from dividing the length (1000 m) by the average linear velocity, $\bar{v}$, where $\bar{v} = q/n_e$. The value of specific discharge, q, at 20°C is equal to $Q/A =$ 1000/80 000 m day^{-1}. The effective porosity, n_e, is given as 0.16. From this information, $\bar{v} = (1000/80\ 000)/0.16$ $= 0.078$ m day .

Therefore, the time taken for a contaminant to travel from Well A to Well B = 1000/0.078 $= 12\,821$ days or about 35 years.

14) Inspection shows that the area of sandstone aquifer between the marshy pools and C has a horizontal water table and, therefore, no flow. The flow pattern must be as shown in Fig. AS.10.4. The flow of spring S must be given by the difference in near-horizontal flow in the first two sections of the aquifer, that is: $Q_S = Q_1 - Q_2$. Darcy's law is used to find Q_1 and Q_2 as follows:

$$Q_1 = K \times (\text{aquifer thickness})$$
$$\times (\Delta h_1/(\text{horizontal distance AS}))$$
$$= 200 \times 125 \times (22.5/5000)$$
$$= 112.5 \ \text{m}^3 \ \text{day}^{-1} \ (\text{per unit width of aquifer}).$$

$$Q_2 = K \times (\text{average aquifer thickness})$$
$$\times (\Delta h_2/(\text{horizontal distance from}$$
$$\text{S to fault below marshy pools}))$$
$$= 200 \times 107.5 \times (12.5/10\ 000)$$
$$= 26.9 \ \text{m}^3 \ \text{day}^{-1} \ (\text{per unit width of aquifer}).$$

a) Hence, the discharge at spring S, $Q_S =$ 112.5 − 26.9 = 85.6 m^3 day^{-1} or about 1 L s^{-1} per unit width of aquifer.

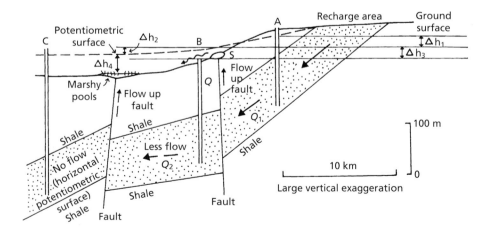

Fig. AS.10.4 Section showing hydraulic head and groundwater flow conditions in a sandstone aquifer with shale confining beds.

b) The entire flow Q_2 is discharged at the marshy pools; hence, its discharge is about 27 m³ day⁻¹ or 0.3 L s⁻¹ per unit width.

c) The transmissivity of the fault at S is given by (flow per unit width of fault plane, $Q_S/(\Delta h_3/(\text{distance along the fault plane}))$ since from Darcy's law, $Q = (\text{width} \times b) \times K \times dh/dl$ and $T = Kb$ such that $T = Q/dh/dl$ per unit width. The hydraulic gradient along the fault plane is equal to Δh_3 divided by the vertical distance of flow along the fault plane from the top of the aquifer to the spring (=20/100). Hence, the transmissivity of the fault at spring, $S = 85.6/(20/100) = 428$ m² day⁻¹.

d) Similarly, the transmissivity of the marshy pools fault is $= 26.9/(\Delta h_4/\text{distance along fault}) = 26.9/(30/120) = 108$ m² day⁻¹.

15) a) Considering that the vertical fissure is 2 m high and 0.1 m wide, the hydraulic radius (characteristic length, d) of the fissure $=$ cross-sectional area/wetted perimeter $= (0.1 \times 2)/4.2 = 0.05$ m. If the water velocity (characteristic velocity), q, is 100 m hour⁻¹ (100/3600 m s⁻¹)

as shown by the tracer movement, then the Reynolds number, $R_e = \dfrac{\rho q d}{\mu}$, can be calculated assuming a temperature of 10°C (with density and viscosity values given in Appendix 2) as follows:

$$R_e = \frac{999.70 \times 100/3600 \times 0.05}{1307 \times 10^{-6}} = 1062$$

Thus, for this Reynolds number, the fissure flow is transitional to turbulent.

b) Assuming that the median grain size of the sand represents the characteristic length, d, and with a specific discharge, q, calculated from the pumping rate divided by the cross-sectional area of flow (Fig. AS.10.5a), then the Reynolds number can be found using the above equation. For values of water density and viscosity chosen from Appendix 2 at an assumed temperature of 10°C, then at a radial distance of 0.25 m from the centre of the well:

$$R_e = \frac{999.70 \times 0.050/(5 \times 2\pi 0.25) \times 1.25 \times 10^{-3}}{1307 \times 10^{-6}}$$
$$= 6.09$$

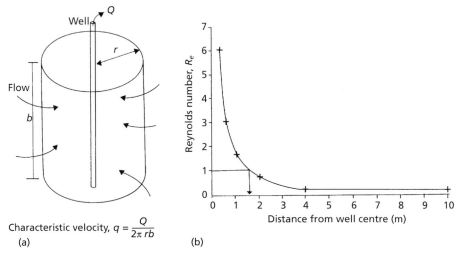

Characteristic velocity, $q = \dfrac{Q}{2\pi rb}$

(a)

(b)

Fig. AS.10.5 (a) Set-up of radial flow towards a pumping well and (b) plot of calculated Reynolds number versus radial distance from a pumping well.

Table AS10.2 Reynolds number in the vicinity of a pumping well.

Distance from well centre line (m)	Reynolds number R_e
0.25	6.09
0.5	3.04
1.0	1.52
2.0	0.76
4.0	0.38
10.0	0.15

For the other radii, Table AS10.2 gives the calculated values of Reynolds number.

When calculated using the above approach, Darcian flow is obeyed up to a value of Reynolds number of about 1. From Fig. AS.10.5 (b), a plot of Reynolds number against radius, Darcy's law only occurs to within about 1.5 m of the well boundary.

17) The specific yield, S_y, of the unconfined alluvial aquifer = (volume of water pumped)/(total volume of aquifer basin drained). The total volume of aquifer basin drained = (valley area) × (water table decline).

Now, $S_y = (500 \times 10^6)/(100\,000 \times 10\,000 \times 3) = 0.166$

Therefore, the specific yield, $S_y = 0.17$

19) a) Assumptions used to construct flow nets are discussed in Box 2.3.

b) Table AS10.3 gives values read from the sketch of the flow net of seepage under the dam (Fig. A10.5).

c) The hydraulic conductivity, K, of the sandstone can be found from the tracer test which showed that the average linear velocity $\bar{v}$ is equal to the distance travelled between points P and Q (400 m) divided by the time taken (5 days) and therefore equal to 80 m day^{-1}. Now, from Darcy's law, $\bar{v} = q/n_e = iK/n_e$. The effective porosity, n_e, of the sandstone is given as 20% and the hydraulic gradient, i, along the length PQ is 50/400. Hence, rearranging and substituting values, $K = \bar{v}n_e/i = (80 \times 0.20)/(50/400) = 128$ m day^{-1}.

d) For the contaminant flowpaths from drums D and E, the hydraulic gradients are, respectively, 50/300 and 50/480. The hydraulic conductivity of the sandstone is 128 m day^{-1} and the effective porosity is 0.20. The time taken for the contaminant to seep out on the downgradient side of the dam is found from (travel distance $\times n_e$)/iK. For drum D, the travel time is $(300 \times 0.2)/(50/300 \times 128) = 2.8$ days. For drum E, the travel time is $(480 \times 0.2)/(50/480 \times 128) = 7.2$ days.

e) The actual breakthrough times of the contaminants may differ from those

Table AS10.3 Head values measured in three piezometer nests.

Relative piezometer position	Head values at piezometer nest A (m)	Head values at piezometer nest B (m)	Head values at piezometer nest C (m)
Upper	122	135.5	165
Middle	124	135	161
Lower	125	135	158

predicted based on advective transport alone due to the effects of hydrodynamic dispersion and attenuation processes such as adsorption and chemical and biological decay. The presence of fractures in the sandstone could potentially provide preferential pathways for the contaminant.

A10.3 Chemical hydrogeology

4) a) From inspection of Table A10.1, the predominant cations and anions in groundwater from the Dammam limestone aquifer are sodium, calcium, chloride and sulphate. The chloride and total dissolved solids (TDS) content show that the water is brackish and could be described as a Na-Cl hydrochemical water type or, considering the second most abundant ions, a Na-Ca-Cl-SO$_4$ water type.

b) The source of Ca^{2+} and HCO$^-$ ions in the groundwater is dissolution of the limestone aquifer lithology which is comprised of calcite (CaCO$_3$). The source of Na$^+$ and SO$_4^{2-}$ (and Cl$^-$) in the Dammam aquifer is the presence of evaporite minerals such as halite (NaCl) and gypsum (CaSO$_4$). Also, the shale horizon at the top of the aquifer is a potential source of sulphate from the oxidative weathering of pyrite.

c) For calcite dissolution in the Dammam aquifer, then the following equation describes the reversible chemical reaction:

$$CaCO_3 = [Ca^{2+}] + [CO_3^-]$$

The calcite saturation index is given by $\Omega_{calcite} = [Ca^{2+}][CO_3^{2-}]/K_{calcite}$
To find $\Omega_{calcite}$ it is necessary to calculate the CO$_3^{2-}$ concentration since a mean value of Ca^{2+} concentration is given as part of the chemical analysis of the groundwater.

Now, consider the dissociation of bicarbonate:

$$HCO_3^- = H^+ + CO_3^{2-}$$

for which the thermodynamic equilibrium constant, K, is

$$K_{HCO3-} = [H^+][CO_3^{2-}]/[HCO_3^-]$$

The following values are given: HCO$_3^-$ = 173 mgL^{-1} (or 173/61 = 2.836 × 10^{-3} mol L^{-1}), pH = 7.4 (or H$^+$ = 10$^{-7.4}$) and K_{HCO3-} = 10$^{-10.33}$. Hence, rearranging the above equation, and assuming that the concentration values are equivalent to the chemical activities, gives

$$[CO_3^{2-}] = (10^{-10.33} \times 2.836 \times 10^{-3})/10^{-7.4}$$
$$= 3.332 \times 10^{-6} \text{ mol } L^{-1}$$

Hence, for $K_{calcite}$ = 10$^{-8.4}$ and Ca^{2+} = 540 mg L^{-1} (or 540/40 = 13.5 × 10^{-3} mol L^{-1}), then the calcite saturation index is found from

$$\Omega_{calcite} = (13.5 \times 10^{-3} \times 3.332 \times 10^{-6})$$
$$/10^{-8.48} = 13.58$$

Therefore, taking the logarithm of the calculated $\Omega_{calcite}$ value gives a calcite saturation index of +1.13 which is *supersaturated*, with calcite expected to precipitate.
For halite dissolution in the Dammam aquifer, then the following equation describes the reversible chemical reaction:

$$NaCl = [Na^+] + [Cl^-]$$

The halite saturation index is given by $\Omega_{halite} = [Na^+][Cl^-]/K_{halite}$
Hence, for K_{halite} = 10$^{-0.773}$, Na$^+$ = 1180 mg L^{-1} (or 1180/23 = 51.3 × 10^{-3} mol L^{-1}) and Cl$^-$ = 2365 mg L^{-1} (or 2365/35 = 67.6 × 10^{-3} mol L^{-1}), then the halite saturation index is found from

$$\Omega_{halite} = (51.3 \times 10^{-3} \times 67.6 \times 10^{-3})/$$

$$10^{-0.773} = 5.85 \times 10^{-4}$$

Therefore, taking the logarithm of the calculated Ω_{halite} value gives a halite saturation index of -3.23 which is *undersaturated*, with halite expected to dissolve.

5) First, for denitrification:

$$\frac{1}{5}NO_3^- + \frac{6}{5}H^+ + e^- = \frac{1}{10}N_{2(g)} + \frac{3}{5}H_2O$$

Multiplying through by 5:

$$NO_3^- + 6H^+ + 5e^- = 1/2N_{2(g)} + 3H_2O$$

From

$$pe = pe^o + 1/n(\log_{10}[\text{oxidants}]/[\text{reductants}])$$

Then, for that denitrification reaction:

$$pe = pe^o + 1/n\left(\log_{10}\left\{(NO_3^-)^1(H^+)^6\right\}/\right.$$

$$\left.\left\{(N_2)^{1/2}(H_2O)^3\right\}\right)$$

Given values are the following: $pe^o = +21.05$ from Table 4.14, $n = 5$, pH $= 8$ (or $H^+ = 10^{-8}$), $NO_3^- = 2 \times 10^{-3}$ mol L^{-1}, $N_{2(g)} = 1$ bar pressure (or 1×10^{-3} mol L^{-1}) and $H_2O = 1$. Substitution in the above equation gives:

$$pe = 21.05 + 1/5\left(\log_{10}\left\{(2 \times 10^{-3})^1(10^{-8})^6\right\}/\right.$$

$$\left.\left\{(1 \times 10^{-3})^{1/2}(1)^3\right\}\right)$$

$$= 21.05 + 1/5(\log_{10}(2 \times 10^{-49.5})) = 11.21$$

Therefore, $Eh = (0.059 \times 11.21)/5 = 0.132$ V or 132 mV

Second, for sulphate reduction:

$$\frac{1}{8}SO_4^{2-} + \frac{9}{8}H^+ + e^- = \frac{1}{8}HS^- + \frac{1}{2}H_2O$$

Multiplying through by 8:

$$SO_4^{2-} + 9H^+ + 8e^- = HS^- + 4H_2O$$

From

$$pe = pe^o + 1/n(\log_{10}[\text{oxidants}]/[\text{reductants}])$$

Then, for that sulphate reduction reaction:

$$pe = pe^o + 1/n\left(\log_{10}\left\{(SO_4^{2-})^1(H^+)^9\right\}/\right.$$

$$\left.\left\{(HS^-)^1(H_2O)^4\right\}\right)$$

Given values are the following: $pe^o = +4.25$ from Table 4.14, $n = 8$, pH $= 8$ (or $H^+ = 10^{-8}$), $SO_4^{2-} = 1 \times 10^{-3}$ mol L^{-1}, $HS^- = 1 \times 10^{-3}$ mol L^{-1} and $H_2O = 1$. Substitution in the previous equation gives:

$$pe = 4.25 + 1/8\left(\log_{10}\left\{(1 \times 10^{-3})^1(10^{-8})^9\right\}/\right.$$

$$\left.\left\{(1 \times 10^{-3})^1(1)^4\right\}\right)$$

$$= 4.25 + 1/8(\log_{10}(10^{-72}))$$

$$= -4.75$$

Therefore, $Eh = (0.059 \times (-4.75))/8 = -0.035$ V or -35 mV

If, during denitrification, ferrous iron (Fe^{2+}) is oxidized to ferric (Fe^{3+}) iron as described by the half-reaction:

$$2Fe^{2+} = 2Fe^{3+} + 2e^- \quad \text{(eq. 4.29)}$$

Then the equation for pe is written as follows:

$$pe = pe^o + 1/n\log_{10}([Fe^{3+}]/[Fe^{2+}])$$

Given values are $pe^o = 13.0$, $n = 1$ (after dividing the redox half-reaction equation through by 2) and $Fe^{2+} = 1 \times 10^{-5}$ mol L^{-1}. As a result of the denitrification reaction, $pe = 11.21$ such that

$$11.21 = 13.0 + 1/1\log_{10}([Fe^{3+}]/[Fe^{2+}])$$

$$11.21 = 13.0 + \log_{10}(Fe^{3+}/1 \times 10^{-5})$$

$$-1.79 = \log_{10}Fe^{3+} - \log_{10}(1 \times 10^{-5})$$

$$-6.79 = \log_{10}Fe^{3+}$$

Therefore, the amount of ferric iron, Fe^{3+}, produced $= 1 \times 10^{-6.79} = 1.62 \times 10^{-7}$ mol L^{-1}.

6) a) When seawater intrudes a coastal freshwater aquifer, the following cation exchange reaction can occur:

$$Na^+ + \frac{1}{2}Ca-X \rightarrow Na-X + \frac{1}{2}Ca^{2+}$$

$$\text{(eq. 4.21)}$$

where X indicates the exchange material. As the exchanger takes up Na^+, Ca^{2+} is released, and the hydrochemical water type evolves from Na-Cl to Ca-Cl.

b) The reverse reaction to (a) above can occur when freshwater flushes a saline aquifer:

$$\tfrac{1}{2}\,Ca^{2+} + Na{-}X$$

$$\rightarrow \tfrac{1}{2}\,Ca{-}X + Na^+ \quad (eq.\ 4.22)$$

where Ca^{2+} is taken up from water in return for Na^+ resulting in a Na-HCO$_3$ water type.

The chemical reactions that occur during freshwater and saline water displacements in aquifers can be identified from a consideration of conservative mixing of fresh and saline water end member solutions and comparing with individual water analyses. For conservative mixing:

$$c_{i,\,mix} = f_{saline} \times c_{i,\,saline} + (1 - f_{saline})c_{i,\,fresh}$$

where c_i is the concentration of ion i; *mix*, *fresh* and *saline* indicate the conservative mixture and end-member fresh and saline waters; and f_{saline} is the fraction of saline water. Any change in the sample composition as a result of reactions, for example cation exchange, other than by simple mixing ($c_{i,react}$) is then simply found from

$$c_{i,react} = c_{i,sample} - c_{i,mix}$$

During seawater intrusion, and for the data shown in Table A10.2, to calculate how much Na^+ has been added to the mixed groundwater sample by cation exchange, then

$$c_{Na,\,mix} = f_{saline}$$
$$\times c_{Na,\,saline} + (1 - f_{saline})c_{Na,\,fresh}$$

Now, using the mixed and saline Cl$^-$ concentration values to indicate the fraction of saline water ($f_{saline} = 440/566$), and with a

freshwater end-member Cl$^-$ concentration value equal to zero, then upon substitution:

$$c_{Na,mix} = (440/566)485 + (1 - 440/566)0$$
$$= 377.0\ mmol\ L^{-1}$$

The amount of Na^+ involved in the cation exchange reaction can be found from

$$c_{Na,react} = c_{Na,sample} - c_{Na,mix}$$

Using the result for $c_{Na,mix}$ found:

$$c_{Na,react} = 374.8 - 377.0$$
$$= -2.2\ mmol\ L^{-1}$$

A similar calculation for Ca^{2+} added to the mixed groundwater sample results in

$$c_{Ca,mix} = (440/566)10.7 + (1 - 440/566)3.0$$
$$= 9.0\ mmol\ L^{-1}$$

and

$$c_{Ca,react} = 10.0 - 9.0 = 1.0\ mmol\ L^{-1}$$

7) In order to complete the trilinear (Piper) diagram, the major ion data need to be presented in terms of chemical equivalence values and then as the percentage that each ion contributes to either the total concentration of cations or anions. The results of these calculations are presented in Table AS10.4 together with the total dissolved solids contents and ionic balance errors ($\{[\Sigma cations - \Sigma anions]/[\Sigma cations + \Sigma anions]\} \times 100$). The ionic balance errors for samples 4, 8, 9, 11 and 13 are greater than 10% with sample 13 greater than 20%. Therefore, these samples should be interpreted with caution. Inspecting the hydrochemical data, it is apparent that all the groundwaters are fresh, even those two boreholes (11 and 13) that are situated at the coast, although these two samples record the highest electrical conductivity values and total dissolved solids content. The trilinear plot (Fig. AS.10.6) shows that all the samples

Table AS10.4 Hydrochemical data for groundwater samples from the Fife and Kinross area, Scotland.

Sample number	Ca^{2+}	Mg^{2+}	$Na^+ + K^+$	HCO_3^-	Cl^-	$SO_4^{2-} + NO_3^-$	Total dissolved solids (TDS)	Ionic balance error
ORS 1	31	53	15	38	20	42	9.6	−3.3
ORS 2	46	39	15	62	18	20	10.5	−4.0
ORS 3	52	37	11	64	10	26	7.5	−6.1
ORS 4	38	46	16	52	20	28	12.8	8.2
ORS 5	46	39	16	52	22	27	10.8	−1.4
ORS 6	64	26	10	53	12	36	7.6	−1.6
ORS 7	63	27	10	63	10	27	6.9	−2.8
Carb 8	47	47	6	75	10	15	17.7	20.1
Carb 9	54	35	11	73	11	17	12.3	14.2
Carb 10	51	36	13	72	9	19	8.8	0.1
Carb 11	32	54	14	83	10	7	28.1	16.8
ORS 12	35	40	25	59	31	11	16.4	6.7
ORS 13	56	12	32	54	33	13	16.7	23.2

Note: All values in %meq L^{-1} except for TDS (meq L^{-1}) and ionic balance error (%).

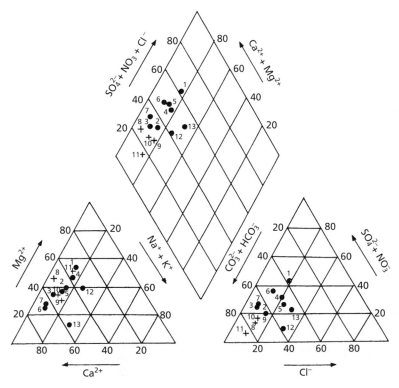

● Old Red Sandstone sample

+ Carboniferous sample

Fig. AS.10.6 Completed trilinear (Piper) diagram of hydrochemical analyses of groundwater samples from the Fife and Kinross area, Scotland.

plot in a similar field and are predominantly Ca-HCO$_3$ in chemical character except for samples 12 and 13, which show the presence of a Na-Cl influence. This is expected in the case of the coastal sample 13 but not for the inland Old Red Sandstone sample 12. This sample appears to be affected by contamination.

The four groundwater samples derived from the Carboniferous rocks differ from the Old Red Sandstone groundwaters in terms of their generally higher Ca^{2+} and HCO$_3^-$ concentrations. As a result, these four samples appear slightly separated from the remainder of the groundwater samples shown in the trilinear plot (Fig. AS.10.6). The groundwaters in the Carboniferous rocks are mildly acidic (pH values < 7.0) as a result of the presence of sandstones, mudstones and shales, although the existence of thin limestone layers contributes Ca^{2+} and HCO$_3^-$ ions from calcite dissolution.

The Carboniferous rocks are poorly permeable with slow groundwater movement and long residence times that promote chemically reducing conditions. Evidence for these conditions is the lower *Eh* values, low dissolved oxygen content and low NO$_3$-N concentrations in groundwaters in the Carboniferous rocks. Two of the samples (samples 8 and 11) have high total Fe and SO$_4^{2-}$ concentrations indicative of oxidative weathering of pyrite (FeS$_2$) with a source in the mudstones, shales and coal strata.

The better permeability developed in the Old Red Sandstone, particularly the Knox Pulpit Formation, means that this unconfined aquifer is more vulnerable to surface-derived pollution. Therefore, it is not surprising that groundwaters in this formation contain NO$_3^-$ in association with modern groundwater recharge (e.g. sample 4 contains 12 mg L^{-1} NO$_3$-N and a tritium content of 27.6 TU).

A10.4 Environmental isotope hydrogeology

4) The plots of δ^2H versus δ^{18}O in Fig. AS.10.7a, δ^{18}O versus Cl$^-$ in Fig. AS.10.7b and Na$^+$ versus Cl$^-$ in Fig. AS.10.7c show the waters to be:

a) meteoric, lying close to the World Meteoric Water Line (WMWL). The Chalybeate spring has the most depleted isotopic composition indicating a possible palaeowater. The Eye Wells 1 and 2 springs lie on an evaporation trend away from the Chalybeate/Magnesium spring waters.

b) all the saline waters (with the exception of Eye Wells 1 and 2) are formed by dilution of the end-member Chalybeate spring water with local meteoric water.

c) the Magnesium spring is anomalous in the sense that it represents a possible palaeowater (similar in isotopic composition to the Chalybeate spring) yet has low salinity suggesting little interaction with saline water at depth.

An annotated conceptual model is presented in Fig. AS.10.8. Saline water exists at depth. Deep descent of recharge water results in either entrainment of a small component of the saline water or a degree of water-rock interaction leading to increased salinity. The depleted isotopic composition of the end-member Chalybeate spring water indicates a possible palaeowater with a long residence time which was recharged during colder climatic conditions than at present. The groundwaters ascend from depth along the Rock Park fault and mix with modern meteoric waters at shallow depths, thus accounting for the range of saline water compositions encountered.

The Eye Wells 1 and 2 have an evaporated Chalybeate spring water composition and probably result from evaporation after ascent

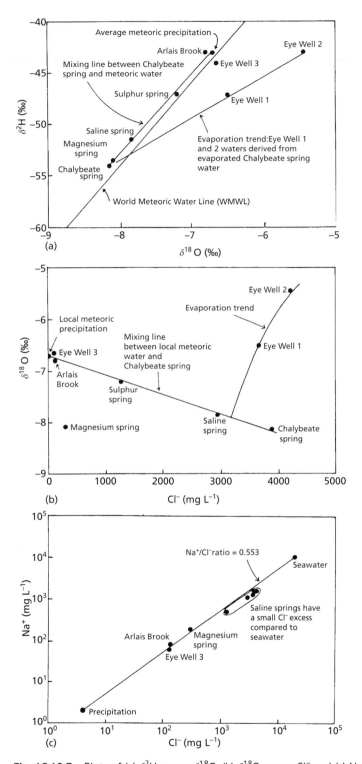

Fig. AS.10.7 Plots of (a) δ^2H versus $\delta^{18}O$, (b) $\delta^{18}O$ versus Cl^- and (c) Na^+ versus Cl^- for spring waters at Llandrindod Wells, mid-Wales.

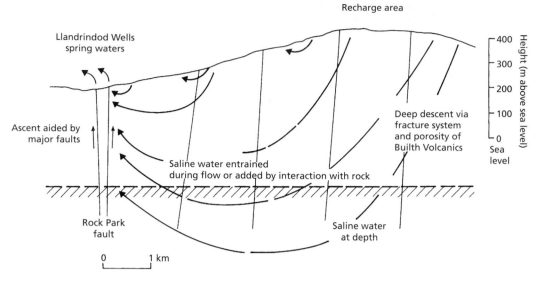

Fig. AS.10.8 Conceptual hydrochemical model of the origin and modification of spring water composition at Llandrindod Wells, mid-Wales.

of the saline water. Eye Well 3 has a modern isotopic composition and low salinity and should not be considered part of the saline waters.

5) a) To demonstrate, consider Site 1 for which there are the following data:

$$A_o = {}^{14}C_{atmos} = 100 \text{ pmc (assumed)}$$
$$A = {}^{14}C_{sample} = 60.8 \text{ pmc}$$
$$\delta^{13}C_{sample} = -13.2\%$$
$$\delta^{13}C_{lithic} = +2\%$$
$$\delta^{13}C_{biogenic} = -26\%$$

Firstly, find the fraction of biogenic carbon, Q, contained in the sample using:

$$Q = \frac{\delta^{13}C_{sample} - \delta^{13}C_{lithic}}{\delta^{13}C_{biogenic} - \delta^{13}C_{lithic}} = \frac{(-13.2) - (+2)}{(-26) - (+2)}$$

$$= \frac{-15.2}{-28} = 0.543$$

Second, the corrected groundwater age can be found from

$$t_c = -8267\log_e\frac{A}{A_o} + 8267\log_e Q$$

$$= -8267\log_e\frac{60.8}{100} + 8267\log_e 0.543$$

Therefore, $t_c = 4113.50 + (-5048.21) = -934.71$ a.

A negative answer is obtained which indicates that the assumed value of A_o is incorrect and that a value greater than 100% is likely as a result of the former detonation of thermonuclear devices in the atmosphere. It can be concluded that Site 1 is modern water.

Similarly, corrected groundwater ages for Sites 2 and 3 are 14 328 a (or 14 ka) and 12 842 a (13 ka), respectively.

b) The distances from outcrop of Sites 2 and 3 are 15.1 and 16.3 km. Given their respective groundwater ages, estimated travel times in the fissured limestone aquifer are $15\,100/14\,000 = 1.08\text{ m a}^{-1}$ and $16\,300/13\,000 = 1.25\text{ m a}^{-1}$. Taking an average of these two values, an indicative average groundwater velocity is 1.17 m a^{-1} or about 1 m a^{-1}.

The calculations are for a porous, fissured limestone aquifer and so the groundwater ages should be interpreted with caution due to potential mixing between younger, fissure water and older, matrix porewater. However, the results for Sites 2 and 3 are consistent in having a low tritium content representative of old water. The groundwater velocities suggest a predominantly slow, intergranular-type flow mechanism, rather than fast, fissure flow, at least at the regional scale.

The stable isotope data show a depletion in values with the old groundwater at Site 3 indicating groundwater recharge under cooler climatic conditions than at present (current conditions are represented by Site 1). Given the age of the groundwater at this site, the recharge probably entered the aquifer during an interstadial period of the last ice age.

A10.5 Stream gauging, infiltration measurements and groundwater recharge estimation

5) The stream discharge calculated using the equation for a thin-plate weir is found from

$$Q = 1.84(L - 0.1 \, nH)H^{3/2}$$

For the given values, then

$$Q = 1.84(1.67 - 0.1 \times 2 \times 0.12)0.12^{3/2}$$
$$= 0.126 \; \text{m}^3 \; \text{s}^{-1}$$

The results of the current metering to measure stream discharge are included in Table AS10.5.

The summation of the stream discharge measurements for all the verticals gives a total stream discharge value of 0.138 m³ s⁻¹. This value compares with 0.126 m³ s⁻¹ obtained by the gauging station and suggests an error of almost 10% in the current metering results. Errors arise because of inaccuracies in measuring the current meter depth below the water surface and also in calculating the contribution to total flow of the vertical sections adjacent to the right and left banks.

6) The slug injection method of dilution gauging provides a value of stream discharge from

$$Q = \int_0^\infty \frac{cv}{C} dt$$

where c is the concentration of injected solution, v is the volume of injected solution and C is the diluted concentration at the downstream station.

This equation can be adapted to

$$Q = \frac{cv}{\sum C \cdot \Delta t}$$

where cv is the mass of tracer injected $= 15$ kg. For the current problem, $\Delta t = 1$ h. Also, note that 1 ppb $= 1 \; \mu\text{g L}^{-1}$. The calculations are given in Table AS10.6 and, from the values obtained, $Q = (15 \times 10^9 \; \mu\text{g})/((32 \; \mu\text{g/L.h})(3600 \; \text{s h}^{-1})(10^3 \; \text{m}^{-3})) = 130.2 \; \text{m}^3 \; \text{s}^{-1}$. The value of stream discharge of 130.2 m³ s⁻¹ from dilution gauging is larger than the value of 110 m³ s⁻¹ obtained from the rating curve. Causes of this discrepancy at the gauging station could include the following: (a) a reduction in the level of the rating curve control section (e.g. a lowered riffle); (b) weed growth that affected the original rating curve but has now been removed; and (c) the stream channel has been dredged/widened.

7) a) From the steady-state (constant rate) dilution gauging method, discharge, Q, is given by $Q = cq/C$, where c = injection concentration, q = injection rate and C = sampled concentration.

After 30 minutes, the tracer is fully mixed across the stream for the following conditions:

$$q = 1.0 \; \text{mL s}^{-1} = 1.0 \times 10^{-6} \; \text{m}^3 \; \text{s}^{-1}$$
$$c = 1000 \; \text{mg L}^{-1} = 1 \, \text{g L}^{-1}$$
$$C = 5 \; \mu\text{g L}^{-1} = 5 \times 10^{-6} \, \text{g L}^{-1}$$

Table AS10.5 Calculation of stream discharge using the velocity-area method.

Distance of object from left bank (m)	Current meter depth below surface (0.6d) (m)	(1) Water depth, d (m)	(2) Area of measurement section (m²)	Stream velocity (m s⁻¹)	(3) Stream discharge (m³ s⁻¹)
0.12	0.30	0.50	$0.50 \times 0.25 =$ 0.125	0.015	0.00188
0.38	0.31	0.52	$0.52 \times 0.25 =$ 0.130	0.035	0.00455
0.62	0.32	0.53	$0.53 \times 0.25 =$ 0.133	0.08	0.01064
0.88	0.31	0.52	$0.52 \times 0.25 =$ 0.130	0.12	0.01560
1.12	0.30	0.50	$0.50 \times 0.25 =$ 0.125	0.23	0.02875
1.38	0.30	0.50	$0.50 \times 0.25 =$ 0.125	0.25	0.03125
1.62	0.32	0.53	$0.53 \times 0.25 =$ 0.133	0.15	0.01995
1.88	0.29	0.48	$0.48 \times 0.25 =$ 0.120	0.10	0.01200
2.12	0.28	0.47	$0.47 \times 0.25 =$ 0.118	0.07	0.00826
2.38	0.26	0.43	$0.43 \times 0.30 =$ 0.129	0.04	0.00516
2.55	–	–	–	–	

Notes: (1) Stream depth $=$ (current meter depth below surface)/0.60.
(2) Area of measurement section = product of the stream depth and the lateral distance bounded by the mid-points between adjacent verticals (see Fig. AS.10.9).
(3) Stream discharge = product of area of measurement and stream velocity.

Table AS10.6 Data for the calculation of stream discharge using the tracer slug injection method.

Concentration (ppb)	Time (h)	Δt (h)	Concentration (ppb) × Δt (h)
0	0	0	0
1	1	1	1
5	2	1	5
10	3	1	10
7	4	1	7
5	5	1	5
3	6	1	3
1	7	1	1
0	8	1	0
			$\sum 32$

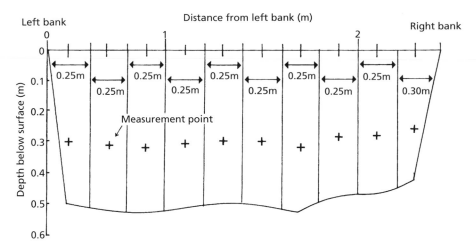

Left bank   Distance from left bank (m)   Right bank

Fig. AS.10.9 Division of channel cross-section into verticals for the purpose of calculating stream discharge using the data shown in Table AS10.5.

Therefore, $Q = (1 \times 1.0 \times 10^{-6})/5 \times 10^{-6} = 0.2 \, m^3 \, s^{-1}$.

b) First, need to establish Manning's n when the flow depth $= 0.3$ m and discharge, $Q = 0.2 \, m^3 \, s^{-1}$. Then, assuming that the value of n is constant, determine velocity and discharge when the flow depth $= 0.5$ m.

Depth $= 0.3$ m

From (a), $Q = 0.2 \, m^3 \, s^{-1}$

Wetted perimeter, $P = 0.3 + 0.3 + 2.4 = 3$ m

Cross-sectional area, $A = 0.3 \times 2.4 = 0.72 \, m^2$

Hydraulic radius, $R = A/P = 0.72/3 = 0.24$ m

Slope of the water surface, $S = 0.001$ and velocity, $V = Q/A = 0.2/0.72 = 0.28$ m s^{-1}

Now, from the Manning formula:

$$V = \frac{R^{\frac{2}{3}} S_f^{0.5}}{n},$$

Rearranging and substituting, Manning's

$$n = \frac{R^{\frac{2}{3}} S_f^{0.5}}{V} = \frac{0.24^{\frac{2}{3}} \times 0.001^{0.5}}{0.28} = 0.044$$

Now, when the flow depth is 0.5 m, the wetted perimeter, $P = 0.5 + 0.5 + 2.4 = 3.4$ m, cross-sectional area, $A = 0.5 \times 2.4 = 1.2 \, m^2$ and hydraulic radius, $R = A/P = 1.2/3.4 = 0.35$ m.

Now, the stream velocity from the Manning formula with a Manning's n value of 0.044 and slope, $S = 0.001$ is, upon substituting:

$$V = \frac{R^{\frac{2}{3}} S_f^{0.5}}{n} = \frac{0.35^{\frac{2}{3}} \times 0.001^{0.5}}{0.044} = 0.36 \, m \, s^{-1}$$

Therefore, the new discharge, $Q = V \times A = 0.36 \times 1.2 = 0.43 \, m^3 \, s^{-1}$

8) a) The soil infiltration characteristics are

f_t, infiltration rate at time, t

f_0, initial infiltration rate at time, $t = 0$ is 53 mm hour^{-1}

f_c, constant or equilibrium infiltration rate after the soil has become saturated (the infiltration capacity) is 5.1 mm hour^{-1}

k, decay constant specific to the soil type is 0.4 hour^{-1}

Therefore, from the Horton equation and upon substituting:

$$f_t = f_c + (f_0 - f_c)e^{-kt}$$
$$= 5.1 + (53 - 5.1)e^{-0.4t}$$

Infiltration rate at time, $t = 2$ hours is:

$f_t = 5.1 + (53 - 5.1)e^{-0.4 \times 2} = 26.6$ mm hour^{-1}

Infiltration rate at time, $t = 6$ hours is:

$f_t = 5.1 + (53 - 5.1)e^{-0.4 \times 6} = 9.4$ mm hour^{-1}

The total volume of infiltration, F, after time, t, is found from

$$F_t = f_c t + \frac{(f_0 - f_c)}{k}(1 - e^{-kt})$$

(b) Therefore, after 6 hours, the total volume of infiltration is

$$F_t = 5.1 \times 6 + \frac{(53 - 5.1)}{0.4}(1 - e^{-0.4 \times 6})$$

$$= 139.5 \text{ mm}$$

9) Rearrangement of the two-component mixing equation $C_T Q_T = C_P Q_P + C_E Q_E$ to express the ratio of Q_P/Q_T, the proportion of pre-event (baseflow) water comprising the total stream discharge, gives

$$\frac{Q_P}{Q_T} = \frac{C_T - C_E}{C_P - C_E}$$

a) With reference to the grassland catchment and using the $\delta^{18}O$ and Cl$^-$ data in turn, application of the above equation provides a method of stream hydrograph separation to calculate the baseflow (Q_P) and storm runoff (event) (Q_E) contributions to the total stream discharge (Q_T) as follows:

For the $\delta^{18}O$ data: $C_T = -7.00\permil$, $C_E = -6.00\permil$ and $C_P = -8.00\permil$

Therefore,

$$\frac{Q_P}{Q_T} = \frac{-7.00 - (-6.00)}{-8.00 - (-6.00)} = \frac{-1}{-2} = 0.5$$

For the Cl$^-$ data: $C_T = 6.0$ mg L^{-1}, $C_E = 2.0$ mg L^{-1} and $C_P = 10.0$ mg L^{-1}

Therefore,

$$\frac{Q_P}{Q_T} = \frac{6.0 - 2.0}{10.0 - 2.0} = \frac{4.0}{8.0} = 0.5$$

Hence, both calculations predict that the baseflow and storm runoff contributions to the total stream discharge are the

same and equal $0.5 \times 2.5 = 1.25$ m^3 s^{-1}.

Similarly, for the forested catchment:

For the $\delta^{18}O$ data: $C_T = -7.00\permil$, $C_E = -6.00\permil$ and $C_P = -7.25\permil$

Therefore,

$$\frac{Q_P}{Q_T} = \frac{-7.00 - (-6.00)}{-7.25 - (-6.00)} = \frac{-1}{-1.25} = 0.8$$

For the Cl$^-$ data: $C_T = 6.0$ mg L^{-1}, $C_E = 2.0$ mg L^{-1} and $C_P = 7.0$ mg L^{-1}

Therefore,

$$\frac{Q_P}{Q_T} = \frac{6.0 - 2.0}{7.0 - 2.0} = \frac{4.0}{5.0} = 0.8$$

In this case, the calculations predict that the baseflow contribution to the total stream discharge is equal to $0.8 \times 1.8 = 1.44$ m^3 s^{-1}. By difference, the storm runoff contribution to the total stream discharge equals $1.8 - 1.44 = 0.36$ m^3 s^{-1}.

b) The response of the two catchments to the same rain event is different. The two catchments are of similar size, but the grassland catchment is developed on a lower permeability, calcareous till soil type and has a greater storm runoff component compared with the forested catchment. The forested catchment is developed on more permeable, acidic soil and exhibits a greater baseflow discharge component. It is also apparent that the forest canopy captures some of the rainfall amount and so explains the smaller total stream discharge from this catchment compared to the grassland catchment.

10) a) A plot of volumetric moisture content against depth below ground level is given in Fig. AS.10.10a. The two profiles for 1 July and 8 July identify a zero flux plane at a depth of 45 cm below ground level. By dividing the profiles into discrete depth intervals (Fig. AS.10.10a), Table AS10.7 shows the evaporation and drainage losses calculated for the 7-day period.

From Table AS10.7, the estimated evaporation losses amount to 8.15 mm

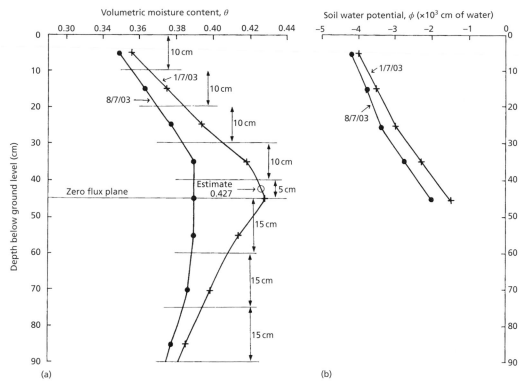

Fig. AS.10.10 Plots of (a) volumetric moisture content and (b) soil water potential versus depth for a clay soil with a cover of short-rooting vegetation using the data given in Table A10.10.

Table AS10.7 Calculation of evaporation and drainage losses from a clay soil profile during a 7-day period.

Depth range (cm)	d (mm)	$\Delta\theta$	$\Delta\theta \times d$ (mm)	Flux OUT = Flux IN + [$\Delta\theta + d$)] (mm)
Evaporation losses				
40–45	50	$0.427^* - 0.390 = 0.037$	1.85	$1.85 = 0 + 1.85$
30–40	100	$0.418 - 0.390 = 0.028$	2.80	$4.65 = 1.85 + 2.80$
20–30	100	$0.394 - 0.378 = 0.016$	1.60	$6.25 = 4.65 + 1.60$
10–20	100	$0.375 - 0.363 = 0.012$	1.20	$7.45 = 6.25 + 1.20$
0–10	100	$0.356 - 0.349 = 0.007$	0.70	$\mathbf{8.15} = 7.45 + 0.70$
Drainage losses				
45–60	150	$0.413 - 0.390 = 0.023$	3.45	$3.45 = 0 + 3.45$
60–75	150	$0.399 - 0.387 = 0.012$	1.80	$5.25 = 3.45 + 1.80$
75–90	150	$0.385 - 0.378 = 0.007$	1.05	$\mathbf{6.30} = 5.25 + 1.05$

* Estimated from graph plot.

and the drainage losses amount to 6.30 mm during the one-week period.

b) From inspection of Fig. AS.10.10b, on 8 July, the difference in soil water potential in the depth range 30–40 cm below ground level is, on average, represented by the difference in soil water potential readings taken at 25 and 45 cm below ground level ($-2.00 - (-3.40)$) = 1.4×10^3 cm of water). Therefore, the hydraulic gradient approximated across 30–40 cm depth is ($1.4 \times 10^3/20 = 70$). The flux of water per square metre of soil lost to evaporation in the depth interval 30–40 cm is equal to 2.80 mm in 7 days (see Table AS10.7). Now, using Darcy's law ($q = -iK$) and rearranging gives an estimate of the unsaturated soil hydraulic conductivity during the measurement period:

$$K = q/i = (2.80 \times 10^{-3}/7)/70$$
$$= 5.71 \times 10^{-6} \text{ m day}^{-1}$$

or 6.61×10^{-11} ms^{-1}.

c) The effect of intense drying on the clay soil would be to produce drying cracks. During a subsequent rain event, water would be able to migrate via these cracks as preferential flow and so disrupting the soil moisture content profile, causing a rise in the position of the zero flux plane. Hence, the amount of water lost by evaporation from the soil profile will be less than if a homogeneous, structureless clay soil existed during the rain event.

11) a) The calculation of the groundwater recharge (hydrological excess, *HXS*) during the period 2002–2003 is presented in Table AS10.8. The start date is 1 April when the catchment soils are assumed to be at field capacity (i.e. the soil moisture deficit $= 0$ mm). *P* is the precipitation amount. *PE* and *AE* are the potential and actual amounts of evapotranspiration with $AE = 0.1(PE)$.

All values are expressed as an equivalent depth of water in mm. A root constant value of 75 mm is assumed for a short-rooted grass cover.

Hence, from the values in Table AS10.8, the quantity of groundwater recharge during the winter period 2002–2003 $= 123 + 19 + 74 + 15 + 45 = 276$ mm.

b) The borehole hydrograph method of groundwater recharge estimation gives

$$\text{Recharge (mm)} = \Delta h \times S$$

where Δh is the amplitude of change in the measured groundwater level during the recharge period (2.75 m) and *S* is the aquifer storage coefficient. For the unconfined sand and gravel aquifer:

$$276 = 2750 \times S$$

and

$$S, \text{ the specific yield } (S_y)$$
$$= 276/2750 = 0.10 \text{ or } 10\%$$

The replenishment of groundwater storage volume in the sand and gravel aquifer during the recharge period is equal to the product of the catchment area (200 km²) and recharge amount $= 200 \times 10^6 \times 0.276 = 5.52 \times 10^7$ m³.

c) If it is assumed that the volume of groundwater recharge contributes to the groundwater discharge component of river flow, then the recharge amount as a fraction of the average annual river flow provides an estimate of the baseflow index for the catchment and is equal to $(5.52 \times 10^7)/(1.843 \times 365 \times 86\,400) = 0.95$. Hence, the baseflow index for this catchment, which is underlain by permeable sand and gravel, is 95%.

d) A fuller assessment of the catchment water balance should be based on longer records of hydrological and hydrogeological data and should also consider other possible flowpaths, for

Table AS10.8 Calculation of hydrological excess water (*HXS*) representing direct groundwater recharge.

Month	*P* (mm)	*PE* (mm)	*P – PE* (mm)	*AE* (mm)	*SMD* (mm)	*HXS* (mm)
2002						
Jan	65	17				
Feb	69	21				
March	34	37				
April	14	90	−76	90	76	
May	29	110	−81	$[29 + (100 - 76) + 0.1(57)] = 59$	$76 + 24 + 6 = 106$	0
June	66	105	−39	$[66 + 0.1(39)] = 70$	$106 + 4 = 110$	0
July	38	112	−74	$[38 + 0.1(74)] = 45$	$110 + 7 = 117$	0
Aug	66	88	−22	$[66 + 0.1(22)] = 68]$	$117 + 2 = 119$	0
Sept	128	54	74	54	$119 - 74 = 45$	0
Oct	73	31	42	31	$45 - 42 = 3$	0
Nov	142	16	126	16	$3 - 126 = -123$	123
Dec	39	20	19	20	0	19
2003						
Jan	90	16	74	16	0	74
Feb	32	17	15	17	0	15
March	78	33	45	33	0	45
April	43	60	−17	60	17	0
May	73	84	−11	84	$17 + 11 = 28$	0
June	23	124	−101	$[23 + (100 - 28) + 0.1(29)] = 98$	$28 + 72 + 3 = 103$	0
July	32	122	−90	$[32 + 0.1(90)] = 41$	$103 + 9 = 112$	0
Aug	18	102	−84	$[18 + 0.1(84)] = 26$	$112 + 8 = 120$	0
Sept	128	54	74	54	$120 - 74 = 46$	0
Oct	28	24	4	24	$46 - 4 = 42$	0
Nov	71	8	63	8	$42 - 63 = -21$	21
Dec	34	7	27	7	0	27

example cross-formational flow of groundwater. The safe yield of any proposed development of groundwater abstractions should consider the desired minimum flow in the river, particularly during drought periods. An ecological assessment is required, although the Q95 statistic (the river flow that is equalled or exceeded for 95% of the time) is a common low flow statistic used when considering the environmental impacts of water resources schemes.

12) Using the chloride budget method of recharge calculation, $R = P \times (C_p/C_s)$, where R is recharge rate, C_s is the mean chloride concentration of the interstitial pore water in the unsaturated zone (8.00

mg L^{-1}) and C_p is the mean chloride concentration in the rainwater (4.33 mg L^{-1}). For a given mean annual precipitation amount of 120.1 mm, then the calculated recharge rate is equal to 120.1/(4.33/8.00) = 65.0 mm a^{-1}.

The aquifer area is 200 km^2 and so the annual volume of groundwater recharge = $200 \times 10^6 \times 65 \times 10^{-3} = 1.3 \times 10^7$ m^3. The drip-feed irrigation water demand is for an area of 26 000 ha or 2.6×10^8 m^2. Hence, the sustainable irrigation rate (assuming no other demands for water) $= (1.3 \times 10^7)/(2.6 \times 10^8)$ m a^{-1} = 0.05 m a^{-1}. This irrigation rate is equivalent to a rate of 0.05 $\times$ 10 000/365 = 1.37 m^3 ha^{-1} day^{-1} or 1370 L ha^{-1} day^{-1}.

The unsaturated zone thickness, l, is 78 m with an average moisture content, θ, of 0.65% (=0.0065). If the recharge rate, i, is 65 mm a^{-1}, then the time of travel (or residence time) of water in the unsaturated zone is found from $1/\bar{v} = l/(i/\theta) = (78)/(65 \times 10^{-3})/(0.0065) = 7.8$ years.

A10.6 Groundwater resources, pumping tests and stream depletion analysis

5) The solution to this problem requires an analysis of steady radial flow to a well using the equilibrium or Thiem equation:

$$T = Kb = \frac{Q}{2\pi(h_2 - h_1)} \log_e \frac{r_2}{r_1}$$

With reference to Fig. AS.10.11, the aquifer transmissivity is found as follows:

$$T = \frac{0.2}{2\pi(30.84 - 29.50)} \log_e \frac{1000}{350}$$

$$= 0.02375 \times 1.0498$$

$$= 0.025 \text{ m}^2 \text{ s}^{-1} \text{ or } 2160 \text{ m}^2 \text{ day}^{-1}$$

From this value of aquifer transmissivity, the hydraulic conductivity is found from $K = T/b$, where the aquifer thickness is 25 m. Hence, $K = 2160/25 = 86.4$ m day^{-1}.

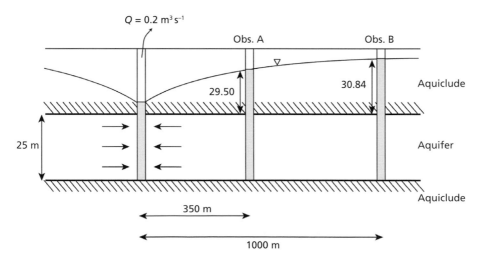

Fig. AS.10.11 Set-up of a pumping test in a confined aquifer with the drawdown response in the abstraction borehole recorded at two observation boreholes.

6) a) A value for hydraulic conductivity, K, is found from the Thiem equation (eq. 7.16) for an unconfined aquifer:

$$K = \left[\frac{Q}{\pi\left(h_2{}^2 - h_1{}^2\right)}\right] \log_e \frac{r_2}{r_1}$$

From the pumping test data:

$Q = 1 \times 10^4 \, \text{m}^3 \, \text{day}^{-1}$

$h_1 = 321.96 \, \text{m}$

$h_2 = 325.00 \, \text{m}$

$r_1 = 50 \, \text{m}$

$r_2 = 1100 \, \text{m}$

On substitution in the Thiem equation:

$$K = \left[\frac{1 \times 10^4}{\pi\,(325^2 - 321.96^2)}\right] \log_e \frac{1100}{50}$$

$$= 5.00 \, \text{m day}^{-1} \text{ or } 5.787 \times 10^{-5} \, \text{m s}^{-1}$$

b) The relationship between hydraulic conductivity, K, and intrinsic permeability, k_i, is

$$K = k_i \frac{\rho g}{\mu}$$

The fluid property values are

$\rho = 999.099 \, \text{kg m}^{-3}$

$\mu = 1.1404 \times 10^{-3} \, \text{kg m}^{-1} \, \text{s}^{-1}$

which yield, for a value of $g = 9.81$ m s^{-2}:

$$\frac{\rho g}{\mu} = \frac{999.099 \times 9.81}{1.1404 \times 10^{-3}}$$

$$= 8\,594\,494.204 \, \text{m}^{-1} \text{s}^{-1}$$

From part (a), $K = 5.787 \times 10^{-5} \, \text{m s}^{-1}$

Therefore,

Intrinsic permeability,

$$k_i = \frac{5.787 \times 10^{-5}}{8\,594\,494.204} = 6.73 \times 10^{-12} \, \text{m}^2$$

or, on conversion: $k_i = \dfrac{6.73 \times 10^{-12}}{9.87 \times 10^{-13}}$

$$= 6.82 \, \text{darcies (6819 millidarcies)}$$

c) The average intrinsic permeability of the sandstone calculated in part (b) from the pumping test is $6.73 \times 10^{-12} \, \text{m}^2$ and compares with an average value from the core samples of 60 millidarcies (or

0.06 darcy), equivalent to $0.06/9.87 \times 10^{-13} = 5.922 \times 10^{-14} \, \text{m}^2$.

Therefore, the average intrinsic permeability of the core material ($5.92 \times 10^{-14} \, \text{m}^2$) is two orders of magnitude lower than the intrinsic permeability of the sandstone calculated from the pumping test ($6.73 \times 10^{-12} \, \text{m}^2$). There are a number of reasons that could explain this difference. For example, the assumed temperature of 15°C of groundwater in the aquifer may have been different to the temperature of the water used to test the core samples. Also, the calculated value of hydraulic conductivity, K, from the pumping test data assumes that the Dupuit assumptions are met for horizontal flow in an unconfined aquifer. However, the principal reason why the value of intrinsic permeability, k_i, based on the pumping test data is higher compared with the average value for core material is the likely presence of fractures in the sandstone aquifer that increases its overall intrinsic permeability.

7) First plot the field data of drawdown (m) versus time (min) on log 3 cycles by 4 cycles graph paper. Next, overlay a copy of the Theis type curve (Fig. 7.14) with the exact same log scales and, keeping the axes parallel, read off the four match points once the best fit is achieved between the type curve and the field curve (see Fig. AS.10.12). Four possible match point values are

$W(u) = 1.0$

Drawdown, $s = 2.8 \, \text{m}$

$1/u = 1.0$

Time, $t = 16.5 \, \text{min}$

The pumping rate $= 295 \, \text{m}^3 \, \text{hour}^{-1}$ and the radial distance, r, from the abstraction borehole to the observation borehole $= 218 \, \text{m}$. Hence, from the Theis equations:

$$s = \frac{Q}{4\pi T} W(u) \quad \text{and} \quad u = \frac{r^2 S}{4Tt}$$

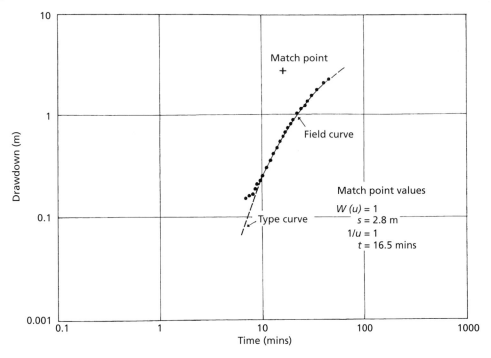

Fig. AS.10.12 Plot showing the overlay of drawdown versus time data (the field curve) with the Theis type curve using data given in Table A10.12. The selection of a match point and associated values used in the pumping test interpretation are also shown.

and rearranging to find T:

$$T = \frac{Q}{4\pi s} W(u) = \frac{295 \times 24}{4\pi 2.8} 1.0 = 201 \ \text{m}^2 \ \text{day}^{-1}$$

and

$$S = \frac{u4Tt}{r^2} = \frac{1.0 \times 4 \times 201 \times 16.5/(60 \times 24)}{218^2}$$

$$= 2 \times 10^{-4}$$

Therefore, the aquifer properties at this confined Chalk aquifer site are transmissivity of $201 \ \text{m}^2 \ \text{day}^{-1}$ and storage coefficient of 0.0002.

8) A plot of drawdown versus time on semilogarithmic graph paper for the data collected at Borehole A is shown in Fig. AS.10.13. The later data fit a straight line with a difference in drawdown $(s_2 - s_1)$ of 3.44 m observed over one log cycle of time. For a drawdown, s, equal to zero the corresponding value of t_o is 470 s. Following the Cooper–Jacob

straight-line method of pumping test analysis, then for a pumping rate of $2000 \ \text{m}^3 \ \text{day}^{-1}$:

$$s_2 - s_1 = \frac{2.3Q}{4\pi T}$$

Thus, transmissivity,

$$T = \frac{2.3 \times 2000}{4\pi \times 3.44} = 106 \ \text{m}^2 \ \text{day}^{-1}$$

For a radial distance of 165 m, the storage coefficient, S, is found from

$$S = \frac{2.25Tt_0}{r^2} = \frac{2.25 \times 106 \times 470/86\,400}{165^2}$$

$$= 4.8 \times 10^{-5}$$

Therefore, the aquifer properties for the confined lower limestone are transmissivity of $106 \ \text{m}^2 \ \text{day}^{-1}$ and storage coefficient of 0.00005. These values for T and S should be regarded with caution, particularly S, since the short duration of the pumping test invalidates the condition that $u < 0.01$.

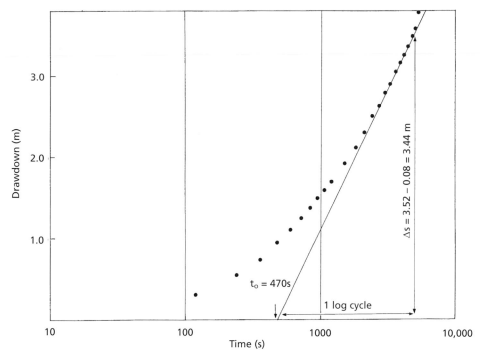

Fig. AS.10.13 Plot of drawdown versus time using given in Table A10.13 for the interpretation of aquifer properties using the Cooper–Jacob method.

The cause of the gradual rise in groundwater level in Borehole B is due to the discharge of the pumped water into the river. The river is in hydraulic contact with the upper limestone aquifer and so recharge is occurring to the shallow water table as a result of the influent river condition.

9) The siting of the new wellfield requires that the river flow is not depleted by more than 2×10^4 m³ day⁻¹ during an operational period, t_p, of 90 days. The wellfield will yield 5×10^4 m³ day⁻¹. To analyse this problem, the method of Jenkins (1968) is used (see Box 6.5).

The allowable volume of stream depletion, v, during time $t_p = 90 \times 2 \times 10^4 = 1.8 \times 10^6$ m³
The net volume of water pumped, Q_t, during time $t_p = 90 \times 5 \times 10^4 = 4.5 \times 10^6$ m³

$$\text{Next compute } v/Q_t = \left(1.8 \times 10^6\right)/ \left(4.5 \times 10^6\right) = 0.4$$

Using Appendix 7, when $v/Q_t = 0.4$, the quotient $t/F = 1.85$, where F is the stream depletion factor.

Now, $tT/L^2S = t/F = 1.85$ and rearranging:

$$L = \left(\frac{tT}{1.85S}\right)^{1/2} = \left(\frac{90 \times 1500}{1.85 \times 0.10}\right)^{1/2} = 854.2 \text{ m}$$

Hence, the wellfield should be located at least 854 m from the river in order to avoid environmental impacts.

Assumptions pertaining to this type of analytical solution are discussed in Section 6.8.2. The assumptions are an oversimplification of reality and only provide a rough estimate of stream depletion effects. By neglecting river bed and river bank sediments and assuming full aquifer penetration by the river, the impact of pumping on the stream flow depletion is over-estimated. The solution can therefore be viewed as conservative. The wellfield is envisaged as a single point of groundwater abstraction, whereas the true effect will be predicted using the principle of superposition of the individual drawdown responses. In practice, the composite cone of depression is too complicated to predict by analytical methods and is better analysed using a numerical model.

A10.7 Contaminant hydrogeology

11) The answer to whether the site should undergo remediation requires the application of the Ogata–Banks equation describing one-dimensional transport of the contaminated groundwater by advection and dispersion:

$$\frac{C}{C_0} = \frac{1}{2}\left[erfc\left(\frac{l - \bar{v}_l t}{2\sqrt{\alpha_l \bar{v}_l t}}\right)\right]$$

The advective velocity, $\bar{v}$, is found using Darcy's law, where $\bar{v} = q/n_e = iK/n_e$. For the given hydraulic parameter values, $\bar{v} = (1 \times 10^{-4} \times 100/86\,400)/(0.115) = 1 \times 10^{-6}$ m s^{-1}.

The problem requires a calculation of whether a relative concentration of $C/C_0 = 0.01$ will be recorded at a distance to the river, l, of 3 km in time, t, of 30 years with an aquifer longitudinal dispersivity, α_l, of 100 m. Substitution of these values in the Ogata–Banks equation yields a predicted value for the travel distance, l, as follows:

$$0.01 = \frac{1}{2}\left[erfc\left(\frac{l - 1 \times 10^{-6} \times 30 \times 365 \times 86\,400}{2\sqrt{\frac{100 \times 1 \times 10^{-6}}{\times 30 \times 365 \times 86\,400}}}\right)\right]$$

$$0.02 = \left[erfc\left(\frac{l - 946.08}{615.17}\right)\right]$$

If $0.02 = erfc(\beta)$, where $\beta = \left(\frac{l-946.08}{615.17}\right)$

Then, using Appendix 8, the value of β to give an $erfc(\beta)$ of 0.02 is 1.65. Hence,

$$1.65 = \left(\frac{l - 946.08}{615.17}\right)$$

Rearranging,

$$l = (1.65)(615.17) + 946.08$$
$$= 1961 \text{ m or about 2 km.}$$

Therefore, with a travel distance of 2 km, the contaminated groundwater will not migrate as far as 3 km, the distance between the Waterside site and the river, and so remediation is not required.

12) The answers to parts (a), (b) and (c) are summarized in Table AS10.9.

(d) The resulting breakthrough curve for the layered aquifer is given in Fig. AS.10.14. The breakthrough time when the concentration of TCE in the borehole is half of the original concentration in the leachate (i.e. $C/C_0 = 0.5$) is read from the graph and equals 60 days. The number of pore volume flushes during this time is found as follows:

Total pore volume of layers 1, 2 and 3 = $(2 \times 100 \times 0.20) + (1 \times 100 \times 0.40) + (3 \times 100 \times 0.15) = 125$ m^3. In 60 days, the total groundwater discharge through the three layers is $7.6 \times 60 = 456$ m^3.

Hence, the number of pore volume flushes in this time = $456/125 = 3.65$.

13) The solution to this problem requires an estimate of the attenuated contaminant transport velocity, $\bar{v}_c$, and from this a calculation of the time of travel, t, from the deep disposal site (P1) to the shallow water table (P2), a distance, l, of 5000 m. The amount of radioactive decay during this time period can then be found from the radioactive decay equation to estimate the concentration of ^{239}Pu reaching the near-surface environment.

From the retardation equation and using the given values:

$$R_d = 1 + K_d\frac{(1-n)\rho}{n} = 1 + 100\frac{(1-0.01)2.65}{0.01}$$
$$= 26\,236$$

Table AS10.9 Calculation of relative concentration breakthrough values of TCE for a layered sand and gravel aquifer.

Layer	K (m day⁻¹)	d (m)	n	ρ_b (g cm⁻³)	f_{oc}	(1) $\bar{v}_w$ (m day⁻¹)	(2) K_d (cm³ g⁻¹)	(3) R_d	(4) $\bar{v}_c$ (m day⁻¹)	(5) t (days)	(6) Q (m³ day⁻¹)	Q/Q_T	C/C_o
1	10	2	0.20	1.60	0.015	5.00	2.25	19.00	0.263	380.2	2.00	7/7.6	0.92
2	50	1	0.40	1.50	0.015	12.50	2.25	9.44	1.324	75.5	5.00	5/7.6	0.66
3	2	3	0.15	1.65	0.015	1.33	2.25	25.75	0.052	1923.1	0.60	7.6/7.6	1.00

Column (1) is found using Darcy's law, where $\bar{v} = q/n_e = iK/n_e$. From inspection of Fig. A10.12, the hydraulic gradient, $i = (17 - 7)/100 = 0.10$.
Column (2) is found from $f_{oc} \times K_{OC}$ where $K_{OC} = 150\,\text{cm}^3\,\text{g}^{-1}$.
Column (3) is calculated using the retardation equation:

$$R_d = 1 + K_d \frac{\rho_b}{n}$$

Column (4) is found from $\bar{v}_c = \bar{v}_w / R_d$
Column (5) is found from time, $t = 100/\bar{v}_c$
Column (6) is found from $Q = AiK$

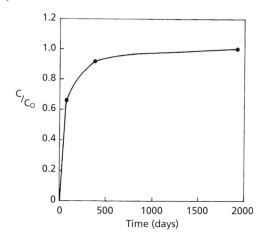

Fig. AS.10.14 Plot of the relative concentration breakthrough curve for TCE in a layered sand and gravel aquifer using data presented in Table AS10.9.

$$\bar{v}_w = q/n = iK/n$$
$$= (((350 - 50)/5000) \times 1 \times 10^{-5})/0.01$$
$$= 6.0 \times 10^{-5}\, \mathrm{m\,s^{-1}}$$
$$\bar{v}_c = \bar{v}_w/R_d = 6.0 \times 10^{-5}/26\,236$$
$$= 2.3 \times 10^{-9}\, \mathrm{m\,s^{-1}}$$

Hence, assuming uniform, one-dimensional groundwater flow in an homogeneous, isotropic aquifer, the time of travel, $t = l/\bar{v}_c = 5000/2.3 \times 10^{-9} = 2.2 \times 10^{12}\,\mathrm{s}$ or 69 328 years.

Using the radioactive decay law $\frac{A}{A_o} = e^{-\lambda t}$ Then, with A_o equal to 100 Beq L^{-1}, the activity, A, of ^{239}Pu at a time of 69 328 years is

$$A = 100 \times e^{-2.84 \times 10^{-5} \times 69\,328}$$
$$= 100 \times 0.14 = 14\ \mathrm{Beq\,L^{-1}}$$

Hence, 14 Beq L^{-1} is a detectable concentration and the shallow groundwater environment is at risk of contamination from radioactive ^{239}Pu. The travel time may be an underestimate since the siltstone contains minor fracturing presenting the possibility of preferential flowpaths through the aquitard between the deep disposal site and the unconfined, shallow aquifer.

14) a) From Darcy's law:
Specific discharge, $q = -iK$
Average linear velocity, $\bar{v} = q/n_e$
Given values are:
Effective porosity, $n_e = 0.25$
Hydraulic conductivity, $K = 120$ m day^{-1}
Hydraulic gradient, $i = 0.005$
Distance between the spill and borehole, $L = 240$ m

The groundwater velocity, $\bar{v}$, is unknown. Once found, the time of travel, t, can be found from $L/\bar{v}$

Hence, for a specific discharge, $q = 0.005 \times 120 = 0.6$ m day^{-1}
Then, $\bar{v} = 0.6/0.25 = 2.4$ m day^{-1}
Therefore, the time taken for the spill to reach the borehole $= 240/2.4 = 100$ days

b) The answer assumes uniform, horizontal groundwater flow and isotropic, homogeneous aquifer properties, which are unlikely given the description of the gravel aquifer as containing sand and clay fractions. Sodium contained in the wastewater is likely to undergo ion exchange or adsorption on the clay fraction making up the aquifer lithology. Chloride is a conservative species and so is not expected to be attenuated and so will arrive at the same time as the groundwater, and ahead of the sodium plume.

15) a) Using Darcy's law, first find the time taken for the groundwater to reach the borehole, a distance of 15 km from the dry cleaning shop:
$q = -iK$
$\bar{v} = q/n_e$ and, therefore, $\bar{v} = iK/n_e$
Given values are:
Effective porosity, $n_e = 0.05$
Hydraulic conductivity, $K = 50$ m day^{-1}
Hydraulic gradient, $i = 0.015$
Distance between the spill and borehole, $L = 240$ m

Substituting the given values, then $\bar{v} = (0.015 \times 50)/0.05 = 15$ m day^{-1}.

Therefore, the time taken for groundwater to reach the borehole from the dry cleaning shop, t_w, = 15 000/15 = 1000 days or 2.74 years.

For the given retardation factor, $R_d = 3.3$, where $R_d = t_c/t_w$, then the time for the PCE contaminant to reach the borehole, t_c, is, on rearranging = 3.3 × 2.74 = 9.04 years.

b) Due to the presence of fissures in the Chalk in which groundwater velocities can be high, the contaminant may arrive at the borehole earlier than calculated based on groundwater flow through the bulk of the Chalk matrix. In addition, the influence of the cone of depression developed around the pumping borehole will act to increase the groundwater velocity compared to transport under a natural hydraulic gradient.

16) a) The maximum concentration of dissolved TCE expected in the aquifer is equal to its solubility (at 20°C) which is 1000 mg L^{-1}.

b) The average linear groundwater velocity is found using Darcy's law, where $\bar{v} = q/n_e = iK/n_e$. The hydraulic gradient, i, is 0.005, the hydraulic conductivity, K, is 2×10^{-3} m s^{-1} and the effective porosity, n_e, is 0.25. Hence, $\bar{v} = (0.005 \times 2 \times 10^{-3})/0.25 = 4 \times 10^{-5}$ m s^{-1}.

c) The spillage of TCE and the release of a dissolved component in the groundwater is likely to be attenuated by hydrophobic sorption of the non-polar organic compound and can be described by the retardation equation, where

$$R_d = 1 + K_d \frac{\rho_b}{n}$$

The retardation factor, R_d is equal to the ratio of the groundwater velocity to the contaminant velocity $\bar{v}_w/\bar{v}_c$. Hence, by finding R_d and knowing $\bar{v}_w$ from part (b), then it is possible to calculate $\bar{v}_c$. The porosity, n, is 0.25 and the bulk density, ρ_b, of the alluvial sand is 2.10 g cm^{-3}. The partition coefficient, K_d, is found using the Schwarzenbach and Westall (1981) equation as follows:

$$\log_{10} K_{OC} = 0.49 + 0.72\log_{10} K_{OW}$$

$\log_{10} K_{OW} = 2.42$ for TCE at 25°C. Substituting in the above equation:

$$\log_{10} K_{OC} = 0.49 + (0.72 \times 2.42) = 2.2324$$

Therefore, $K_{OC} = 170.765$.

From the relationship $K_d = K_{OC} \times f_{OC}$, then for a fraction of organic carbon in the aquifer, f_{OC}, of 0.5% or 0.005, then $K_d = 170.765 \times 0.005 = 0.8538$ cm^3 g^{-1}. The retardation factor is now found using the retardation equation:

$$R_d = 1 + 0.8538\frac{2.10}{0.25} = 8.172$$

Therefore, the contaminant velocity $\bar{v}_c = \bar{v}_w/R_d$
$= 4 \times 10^{-5}/8.172 = 4.895 \times 10^{-6}$ m s^{-1}. Ignoring dispersion and assuming uniform groundwater flow, the time taken for the dissolved plume of TCE to reach the river at a distance of 400 m is equal to 400/($4.895 \times 10^{-6} \times 86\,400$) = 946 days or about 2.6 years.

17) The solution requires the application of single-well capture-zone type curves (Fig. A10.15) overlain on the site map (Fig. A10.14) having made sure that the scale bars are the same. The choice of capture-zone type curve is found from calculating Q/qb, where Q is the total spring discharge of $(2.5 + 0.5) \times 10^3$ m^3 day^{-1}, b is the aquifer

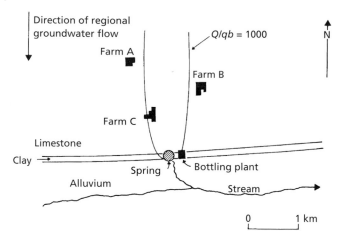

Fig. AS.10.15 Capture-zone type curve for a bottling plant supplied by water from a spring emerging from a limestone aquifer.

thickness of 20 m and q is the specific discharge found using Darcy's law ($q = -iK$), where i is the hydraulic gradient of 6×10^{-4} and K is the hydraulic conductivity. The hydraulic conductivity, K, is found from T/b where T is the aquifer transmissivity value of 5000 m^2 day^{-1}. From this information, $q = (5000/20) \times 6 \times 10^{-4} = 0.15$ m day^{-1}. Now, $Q/qb = (3.0 \times 10^3)/(0.15 \times 20) = 1000$. For a value of $Q/qb = 1000$, the associated capture-zone type curve encompasses Farm C but not farms A or B (see Fig. AS.10.15). Hence, of the dairy farms, Farm C requires further investigation as a potential source of the ammonium and biological oxygen demand in the spring water at the bottling plant.

As a limestone aquifer, the possibility of anisotropy and heterogeneity caused by the solutional weathering of bedding planes and joints to produce fissuring is a problem since all three farms are either within or close to the chosen capture-zone type curve. Therefore, at this stage of the investigation, further work is required to characterize the aquifer properties, potentially involving tracer tests to determine which of the farms, if any, is causing deterioration in groundwater quality.

References

Chebotarev, I.I. (1955) Metamorphism of natural water in the crust of weathering. *Geochimica et Cosmochimica Acta* **8**, 22–48, 137–170, 198–212.

Jenkins, C.T. (1968) *Computation of rate and volume of stream depletion by wells.* In: *Techniques of Water-Resources Investigations of the United States Geological Survey, Chapter D1. Book 4, Hydrologic Analysis and Interpretation,* Series 04-D1. United States Department of the Interior, Washington, 17 pp.

Merrin, P.D. (1996) *A geochemical investigation into the nature and occurrence of groundwaters from the Llandrindod Wells Ordovician Inlier, Wales.* MSc Thesis, University of East Anglia, Norwich.

Robins, N.S. (1986) Groundwater chemistry of the main aquifers in Scotland. *British Geological Survey Report* **18**, 2.

Saleh, A., Al-Ruwaih, F. and Shehata, M. (1999) Hydrogeochemical processes operating within the main aquifers of Kuwait. *Journal of Arid Environments* **42**, 195–209.

Schwarzenbach, R.P. and Westall, J. (1981) Transport of nonpolar organic compounds from surface water to groundwater. Laboratory sorption studies. *Environmental Science & Technology* **15**, 1360–1367.

Index

Page numbers in *italics* refer to figures, those in **bold** to tables.

Hydrogeology: Principles and Practice, Third Edition. Kevin M. Hiscock and Victor F. Bense.
© 2021 John Wiley & Sons Ltd. Published 2021 by John Wiley & Sons Ltd.
Companion website: www.wiley.com/go/hiscock/hydrogeology3e

(a)

(b)

Plate 1.1 Examples of now redundant village pumps once widespread in their use in Britain: (a) the wooden pump on Queen's Square, Attleborough, Norfolk, enclosed 1897; and (b) the large Shalders pump used in the days before tarmac for dust-laying on the old turnpike (Newmarket Road) at Cringleford, near Norwich, Norfolk.

Hydrogeology: Principles and Practice, Third Edition. Kevin M. Hiscock and Victor F. Bense.
© 2021 John Wiley & Sons Ltd. Published 2021 by John Wiley & Sons Ltd.
Companion website: www.wiley.com/go/hiscock/hydrogeology3e

Plate 1.2 Arcade of the aqueduct Aqua Claudia situated in the Parco degli Acquedotti, 8 km east of Rome. The aqueduct, which is built of cut stone masonry, also carries the brick-faced concrete Anio Novus, added later on top of the Aqua Claudia.

Plate 1.3 The baroque *mostra* of the Trevi Fountain in Rome. Designed by Nicola Salvi in 1732, and fed by the Vergine aqueduct, it depicts Neptune's chariots being led by Tritons with sea horses, one wild and one docile, representing the various moods of the sea.

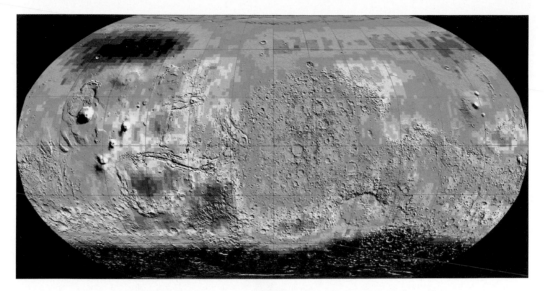

Plate 1.4 Global map of epithermal neutron currents measured on the planet Mars obtained by the NASA Odyssey Neutron Spectrometer orbiter. Epithermal neutrons provide the most sensitive measure of hydrogen in surface soils. Inspection of the global epithermal map shows high hydrogen content (blue colour) in surface soils south of about 60° latitude and in a ring that almost surrounds the north polar cap. The maximum intensity in the northern ring coincides with a region of high albedo and low thermal inertia, which are both required for near-surface water ice to be stable. Also seen are large regions near the equator that contain enhanced near-surface hydrogen, which is most likely in the form of chemically and/or physically bound water and/or hydroxyl radicals since water ice is not stable near the equator. (Source: Reproduced from Los Alamos National Laboratory. © Copyright 2011 Los Alamos National Security, LLC. All rights reserved.)

Plate 1.5 Hose reel and rain gun irrigation system applied to a potato field in Norfolk, eastern England on 30 May 2020 and supplied by groundwater from the underlying Cretaceous Chalk aquifer.

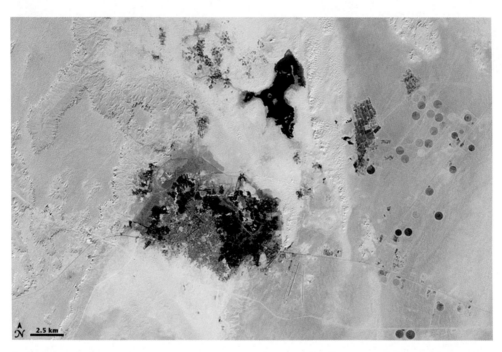

Plate 1.6 NASA Landsat-7 satellite image of the Ouargla Oasis, Algeria on 20 December, 2000. In this false-colour image, red indicates vegetation (the brighter the red, the more dominant the vegetation). Pale pink and orange tones show the desert landscape of sand and rock outcrops. The satellite image shows date palms surrounding the urban area of Ouargla and Chott Aïn el Beïda in the southwest, a saline depression that has traditionally collected irrigation runoff, as well as the proliferation of irrigated land to the north and east of Ouargla in the vicinity of Chott Oum el Raneb. The width of the image shown is approximately 40 km. (Source: Reproduced from http://earthobservatory.nasa.gov/IOTD. NASA images created by Jesse Allen and Rob Simmon, using Landsat data provided by the United States Geological Survey. Caption by Michon Scott.)

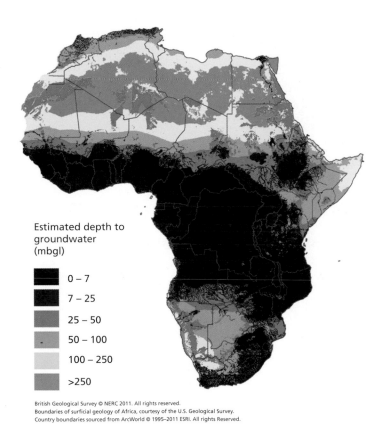

Estimated depth to
groundwater
(mbgl)

■	0 – 7
■	7 – 25
■	25 – 50
■	50 – 100
□	100 – 250
■	>250

British Geological Survey © NERC 2011. All rights reserved.
Boundaries of surficial geology of Africa, courtesy of the U.S. Geological Survey.
Country boundaries sourced from ArcWorld © 1995–2011 ESRI. All rights Reserved.

Plate 1.7   Estimated depth in metres below ground level (m bgl) to groundwater in Africa (Bonsor and MacDonald 2011). (Source: Bonsor, H.C. and MacDonald, A.M. (2011). *An initial estimate of depth to groundwater across Africa, 2011*, British Geological Survey. © 2011, British Geological Survey.)

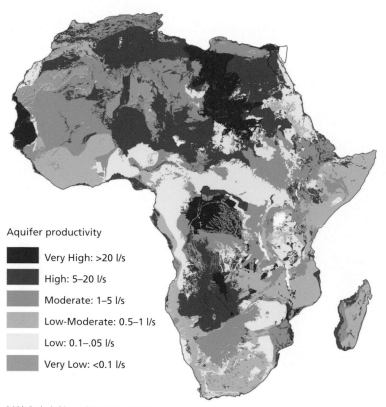

Aquifer productivity

- ■ Very High: >20 l/s
- ■ High: 5–20 l/s
- ■ Moderate: 1–5 l/s
- ■ Low-Moderate: 0.5–1 l/s
- ■ Low: 0.1–.05 l/s
- ■ Very Low: <0.1 l/s

British Geological Survey © NERC 2011. All rights reserved.
Boundaries of surficial geology of Africa, courtesy of the U.S. Geological Survey.
Country boundaries sourced from ArcWorld © 1995–2011 ESRI. All rights Reserved.

Plate 1.8   Aquifer productivity in litres per second (L s⁻¹) for Africa showing the likely interquartile range for boreholes drilled and sited using appropriate techniques and expertise (Bonsor and MacDonald 2011). (Source: Bonsor, H.C. and MacDonald, A.M. (2011). *An initial estimate of depth to groundwater across Africa, 2011*, British Geological Survey. © 2011, British Geological Survey.)

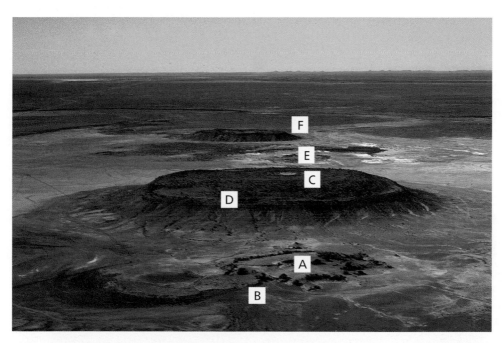

Plate 2.1   Aerial view of artesian springs and spring mounds west of Lake Eyre South (137°E, 29°S) in the Great Artesian Basin in northern South Australia (see Box 2.11) showing the flowing artesian Beresford Spring (A in foreground), the large, 45 m high Beresford Hill with an extinct spring vent (C), the flowing artesian Warburton Spring (E), and the flat topped hill (F) capped by spring carbonate deposits (tufa) overlying Bulldog Shale. The diameter of the upper part of the circular Beresford Hill, above the rim, is about 400 m. Luminescence ages of 13.9 ± 1 ka were determined for samples from the carbonate mound of the actively flowing Beresford Spring (B) and of 128 ± 33 ka from the northwest side of the dry extinct Beresford Hill spring carbonate mound deposits (D) (Prescott and Habermehl 2008). (Source: Prescott, J.R. and Habermehl, M.A. (2008) Luminescence dating of spring mound deposits in the southwestern Great Artesian Basin, northern South Australia. *Australian Journal of Earth Sciences* **55**, 167–181. Reproduced with permission from Taylor & Francis.)

Plate 2.2 Big Bubbler Spring, with its spring outlet on top of an elevated mound, located west of Lake Eyre South in the Great Artesian Basin in northern South Australia (see Box 2.11). The spring outflow runs into a small channel and forms small wetlands (to the right). Hamilton Hill and its cap of spring carbonate deposits is visible in the background and is similar to Beresford Hill (Plate 2.1) located approximately 30 km to the north-west (Prescott and Habermehl 2008). (Source: Prescott, J.R. and Habermehl, M.A. (2008) Luminescence dating of spring mound deposits in the southwestern Great Artesian Basin, northern South Australia. *Australian Journal of Earth Sciences* **55**, 167–181. Reproduced with permission from Taylor & Francis.)

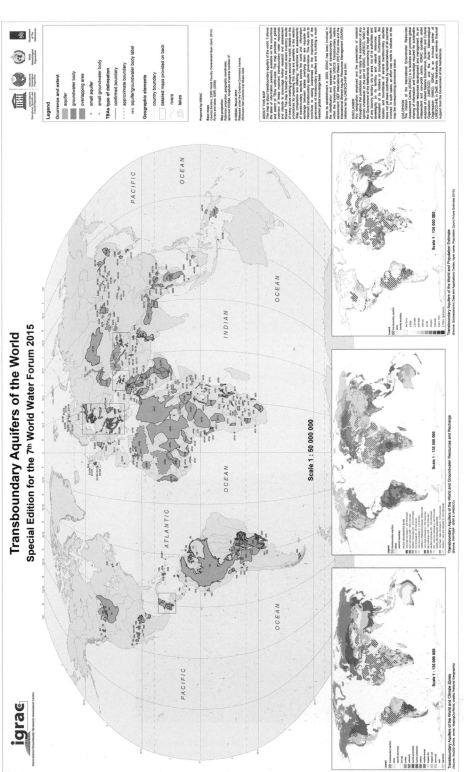

Plate 2.3 1 : 50 000 000 Transboundary Aquifers of the World (Special Edition for the 7th World Water Forum 2015) map. There are 592 identified transboundary aquifers, including transboundary 'groundwater bodies' as defined by the European Union Water Framework Directive, underlying almost every nation. Areas of transboundary aquifer extent are shown with brown shading and areas of transboundary groundwater body extent are shown with green shading, with overlapping aquifers and groundwater bodies shown in gold shading. Individual blue squares and green circles, respectively, indicate small aquifers and groundwater bodies (<6000 km²). The thematic inset maps combine, from left to right, respectively, the delineations of transboundary aquifers of the world with maps of climate zones, groundwater resources and recharge, and population at 1 : 135 000 000. For more information on individual transboundary aquifers and groundwater bodies and an extended view of the small aquifers and groundwater bodies, visit IGRAC's online Global Groundwater Information System: (https://ggis.un-igrac.org/ggis-viewer/viewer/tbamap/public/default). (Source: Transboundary Aquifers of the World, Special Edition for the 7 World Water Forum 2015, IGRAC. © 2015, IGRAC.)

Plate 2.4 Groundwater discharge in the intertidal zone of Kinvara Bay on 15 September 2010 at Dunguaire Castle, County Galway, Ireland.

Plate 2.5 Lake Caherglassaun (for location see Box 2.14, Fig. 2.66) responding to high tide as observed at 14.53 h on 13 September 2006 in the karst aquifer of the Gort Lowlands, County Galway, Ireland.

Plate 2.6 The Carran Depression and turlough (a fluctuating, groundwater level-controlled ephemeral lake) on 13 September 2006, The Burren, County Clare, Ireland.

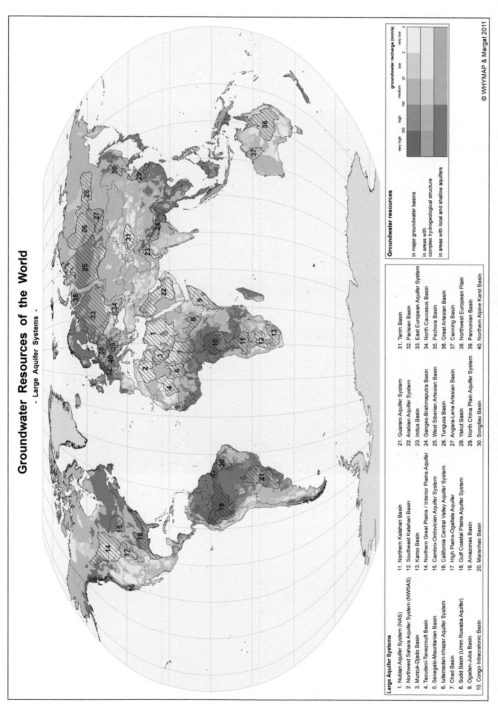

Groundwater Resources of the World
- Large Aquifer Systems -

Large Aquifer Systems

1. Nubian Aquifer System (NAS)	11. Northern Kalahari Basin	21. Guarani Aquifer System
2. Northwest Sahara Aquifer System (NWSAS)	12. Southeast Kalahari Basin	22. Arabian Aquifer System
3. Murzuk–Djado Basin	13. Karoo Basin	23. Indus Basin
4. Taoudeni–Tanezrouft Basin	14. Northern Great Plains / Interior Plains Aquifer	24. Ganges–Brahmaputra Basin
5. Senegalo–Mauritanian Basin	15. Cambro–Ordovician Aquifer System	25. West Siberian Artesian Basin
6. Iullemeden–Irhazer Aquifer System	16. California Central Valley Aquifer System	26. Tunguss Basin
7. Chad Basin	17. High Plains–Ogallala Aquifer	27. Angara–Lena Artesian Basin
8. Sudd Basin (Umm Ruwaba Aquifer)	18. Gulf Coastal Plains Aquifer System	28. Yakut Basin
9. Ogaden–Juba Basin	19. Amazonas Basin	29. North China Plain Aquifer System
10. Congo Intracratonic Basin	20. Maranhao Basin	30. Songliao Basin
		31. Tarim Basin
		32. Parisian Basin
		33. East European Aquifer System
		34. North Caucasus Basin
		35. Pechora Basin
		36. Great Artesian Basin
		37. Canning Basin
		38. Northwest European Plain
		39. Pannonian Basin
		40. Northern Alpine Karst Basin

Groundwater resources

in major groundwater basins

in areas with
complex hydrogeological structure

in areas with local and shallow aquifers

groundwater recharge (mm/a)

very high	high	medium	low	very low
300	100	20	2	0

© WHYMAP & Margat 2011

Plate 2.7 1 : 25 000 000 Groundwater Resources of the World (2008 edition) map showing the distribution of large aquifer systems (excluding Antarctica). Blue shading represents major groundwater basins, green shading areas with complex hydrogeological structure and brown shading areas with local and shallow aquifers. Darker and lighter colours represent areas with high and low groundwater recharge rates, respectively, generally above and below 100 mm a^{-1}. For further discussion see Section 2.17. (Source: Wall map "Groundwater Resources of the World", Global groundwater wall map, 2008. © 2008, WHYMAP.)

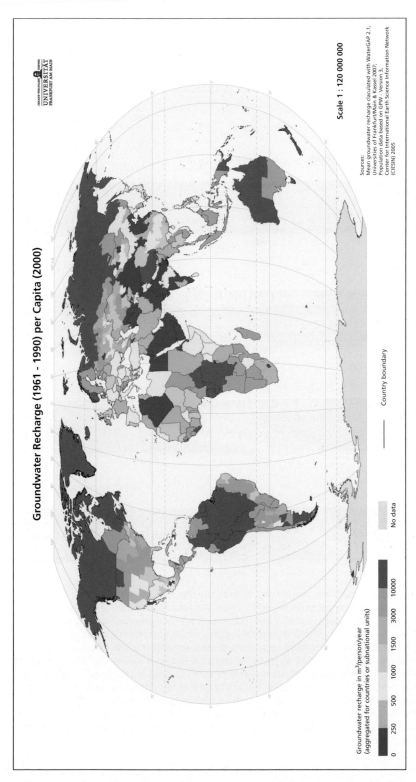

Plate 2.8 1 : 120 000 000 Groundwater Recharge (1961 – 1990) per Capita (2000) map showing groundwater recharge in m³ capita⁻¹ a⁻¹ aggregated for countries or sub-national units (excluding Antarctica). (Source: Wall map "Groundwater Resources of the World", Global groundwater wall map, 2008. © 2008, WHYMAP.)

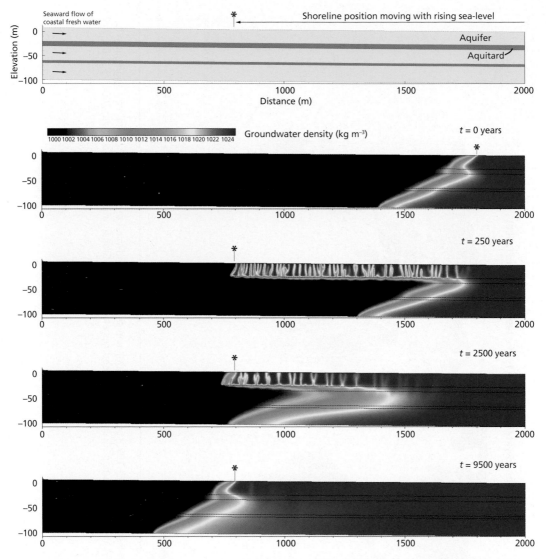

Plate 3.1 Variable-density groundwater flow simulations to evaluate the efficiency of different styles of salinization processes in layered aquifer systems on the continental shelf during and after transgression of the sea (see Section 3.6.1 and Fig. 3.12). The upper panel shows the model set-up representing a slightly seaward dipping layered aquifer system in which the left-hand boundary represents fresh, meteoric water originating as recharge in the hinterland. The right-hand boundary represents coastal seawater. In the initial steady-state situation ($t = 0$ years) the sideways sag of the saline water underneath the sea floor results in a tongue of saline water in the deeper inland aquifers. When sea-level rises, seawater starts to sink into the upper aquifer with a characteristic finger pattern indicative of free-convection replacing fresh water. This process of salinization is rapid compared to the salinization process in the deeper aquifer which only proceeds slowly by transverse movement of the saline-fresh interface. The simulation shows that it takes millennia for these processes to result in complete salinization of sub-seafloor aquifers which explains the current occurrence of fresh water in many parts of the continental shelf (e.g. Fig. 3.13).

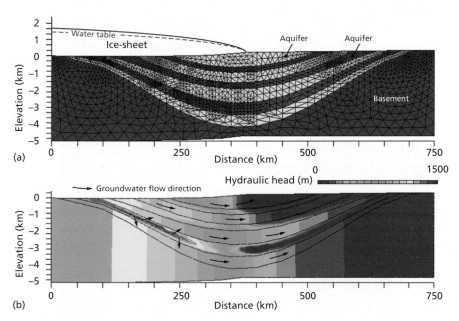

Plate 3.2 Example of a numerical simulation illustrating aspects of the hydrodynamics within sedimentary basins during glaciation. (a) A bowl-shaped sedimentary basin is conceptualized consisting of several thick aquifers and aquitards. This basin is overridden by an ice-sheet, which results in a complex hydrodynamic response. A deformation of the finite-element mesh accommodates the flexure of the sedimentary basin caused by the weight of the ice-sheet. (b) The high hydraulic head at the ice-sheet base is propagated into the aquifer units in the basin and results in a strong groundwater flow component away from the base of the ice-sheet. At the same time, the increasing weight exerted as the ice-sheet advances results in a build-up of hydraulic head in the aquitard units in the basin which are considerably more compressible than the aquifers. Consequently, groundwater is moving away from these aquitard units. In this model simulation the lower aquitard is more compressible than the upper aquitard (Bense and Person 2008). (Source: Adapted from Bense, V.F. and Person, M.A. (2008) Transient hydrodynamics in inter-cratonic sedimentary basins during glacial cycles. *Journal of Geophysical Research* **113**, F04005.)

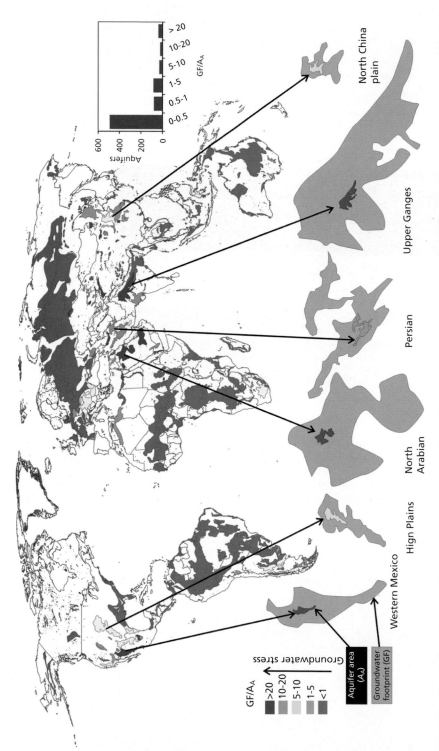

Plate 6.1 Global map of the groundwater footprint of aquifers. Six aquifers that are important to agriculture are shown at the bottom of the map (at the same scale as the global map) with the surrounding grey areas indicating the groundwater footprint proportionally at the same scale. The ratio GF/A_A indicates widespread stress of groundwater resources and/or groundwater-dependent ecosystems. The inset histogram shows that GF is less than A_A for most aquifers (Gleeson et al. 2012). (Source: Gleeson, T., Wada, Y., Bierkens, M.F.P. and van Beek, L.P.H. (2012) Water balance of global aquifers revealed by groundwater footprint. *Nature* **488**, 197 – 200.)

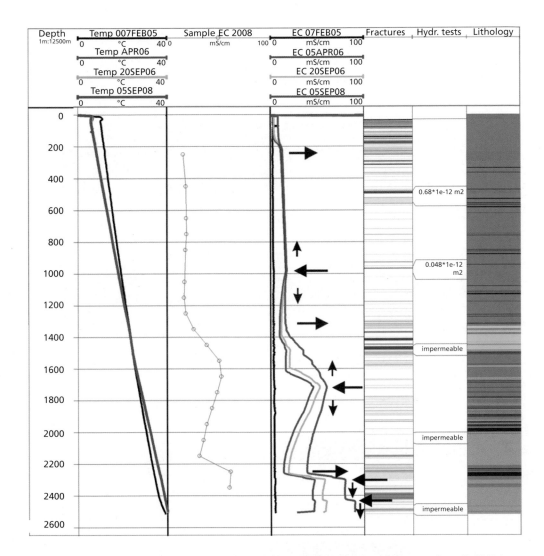

Plate 7.1 Temperature and fluid electrical conductivity (EC) logs in the Outokumpu Deep Drill Hole, eastern Finland. The 2516 m deep research borehole was drilled in 2004–2005 into a Palaeoproterozoic metasedimentary, igneous and ophiolite-related sequence of rocks in a classical ore province with massive Cu-Co-Zn sulphide deposits. The 'Sample EC' column shows the results of drill borehole water sampling in 2008. Arrows pointing to the left indicate interpreted depths of saline formation fluid flowing into the borehole and arrows to the right indicate fluid flowing out of the borehole. Arrows pointing up and down indicate the flow direction in the borehole. The 'Fractures' column indicates the interpreted fractures from sonic, electrical potential and calliper logs. The 'Hydraulic tests' column shows the test intervals and hydraulic permeabilities from packer experiments during drilling breaks. The 'Lithology' column shows the rock types (blue: metasediments; green and orange: ophiolite-derived serpentinite and skarn rocks; pink: pegmatitic granite). (Source: Adapted from Ahonen *et al.* 2004.)

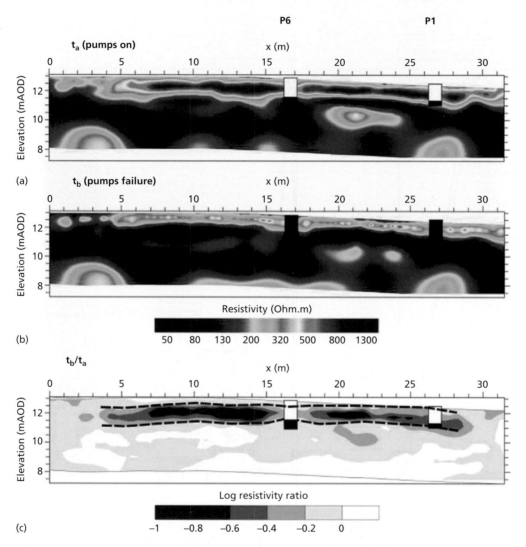

Plate 7.2 Automated time-lapse electrical resistivity tomography (ALERT) monitoring results during an interruption in groundwater pumping in an operational Lower Cretaceous sand and gravel quarry in West Sussex, England. Two times are shown: (a) t_a and (b) t_b imaged 15 days apart, as well as (c) the log resistivity ratio (t_b/t_a) plot showing sub-surface change. Water levels shown are for piezometers P1 and P6. Dashed lines show the minimum and maximum water levels estimated from the log resistivity ratio section (Chambers *et al.* 2015). (Source: Adapted from Chambers, J.E., Meldrum, P.I., Wilkinson, P.B. *et al.* (2015) Spatial monitoring of groundwater drawdown and rebound associated with quarry dewatering using automated time-lapse electrical resistivity tomography and distribution guided clustering. *Engineering Geology* **193**, 412–420.)

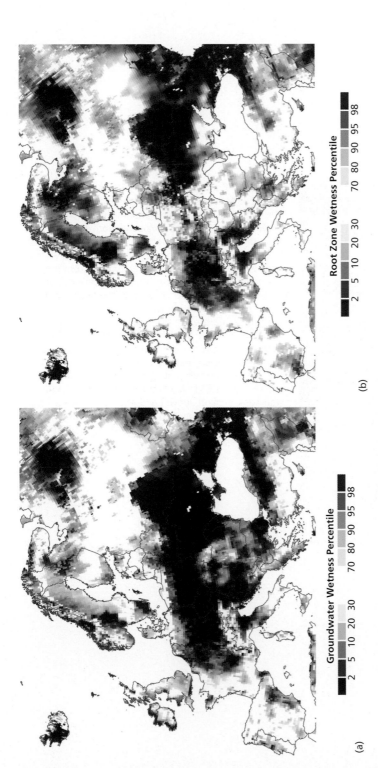

Plate 7.3 Satellite-derived images of (a) shallow groundwater storage and (b) root zone soil moisture content in Europe on 22 June 2020 as measured by the Gravity Recovery and Climate Experiment Follow On (GRACE-FO). GRACE-FO employs a pair of satellites that detect the movement of water based on variations in the Earth's gravity field by measuring subtle shifts in gravity from month to month. Variations in land topography, ocean tides and the addition or subtraction of water change the distribution of the Earth's mass and gravity field. Measurements are integrated with data from the original GRACE mission (2002–2017), together with current and historical ground-based observations using a sophisticated numerical model of water and energy processes at the land surface. The colours depict the wetness percentile to illustrate the status of groundwater storage and soil moisture content compared to long-term records for the month. Blue areas have more abundant water than usual, while orange and red areas have less. The darkest red areas represent dry conditions that should occur only 2% of the time (a return period of about once every 50 years). Much of Europe experienced drought in the summers of 2018 and 2019, followed by little snow in the winter of 2019–2020, the warmest on record. As a consequence, much of the continent began 2020 with a significant water deficit, with the threat of a groundwater drought and implications for maize and wheat yields compared to the five-year average in a number of countries. (Source: Signs of Drought in European Groundwater, NASA Earth Observatory, https://earthobservatory.nasa.gov/images/146888/signs-of-drought-in-european-groundwater?src=eoa-iotd.)

Plate 8.1 Replica pump with missing handle (see Plate 1b for comparison) in present-day Broadwick Street, Soho, London. The handle from the original Broad Street pump was famously removed on 8 September 1854 on the recommendation of Dr John Snow (1813–1858) who had concluded that the outbreaks of deaths from cholera among residents of the parishes of St James and St Anne were due to drinking contaminated water from the Broad Street well. From his investigation into the epidemiology of the cholera outbreak around the well, Snow gained valuable evidence that cholera is spread by contamination of drinking water. Subsequent research by others showed that the well was contaminated by sewage from an adjacent cess pool at 40 Broad Street entering the 1.83 m diameter, 8.8 m deep, brick-lined well sunk in sand above London Clay. This case represents one of the first, if not the first, study of an incident of groundwater contamination in Great Britain (Price 2004).

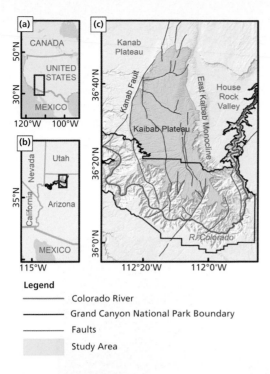

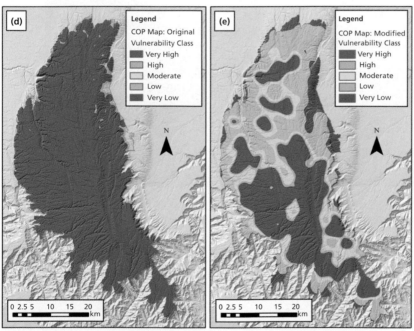

Plate 9.1 (a) and (b) Location of the Kaibab Plateau in the Colorado Plateau physiographic province (maximum elevation of 2807 m) north of the Grand Canyon, Arizona, United States, including the outline of the Grand Canyon National Park. (c) Shaded relief image of the Kaibab Plateau and surrounding region with approximate locations of major faults in the area. (d) and (e) Two karst aquifer vulnerability maps of the deep (approximately between 650 and 1000 m below ground surface), semi-confined Kaibab Plateau R (Redwall-Muav) aquifer system created, respectively, with the original concentration-overburden-precipitation (COP) method described by Vías *et al.* (2006) and the modified COP method of Jones *et al.* (2019) that uses sinkhole density as well as the location of faulted and fractured rock to model intrinsic vulnerability. Note that the modified model has a reduced overall intrinsic vulnerability to contamination and greater spatial variation of vulnerability (Jones *et al.* 2019). (Source: Adapted from Jones, N.A., Hansen, J., Springer, A.E. *et al.* (2019) Modeling intrinsic vulnerability of complex karst aquifers: modifying the COP method to account for sinkhole density and fault location. *Hydrogeology Journal* **27**, 2857–2868.)

Plate 10.1 An example of a dune slack at Winterton Dunes National Nature Reserve on the east coast of Norfolk, eastern England, observed in September 2020. Dune slack (or pond) habitats are a type of wetland that appear as damp or wet hollows left between sand dunes where, as here, the groundwater reaches or approaches the surface of the sand. The unusual acidic dunes and heaths at Winterton are internationally important for the rare groups of plants and animals which they support. The temporary pools in the dune slacks provide breeding sites for nationally important colonies of natterjack toads. The natterjack toad *Epidalea calamita* is often associated with dune slacks. To breed successfully, natterjacks require warmer water such as found in shallow dune slacks.

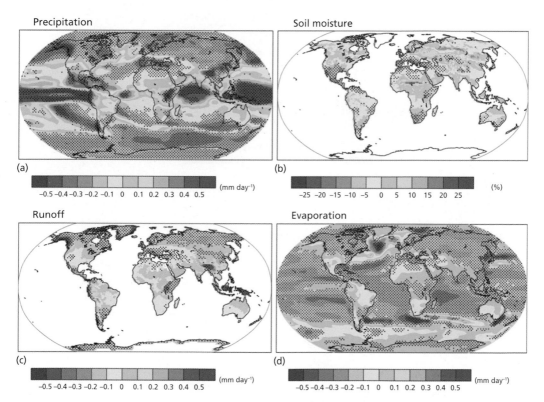

Plate 10.2 Multi-model mean changes in: (a) precipitation (mm/day), (b) soil moisture content (%), (c) runoff (mm/day) and (d) evaporation (mm/day). To indicate consistency in the sign of change, regions are stippled where at least 80% of models agree on the sign of the mean change. Changes are annual means for the medium, A1B scenario 'greenhouse gas' emissions scenario for the period 2080–2099 relative to 1980–1999. Soil moisture and runoff changes are shown at land points with valid data from at least 10 models (Collins *et al.* 2007). (Source: Collins, W.D., Friedlingstein, P., Gaye, A.T. *et al.* (2007) Global climate projections, Chapter 10. In: *Climate Change 2007: The Physical Science Basis. Contribution of Working Group I to the Fourth Assessment Report of the Intergovernmental Panel on Climate Change* (eds S. Solomon, D. Qin, M. Manning *et al.*). Cambridge University Press, Cambridge, pp. 747–846. © 2007, Cambridge University Press.)

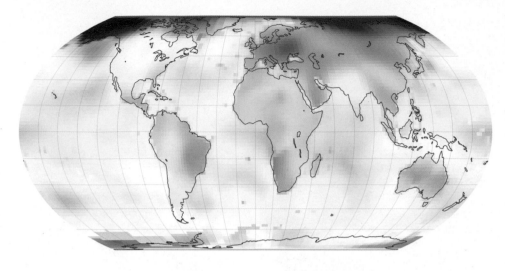

Temperature Anomaly (°C)

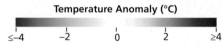

≤−4   −2   0   2   ≥4

(a)

Global Temperature Anomaly
Difference from 1951 – 1980 average, °C

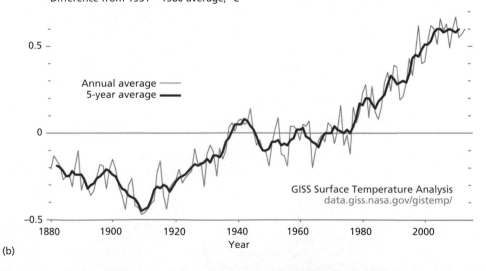

(b)

Plate 10.3   (a) Map showing global land-ocean temperature anomalies in 2019. Regional temperature anomalies are compared with the average base period (1951–1980). (Source: NASA (2020). *2019 was the second warmest year on record.* https://earthobservatory.nasa.gov/images/146154/2019-was-the-second-warmest-year-on-record (accessed 13 September 2020).) (b) NASA Goddard Institute for Space Studies (GISS) graph showing global surface temperature anomalies from 1880 through to 2013 compared to the base period from 1951 to 1980. The thin red line shows the annual temperature anomaly, while the thicker red line shows the five-year running average. (Source: Global Temperature Anomaly, 1880–2013, NASA Earth Observatory, NASA.)